Seeds Handbook

Seeds Handbook

Biology, Production, Processing, and Storage

Second Edition, Revised and Expanded

Babasaheb B. Desai

Mahatma Phule Agricultural University
Rahuri, India

MARCEL DEKKER, INC.   NEW YORK · BASEL

FIRST INDIAN REPRINT 2009

The first edition of this book was published as *Seeds Handbook: Biology, Production, Processing, and Storage*, B. B. Desai, P. M. Kotecha, and D. K. Salunkhe, 1997 (Marcel Dekker, Inc.), ISBN: 0-8247-0042-2.

Library of Congress Cataloging-in-Publication Data
A catalog record for this book is available from the Library of Congress.

ISBN 10: 0-8247-4800-X
ISBN 13: 978-0-8247-4800-5

Headquarters
Marcel Dekker, Inc.
270 Madison Avenue, New York, NY 10016, U.S.A.
tel: 212-696-9000; fax: 212-685-4540

Distribution and Customer Service
Marcel Dekker, Inc.
Cimarron Road, Monticello, New York 12701, U.S.A.
tel: 800-228-1160; fax: 845-796-1772

Eastern Hemisphere Distribution
Marcel Dekker AG
Hutgasse 4, Postfach 812, CH-4001 Basel, Switzerland
tel: 41-61-260-6300; fax: 41-61-260-6333

World Wide Web
http://www.dekker.com

The publisher offers discounts on this book when ordered in bulk quantities. For more information, write to Special Sales/Professional Marketing at the headquarters address above.

Printed and bound in India by Replika Press Pvt. Ltd.

FOR SALE IN SOUTH ASIA ONLY

To the memories of
Vikas and Varsha

Preface

Seed is the most important product of a plant's life cycle, and seed germ plasm constitutes the evolutionary continuum of the plant species. The essential role of seeds in crop establishment and improvement has long been recognized. As a prime dispersal unit, seeds can be stored for long periods of time in a suspended metabolic state without losing the essential genetic attributes of the crop. Considering the vital role of agriculture in a country's economy, the continuing development of improved seeds of crop plants has taken on increasing importance. Today seed industries and technologies are being evolved to ensure the continuous supply of high-quality seeds of crop plants to farmers.

The advent of hybrid and high-yielding varieties of crops with economic impact has necessitated greater care in the maintenance and preservation of seeds as a source of foods, feeds, and fibers. Seed production has evolved as a prime enterprise in modern agriculture and can no longer be regarded as merely a by-product of the agricultural industry. One of the quickest and easiest ways of increasing agricultural and horticultural production is to harness the higher yields and better quality offered by improved crop seeds. The increasing demand for high-quality seed for agricultural crops has remained unsatisfied and is expected to continue to expand.

Seeds Handbook is intended to present the basic information and update all areas of seed biology, production, and processing. Treated in detail are flowering, fruiting, seed morphology and development, biochemistry of seed dormancy, germination, viability, and longevity. Also covered is the production of seeds of all important agronomic and horticultural crops, including cereals, pulses, oilseeds, fibers, forages, vegetables, flowers and ornamental crops, as well as seed processing, seed testing and certification, seed storage, transportation, distribution, and marketing. Altogether, the volume deals with the practical methods of producing seeds and improving the genetic and physiological characteristics of over 80 major and minor crops. In addition, it addresses modern biotechnological aspects, such as production of synthetic seeds, loss-reduction biotechnologies of seeds, and new developments in the seed production business and industry. Future challenges and research needs are also discussed.

The second revised and expanded volume of *Seeds Handbook* elaborates in more detail every aspect of seed science and technology, starting from seed biology, seed production, processing, storage, and distribution. Seed biotechnological aspects have been discussed at length, including three new chapters on micropropagation, genetically modified (GM) or transgenic seeds and crops, and world food security. The chapter on synthetic seed biotechnology has been revised bringing in newer information on this subject as have all the remaining chapters of the earlier volume. The latest advances in the diverse and rapidly expanding field of seed science and technology have been updated and revised.

With over 1000 references, *Seeds Handbook* is designed to serve as an indispensable resource for upper-level undergraduate and graduate students, and teachers in the disciplines of agronomy, horticulture, animal husbandry, botany, cytogenetics and breeding, physiology and biochemistry, and seed technology. It also serves as a reference for researchers, planners, government agencies, and others in the seed industry.

I express my deep sense of gratitude to my wife, Vilasini, and daughter, Savita, for their assistance and patience during the preparation of the manuscript. The help received from Dr. Shivaputra Kotur and his colleagues, Dr. B. N. S. Murthy and Dr. Ashok of the Indian Institute of Horticultural Research, Hessergatta, Bangalore, is highly appreciated. Finally, it was a delightful experience to have associated with Ms. Theresa Stockton of Marcel Dekker, Inc. Without her cooperation and assistance this work would not have been completed.

Babasaheb B. Desai

Contents

Part IV Newer Seed Biotechnologies

Seeds Handbook

1
Introduction

Life is death—death is life. 'form' and 'no-form' are same, 'Tree and a Seed'—
seed is disappearance of the tree, tree has moved in 'no-form,' Nothingness has
tremendous potential—immense, enormous, vast, One single seed can fill all the
planets with greenery.

Osho

This was the goal of the leaf and the root, for this is the source of the root and
bud. This is the seed, compact of God wherein all mystery is enfolded.

G. S. Galbraith

Despite significant advances in the understanding of seeds and various aspects of seed
science and technology, the mystery surrounding seeds continues to grow. Humankind
has thrived on and sustained life with the help of seeds.

I. SEEDS: A SOURCE OF FOOD SECURITY

Humankind's search for food dates back to its history on this planet. Plants and
animals, including marine species, constitute the major sources of food. Soil is the
main source of food and feed for human beings and animals. Agricultural and
animal products from the land and marine products from the ocean are capable of
meeting the energy and protein needs of the world's population. By far the greatest
portion of the food consumed by people worldwide consists of the seeds of
agricultural crops.

Agriculture and civilization have progressed simultaneously along with seed
husbandry and the history of the development of new crops and their varieties. Many
people ended nomadic lives to settle permanently as they learned to plant, harvest, and
preserve seeds of certain grasses over the winter. The development of major human
civilizations had its basis in the culture of three cereal staple grains: wheat, rice, and
maize. The Mesopotamians planted wheat along the banks of the Tigris and Euphrates
rivers, the Chinese grew rice on the banks of the Hwang Ho and Yangtze rivers, and
the Mayans cultivated maize along the dry flat plains of the Yucatan (1).

Seeds of agricultural crops have been and will continue to be the major source of
food worldwide. The large-seeded grasses or cereals belonging to the plant family
Poaceae contribute more food than any other family, followed by the Fabaceae,
constituting legumes and pulses, such as groundnuts, peas, and beans. About nine-
tenths of all seeds cultivated are cereal grains forming the breadstuffs of the world.
Whereas the cereals provide the most important food component, carbohydrate, as
well as some protein and other vital constituents, legume seeds and pulses are usually

1

richer in protein than are cereal seeds both in terms of quantity and quality. Oilseeds like sunflower, groundnut, safflower, rapeseed, palm, cottonseed, and soybean constitute sources of edible oils and proteins. The seeds of fruits and vegetables are important sources of vitamins and minerals. Seeds are also used as spices and to make beverages and drugs. In addition to their food and medicinal value, seeds of agricultural crops are used in many other products, such as fiber (cotton, jute, flax), soaps, detergents, cosmetics, and a range of industrial products like resins, paper, paints, varnishes, linoleum, jewelry, and ornamentals.

Plants begin their growth within the buds in the meristematic regions, where cell division and elongation processes cause plant tissues to differentiate into specific reproductive organs like flowers, fruits, and seeds. A seed is a ripened ovule containing an embryo, and a fruit is a ripened ovary containing the seeds.

The life of a seed begins deep within the burgeoning flower, where the tiny embryo starts growing, forming an outer integument or seed coat. The latter protects the delicate embryo and also provides nutrition and an ideal environment for the growth, maturation, and development of the embryo and endosperm of the young seed.

Seed size and shape vary greatly within plant species and types ranging from the smallest dustlike seeds of orchids, tobacco, and some grasses (Kingston velvet, bent grass) to the largest seeds of acorns, walnuts, and coconuts. Seeds may be round, oval, triangular, elliptical, elongated, or irregular in shape. Seed color varies from the most common black or brown to the hues of red, yellow, purple, green, white, or multicolored seeds. Seeds may be smooth surfaced or rough textured with silky hairs, cottony masses, hooks, bristles, or winglike structures, which have important functions in seed dispersal and survival. Seeds are dispersed via wind, water, and birds and other animals, with their external surface features aiding in the dissemination process. Seeds are also well equipped to germinate and resume their growth at an appropriate place and time to ensure their survival. While some seeds are able to germinate and grow immediately after their dispersal, others, including weed seeds, can remain in a dormant state for long periods until conditions are suitable for their germination and growth.

As a microcosm of life, seeds represent a neatly wrapped package containing a living organism capable of exhibiting almost all processes found in the mature plant, such as respiration, cell division, morphogenesis, and photosynthesis. While important strides in the understanding of seeds have been made in the past two or more decades, seeds remain mysterious structures containing wonders of life yet to be revealed (2).

Agriculture and human civilization have progressed together from early crop improvements through the cultivation of indigenous useful crop plants and the introduction of crop varieties into new areas. This was followed by the selection of superior types of cultivated plants. Soon the now well-known scientific techniques of selection, hybridization, and polyploidization were developed, which paved the way to significant improvement in both yield and quality of agricultural crops. Although the pace of progress remained slow for a long time, the availability of improved techniques of plant breeding and genetics during the last two to three decades has enabled agricultural scientists and crop specialists to improve the yield potential of major cereals and millets on a revolutionary scale. The discovery of dwarfing genes in crops like wheat and the response of self-fertilized cereals to increased doses of fertilizers as well as the exploitation of hybrid vigor in cross-fertilized crops formed the basis of major advances in crop yields. The introduction of fertilizer-responsive dwarf varieties of paddy and wheat and promising hybrids of maize, sorghum, and pearl millet in recent years have more than doubled the yields of cereal crops.

Although seed is a basic input and forms only a small part of the total cost of cultivation, unless farmers use a good seed of assured quality, the investment in other inputs like fertilizer, irrigation, and plant protection will not be worthwhile. Only high-quality seeds of assured genetic purity can be expected to respond fully to all other inputs (3).

The hitherto indifference of farmers in Third World Countries to quality seeds epitomizes the more general indifference toward scientific agriculture. However, this picture has changed during recent times, as evidenced by significant increases in crop yields in developing countries. Because agriculture is a biological industry, its success depends upon the use of good-quality seed, and the pace of progress in food production will largely depend upon the speed with which high-quality seeds of agricultural crops are generated and made available to farmers when required.

Seed propagates around 90% of all crops grown on the earth. Modern agricultural crop seeds are developed with desired characters by altering their genes. Seed quality control systems are fundamental in seed production. Seed is an efficient means of spreading infection over long distances, and it can harbor pathogens of considerable longevity. Losses caused by seedborne and seed-transmitted diseases have been widely reported and are considered of great economic importance, especially in tropical and subtropical countries like India. Diseases exclusively seedborne can be combated, indeed eliminated, by using healthy seed (4).

The seed represents a living embryo embedded in the supporting food storage tissue. Seed technology, therefore, aims to protect this biological entity, looking after its "welfare" during various stages of seed production, processing, handling, transportation, storage, and marketing. Food technology, on the other hand, is concerned only with the supporting food tissue. Seed technology is an interdisciplinary science, encompassing a broad range of subjects, and includes development of superior crops and their varieties through breeding and cytogenetics, their evaluation and release, seed production, handling, processing, storage, testing, certification, quality control, marketing, and distribution. It also includes research into seed growth and development, seed physiology, seed dormancy and germination, seed viability and longevity, seed pathology and microbiology, and seed enhancement based on modern botanical and agricultural sciences.

According to Feistritzer (5), "seed technology" plays the following major roles:

1. A carrier of new technologies
2. A basic tool for secured food supply
3. The principal means to secure crop yields in less favorable production areas
4. A medium for rapid rehabilitation of agriculture during times of natural disaster

National seed stocks should provide improved seeds in emergency periods to production areas for the rapid production of food grains and should supply seeds to disaster regions for resowing.

The major goal of seed technology is to increase the production of high-quality agricultural crops through the effective distribution of seeds of high-yielding varieties. The success of seed technology depends on the following four functions being carried out effectively (3):

1. *Rapid multiplication*: The time taken to make available the desired quantities of high-quality seeds of improved varieties of crops to farmers is a measure of efficiency in the successful development of a country's seed industry.
2. *Timely supply*: The improved seeds of new crops and their varieties must be supplied to farmers in time so as not to disturb the planting schedule.
3. *Assured high quality of seeds*: The expected dividends and returns are achieved only by using good-quality seeds of improved crop varieties.
4. *Reasonable price*: The average farmer should be able to afford high-quality seed.

Seeds thus constitute an important source of food security for humans.

Recent developments in seed production-sowing to harvest, postharvest seed technology, meeting seed quality standards, assessing seed lot viability and potential performance, seed lot hygiene, seed testing and technology; and advances in seed physiology were reviewed at the Seed Symposium of the 25th ISTA Congress held in Pretoria, South Africa, in April 1998 (6).

II. WHAT DOES THE BOOK ENCOMPASS?

This second revised and expanded volume of *Seeds Handbook* deals with all aspects of seed production and technology, including morphological aspects of seed biology, such as flowering, fruit and seed development, seed dormancy, germination, viability, and longevity. These aspects are dealt with in detail in Part I: Seed Biology (Chapters 2–5). Part II: Seed Production (Chapters 6–13) deals with the basic principles of seed production, describing both the genetic and agronomic principles and the technology of seed production of all important major and minor agricultural and horticultural crops, including cereals, millets, pulses, oilseeds, vegetables, flowers and ornamentals, sugar and fiber crops, and grasses and forage legumes. The soil and climatic requirements of these crops, land preparation and isolation requirements for seed production, irrigation, fertilization and weed control, plant protection, rouging, harvesting, and postharvest operations such as drying, threshing, cleaning, seed treatment, storage, and seed enhancement are dealt with in detail, updating the information available. Part III: Seed Processing and Storage Technology (Chapters 14–20) is devoted to various aspects of seed processing and storage technology for all agricultural crops in general, namely, drying, cleaning, upgrading, seed treatment, seed packaging, and handling; seed storage, transportation, and marketing; seed testing, certification, and legislation. Chapter 21 describes modern developments in the seed production industry. Recent advances in seed biotechnology are described in Part IV, Newer Seed Biotechnologies, including new chapters, namely on micropropagation, genetically modified (GM) or transgenic seeds and crops, and world food security. The artificial seed biotechnology through somatic embryogenesis (synseeds) and loss reduction biotechnology of seeds during their production, processing and storage have been updated and included in Part IV (Chapters 23 and 25). The latest advances in the diverse and rapidly expanding field of seed science and technology have been integrated and updated, delineating various aspects of seed, starting from its biology, biotechnology, production, processing, storage, and distribution.

REFERENCES

1. L. O. Copeland and M. B. McDonald, *Seed Science and Technology*, 3rd ed., Chapman & Hall, New York, 1995.
2. V. R. Boswell, What seeds are and do. *Seeds: The Yearbook of Agriculture*. U.S. Department of Agriculture, Washington, DC, 1961, p. 1.
3. R. L. Agarwal, *Seed Technology*. Oxford and IBH, New Delhi, 1980.
4. T. Singh and P.C. Trivedi (eds.), *Seed Pathology*. Printwell, Jaipur, India, 2000.
5. W. P. Feistritzer, The role of seed technology for agricultural development, *Seed Sci. Technol.* 3:415 (1975).
6. J. G. Hampton, A. H. Martinelli, J. M. Farrant, H. M. J. Schmiermann, A. A. Powell, A. M. Abdel-Monem, K. Mtindi, D. Come and A. B. Ednie, Seed technology—past, present and future. *Seed Sci. Technol.* 27:681 (1999).

2
Seed Morphology and Development

I. INTRODUCTION

A. What Are Seeds?

Seeds provide a way for species to survive in addition to protecting and sustaining other life. By way of seeds, embryonic life can be practically suspended and then revived to develop again. Like highly organized fortresses supplied with specialized foods, seeds form vehicles for the spread of new life from one place to another. They are raw materials for innumerable products important to human beings and animals. They are wealth, beauty, and a symbol of friendship and goodwill. Seeds are a source of wonder and objects of earnest inquiry in our ceaseless search for an understanding of living things. Unwanted kinds of seeds, such as weeds, can be enemies and a source of trouble. As containers of embryos of a new generation, seeds ensure their main purpose—continuity of life through their numbers, forms, and structures.

Victor R. Boswell in 1961 (1) wrote:

> Some dry seeds become so well protected and insensitive that they can tolerate sharp, deep-freeze temperatures for years, with no harm or loss of vigor. A light-sensitive seed, while dry, may be so well protected and so insensitive that it is quite unaffected by daylong exposure to sunlight, yet, after it becomes moist, it may respond to a light exposure from a flash lamp as short as one one-thousandth of a second. We still do not know exactly what chain of events is set in motion by that flash and how? Some seeds require alternating temperatures in order to grow, while others do not. Some seeds live for decades and scores of years, while others, apparently as well protected, die in two or three years. Some small plants produce seeds that are much larger than the seeds of some much larger plants. One kind of seed develops completely in a few days while another takes years. Seeds are so wondrously different among species and yet all are quite evidently evolved to accomplish exactly the same thing.

Seeds remain a source of wonder.

Seeds form by the combination of mature male and female gametes coming from the stamen and pistil of the flower, respectively, in a process known as fertilization, or *syngamy*. In addition to formation of a seed, the process of fertilization is responsible for the level of genetic variation present in the zygote. Fertilization in angiosperm typically occurs by either self- or cross fertilization. Plants can also reproduce asexually either by vegetative propagation through stolons, rhizomes, tubers, tillers, bulbs, bulbils, or corms or through apomixis by producing seeds and vegetative propagules by asexual methods.

7

Botanically seeds are essentially young plants whose life activities are going on at a minimum rate. They represent the most critical phase of a plant's life cycle and are responsible for the evolutionary continuum of plant species. Two botanical families contain almost all of the food and feed crops comprising of the world's diet. The Poaceae, consisting of large-seeded grasses or cereals, contributes more food crops than any other plant family. Cereals comprise approximately 90% of all cultivated seed crops, providing the most important source of carbohydrate as well as some protein and other vital food nutrients. Throughout human history, cereals in the form of rice, wheat, and maize have been major sources of food, followed by other plants like sorghum, oat, barley, rye, and minor millets. Fabaceae is the second most important food family, consisting of legume food crops or pulses and oil seeds like groundnut, soybean, chickpea, pea, lentil, and beans. These crops contain more protein than cereal seeds, and pulse protein has a better balance of amino acids for human nutrition than do cereal grains (2).

Two classes of plants produce seeds, namely, gymnosperms and angiosperms, the latter being more highly developed than the former. Gymnosperms, or "nacked seed" plants, produce seeds without ovaries, flowers, or fruits; for example, conifers in which seeds are borne in pairs at the base of the scales of cones. In the highly developed angiosperms, the ovule and the seed develop within an ovary or the seed vessel. Kozlowski and Gunn (3) grouped seed-producing plants into three taxa—Angiospermae, Gymnospermae, and Pteridospermae—the last group being represented only by fossils from the early periods. The pteridosperms were among the first plants to produce true ovules and seeds. The ovary is the part of the flower containing the ovule with its egg or female sex cell. The ovary later turns into a fruit with the developed ovule or ovule seeds inside.

The embryo sac containing the tiny female cells is present in the ovule, which lies deep within the ovary of the mother flower. The embryo develops after the egg is fertilized by a sperm cell from the pollen tube, leading to perpetuation of the parent's life. Along with the embryo, a special store of food also develops in the forms of carbohydrates, proteins, lipids, and minerals to nourish the embryonic plant within the seed after it is separated from its mother plant. The ovary and the attached tissue with stored food thus become the fruit of the plant. Botanically a seed is a ripened ovule containing an embryo, and the fruit is a ripened ovary containing the seeds. Seeds of many vegetable crops are actually fruits. Most of them such as sweet corn, lettuce, and spinach contain only one seed, while others like members of the carrot family produce two-part fruits, each with one seed. Still other fruits, such as beets, have one or several seeds.

Although the seeds of most species of agronomic crops, vegetables, and flowers come reasonably true to variety with appropriate precautions to keep the pollen of undesired types from reaching the flowers of desired types, seeds of some crop plants like potatoes, cultivated fruit trees, grapes, berries, and ornamental plants fail to come true to their type because their sex cells carry random assortments of characteristics (1). Such crop plants are preferably propagated asexually or vegetatively by means of plant parts such as buds, roots, stems with attached buds, bulbs, and tubers.

B. What is Biotechnology?

Biotechnology (biology + technology) may be regarded as the science of improving living organisms through various technological changes. It involves molecular biology, biochemistry, and engineering. Biotechnology involves the manipulation of information flow in living organisms and tailoring or modifying the existing genetic and biochemical processes in organisms to the requirements of society. Through biotechnology, "bad" genes can be replaced with "good" ones to help treat genetic diseases. In agriculture, biotechnology has already led to a minor revolution in plant and seed production technology by giving us the means to tailor crops to meet specific human needs. Some of the most exciting possibilities brought about through biotechnology and high-technology agriculture are as follows:

1. Making food, feed, and fiber crops more resistant to diseases, insect pests, and various types of stresses such as extreme temperatures, acidity, and alkalinity. A single gene often confers disease resistance, which may be relatively easy to move from plant to plant. One way of achieving this goal is through the production of more desirable artificial or synthetic seeds by direct manipulation of the embryo (somatic embryogenesis) (see Chapters 22–24).
2. Creating hardy new crop plants capable of thriving in saline soils or withstanding periods of drought (see Chapter 24).
3. Giving staple food crops like wheat, rice, and maize the ability to make their own nitrogen-rich fertilizers, which would entail using solar energy to make ammonia directly from the air.
4. Increasing crop yields by improving the way plants use the sun's energy during photosynthesis.

The ultimate importance of biotechnology for plant science is that it provides a tool for the transfer of any gene from any source into plants (4).

II. FLOWERING, FRUIT, AND SEED DEVELOPMENT

Fruits and seeds in angiosperms are derived from flowers in a sequential process involving flower bud induction and initiation; flower differentiation and development, pollination, and fertilization; fruit set and seed formation, growth, and development of fruit; and seed and fruit senescence. The flowers in angiosperms are the specialized modified leaves borne or arranged on the stem primarily for sexual reproduction. Such plants grow vegetatively until they are fully mature to produce flowers, at which stage certain environmental stimuli such as light and temperature can trigger the transformation of vegetative meristems into reproductive meristems. This transformation takes place well ahead of actual flowering. Depending upon the plant species, the time required to attain vegetative maturity varies from a few days (flowering annuals) to several days (forest trees). However, flower bud initiation in some bulb crops takes place before sprouting. At maturity, most flowering plants become sensitive to environmental stimuli.

A. Photoperiodism and Vernalization

Flowering in some crop plants is induced by changes in day length (photoperiodism) and/or low temperatures (vernalization). The photoperiodic effects on flowering were first discovered in 1920 by W. N. Garner and H. A. Allard, plant physiologists in the U.S. Department of Agriculture, who observed that the Maryland Mammoth tobacco and Biloxi and other varieties of soybean, which normally bloom in autumn, could be made to flower in June or July by subjecting them to artificially shortened days and lengthened nights. This observation led to one of the most significant discoveries in botany, marking the recognition of a hitherto unsuspected feature of the environment—the daily duration of light and darkness as an important factor regulating plant growth and development. Gamer and Allard rightly called the phenomenon *photoperiodism*, because it recognized the importance of both a light-requiring (*photo*) reaction and a time-measuring one (*period*) in the response. Photoperiodism is thus a response to the relative lengths of day and night or, more specifically, a response to the duration and timing of light and dark periods. This phenomenon is an integral part of the overall adaptation of many kinds of organisms to their natural environment (5). Countless investigators confirmed the discovery promptly, and the phenomenon was found to be common to many flowering plants.

Plants like spinach, sugar beet, *Hibiscus*, coneflower, dill, *Fuschia*, henbane, and *Sedum* require long days for flower induction (long-day plants), while others like *Bryophyllum*, *Chrysanthemum*, cocklebur, *Cosmos*, *Kalanchoe*, poinsettia, strawberry, tobacco, and *Viola* require short days for flowering (short-day plants). In contrast, several plants like artichoke, balsam, *Gardenia*, cucumber, *Comphrena*, lima bean, cineraria, and tomato, which are not influenced by day length, are called day-neutral plants (6). Later it was established that it is the critical day length that induces the transformation of vegetative buds into flower buds. Long-day plants are those that require day lengths longer than their critical day length, while short-day plants require days shorter than their critical day length. This critical day length can vary slightly with the prevailing temperatures. Figure 1 depicts the photoperiodic control of short-day and long-day plants, as influenced by the light regimen. According to Moore (7), short-day plants require exposure to a minimum number of photoinductive cycles (which must be consecutive for maximum effectiveness) during which an uninterrupted dark period equals or exceeds a critical duration after attaining "ripeness to

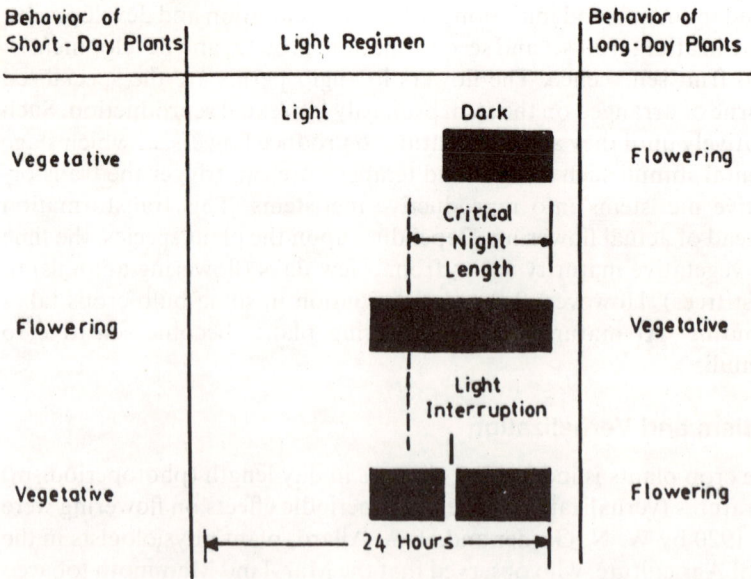

Figure 1   Photoperiodic control of flowering. (From Ref. 6.)

flower," whereas long-day plants require exposure to a certain minimum number of cycles (not necessarily consecutive) during which the photoperiod equals or exceeds some critical duration. After attaining ripeness to flower, long-day plants do not require a dark period.

Based on their floral induction responses to day length (photoperiod), crop species have been classified as short-day, long-day, intermediate-day, or day-neutral, although it is really the length of the night or dark period that is the critical factor influencing flowering. Table 1 gives examples of some agricultural crop species requiring photoperiod and vernalization to induce flowering.

The flowering stimulus is formed in the leaves in response to photoperiod and is transported to the meristematic region where it exerts its influence. Research conducted by the U.S. Department of Agriculture established the involvement of a natural plant pigment (phytochrome) in the photoperiodic response of plants. This pigment exists in two forms. The P_r form absorbs red light (660 nm) and is converted to a far-red (730 nm)–absorbing form, P_{fr}. The P_{fr} form absorbs far-red light and is converted into the P_r form. It can also be converted to the P_r form, although slowly during long uninterrupted dark periods. The conversion of P_r (inactive) to P_{fr} (active) in light and how long it remains in the P_{fr} form usually determines the plant's flowering response. Photoperiodic and nonphotoperiodic plants' responses were interpreted to conclude that there is a single photoreceptor pigment involved which exists in two photo-interconvertible forms:

$$P_r \underset{\text{far-red light}}{\overset{\text{red light}}{\rightleftharpoons}} P_{fr}$$

Thus, the response of a plant to red light (i.e., in the red, far-red reversible responses) results from the action of P_{fr} thereby generated, and response to far-red results from the absence of action of P_{fr} (7). The exact mechanisms of photoperiodism remain to be elucidated fully.

Some plants (biennials and perennials) require low temperatures for flower induction. This phenomenon, known as vernalization, was first demonstrated in winter wheat. Vernalization is an induction or acceleration of flowering by low

Table 1 Photoperiodic and Vernalization Responses of Some Agricultural Crop Species

Type of response	Short-day plants	Day-neutral plants	Long-day plants
Obligate photoperiodic response	Soybean, rice, dry bean, coffee	Soybean, cotton, potato, rice, sunflower, tobacco	Oat, annual rye grass, canary grass, red clover, timothy grass, spinach, radish
Facultative photoperiodic response	Soybean, cotton, sugarcane, rice, potato, sunflower		Cabbage, spring barley, spring rye, potato, sunflower, red clover
Positive vernalization requirement	Onion	Onion, carrot, broad bean	Winter oat, winter barley, perennial rye grass, winter wheat, sugarbeet

Source: Ref. 2.

temperature. The temperature and the length of vernalization required to induce flowering vary with plant species, and cultivars within the species may also have different requirements. The crop plants are classified into (a) those that require low temperatures for flowering (e.g., beets, Brussels sprouts, carrots, celery, cauliflower, and cabbage) and (b) those in which low temperatures can induce early flowering (e.g., peas, lettuce, and spinach). In certain cases, even the seeds can be vernalized (e.g., beets and kohlrabi), while others need a specified number of leaves before they can be vernalized. Garden plants like gladiolus, hyacinth, narcissus, and tulip require low temperatures to promote flower development, while some perennials require them for breaking the rest period. It is important to note that many fruit trees, garden plants (e.g., roses, carnations, and gerbera), and vegetable crops are self-inductive to flowering.

Fruits of Brazil nut (*Bertholletia excelsa*) were collected from the tree and the morphological development of the embryo was described and diagrammatic drawings were prepared. In a very early maturation phase, the ovule presents a liquid nuclear endosperm filling the embryo sac. Later, a cellular endosperm is formed lining the wall of the embryo sac. After digesting the endosperm and completing its development, the embryo fills the embryo sac cavity. The embryo comprises a massive hypocotyl without being differentiated into radicle, plumule, and cotyledons. Although undifferentiated, the *B. excelsa* embryo presents shoot and root poles lying at its opposite extremities (7a).

B. Stimulants of Flower Initiation

Transformation of a vegetative bud into a reproductive one follows a definite sequence of biochemical events. This transformation is triggered by biochemical changes in relative proportions of various phytohormones such as auxins, gibberellins, and abscisic acid (ABA) induced by external stimuli like photoperiod and/or vernalization. Flower initiation takes place in meristems, which have high levels of gibberellins as compared to auxins and ABA. It is known that the conditions that promote flower initiation enhance the relative proportion of gibberellins in apical meristems, while decreasing the levels of inhibitors, particularly ABA. This hypothesis is supported by the fact that gibberellic acid (GA) is the only known chemical to induce flowering during noninductive photoperiodic conditions in long-day plants, and it can also substitute for low temperatures (5).

Flower initiation can occur only in apices in which DNA and RNA synthesis is taking place. RNA may result in the production of new enzyme systems not previously present. This is evident from the fact that application of GA and indole acetic acid (IAA) increases RNA synthesis in apical meristems. The relative GA levels in meristems were also shown to increase just before flower initiation (Fig. 2). Evidence also indicates that GA increases the conversion of tryptophan to IAA or promotes IAA transport to the meristematic region which is required for the differentiation of the pistil. High GA levels have been known to promote development of staminate flowers, particularly in monoecious plants. The levels of these hormones decline at the time of anthesis.

After the discovery of photoperiodism, a hypothesis was developed that one or more specific flowering hormones is responsible for floral initiation in plants. Such hypothetical hormones or stimuli were called florigen, vernalin, or anthesin (8); the

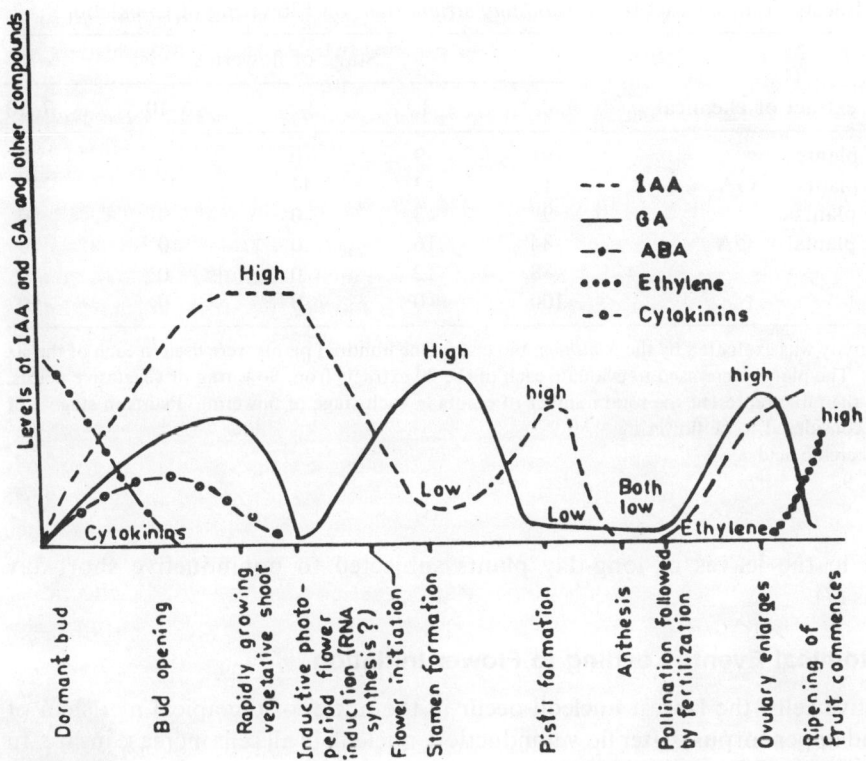

Figure 2 Levels of IAA, GA, and other compounds in stem apex, flower, and fruit. (From Ref. 11.)

latter emphasized the role of GA in the flowering of many plants. Attempts to isolate and characterize the flowering stimulus (or stimuli) ended in failures. The extracts prepared from the leaves and buds of flowering cocklebur (*Xanthium strumarium*) were found to stimulate flowering in the short-day plant *Lemma perpalsilla* under noninductive conditions if supplemented with GA (Tables 2 and 3). Hodson and Hammer (9) named the unidentified active substance in the extracts *florigenic acid.*

Circumstantial evidence points to the conclusion that flower initiation is controlled by hormones (6); one or more positively acting florigens and negatively acting inhibitors, however, remain to be identified. Some known hormones definitely appear to be involved in the initiation and regulation of flowering.

Exogenous application of GA is known to cause flowering in several species of long-day and vernalization-requiring plants under noninductive environmental conditions (10). According to Moore (7), GA is not the long sought after florigen, because GAs do not cause flowering of short-day plants under noninductive conditions or even in all long-day plants. If GA is not a florigen, it conceivably might be a vernalin.

Although ABA is known to cause flowering of certain short-day plants (e.g., *Chenopodium rubrum*) under long-day conditions, it is also not a florigen, because some short-day and long-day plants fail to flower with exogenous application of ABA. It may be at most one of the postulated flowering inhibitors long thought to be

Table 2 Effects of an Extract from *Xanthium strumarium* on Flowering of Cocklebur[a]

	Stage of flowering				
Treatment extract or chemical	0	1	2–5	6–10	7–10
Flowering plants	91	9	0	0	0
Flowering plants + GA	11	13	44	21	11
Vegetative plants	97	3	0	0	0
Vegetative plants + GA	84	16	0	0	0
GA	88	12	0	0	0
Water	100	0	0	0	0

[a] Extract activity was evaluated by the *Xanthium* bioassay. One hundred plants were used in each of the six treatments. The plants were used to evaluate each of the 20 extracts from flowering or vegetative plants. Figures in the table represent the total number of plants in each stage of flowering. Plants in stage 2 or above are considered to be flowering.
GA = Gibberellic acid.
Source: Ref. 9.

produced in the leaves of long-day plants subjected to noninductive short-day conditions (7).

C. Anatomical Events Leading to Flower Initiation

In vegetative cells, the largest nucleoli occur in the nuclei of the apical meristem of mantle and upper corpus. After flower induction, nucleoli in all cells increase in size. In the initial stages of flower initiation, the mantle cells are larger than the corpus cells. Their nuclei and cytoplasm stain lighter than those of corpus cells. They also contain larger vacuoles. Later on in the induction process, the number of mantle cells increases. This is followed by a distinct increase in the rate of cell division in the apical region. The cells on the surface elongate and divide anticlinally, while those in second and third layers divide periclinally. The cells in the fourth layer divide both anticlinally and periclinally, resulting in an increase in the height of pith cells and in the length of upper internodes. This is followed by a decrease in foliar initiation. The foliar primordia are differentiated into bracts instead of leaves with an increase in the

Table 3 Effects of Extract from *Xanthium strumarium* on Flowering of *Lemma purpusilla*[a]

Treatment	Plant flowering (%)									
Flowering plants	45	31	54	37	58	46	45	51	39	60
Flowering plants + GA	0	0	0	0	0	0	0	0	0	0
Vegetative plants	0	0	9	0	7	0	0	0	2	0
Vegetative plants + GA	0	0	0	0	0	0	0	0	0	0
GA	0	0	0	0	0	0	0	0	0	0
Water	0	0	5	0	11	3	0	0	1	5

[a] Extract activity was evaluated by the *Lemma* bioassay. A total of 30,000 plants were used to evaluate the activity of 20 extracts. The extracts were tested with and without the addition of GA. Each figure in table represents the average response of five lots of 100 plants each.
GA = Gibberellic acid.
Source: Ref. 9.

inductive cycle. The apical dominance is reduced, and as a result more buds are differentiated in the axils of leaves. The apex becomes about two to three times taller than its diameter, and the flower primordia become visible. After awhile, the primordia grow and differentiation of corolla tube, stamen, pistil, and ovule takes place depending upon the hormonal balance. Exposure of flower buds to unfavorable conditions can prevent further development with the result that production of a crown bud or deformed flower takes place. Depending upon the species, these changes may take place with only one inductive cycle or with several consecutive inductive cycles.

An ovary may contain one ovule or several ovules. Each ovule arises separately from the placenta, a minute dome-shaped projection called the nucellus. Each ovule has one large cell, a megaspore mother cell (megasporophyte), which by a meiosis consisting of two successive divisions produces four haploid megaspores. Of these four megaspores, only one develops into an embryo sac, while the others degenerate. This megaspore undergoes mitosis three times to give rise to eight nuclei, three of which move to either end forming polar nuclei, while two remain in the center. The nuclei located opposite to the micropilar end are called antipodal cells, while those in the center fuse together to form fusion nuclei containing 2n chromosomes. Of the three nuclei on the micropilar end, one forms a central egg nucleus, while the other two form the synergids on either side of the egg nucleus.

Simultaneously with the formation of megaspores in the embryo sac, the differentiation of microspores takes place in the stamen. The mature pollen grains are released from the anther and carried to the stigma of the same flower or different flowers by insects, wind, water, gravity, or other pollinating agents (11).

The structure of a complete normal flower (Fig. 3) consists of sepals, petals, stamen, and pistil. Only the latter two organs (stamen and pistil) are directly involved in sexual propagation. The petals in some plant species are colored, and flowers may contain the nectar and scent of essential oils to attract insects and other pollinating agents.

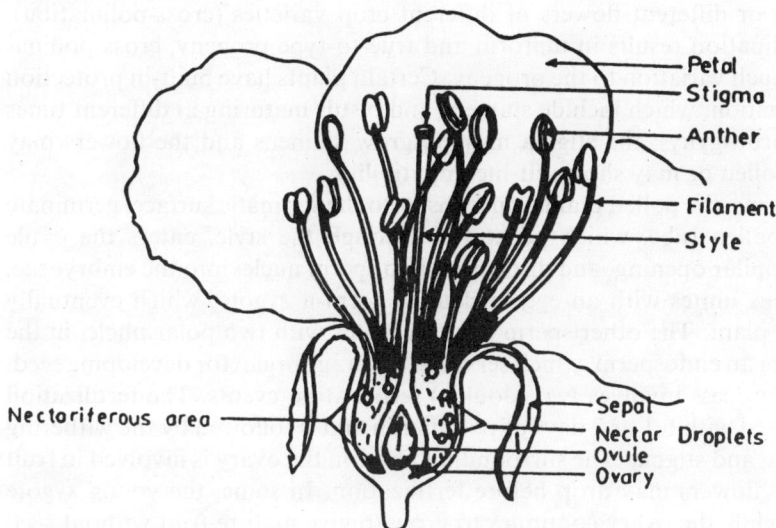

Figure 3 Structure of an angiosperm flower. (From Ref. 11.)

The male and female reproductive organs may be present on the same flower (hermaphrodite), or on different flowers on the same plant (monoecious), or on different plants (dioecious). In monoecicus plants, the male flowers are produced first and the ratio between male and female flowers is usually controlled genetically. However, spraying certain growth hormones or regulators in the initial stages can alter the sex ratio.

D. Pollination and Fruit and Seed Development

Salunkhe et al. (11) discussed the formation of seed in higher plants under the following six developmental stages of plant growth:

1. Formation of stamens and pistils in the flower
2. Anthesis, or opening of the flower, which signals the maturity of stamen and pistil
3. Pollination, consisting of transfer of pollen grains from stamen to the pistil, germination of pollen grains, and formation of the pollen tube, which penetrates the embryo sac in the ovary
4. Fertilization of the egg cell and polar nuclei with sperm nuclei from the pollen tube
5. Growth of a differentiated egg and its differentiation into embryo and surrounding endosperm and seed coat
6. Maturation of seed and accumulation of stored food

In insect-pollinated crops, the pollination process requires conditions favorable for the maximum activity of insects. The release of pollen grains from the anther or stamen is influenced by temperature and humidity; low-temperature and high-humidity conditions greatly influence its dehiscence.

Pollination can be accomplished by the transfer of pollen grains of one anther to the stigma of the same flower (self-pollination), different flowers of the same plant (self-pollination), or different flowers of different crop varieties (cross-pollination). Whereas self-pollination results in uniform and true-to-type progeny, cross-pollination introduces much variation to the progeny. Certain plants have built-in protection against self-pollination, which include stamens and pistils maturing at different times (protoandry or protogyny); the stigma may outgrow stamens and the flowers may have nonviable pollen or may show self-incompatibility.

In angiosperms, the pollen grains, on transfer to the stigmatic surface, germinate to produce the pollen tube, which penetrates through the style, enters the ovule through the micropilar opening, and discharges two sperm nuclei into the embryosac. One sperm nucleus unites with an egg nucleus to form a zygote, which eventually becomes the new plant. The other sperm nucleus unites with two polar nuclei in the embryosac to form an endosperm, which serves as a storage organ for developing seed. Thus, the whole process involves two (double) fertilization events. The fertilization process is completed within 1 or 2 days of pollination and is followed by the withering of corolla, stamen, and stigma. The surrounding tissue in the ovary is involved in fruit formation. Many flowers may drop before fertilization. In some, the young zygote may be aborted while the ovary continues to grow to give mature fruit without seed (e.g., Thompson seedless grape) (11).

The seed and fruit usually do not grow unless pollination and fertilization have taken place. In some plants, however, fruit set takes place without involving pollination and fertilization (parthenocarpy). The parthenocarpic fruits are thus seedless or have nonviable seeds (e.g., banana, Washington naval orange, fig, Oriental persimmon). In parthenocarpy, the relative proportion of certain phytohormones is thought to be altered during the pollination and fertilization processes. The exogenous application of auxins (tomato, pepper, and figs), gibberellins, and cytokinins (grapes) can replace pollination and/or fertilization stimulus and produce parthenocarpic fruits or seeds.

In several angiosperms (citrus, mango), more than one embryo develops per seed (polyembryony). In such cases, only one embryo arises sexually (fertilization), while others called nucellar embryos develop from cells (2n) of nucellus. In polyembryonic varieties of crops, the induction of nucellar embryos in vivo appears to depend upon the stimulus provided by pollination and/or fertilization and early development of the zygotic embryo. Polyembryony is thought to be a recessive hereditary characteristic controlled by multiple genes, which regulate the synthesis of a potent inhibitor of embryogenesis in nucellar cells of monoembryonic varieties (12). The frequency of nucellar embryo development in polyembryonic varieties is influenced by the nutritional status of the fruit, pollen, parent, and environmental factors.

Environmental conditions like temperature, light, and soil moisture influence seed and fruit set. The level of fruiting or crop load also affects the development of fruits and seeds. While a heavy crop load affects the quality of the finished product, a low fruit set gives an unprofitable seed crop.

Following the fruit set, the fruits and seeds begin to grow and act as a strong sink for the food material. Figure 4 shows the development of different tissues in lettuce fruit (13). The hormonal status of auxins, gibberellins, cytokinins, and ethylene influence the developmental process in fruits and seeds.

The development of the zygote and embryo follows a definite pattern. First, cell division in the zygote results in two cells, an axial (distal or apical) and a basal cell. In dicot (dicotyledonous) plants, the basal cell undergoes limited cell division to produce

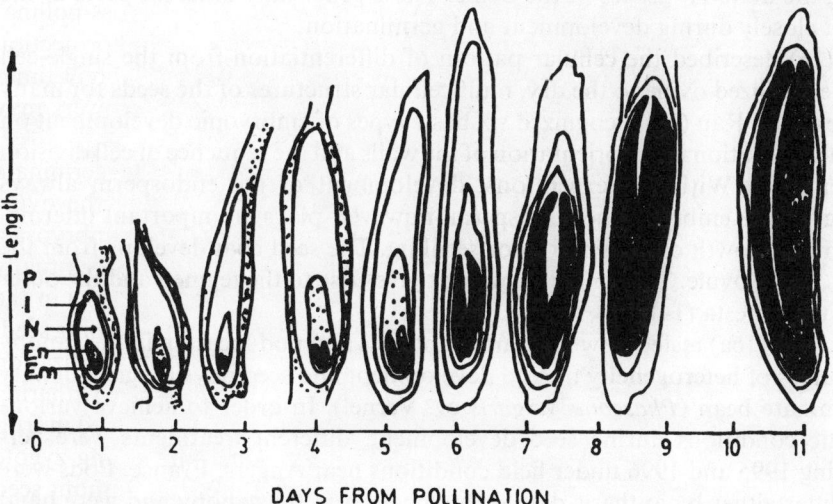

Figure 4 Development of lettuce fruit (P = pericarp, I = integument, N = nucellus, En = endosperm, Em = embryo). (From Ref. 13.)

the suspensor and may even contribute to the embryo, while the axial cell divides and then differentiates into the embryo. In contrast, in monocot (monocotyledonous) plants, the basal cell forms the terminal cell (houstoria) of the suspensor, while the embryo and other two cells of suspensor arise from the axial cell. Dicot plants contain two cotyledons, while in the monocots, one cotyledon is reduced to an absorptive scutellum, which in mature seeds lies adjacent to nutritive endosperm. Monocot plants also possess a coleorhiza (which covers the radical) and a coleoptile (first foliage leaf covering the plumule) (11).

The true endosperm, which is a unique characteristic of angiosperms, arises by a triple nucleus fusion (two polar and one sperm nucleus) to give triploid tissue. Some plants have a diploid, tetraploid, pentaploid, or polyploid endosperm. The endosperm of angiosperms may be nuclear (endosperm nucleus undergoes nuclear division before cellularization), cellular (endosperm nucleus does not undergo free nuclear division), or of an intermediate, holobial type (nuclear division proceeds and is followed by cell formation). The endosperm draws some of its nutrients directly from adjacent cells, while the major portion is transported from leaves and other sources. The endosperm acts as a food storage compartment for the developing embryo. While monocot embryos draw only a limited amount of food materials from the endosperm during maturation, most dicot embryos deplete the endospermic food materials during maturation and store them in cotyledons. The remnants of the depleted endoplasmic tissues in monocots lie between the starchy endosperm and scutellum. In lettuce seed, the depleted endosperm persists as a two- to three-cell-layer–thick structure surrounding the embryo and acts as a barrier for radical emergence. However, in endospermic dicot seeds, the endosperm remains intact. The integument tissues in the ovule undergo changes during seed maturation to form a seed coat. The structure of the seed coat varies considerably between the species (11).

According to Boesewinkel and Bowman (14), angiosperm seeds consist of three genetically different components: (a) the embryo developed from a zygote, (b) the usually triploid endosperm formed by the fusion of the two polar nuclei with the second spermatic nucleus, and (c) the seed coat formed out of the integuments, representing the material tissues of the ovule. These genetically different parts of the seed interact closely during development and germination.

Johri (15) described the cellular pattern of differentiation from the single-cell zygote of the fertilized ovule to the dry, multicellular structures of the seeds for many plants. Natesh and Rau (16) recognized six basic types of embryonic development on the basis of the variation in the orientation of the walls and the sequence of cell division in the proembryo. With few exceptions, development of the endosperm always precedes that of the embryo. The endosperm, however, plays an important intermediary role in the growth of the embryo or seedling. The seed coat develops from the integuments of the ovule. The inner integument gives rise to the tegmen and the outer integument to the testa (14).

Coste et al. (16a) tested how sequential anthesis and pod location in the canopy could be sources of heterogeneity of seed development and seed physiological quality for a determinate bean (*Phaseolus vulgaris* cv. Vernel). In order to achieve various microclimatic conditions during seed development, different treatments were performed during 1995 and 1996 under field conditions near Angers, France. Pods were chosen to differ either by anthesis date or by location in the canopy and were hand harvested at regular intervals. After moisture measurement on a subsample and slow drying, seed quality was assessed by standard germination, conductivity, and controlled deterioration tests. During seed filling and desiccation, seed moisture was found to be a good indicator of seed development. During the end of seed filling and the beginning of seed desiccation, seed quality increased as seed moisture decreased and leveled off at about 0.4 g per gram with no differences between years and treatments or between anthesis date and location of pods. Thus, it was concluded that during this phase, seed quality depended only on seed developmental stage. After the end of desiccation, seed vigor decreased linearly with increasing time from the end of desiccation; rate of deterioration differed between years and treatments but not between anthesis date and location of pods.

III. SEED PHYSIOLOGY AND COMPOSITION

Figure 5 depicts the major events that take place during seed maturation. Significant physiological discoveries have arisen from the research conducted on leguminaceous crops like peas and beans. In these crops, pods arising from ovary wall first increase in length, followed by an increase in wall thickness. The seeds start depositing storage reserves only after the pods have attained their maximum fresh weight. The leaves and pods at the lowest reproductive node of the field pea provide approximately 70% of the carbon required by the developing seed at that node. The leaflets serve as a major source for food reserve throughout seed development, while the stipules provide photosynthates during the early stage. About 85% of the carbon from leaflets and

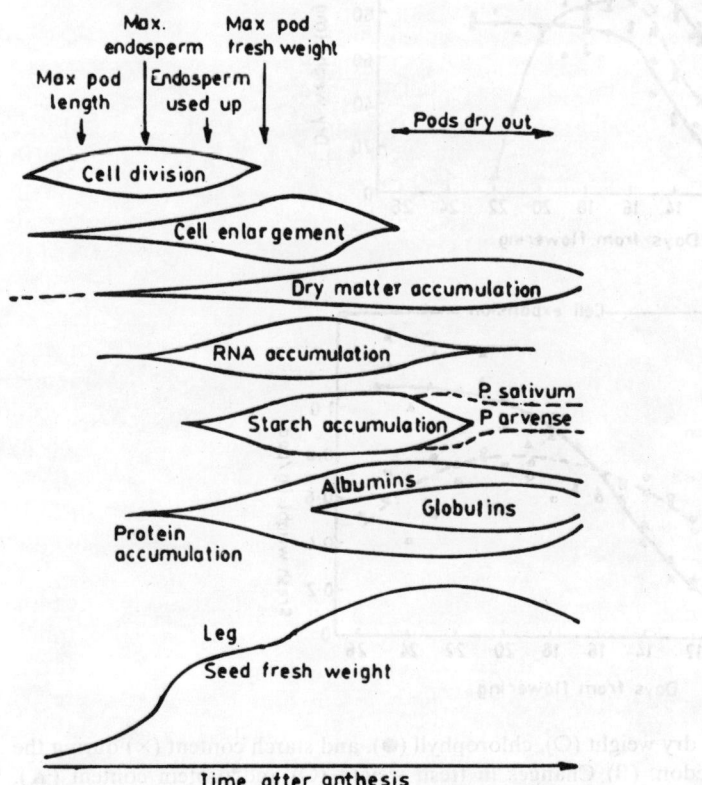

Figure 5 The events during seed maturation in pea. (Ref. 48.)

stipules come from diurnal changes in translocatary system. The translocation of sugars into the pod increases until noon, when leaf photosynthesis and phloem sugar levels are high. The phloem amino acids were also found to fluctuate with levels of sucrose. The carbon provided by CO_2 fixation is mostly utilized for recycling of respired carbon from the developing seed. The pod and seed coat have higher phosphoenol pyruvate (PEP) carboxylase (a key enzyme in dark CO_2 fixation) during the first 10–15 days of anthesis. PEP carboxylase levels decrease slowly until the seed matures (17). This enzyme is synthesized and stored in the cotyledons in the later stages of development. In some plants, the photosynthetic contribution of the developing pod is much higher than it is in peas and beans (18).

Amino acids in dicot plants are derived from a variety of sources, including leaves. In most annual plants, fruiting and seed development are closely followed by leaf senescence, which yields amino acids on hydrolysis of protein. The amino acids liberated in senescing leaves are translocated to the developing seeds.

The changes in protein, starch, and nucleic acid content during the development of pea cotyledon are depicted in Figure 6 (19). The cotyledons of developing seed in

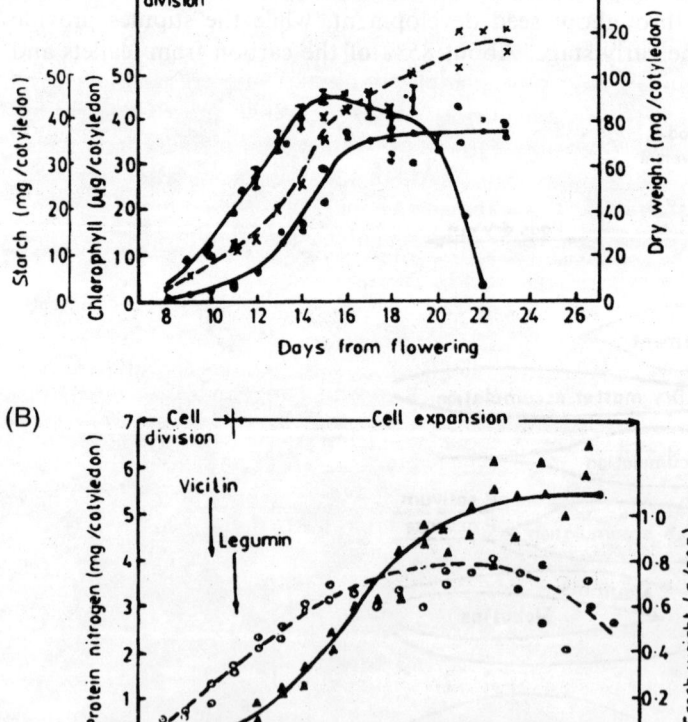

Figure 6   (A) Changes in dry weight (O), chlorophyll (●), and starch content (×) during the development of pea cotyledon. (B) Changes in fresh weight (O) and protein content (▲). (From Ref. 19.)

legume crops pass through a brief cell division phase, followed by an extended cell expansion phase. In nonendoplasmic seeds, the endosperm is occluded in the initial stages of cell expansion phase. During this phase, the major portion of storage reserves is laid down in cotyledons. In peas, starch accounts for about 35–45% of the total food reserve, while protein contributes approximately 25% of the dry weight of mature seed. The starch grains are located in chloroplasts and are visible 10 days after anthesis. In the initial stages, the photosynthetic products are used for starch formation, while later on in development, translocated sugars are also employed for synthesizing storage starch. Bulk protein synthesis occurs during cell expansion, when there is an increase in DNA and RNA synthesis (Fig. 7). The protein in seeds appears as small protein bodies dispersed in cytoplasm among larger starch molecules. In legume seeds, legumin and globulin (1: 1.4) are the major forms of seed proteins. Several legumes including fenugreek (*Trigonella foenumgraecum*) have endospermic seeds, which store galactomannans in this tissue (20). Galactomannans begin to be deposited in the early stages of development, which continues until the seed fully matures. Other low molecular weight carbohydrates like raffinose, stachyose, and sucrose are also found in these seeds.

In monocot seeds, starch accounts for the major portion of dry weight. Another major difference between monocot and dicot seeds is the presence of an aleurone layer covering the endosperm in the monocot seeds (Figs. 8–11) (21).

Seeds vary considerably in their contents of carbohydrates, lipids, and proteins and form a distinct biochemical contrast to their parent plants. They are relatively dry, usually containing about 10–20% moisture at maturity. While most of the seed material is present as one of the common types of polymer, seed polymers are almost certainly different from those of the parent plant. In most seeds, the carbohydrate predominates, which is starch with β-glucans and hemicelluloses as minor constituents. In some seeds, the major component is lipid, usually of the triglyceride type, and

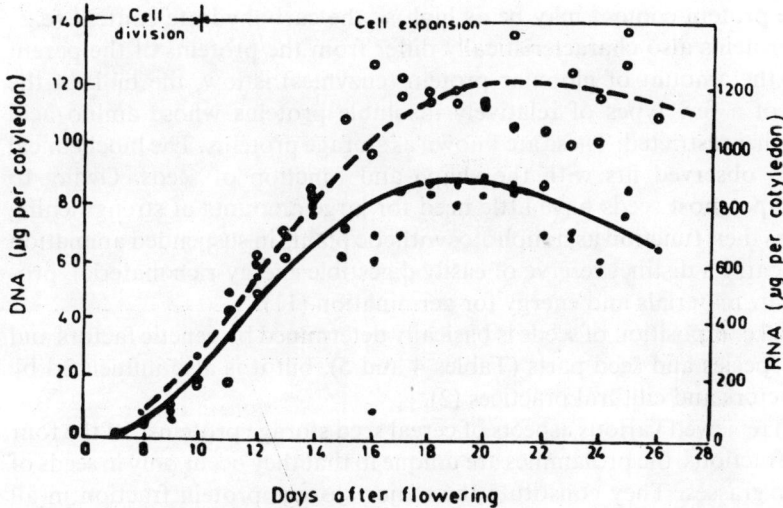

Figure 7 Changes in DNA (O) and RNA (●) during the development of pea cotyledons. (From Ref. 19.)

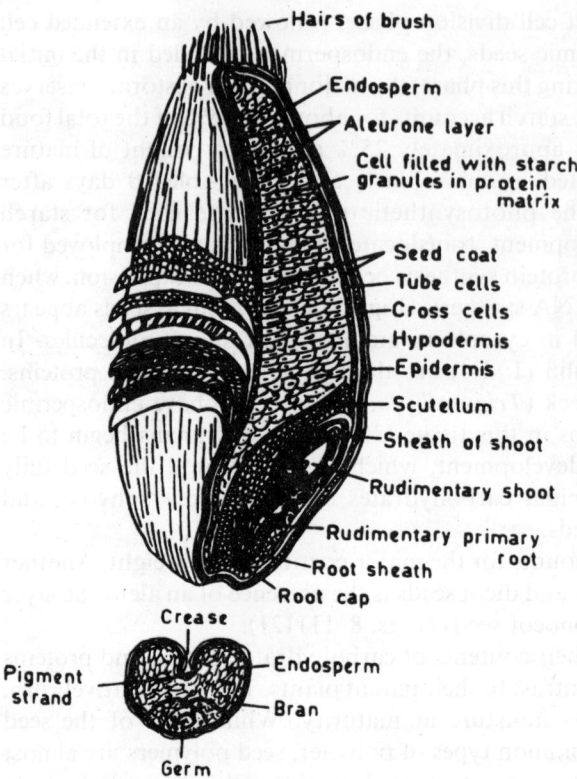

Hairs of brush

Endosperm

Aleurone layer

Cell filled with starch granules in protein matrix

Seed coat

Tube cells

Cross cells

Hypodermis

Epidermis

Scutellum

Sheath of shoot

Rudimentary shoot

Rudimentary primary root

Root sheath

Root cap

Crease

Endosperm

Pigment strand

Bran

Germ

Figure 8 Longitudinal and cross section through a wheat kernel. (From Wheat Flour Institute, Washington, DC.)

in such seeds the protein content may be as high as the carbohydrate content (e.g., soybean). Seed proteins also characteristically differ from the proteins of the parent plant. Although the amount of globular protein (enzymes) is low, the bulk of the protein consists of a few types of relatively insoluble proteins whose amino acid composition is rather restricted. These are known as storage proteins. The biochemical composition thus observed fits with the shape and function of seeds. Owing to their compact shape, most seeds have little need for large amounts of strengthening materials, whereas their function as nonphotosynthetic plants in suspended animation requires them to carry a distinct reserve of easily digestible energy-rich material, providing both the raw materials and energy for germination (11).

The chemical composition of seeds is basically determined by genetic factors and varies with crop species and seed parts (Tables 4 and 5), but it is also influenced by environmental factors and cultural practices (2).

Shewry (24) reviewed various aspects of cereal seed storage proteins. Of the four Osborne protein fractions, the prolamines are unique in that they occur only in seeds of cereals and other grasses. They constitute the major storage protein fraction in all cereals except oats and rice, where the major storage proteins are globulins and glutelins, respectively. According to Vitale and Bollini (25), deficiency of methionine and cysteine in legume seeds is mostly due to the low content of these amino acids in

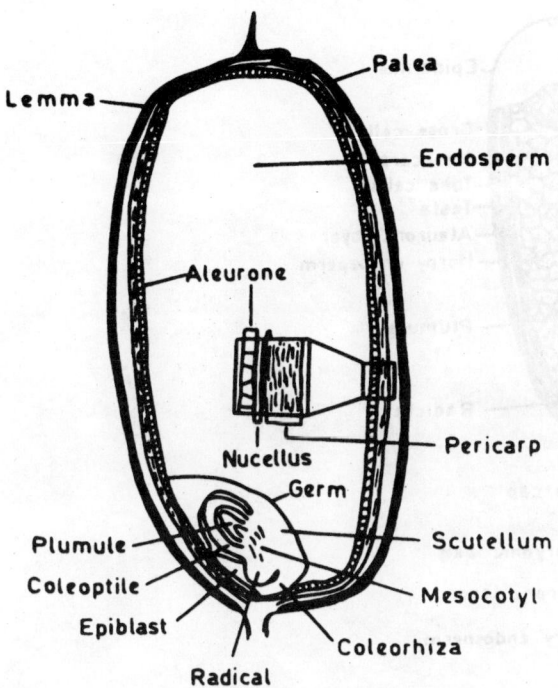

Figure 9 Schematic diagram of a midlongitudinal section of rice caryopsis. (From Ref. 48.) (From Y. Pomeranz, and R. L. Ory, *Handbook of Processing and Utilization in Agriculture*, Vol. 2, Part 1: Plant Products (I. A. Wolff, ed.) CRC Press, Boca Raton, Fla., 1982, p. 139.

storage proteins and lectins. A more favorable amino acid composition for legume seeds may be achieved by means of gene manipulation and transfer among plant species for the improvement of protein quality for animal consumption. One must, however, take into account the extent of change in storage proteins that could be tolerated by the plant, since a balance among different storage proteins may exist in the seed. Osborne and Bliss (26), for example, showed that in bean a decrease in phytohemagglutinin (PHA) induced a higher accumulation of phaseolin. In pea, conditions of sulfur deficiency negatively affect accumulation of legumin, which is higher in sulfur amino acids and stimulates the accumulation of vicilin (27,28). Nutritional improvement of legume seeds can also be achieved by modifying the lectin content of the seed (25). Morton et al. (29), describing recent developments in seed storage protein gene expression, indicated that the changes in seed storage protein gene expression in response to sulfur deficiency could be regulated depending on the gene and species.

While a few polymers make up the bulk of the dry weight of seeds, they may also contain certain compounds in low concentrations that have effects on humans and animals; for example, alkaloids like caffeine from coffee or cocoa and morphine from poppy seeds. Seeds may also contain some toxic nonprotein amino acids and antinutritional factors such as agglutinins and lectins. In addition, plant seeds have a capacity to synthesize a variety of dietary essential components required by humans and animals; for example, vitamins, essential amino acids, and essential fatty acids (11).

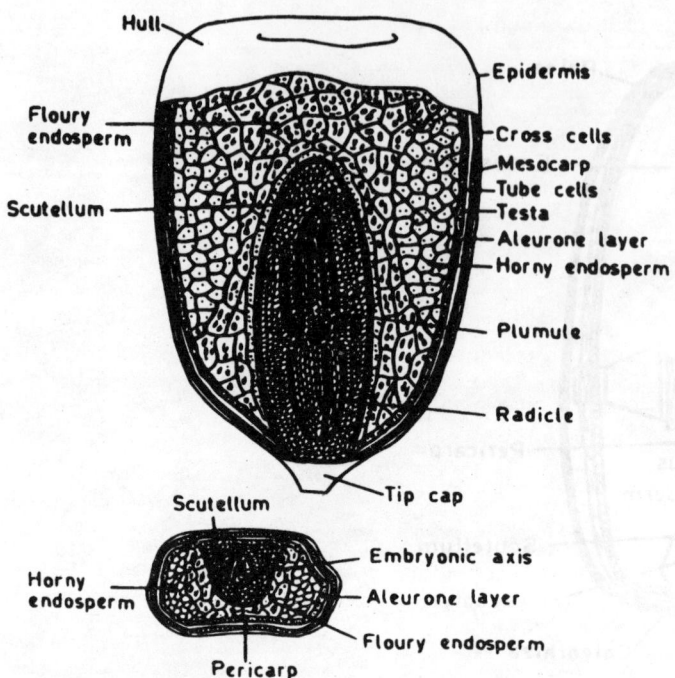

Figure 10 Longitudinal and cross section through a corn kernel. (From Wheat Flour Institute, Washington, DC.)

Ultrathin-layer isoelectric focusing (UTLIEF) of seed proteins soluble in 4-mmol l-1 NaCl was used to discriminate tomato varieties. Eight polymorphic protein bands were found on the gel with a pH range of 4–8, of which six bands were reported for the first time. About 87% of pair-wise comparisons of the 11 varieties tested showed at least one different protein band. Almost all varieties were found to be not homozygous genetically (29a).

Zamski (30) reviewed the physiological and biochemical aspects of developing seeds. From a biochemical point of view, seeds exhibit heterogeneous sink character-

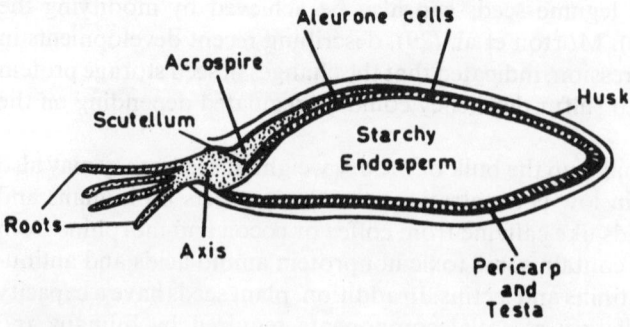

Figure 11 Longitudinal section of a germinated barley kernel. (From Ref. 49.).

Table 4 Average Percentage of Protein and Fat in Seeds

Crop/Plant	Protein (%)	Fat (lipid) (%)
Acorn (red oak)	3.2	10.7
Barley	8.7	1.9
Mung bean	23.6	0.2
Navy bean	22.9	1.4
Pinto bean	22.5	1.2
Beechnuts	15.0	30.6
Buckwheat	10.3	2.3
Chickpeas	20.3	4.3
Cotton seed kernel (without hull)	38.4	33.3
Flax seed	24.0	35.9
Kafir grain	11.0	2.9
Mustard, wild	23.0	38.8
Oats	12.0	4.6
Pea	23.4	1.2
Peanut (without hulls)	30.4	47.7
Rapeseed	20.4	43.6
Rice (rough grain)	7.9	1.8
Rye	12.6	1.7
Soybean	37.9	18.0
Vetch	29.6	0.8
Wheat	13.2	1.9

Source: Ref. 22.

istics. Angiosperm seeds store a high percentage of polysaccharides, proteins, lipids, and sometimes hemicelluloses and organic phosphate (31), whereas cotyledons of *Vicia faba* store about 35% of starch and 36% of protein (32). In addition to starch, most seeds have been found to accumulate oligosaccharides and other polysaccharides (33). Seeds consume only a fraction of the imported sugars as an energy source during their development. Sivak and Preiss (34) reviewed the structure and mode of synthesis of starch in developing seeds.

Zamski (30) concluded that plants invest huge amounts of carbohydrates in fruits and seeds, which are irreversible sinks but are important for the long-term

Table 5 Chemical Composition of Different Parts of Maize Seed[a]

Component	Entire seed	Endosperm	Embryo	Pericarp test[a]
Starch	74.0	87.8	9.0	7.0
Sugars	1.8	0.8	10.4	0.5
Oil (lipid)	3.9	0.8	31.1	1.2
Protein	8.2	7.2	18.9	3.8
Ash	1.5	0.5	11.3	1.0

[a] Includes only selected chemical components.
Source: Ref. 23.

survival of the plant species. The developing fruits and seeds constitute strong sinks and exhibit a very high amount of assimilate import.

Holubowicz et al. (34a) studied the internal seed structure of *Amaranthus caudatus, Celosia argentea* var. *plumosa, Gomphrena globosa, Calendula officinalis, Callistephus chinensis, Dahlia pinnata, Rudbeckia hirta, Tagetes erecta, Zinnia elegans, Cheiranthus cheiri, Iberis amara, Matthiola bicornis* (*Matthiola longipetala subsp. bicornis*), *Dianthus chinensis, Nigella damascena, Reseda odorata*, and *Viola wittrockiana* and described these structures based on microphotographs taken with a binocular microscope. They have also presented the preparation procedures, described imbibition and staining of the seeds, and tabulated data of some geometrical characteristics of embryos and seeds.

Poornima and Udaiyan (34b) estimated simple sugars and polyphenols in fresh neem seeds and in seeds stored for 15 and 30 days at room temperature (28–30°C). A gradual increase in simple sugars and polyphenols was observed in the seeds following storage, indicating deterioration of seed reserves and production of toxic substances, which may be some of the causes for loss of seed viability.

Lahuta et al. (34c) conducted comparative studies on the accumulation of carbohydrates in maturing seeds of two field bean (*V. faba* L. var. minor) cultivars Dino (classic morphotype) and Tibo (self-determinate growth habit) in 1990–1994. The plant growth conditions exerted a significant effect on the dynamics of the accumulation of carbohydrates and their final levels in mature seeds. The crops were grown in the field for 5 years, in 2 of which weather conditions were unfavorable for *V. faba*. Drought in 1992 and 1994 shortened the period of embryo development by several weeks, giving smaller seeds, which contained less starch but more oligosaccharides, sucrose, and reducing monosaccharides. Water stress stimulated the accumulation of raffinose family oligosaccharides. The qualitative and quantitative composition of oligosaccharides changed to a greater extent in the seeds of cv. Tibo than in those of cv. Dino.

The biochemistry of triacylglycerols (TAG) and their synthesis in developing seeds has been extensively reviewed (35–39). The composition of TAG varies widely among plant species, often characterized by unusual fatty acids. Different aspects of TAG synthesis are regulated by the acyl specificity and selectivity of enzymes dealing with acyl donors and acyl acceptors. Active research in this area has improved our understanding of different mechanisms of lipid biosynthesis (36,38). The differences in the biochemistry of TAG synthesis in the production of TAGs containing unsaturated, oxygenated very long-chain, and medium-chain fatty acids were described recently by Miquel and Browse (39).

Because of the commercial importance of plant seed oils, enzymes responsible for the steriospecific distribution of fatty acids on TAG have been particularly emphasized, as they are important targets for genetic engineering (40,41). Many of the steps of TAG synthesis have received increasing attention by research workers, revealing the intricacy of different synthetic reactions. Advances in protein purification and cloning techniques have made available several genes encoding some of the enzymes of lipid metabolism, and attempts to use these genes to manipulate seed oil fatty acid composition have been successful (39). The increasing availability of cloned genes is expected to be profitable for oil seed biotechnology, and an array of newer techniques will deepen our knowledge of lipid biosynthesis and understanding of its regulatory mechanisms. However, much research remains to be accomplished in order to elucidate the mechanisms of oil body assembly, the regulation of the genes involved, and the process of oil body mobilization (42).

In addition to carbohydrates, lipids, and proteins, several mineral nutrients such as phosphorus, magnesium, potassium, calcium, and other elements are also sequestered within mature seeds during their development in the form of phytin or phytate. Lott et al. (43) dealt with the mechanisms and regulation of mineral nutrient storage during seed formation. More pertinent information can be gathered from reviews on phytate biosynthesis (44,45), inositol formation (46), and transport of nutrients into developing seeds (47). On the basis of current research concerning phytin and mineral storage in seeds, Lott et al. (43) have posed several questions regarding the roles, mechanisms, and reasons for patterns of accumulation of phytate and pathways of phytin biosynthesis. The synthesis, deposition, physiological role, and mobilization of phytin in seeds need to be investigated further.

REFERENCES

1. V. R. Boswell, What seeds are and do. An Introduction, *Seeds: The Yearbook of Agriculture*, U.S. Department of Agriculture, Washington, DC, 1961.
2. L. O. Copeland and M. B. McDonald, *Principles of Seed Science and Technology*, 3rd ed., Chapman & Hall, New York, 1995, p. 17.
3. T. T. Kozlowski and C. R. Gunn, Importance of characters of seeds, *Physiological Ecology: Seed Biology*, Vol. I, 1972, p. 1.
4. R. Cook, Engineering a new agriculture, *Technology Review* (May/June), Alumni Association of the Massachusetts Institute of Technology, Cambridge, MA, 1982, p. 22.
5. T. C. Moore and P. R. Eckland, Role of gibberellins in the development of fruits and seeds, *Gibberellins and Plant Growth* (H. N. Krishnamoorthy ed.), Wiley Eastern, New Delhi, 1975, p. 145.
6. F. B. Salisbury and C. W. Ross, *Plant Physiology*, 2nd ed., Wadsworth, Belmont, CA, 1978.
7. T. C. Moore, *Biochemistry and Physiology of Plant Hormones*, Narosa, New Delhi, 1980, p. 230.
7a. R. Cunha, M. A. do Prado, J.E.U. de Carvalho, and M. de Goes, Morphological studies on the development of the recalcitrant seed of *Bertholletia excelsa* H.B.K. (Brazil nut), *Seed Sci. Technol.* 24:581 (1996).
8. M. K. Chailakhyan, Flowering and photoperiodism of plants, *Plant Sci. Bull.* 3(3):1 (1970).
9. H. K. Hodson and K. C. Hammer, Floral inducing extract from *Xanthium*, *Science*, 67:384 (1970), p. 1.
10. A. Long, The effect of gibberellin upon flower formation, Proceedings of National Academy of Sciences. Washington, D.C., 1957, p. 709.
11. D. K. Salunkhe, N. R. Bhat and B. B. Desai, *Vegetable and Flower Seed Production*, Agricole, New Delhi, 1987.
12. K. Esau, *Anatomy of Seed Plants*, 2nd ed., John Wiley; New York, 1973.
13. H. A. Jones, Pollination and life history of lettuce, (*Lectuca sativa* L.), *Hilgardia* 2:425 (1927).
14. F. D. Boesewinkel and F. Bowman, Seed morphology and development, *Seed Development and Germination* (J. Kigel, G. Galili eds.), Marcel Dekker, New York, 1995, p. 1.
15. B. M. Johri, *Embryology of Angiosperms*, Springer-Verlag, Berlin, 1984.
16. S. Natesh and M. A. Rau, The embryo, *Embryology of Angiosperms* (B. M. Johri ed.), Springer-Verlag, Berlin, 1984, p. 377.

16a. F. Coste, B. Ney and Y. Crozat, Seed development and seed physiological quality of field grown beans (*Phaseolus vulgaris* L.), *Seed Sci. Technol.* 29:121 (2001).

17. C. L. Hedley, C. M. Harvey, and R. S. Keely, Role of PEP carboxylase during seed development in *Pisum sativum*, *Nature*, 258:352 (1975).

18. B. Quebedeaux and R. Cholet, Growth and development of soybean (*Glycine max* [L] Morr.) pods. *Plant Physiol*, 55:745 (1975).

19. A. Millerd and D. Spencer, Changes in RNA synthesizing activity and template activity in nuclei from cotyledons of developing pea seeds, *Austr. J. Plant Physiol.* 1:331 (1974).

20. J. S. G. Reid and Meier, Formation of reverse galactomanan in the seed of *Trigonella foenumgraecum*, *Phytochemistry* 9:513 (1970).

21. S. S. Deshpande, B. Singh and U. Singh, Cereals, *Foods of Plant Origin: Production. Technology and Human Nutrition* (D. K. Salunkhe, S. S. Deshpande eds.), Van Nostrand, Reinhold, New York, 1991, p. 6.

22. F. B. Morrison, *Feeds and Feeding*, Morrison, Ithaca, NY, 1961.

23. F. R. Earle, J. J. Curtice and J. E. Hubbard, Composition of the component parts of the corn kernel, *Cereal Chem.* 23:507 (1956).

24. P. R. Shewry, Cereal seed storage proteins, *Seed Development and Germination* (J. Kigel, G. Galili eds.), Marcel Dekker, New York, 1995, p. 45.

25. B. A. Vitale and R. Bollini, Legume storage proteins; *Seed Development and Germination*, (J. Kigel, G. Galili, eds.), Marcel Dekker, New York, 1995, p. 73.

26. T. C. Osborne and F. A. Bliss, Effects of genetically removing lectin seed protein on horticultural and seed characteristics of common bean, *J. Am. Soc. Hort. Sci.* 110:484 (1985).

27. R. J. Blagrove, J. M. Gillespie, and P. J. Randall, Effects of sulfur supply on the seed globulin composition of *Lupinus angustifolius. Austr.J. Plant Physiol.* 3:173 (1976).

28. P. M. Chandler, T. J. V. Higgins, P. J. Randall and D. Spencer, Regulation of legumin levels in developing pea seeds under conditions of sulfur deficiency, *Plant Physiol.* 71:47 (1983).

29. R. L. Morton, D. Quiggin and T. J. V. Higgins, Regulation of seed storage protein gene expression, *Seed Development and Germination* (J. Kigel, G. Galili, eds.), Marcel Dekker, New York, 1995, p. 103.

29a. X. F. Wang, R. Knoblauch, and N. Leist, Varietal discrimination of tomato (*Lycopersicon esculentum* L) by ultrathin-layer isoelectric focusing of seed protein, *Seed Sci. Technol.* 28:521 (2000).

30. E. Zamski, Transport and accumulation of carbohydrates in developing seeds, *Seed Development and Germination* (J. Kigel, G. Galili, eds.), Marcel Dekker, New York, 1995, p. 25.

31. A. M. Mayer and A. Poljakoff-Mayber, *The Germination of Seeds*, 3rd ed., Pergamon Press, Oxford, 1982.

32. H. A. Ross and H. V. Davis, Sucrose-degrading enzymes in developing cotyledons of *Vicia faba. J. Exp. Botany* 43(Suppl.):14 (1992).

33. N. K. Matheon, The synthesis of reserve oligosaccharides and polysaccharides in seeds, *Seed Physiology, Vol. 1, Development* (D. R. Murray, ed.), Academic Press, Orlando, FL, 1984.

34. M. N. Sivak and J. Preiss, Starch synthesis in seeds, *Seed Development and Germination* (J. Kigel, G. Galili, eds.), Marcel Dekker, New York, 1995, p. 139.

34a. R. Holubowicz, J. Rutkowska, and T. W. Bralewski, Internal seed structure of selected ornamental species, *Folia-Hortic.* 12:93 (2000).

34b. S. Poornima and K. Udaiyan, Estimation of simple sugars and polyphenols in neem seeds during storage, *Seed Sci. Technol.*, 28:517 (2000).

34c. L. B. Lahuta, A. Login, A. Rejowski, A. Socha and K. Zalewski, Influence of water deficit on the accumulation of sugars in developing field bean (*Vicia faba* var. minor) seeds, *Seed Sci Technol.* 28:93 (2000).

35. C. R. Slack and J. A. Browse, Synthesis of storage lipids in developing seeds, *Seed Physiology* (D. R. Murray, ed.), Academic Press, Orlando, FL, 1984, p. 209.
36. S. Styme and A. K. Stobart, Triacylglycerol biosynthesis, *The Biochemistry of Plants*, Vol. 9 (P. K. Stumpf, E. E. Conn, eds.), Academic Press, New York, 1987, p. 175.
37. J. Browse and C. Somerville, Glycerolipid synthesis: biochemistry and regulation, *Ann. Rev. Plant. Physiol. Plant Mol. Biol.* 42:467 (1991).
38. M. Frentzen, Acyltransferase and triacylglycerol, *Lipid Metabolism in Plants* (T. S. Moore, ed.), CRC Press, Boca Raton, FL, 1993, p. 195.
39. M. Miquel and J. Browse, Lipid biosynthesis in developing seeds, *Seed Development and Germination* (J. Kigel, G. Galili, eds.), Marcel Dekker, New York, 1995, p. 169.
40. J. F. Battey, K. M. Schmid and J. B. Ohlrogge, Genetic engineering for plant oils: potential and limitations, *TIBTECH* 7:122 (1989).
41. M. J. Hills and D. J. Murphy, Biotechnology of oilseeds, *Biotech. Gen. Eng. Rev.* 9:1 (1991).
42. E. M. Herman, Cell and molecular biology of seed oil bodies, *Seed Development and Germination* (J. Kigel, G. Galili, eds.), Marcel Dekker, New York, 1995, p. 195.
43. N. A. Lott, J. S. Greenwood and G. D. Batten, Mechanisms and regulation of mineral nutrient storage during seed development, *Seed Development and Germination* (J. Kigel, G. Galili, eds.), Marcel Dekker, New York, 1995, p. 215.
44. E. Graf, ed., *Phytic Acid: Chemistry and Application*, Pilatus Press, Minneapolis, 1986.
45. J. S. Greenwood, Phytin synthesis and deposition, *Recent Advances in the Development and Germination of Seeds* (R. B. Taylorson, ed.), Plenum Press, New York, 1989, p. 109.
46. F. A. Loewus and M. W. Loewus, Myo-inositol: its biosynthesis and metabolism, *Ann. Rev. Plant Physiol.* 34:137 (1983).
47. P. Wolswinkel, Transport of nutrients into developing seeds: a review of physiological mechanisms, *Seed Sci. Res.* 2:59 (1992).
48. Y. Pomeranz and R. L. Ory, *Handbook of Processing and Utilization in Agriculture*, Vol. 2, Part I, *Plant Products* (I. A. Wolff ed.), CRCPress, Boca Raton, FL, 1982, p. 139.
49. W. C. Burger, *Handbook Processing and Utilization in Agriculture*, Vol. 2, Part I, *Plant Products* (I. A. Wolff, ed.), CRC Press, Boca Raton, FL, 1982, p. 187.

3
Seed Dormancy

I. INTRODUCTION

After seed matures, germination follows, but these two processes are normally separated both by time and space. The interval between the two events may·vary from a couple of hours to many years and from a few centimeters to thousands of kilometers. According to Pollock and Toole (1), the function of the seed is to carry its embryo plant through the hazards of time and space to a time and space where the new plant can grow, flower, and in its turn produce seeds. Delayed germination is not accidental—it represents the physiological mechanisms that keep the seed in a nongerminating state. Since young plants are vulnerable to the hazards of drought and extremes of heat and cold, it is advantageous for the seed to remain in an inactive condition until it reaches a time and space appropriate for germination. In the nongrowing condition, the seed moisture content is low and cell protoplasm is protected from damage owing to its low metabolic rate. The seed thus survives on its own nutrient reserves for a long period.

II. RESTING, GERMINATION BLOCKS, AND AFTER-RIPENING

The term *seed dormancy* has been used to describe two inactive conditions: one resulting from unfavorable environmental conditions and the other due to internally imposed germination blocks. Seed germination may be delayed, for example, by inadequate water supply or unfavorable temperatures. In some seeds, germination may be prevented by the presence of blocking mechanisms within the seed, which must be removed before germination can occur. The terms *rest*, *resting*, and *rest periods* have been used to describe seeds or buds that are inactive because of such internal blocks. It is necessary to distinguish seed dormancy resulting from an unfavorable environment from that due to the presence of inherent blocks.

Seeds possess remarkably complex and effective protective mechanisms to ensure their long survival. Such prolonged survival of the embryo within the seed ensures dispersal, distribution, or spread of the plant species over long distances. Many seeds have the ability to adapt to seasonal rhythms, that is, they have a natural time clock or a delayed-action mechanism that ensures that the seeds will remain dormant until another growing season rolls around, which may be long enough to permit another generation of seeds to mature (2).

The length of dormancy and the nature of delaying or resting mechanisms differ greatly among various species and cultivars of crops. Karseen and Hilhorst (3) described annual cycles of dormancy undergone by the seeds of some weed species.

Citing the work of Karseen (4), these investigators schematically represented various changes in primary dormancy, its release, germination, quiescence, and secondary dormancy (Fig. 1). Primarily dormant seeds, either dry or imbibed, require exposure to a certain temperature for a certain period to relieve dormancy (step 1 in Fig. 1). Only then will seeds become sensitive to factors that stimulate germination. However, germination will occur only when the complete set of required factors is present (step 2 in Fig. 1). If not, germination will be inhibited (step 4 in Fig. 1), the state being called quiescence. When quiescence is extended for longer periods, the seed may enter a new state of dormancy, called secondary dormancy (step 3 in Fig. 1), which can be broken under suitable temperature conditions. The cycle appears to depend on seasonal fluctuations of the soil temperature (5).

Bewley and Black (6) distinguished the term *quiescence* from dormancy; the former refers to nongermination of seed due to either lack of water or unsuitable environmental conditions, whereas the response of the seed to fail to germinate under favorable conditions points to dormancy of the seed due to its internal germination blocks. The block may disappear slowly from the dry seed (after-ripening). The dormancy can be terminated under the influence of some factor not required for germination itself but only to prime the seed subsequently to respond to the conditions that support germination. Thus, a dormant seed may be promoted to germinate by a discontinuity of the conditions, with one condition being necessary to prime or potentiate seeds to germinate and another adequate for germination itself. Seeds of many plant species, for example, require chilling for several weeks at 4°C for germination. These seeds normally do not germinate at low temperatures or do so very sluggishly, but germinate readily when they are transferred to warmer conditions; that is, in order to germinate, these dormant seeds must experience a discontinuity in conditions—a low temperature to break dormancy and a higher one to support germination. Similarly, some seeds germinate only after they have been exposed to light even if only for a few seconds. The dormancy of these seeds is broken by another environmental factor, light, but seeds germinate perfectly under dark conditions. Bewley and Black (6) pointed out that chilling and light are not factors strictly required continually during germination but only to trigger or potentiate the process. Seed dormancy varies with external conditions like temperature and with time both in the developing and in the mature seed.

The term *dormancy*, or resting, cannot be applied rigidly to describe the response of certain seeds. Lettuce, for example, germinates promptly in total darkness if it is planted in moist soil at 57°F (13.9°C), since it does not contain a germination block.

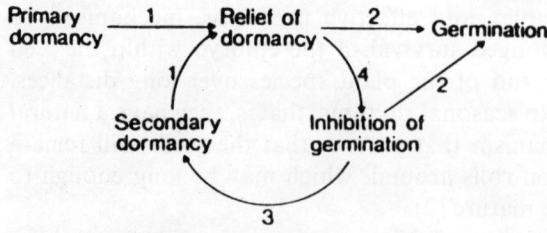

Figure 1   Effects of chemical environment on seed germination. Schematic presentation of changes in dormancy. (From Ref. 3.)

If it is planted at 84°F (28.9°C), however, it remains inactive. After a few days at 84°F, if the temperature drops to 57°F (13.9°C), the seed still does not germinate. Exposure of the seed to high temperature induces the formation of a germination block that did not exist previously. This block can be removed by exposing the lettuce seed to red light. It is, therefore, necessary to list carefully the conditions under which seed germination is attempted. The variety of the seed crop and its previous history are also important.

Germination blocks are not absolute but rather relative. The growth of some seeds, for example, sour cherry *(Prunus cerasus)*, is not completely stopped even in a blocked seed at low temperature. Root and shoot cell division continues and the whole embryonic axis grows slightly at a time when the seed does not germinate even under the most favorable conditions. Something obviously does occur during after-ripening to permit the subsequent germination, indicating that the blocks are only relative. A blocked seed is like a motionless automobile with its motor running at idling speed with its gears disengaged (1).

It is common knowledge that all viable seeds when planted do not germinate; the proportion of the seeds that do germinate varies with the conditions of germination, because all seeds are not genetically identical. The expression of a seed's heredity in the form of germination depends on the environment during seed formation, maturity, and the variable germination blocks. The genetic and environmental factors interact to produce greater variability in the magnitude of germination of different kinds of seeds.

Germination blocks act through a number of different physiological mechanisms; some may be simple and well understood, whereas others are complex and unknown. The presence of a hard seed coat impermeable to water may completely block germination in some seeds such as water lily, cotton, okra, clover, and some beans. During ripening and drying processes, the seed coats of these species become impermeable to water, with the seed moisture content reaching a low level. Some germination blocks may be localized to affect the embryo. Blocked apple seeds germinate only after undergoing a period of after-ripening; that is, remaining moist at a temperature around 40°F (4.4°C). The imbibed seeds chilled for 2–3 months germinate promptly, producing normal seedlings.

Germination blocks can be removed by a natural combination of time and exposure to the elements. A hard seed coat may be softened in the soil by alternate freezing and thawing or wetting and drying. Microorganisms using the seed coat as a source of organic carbon may rot it off in the soil, and a low soil temperature may meet the chilling requirement for seed germination. Drying for a period of time constitutes after-ripening in the dry condition, which induces favorable changes in the seed coat to permit germination. Mechanical scarification consisting of blowing seeds against abrasive points or rubbing them over an abrasive surface may be employed to remove artificially the hard seed coats. Hot water or acids also are used to scarify the seed coats chemically (e.g., dropseed grass).

The leaching action of water and light initiates changes in seeds, permitting their germination. Some desert plant seeds require a large amount of water to remove the inhibitors blocking germination. The conditions against which blocks protect the seed often are the same ones that serve to remove the blocks, which is closely related to the seasonal changes that seeds encounter (1).

Both germination blocks and after-ripening are reversible; after-ripened seeds are ready to germinate. However, blocking may again be induced in such seeds if they are exposed to conditions such as oxygen deficiency, an excess of water, or a temperature too high for germination. After-ripening is a major factor in the longevity of some weed seeds. The carefully after-ripened seeds in a refrigerator may revert to a blocked condition if planted at higher temperatures (above 70°F or 21.10°C), resulting in a second dormancy (e.g., seeds of pine, rose, apple, cocklebur).

Germination blocks may be physical or chemical, the former being caused by structural changes in seed coats surrounding the embryo. Chemical blocks are of two types:

1. Those inhibiting chemicals in tissues surrounding the seed
2. Chemical inhibitors within the embryo itself.

Many seeds may have more than one type of block (clover). Physical blocks are associated with the structure of the seed coats and other tissues surrounding the embryo. In addition to acting as germination blocks, these tissues also protect the embryo from mechanical damage and from attack by microorganisms. Some blocks act by preventing the entry of water and exchange of oxygen and carbon dioxide through the seed coats. Most seeds need an abundant supply of oxygen for enhanced metabolism during germination. A restricted supply of oxygen may induce changes in seed metabolism, imposing germination blocks.

Chemical blocks may be present in tissues surrounding the embryo or in the outermost covering of the ovary or fruit wall. Failure of seeds to germinate in the fruit may be attributed to such chemical blocks. Inhibitors are also known to be present in the seed coats and other membranes around the embryo. Hundreds of inhibitors of seed germination have been isolated and chemically characterized. Most of these are nonspecific, blocking germination in many kinds of seeds besides those from plants in which they are found. Blocks in seed coats and membranes are simpler and better understood than those present within the embryo. Although most growth phenomena are chemical processes, it may be misleading to suggest that all embryo blocks are chemical (1). The embryo blocks are not necessarily caused by growth-inhibiting chemicals and may result from a deficiency of some essential compound. After-ripening may permit accumulation of the missing compound to a level permitting germination (e.g., loblolly pine). It is rather difficult to isolate and identify an inhibitor from an embryo causing germination blocks. After isolation, the active compound must be measured outside of the plant using some bioassays. Cyanide is a powerful inhibitor controlling the germination of many seeds. It occurs as a glycoside in the embryo, having no growth-inhibiting properties. Free cyanide is formed only when the cells are damaged. However, evidence is not available to show that cyanide actually functions as a germination inhibitor. Pollock and Toole (1) described the following rigorous test to decide whether an extracted compound functions in the seed as an inhibitor: The true inhibitor should change in quantity parallel to changes in the physiological condition of the seed. If the seed is strongly blocked, a germination inhibitor should be present in high concentration. As the block is removed, the concentration of the inhibitor should decline to a minimum, leading to prompt germination. If after-ripening is reversed by high temperature or lack of oxygen, the concentration of the inhibitor should increase. It was thus concluded that although growth-inhibiting chemicals may very well control germination, they are not the only possible mechanism(s). Changes in cellular organization and processes of respiration, photosynthesis, and protein synthesis occurring in microsomes and mitochondria may also account for certain mechanisms controlling germination, which are yet to be investigated fully.

Whereas some germination blocks can be removed easily, others require more strenuous efforts. The mechanism of stimulation of germination by light also controls several other growth responses of crop plants, including flowering. Temperature may influence one type of chemical reaction in plants differently than others. The low-temperature requirement of some plant seeds for germination is not fully understood. Some physical and/or chemical blocks may be interdependent. Artificially applied chemicals and growth regulators have been known to interact with light and temperature. Since interactions exist and such interacting blocks control germination; germination appears to proceed by a number of alternate pathways. Blocks imposed and removed by various environmental conditions may close these pathways (light and temperature) as well as by chemical treatments. The activity of cellular reactions catalyzed by enzymes is regulated by many factors, including temperature, which may reflect alternate pathways of germination. According to Yogeesha et al. (6a), seed dormancy in eggplant was due to the presence of high levels of abscisic acid, a germination inhibitor in eggplant. They noticed low levels of cytokinins in the dormant variety compared to that of the nondormant type. This dormancy could be overcome successfully by treating the seeds with 200 ppm GA_3 (gibberellic acid).

III. BIOCHEMICAL MECHANISMS OF DORMANCY

Seed dormancy varies with external conditions, especially with time and temperature changes, both in the developing and mature seed. Vegis (7) adduced that seeds change in their temperature tolerance in the following way: As dormancy sets in (predormancy) and develops, the range of temperature over which seeds can germinate becomes narrower, until eventually it does not occur at any temperature (full dormancy). Alternatively, germination can occur over a narrow range of temperature. There is a widening of the temperature range over which germination can proceed as dormancy is lost (postdormancy). In certain cases, the ability to germinate may be retained at lower temperatures and dormancy is manifested only at higher temperatures, the seeds thus showing relative dormancy. An example of the relative dormancy of wheat in contrast to the complete dormancy of wild oat *(Avena fatua)* is shown in Figure 2. Bewley and Black (6) stated that in freshly harvested wheat, dormancy is manifested only at temperatures above about 18°C, whereas in *A. fatua*, it is present at all temperatures. Thus, dormancy may completely prevent germination at some temperatures, but only slow it down at others. The failure of a seed to germinate owing to unsuitable conditions should not be confused with the seed's inability to germinate owing to relative dormancy. The beneficial effects of priming treatments like chilling and light can easily demonstrate the relative dormancy of seeds.

Various categories of seed dormancy have been described according to the manner of their origin. Some investigators distinguish only between primary dormancy (the state of the seed as shed from the mother plant) and secondary dormancy, which is induced in a mature, imbibed seed by certain unfavorable environmental conditions for the germination. Primary dormancy (or dormancy) is divided into *innate* and *enforced* dormancy, which are inherent in the seed as it develops on the mother plant. Secondary dormancy is manifested only under certain environmental

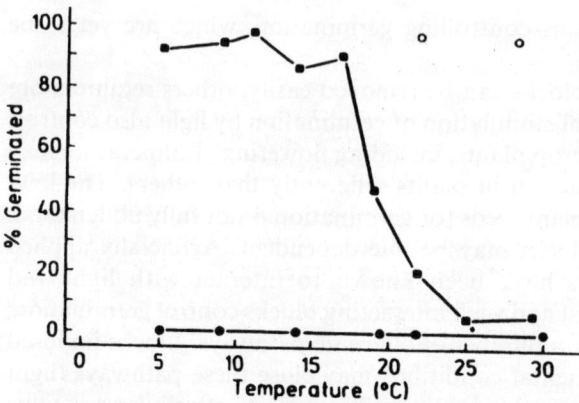

Figure 2 Relative and complete dormancy exemplified by wheat [*Triticum aestivum* (■)] and wild oat [*Avena fatua* (●)]. Germination at different temperatures was determined 16 days after the start of imbibition. To show that *A. fatua* is viable, naked caryopses were stimulated to germinate by 0.33 mM gibberellic acid (GA_3) at two temperatures (○). (*A. fatua*, Montana strain. *T. aestivum* cv. Capelle Desprez.) (From Ref. 6.)

conditions, which may be a case of relative dormancy. Bewley and Black (6) described a variety of such terms equivalent in their meaning, citing examples of clarification.

Two types of dormancy mechanisms are known: (a) coat-imposed dormancy and (b) embryo dormancy. The mechanism(s) of dormancy should be able to explain the nature of the block(s) within the seed preventing germination under apparently favorable conditions and how they operate. A full appreciation of the mechanism of dormancy eventually should also consider which environmental factors cause dormancy and what is required to break dormancy. Nikolaeva (8) and, more recently, Bewley and Black (6) have reviewed the dormancy mechanisms at length in several subdivisions. According to Bewley and Black (6), the basic dichotomy of embryo dormancy and coat-imposed dormancy is still retained, and in some species, both types may exist simultaneously or consecutively.

A. Coat-Imposed Dormancy

In many plant species, seed dormancy is imposed by the structures surrounding the embryo (seed coat), which may include glumes, palea and lemmae (grasses), the pericarp, testa, perisperm, and endosperm. The embryo in these cases when isolated experimentally from the enclosing structures is not dormant. Certain physical and chemical treatments of the seed coat also permit the embryo to germinate; for example, abrasion (scarification) perforation, or treatment with mineral acids.

Bewley and Black (6) stated the following possible mechanisms of prevention of embryo germination by the covering structures:

1. Interference with water uptake
2. Interference with gaseous exchange
3. Presence of chemical inhibitors
4. Seed coat as barrier for release of inhibitors from the embryo

5. Modification of light reaching the embryo
6. Seed coats exerting mechanical restraint

One or the combined action of two or three of these effects may be responsible for the maintenance of dormancy in seeds.

Impermeability of the seed coat to water can be an overriding cause of dormancy and delayed germination in most hard seeds of leguminous crops as well as in some members of the Cannaceae: Chenopodiaceae, Convolvulaceae, Convallariaceae, Geraniaceae, Graminae, Liliaceae, Malvaceae, and Solanaceae (9–11). In most hard-coated leguminous and other seeds, germination may be delayed owing to inadequate uptake of water by the embryo, although these seeds may be generally described as dormant. Seeds of *Cassia sericea* subjected to hot water treatment for 5 minutes or acid scarification for 3 minutes showed maximum germination (88.0 and 87.0%, respectively) as compared to the control (14.0%) (11a).

In leguminous seeds, the testa is the main barrier to the entry of water, and the outermost region of the testa, particularly the waxy cuticle, is the major impermeable layer, which can be washed out with ethanol to confer water permeability. In the species of Trifolieae, the inner regions of the testa may be responsible for restricting the entry of water (9). The testa of leguminous seeds is composed of hilum, strophiole, and micropyle; the latter often is occluded in impermeable seeds of *Phaseolous lunatus* (12) and other plant species. The hilum is also impervious to water, and under moist conditions, the fissure in the hilum is closed but tends to open only in a low relative humidity. This device plays an important role in the drying of the seed, including its testa, resulting in the establishment of its impermeability to water. The strophiole also does not conduct water unless the seed has experienced certain conditions, at which point it becomes the major site for the entry of water.

It has been observed that the removal of the covering tissues allows the embryo to germinate even in seeds in which the seed coats do not prevent entry of water. Clearly, the coat that imposes dormancy, when removed, allows the embryo to come in contact with oxygen. The covering tissues in such cases probably hinder exchanges of gases; that is, intake of oxygen and escape of carbon dioxide from the embryo. Some seeds possess mucilaginous coats (e.g., *Sinapis arvensis*), and the mucilage of the testa in such seeds may be responsible, at least partially, for the impermeability of the gaseous exchange. The testa of *Blepharis persica* has abundant mucilage, and when it is exposed to excessive amounts of water, it swells greatly to occupy completely the spaces between multicellular hairs protruding from the coat's surface. This forms an effective barrier against oxygen, leading to delayed germination. An atmosphere of oxygen overcomes the inhibitory effect of the overwetted mucilage, allowing germination (6).

The presence of phenolics in the hull has been implicated in the coat-imposed dormancy with oxygen consumption (barley). Barley testa has enzymes like peroxidase, cytochrome oxidase, and phosphoglycerate dehydrogenase, which could account for some oxygen consumption. In rice *(Oryza sativa)*, dormancy is also imposed by the hull. The naked caryopsis is known to germinate readily, and therefore intact rice grains (with hull present) can be made to germinate by subjecting them to an atmosphere of oxygen or by puncturing the glumes over the embryo. The imbibed hull takes up oxygen because of its peroxidase activity. The oxygen consumption of dormant grain hulls is about double that of nondormant seeds, which coincides with the twofold activity of peroxidase in the dormant hulls. Also, the oxygen uptake of the intact dormant grain is more than 50% higher than that of the nondormant grain, apparently due to the hull effect. Bewley and Black (6) suggested that the coat-imposed dormancy in rice and other cereals is caused by the consumption of oxygen by the peroxidases present in the hull, limiting the amount of oxygen reaching the embryo. The finding that peroxidase activity declines during the natural loss of dormancy (after-ripening) appears to be consistent with this concept.

As a consequence of the effect of the seed coats in limiting the amount of oxygen available to the embryo, the respiration needed for germination would also be affected, enforcing the embryo to remain dormant. The isolated embryos of some seeds are, however, capable of germinating under extremely low oxygen tensions and even under nitrogen. The deleterious effects of low oxygen levels (because of coat impedance) may also be attributed to factors other than reduced respiration, such as the production of inhibitors or the failure to oxidize inhibitors already present.

The seeds of many plant species contain germination inhibitors. Bewley and Black (6) cited several examples of inhibitors known to be present in the coat (Table 1). Limited evidence is available, however, to support the contention that these inhibitors present in the covering tissue might actually account for the coat's imposition and maintenance of dormancy. More rigorous tests of a correlation between the magnitude of dormancy and the level of inhibitor are needed to prove that the inhibitor is an important component of the dormancy-regulating system.

The inhibitors present in the outer coats of rice and wheat have been known to decrease as dormancy is lost during after-ripening process (13). Wheat testa contains catechins (tannins), which are precursors of the red pigment phlobaphene (14). The red-coated wheats are, therefore, thought to have high levels of inhibitory catechins and their derivatives, whereas the white wheats have low levels of these compounds. The white wheats are also highly susceptible to preharvest sprouting as compared to the red wheats. There appears to be persuasive evidence to support the thesis that dormancy in some wheat cultivars may actually be controlled by the inhibitors present in their seed coats.

Seed coats may also impair the outward diffusion of inhibitors present in the embryo, forming an impermeable barrier. Thus, the embryo is not able to germinate because of its retention of high levels of inhibitor. The isolated embryos, when wetted with water, lose their inhibitors and regain germination capacity. Dormancy in wild oats *(A. fatua),* for example, is normally imposed and maintained by the hull (lemma and paleae). It is lost when naked caryopses are held on a wet substratum, ensuring rapid germination. In contrast, if they are held under high humidity, and caryopses are able to take up the same quantity of water, dormancy is maintained (6). In the latter condition, since the hull is removed, constraints due to either mechanical or gaseous exchange cannot operate. The inhibitor present in the caryopsis, however, cannot diffuse out through the hull. The absence of surrounding water thus prevents the outward diffusion of inhibitor from the naked caryopsis. The hull in these seeds appears to form a barrier against the escape of the inhibitor.

Many light-sensitive seeds possess coat-induced dormancy; naked embryos germinate irrespective of the light/dark conditions, with only a few showing true embryo dormancy. Intact light-requiring seeds can be stimulated to germinate when a certain ratio of P_{fr} (active) and P_r (inactive) forms of phytochrome (P) is established within the embryo by the combined action of red and far-red lights; the ratio needed

Table 1 Inhibitors Present in the Coat of Some Seeds

Species	Location of inhibitor	Nature of inhibitor[a]
Acer nequndo	Pericarp	Possibly ABA
Aegilops ovata	Hull	A monoepoxylignanolide
Avena sativa	Hull	Unknown
Beta vulgaris	Pericarp	Various phenolic acids, possibly ABA, *cis*-4-cyclohexene-1, 2-dicarboxymide, high concentration of inorganic ions
Betula pubescens	Pericarp	Unknown
Bouteloua curtipendula	Glumes, lemma, palea	Possibly coumarin and its derivatives
Comptonia peregrina	Testa	Possibly ABA
Corylus avellana	Testa	ABA
Elaeagnus angustifolia	Pericarp, testa	Possibly coumarin
E. umbellata	Pericarp, testa	Possibly coumarin
Fraxinus americana	Pericarp	ABA
F. ornus	Pericarp	ABA
Hordeum vulgare	Hull	Coumarin, phenolic acids, scopoletin
Iris spp.	Endosperm	Unknown
Oryza sativa	Hull	Probably ABA
Prunus persica	Testa	ABA
Rosa canina	Pericarp, testa	ABA
R. arvensis	Pericarp, testa	ABA
Sinapsis arvensis	Testa	Unknown
Triticum spp.	Pericarp, testa	Catechin, catechin tannins, several unknowns
Zilla macroptera	Pericarp	Unknown

[a] Note all named chemicals were rigorously characterized.
Source: Ref. 6.

depends upon the plant species. Since the light must pass through the surrounding tissues (coat) enclosing the embryo, they could act as a filter, altering the proportion of P_{fr} and P_r radiation reaching the embryo. Thus, in addition to coat- induced dormancy, the embryo of light-sensitive seeds may also suffer from an inadequate light environment, resulting in dormancy (6), The coat thickness and pigmentation, as affected by the environment during the development of seeds on the mother plant, modify the coat to act as a light filter. Seeds with thick, dark coats are comparatively less responsive to light than those with thin, light-coated seeds (e.g., *Chenopodium album*).

In addition to the foregoing mechanisms, the coat might also act by exerting a mechanical restraint. The coats of many seeds and nuts (indehiscent fruits) consist of hard, tough tissues offering mechanical resistance to the growth of embryo, The endosperm of lettuce, for example, imposes dormancy by acting as a mechanical barrier against germination unless hydrolases (cellulose and pectin-hydrolyzing enzymes) are newly synthesized to weaken the mechanical barrier of these polymers, Copeland and McDonald (15) have described various methods of breaking exogenous dormancy.

Hard seededness is a serious problem in the multicut leafy vegetable *Trigonella corniculata*, which has around 50% hard seeds at the time of sowing. Different methods such as chemical scarification by immersing in concentrated sulfuric acid for 2, 5, 10, 20, 30, and 40 minutes, hot water treatment at 90°C for 5 min, and sandpaper scarification for 5 min were used to break the dormancy due to hard seededness in this species. The treated seeds were evaluated for their quality parameters to assess the impact of various dormancy-breaking treatments. Acid scarification by immersing the seeds for 10 min in concentrated H_2SO_4 was most effective in eliminating hard seededness while maintaining a seed viability of 96.5%. In this treatment, the seed quality characters were also superior as measured by seedling fresh weight, seedling vigor, coefficient of variation in seedling, fresh weight, and electrolyte leakage from treated seeds. Sandpaper scarification for 5 min was less effective, as about 25% hard seeds remained, but this treatment was able to maintain high seed quality and seed viability of 96.5%. Hot water treatment was ineffective in breaking the hard seed dormancy (15a).

B. Embryo Dormancy

The failure of an isolated (viable and mature) embryo to germinate constitutes a case of embryo dormancy. Naked embryos, when placed on a wet substratum, remain dormant under the conditions suitable for germination. This dormancy may be due to either a deficiency in the axis or some metabolic blocks within cotyledons. The embryos of the freshly harvested seeds of *Xanthium pennsylvanicum* are dormant, whereas 1-year-old seeds of the same species are not, indicating the existence of some metabolic deficiency in the cotyledons and in the axes of the dormant embryos (16). Seeds of the plant species from Rosaceae (woody trees) suffer greatly from embryo dormancy. In apple *(Malus sylvestris)*, the intensity and magnitude of embryo dormancy may depend upon variety, provenance, year of harvest, and other factors.

Physiological dwarfism is experienced by some species in which the isolated embryo can germinate but does so very sluggishly to produce slowly growing dwarf seedlings. Chilling or gibberellins may completely break embryo dormancy and also convert physiological dwarfs into normal plants (6).

The control of embryo dormancy involves both the cotyledons and germination inhibitors. In many plant species, the cotyledons may be responsible for inhibiting the growth of the axis in dormant embryos. The dormancy of such seeds is broken by removing one or both the cotyledons from the isolated dormant embryos. The embryonic axis of the dormant isolated embryo can be made to germinate by excising its scutellum (modified cotyledon). In apple, the degree of dormancy appears to be a function of the amount of cotyledon left attached to the embryo, suggesting that the dormancy of the axis in the intact embryo, in these cases, is maintained by some inhibitory action of the cotyledon(s). In both epicotyl dormancy and physiological dwarfism, the cotyledons exert a similar inhibitory influence over the axis. The cotyledons also seem to be inducing secondary dormancy. When isolated embryos of nondormant apple seeds, for example, are held under unfavorable conditions for germination, a secondary dormancy is induced, which does not occur in embryos from which portions of cotyledons have been removed (6).

The physiological and biochemical basis of cotyledon action in embryo dormancy is not fully understood. The presence of inhibitors such as abscisic acid in cotyledons is thought to be involved. These chemical inhibitors are leached out of cotyledons resting on the wet substratum, inducing germination. Enlargement and greening are prevented in cotyledons wetted with a solution of abscisic acid. This observation suggests that abscisic acid–type inhibitors present in cotyledons may be controlling the growth of the axis.

Experimental evidence exists to show that:

1. Inhibitors are found in embryos of many plant species possessing embryo dormancy.
2. Leaching out of the inhibitors promotes germination of the isolated, dormant embryos.
3. Embryo dormancy can be induced by treating nondormant embryos with a known inhibitor.
4. Treatments that break dormancy in some cases can cause a drop in the inhibitor level in the embryo.

Abscisic acid (ABA) has been found to be the most predominant inhibitor in several species. The extent of embryo dormancy can be correlated with the concentration of ABA present. Prolonged washing (leaching) can induce germination of the dormant embryos, probably by leaching out of the inhibitors (ABA) from the dormant embryos.

Embryo dormancy may also be related to its morphological immaturity. Such embryos do not germinate when isolated, but may require a period of further development to be able to germinate. Immature embryos are characterized by their small size and poor differentiation, needing further growth and development before being ready for germination. The developmental period may, in some cases, be followed by a period of embryo dormancy broken by chilling. Development of immature embryos within the seeds of *Heracleum sphondylium* best occurs at low temperatures (2–3°C) over a period of 12–15 weeks. Light may also support embryo maturation in some species (alpine), wherein continuous light or several long daily photoperiods encourage embryo growth and its subsequent emergence from the seed (6).

Environmental conditions during seed development and maturation such as day length experienced by the mother plant, position of the seed on the mother plant, and age of the mother plant at the time of flower induction influence seed dormancy. Plants differ in their response to a photoperiod and their expression of seed dormancy (17–19). For example, longer days seem to promote the germination of *Polypogon mospeliensis* (Table 2). The data presented in Table 3 show that seeds from primary umbels of three cultivars of celery *(Apium graveolens)* are heavier, more mature, and more dormant than those produced elsewhere on the plant (20).

Kermode (21) described the regulatory mechanisms in the transition from seed development to germination. Embryo development occurs under the controlling influence and protection of the maternal environment. The interactions between the embryo and the surrounding tissues (seed environment), which are in intimate contact with the maternal sporophyte (environment), modulate and control the course of embryogeny. Important cues in regulating embryogenesis may be generated both

Table 2 Influence of Day Length Under Greenhouse and Outdoor Conditions on *Polypogon mospeliensis* During Growth and Seed Maturation and Germinability

Day length during growth of mother plant and seed maturation (h)	Seeds from greenhouse plants, germination (%)		Seeds from outdoor plants, germination (%)	
	3 days	7 days	3 days	7 days
9.0	0.0	0.0	1.5	10.0
11.0	0.0	0.0	9.5	17.0
12.0	17.5	19.0	64.0	86.5
13.5	38.0	44.5	98.5	98.5
15.0	60.5	66.0	97.0	98.5
18.0	90.5	91.0	98.0	99.0
Control (natural day length)	93.0	96.0	91.5	93.0

Source: Ref. 17.

internally by the embryo and externally from the surrounding seed tissues and other plant parts.

IV. INVOLVEMENT OF HORMONES AND GROWTH REGULATORS

It is hypothesized that seed dormancy is regulated by the interaction of growth promotors and inhibitors, with the latter in particular controlling the onset and maintenance of dormancy. ABA is one of the most important naturally occurring

Table 3 Seed Position of Umbel, Weight (g), and Germination (%) After 21 Days at 18°C in Light in Three Celery Cultivars

Cultivar	Umbel position	Mean seed weight (mg)	Germination (%)
Green snap	Primary	0.590	51
	Secondary	0.440	85
	Tertiery	0.386	94
	Quaternary	0.382	80
		(0.069)[a]	(9.8)
Lathom blanching	Primary	0.474	50
	Secondary	0.438	72
	Tertiery	0.380	94
	Quarternary	0.348	82
		(0.069)	(9.2)
Ely white	Primary	0.590	59
	Secondary	0.468	62
	Tertiery	0.490	80
	Quaternary	0.520	87
		(0.086)	(7.3)

[a] LSD at 5% in parenthesis.
Source: Ref. 20.

inhibitors inducing dormancy. It is specifically involved with the onset and maintenance of dormancy. The kinetics of the production and disappearance of the inhibitor
during seed development of dormant and nondormant plant species have established
the role of ABA in dormancy. ABA, which appears in grains during their development,
is thought to be retained into maturity by dormant varieties but not by nondormant
ones. Balboa-Zavala and Dennis (22) monitored the germination ability of isolated
embryos of apple seeds during seed development and found that neither free nor
bound ABA levels in the embryonic axes and cotyledons showed any consistent
correlation with the acquisition of dormancy (Fig. 3). The germination capacity of the
embryos dropped to zero before an increase in ABA concentration occurred (see
Fig. 2). Such studies question the universal role of ABA in the induction of dormancy.
Bewley and Black (6) have cautioned about drawing general conclusions based on

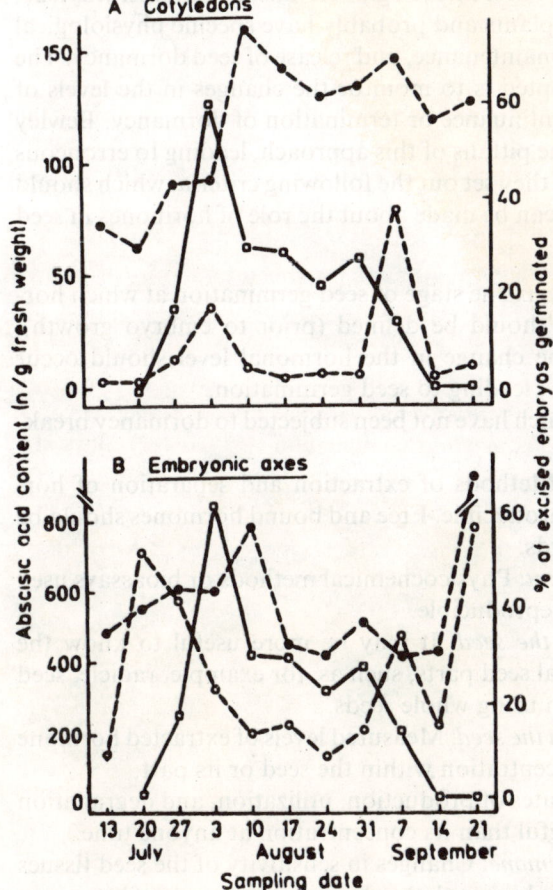

Figure 3 Extractable levels of ABA and the development of dormancy in apple embryos.
Seeds were collected from McIntosh apples during their development. Free (○) and bound (●)
abscisic acid was extracted from (A) cotyledons and (B) embryonic axes and determined by gas
liquid chromatography and electron capture detection. Germination (□) of isolated embryos
of different ages was tested at 20°C. (From Ref. 22.)

these experiments, which reveal no significant changes involving the action of endogenous inhibitors. The dormancy may be induced by an increase in the embryo's sensitivity to the inhibitor. On this basis, the total and extractable inhibitor need not change at all, yet it could play a role in the inception of dormancy. Attempts to establish the role of inhibitors and other hormones in the induction of dormancy by correlating total extractable growth substance with the plant's physiological behavior have not been proved to be rewarding. Promotive hormones such as gibberellins and cytokinins counteract the action of inhibitors. It is, therefore, possible that dormancy in the developing seeds decreases. The onset of dormancy in some plant species (*A. fatua* and *A. ludoviciana*) has been prevented by treating mother plants with gibberellic acid (GA) during seed development. The treatment presumably obviates dormancy by artificially elevating the level of promoting hormones (GA) in the maturing caryopses.

The mechanisms of dormancy and its release involve the action of inhibitor(s) and the growth-promoting hormones. These include gibberellins, cytokinins, ethylene, and ABA, which occur commonly in plants and probably have specific physiological and biochemical effects on induction, maintenance, and release of seed dormancy. The experimental approach generally adopted is to monitor the changes in the levels of extractable hormone in relation to continuance or termination of dormancy. Bewley and Black (6) have cautioned about the pitfalls of this approach, leading to erroneous interpretations of the results. Further, they set out the following criteria, which should be met before acceptable conclusions can be made about the role of hormones in seed dormancy.

1. *The time course of germination*: The stage of seed germination at which hormone contents are determined should be defined (prior to embryo growth), excluding germinated seeds. The change in the hormonal level should occur during termination of dormancy, leading to seed germination.
2. *Controls*: Dormant seeds, which have not been subjected to dormancy breaking treatment, must be included.
3. *Extraction and separation*: Methods of extraction and separation of hormones should be reliable and reproducible. Free and bound hormones should be examined using internal standards.
4. *Identification and measurement*: Physicochemical methods or bioassays used must be specific, sensitive, and reproducible.
5. *Occurrence of hormones in the seed*: It may be more useful to know the hormonal status of the individual seed parts, such as, for example, radicle, seed coat and cotyledons, rather than using whole seeds.
6. *Concentration of hormones in the seed*: Measured levels of extracted hormone should be transformed into concentration within the seed or its part.
7. *Turnover of hormones*: The rates of production, utilization, and degradation of a hormone are more meaningful than its concentration at anyone time.
8. *Sensitivity to endogenous hormone*: Changes in sensitivity of the seed tissues to its own endogenous hormone, along with the changes in amounts of hormone, will be more meaningful.

According to the generally accepted hormonal theory of seed dormancy, the latter is imposed and maintained by the inhibitor(s), which may decline during the disappearance of dormancy. Promoters are responsible for the release of dormancy. Thus, light, chilling, and other factors may break seed dormancy by causing a drop in the inhibitor level and a rise in the level of promoters (6).

Investigators have tried to elucidate the mechanisms of seed dormancy by studying the application of growth substances to seeds to release them from dormancy and promote germination. Since all active growth regulators (gibberellins, cytokinins, and ethylene) are found endogenously in seeds, it is argued that if these chemicals are active when supplied to the seed, the endogenously present hormones should act in a similar manner. It has been suggested that seed dormancy is controlled by the balance of promoters and inhibitors (23–25).

Gibberellins have been found to release dormancy in seeds of several plant species. They also accelerate germination of nondormant seeds, and even among dormant seeds, both types (i.e., seeds with coat-imposed dormancy [e.g., barley, lettuce] and embryo-imposed dormancy [wild oats]) are promoted. The effective physiological concentrations vary from 10^{-5} to 10^{-3} M and a mixture of GA_4 and GA_7 rather than GA_3 (gibberellic acid) has been found to be effective.

Miller (26) investigated the effects of cytokinins on the germination of lettuce seeds. The dormancy of lettuce seeds could be overcome by supplying kinetin in darkness. However, kinetin was found to be effective only in the presence of low levels of light, which in themselves were ineffective to promote germination. In complete darkness, kinetin and other cytokinins could break the dormancy of only a small percentage of lettuce seeds, with abnormal emergence of embryo and cotyledons growing first. According to Thomas (27), cytokinins act by counteracting the effect of inhibitors, especially ABA, when applied simultaneously (Fig. 4). The dormancy-breaking

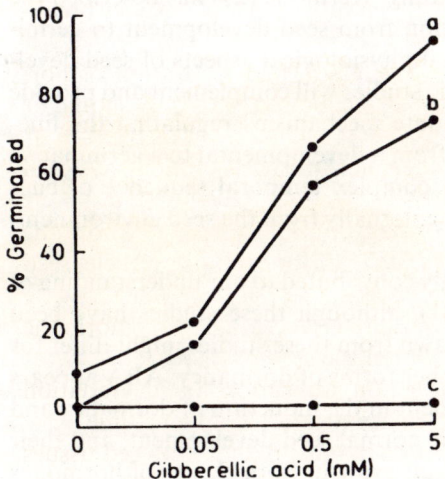

Figure 4 Effects of applied kinetin, abscisic acid, and gibberellic acid on germination of lettuce (cv. Grand Rapids) in darkness at 25°C [a = gibberellic acid; b = gibberellic acid + abscisic acid (0.04 mM) + cytokinin (0.05 mM); c = gibberellic acid + abscisic acid (0.04 mM)]. (From Ref. 27.)

effect of GA is prevented by ABA, whose inhibition is overcome by kinetin. Cytokinins are thus thought to have a "permissive" role, allowing a second promoter (GA) to act. Cytokinins also reverse the action of ABA on isolated lettuce embryos in the absence of gibberellins, which are required only by the intact seed.

Ethylene effectively breaks the dormancy of many seeds when supplied either directly as a gas or in the form of ethylene-releasing substances like Ethrel. The effective concentrations range from 0.1 to 200 µL/L. Taylorson (28) examined 43 species of weed seeds and found that 32 species showed no response to ethylene. Several responsive species demonstrated interactions between light and ethylene.

Kermode (21) pointed out limitations in our understanding of the precise roles played by ABA and the osmotic potential within the developing seed. It is difficult to conclude that control over maturation and germinability rests universally with these "cues." Important advances have resulted from a movement away from traditional approaches to a study of regulatory factors in seed development. The pathways of ABA biosynthesis and degradation and developmental changes in metabolism both in embryo and maternal seed tissues are being elucidated. The factors involved in the partitioning of ABA, particularly in the control of ABA movement from the maternal tissues (seed coat) to the embryo, are also becoming understood.

The relationship between the regulation by ABA and a low osmotic potential is complex. ABA and osmotic potential in some cases may act synergistically, whereas in other instances, they seem to operate through separate pathways eliciting different metabolic effects (21). Desiccation, whether natural or imposed, plays an important role in switching seeds from a developmental mode to one that is essential to promote germination. Premature drying may permanently redirect metabolism from a developmental to a germination mode, effecting a normal temporally regulated "switch." It is not known how, at the molecular level, drying leads to suppression of the synthesis of developmental proteins and the induction of synthesis of proteins required for the germination and subsequent growth of the seedling. Kermode (29) has described the regulatory mechanisms involved in the transition from seed development to germination. Bewley (30) reviewed some fundamental physiological aspects of seed development. According to Kermode (21), these basic studies will complement and provide the foundation for molecular studies to elucidate mechanisms regulating the fine-tuned temporal changeover in gene expression from a developmental to a germinative mode. A process, which probably involves a complex temporal sequence of cues generated both internally from the embryo and externally from the seed environment, remains to be unraveled.

The study of hormone mutants has strongly contributed to our understanding of the control of dormancy and germination (31), although these studies have been restricted only to a few species. Conclusions drawn from these studies might differ for seeds with embryo dormancy and those with other forms of dormancy; ABA appears to play an important role during seed development in the induction of dormancy and desiccation tolerance. GAs are not required for normal seed development, and their role appears to be restricted to the control of fruit growth. Both classes of hormones are directly involved in the regulation of dormancy in the mature seed. Dormancy is actually controlled at the level of availability or activity of the receptor for P_{fr}. Nitrate might be a cofactor at that level. An active P_{fr}-receptor complex is required to start the signal transduction pathway leading to germination. GA synthesis and sensitivity to GAs are essential parts of this pathway. Seed germination depends on the presence of endogenous or exogenously applied GA and may be inhibited by the presence of ABA (31).

V. THE BIOLOGICAL SIGNIFICANCE OF DORMANCY

The understanding of the complexities of the mechanisms of seed dormancy is of great biological significance. An understanding of the phenomenon will lead to control of seed germination, which is of enormous practical importance (e.g., control of weeds). Seed germination is a critical stage in the life of each crop plant, affecting both the vigor and stand of the crop and the final crop yield. Understanding the natural mechanism(s) controlling germination would provide a scientific basis for raising better crops. Knowledge of dormancy mechanisms would enable us to cause or prevent germination at will and possibly control the storage life of seeds. An understanding of the mechanisms of germination blocks in seeds could also contribute to a larger understanding and control of growth and development in other organisms.

According to Bewley and Black (6), dormancy is a device for optimizing the distribution of germination in time or space, having great ecological importance. Extension of germination over an extended period of time occurs because seeds of many species show variability in the depth of dormancy, with the population exhibiting sporadic release from dormancy and irregular germination. The temporal dispersal thus enhances the spread and survival of the species. Seasonal environmental changes (light and temperature) ultimately regulate seed dormancy and germination, controlling the seed dispersal.

Four basic patterns of germination have been recognized in relation to its temporal distribution (32):

1. Quasisimultaneous, when germination of all the seeds occurs over a relatively brief period
2. Continuous, in which seeds germinate over an extended period of time
3. Intermittent, irregular germination over long periods of time (multimodal distribution)
4. Periodic multimodal germination showing more regular periodicity

These patterns result from the dormancy characteristics and from their interplay with various environmental factors like temperature. Dormancy is also important for preventing vivipary and precocious germination; that is, germination while the seed is on the mother plant, which is important in certain crop species like cereals. Eira and Caldas (33) presented an overview of dormancy in seeds from physiological and ecological viewpoints and discussed the intimate connection between dormancy and seed development and germination.

The development of primary dormancy in seeds of *P. avium* during maturation was studied in selected trees in 2 consecutive years (1994 and 1995). Seeds without endocarp were prechilled and germinated at 4°C, and the mean time to complete germination was recorded as a relative, quantitative measure of dormancy level. Corresponding levels of germination capacity, dry weight, and moisture content of the seeds were determined. Changes in dry weight and moisture content followed the behavior of general seed development. Seed acquired a maximum ability to germinate 4–6 weeks before full maturity, at which point they began to germinate after a few weeks at 4°C, thus displaying a shallow dormancy. Generally, dormancy levels increased with increasing maturity at rates of up to 4 weeks deeper dormancy per week during maturation. Dormancy levels are suggested to reach a final plateau. Both induction rate and final level depend on the tree and year. The levels of dormancy were different in the 2 years for the same dry weight and moisture content of seeds, which suggests that other factors affect dormancy induction significantly. Environmental factors, especially high temperatures, are hypothesized to be important for dormancy induction (34).

Shoenia filifolia subsp. *subulifolia* has an innate dormancy period that prevents germination during summer thundershower events. High-temperature dry storage or applications of nitrate or gibberellic acid (GA_3 at 30 µM) shortened the required after-ripening period. Combinations of KNO_3 (10 mM) and GA_3 had an additive effect on germination. In addition to KNO_3, Ca $(NO_3)_2$ (5 mM) and K_3 [Fe $(CN)_6$] (10 mM) also increased germination percentages. However, NH_4Cl (10 mM), KNO_3 (10 mM), $NaNO_3$ (0.5 mM), and a 1% concentration of smoke water (Kings Park and Botanic Garden Smoky Seed Starter) showed no stimulatory effect on germination in this arid-zone ephemeral. The solution of combustion products did not affect 27-week-old seeds when grown in the light, but the light requirement tended to be overcome by gradually increasing concentrations of smoke water (0, 0.02, 0.1, 0.2, 1, 10, or 100% v/v). The after-ripening requirement, nitrate, GA, and the P_{fr} form of phytochrome may be linked in *S. filifolia* subsp. *subulifolia* via their association with cellular membranes. The capacity of smoke water also to stimulate germination in the dark may indicate that the active ingredient in smoke could be a nitrous oxide–generating compound. Seeds with sufficient levels of GA_3 to sense light and/or increased levels of water-soluble nitrogenous compounds can germinate with the first winter rains in open-soil sites of this Mediterranean-type climatic region of western Australia (35).

Hitchmough et al. (36) investigated the efficacy of a range of presowing treatments in overcoming dormancy in wild collected and cultivated genotypes of *Trollius europaeus* (Globeflower) and the cultivated garden hybrid T. Golden Queen to identify effective, yet practical treatments for pretreating *T. europaeus* seed prior to sowing with other native forbs and grasses in wet wildflower meadow reconstruction projects. T. Golden Queen was included in initial experiments as a contrast genotype whose dormancy and germination are better understood than that of *T. europaeus*. Seed of wild collected *T. europaeus* proved to be the most dormant genotype, with T. Golden Queen the least dormant. The only genotype to germinate following sowing of untreated seed at 20°C was T. Golden Queen. Following initial intergenotype comparisons, a range of dormancy breaking treatments including wet prechilling, prewashing, chilling, freezing, warm storage, and GA presoaks were applied to wild collected *T. europaeus* seed. The most effective treatment for breaking dormancy was to prewash the seed in running tap water for 7 days then presoak for 24 hr in GA_{4+7} or GA_3 at 10^{-4} M and 10^{-1} M, respectively. With these latter treatments, germination commences 3–6 days after sowing at 20°C. By day 12, 75–85% of sown seed had germinated. Given its low cost and the ease of undertaking, prewashing followed by soaking in GA_3 at between 10^{-2} M and 10^{-1} M is proposed as an effective dormancy-breaking treatment for *T. europaeus* seed prior to field sowing in spring or early summer.

REFERENCES

1. B. M. Pollock and V. K. Toole, After-ripening, rest period and dormancy, *Seeds: The Yearbook of Agriculture*, U.S. Department of Agriculture, Washington, DC, 1961, p. 106.

2. V. R. Boswell, Flowering habit and production of seeds, *Seeds: The Yearbook of Agriculture*, U.S. Department of Agriculture, Washington, DC, 1961, p. 57.
3. C. M. Karseen and H. W. M. Hilhorst, Effect of chemical environment on seed germination, *Seeds: The Ecology of Regeneration in Plant Communities* (M. Fenner, ed.), C.A.B. International, Wallingford, UK, 1992, p. 327.
4. C. M. Karseen, Seasonal patterns of dormancy in weeds, *The Physiology and Biochemistry of Seed Development Dormancy and Germination* (A. A. Khan, ed.), Elsevier, Amsterdam, 1982, p. 243.
5. H. J. Bouwmeister, The effect of environmental conditions on the seasonal dormancy and germination of weed seeds, Ph.D. dissertation, Agricultural University, Wageningen, The Netherlands.
6. J. D. Bewley and M. Black, *Physiology and Biochemistry of Seeds in Relation to Germination*, Springer-Verlag, Berlin, 1982.
6a. H. S. Yogeesha, K. K. Upreti, K. Bhunuprakash and G. S. R. Murti, Seed dormancy in eggplant (*Solanum melongena* L.) var. Arka Neelam, XI National Seminar on Quality Seed to Enhance Agricultural Productivity, UAS, Dharwar, Jan 18–20, 2002, *Seed Tech News* 32: (A. Gaur, A.K. Vari, J.L. Varshney, K. Kant, eds.), Indian Society Seed Technology, Division of Seed Science and Technology, IARI, New Delhi, March 2002, p. 113.
7. A. Vegis, Dormancy in higher plants, *Ann. R. Plant Physiol.* 15:185, (1964).
8. M. G. Nikolaeva, *Physiology of Deep Dormancy in Seeds*. Leningrad, Izdakel 'Stvo Nauka, Isr. Prog. Sci. Transl. Jeruslem, 1969.
9. L. A. T. Ballard, Physical barriers to germination, *Seed Sci. Technol.* 1,285, (1973).
10. L. V. Barton, Seed dormancy: General survey of dormancy types in seeds and dormancy imposed by external agents, *Encyclopedia of Plant Physiology*, Vol. 15/2 (W. Ruhland, ed.) p. 909, Springer-Verlag, Berlin, 1965, p. 909.
11. M. P. Rolston, Water impermeable seed dormancy, *Bot. Rev.* 44:365, (1978).
11a. D. S. Uppar, N. K. Biradar Patil, M. Shekhargouda and P. N. Umapathy, Studies on seed dormancy in *Cassia sericea*, XI National Seminar on Quality Seed to Enhance Agricultural Productivity, UAS, Dharwar, Jan 18–20, 2002, *Seed Technol. News* 32 (A. Vari, A. K. Vari, J. L. Varshney and K. Kant, eds.), Indian Society Seed Technology, Division of Seed Science and Technology, IARI, New Delhi, March 2002, p. 121.
12. W. Steinswat, L. H. Pollard and W. F. Campbell, Nature of hard seediness in lima beans *(Phaseolus lunatus* L.), *J. Am. Soc. Hort. Sci.* 96:312, (1971).
13. T. Miyamoto, N. E. Tolbert and E. H. Everson, Germination inhibitors related to dormancy in wheat seeds, *Plant Physiol.* 36:739, (1961).
14. T. Miyamoto and E. H. Everson, Biochemical and physiological studies of wheat seed pigmentation, *Agron. J.* 50:733, (1958).
15. L. O. Copeland and M. B. McDonald, *Principles of Seed Science and Technology*, 3rd ed. Chapman & Hall, New York, 1995.
16. Y. Esashi, H. Katoh, Y. Hafa and N. Goto, Dormancy and impotency of cocklebur seeds, *Plant J Physiol.* 59:122, (1977).
17. Y. Gutterman, Phenotypic maternal effect of photoperiod on seed germination, *The Physiology and Biochemistry of Seed Development, Dormancy and Germination* (A. A. Khan, ed.), Elsevier, Amsterdam, 1982, p. 67.
18. Y. Gutterman, Maternal effects on seeds during development, *Seeds: The Ecology of Regeneration in Plant Communities* (M. Fenner, ed.), C.A.B. International, Wallingford, UK, 1992, p. 27.
19. Y. Gutterman, *Seed Germination in Desert Plants*, Springer-Verlag, Berlin, 1993.
20. T. H. Thomas, N. L. Biddington and D. F. O'Toole, Relationship between position on the parent plant and dormancy characteristics of seeds of three cultivars of celery (*Apium graveolens*), *Physiol. Plant.* 45:492, (1979).
21. A. R. Kermóde, Regulatory mechanisms in the transition from seed development to

germination: Interactions between the embryo and the seed environment, *Seed Development and Germination* (J. Kigel and G. Galili, eds.), Marcel Dekker, New York, 1995, p. 273.

22. O. Balboa-Zavala and F. G. Dennis, Abscisic acid and apple seed dormancy, *J. Am. Soc. Hort. Sci.* 102:633, (1977).

23. R. D. Amen, A model of seed dormancy, *Bot. Rev.* 34:1, (1968).

24. R. C. Jann and R. D. Amen, What is germination? *The Physiology and Biology of Seed Dormancy and Germination* (A. A. Khan, ed.), North Holland, Amsterdam, 1982.

25. P. F. Wareing and P. F. Saunders, Hormones and dormancy, *Ann. R. Plant Physiol.* 22:261, (1971).

26. C. O. Miller, Effects of cytokinins on germination of lettuce seeds, *Plant Physiol.* 31:318, (1956).

27. T. H. Thomas, Cytokinins, cytokinin-active compounds and seed germination, *The Physiology and Biochemistry of Seed Dormancy and Germination* (A. A. Khan, ed.), North Holland, Amsterdam, 1977, p. 3.

28. R. B. Taylorson, Response of weed seeds to ethylene and related hydrocarbons, *Weed Sci.* 27:7, (1979).

29. A. R. Kermode, Regulatory mechanisms involved in the transition from seed development to germination, *CRC Crit. Rev. Plant Sci.* 9:155, (1990).

30. J. D. Bewley, Challenges in seed physiology, challenges for the second decade, *Proceedings of the Tenth Annual Seed Technology Conf.* (1. S. Bums ed.), Iowa State University Press, Ames, 1955, p. 51.

31. C. M. Karseen, Hormonal regulation of seed development, dormancy and germination studied by genetic control, *Seed Development and Germination* (J. Kigel and G. Galili, eds.) p. 333, Marcel Dekker, New York, 1995, p. 333.

32. E. Salisbury, *Weeds and Aliens*, Collins, London, 1961.

33. M. T. S. Eira and L. S. Caldas, Seed dormancy and germination as concurrent process, Seed Physiology Papers—Presented at the VII Brazilian Plant Physiology Congress, Brasilia, July 1999. *Seed Abstr.* 25:211 (2002).

34. M. Jensen and E. M. Eriksen, Development of primary dormancy in seeds of *Prunus avium* during maturation, *Seed Sci. Technol.* 29:307, (2001).

35. J. A. Plumer, A. D. Rogers, D. W. Turner and D. T. Bell, Light, nitrogenous compounds, smoke and GA3 break dormancy and enhance germination in the Australian everlasting daisy, *Shoenia filifolia*, subsp. subulifolia, *Seed Sci. Technol.* 29:321, (2001).

36. J. D. Hitchmough, J. Gough and B. Corr, Germination and dormancy in a wild collected genotype of *Trollius europaeus*, *Seed Sci. Technol.* 28:549, (2000).

I. INTRODUCTION

Germination may be defined as an emergence of embryo from the seed by starting a variety of anabolic and catabolic activities, including respiration, protein synthesis, and mobilization of food reserves after it has absorbed water. To the seed analyst, germination means the emergence and development from the seed embryo of the essential structures that indicate the seed's ability to produce a normal plant under favorable conditions (1). The presence of oxygen is necessary to allow some aerobic respiration, as is a temperature suitable to permit various metabolic processes to proceed. Seeds of many plant species, however, fail to germinate in spite of favorable conditions owing to dormancy.

The mobilization of food reserves is not strictly a component of germination, but only a uniquely associated aspect. An individual seed may or may not have the ability to germinate, but for a seed population, it is possible to express its capacity to germinate (germinability) in terms of the percentage of seeds that germinate under favorable conditions. Bewley and Black (2) distinguished seed germinability from germination rate; the latter is defined as the germination percentage obtained after a certain time under stipulated conditions, which may or may not be optimal (e.g., in the presence of certain chemicals or at a certain temperature). This is the way in which germination is commonly measured. The germination rate is the reciprocal of the time (t) to germination, which can be expressed for a single seed, for a population, or for a certain fraction of that population (50%). Germination thus can be divided into two parts: (a) biochemical preparative processes and (b) emergence of the embryo itself. The expression of the final germination percentage indicates the proportion of seeds reaching the final stage of emergence, but it reveals nothing about the time taken to reach this stage.

II. BIOCHEMISTRY AND PHYSIOLOGY OF SEED GERMINATION

In 1968, Amen put forth the first model of the sequence of biochemical events occurring during seed germination. This was only a generalized scheme because of the specific needs of individual plant species, with some requiring light, whereas others need cold temperatures to induce germination. Amen's hypothetical model highlights the many events that must occur during germination. According to Amen (3), seed germination is controlled by a balance between growth inhibitors and growth promotors. If inhibitors are present in physiologically greater concentrations than promotors, dormancy is ensured; a triggering agent like light or temperature is required either to inactivate or decrease the level of inhibitors in seeds. When this occurs, a germination stimulant such as gibberellic acid can exert its promotive influence to initiate and bring about the process of germination. Gibberellins increase the de novo synthesis of hydrolytic enzymes like (α-amylase and ribonuclease and activate release of β-amylases, which in turn hydrolyze starch, providing the necessary respiratory substrates to yield the energy required for germination. Ribonucleases hydrolyze nucleic acids, the products of which are used to recode new RNA species needed during the later stages of seed germination. The proteolytic enzymes and cellulases together degrade cell walls, a step essential in the loosening of the seed coat prior to radicle protrusion (Fig. 1). Muscolo et al. (3a) recently hypothesized that humic substances inhibit seed germination by affecting metabolic processes, and in particular the tricarboxylic acid (TCA) cycle that provides the carbon skeletons necessary to synthesize amino acids. Soltani et al. (3b) determined the effects of seed size, salinity, and their interaction effect on germination and heterotrophic seedling growth of chickpea cultivars. The use of large seed would be an advantage in producing more vigorous seedlings of chickpea in a nonsaline condition. Under saline conditions, however, seed grading does result in more vigorous seedlings.

Schuab et al (3c) conducted an experiment to evaluate the effect of p-coumaric acid (p-CA) on germination and vigor of soyabean seeds and to quantify total lipids and total soluble proteins of Embrapa-58 variety. The seeds were laid out on three sheets of towel paper and treated with different phenolic acid concentrations (0, 0.1, 1.0, 5.0, and 10.0 mM). These sheets were then made into rolls and placed in a Mangelsdorf germinator at 25°C for 8 days. The seeds were evaluated by three different means: a germination test, radicle and hypocotyl lengths, as well as their fresh and dry biomass. For the biochemistry evaluations, the seeds were germinated for a 7-day period, after which, the seedlings were dried at 80°C for 24 hr, and then, the lipid content of cotyledons and proteins in the three components of seedlings were assessed. The results showed that there were no significant differences in terms of cumulative seed germination in the presence of the compound studied, although a delay was

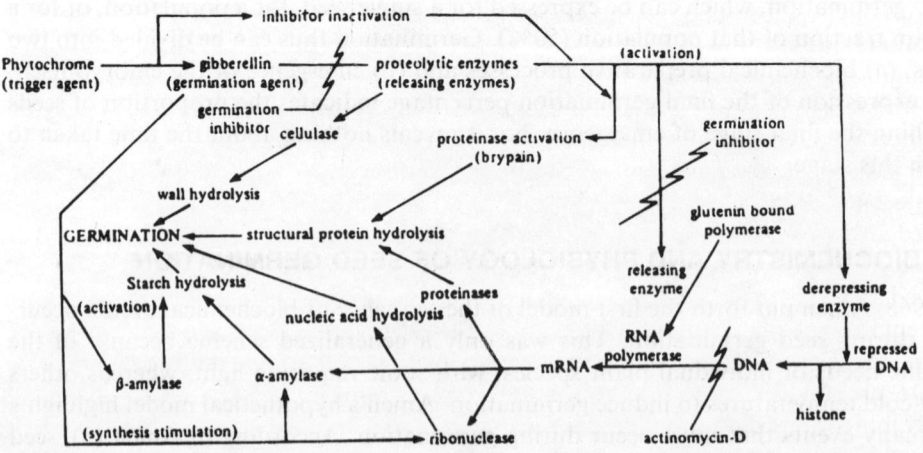

Figure 1 Possible biochemical pathways in the initiation of seed germination. (From Ref. 3.)

observed in the germinative process. However, in the seed germination test, a great reduction in normal seedlings was observed. In relation to the growth evaluation of seedlings, a great reduction in the length of radicles was verified. As for the biomass evaluation, a decrease of fresh and dry biomass, mainly of radicles, was observed. In the biochemistry evaluations, an increase of the lipids and protein content was found out in the presence of p-CA. Although the action mechanisms of these allelochemicals are not completely elucidated, the results showed alterations not only in the seedlings morphology, especially in the roots, but also in the metabolism of proteins and in the mobilization of lipids.

Some glucosinolate enzyme hydrolysis–derived products were chemically characterized and studied to investigate their inhibitory activity on seed germination (3d). The glucosinolates epi-progoitrin, glucoerucin, and glucoraphenin were isolated from ripe seeds of *Crambe abyssinica*, *Eruca sativa* (*E. vesicaria*), and *Raphanus sativus*, respectively. The effects of hydrolysis products on seed germination of the weeds *Chenopodium album*, *Portulaca oleracea*, and *Echinochloa crus-galli* and lettuce (*Lactuca sativa*) cv. Romana (control) were evaluated under controlled conditions. A total inhibition of seed germination produced by the hydrolysis products of glucoerucin at a concentration of 10 mg/mL^{-1} of native glucosinolate was evident. The hydrolysis products of glucoraphenin showed similar activity. The only exception was *E. crusgalli*, which demonstrated germination values significantly higher than those of the other species. The hydrolysis products of epi-progoitrin, mainly 5-vinyloxazolidine-2-thione, gave a high percentage of abnormal seedlings even at a concentration of 1 mg/mL^{-1} of native glucosinolate.

Catharanthus roseus cv. Alba seeds exhibited only 40% germination in dark at 25°C and removal of the seed coat did not produce any significant effect; 100% germination was found in decoated seeds at a GA$_3$ concentration as low as 0.144 mM, whereas 2.886 mM was necessary for coated ones (3e). Concentrations as high as 2.886 mM inhibited the germination when made available to decoated seeds. A similar effect was noted in experiments with different concentrations of KNO$_3$. H$_2$SO$_4$ of 9(N) produced 81% germination in coated seeds with only 5 min of pretreatment, but inhibition was noticed with a further increase of pretreatment time. As revealed from scanning electron microscopy (SEM), H$_2$SO$_4$ produced cracks at the hilar region and degraded the cuticular layer. But such an effect did not appear to be favorable for increasing germination, since removal of the seed coat was of no effect. It therefore appears that GA$_3$ and KNO$_3$, because of their acidic property, lowered the internal seed tissue pH and increased the germination percentage. Different buffers, which are nonconventional in the field of germination as promoters, produced 78% (acetate, pH 4.5), 59% (citrate, pH 5), and 69% (phosphate, pH 5) germination, respectively. Both respiratory (dehydrogenases) and hydrolytic (amylases and proteases) enzymes showed enhanced activity either in promoters or in acid immersion or in buffer, all of which showed a pH in between the acidic range the neutral (control). 10^{-5} M of BAS 111.W, a potent inhibitor of GA biosynthesis, inhibited germination but was not inhibitory to normal seedling growth. Retardant application within 24 hr in acid immersion, 48 hr in KNO$_3$, phosphate and acetate buffer pretreatment, and 120 hr in control (DW) distinctly inhibited germination. It therefore appears that maintenance of an acidic pH of the seed tissue for a specific duration and concomitant synthesis of GA$_3$ was a prerequisite for visible germination of *C. roseus* cv. Alba seeds.

A. Water Uptake

Rehydration of seeds is an initial essential step toward germination. The quantity of water taken up by seeds is generally small (not exceeding two to three times the seed dry weight) and depends, for example, on the size of the seed and the hydratability of the contents. Substantial quantities of water are required for the establishment and subsequent growth of the seedling. Bewley and Black (2) described seed germination and growth, considering two major factors: (a) water requirements of the seed and (b) the relationship between the seed and its substrate (the soil).

The magnitude of rehydration depends on the force with which water is held in the soil and the amount of work to be performed by the seed to remove it. The driving force for the movement of water is expressed in water potential (Ψ). The cell (seed) will absorb water only if the water potential of its surroundings is higher or less negative than that of the cell. Both the cell and soil have their own water potentials, which are the sum of (a) $\Psi\pi$, the osmotic or solute concentration effect; (b) Ψm, the matric (or hydrational) potential, which is contributed by matrices such as cell walls, protein bodies to be hydrated to bind water; and (c) Ψp, the turgor (hydrostatic) pressure resulting from the entry of water into the cell. In the case of soil water potential, only Ψm, plays an important role. The difference between the water potential of seed and soil is only one of the factors influencing the rate of flow of water from the soil environment to the seed. Other factors include internal impedance of the soil matrix and external impedance (degree of contact of seed with water), as well as internal impedance (seed coat and air spaces) (2).

Owing to rapid water uptake, the expanding cells increasingly vacuolate until they acquire the large central vacuole characteristic of elongated cells. Water uptake increases because of changes in the water potential in the cells, which is in turn is attributed to decreased (more negative) solute potential ($\Psi\pi$).

B. Expression of Germination

A number of reports suggest that cell division precedes expansion of the axis through cell enlargement (4,5). Studies on *Prunus cerasus* (sour cherry) indicated that the number of cells per axis increases by 4 weeks of after-ripening at 5°C, before any increase in the axis length occurs. The embryos of sour cherry seeds grow to maturity during after-ripening at low temperature (stratification). This phenomenon should not be related to radicle expansion during germination. The radicle elongation occurs in two phases: an initial phase of slowly accelerating elongation followed by a more rapid one (2). The first phase is not accompanied by changes in dry weight, and the second phase is characterized by mobilization of reserves and translocation of catabolites to the axes (e.g., *Vicia* seeds). In *Vicia*, the radicle elongates prior to its emergence and penetration through the seed coat, which is accompanied by mitosis. A similar biphasic mode of radicle elongation has been observed in barley and peas, although in barley, the radicle emerges from the seed prior to the onset of the second phase of elongation; that is, before cell division begins. In some seeds, the radicle has been found to elongate prior to its puncturing of the seed coat, and the emergence of the radicle from the seed is the result of cell elongation (2).

Two major aspects controlling seed germination are (a) specific events essential for the occurrence of visible germination and (b) internal controls, if any, responsible for setting these events in motion.

Since germination culminates in radicle emergence, in most cases accompanied only by cell elongation and not necessarily cell division, specific events concerning germination are related to phenomena connected with and leading to cell extension. Loosening of the cell wall as a consequence of the increased uptake of water by the cell may be a critical factor in the expansion of radicle cells. Cell wall plasticization by invoking hydrogen ion secretion is another possible mechanism by which radicle cells expand. It has been adduced that radicle emergence is stimulated by the acidification of the cell walls (H^+), thus necessitating proton extrusion as a prelude to visible germination. Continued synthesis of proteins (critical enzymes or cell wall components) and respiration are also known to be essential for cell expansion in stems and coleoptiles (2).

Growth hormones, especially auxins and gibberellins, are known to be involved in plant cell elongation. The requirements of some dormant seeds of the specific triggering factors like light, chilling, or after-ripening have been eliminated wholly or partially by supplying gibberellins, cytokinins, or ethylene to seeds. In a few cases, amounts of promotive hormonal regulators have been found to increase in treated seeds before radicle emergence. There is, however, very limited evidence to show that these growth hormones are synthesized during germination of nondormant seeds. In barley seeds, the gibberellin levels increase as the embryo begins growing as a visible sign of germination, but there is no evidence of hormone production before growth takes place. However, the rate of germination of embryos in barley has been observed to respond positively to applied gibberellin. In other seeds, substances such as CCC and AMO 1618, which inhibit gibberellin biosynthesis, prevent germination. The production of gibberelins may therefore be a prerequisite for radicle emergence in these seeds.

According to Khanna et al. (5a), crotalaria seeds attain physiological maturity at 30 days after anthesis (DAA), and thus pods should be harvested at 45–50 DAA when moisture content of seeds are around 8.5% for their optimum germination.

Seeds of *Pistacia mutica*, a potential rootstock for pistachio production, have poor germination and emergence due to seed endocarp impermeability. The objectives of this research were to evaluate the effect of chemical scarification with sulfuric acid on germination and emergence, to analyze the change in endocarp permeability of scarified seeds, and to identify the best predictor of the percentage and rate of emergence in field conditions. The chemical scarification significantly increased the percentages and rates of emergence and germination compared to nontreated seeds. No statistical differences were observed among 2, 3, 4, and 5 hr of treatment. However, soaking seeds in sulfuric acid for 5 hr resulted in the highest germination percentage and rate, and the highest emergence percentage and rate were obtained in the seeds treated for 4 and 5 hr, respectively. The longer the scarification time, the greater the increase in seed endocarp permeability and imbibition rate. This was the main cause of the emergence and germination improvement in relation to the control. The reduction of seed weight during scarification was the best predictor of the percentage and rate of emergence (5b). Naidu et al. (5c) studied the effect of GA_3, IBA, or IAA each at 250, 500, 1000, 1500, or 2000 ppm on soapnut seed germination in laboratory experiments. Seeds were soaked at room temperature for 10, 50 hr GA_3 was more effective in improving seed germination in soapnut than either IBA or IAA. The effectiveness increased with a concentration up to 1500 ppm and with duration of soaking.

Singh et al. (5d) osmoconditioned seeds of muskmelon (*Cucumis melo*) cultivars Pb. Hybrid and Pb. Sunehri with polyethylene glycol (PEG) and KNO_3 solutions to enhance their performance at low temperature. Osmoconditioning both with PEG 6000 and KNO_3 increased the percentage germination, speed of germination, vigor in terms of dry weight and length of the seedlings, and root/shoot ratios. KNO_3 priming had a greater effect than PEG priming. Other vigor parameters such as electrical conductance of seed leachates were decreased by osmoconditioning, whereas dehydrogenase activity was enhanced. Osmoconditioning resulted in increased amylase activity.

Studies were conducted with freshly harvested *G. sylvestre* seeds to assess the influence of seed treatment and containers on storability of seeds. The seeds treated with thiram at $2 \, g \, kg^{-1}$, stored both in cloth bags and 700-gauge polythene-lined cloth bags at 8% moisture content recorded 48% germimation after 6 months of storage. The untreated seeds recorded lower germination and vigor than the treated seeds. But vigor index values were significantly higher in seeds stored in 700-gauge polythene bags than in those stored in cloth bags. The electrical conductivity of seed leachate was also of a lower order in treated seeds stored in moisture-proof containers. The protein and oil content of stored seeds showed significant differences (5e).

Aconitum heterophyllum and *A. balfourii* are important medicinal herbs of the Himalayan region. Pandey et al. (5f) examined the effect of plant growth substances (PGSs; i.e., abscisic acid, 6-benzylaminopurine, gibberellic acid, and zeatin riboside) and two nitrogenous compounds (thiourea and potassium nitrate) for enhancing and synchronizing uniform germination. The tetrazolium (Tz) staining pattern indicated that freshly collected seeds had high viability, which decreased following storage at 4°C for 6 and 12 months. The treatments and time of seed germination were significantly different ($P < .01$). Gibberellic acid (GA_3; 250 µM) significantly enhanced seed germination (42.5% compared to 27.5% in control) in *A. balfourii* within 15 weeks but was inhibitory in *A. heterophyllum*. 6-Benzylaminopurine (BAP; 25 and 250 µM) and zeatin riboside (ZR; 5 and 250 µM) did not enhance germination in *A. balfourii*; 50 µM ZR was actually inhibitory. In *A. heterophyllum*, the lower concentration of BAP was inhibitory (7.5% compared to 25.0% in control), whereas 250 µM BAP enhanced germination (42.5% compared to 25.0% in control); the higher concentration of ZR was inhibitory. The combined treatments of gibberellin and cytokinin in general resulted in reduced germination in both species. Among the nitrogenous compounds, thiourea increased the rate and germination percentage in both species, but potassium nitrate enhanced germination in *A. balfourii* only. Seed germination was first detected in *A. balfourii* in the 5th week (2.5%) following treatment with 65 mM thiourea, and this value increased to 22.5% in the 7th week and reached as high as 75% (compared to 27.5% in control) in the 15th week. A higher dose of thiourea (130 mM) resulted in a rapid and high germination rate (40% compared to 0% in control) in the 7th week, reaching 75% in 10th and 12th weeks and a maximum 80% (compared to 27.5% in control) in the 15th week. In *A. heterophyllum*, however, thiourea only marginally enhanced germination even up to the 15th week. In *A. balfourii*, KNO_3 (50 and 100 mM) significantly enhanced germination (62–70%) within 15 weeks.

S. trifoliatus seeds were exposed to temperatures of 30–100°C for periods ranging from 5 min to 5 days. The enhancement in germination was dependent on temperature and period of exposure. Incubation at 60°C for 1–5 hr resulted in 84–88% germination. The scarification of seeds was also carried out with different concentrations of inorganic acids, such as sulfuric, nitric, and hydrochloric acids. Concentrated or diluted sulfuric acid gave the best results in terms of the germination percentage followed by nitric and hydrochloric acids (5g).

Kattimani et al. (5h) studied the effect of presowing seed treatments on seed vigor, root length, and dry root yield of the medicinal plant *Withania somnifera* under the agroclimatic conditions of Andhra Pradesh, India. Soaking in 1% sodium nitrate gave the most rapid germination and highest percentage germination (92% compared with 26% in untreated controls). Seeds soaked with nitrates of sodium and potassium at 100% for 24 hr produced more vigorous seedlings, higher dry matter accumulation, and root length as compared to unsoaked and water-soaked seeds. Unsoaked seeds had the lowest root length throughout the different stages of crop growth except 15 days after sowing, whereas zinc nitrate (0.5%) increased root length compared with most other treatments. Seeds soaked with nitrates of sodium and potassium at 10% recorded maximum dry root yield at harvest of 3.93 and 3.43 g/plant, respectively.

Pink (light) and dark gray to black (dark) seed fractions were selected from three samples of radish cv. Saxa. Seed mycoflora, germination capacity, the incidence of disease symptoms on germs, and ungerminated seeds were assessed in color fractions and nonfractionated samples. The color of seed coat was considered as an indicator of seed quality. The presence of *A. brassicae* was correlated with the color change (5i).

C. Respiration

In addition to enzyme and organelle activity and protein and nucleic acid (RNA) synthesis, respiration constitutes one of the important biochemical processes occurring during germination. Both anaerobic and aerobic respiratory pathways, namely, glycolysis, the pentose phosphate pathway (PPP), and the TCA cycle, are thought to be activated in the imbibed seed. These oxidative pathways are responsible for the production of the key intermediates in cellular metabolism: energy in the form of ATP and reducing power as the reduced pyridine nucleotides (NADH and NADPH). Glycolysis is catalyzed by cytoplasmic enzymes and is operative under both aerobic and anaerobic conditions to convert glucose to pyruvate, which may be further reduced to ethanol (in the absence of oxygen) or lactate if no decarboxylation occurs. Anaerobic fermentation of glucose yields only two net ATP molecules per glucose molecule respired in contrast to six ATPs produced under the aerobic conditions. Under the aerobic conditions, pyruvate is further oxidized within mitochondria after the oxidative decarboxylation of pyruvate to acetyl coenzyme A (acetyl CoA). The latter is completely oxidized to CO_2 and water through the TCA cycle, yielding a net 28 ATP molecules per glucose molecule respired.

The reducing power, NADPH, required for the reductive biosynthesis of lipids and steroids is made available by the pentose phosphate pathway. The intermediates of this pathway, such as 5-C and 7-C compounds, serve as precursors for the synthesis of aromatics and nucleic acids.

Moringa (6) demonstrated that seeds of 43 of 78 land plant genera, representing 24 families, germinated under the conditions of reduced oxygen tension (i.e., under water). Of 43 genera that germinated, 18 also germinated in the presence of oxygen on moist paper. According to Bewley and Black (2), seeds of some aquatic species actually germinate better under partially anaerobic conditions, probably by adapting to their habitat. The seeds of cattail or bullrush *(Typha latifolia)* germinate rather poorly in air unless decoated but germinate well when intact in a mixture of 99% hydrogen in air. It has been shown that rice grains germinate about 80% in 0.3% oxygen, whereas wheat grains require a minimum of 5.2% oxygen for an equivalent amount of germination. In these cereals, germination at reduced oxygen tension was shown to inhibit root growth more than shoot growth.

Seeds of many land species fail to germinate under water by losing their viability. In contrast, air drying of wild rice grains *(Zizania aquatica)* for 90 days resulted in loss of viability, but their after-ripening at low temperatures in stagnant water accelerated the rate of germination, representing an adaptation to habitat. Seeds of Indian lotus germinate fully in 100% N_2, hydrogen, or CO_2, probably because oxygen is made available to the embryo through an internal cavity and from intercellular spaces of the seed tissues. Analysis of 0.2 mL of the trapped internal gas showed 18.3% O_2, 0.74% CO_2, and 80.93% N_2 (2).

Forward (7) experimentally demonstrated the requirement of oxygen for seed germination (Table 1). CO_2 concentrations higher than 0.03% retard germination, whereas nitrogen gas has no influence on it.

Kolloffel (8) studied the pattern of O_2 consumption by imbibed pea seeds in two previsible and two postvisible germination phases (Fig. 2A). The germinating axis may contribute to only about 10% of the total seed respiration, with the rest being attributed to the nongrowing cotyledons. Phase I is characterized by a sharp rise in respiration lasting about 10 hr, which is associated with the activation and hydration of mitochondrial enzymes of the TCA cycle and electron transport chain. The respiration during this phase rises linearly with the degree of swelling of the cotyledon tissue (Fig. 2B). Phase II is characterized by a lag in respiration between 10 and 25 hr after the beginning of imbibition. Now cotyledons are hydrated completely and all preexisting enzymes may be activated. The respiratory quotient (RQ) rises above 3.0, indicating some anaerobic respiration. At this stage, germination may be hindered for want of oxygen, which is restricted by the intact seed coat. The removal of the seed coat in pea seeds increases both the rate of water uptake and swelling (Fig. 2C) and decreases the lag phase (Fig. 2A,D). The rapid uptake of oxygen by intact seeds (with testae) during the early stages of imbibition (phase I) and the restriction of oxygen by testae in phase II has not been explained. Pea seed testae appear to become less permeable to oxygen as imbibition proceeds. The radicle penetrates the testa between

Table 1 Effect of CO_2/O_2 Ratios on Germination of Oat Seeds

Gas mixture		
% CO_2	% O_2	% Germination
0.0	20.9	100
16.9	17.4	93
30.0	14.7	50
35.0	13.6	31
36.8	13.2	10
38.7	12.8	1

Source: Ref. 7.

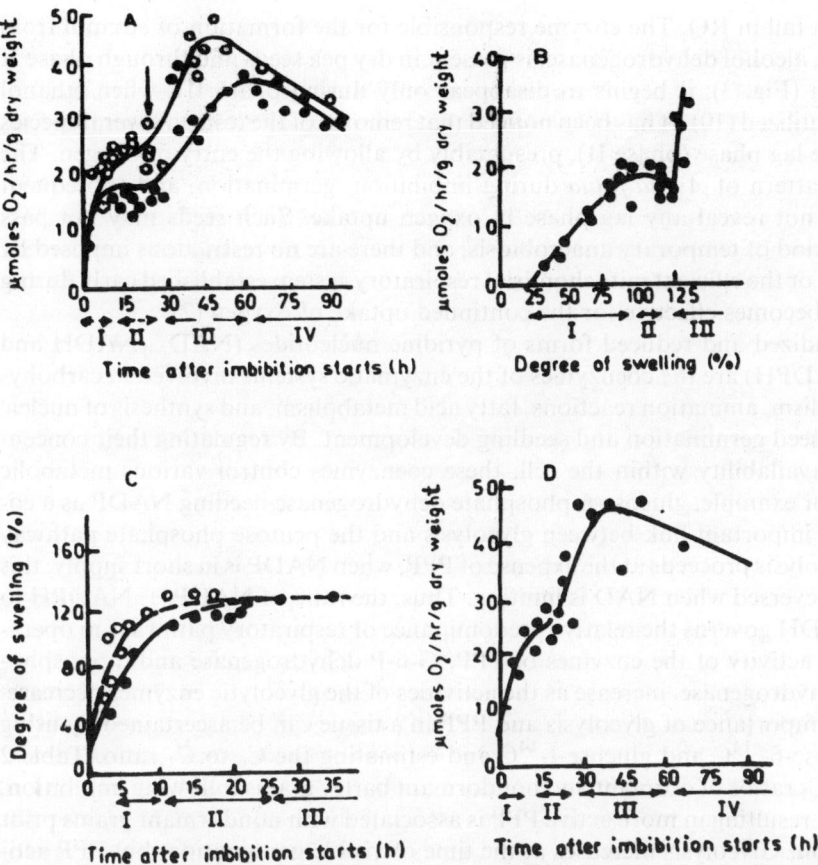

Figure 2 (A) The course of respiration of intact dark-germinated *Pisum sativum* seeds (●) cv. Rondo and of cotyledons with seed coat dissected away (○). Arrow indicates approximate time of visible germination. (B) The relation between the degree of swelling and the respiration rate of cotyledons in intact germinated seeds. (C) The swelling of cotyledons from intact seeds (●) and of cotyledons from seeds imbibed without a seed coat (○). (D) Respiration of excised cotyledons. See text for explanation of phases I–IV. (From Ref. 2.)

phases II and III. Phase III is characterized by a second respiratory burst, ascribed partially to an increased oxygen supply through the newly punctured testa. The activity of the newly synthesized mitochondria and respiratory enzymes in the dividing cells of the growing axis also contribute to this respiration. The RQ falls to about 1.0, suggesting that sugars are being catabolized through aerobic respiration. Phase IV is marked by significant fall in respiration coinciding with disintegration of cotyledons following depletion of reserve food. The lengths of the respiration phases in seeds may vary from species to species (2).

Anaerobiosis appears to be the possible cause of the lag phase in the respiration of some, if not all, seeds. The products of anaerobic respiration, ethanol and lactate, which accumulate in pea seed prior to the penetration of the radicle through the enclosing testa (9), decrease in quantity following radicle emergence, because they are metabolized by the aerobic conditions, as evidenced by an increase in O_2 uptake, CO_2

output, and a fall in RQ. The enzyme responsible for the formation of ethanol from acetaldehyde, alcohol dehydrogenase, is present in dry pea seeds and through phase II of respiration (Fig. 3); it begins to disappear only during phase III, when ethanol begins being utilized (10). It has been noticed that removal of the testa in several species eliminates the lag phase (phase II), presumably by allowing the entry of oxygen. The respiratory pattern of *Avena fatua* during imbibition, germination, and subsequent growth does not reveal any lag phase in oxygen uptake. Such seeds may not pass through a period of temporary anaerobiosis, and there are no restrictions imposed by the seed coat, or the efficient mitochondrial respiratory system established early during germination becomes effective for the continued uptake of oxygen (2).

The oxidized and reduced forms of pyridine nucleotides (NAD^+/NADH and $NAPP^+$/ NADPH) are the coenzymes of the enzymatic systems involved in carbohydrate metabolism, amination reactions, fatty acid metabolism, and synthesis of nucleic acids during seed germination and seedling development. By regulating their concentration and availability within the cell, these coenzymes control various metabolic pathways. For example, glucose-6-phosphate dehydrogenase needing NADP as a coenzyme is an important link between glycolysis and the pentose phosphate pathway (Fig. 4). Glycolysis proceeds at the expense of PPP, when NADP is in short supply; this condition is reversed when NAD is limiting. Thus, the ratio of NADP + NADPH to NAD + NADH governs the relative predominance of respiratory pathways in operation (2). The activity of the enzymes of PPP, G-6-P dehydrogenase and 6-phosphogluconate dehydrogenase, increase as the activities of the glycolytic enzymes decrease. The relative importance of glycolysis and PPP in a tissue can be ascertained by using labeled glucose-6- ^{14}C and glucose-I-^{14}C and estimating the C_6 to C_1 ratio. Table 2 shows C_6 to C_1 ratios of dormant and nondormant barley grains following imbibition. A lower ratio resulting in more active PPP is associated with nondormant grains prior to germination. Glycolysis increases at the time of visible germination, but PPP activity may enhance it again as the seedlings begin to grow (11). Roberts (12) adduced

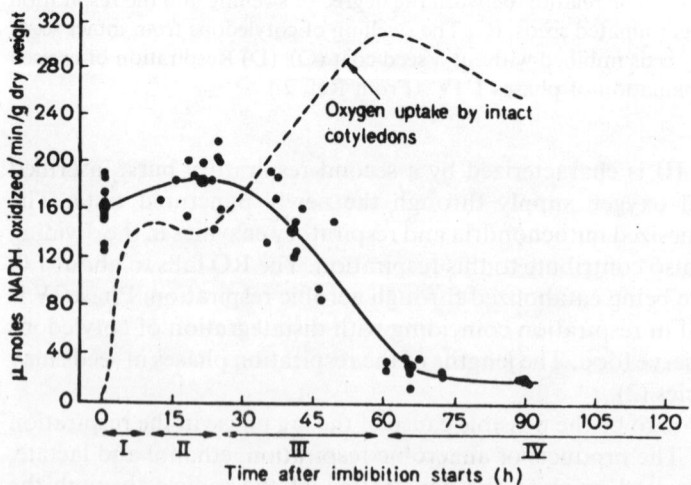

Figure 3 Alcohol dehydrogenase activity (●) in, and oxygen uptake by cotyledons of dark-germinated intact pea seeds cv. Rondo. Germination time as in Figure 2a. (From Ref. 10.)

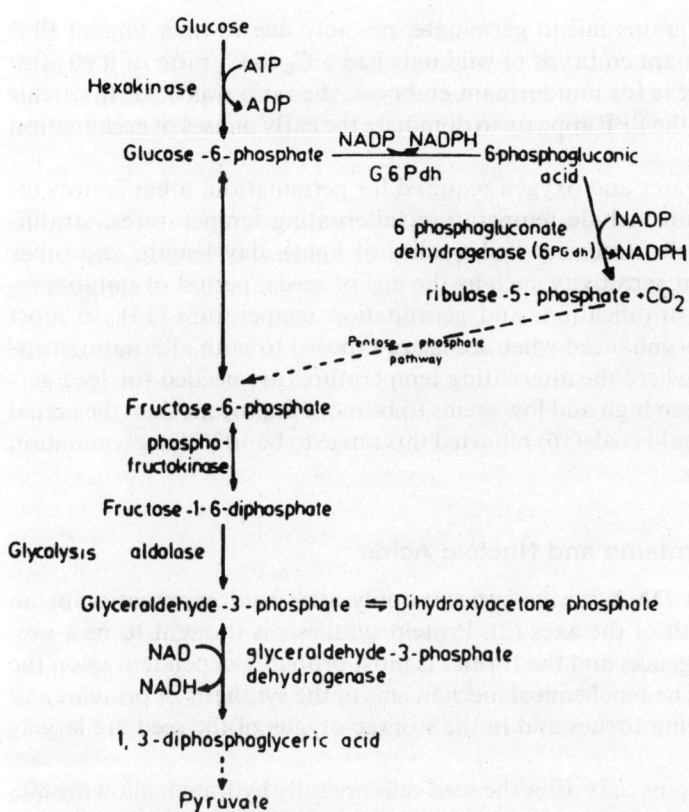

Figure 4 The link between glycolysis and pentose phosphate pathway. (From Ref. 2.)

Table 2 Evolution of $^{14}CO_2$ from Glucose-6-^{14}C and from Glucose-1-^{14}C (C-6/C-1 ratios) When Applied for 3 hr Periods to Pellas Barley Grains at Different Times

Time from start of imbibition (h)	C-6/C-1 ratio	
	Dormant	Nondormant
0.5–3.5	0.18	0.13
3.5–6.5	0.29	0.15
6.5–9.5	0.28	0.13
12.5–15.5	0.42	0.14
21.5–24.5	0.33	0.12
33.5–36.5[a]	0.35	0.28

[a] Radicle appeared during this period in nondormant grains.
Source: Ref. 12.

that the dormant barley grains fail to germinate, possibly due to their limited PPP activity. The excised dormant embryos of wild oats had a C_6 to C_1 ratio of 0.90 after 10 hr of imbibition, whereas for nondormant embryos, the ratio was 0.55. In cereals like barley, oats, and rice, the PPP appears to dominate the early phases of germination after water uptake.

In addition to the water and oxygen required for germination, other factors influencing seed germination include temperature (alternating temperatures, stratification, or prechilling), light (intensity and quality of light), day length, and other factors that influence light sensitivity such as the age of seeds, period of imbibition, imbibition temperature, stratification, and germination temperature (13). In most seeds, total germination is enhanced when seeds are exposed to both alternating temperatures and light (14); where the alternating temperatures are needed for seed germination, the range between high and low seems to be more important than the actual temperatures (15). Mcdonald et al. (16) reported this range to be 10°C for germination of most cool season seeds.

D. Biosynthesis of Proteins and Nucleic Acids

It is well established that DNA synthesis occurs only after seed germination as an integral part of the growth of the axes (2). Protein synthesis is thought to be a prerequisite for radicle emergence, and the former is most probably dependent upon the prior synthesis of RNA. The biochemical mechanisms of the synthesis of proteins and nucleic acids within growing tissues and in the storage organs of the seed are largely identical.

Protein synthesis begins only after the seed cells are fully hydrated, allowing 80S ribosomes to associate with mRNA. In wheat embryos, protein synthesis is initiated with small 40S ribosomal units becoming attached to the initiating tRNA; namely, methionyl tRNA. The large (60S) initiation complex is established at the starting point on the mRNA for the commencement of protein synthesis (17). Formation of the initiation complex requires several factors, including enzymes, which are present in the cytoplasm of dry wheat embryo, and expenditure of energy in the form of GTP and ATP. Other factors such as elongation factors, termination factors, and organization of polypeptide chain(s) synthesized are also essential.

The time required for starting protein synthesis after imbibition varies with species from a few minutes to a couple of hours. Dissected axes of bean seeds (*Phaseolus lunatus, P. vulgaris*) and isolated embryos of cereals such as wheat, rice, and rye begin the protein synthesis within 30–60 min of imbibition (2). Polyribosomes comprising several ribosomes attached to mRNA constitute the active protein-synthesizing complex, which is formed within 15 min of imbibition in wheat embryos. Ribosomes present in dry whole seeds or in dry embryos and dry dicot storage organs retain their potential for the expression of protein synthesis during seed germination. Dry seeds do not contain mRNA associated with the ribosomes, but soon after imbibition ribosomes are associated with mRNA to form a polysomal complex. Marcus (18), employing supernatant (cytoplasmic or postribosomal) fractions obtained from dry wheat embryos, demonstrated in vitro protein synthesis in the presence of added ribosomes and mRNA (or polysomes) and ATP/GTP. It is presumed that the cytoplasmic components essential for protein synthesis, namely, initiation and elongation factors, tRNAs, amino acids, and enzymes (aminoacyl-tRNA synthetases), are present in the dry seed in sufficient quantities to permit resumption of protein synthesis in the seed upon imbibition. It is believed that mRNAs are stored in seeds for later utilization during imbibition, germination, and growth to enable the seed to begin protein synthesis readily after rehydration. The emerging pattern of RNA and protein synthesis in the embryo and axes is capable of synthesizing all types of RNAs as soon as the tissue is hydrated, long before the radicle is expanded. Synthesis of mRNA is not an essential prelude to the synthesis of specific proteins for germination. However, synthesis and translation of mRNA does appear to be a prerequisite for germination. It is possible that the components present in the dry seed might not be in sufficient quantity to support protein synthesis during germination (2). It is not known which specific enzymes (and proteins) are required for the synthesis of proteins necessary for germination to take place.

Bewley and Black (2) differentiated between the concepts of "DNA synthesis" and "mitotic cell division," which have been used synonymously. Whereas some cells that never divide synthesize DNA, others may synthesize DNA many hours before cell division begins. In *A. cera* seeds, following germination nuclei become visibly enlarged, increasing from about 7 μm (in quiescent state) to 15 μm when activated. Mitosis in these seeds is separated from DNA synthesis by at least 9 hr. The onset of DNA synthesis in a germinated seed varies with the species. Whereas in *Zea mays* it may start 30 hr after imbibition, rapid synthesis of DNA (6 hr) following imbibition occurs in barley embryos. Also, DNA synthesis in the cytoplasm appears to precede that in the nucleus. In the activity of both soluble and a chromatin-bound DNA polymerase in germinating pea, the axis increases prior to DNA synthesis (18). The increase in polymerase activity, however, may not be a prerequisite for DNA replication. The levels of kinase and polymerase may or may not control DNA synthesis, and hence subsequent mitosis may not control germination mechanisms, because the latter event does not necessitate either DNA synthesis or cell division.

III. MOBILIZATION OF FOOD RESERVES

After the emergence of the radicle, the growing axis of germinated seed maintains its growth using the stored food reserves. The storage organs of seeds contain substantial quantities of at least two major food reserves in the form of complex polymers: carbohydrates, lipids, proteins, and phosphorus-containing compounds. These food reserves must be hydrolyzed or degraded to their simpler monomers, which are eventually catabolized enzymatically for the production of energy (ATP) and other essential metabolites for the growth and development.

The embryonic axis obtains energy for its growth from the hydrolyzed soluble forms of the storage compounds, the former being translocated from the endosperm to the embryo and transformed to energy molecules that can be readily utilized by the embryonic axis. The endosperm initially becomes enriched with soluble products like glucose and maltose, which are then absorbed by the scutellum and transformed enzymatically to sucrose. The sucrose is then translocated to the adjacent embryonic axis to serve as a source of energy (Fig. 5). Because of the absence of an aleuronelike tissue synthesizing hydrolytic enzymes in dicots, the role of hormones in dicot seed germination is a subject of debate (13). Gibberellins are known to trigger hydrolytic enzymes in some cases, but the degree of activation is much less than that observed in cereals. It is postulated that dicot seed germination is mediated by the growing

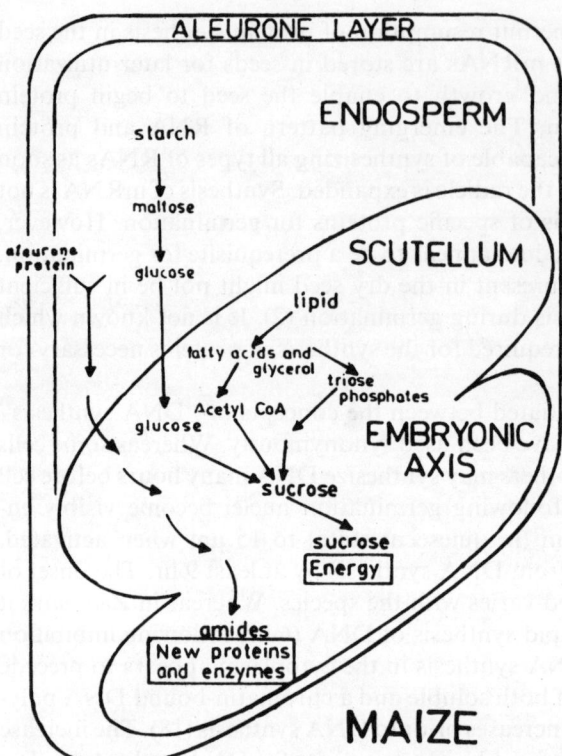

Figure 5 Breakdown of storage compounds and mobilization of their products during the germination of maize seeds. (From Ref. 14.)

embryonic axis. As the latter continues to grow, it incorporates breakdown products into the synthesis of new compounds, which reduces the concentration of compounds in the cotyledons. This in turn stimulates the hydrolysis of other storage reserves for use by the embryonic axis through a feedback mechanism (Fig. 6).

A. Carbohydrates

Starch is the major reserve carbohydrate present in most seeds along with other polysaccharides such as hemicellulose, galactomannan, and certain oligosaccharides. Amylose and amylopectin are hydrolytically degraded by α- and β-amylases in the following manner (2):

$$\text{Amylose} \xrightarrow{\text{α-amylase}} \text{glucose} + \text{α-maltose} + \text{α-maltotriose}$$

$$\text{Amylopectin} \xrightarrow{\text{α-amylase}} \text{glucose} + \text{α-maltose} + \text{α-maltotriose} + \text{α-limit dextrin}$$

$$\text{α-Maltotriose} \xrightarrow{\text{α-amylase}} \text{maltose} \xrightarrow[\text{(β-glucosidase)}]{\text{maltase}} \text{glucose}$$

$$\text{Amylose} \xrightarrow{\quad \beta\text{-amylase} \quad} \beta\text{-maltose}$$

$$\text{Amylopectin-} \xrightarrow{\quad \text{-amylase} \quad} \beta\text{-maltose} + \beta\text{-limit dextrin}$$

$$\text{Limit-dextrins} \xrightarrow[\alpha\text{-glucosidase, limit dextrinase}]{\text{debranching enzymes}} \text{glucose}$$

Whereas in cereals starch is degraded to yield more maltose, dicot seeds yield more glucose and maltotriose, which is due to the difference in the relative activity of â-amylase in these two types of seeds. Starch may also be degraded phosphorolytically:

$$\text{Amylose} + \text{amylopectin} + \text{Pi} \xrightarrow[\text{phosphorylase}]{\text{starch}} \text{glucose-1-P} + \text{limit dextrin}$$

α-Amylase hydrolyzes α-1,4- glycosidic bonds of both amylose and amylopectin; however, it cannot catalyze the α-1,6 bonds of branch points of amylopectin, thus producing limit dextrins. The latter are utilized for the synthesis of amylopectin or may be catabolized by debranching enzymes to produce oligosaccharides and glucose. β-Amylase cleaves successive maltose units from both amylose and amylopectin from the nonreducing end. β-Amylases are capable of degrading only the larger dextrins released from the native starch grains by prior attack of α-amylases. Maltose is hydrolyzed to glucose by α-glucosidase. Starch phosphorylase cleaves to glucose from the nonreducing end of polysaccharide chains by introducing phosphate rather than

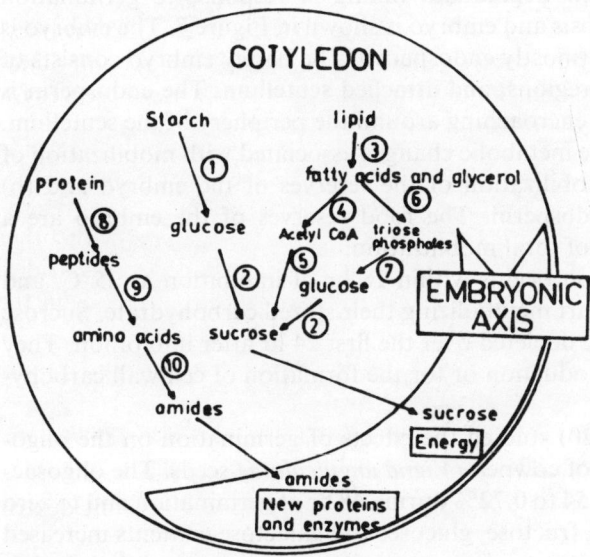

Figure 6 Breakdown of storage compounds and mobilization of their products during the germination of soybean seeds. (From Ref. 14.)

water, releasing glucose-1-phosphate molecules of amylose and amylopectin within two or three glucose residues of a α-1,6 branch point.

The plant world has selected sucrose to be the major form of carbohydrate transport, because sucrose is uniquely resistant to hydrolysis by several enzymes (except invertase). Glucose-1-phosphate, the product of starch phosphorylation, is a direct substrate for sucrose biosynthesis. Free glucose, however, must first be phosphorylated, namely to glucose-6-P, and then isomerized to G-1-P to be combined with uridine diphosphoglucose (UDPG), which transfers glucose to free fructose or to fructose-1-P as follows (2):

UDPG phosphorylase

G-1-P + UTP $\xrightarrow{\hspace{5cm}}$ -UDPG + PPi

Sucrose synthetase

UDPG + fructose $\longleftarrow$============$\longrightarrow$ sucrose + UDP

or

Sucrose-6-P synthetase

UDPG + fructose-1-P $\longleftarrow$============$\longrightarrow$ sucrose-6-P + UDP

Sucrose-6-phosphatase

Sucrose-6-P $\xrightarrow{\hspace{4cm}}$ sucrose + Pi

Sucrose can be hydrolyzed to glucose and fructose by (α-fructofuranosidase [invertase] whenever needed.

Among cereal grains, barley *(Hordeum* sp.) has been thoroughly investigated as to its mobilization of carbohydrate reserves during germination. Barley varieties exhibit significant variation in the degree and timing of response to germination stimuli. Structure of wheat caryopsis and embryo is shown in Figure 7. The embryo is situated at the base of the kernel (mostly endosperm). The barley embryo consists of the acrospire, nodal and rootlet regions, and attached scutellum. The endosperm is surrounded by the aleuron layer, encroaching around the periphery of the scutellum. Bewley and Black (2) reviewed the metabolic changes associated with mobilization of food under two headings: (a) mobilization of the reserves of the embryo and (b) utilization of the reserves of endosperm. The food reserves of the embryo are a quantitatively minor component of total mobilization.

In barley grains, germination begins within 18 hr of imbibition at 25°C, and within 24 hr detached embryos start metabolizing their stored carbohydrate. Sucrose and the trisaccharide raffinose are depleted over the first 24 hr after imbibition. They may be used for energy (ATP) production or for the formation of cell wall carbohydrates (2).

Akinlosotu and Akinyele (20) studied the effects of germination on the oligosaccharide and nutrient contents of cowpea *(Vigna unguiculata)* seeds. The oligosaccharide content decreased from 9.54 to 0.72% during 48 hr of germination and to zero after 72 hr, whereas total sucrose, fructose, glucose, and galactose contents increased from 2.03 to 7.99% after 72 hr. Protein and energy contents also increased slightly, but Ca and Fe contents decreased. The ascorbic acid and niacin contents increased and the thiamine content of seeds decreased.

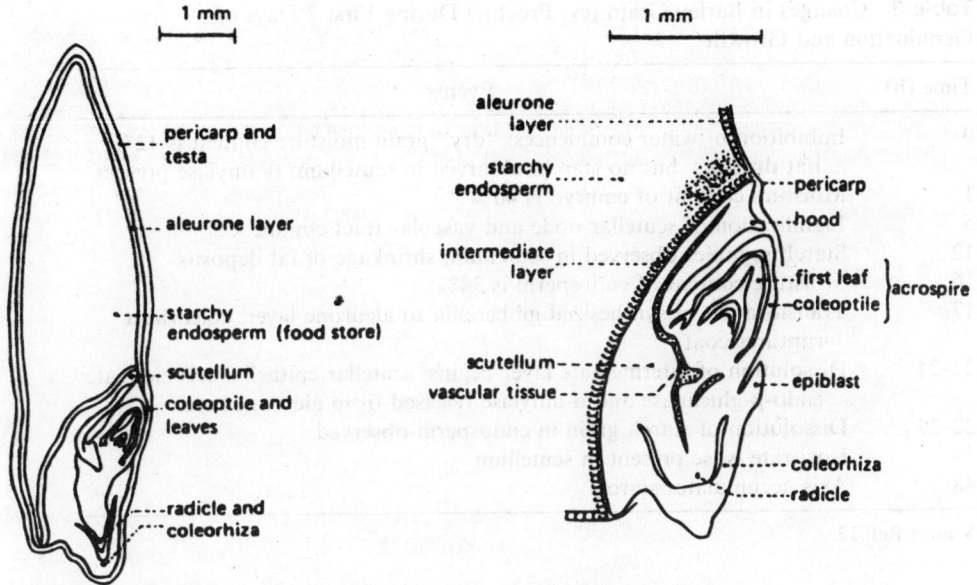

Figure 7 Structure of wheat caryopsis and embryo. (From Ref. 2.)

Gibberellins are synthesized by the seed embryo in either the nodal region of the axis or the scutellum. Indole acetic acid is also synthesized by the coleoptile tip soon after imbibition and then diffuses to the scutellar node. This auxin promotes the lignification of the cell walls in the center of the node and subsequently plays a role in depositing lignin in the elongated provascular strand of the scutellum. Bewley and Black (2) listed the chronological sequence of the events leading up to carbo-hydrate breakdown during the first two days of germination and growth of barley seeds (Table 3). These events pertain to barley seeds imbibed and germinated under nonmalting conditions.

These events (Table 3) begin after 5 hr of imbibition, at which time the vascular system is ready for translocating metabolites from the scutellum to the embryo and possibly for the transport of gibberellins synthesized at the node to the scutellum for their release. The gibberellin thus released diffuses into the aleurone layer to stimulate the synthesis and release of a number of hydrolytic enzymes into the endosperm. The GA released from the embryo induces the aleurone layer to undergo a series of metabolic changes, which result in the release of α-amylase and other hydrolases into the endosperm for the degradation of food reserves stored in it. Bilderback (21) noted that seven isoenzymes of α-amylase (MW 45,000) appear in GA-treated isolated aleurone layers. The participation of β-amylase in starch degradation has also been adduced. Nine isoenzymes of β-amylase have been separated from the starchy endo-spenn of barley, four of which had a molecular weight of 43,000, and the remaining were the multimers of these isoenzymes. In vivo, gibberellins stimulate the activation of α-amylase, probably by increasing proteinase production and release from the aleurone layer. α-Amylase and proteinase may be simultaneously produced and released from GA-stimulated isolated aleurone layers. There is no de novo synthesis

Table 3 Changes in Barley Grain (cv. Proctor) During First 2 Days of
Germination and Growth

Time (h)	Events
0	Imbibition of water commences: "dry" grain moisture content is 12%. Fat droplets, but no starch, observed in scutellum; β-amylase present.
1	Moisture content of embryo is 40%
5	Lignification of scutellar node and vascular tract commences
12	Starch granules observed in scutellum; shrinkage of fat deposits
16	Moisture content of endosperm is 34%
17	Translocating of synthesized gibberellin to aleurone layer; coleorhiza ruptures coat
22–24	Dissolution of intermediate layer begins; scutellar epithelial cels elongate: endo-β-glucanase and α-amylase released from aleurone layer
22–29	Dissolution of starch grain in endosperm observed
24	Isocitrate lyase present in scutellum
48	Axis accumulates starch

Source: Ref. 12.

of β-amylases during mobilization of the starchy reserves. β-Amylases alone are not able to degrade native starch grains and can only digest starch that is previously solubilized by α-amylase (2).

B. Lipids

Adams and Novellie (22) isolated a number of hydrolytic enzymes such as α-amylase, β-amylase, α-glucosidase, proteinase, phosphatases, and ribonuclease from ungerminated sorghum grains, indicating the presence of several carbohydrates in cereals even before germination. In maize (and perhaps also in sorghum), the hydrolysis of food reserves after germination is initiated by the release of hydrolases preformed in the endosperm during maturation rather than the de novo synthesis of hydrolases, which normally occurs in other cereals like barley, wheat, rice, and oats. It is, however, possible that the endosperms of maize and sorghum may contain sufficient endogenous GA (produced during seed development) to stimulate the production of hydrolases without requiring more GA from the embryo or any additional effect of exogenously supplied GA (2).

The scutellum (a modified cotyledon) absorbs the products of starch hydrolysis, which are further converted into sucrose and transported to the growing embryo. The scutellum can also absorb glucose and maltose.

Triglycerides, which are the major form of stored lipids in seeds, are hydrolyzed by lipases to diglycerides, monoglycerides, and then to glycerol and fatty acids. Glycerol is metabolized through glycosis after its oxidation to triose phosphates. The fatty acids are further oxidized through the β-oxidation pathway by sequentially removing two carbon atoms in the form of acetyl CoA. The latter enters the TCA cycle for its complete oxidation to CO_2 and water. The lipolysis of fats to fatty acids and glycerol in seeds takes place in fat-storing oil bodies, whereas fatty acid oxidation and synthesis of succinate via glyoxylate cycle occurs in glyoxysomes. Succinate is converted to oxaloacetate in the mitochondria. Oxaloacetate is then metabolized to sucrose in the cytoplasm.

Muto and Beevers (23) investigated mobilization of stored lipids in the castor bean, which begins on the third day after imbibition and ends on the seventh day. By this time, the endospenn is liquefied and its contents are absorbed by the expanding cotyledons. Acid lipase (pH 5) is activated on the second day of imbibition, reaches its peak of activity, and declines sharply until day 4, at which time an alkaline lipase (pH 9) appears and reaches its maximum activity. Whereas the acid lipase is associated with the fat, the alkaline lipase remains associated with the particulate fraction. Huang (24) showed that the glycerol released by lipase in the castor bean endosperm is phospho rylated by glycerol kinase in the cytosol to give á-glycerol phosphate, which is oxidized to dihydroxyacetone phosphate in the mitochondria, and is released into the cytosol and converted to hexose. Free fatty acids in many oil seeds are rapidly degraded to carbohydrate within the endosperm.

C. Proteins

Proteinases or proteases hydrolyze stored proteins into their constituent amino acids. The enzymes consist of endopeptidases, which catalyze the internal bonds of polypeptides to yield smaller polypeptides, amino peptidases, and carboxypeptidases, which sequentially cleave the terminal amino acids from the free amino and carboxyl ends of the polypeptide chain, respectively. The peptide hydrolases degrade various smaller peptides into amino acids. The liberated free amino acids are either used for protein synthesis or oxidized for energy after their deamination. The ammonia is absorbed into glutamine and aspargine. In cereals, the reserve protein foods are stored in the aleurone bodies of the aleurone layer and in the protein bodies of the endosperm. The aleurone grain proteins are mobilized to provide amino acids for the synthesis of important hydrolases such as α-amylase. In barley, aleurone layers release several different proteinases for the mobilization of reserve proteins of the starchy endosperm following germination (25). In barley, three different proteolytic systems exist (2):

1. Hydrolysis of aleurone layer proteins to provide amino acids for the synthesis of hydrolytic enzymes. GA produced by the embryo probably controls this system.
2. The mobilization of stored reserves in starchy endosperm for use by the developing seedling. This system consists of two components: (a) proteinase synthesized and secreted by the aleurone layer and controlled by GA and (b) proteinase preformed in the endosperm itself and activated therein.
3. A system in the axis, which may be important for protein turnover in the growing embryo and/or for hydrolysis of any small peptides transported into the scutellum from the starchy endosperm.

Harvey and Oaks (26) found maize endosperm to contain zein (stored in protein bodies) and glutelin (stored in the cytoplasmic matrix). Both storage proteins are mobilized soon after germination (about 20 hr after imbibition), with the peak of hydrolysis taking place between third and eighth day after imbibition, which coincides with the appearance of an acid endopeptidase synthesized de novo in the endosperm independent of any control by the embryo or added GA.

Glutamine is the most abundantly available amino acid in the free amino acid pool of partially digested maize endosperm, which is taken up by the scutellum for its transportation to the growing axis. Other amino acids also may be taken up by the scutellum, although there is limited competition between these amino acids and glutamine. The latter is believed to be transported by a carrier protein. In the growing maize seedling, uptake of glutamine by the scutellum coincides with the uptake of sugars (2).

In *Phaseolus vulgaris,* proteinase activity is at its peak about 5 days after imbibition, probably by activating the enzyme present in the seed rather than its de novo synthesis (27). Dry pea seeds *(Pisum sativum* cv. Alaska) also contain low levels of soluble proteinase, which increase about four times after 5 days of imbibition, and the peak lasts for about 10 days (28). It has also been noticed that in pea seeds protein hydrolysis begins even before there is any detectable activity of the major proteinase (acid sulfhydryl proteinase). It is suggested that the early protein degradation may be attributed to the combined activity of preformed soluble peptidases (amino peptidases) present in dry dicot seeds (2).

D. Phytic Acid

Many seeds possess their reserve phosphates in the form of phytic acid or myoinositol hexaphosphate, which may account for more than 50% of the total stored phosphate. Since phytic acid is usually present in the form of its potassium, magnesium, or calcium salt (as phytin or phytate), the latter also serves as a major source of these macronutrients. Enzyme phytase degrades phytin to release its phosphate and associated mineral cations and myoinositol. According to Roberts et al. (29), myoinositol derived from phytin may be important to the growing seedling as a precursor of all pentosyl and uranosyl sugar units associated with pectin and other polysaccharides of the cell wall.

Hall and Hodges (30) monitored the fate of various phosphate fractions in different parts of oat grains during the first 8 days after imbibition. The dry seeds had 53% phytic acid and 27% phosphate in the form of lipid, protein, and nucleic acid. Seed germination was visible on the second day; subsequent growth was marked by an increasing fresh weight of shoots and roots. The rise in total phosphate levels in the growing shoot/root axis was accompanied by a fall in the nonaxial parts of the seed. The phosphate from phytic acid is released from its storage form and is transported as free inorganic phosphate to the growing regions. The nucleic acid phosphate found abundantly in dry grain, along with lipid and protein phosphates, decreases in the nongrowing regions of the grain and increases in the shoot and root regions soon after germination. A phytate in barley is associated with the protein bodies of the aleurone layer, which is the storage site for phytin (31). However, in wheat and rice, neither the phytin nor the phytase is associated with the protein bodies of the starchy endosperm (32). The phytin is degraded by phytase to myoinositol, phosphate, potassium, magnesium, and calcium. These catabolites move from the aleurone layer and are distributed into the developing seedling by diffusion through the endosperm.

Nyctanthes arbortristis, cultivated in Indian gardens for its fragrant flowers and a source of ethnic medicines, is a seed-propagated plant. The seeds exhibit a poor germination rate. Phenolic compounds, leached out of imbibed seeds, interfer in germination. The inhibitory compounds were present in the pericarp and seed coat, and removal of these layers increased germination; germination was 24% in intact seeds and 94% with both the pericarp and seed coat removed. Leaching of the germination inhibitors was restricted by treating seeds prior to germination with solutions of antioxidants like polyvinylpyrrolidone (PVP) and polyvinylpolypyrrolidone (PVPP), resulting in better germination (32a).

IV. INVOLVEMENT OF HORMONES AND GROWTH REGULATORS

Various plant hormones regulate the metabolic processes in the mobilization of stored food reserves in germinating seeds. These aspects have been investigated in more detail using cereal grains such as barley and wheat. The stored reserves in the endosperm of cereals are mobilized largely by the hydrolytic enzymes synthesized in the aleurone layer. The gibberellin coming from the embryo stimulates the activity of these hydrolases. Thus, control over the mobilization of food reserves is exerted by the embryo (2).

Paleg (33) demonstrated that embryoless barley grains synthesized more α-amylase when they were incubated with isolated embryos than they did when incubated alone. The promotive factor was partially purified and identified as a gibberellin. It was also shown that chemically characterized gibberellic acid stimulated degradation of the endosperm in embryoless grains, which is similar to that occurring in intact kernel. It was thus established that the gibberellin produced by the embryo induces the aleurone layer to synthesize food-mobilizing enzymes. Two gibberellic acids (GA_1 and GA_3) are the major gibberellins produced by the germinating embryo, although other acids (GA_2, GA_4, GA_7, and GA_{22}) have been found to play certain roles such as activation of aleurone cells (GA_3 and GA_7) and control of embryo growth (GA_1 and GA_4).

In barley endosperm, α-amylase is synthesized and secreted by the aleurone cells under the influence of GA. The latter also stimulates synthesis of other hydrolyases like proteinase, pentosanase, limit dextrinase, and α-glucosidase. Three enzymes, namely, ribonuclease, β-1,3-glucanase, and phosphatase, which are secreted by GA-treated aleurone cells, do not depend on gibberellin for their synthesis (2).

Several reviews have pointed out that the aleurone layer of barley has the following unique features that make it an ideal experimental material to study the control of mobilization of food reserves during seed germination (34–40):

1. GA triggers production of α-amylase in vivo.
2. Enzyme response is virtually confined to GA.
3. Aleurone cells do not undergo division.
4. Target tissue consists of only one type of cells.
5. Cells can be isolated free of other tissues and still respond to GAs.
6. Substrate does not influence the response of isolated aleurone cells.

By using a density-labeling technique, Filner and Varner (41) demonstrated that GA stimulates de novo synthesis of α-amylase in barley aleurone tissue. Chrisppels and Varner (42) showed that isolated barley aleurone tissue requires about 8 hr of exposure to GA for the commencement of α-amylase secretion (Fig. 8). Bewley and Black (2) summarized the events taking place in isolated barley aleurone layers in response to GA (Table 4).

Kermode (43) reviewed the current knowledge of seed development and germination, discussing the roles of ABA and osmotic potential, with special reference to the deposition of storage reserves, desiccation tolerance, and prevention of germination. Gibberellins have also been found to induce synthesis and secretion of hydrolases in other cereals like wheat and wild oat (*A. fatua*). The events taking place in the wheat

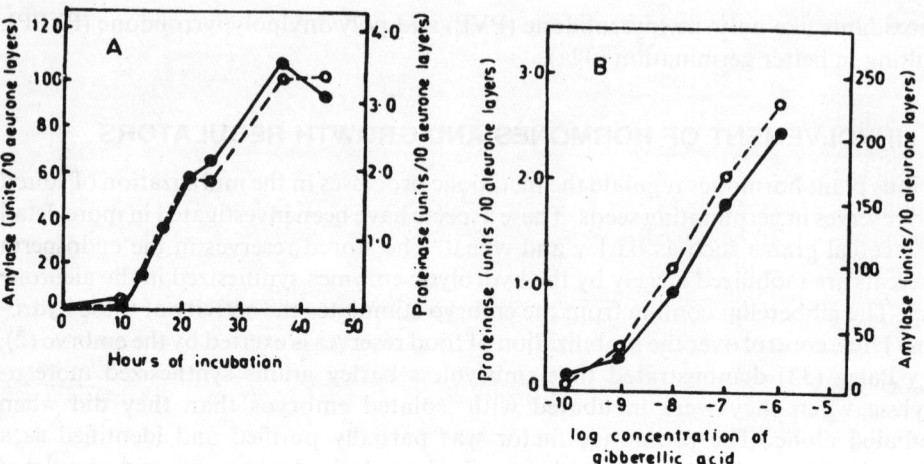

Figure 8 (A) Time course of release of amylase and proteinase by aleurone layers of Himalaya barley in the presence of 1 μM GA$_3$. (B) Release of these two enzymes by aleurone layers in response to various GA$_3$ concentrations: amylase (O), proteinase (●). (From Ref. 2.)

aleurone layer that are stimulated by the embryo include induction of triglyceride metabolism by neutral lipase, stimulation of activities of the enzymes of β-oxidation and isocitrate lyase and malate synthetase of the glycoxylate cycle, stimulation of phytase followed by degradation of phytin, and induction of α-amylase synthesis (32,44,45). Other events, such as stimulation of enzymes of glycolysis and gluconeogenesis, PPP, TCA, and the electron transport chain, induction of nucleic acid and phospholipid synthesis, and proliferation of endoplasmic reticulum take place independent of the embryo (46).

IAA plus glutamine can induce triglyceride metabolism in the aleurone tissue of wheat half-grains (Table 5). It is not known whether these controlling factors are transmitted from the embryo. The induction of α-amylase in aleurone tissue is enhanced by the presence of either cytokinin or the attached starchy endosperm (47). These investigators suggested that GA-stimulated α-amylase induction in the aleurone layer of the intact wheat grain needs prior "sensitization" by a cytokinin released from the endosperm.

The deembryonated kernels of some maize seeds (cvs. Senaca Chief sweet corn Pride, and some hybrids), unlike those of barley, wheat, and oat, do not require added GA to stimulate α-amylase activity (48). Any GA in the aleurone cells and/or starchy endosperm deposited there during seed development may stimulate enzyme synthesis. Alternatively, the hydrolases might be preformed in the endosperm during seed maturation and then released, rather than synthesized, following germination.

Applied hormones also influence plant growth. Cytokinins, for example, promote cotyledon expansion, which requires synthesis of new cell wall material. In cytokinin-treated cotyledons, fats are hydrolyzed rapidly without decreasing their dry weight. These observations indicate a probability of transfer of carbon from triglyceride to cellulose; that is, from source to sink. The mode of action of cytokinin on the acceleration of fat hydrolysis and isocitrate lyase activity is not known. The cytokinin may be acting simply to create an internal sink through its effect on cell expansion (2).

Table 4 Response of Isolated Barley Aleurone Layers to GA

Event	Status of knowledge
1. GA induction	
GA induces de novo synthesis of α-amylase	Undisputed
2. During lag phase	
Amino acids must be provided for de novo hydrolase synthesis	Undisputed
GA induces ribosome and membrane synthesis and PER (proliferation of endoplasmic reticulum)	Disputed
GA enhances polysome formation and protein synthesis	Disputed
GA causes qualitative changes in protein synthesis	Undisputed
GA enhances synthesis of poly(A) containing RNA, including mRNA for α-amylase	Possible, but GA-enhanced activation or protection of m-RNA not eliminated
GA acts at the translational level of protein synthesis	No firm evidence available
GA action is mediated via cyclic AMP	Highly unlikely
3. α-Amylase release	
Mode of enzyme release is soluble	Possible
Mode of enzyme release is vascular	More likely, but requires confirmation
Release is an energy-dependent process	Undisputed
4. Other hydrolases	
GA induces de novo synthesis of proteinase, limit dextrinase, pentosanase, and α-glucosidase	Undisputed
Acid phosphatase, ribonuclease, β-1,3-glucanase, and peroxidase synthesized in absence of GA, but GA enhances their release	Undisputed
Major enzyme releases after release channels in aleurone tissue are formed	Undisputed
Pentosan biosynthesis ceases when cell walls are degraded	Undisputed
GA induces ion release	Undisputed, but mechanism unknown

Source: Ref. 2.

V. ENVIRONMENTAL STRESSES

Temperature, light, oxygen, carbon dioxide, and factors influencing the availability of water constitute the main environmental factors controlling seed germination. Any of these factors can favor or inhibit germination in the natural environment. The inhibiting factors and stresses may have profound long-term effects on seed physiology to induce secondary dormancy in seeds, requiring another set of factors to release this dormancy.

Table 5 Sensitivity of Embryo-Dependent Responses of Wheat Aleurone Tissue to GA and Other Growth Promoters

Event	Response to GA	Alternative promoting substances
Induction of triglyceride metabolism	None	Starchy endosperm factor (cytokinin?) induces about 20% activity Cytokinin: (100 mM) induces about 10% activity IAA (10 μM) induces about 10% activity IAA (10 nM) + 1 mM glutamine induces full activity
Neutral lipase activity	None	Embryo factor (including GA?) alone induces about 15% activity IAA (10 μM) + 1 mM glutamine induces about 70% activity Nitrogenous compounds induce 0–55% activity
Isocitrate lyase and malate sythetase activity	Stimulated	None
Release of inorganic phosphate and other ions	Stimulated	Starchy endosperm factor (cytokinin?) might reduce ion release
Induction of α-amylase synthesis	Stimulated	Preinduction of aleurone tissue with cytokin enhances subsequent GA stimulation

Source: Ref. 2.

A. Secondary Dormancy

Whereas primary dormancy in seeds occurs during their development and maturation, secondary dormancy arises in mature seeds as a result of certain environmental stresses causing inhibition of germination. Bewley and Black (49) listed several dormancy-inducing factors, giving examples of each; namely, temperatures above maximum for germination (thermodormancy), temperatures below minimum for germination, darkness (skotodormancy), prolonged white light (photodormancy), prolonged far-red light, anaerobic conditions, carbon dioxide, water stress, gamma-irradiation, drying, and chemicals such as coumarin, naringen, and abscisic acid.

Thermodonnancy caused by above-maximum temperatures is very common (e.g., lettuce). Such seeds do not germinate even when exposed to moderate temperatures favorable for seed germination. The time and temperature relationship appears to be involved in inducing secondary dormancy. Secondary dormancy caused by temperatures below the minimum for germination is less common than thermodormancy. However, both lower and higher temperatures inhibit some seeds like those of *Taraxacum megalorhizon*. The seeds of many winter annuals exhibit low-temperature–induced dormancy.

Skotodormancy is a type of thermodormancy occurring when light-requiring seeds (with primary dormancy) are held imbibed in darkness for several days. The seeds lose their sensitivity to light and gibberellin and require other kind of treatment to release them from dormancy (50,51). Skotodormancy is temperature dependent and can occur only at those temperatures at which there is a light requirement (e.g., above 23°C in Grand Rapids lettuce). In these seeds, skotodormancy appears to be a coat-imposed dormancy (51).

Photodormancy is induced by prolonged white light in some species (*Nemophila insignis*) and by far-red light in many others. The rate of photochrome cycling $(P_r \leftarrow P_{fr})$ and blue light photoreceptor are responsible for the inhibition. The secondary dormancy thus induced can be released by substantially lowering temperature or by isolating the embryo. Inhibition by prolonged far-red light can be removed by irradiation with light effective in establishing a favorable phytochrome photoequilibrium.

High osmotic pressure (−9 bar) can prevent germination of seeds that cannot be recovered by transferring them to water. This secondary dormancy imposed by the seed coat in seeds like those of lettuce can be relieved by puncturing or removing the seed coat (52).

Anaerobic conditions may induce secondary dormancy in seeds (e.g., *Xanthium pennsyl vanicum*) especially at high temperatures such as 27°C. Partial anaerobiosis may be responsible for developing secondary dormancy in buried seeds. Chemicals such as coumarin and naringenin can inhibit germination of dark-germinating lettuce seeds, with the affected seeds becoming light requiring (53,54). ABA induces secondary dormancy in lettuce seeds whose primary dormancy has been broken by light. These seeds do not germinate in the presence of ABA and require further exposure to light in the absence of the inhibitor to promote germination (55).

The environmental stresses imposing secondary dormancy can be counteracted by certain treatments to render them ineffective. Karssen (56) reported that when seeds of *Sisymbrium officinale* and *Polygonum persicaria* are treated with nitrate, they evade dormancy. In many species of seeds, treatments with light or growth regulators (GA) antagonize dormancy-inducing factors like high and low temperatures and osmotic pressures (57). Higher concentrations of oxygen and ethylene in the environment have been found to prevent dormancy in *Xanthium* sp. In contrast to this, anaerobiosis allows some seeds (lettuce, *Rumex crispus,* and S. *officinale*) to escape the effects of higher temperatures and darkness (55).

B. Factors Influencing Seed Germinability

The factors imposing primary dormancy are sometimes found to terminate secondary dormancy. Thus, thermodormancy in lettuce can be overcome by low temperature, or increased oxygen, or carbon dioxide (49). Similarly, dormancy of lettuce imposed by water stress can be relieved by low temperatures and light, whereas osmotically induced dormancy of *Chenopodium* sp. can be broken by light and gibberellins (GA_4 and GA_7). Skotodormancy in *Verbascum blattaria* and *Polygonum persicaria* can be released by chilling and in lettuce by a combination a of light, gibberellin, cytokinin, thiourea, and ethylene (51,58,59). The mechanism of secondary dormancy is not fully understood. A rhythm of dormancy can be traced in nature when seeds in the field pass through cycles of increasing and decreasing dormancy. The seeds dispersed from the mother plant may or may not have primary dormancy, but when either type encounters field conditions unfavorable for germination, a secondary dormancy may set in. Subsequently, this may slowly dissipate and some seeds may succeed in germinating, but those that do not germinate may again encounter unfavorable conditions resulting in secondary dormancy (49).

1. Water Stress

Seed germinability is significantly influenced by the water stress during seed development and maturation. The organs growing most rapidly at the time of water stress suffer the greatest loss of their growth and development (e.g., floral initiation, inflorescence development, anthesis, and fertilization and grain filling).

Enhanced water stress may delay radicle emergence, slow down completion of germination, or reduce total germination. Temperature extremes may accentuate the adverse effects of osmotica; the specific effects vary with the species or cultivars. The initiation of radicle elongation is prevented in osmotically stressed embryos, although the metabolic events associated with germination may continue. The initiation of radicle elongation, indicating termination of the germination process, is suppressed by osmotic potentials, which obviously have no inhibitory effect on the subsequent plant growth (60,61).

The effects of physical treatment and the presence of the pericarp and sarcotesta on seed germination were studied in sago palm (*Metroxylon sagu*). Germination counts were made for seeds submerged completely in water at 30°C in the dark. Seeds from which the pericarp (exocarp and mesocarp) and sarcotesta were removed, that is, cleaned seeds, germinated. In contrast, seeds with the pericarp and/or sarcotesta still attached, that is, uncleaned seeds, did not germinate at all within 100 days, because water absorption was restricted mainly by the pericarp. When cleaned seeds were placed in water with the pericarp, germination was delayed. Soaking cleaned seeds with the pericarp and sarcotesta resulted in no germination in 100 days. However, germination was not hindered when the cleaned seeds were mixed together with only the sarcotesta. The presence of the pericarp and sarcotesta were major factors limiting germination of sago palm seeds. The investigators have discussed the mechanism by which this inhibition of germination is effected (61a).

Chachalis and Smith (61b) applied a hydrophobic polymer to soyabean (*Glycine max*) seeds to regulate imbibition, reduce imbibition damage and improve germination and seedling emergence. These studies were conducted with genotypes prone to imbibition damage due to either seed coat splitting or low seed vigor. The effect of polymer application on seedling emergence and growth was tested under normal (75% soil water-holding capacity; SWHC) or short-term flooded (100% SWHC for the first 4 days and 75% SWHC subsequently) soil conditions. Coating with 24 mg of polymer per seed (Vinamul 3650) regulated the rate of water uptake, reduced imbibition damage, and improved the percentage of germination and seedling emergence. In short-term flooding soil conditions, low seedling emergence was observed regardless of seed aging and was particularly evident in seeds that possessed a high proportion of split testae. There was little difference between uncoated and coated seeds in the time to 50% emergence and seedling growth. The coating with a hydrophobic polymer could reduce rates of water uptake, lower solute leakage, and improve vital tetrazolium chloride staining and partially improve germination or emergence of soyabean seedlings.

Yanping et al. (61c) investigated the effects of storage temperature and seed moisture content on the vigor of Welsh onion (*Allium fistulosum* L.) seeds. After 2 years' storage, several indices of seed vigor, including dehydrogenase activity, germination energy, vigor index, and the emergence percentage of the seeds, declined as storage temperature and seed moisture content increased. When the seeds were stored at room temperature or 6°C for 2 years, the seed moisture content significantly influenced the indices. Increasing seed moisture content resulted in a significant decrease of the indices, some of which were reduced to zero. Low-temperature storage at −6°C reduced the effect of seed moisture content on seed vigor. Seedling emergence from seeds stored for 2 years was closely associated with the storage temperature and seed moisture content. The low-temperature storage of the seeds with low moisture content maintained high seed vigor. If seeds in which the moisture content had been reduced to less than 6% were stored at −6°C, the germination percentage was above 90% after 3 years' storage.

2. Salinity Stress

Salinity stress affects germination by preventing the uptake of water owing to the osmotic potential, as well as by allowing entry of toxic ions into the developing embryo or seedling (49). Factors such as soil type, amount of organic matter present, species and cultivars, and temperature also influence the salt-tolerance limits of the plant species. A plant's sensitivity to salt stress is much greater during seed germination than during subsequent phases, such as seedling growth and development. Radicle growth has been observed to proceed at a lower water potential than the initiation of radicle elongation (61). The seeds of plants adapted to higher salt stresses are found to germinate only when the salt content of their habitat reaches a low level; for example, after heavy rainfall. Although the seeds of most halophytes germinate best in freshwater, they tolerate prolonged soaking in saltwater (62,63). A saltwater treatment prior to transfer of freshwater even promotes germination (64). In contrast, most nonhalophytes are far more sensitive to a presoaking treatment in saline solutions. Alfalfa seeds, for example, germinate very poorly in salt solutions, exceeding a concentration of 0.3–0.4 M, which is attributed to the probable accumulation of toxic ions.

Ruan et al. (64a) investigated effects of various priming treatments, using $CaCl_2$, NaCl, and GA_3. Priming was not found to improve rice seed germination, but significantly increased germination energy and germination index and seedling stand establishment in flooded soil. The addition of GA_3 to $CaCl_2 + NaCl$ did not significantly increase either the speed of emergence or stand establishment as compared with $CaCl_2 + NaCl$ priming solution. The priming of rice seeds might contribute to improved seedling establishment in anaerobic soil.

3. Dehydration and Desiccation

Seed hardening associated with dehydration-rehydration cycles affects germination and seedling growth. The cells that are elongating appear to be less sensitive than those that are dividing during germination. Dehydration of cereal seedlings may permanently injure the primary root. A loss of vigor due to desiccation has been correlated with the amount of leakage of substances (sugars, organic acids, amino acids) during initial imbibition. Desiccation below a critical moisture level results in membrane damage. Although desiccation of seeds or seedlings during their desiccation-insensitive stage has no severe irreversible effects upon membrane integrity; during the sensitive phase, membrane leakiness is increased by desiccation. Desiccation, however, does not adversely affect grains' ability to carry out protein synthesis. Since germination is not delayed by an interrupting period of desiccation, the grains seem to resume protein synthesis at a point close to where it stopped when desiccation was imposed. This reflects the ability of desiccated grains to resume protein synthesis at a more advanced stage of germination. Bewley and Black (49) concluded that germinating seeds tend to be insensitive to moderate water stress or to desiccation stress. These seeds are capable of resuming cellular processes more or less at the point of interruption by water loss. The wetting and drying cycles during the early stages of seedling development do not seem to have any permanent injurious effects on subsequent metabolism. As the seedlings develop, however, they become more sensitive to desiccation, resulting in permanent damage due to drying.

4. Seasonal Fluctuations and Cycles

Egley (65) described the effects of environmental and biological factors that may regulate seed dormancy transitions and the timing of seed germination in soil. The environment and the mother plant influence the dormancy and germination of some seeds during seed development (66). Both the seed's genetic make-up and environmental experience play important roles in programming dormancy and germination. Different environmental experiences of seeds from the same mother plant may result in variations in their time of germination, increasing the chances for establishment and growth to maturity of some seedlings having favorable experiences. Seasonal temperature fluctuations influence dormancy cycles and time of germination of seeds of annual plants (67), with germination being restricted to a particular period of the annual cycle when the probability of a plant's survival and reproduction is high. Temperature is the key signal indicating optimum time for germination. Other environmental and soil factors also interact to influence germination time, but their roles remain secondary to that of temperature (65). Cyclic changes in hormone synthesis or changes in sensitivity to hormones or environmental germination stimuli may be involved in the loss or acquisition of dormancy (68,69). Thus, the changing sensitivity of seeds to either environmental factors or hormonal signals may serve as important sensors in the regulation of seed dormancy and germination. Cycles of varying sensitivity (i.e., number or availability of receptor sites) can be critical in seed dormancy cycles.

Menaka et al. (70) noted that within the inflorescence, the amaranthus seeds' middle position possessed higher germination and vigor than the seeds of proximate and distal positions, which may be collected for use as basal seeds for further multiplication.

Seed germination in cinchona does not improve after acid treatment, where 20 days' washing in water improved the germinability. Cinchona seeds possessing 9.0–19.6 % moisture content were successfully cryopreserved and showed more than 90% germination when tested after 1 year of storage in the vapor phase of liquid nitrogen (at -150°C) (71). Yucel (72) examined seed germination characteristics of six *Salvia* species (*S. cryptantha, S. cyanescens, S. dichroantha, S. tchihatcheffii, S. aethiopis, S. virgata*), and the effects of various concentrations of salt (NaCl), nitrate (KNO_3), and acid (H_2SO_4) on the germination rates of these species were examined. It was observed that, although all salt concentrations reduced the germination percentage, low salt concentrations (0.5–1.0%) had no effect on the germination speed, whereas high concentrations (1–3%) had an inhibiting effect. Increasing the concentration of potassium nitrate reduced the germination percentage for all six *Salvia* species, but two species (*S. cyanescens* and *S. dichroantha*) gave a higher germination speed. Sulfuric acid inhibited both the germination percentage and speed, inhibiting or preventing germination altogether. Significant differences in sensitivity to salt and acid were also determined among the six *Salvia* species.

To enhance seed germination, seeds were subjected to (a) rupture of the coat by piercing, scarification with concentrated H_2SO_4 (exposure for 1, 5, 10, 15, and 20 min) or sandpaper (P 180), ultrasound, dry (dry, hot air at 80, 100, 110, or 120°C [±2°C] for 1, 2, or 3 min), and wet (dipping in boiling water for 1, 2, or 3 min and then cooled in distilled water or dipped in 100 mL of hot water (90 or 70°C) and then left to cool at 20–22°C) heat; (b) softening of the coat or removal of possible physiological dormancy by prechilling at 5–7°C for 60 days, soaking in water for 24 and 48 hr, in ethanol for 8 hr, and in sodium hypochlorite for various times (5, 10, 30, 60, 90, 120, and 180 minutes); and (c) embryo stimulation with GA_3 (50 and 500 µg/Li). The only effective treatments were seed coat piercing, sand paper scarification (final germination of 82–85% and a germination rate of 72–74), and to a lesser extent prechilling. Acid scarification was ineffective if performed for 1 min and harmful if performed for 5 min or more. All the heat treatments, the soaking in 70% ethanol and leaving the seeds to germinate in 1–4% ethanol-water solutions were harmful to the seeds. Soaking in sodium hypochlorite for 5 min had no effect, whereas soaking for 10 min or more was harmful. Ultrasound treatment had no effect if applied for 30 min and was harmful if applied for 60 min. Soaking in water and the addition of GA_3 were ineffective (73).

Welbaum et al. (74) have traced the development of understanding of the forces driving and/or constraining radicle emergence from the seed from its conceptual and physiological origins to its cellular and molecular mechanisms. The driving forces and constraints on expansion by the embryo are examined, particularly for seeds in which the embryo is surrounded by endospermic and testae. Tissues that restrict growth, and using mainly horticultural and agricultural species as examples. Models have been developed to predict germination based on thermal time, hydrotime, and combined hydrothermal time. These population-based models indicate that the timing of germination is closely tied to physiologically determined temperature and water potential thresholds for radicle emergence, which vary among individual seeds in a population. The restraint imposed by tissues surrounding the radicle is a major determinant of the threshold water potential. Enzymatic weakening of these tissues is a key event regulating the timing of radicle emergence. Considerable evidence suggests that endo-β-mannanase is involved in this process in a number of species, although it is doubtful that it is the sole determinant of when radicle emergence occurs. Molecular and biochemical studies are revealing the complexity of events occurring in endospermic and embryonic cells associated with the completion of germination. Unique permeability properties and the presence of enzymes associated with pathogen resistance suggest additional functional roles for the tissues enclosing the embryo. The insights gained from physiology and modeling are being extended by the application of molecular techniques to identify and determine the function of genes expressed in association with germination. Single-seed assay methods, in vivo reporters, specific modification of gene expression, and mutagenesis will be critical technologies for advancing the understanding of germination.

Brits et al. (75) quantified the newly proposed seed biological phenomenon of desiccation-scarification of different testal layers found in *Leucospermum* spp. in Cape fynbos. Intact seeds of *L. cordifolium* were exposed for varying periods to different temperatures, and breakage tests were then performed on both the exotesta and endotesta, the latter following immersion in water. Heat intensity and duration correlated positively with the degree of breakage of both testal layers. The viability of embryos, tested with tetrazolium, was not affected. The effects were furthermore determined on germination itself, of desiccation-scarification by means of heat, and other oxygenation treatments. These treatments distinguished between the effects of oxygenation, desiccation, and heat in the germination of deeply dormant seeds of *L. cordifolium, L. glabrum,* and *L. reflexum.* The promotive effects of heat desiccation on germination were similar for the three species and could not be linked to possible physiological changes caused by high temperature per se. Combined heat desiccation pulsing (scarification) of the exotesta and endotesta resulted in strongly increased germination compared with controls; however, the exotesta appears to act as the primary barrier to oxygen diffusion to the embryo. In natural fynbos, desiccation-mediated scarification of the testal layers was the main means of regulating oxygenation, and thus synchronous germination of dormant *Leucospermum* seeds after exposure to fire.

Seeds of seven field bean cultivars stored for 9 years at -14, $+4$, or $+18°C$ and a water content of 8–10%, were analyzed for viability/vigor level and the content and composition of phospholipids. Seeds retained higher vigor and viability when stored at lower temperatures. The major phospholipids in high-viability seeds (germinability 87–97%) were phosphatidylcholine (PC), phosphatidylethanolamine (PE), and phosphatidylinositol (PI). In deteriorated seeds, PE and PC declined almost completely, whereas the PI level was reduced by 50%. Seeds with intermediate-viability levels (germinability 77–90%) contained intermediate amounts of these phospholipids. The above pattern was observed in both cotyledons and embryonic axes. The extent of phospholipid degradation was correlated with the level of seed exudate conductivity and viability depression due to prolonged storage (76).

Brown and Staden (77) have shown that smoke derived from burning plant material stimulated seed germination in wildflower species from fire-dependent plant communities in South Africa and Australia. In a laboratory study, seeds of *Syncarpha vestita* and *Rhodocoma gigantea* presoaked in plant smoke extract had a significantly higher germination percentage and maintained better viability under storage than the untreated seeds. Kochankov et al. (78) studied the effects of temperature, growth regulators, and other chemicals on the germination rate, emergence, and seedling survival of *Echinacea purpurea* under normal and stress conditions. Seeds of *E. purpurea* germinated from 5 to 35°C. The greatest final germination percentage and the shortest germination time were observed at 25°C. Lowering the osmotic potential from 0 to -0.5 MPa decreased germination at 20 and 25°C. Osmotic potential below -0.5 MPa completely inhibited germination at these temperatures. Ethephon (1445 mg/L) and GA_3 (60 mg/L) enhanced germination rates and increased the final germination percentage of water-stressed seeds. Ethephon and GA_3 also increased emergence and seedling development. The seedlings survived exposure to 0 and -2°C for 17 hr, but 17 hr exposure to $-5°C$ killed all seedlings irrespective of treatments.

North American pawpaw is a temperate species in the mostly tropical Annonaceae. In a germination study, a small proportion (12%) of the seed population germinated after removal from the fruit. The remaining seeds required 8 weeks of chilling stratification to satisfy dormancy. In addition, pawpaw seeds displayed a moderate form of recalcitrance. Seeds lost 50% viability when they were dried from their initial 37–25% moisture. The critical value for total loss in viability was between 15 and 5% moisture. Also, following stratification, pawpaw seeds germinated at 45% without additional water in the germination medium. There was no significant effect of light or temperature (25, 30, or 20/30°C) on standard germination. Seeds showed the best germination with 1 mL of water added to the germination medium in Petri dishes. For standard germination testing, it is suggested that pawpaw seeds should be stratified for 100 days at 5°C followed by germination in rolled towels at constant 25 or alternating 20/30°C. Light was not required and final germination counts should be after 5 weeks (79).

Mature intact pawpaw seeds exhibit delayed germination in the natural environment. Removal of the sarcotesta, water leaching, or drying the seed relieves inhibition, whereas soaking the seed from which the sarcotesta has been removed in the expressed juice of the sarcotesta causes inhibition. High-performance liquid chromatography (HPLC) and bioassays showed that the levels (7.8 ng/g FW) of ABA in pawpaw seeds were insufficient to cause the degree of germination inhibition observed. Acid hydrolysis of pawpaw seed sarcotesta released a number of ether-soluble phenolics. These phenolic compounds were subjected to solvent partitioning and thin-layer chromatography (TLC); five major phenolics were visible. These phenolic fractions inhibited germination and seedling growth of lettuce seeds. The most dominant fraction among the phenolics was collected, crystallized, and identified as p-hydroxybenzoic acid by nuclear magnetic resonance (NMR). The phenolics in the sarcotesta appear to play an important role during dormancy and germination of pawpaw (80).

Grzesik and Nowak (81) investigated the effects of seed matriconditioning and storage or hydropriming with chemicals on *Helichrysum bracteatum* germination, emergence and seedling survival at different temperatures or water-stress conditions. For matriconditioning of *Helichrysum* at 15°C, the best ratio of seeds to Micro-Cel E to water was 1.0:0.4:1.0 for a 6-day treatment, although other ratios such as 1.0:0.2:0.6 and 1.0:0.3:0.8 were also effective. Matriconditioned seeds could be stored in room conditions for 5 months without adverse effect on the final germination percentage and mean germination time. Matriconditioning of seeds for 6 days with Micro-Cel E in the presence of kinetin (20 mg/L), Busz (15 000 mg/L; a mixture of N, P, and K), ethephon (1500 mg/L), and GA$_3$ (300 mg/L) improved seed performance and seedling emergence. Such treatment also increased seedling frost resistance and decreased the harmful effect of water stress on seed performance. Hydropriming for 6 days at 15°C, in the same ratios of chemicals (which kept seeds ungerminated during priming), accelerated germination at 5–30°C after conditioning and did not inhibit germination percentage. Grzesik and associates (82) further examined the yield and germination of *Callistephus chinensis* cv. Aleksandra seeds harvested twice at 2-week intervals in 1993 and 1994 from primary, secondary, and tertiary capitula and then stored for 1 year. The yield of seeds, independently of year, was highest when taken from secondary capitula. The mature seeds harvested in early autumn from the primary or secondary capitula were heavier and germinated better than those collected in late autumn and from the tertiary ones. The range of optimal temperatures for germination of unstored seeds was 5–15°C and it broadened during storage, indicating release from dormancy. The yield and germination of Verbena X hybrida cv. Amulet and Malgosia seeds, collected at different dates in 1993 and 1994, were also examined immediately after harvest or after 6 or 12 months' storage. Mature seeds harvested at the beginning of October were heavier and germinated better than those obtained later. The best temperature for germination of seeds immediately after harvest was 5–15°C. Germination increased when seeds were stored for 6 months in conditions favorable for preservation. The carbohydrate content was higher in the seeds harvested in early autumn and was lower in the seeds collected later. Seed composition depended on time of seed maturation (83).

Bhattacharjee and Mukherjee (84) studied the effects of high-temperature treatment on *Amaranthus lividus* seeds during the early imbibitional phase in terms of germination behavior, leakage of ultraviolet (UV)–absorbing substances, electrolyte leakage, and free radical–mediated membrane deterioration. Transfer of seeds during the early imbibitional period from 25 to 45°C for 4, 8, 12, and 16 hr resulted in the leakage of UV-absorbing substances and electrolytes up to 72 hr. Leakage of α-NH_2 and soluble carbohydrates was measured and the relative leakage ratio and membrane injury index were calculated. High-temperature treatment decreased ethylene formation in germinating seeds. The high temperature caused greater membrane damage by membrane lipid peroxidation in germinating seeds. The elevated level of malondialdehyde and higher lipoxygenase activity in high-temperature–treated germinating seeds substantiated this. Involvement of free oxygen radicals in membrane deterioration in the germinating seeds under heat shock could be indirectly inferred by the reduced activities of free radical scavengers like peroxidase, catalase, and superoxide dismutase with a concomitant rise in free H_2O_2 level.

REFERENCES

1. Association of Official Seed Analysts, Rules for testing seeds. *J. Seed Technol. 12*(3): 109 (1991).
2. J.D. Bewley and M. Black, *Physiology and Biochemistry of Seeds in Relation to Germination*, Vol. I, Springer-Verlag, Berlin, 1978.
3. R.D. Amen. A model of seed dormancy, *Bot. Rev. 34*:1 (1965).
3a. A. Muscolo, M.R. Panuccio, M. Sidari, E. Sessi and S. Nardi, Alternation of amino acid metabolism by humic substances during germination of *Pinus laricio* seeds, *Seed Sci. Technol. 30*:1 (2002).
3b. A. Soltani, S. Galeshi, E. Zeinali and N. Latifi, Germination, seed reserve utilization and seedling growth of chickpea as influenced by salinity and seed size. *Seed Sci. Technol. 30*:51 (2002).
3c. S.R.P. Scuab, A.L. Braccini, O. Ferrarese-Filho, C.A. Scapim and M.C.L. Braccini, Physiological seed quality evaluation and seedling lipid and protein content of soybean (*Glycine max* L. Merril), in the presence of p-coumaric acid, *Seed Sci. Technol. 29*:151 (2001).
3d. L. Angelini, L. Lazzeri, S. Galletti, A. Cozzani, M. Macchia and S. Palmieri, Antigerminative activity of three glucosinolate-derived products generated by myrosinase hydrolysis, *Seed Sci. Technol. 26*:771 (1998).
3e. S. Choudhury and K. Gupta, Studies on the germination mechanism of Catharanthus roses L., G. DON cv. Alba seeds: effect of promoters and pH. *Seed Sci. Technol. 26*: 719 (1998).
4. B.M. Pollock and H.O. Olney, Studies of the rest period. 1. Growth, transportation and respiratory changes in the embryonic organs of the after-ripening cherry seed, *Plant Physiol. 34*(131): (1959).

5. G.P. Berlyn, Seed germination and morphogenesis, *Seed Biology* (T.T. Kozlowski ed.), Academic Press, New York, 1972, p. 223.

5a. M. Khanna, A.K. Vari, S. Baranwal, S.P. Sharma and I. Jethani, Seed germination in relation to stage of maturity, XI National Seminar on Quality Seed to Enhance Agricultural Productivity, UAS, Dharwar, Jan 18–20, 2002, *Seed Tech News* 32: (A. Gaur, A.K. Vari, J.L. Varshney, and K. Kant, eds.), Indian Society Seed Technology, Division of Seed Science and Technology, IARI, New Delhi, March 2002, p. 84.

5b. S. Caloggero and C.A. Parera, Improved germination and emergence of Pistacia mutica by presowing chemical scarification, *Seed Sci. Technol. 28*: 253 (2000).

5c. C.V. Naidu, G. Rajendradu and P.M. Swamy, Effect of plant growth regulators on seed germination of *Sapindus trifolitus* Vahl., *Seed Sci. Technol. 28*: 249 (2000).

5d. G. Singh, S.S. Gill and K.K. Sandhu, Improved performance of muskmelon (Cucumis melo) seeds with osmoconditioning, *Acta-Agrobotanica 52*: 121 (1999).

5e. C. Harakumar, P. Srimathi and K. Malarkodi, Seed storage studies in *Gymnema sylvistre, Madras Agric. J. 86*: 323 (2000).

5f. H. Pandey, S.K. Nandi, M. Nadeem and L.M.S. Palni, Chemical stimulation of seed germination in *Aconitum heterophyllum* Wall. and *A. balfourii* Stapf.: important Himalayan species of medicinal value, *Seed Sci. Technol. 28*: 39 (2000).

5g. C.V. Naidu, G. Ranjendradu and P.M. Swamy, Effect of temperature and acid scarification on seed germination of *Sapindus trifoliatus, Seed Sci. Technol. 27*: 885 (1999).

5h. K.N. Kattimani, Y.N. Reddy and B.R. Rao, Effect of pre-sowing seed treatment on germination, seedling emergence, seedling vigor and root yield of Ashwagandha (*Withania somnifera* Daunal.), *Seed Sci. Technol. 27*: 483 (1999).

5i. K. Tylkowska, Relationships between the color of radish seed coat, the occurrence of *Alternaria* spp and seed germination. *Phytopathol.Polonica, 22*: 107 (1995).

6. T. Moringa, Germination of seed under water, *Am. J. Bot. 13*: 126 (1926).

7. B.F. Forward, Studies on germination of oats, *Proc. Int. Seed Testing Assoc. 23*: 23 (1958).

8. C. Kolloffel, Respiration rate and mitochondrial activity in the cotyledons of *Pisum sativum* L. during germination, *Acta Bot. (Neerl.) 16*: 111 (1967).

9. E.A. Cossins, Formation and metabolism of lactic acid during germination of pea seedlings, *Nature 203*: 989 (1964).

10. C. Kolloffel, Activity of alcohol dehydrogenase in the cotyledons of peas germinated under different environmental conditions, *Acta Bot. (Neerl.) 17*:70 (1968).

11. Y. Yamamoto, Pyridine nucleotide content in the higher plant: effect of age of tissue. *Plant Physiol. 38*: 45 (1963).

12. E.H. Roberts, Seed dormancy and oxidation process, *Dormancy and Survival* (H.W. Woolhouse ed.), Cambridge University Press, Cambridge, UK, *Symp. Soc. Exp. Biol.* 23: 143 (1969)

13. L.O. Copeland and M.B. McDonald, *Principles of Seed Science and Technology,* 3rd ed., Chapman & Hall, New York, 1995.

14. M.B. McDonald, Seed germination and seedling establishment, *Physiology and Determination of Crop Yield* (K.J. Boote, T.R. Sinclair, eds.), Crop Science Society of America, Madison, WI, 1994.

15. A.J. Murdoch, E.H. Roberts, and C.O. Goedert, A model for germination responses to alternating temperatures, *Ann. Bot. 63*: 97 (1989).

16. M.B. McDonald, L.O. Copeland, A.D. Knapp, and D.F. Grabe, Seed development, germination and quality, *Cool-Season Grass Monograph* (L. Moser ed.), American Society of Agronomy, Madison, WI, 1994.

17. D.P. Weeks, D.P.S. Verma, S.N. Seal, and A. Marcus, Role of ribosomal subunits in Eucaryotic protein chain initiation, *Nature 236*: 167 (1972).

18. A. Marcus, Seed germination and capacity for protein synthesis, *Symp. Soc. Exp. Biol.* 23: 143 (1969).

19. N.E. Robinson and J.A. Bryant, Development of chromatin bound and soluble DNA polymerase activated during germination of *Pisum sativum* L., *Planta (Berl.) 127*: 69 (1975).

20. A. Akinlosotu and I.O. Akinyele, The effect of germination on the oligosaccharide and nutrient content of cowpeas (*Vigna linquiculata*), *Food Chem. 39*(2): 157 (1991).

21. D.E. Bilderback, Amylases from aleurone layers and starchy endosperm of barley seeds, *Plant Physiol. 53*: 480 (1974).

22. C.A. Adams and L. Novellie, Acid hydrolysis and autolytic properties of protein bodies and spherosomes isolated from ungerminated seeds of *Sorghum bicolor*, Linn. Moench. *Plant Physiol. 55*: 7 (1975).

23. S. Muto and H. Beevers, Lipase activities in castor bean endosperm during germination. *Plant Physiol. 54*: 23 (1974).

24. A.H.C. Huang, Enzymes of glycerol metabolism in the storage tissues of fatty seedlings, *Plant Physiol. 55*: 555 (1975).

25. N.O. Sundblom and J. Mikola, On the nature of the proteinases secreted by the aleurone layer of barley grain, *Physiol. Plant. 27*: 281 (1972).

26. B.M.R. Harvey and A. Oaks, The role of gibberellic acid in the hydrolysis of endosperm reserves in *Zea mays*, *Planta (Berl.) 121*: 67 (1974).

27. H. Yomo and K. Srinivasan, Protein breakdown and formation of protease in attached and detached cotyledons of *Phaseolus vulgaris* L., *Plant Physiol. 52*: 671 (1973).

28. H. Yomo and J.E. Varner, Control of the formation of amylases and proteases in the cotyledons of germinating peas, *Plant Physiol. 51*: 708 (1973).

29. R.M. Roberts, J. Deshusses and F. Loewus, Inositol metabolism in plants. IV. Conversion of myoinositol to uronic acid and pentose units of acidic polysaccharides in root tips of *Zea mays*, *Plant Physiol. 43*: 979 (1968).

30. J.R. Hall and T.K. Hodges, Phosphorus metabolism of germinating oat seeds, Plant Physiol. *41*: 1459 (1966).

31. B. Tronier, R.L. Ory and K.W. Henningsen, Characterization of fine structure and proteins from barley protein bodies, *Phytochemistry 10*: 1207 (1971).

32. D. Eastwood and D.L. Laidman, The mobilization of macronutrient elements in the germinating wheat grain, *Phytochemistry 10*: 1275 (1971).

32a. S. Bhattacharyya, B. Das and T.K. Bose, Investigation on seed germination of *Nyctanthes arbortristis* (*Oleaceae*) in relation to the total phenol content, *Seed Sci. Technol. 27*: 321 (1999).

33. L.G. Paleg, Physiological effects of gibberellic acid. I. On carbohydrate metabolism and amylase activity of barley endosperm, *Plant Physiol. 35*: 293 (1960).

34. H. Yomo and J.E. Varner, Hormonal control of a secretary tissue. Current Topics in Developmental Biology, Vol. 1. (A.A. Moscona and A. Monroy, eds.), Academic Press, New York, 1971, p. 3.

35. D.E. Briggs, Hormones and carbohydrate metabolism in germinating cereal grains, *Biosynthesis and its Control in Plants (B.V. Milborrow, ed.)*, Academic Press, London, 1973, p. 219.

36. R.L. Jones, Gibberellins: Their physiological role, *Ann. R. Plant Physiol. 24*: 571 (1973).

37. J.E. Varner, Gibberellin control of a secretary tissue, *The Chemistry and Biochemistry of Plant Hormones: Recent Advances in Phytochemistry*, Vol. II (V.C. Runeckles, E. Sondheimer, and D.C. Walton, eds.), Academic Press, New York, 1974, p. 123.

38. G.H. Palmer, The industrial use of gibberellic acid and its scientific basis-a review, *J. Inst. Brew. 80*: 13 (1974).

39. J.D. Mann, Mechanism of action of gibberellins. *Gibberellins and Plant Growth* (H.N. Krishnamoorthy, ed.), Wiley Eastern, New Delhi, 1975.

40. G.H. Palmer and G.N. Bathgate, Malting and brewing, *Advances in Cereal Science and*

Technology (Y. Pomeranz, ed.), American Association of Cereal Chemists, St. Paul, MN, 1976, p. 237.

41. P. Filner and J.E. Varner, A test for de *novo* synthesis of enzymes density labeling with H_2O^{18} of barley α-amylase induced by gibberellic acid. *Proc. Natl. Acad. Sci. USA. 58*: 1520 (1967).

42. M.J. Chrispeels and J.E. Varner, Gibberellic acid enhanced synthesis and release of α-amylase and ribonuclease by isolated barley aleurone layers, *Plant Physiol. 42*: 398 (1967).

43. A.R. Kermode, Regulatory mechanisms involved in the transition from seed development to germination, *Crit. Rev. Plant Sci. 9*(2): 155 (1990).

44. L.G. Paleg, B.G. Coombe, and M.S. Buttrose, Physiological effects of gibberellic acid. V. Endosperm responses of barley, wheat and oats, *Plant Physiol. 37*: 798 (1962).

45. R.J.A. Tavener and D.L. Laidman, The induction of lipase activity in the germinating wheat grain, *Phytochemistry 11*: 989 (1972).

46. K. Varty and D.L. Laidman, The pattern and control of phospholipid metabolism in wheat aleurone tissue, *J. Exp. Bot. 27*: 748 (1976).

47. D. Eastwood, R.J.A. Tavener, and D.L. Laidman, Sequential action of cytokinin and gibberellic acid in wheat aleurone tissue, *Nature 221*: 1267 (1969).

48. L.D. Goldstein and P.H. Jennings, The occurrence and development of amylase enzymes in incubated deembryonated maize kernels, *Plant Physiol. 55*: 893 (1975).

49. J.D. Bewley and M. Black, *Physiology and Biochemistry of Seeds*, Vol. 2, Springer-Verlag, Berlin, 1982.

50. R.B. Tayorson and S.B. Hendricks, Phytochrome transformation and action in seeds of *Rumese crispus* L. during secondary dormancy, *Plant Physiol. 52*: 475 (1973).

51. J.D. Bewley, Secondary dormancy (Skotodormancy) in seeds of lettuce (*Lactuca sativa* cv. Grand Rapid) and its release by light, gibberellic acid and bezyladenine, *Physiol. Plant. 49*: 277 (1980).

52. A. Kahn, An analysis of "dark osmotic inhibition" of germination of lettuce seeds, *Plant Physiol. 35*:1 (1960).

53. G. Nutile, Inducing dormancy in lettuce seeds with coumarin, *Plant Physiol. 20*: 433 (1945).

54. J.C. White, J.R. Hillman and I.D.I. Phillips, Studies on the chemical induction of a light requirement for germination in seeds of lettuce, *Lactuca sativa* L. cv Great Lakes, *J. Exp. Bot. 23*: 987 (1972).

55. C.M. Karseen, Environmental conditions and endogenous mechanisms involved in secondary dormancy of seeds, *Isr. J. Bot. 29*: 45 (1981).

56. C.M. Karseen, Patterns of change in dormancy during burial of seeds in soil, *Isr. J. Bot. 29*: 65 (1981).

57. D.N. Kristie, P.K. Bassi and M.J. Spencer, Factors affecting the induction of secondary dormancy in lettuce, *Plant Physiol. 67*: 1224 (1981).

58. H.L. Speer, A.L. Hsiao and W. Vidaver, Effects of germination promoting substances given in conjugation with red light on the phytochrome mediated germination of dormant lettuce seeds (*Lactuca sativa* L.), *Plant Physiol. 54*: 852 (1974).

59. W. Vidaver and A.L. Hsiao, Actions of gibberellic acid and phytochrome on the germination of Grand Rapids lettuce seeds, *Plant Physiol. 53*: 266 (1974).

60. T.W. Hegarty, The physiology of seed hydration and dehydration and the relation between water stress and the control of germination: a review, *Plant Cell Environ. 1*: 101 (1978).

61. T.W. Hegarty and H.A. Ross, Some characteristics of the water sensitive process in the inhibition of germination by water stress, *Ann. Bot. (Lond.) 42*: 1223 (1978).

61a. H. Ehara, O. Morita, C. Komada and M. Goto, Effect of physical treatment and presence of the pericarp and sarcotesta on seed germination in sago palm (*Metroxylon sagu* Rottb.), *Seed Sci. Technol. 29*: 83 (2001).

61b. D. Chachalis and M.L. Smith, Hydrophobic-polymer application reduces imbibition rate and partially improves germination or emergence of soybean seedlings, *Seed Sci. Technol. 29*: 91 (2001).

61c. Y. Yanping, G. Ronggi, S. Qingquan and L. Shengfu, Vigor of Welsh onion seeds in relation to storage temperature and seed moisture content, *Seed Sci. Technol. 28*: 817 (2000).

62. V.J. Chapman, *Salt Marshes, Salt Deserts of the World*. International Science Publication, New York, 1960, p. 315.

63. I.A. Ungar, Holophyte seed germination, *Bot. Review, 44*: 233 (1978).

64. L.A. Boorman, Some aspects of the reproductive biology of *Limonium vulgare* Mill and *Limonium humile* Mill, *Ann. Bot. (Lond.) 32*: 803 (1968).

64a. S. Ruan, Q. Xue and K. Tylkowska, The influence of priming on germination of rice (*Oryza sativa* L.) seeds and seedling emergence and performance in flooded soil, *Seed Sci. Technol. 30*: 61 (2002).

65. G.H. Egley, Seed germination in soil: Dormancy cycles. *Seed Development and Germination*, (J. Kigel, G. Galili, eds.), Marcel Dekker, New York, 1995, p. 529.

66. Y. Gutterman, Material effects on seeds during development, *Seeds: The Ecology of Regeneration in Plant Communities* (M. Fenner, ed.), C.A.B. International, Wallingford, UK, 1992, p. 27.

67. J.M. Baskin and C.C. Baskin, Physiology of dormancy and germination in relation to seed bank ecology, *Ecology and Soil Seed Banks,* Academic Press, San Diego, 1985, p. 53.

68. H.W.M. Hilhorst and C.M. Karseen, Seed dormancy and germination: The role of abscisic acid and gibberellins and the importance of hormone mutants, *Plant Growth Regulation 11*: 225 (1992).

69. C.M. Karseen and H.W.M. Hilhorst, Effect of chemical environment on seed germination, *Seeds: The Ecology of Regeneration in Plant Communities* (M. Fenner, ed.), C.A.B. International, Wallingford, UK, 1992, p. 327.

70. C. Menaka, P. Balmurugan, and K. Natrajan, Position of seeds in the inflorescence on quality in Amaranthus, XI National Seminar on Quality Seed to Enhance Agricultural Productivity, UAS, Dharwar, Jan 18–20, 2002, *Seed Tech News* 32: (A. Gaur, A.K. Vari, J.L. Varshney, K. Kant, eds.), Indian Society Seed Technology, Division of Seed Science and Technology, IARI, New Delhi, March 2002, p. 29.

71. J. Som, Studies on germinabiliy and cryopreservation of Cinchona (*C. ledgeriana*) seeds, *Seed Sci Technol. 28*: 865 (2000).

72. E. Yucel, Effects of different salt (NaCl), nitrate (KNO₃) and acid (H₂SO₄) concentrations on the germination of some. Salvia species seeds, *Seed Sci. Technol. 28*: 853 (2000).

73. A. Boscagli and B. Sette, Seed germination enhancement in *Satureja montana* L, *Seed Sci. Technol. 29*: 347 (2001).

74. G.E. Welbaum, K.J. Bradford, K.J. Yim-Kyuock, D.T. Booth, M.O. Oluoch and K.O. Yim, Biophysical, physiological and biochemical processes regulating seed germination, *Seed Sci. Res. 8*: 161 (1998).

75. G.J. Brits, F.J. Calitz and N.A.C. Brown, Heat desiccation as a seed scarifying agent in *Leucospermum* spp. (*Proteaceae*) and its effects on the testa, viability and germination, *Seed Sci. Technol. 27*: 163 (1999).

76. D.J. Michalczyk, R.J. Gorecki, K. Kulka and A. Rejowski, Changes in phospholipid fractions of field bean (*Vicia faba* var. Minor Harz) seeds stored at different temperatures, *Plant-Breeding & Seed Sci. 42*: 37 (1998).

77. N.A.C. Brown and J. van Staden, Plant-derived smoke: an effective seed pre-soaking treatment for wildflower species and with potential for horticultural and vegetable crops, *Seed Sci. Technol. 26*: 669 (1998).

78. V.G. Kochankov, M. Grzesik, M. Chojnowski and J. Nowak, Effect of temperature,

growth regulators and other chemicals on *Echinacea purpurea* L. Moench seed germination and seedling survival, *Seed Sci. Technol. 26*: 547 (1998).

79. C.H. Finneseth, D.R. Layne and R.L. Geneve, Requirements for seed germination in North American pawpaw (*Asimina triloba* L Dunal), *Seed Sci. Technol. 26*: 471 (1998).

80. Y.J. Chow and C.H. Lin, p-Hydroxybenzoic acid as the major phenolic germination inhibitor of papaya seed, *Seed Sci. Technol. 19*: 167 (1991).

81. M. Grzesik and J. Nowak, Effects of matriconditioning and hydropriming on *Helichrysum bracteatum* L. seed germination, seedling emergence and stress tolerance, *Seed Sci. Technol. 26*: 363 (1998).

82. M. Grzesik, K. Gornik and M.G. Chojnowski, Effect of harvest time on the quality of *Callistephus chinensis* Nees cv. Aleksandra seeds collected from different parts of plant, *Seed Sci. Technol. 26*: 263 (1998).

83. M. Grzesik, K. Gornik and M.G. Chojnowski, Effect of environmental conditions and the harvest time on the seed quality of Verbena X hybrida Voss, *Seed Sci. Technol. 26*: 131 (1998).

84. S. Bhattacharjee and A.K. Mukherjee, The deleterious effects of high temperature during early germination of *Amaranthus lividus*; *Seed Sci. Technol. 26*: 1 (1998).

5
Seed Viability and Longevity

I. INTRODUCTION

How long a seed can remain alive varies greatly from the short-lived seeds of food crops to the long-lived seeds of weeds and wild plants. According to Quick (1), seed longevity is an ecological characteristic of a plant as well as a morphological and biochemical one. Plants (particularly woody plants) characteristic of arid climates have in general longer lived seeds than do plants from tropical or humid-temperate habitats. Seeds of Leguminosae, including those of the species of Mimosaceae, are long lived with a marked tendency for longevity. Other plant families that have greater than average proportions of species with exceptionally long-lived seeds include the palms, cannas, waterlilies (lotuses), spurges, soapberries, buckthorns, mallows, and morning glories. Within a single plant species, varieties and strains with somewhat different genetic constitutions may vary in germination and longevity.

A. Life Span of Seeds

On the basis of life span, seeds can be conveniently but rather arbitrarily divided into three categories:

1. *Microbiotic seeds,* which have a life span that does not exceed 3 years
2. *Mesobiotic seeds,* which have a life span varying from 3 to 15 years
3. *Macrobiotic seeds,* which have a life span ranging from 15 to more than 100 years

These categories, however, mean little until information on a favorable storage environment of these seeds is available (2). Improving storage conditions for any given seed may change its categorization from microbiotic to mesobiotic or even macrobiotic. Harrington (3) cited a few examples of viability spans of some seeds in certain storage environments (Table 1).

Some seeds are genetically and chemically well equipped for longer storability than others under similar conditions; for example, *Canna, Lotus,* and *Lupins,* which are viable after even 500 years. Harrington (3) reported that seeds of *Albizia, Cassia, Goodia,* and *Trifolium* can retain their germinability as long as 100 years. Seeds of most species of agricultural crops are relatively short-lived, including vegetables and agronomic crops. Justice and Bass (4) classified the crop species on the basis of a relative storability index (1 = 50% of the seeds expected to germinate after 1–2 years' storage; 2 = 50% of the seeds expected to germinate after 3–5 years' storage; 3 = 50% of the seeds expected to germinate after 5 or more years' storage), as shown in Table 2.

Table 1 Viability Span of Some Seeds with Specified Storage Environment

Plant species	Longevity (y) and % germinated[a]	Storage environment
Corn poppy (*Papaver rhoeas*)	10 (53)	Laboratory
Rape (*B. napus*)	10 (12)	Laboratory
Squash, marrow (*C. pepo*)	10 (55)	Laboratory, sealed
Soybean (*Glycine max*)	13	Storage
White spruce (*Picea glauca*)	15 (40)	−4°C, sealed
Hemp (*Cannabis sativa*)	19	Laboratory, sealed
Tobacco (*N. tabacum*)	20 (92)	Laboratory, sealed
Lettuce (*L. sativa*)	20 (86)	−4°C, 8% RH
Dwarf bean (*P. vulgaris*)	22 (30)	Laboratory
Onion (*A. cepa*)	22 (33)	Dry, laboratory
Strawberry (*Fragaria sp.*)	23 (89)	5°C, dry
Barley (*Hordeum vulgare*)	32 (96)	Storage
Alfalfa (*Medicago sativa*)	78 (22)	Laboratory
Kidney vetch (*A. vulneraria*)	90 (4)	Laboratory

[a] Figures in parentheses indicate percentage germinated.
Source: Ref. 3.

The longevity of seeds in storage is influenced by factors such as temperature, moisture, and oxygen pressure. Lower temperatures and moisture content enhance the period of longevity of seeds, whereas the higher oxygen pressure for many species shortens it. Exceptions to these generalizations are the recalcitrant (unorthodox) seeds of some plant species, which must retain relatively high moisture content in order to maintain maximum viability (2). According to Berjak et al. (5) and Smith and Berjak (6), the recalcitrant seeds are shed at high water content and are intolerant of dehydration and often also of chilling. There are, however, degrees of recalcitrant behavior ranging from seeds tolerating both relatively substantial water loss and low temperature to those that rapidly lose viability when dehydrated even slightly and/or are exposed to chilling.

Recalcitrant and orthodox seeds differ greatly in their ecology and seed morphology (7). Most recalcitrant seeds mature and exist in their fruits and are covered with fleshy or juicy layers and on impermeable testa. At physiological maturity, the recalcitrant seeds have a much higher moisture content (50–70%) than orthodox seeds (30–50%). They are also larger than most orthodox seeds, although their embryos are only about 15% the size of an orthodox seed embryo (8). The recalcitrant seeds are thus difficult to store. They probably do not experience dormancy and continue to develop and progress toward germination (9). Chin and Roberts (10) have reviewed various aspects of the handling and storage of recalcitrant seeds. Some seeds now appear to fit neither an orthodox nor a recalcitrant category. Ellis (11) suggested an intermediate category for these seeds; for example, citrus and coffee seeds (8).

Ferrant et al. (12) suggested that seed behavior should be considered in the context of the natural habitat of the individual plant species in order to understand the phenomenon of recalcitrance. The large-seeded species of hardwoods (e.g., *Corylus*,

Table 2 Classification of Seeds of Crop Species on the Basis of Relative Storability Index

	Relative storability index	
1	2	3
Reed canary grass	Barley	Alfalfa
Field corn	Kentucky grass	Sugarbeet
Cow pea	Mountain brome	Velvet bent grass
Harding grass	Buck wheat	Atsike clover
Yellow lupine	Chrimson clover	White sweet clover
Pearl millet	Chewing fescue	Tomato
Orchard grass	Flax	Creeping bent grass
Soybean	Rope	Buffalo grass
Lettuce	Perennial rye grass	Carpet grass
Parsley	Bird foot trefoil	Hairy vetch
Bermuda grass	Wheat	
Cotton	Cabbage	
Dollis grass	Sweet corn	
Korean lespedeza	Pumpkin	
Foxtail millet	Bahia grass	
Peanut	Field bean	
Red top	Smooth brome	
Rye	Berseem clover	
Sorghum	Subterranean clover	
Sunflower	Tall fescue	
Onion	Oats	
Parsnip	Trimothy	
	Broccoli	
	Carrot, turnip	
	New Zealand spinach	

Source: Ref. 4.

Castanea, Quercus, Aescujus, Salix, and *Jugjans)* and plantation crops *(Coffee, arabica, Cola nitida, Theobroma cacao,* and *Hevea brasiliensis*), as well as the seeds of most aquatic species (e.g., *Zizania aquatica*), rapidly lose their viability in dry conditions. The recalcitrant seeds of some tropical species do not sustain low temperatures during storage. Seeds of *Z. aquatica* must be stored moist at low temperatures, a condition that makes them lose their dormancy and sprout. Deterioration of moist seeds due to contaminating microorganisms must also be prevented. The optimal conditions for the storage of recalcitrant seeds need to be determined to establish the quantitative relationship between the environmental parameters and viability. Chin and Roberts (10) have reviewed the characteristics, germination, processing, and storage of calcitrant seeds.

Merlin and Palanisamy (12a) evaluated the viability and storability of jackfruit (cv. Singapore) seeds. Under ambient dry conditions with free air circulation, the loss of seed moisture results in a significant reduction of seed germination. Seeds treated with carbendazim at 2 g/kg and stored in a polythene bag at 10°C recorded 24%

germination after 56 days of storage, whereas seeds stored in a cloth bag under ambient conditions lost viability within 21 days.

Euterpe edulis, an economically important palm species in southern Brazil, has been reported to present recalcitrant seeds. Andrade (12b) investigated seed desiccation sensitivity under different drying regimens and evaluated seed storage during 12 months. Germination and moisture content (fresh weight basis) were assessed periodically during desiccation methods. Viability was maintained when fresh seed was dehydrated from 45 to about 35–30% moisture content; no germination occurred after drying to 24–18% moisture content. Desiccation sensitivity was independent of drying method. These characteristics substantiate previous indication of a "recalcitrant" behavior. In the storage trials, seeds were tested for germination following storage in polythene bags at 15, 12, and 5°C and 44, 40, and 36% moisture content. At 15 and 12°C and 40% moisture content, no further reduction in germination occurred for 12 months. In contrast, seeds stored with 36% moisture content lost germination independent of the storage temperature used. Chilling damage occurred at 5°C for all seed moisture contents. Partial drying to 40% moisture content before storage at 12–15°C is recommended in the short term.

Seed aging is natural phenomenon that occurs in all seeds, including those stored in dry and low-temperature rooms. *Phaseolus vulgaris* cv. Carioca seeds, harvested in five different years (from 1993 to 1997) in São Paulo, Brazil, were stored at 15°C. Seed samples were germinated and evaluated according to The Brazilian Rules for Testing Seeds. The newest seed sample was submitted to artificial aging. Small (1 g) samples of all materials (naturally and artificially aged) were ground and proteins extracted. Equal quantities of protein were loaded onto the gels and electrophoresis was carried out. Although seeds submitted to 24, 48, 72, and 96 hr of artificial aging did not show any statistical difference (5%) in relation to unaged seeds in physiological parameters related to emergence, some statistically different parameters related to dry matter and length of shoots and roots occurred. Twenty-four hours of artificial aging at 41°C improved these factors. Protein patterns were changed after 72 hr of artificial aging and naturally aged seeds showed alterations after 2 years' storage at 15°C. Aged seeds, naturally and artificially, had decreasing germination, vigor, and changes in the characteristic banding pattern of proteins. Physiological parameters and electrophoresis examination showed that for naturally aged seeds physiological parameters were more sensitive, whereas artificially aged seeds (41°C/100% RH) had electrophoretic profiles that were more efficient for seed lot discrimination (12c).

Aisandra butyracea (*Diploknema butyracea*) seeds are recalcitrant. A moisture content below 25% resulted in loss of viability. Rapid decline in seed viability due to moisture loss was accompanied by a reduction in soluble protein and carbohydrate content. The storage of seed with pulp at 3°C was effective in slowing the metabolic activity of seeds, but viability could not be extended beyond a term of 20 days (12d).

Kumar et al. (12e) studied variability for seed longevity among 10 pea varieties under accelerated aging conditions (80% relative humidity and 40 ± 1°C temperature). Germination and seedling vigor decreased in all the varieties with an increase in the duration of storage. However, varieties differed in the degree of germination decline even when stored under similar conditions and duration. On the basis of LD_{50} of germination, the varieties JP888, T163, Pant P9, and EC33866 were categorized as good storers; Rachna, Hans, PG3, and KFPD8 as moderate storers; and HFP4 and Pusa 10 as poor storers. Seed size was not correlated with seed viability in storage;

however, seed size within a genotype was found to affect longevity in some varieties. Seed longevity is genotypically controlled.

The longevity of large and small seeds of *Amaranthus gangeticus* (*A. tricolor*) and *A. hybridus* stored at different moisture levels and temperatures was studied. Small seeds (<1 mm) could be stored for longer than large (>1.41 mm) ones. Within temperature ranges of 25–40 and 22–40°C for *A. hybridus* and *A. tricolor*, respectively, seeds retained viability better at lower moisture contents and storage temperatures. Time required for germination increased with seed age, with the rate of increase being higher in more severe environments. Before storage, large seeds germinated faster than small ones, but this trend was reversed as seeds aged. Large seeds produced taller seedlings than did small seeds before storage but not after storage. The frequency of abnormal seedlings increased with the storage period, with the higher frequency generally being in large seeds (12f).

Choudhuri et al. (12g) subjected seeds stored for up to 15 months under ambient laboratory conditions to hydration-dehydration treatments by (a) soaking in water for 6 hr, (b) dipping for 5 min and then keeping covered for 2 hr, and (c) moisture equilibration with a water-saturated atmosphere for 48 hr, followed in all cases by slow drying back to the original weight (7.8% moisture content). Seeds were examined for loss of vigor and viability during subsequent storage under accelerated or natural aging conditions. Hydration-dehydration treatments effectively slowed down physiological deterioration, but the effects depended on seed vigor at the time of treatment. The use of monosodium hydrogen phosphate (10^{-4} M) and sodium sulfate (10^{-4} M) in the soak water caused only minor additional advantages. The lower electrical conductance of seed steep water and reduced leaching of sugars and amino acids of moisture equilibrated–dried seeds following aging indicated better membrane integrity maintenance than that in untreated seeds. The improved storability of hydrated-dehydrated seeds was also associated with greater dehydrogenase activity and appreciably lowers lipid peroxide formation in the cells.

II. CAUSES OF LOSS OF SEED VIABILITY

Harrington (13,14) established the pattern of loss of viability of seeds related to their storage environment and classified the majority of seed species as "orthodox seeds." These seeds could conform to the following rules of thumb:

1. The storage life of the seeds doubles for each 1% decrease in seed moisture content. For several seed species, the storage life is doubled for each 2% decrease in moisture content (2).
2. The storage life of seeds doubles for each 10°F (5.6°C) decrease in storage temperature.
3. The arithmetic sum of the storage temperature in °F and % relative humidity (RH) should not exceed 100, with no more than half being contributed by the temperature.

The seed moisture content is generally calculated on a wet weight basis, according to International Seed Testing Association (ISTA) practices:

$$\% \text{ Moisture content} = \frac{\text{Fresh wt. of seed } - \text{ Dry wt. of seed}}{\text{Fresh wt. of seed}} \times 100$$

whereas many biologists calculate seed moisture content on a dry weight basis:

$$\% \text{ Moisture content} = \frac{\text{Fresh wt. of seed} - \text{Dry wt. of seed}}{\text{Dry wt. of seed}} \times 100$$

This makes a significant difference at high seed moisture content values.

A. Seed Moisture

At moisture contents exceeding 30%, nondormant seeds may germinate, and at moisture contents from 18 to 30%, rapid deterioration by microorganisms may set in, especially in aerobic conditions. Christensen (15) observed that fungi can grow and destroy many seeds stored at a 10–18% moisture content. Respiration of seeds enhances at a moisture content in excess of 18–20%. Insect activity is negligible below 8–9% moisture content. However, seeds dried below 4–5% moisture content, although immune from attack by insects and fungi, may deteriorate faster than those maintained at a moisture content 1–2% higher (2).

According to Copeland and McDonald (8), the hygroscopic nature of seeds allows them to maintain their equivalent moisture content at a given relative humidity, equilibrium being attained when seeds have no further tendency to either absorb or lose moisture. Hygroscopic equilibrium curves (absorption isotherms) graphically express the relationship between the seeds' moisture content and their ambient relative humidity at constant temperatures. These are established by measuring the absorption or desorption at successive relative humidities and can be used to predict the seed moisture content at any given relative humidity. The hygroscopic moisture equilibrium curve in Fig. 1 is a sigmoidal curve with three distinct phases of water absorption or desorption. Phase one represents tightly held water, which may be part of the chemical structure of seeds and cannot be removed without destroying the seed tissue. The second phase represents water held more loosely. For most seeds, this represents a straight-line relationship between humidity and moisture content. Water held by the third (upper) phase is most easily removed by drying and contributes significantly to seed deterioration during storage (Fig. 1) (8).

The activities of the pests of stored seeds are closely related to the RH of the interseed atmosphere rather than to the seed moisture content per se. The moisture content of the oilseeds may be different than that of starchy cereal seeds even if the equilibrium RH for both is the same. Thus, seeds having higher starch content (grasses, cereals, and most legumes except soybean and groundnut) have a moisture content of about 10% at 45% RH, whereas oilseeds like sunflower and castor have about 6–7% moisture at this RH. Seeds to be cold stored (0–5°C) need to be sealed in moisture-proof containers or stored in a dehumidified atmosphere to avoid deterioration.

Agrawal (16) reported deterioration of orthodox seeds during storage. In these seeds, a decrease in the activity of PEP (pyruvate-enol-phosphate) carboxylase and RuBP (ribulose-biphosphate) carboxylase and an increase in protease activity was observed during storage. The increased protease activity was thought to be responsible for the decrease in the activities of other enzymes. Leaching of water-soluble sugars and leucine-^{14}C increased with seed deterioration. The enhanced leaching was attributed to membrane deterioration during seed storage. The Embden-Meyerhof pathway (EMP) before seed deterioration exclusively oxidized glucose, whereupon the pentose phosphate oxidized a part of the glucose pathway. These changes were found

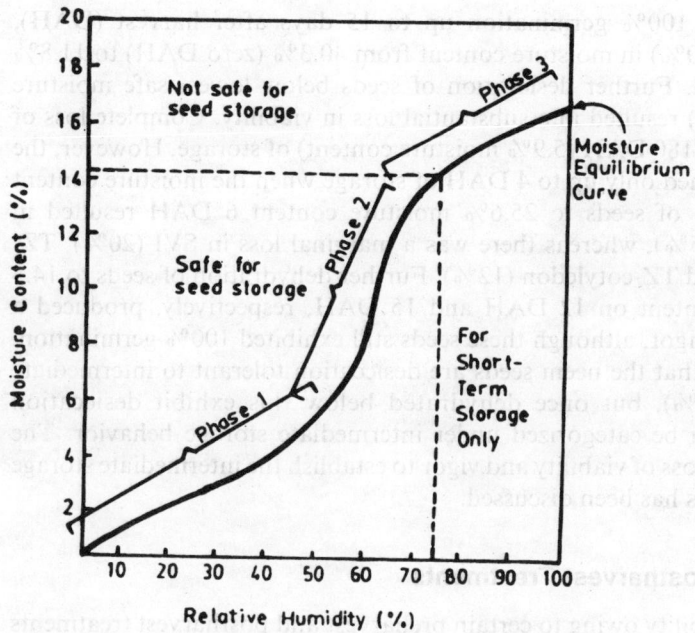

Figure 1 An absorption isotherm showing relationship of seed moisture content with relative humidity of air at a given temperature. (From Ref. 8.)

to precede the loss of seed germination. The freeze-drying of certain seeds improves their longevity in storage if their initial moisture content before treatment is less than 15% (2).

Roberts (17–19) developed mathematical formulas to predict with reasonable accuracy the percentage of viability of a seed population in relation to storage temperature and seed moisture content. A study of the survival curves of seven species of seeds (wheat, rice, peas, barley, broad bean, tomato, and onion) showed that under a given set of constant conditions, a given sample of seeds has a mean viability period, and that the viability periods of the individual seeds in a population are randomly distributed around the mean value. The survival curves of percentage of viability plotted on a probability scale indicate that under constant storage conditions, the frequency distribution of seed deaths in time is normal even in adverse storage environments (accelerated aging) where seeds die rapidly.

The nomographs, however, can be used only as a rough guide to predict the viability of any one species. Bewley and Black (2) described the major limitations·in the use of these nomographs, outlining the problems faced in determining optimum storage conditions. Different cultivars and harvests of a species may show different viability characteristics under the same storage conditions. The differences between harvests according to storage conditions (32°C; 70–90% RH) can be very large. Similarly, cultivar differences in viability can be great when the seeds are stored under uniform conditions (20).

Varghese and Naithani (20a) examined the response of viability and vigor to desiccation and storage period in neem seeds in ambient conditions. Freshly harvested neem seeds maintained 100% germination up to 15 days after harvest (DAH), although massive loss (70%) in moisture content from 40.3% (zero DAH) to 11.8% (15 DAH) was recorded. Further desiccation of seeds below lowest safe moisture content (LSMC) (11.8%) resulted in a substantial loss in viability. Complete loss of viability was recorded at 180 DAH (5.9% moisture content) of storage. However, the vigor index was maintained only up to 4 DAH of storage when the moisture content was 29.5%. Desiccation of seeds to 25.6% moisture content 6 DAH resulted in significant loss in GI (75%), whereas there was a marginal loss in SVI (20%), TZ-embryonic axes (4%) and TZ-cotyledon (12%). Further dehydration of seeds to 14.4 and 11.8% moisture content on 12 DAH and 15 DAH, respectively, produced a substantial loss in seed vigor, although these seeds still exhibited 100% germination. These results suggested that the neem seeds are desiccation tolerant to intermediate moisture content (11.8%), but once dehydrated below this exhibit desiccation sensitivity, and thus can be categorized under intermediate storage behavior. The significance of drying in loss of viability and vigor to establish the intermediate storage physiology of neem seeds has been discussed.

B. Harvesting and Postharvest Treatments

Seeds may lose their viability owing to certain preharvest and postharvest treatments (i.e., threshing, drying). Mechanical injuries inflicted during harvesting severely decrease the viability of some seeds (21). Dickson and Boettger (22) reported that certain lines of large-seeded legumes were more susceptible to mechanical damage than other cultivars. Cereal grains such as wheat, rice, and barley are comparatively more immune to mechanical damage because of the presence of a protective palea and lemma. Small-seeded species tend to escape harvest injury, and spherical seeds are better protected against injury than the elongated or irregularly shaped seeds. The presence of surrounding structures (e.g., cereal grains with intact hulls) enhances viability during storage. The chaff or hulls are also known to protect seeds from mold attack. Scarification, mechanical dehulling, and delinting (cotton) with mineral acids significantly decrease the storage life of seeds.

The injured and deeply bruised areas on seeds serve as centers for infection by microorganisms, resulting in accelerated aging. Injuries close to the embryonic axis bring a more rapid loss of viability during storage than the injuries located at other places.

Quick or excessive drying as well as drying at elevated temperatures can reduce seed viability. Poor ripening and harvesting conditions also lower the viability of seeds. The viability values of seeds predicted by using nomographs may be overestimated if the seeds are subjected to severe harvesting and postharvest treatments (2).

C. Storage Conditions

Roberts (17) calculated values of viability constants for seeds under hermetic (airtight) storage conditions; that is, by maintaining temperature and moisture constant. However, respiration by both seeds and the associated microorganisms changes the gaseous atmosphere of the storage room. Roberts and Abdulla (23) observed that pea seeds stored at 18.4% moisture at 25°C decreased the O_2 content from 21.0 to 1.4%

and increased the CO_2 content from 0.03 to 12.0% within 11.3 weeks; the time taken for 50% loss of viability of pea seeds under these conditions. The viability of seeds stored in an atmosphere of nitrogen was considerably higher than those stored in aerobic conditions for similar periods of time, but not much greater than those hermetically sealed in air.

Most seeds retain their viability better in sealed containers than in open storage (3,20). Bewley and Black (2) attribute this improvement to the fact that sealed containers provide a simple and convenient method of controlling the seed moisture content after their drying. A decrease in O_2 concentrations, whether obtained by vacuum or by raising the partial pressure of CO_2, N_2, argon, or helium, increases viability of seeds. The deleterious effects of oxygen were found to be more marked for barley stored at higher moisture content. Also, the advantage of an oxygen-free atmosphere decreases with an improvement in other storage conditions.

Ellis and Roberts (24) modified the basic viability equations, taking into account variations in potential longevity between seed lots within a species due to genotype and prestorage conditions. The new equation describes the relationship between the standard deviation of the distribution of seed deaths in time and the storage environment. Ellis and Roberts (25) tested viability formulas and nomographs on barley. The two seed lots of cv. Golden Promise and Julia, which had been produced in the same field environment, did not differ in viability. Their longevity in similar storage conditions, however, differed considerably. The observed and predicted half-viability periods of these two seed lots of barley agreed reasonably, with an average predictive error of about 14%. According to Bewley and Black (2), further studies are needed to establish a more accurate prediction of the viability of seeds in storage, especially for application of viability equations to more species, including recalcitrant species, and to incorporate the quantitative effects of oxygen into the viability equations.

In addition to these causes, other factors influencing the life span of seeds include internal factors like the physical state and physiological condition of the seed, mechanical damage during harvesting, cleaning and handling, the presence of microflora, and seed maturity. Factors such as temperature, moisture, variety, and nutrient status influence seed maturity and consequently seed storability (8).

Seeds of 11 cultivars of six-grain leguminous species were placed in cloth bags and stored under room conditions (25a). Standard germination tests were conducted at 30, 162, 298, 443, 610, 960, 1099, 1250, 1401, 1567, and 1749 days after seed harvest. Germination started to decline at 183 days after harvest for *Phaseolus vulgaris* cv. Milionario; at 486 days for *P. vulgaris* cv. Fortuna; at 1012 days for *P. lunatus*; at 419 days for *Vigna angularis* cv. Kintoki; at 490 days for *V. angularis* cv. Dainagon; at 720 days for *V. unguiculata* cv. EPACE-6; at 964 days for *V. unguiculata* cv. CNC 0434; at 828 days for *V. umbellata* introduction E-7; at 906 days for *V. umbellata* introduction E-18; at 1188 days for *V. radiata* introduction GL 388; at 1341 days for *V. radiata* introduction KY 2013. All species, except *V. angularis*, maintained germination of more than 80% for a longer period than did common beans. Loss of viability occurred approximately at 1160 days after harvest for *P. vulgaris* (both cultivars); at about 1290 days for *V. angularis* (both cultivars); at 1680 days for *V. unguiculata* (both cultivars); at 1684 days for *P. lunatus*; at 1734 days for *V. umbellata* introduction E-18. More than 30% of the seeds from *V. umbellata* introduction E-7 were viable 1749 days after harvest, when most of *V. radiata* seeds were also viable.

III. SEED PATHOLOGY

Copeland and McDonald (8) have described the following causes of seed deterioration:

1. Lipid peroxidation (free-radical chain reactions)
2. Degradation of functional structures
3. Inability of ribosomes to dissociate
4. Enzyme degradation and inactivation
5. Formation and activation of hydrolytic enzymes
6. Breakdown in mechanisms for triggering germination
7. Genetic degradation (random somatic mutations)
8. Depletion of food reserves
9. Starvation of meristematic cells
10. Accumulation of toxic compounds

Christensen (15,26) and Christensen and Kaufmann (27) have reviewed the work carried out on deterioration of seeds during storage due to different microflora and other pests. Of the various pests, bacteria play an important role in the deterioration of seeds during storage. Seedborne pathogenic bacteria do not affect seed germination unless the infection has progressed to the point of obvious decay. Bacteria require free water to grow in stored seeds, conditions that would induce most seeds to germinate. Such moist conditions would also encourage growth of fungi, which would tend to suppress the growth of bacteria.

Both field and storage fungi can deteriorate seeds during storage. The field fungi invade developing seeds in the field or during subsequent preharvest operations. These fungi need a high moisture content for growth (e.g., 30–33% in cereals on a dry weight basis). The unusual wet conditions at harvest can result in fungal infestation in the field, followed by the rapid deterioration, including seed discoloration, death of ovules and embryos, seed shriveling, and production of toxic substances. Wheat and barley are reported to be particularly associated with *Alternaria, Fusarium,* and *Helmintho-sporium* spp. Plant seeds such as pea, tomato, melon, and maize are generally less contaminated by field fungi than the seeds of wheat, oat, or barley, which are more freely exposed to airborne fungi (2).

Aspergillus and *Penicillium,* which constitute the major storage fungi, thrive only under storage conditions and are usually absent in seeds of standing plants in the field. Storage fungi have a sharply defined lower limit of seed moisture content below which they do not grow (Table 3). Other factors influencing the fungal prevalence include their ability to penetrate seeds, seed conditions, nutrient availability, temperature, and time. Storage fungi decrease seed viability, cause discoloration, produce mycotoxins and heat, and develop mustiness and caking (2). The microorganisms accelerate the loss of seed viability at high moisture contents. The seed moisture contents in equilibrium with RH below 68% are not sufficient for the growth of microorganisms. Microorganisms are not involved in seed deterioration occurring at moisture contents less than about 13% in starchy seeds and less than about 9% in oilseeds (Table 3).

Sinclair (28) described seeds as a microcosm of microbes, with the potential for carrying a wide variety of fungi, bacteria, viruses, and often nematodes causing diseases in seedlings and plants. Some live on the seed surface without visibly affecting its appearance, but they may become harmful when environmental conditions favor their growth and reproduction. Seed microflora often cause problems like seed rots

Table 3 Minimum Relative Humidities for the Growth of Common Storage Fungi at Their Optimum Temperature (27–30°C) and the Appropriate Moisture Content of Various Seeds in Equilibrium with RH

		% Moisture content				
			Rice			
Fungus	Minimum RH (%)	Wheat, maize, sorghum	Rough	Polished	Soybean	Sunflower
Aspergillus holoptilicus	68	13	12.5	14	12.5	8.5
A. restrictus	70	14	13.5	15	13	9.5
A. glaucus	73	15	14.5	15.5	14	10.5
A. candidus, A. ochraceus	80	16	15	16.5	16	11.5
A. flavus	85	18	16.5	17.5	18	13.5
Penicillum spp.	80–90	16	15	16.5	16	11.5

Source: Ref. 15.

and seedling blight, with resultant variability in germination. Some microorganisms live in nonliving outer seed tissue (bracts, pericarps, or seed coats) and attack the germinating seed under favorable conditions. Still others are carried inside the seed on or inside of the embryo or endosperm. These may not kill the seed but can delay germination and weaken seedlings. Some microbes survive in the embryo and the resulting seedling to cause infection to the seed crop.

Fungi cause the largest number of plant diseases and occur more commonly in or on seeds than bacteria or viruses. More than 8000 species of fungi have been identified as plant pathogens. Whereas saprophytic fungi are not specific to any particular host and are found on seeds of different plants, the pathogenic fungi are normally confined to a limited host range. Both types may occur on the seed surface in cracks or inside the seed coat, but pathogenic fungi can also occur within the seed tissue. Seeds infected with pathogenic fungi may not germinate normally and may provide inoculum for further spread of disease, causing a reduction in crop yield and quality (8).

The occurrence of fungi on 57 onion Wolska seed samples from 1993 to 1995 was determined on potato dextrose agar and prune lactose agar media. Moreover, seedling and bulb infection and seed germination were assessed. From the onion seedborne pathogens, *Botrytis allii*, *B. cinerea*, *Fusarium* spp., and *Pleospora herbarum* were found on seeds, seedlings, and bulbs, whereas *Alternaria porri* was found only on seeds. The level of seed infection by *B. allii* was highest and seed germination was lowest in 1994. Surface sterilization resulted in a considerable reduction in seed infection by this pathogen. The highest infection of seedlings was observed in the season 1995 and 1996. However, the bulb infection by this pathogen in this season as determined before storage, in store, and after storage was the lowest. On the other hand, in the season 1993 and 1994, the level of seedling infection by *B. allii* was lowest, whereas bulb infection in store and after storage was highest. The relationships between the level of inner seed infection and seed germination and infection of seedlings and bulbs were investigated too. Inner seed infection lower than 5% did not guarantee healthy onion bulbs at a satisfactory level under meteorological conditions favorable for the disease development. In years less favorable for its development, only at a seed infection level higher than 10%, the tendency to deterioration of some seed parameters and plant characters was observed (28a). Tylkowska and Dorna (28b) studied the effects of cinnamon, garlic, greater celandine, ginger, and some fungicides on the growth of pathogenic fungi isolated from onion, cabbage, and carrot seeds. Among the plant materials applied, ginger most effectively inhibited growth of *A. brassicae, A. brassicicola, B. allii,* and *S. botryosum,* whereas garlic was the best in controlling growth of *A. alternata, A. dauci, A. radicina, F. avenaceum,* and *P. lingam.* None of them diminished the growth of *B. cinerea.*

Abdelmonem (28c) discussed the health of seeds from Egyptian seed crops, and described methods used for detection of seedborne pathogens and listed seedborne diseases found on 23 different crops or groups of crops (including field, horticultural, and forage crops) in Egypt.

Insects and mites can also cause seeds to deteriorate during storage (29). However, at <8% moisture content, there is little or no activity of a weevils, flour beetles, and seed borers, whose activity increases considerably above 15% moisture and is optimal at 30–35°C. Mites are found to survive below 60% RH, although they can tolerate a wide range of temperature from 3 to 31°C (2).

IV. BIOCHEMISTRY OF SEED VIABILITY

A. Aging of Seeds

Bewley and Black (2) listed the following symptoms of deterioration exhibited by aging seeds:

1. Change in seed color, seed coat, and/or embryo
2. Delay of radicle emergence and seedling growth
3. Decreased total germination by a seed population
4. Decline in tolerance to suboptimal conditions during germination and/or growth
5. Increase in the number of abnormal seedlings
6. Lower tolerance to adverse storage conditions
7. Higher sensitivity to radiation treatments
8. Mustiness
9. Increased heat production during storage

Various postharvest treatments, including adverse storage conditions, produce "aged seeds" with the foregoing symptoms of deterioration, leading to reduced viability and germinability. Viability responses of seeds vary from zero germination through reduced seedling vigor to no apparent visible change. The metabolic changes occurring within the seeds as manifested by the morphology and physiology of seed deterioration constitute "aging" of seeds. These changes are not caused by any contamination by microorganisms or other external agents.

Aging of seeds under both natural and artificial conditions led to a decline in the viability of onion seeds (29a). Reduction in germination percentage shoot and root length, total seedling weight, and vigor index was observed during aging, more in the case of artificially aged seeds (42°C at 100% RH for 72 hr). Aged seeds also showed higher electrical conductivity and lipid peroxidation values and lower α-amylase, catalase, and peroxidase activity compared to the control seeds.

B. Biological Oxidation of Viable Seeds

Biological oxidation in the form of mitochondrial activity of viable embryos and axes increases with time after the start of imbibition with a progressive increase in oxidative phosphorylation and ultrastructural integrity of mitochondria. Unlike viable seeds, nonviable seeds are characterized by swollen and disrupted mitochondria, which become increasingly disorganized after imbibition resulting in complete lysis after 2–3 days. This is accompanied by extensive disruption of the cytoplasm. Mitochondria in viable seeds, however, develop following imbibition (2).

The ATP content of aged, nonviable crimson clover (*Trifolium incarnatum*) seeds was found to be less than 1% of that in viable controls (30). In embryonic axes from rapidly aged soybean seeds, the ATP content was as high as 35% of the controls, but still insufficient to support germination (31). A decline in respiratory enzymes like cytochrome oxidase, malic and alcohol dehydrogenase, succinic and glutamic dehydrogenase, catalase, and peroxidase in nonviable seeds may contribute to respiratory failure. Since the production of some respiratory enzymes and isoenzymes is an indication of embryo viability, the inability of the seed to correct these enzyme deficiencies incurred during development or storage can be associated with loss of seed viability (2).

C. Seed Viability Tests

Seed viability can be easily detected by employing the commonly used tetrazolium test, which involves submerging a seed in a 1% solution of colorless 2,3,5-triphenyltetrazolium chloride or bromide. Reduction of this compound to red-colored formazan by the dehydrogenases of viable seeds is an indication of living tissue (Fig. 2).

The tetrazolium test differentiates between viable and dead embryonic tissues on the basis of their relative respiration rate in the hydrated state, using dehydrogenase activity as an index of respiration rate and seed viability (8). Seed viability can also be measured by employing several other tests such as the hydrogen peroxide (H_2O_2) test, the ferric chloride test, the indoxyl acetate test, the fast green test, the sodium hypochlorite test, the conductivity test, excised embryo tests, and radiographic examinations (32–36).

The test correlates significantly with respiration and seedling vigor. Aging treatments, especially accelerated aging, subjecting seeds to elevated temperatures, sometimes at high RH, lead to a loss of viability without concomitant decrease in

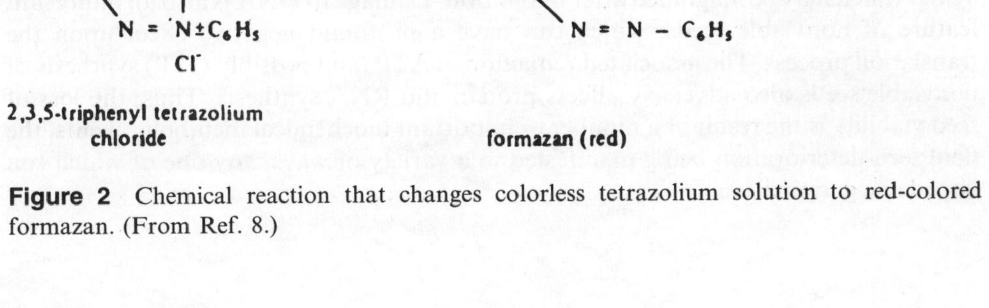

2,3,5-triphenyl tetrazolium chloride   formazan (red)

Figure 2 Chemical reaction that changes colorless tetrazolium solution to red-colored formazan. (From Ref. 8.)

dehydrogenase activity, resulting in an overestimation of viability. Woodstock (37) reviewed the physiological and biochemical aspects of the viability tests in detail.

The degree of correlation between oxygen uptake and seedling vigor varies with the time at which respiration is measured and the number of days elapsed before which, seedling vigor is measured. According to Bewley and Black (2), more research is needed to understand the respiratory metabolism of deteriorating seeds and seed parts. Studies designed to test the activity and integrity of mitochondria, their enzymes, the capacity for oxygen uptake, and ATP production in variously aged seeds at different stages of deterioration are desirable. The vigor of barley seedlings can be related to their mitochondrial content and to the rate of mitochondrial oxidase activity (38). Rapidly growing seedlings contain higher quantities of mitochondrial protein, which accounts for the enhanced respiration rates of seedlings produced from heavier seeds. Also, hybrid vigor in maize seems to be associated with the enhanced activity of NAD-linked mitochondrial enzymes, leading to more efficient electron transport and oxidative phosphorylation. In lettuce, highly significant correlations were noticed between the ATP content of tested seeds and seed vigor; that is, seed weight, seedling weight, and hypocotyl length. In other plant species, however, the correlation between ATP content and seed vigor was less than striking (39).

D. Other Biochemical Events Related to Seed Viability

The protein-and RNA-synthesizing apparatus in the embryos and embryonic axes of nonviable seeds is highly disorganized, as evidenced by the inability of these organelles to incorporate exogenously supplied radioactive amino acids into protein upon hydration (40). Even the viable embryos of rye from populations showing reduced viability exhibited signs of deterioration in protein synthesis. The decreased protein synthesis of nonviable seeds is usually accompanied by an extensive loss of the capacity of embryos to synthesize RNA (2), which may be totally absent in completely nonviable tissues. Sen and Osborne (41), using rye embryos, observed that a gradual loss of viability was associated with a decline in incorporation of radioactive precursors into all the major classes of RNA. The processing of precursor rRNA into 255 and 185 RNA of the ribosomal subunits was also retarded. An impairment in the transcription mechanism as a consequence of damaged nuclear DNA may be associated with the loss of capacity for protein synthesis and seed viability (2). However, protein synthesis may also be impaired at the level of translation.

The capacity of protein synthesis of the populations of seeds showing various degrees of viability also decreases with reduced germinability and vigor; for example, in sun-bleached lima bean axes (42) and cereal grains and embryos (37). The loss of seed viability associated with a decreased capacity for protein synthesis can be attributed to lesions in the ribosomes and some supernatant factors occurring during aging, which may be magnified after imbibition. Damage to DNA is also an important feature of nonviable seeds, which can have a profound negative effect upon the translation process. The associated reduction in ATP (and possibly GTP) synthesis of nonviable seeds also adversely affects protein and RNA synthesis. Thus, the loss of seed viability is the result of a number of important biochemical metabolic events, the final seed deterioration being manifested in a variety of ways, any one of which can impede its germination.

According to Roberts (17,18), almost any combination of time, temperature, and moisture content leading to a loss of seed viability during storage will also lead to a predictable amount of genetic damage in the survivors. Chromosomal breakdown was noticed during the anaphase of the first mitotic division of the root tips of surviving populations of garden pea, broad bean seeds, and barley grains aged by various combinations of temperature and moisture content to produce different rates of loss of viability (17). Since DNA changes can lead to impaired transcription and translation processes, upon which seed germination depends, it is likely that DNA lesions play an important role in viability loss. An early event after the imbibition of stored seeds might be the repair of damage to macromolecules or macromolecular structures suffered during storage. Lesions to DNA segments containing the genes for repair enzymes could be more deleterious than lesions to less essential DNA segments (2).

A buildup of chemical mutagens during the aging of seeds may lead to chromosomal aberrations, resulting in partially degraded DNA molecules. A variety of seed extracts and common metabolites are thought to induce chromosomal fragmentation. However, further research is needed to prove this hypothesis.

Many seeds remain in the soil in the imbibed state for long periods of time before the right conditions exist for their germination. Seeds obviously become adapted to such conditions. The loss of viability of seeds with low moisture content (9,7%) is due to the inactivation of the enzymes necessary to repair damage to DNA and other essential macromolecules in the cytoplasm incurred during storage. At still lower water contents, there is not enough damage to decrease germination. If the damage is repaired only in the imbibed state, then it will occur in wet-stored seeds and will be continuously repaired without accumulation. Thus, such seeds will remain viable. In dry-stored seeds, however, the repair mechanisms will become activated only after poststorage imbibition, by which time damage may be too extensive to be repaired, resulting in loss of seed viability. The survival of long-stored seeds is related to their capacity to avoid lethal damage during storage and carry on essential repairs during germination (2).

Edwards (43) noted that the extent of metabolism of air-dried seeds is a function of their water content. Enzymatic decarboxylation and transamination of glutamate with an 18% moisture content and transamination of α-ketoglutarate with alanine at a 15% moisture content have been reported in wheat grains (44). Enzyme activity occurs in dry seeds at low moisture contents, but other factors, including protein content, might influence enzyme activity. According to Bewley and Black (2), localized enzymatic reactions within dry cells could lead to the depletion of substrates essential to germination. It is not known whether the activities of certain enzymes lead to loss of viability during storage (through depletion of substrates or production of toxic substances) or whether lack of enzyme activity, such as membrane repair enzymes, is lethal to the germinating tissue.

V. DEPLETION OF FOOD RESERVES

The mobilization of food reserves in the germinating seed is controlled by the embryo or embryonic axes of viable seeds. Since germination takes place well before the mobilization of the food reserves, seedling vigor rather than seed viability is affected by the deterioration of these food reserves, Loss of seed viability is not usually accompanied by any dramatic changes in stored food reserves. During unfavorable storage conditions, fungi and other microbes may degrade these compounds to thrive themselves on the products of hydrolysis. Activated enzymes in some cases also might bring about these changes.

Deterioration of seeds during storage can result in marked changes in food reserves and the activity of enzymes capable of degrading them (2). Although de novo synthesis of such enzymes does not occur in nonviable seeds, preformed enzymes may remain active for many years after seeds lose their germinability. Depending upon seed species, the loss of capacity to synthesize hydrolases can either accompany the loss of viability or precede it. Whereas in barley and rye grains deterioration of the embryo and endosperm (aleurone layer—site of hydrolase synthesis) occurs independently, in wheat it occurs at identical rates (45). Anderson (46), however, noted that germinated, aged barley grains exhibited about 50–70% lower α-amylase activity as compared to their unaged controls.

Response of α-amylase to gibberellins in the barley aleurone layers changes during storage; it increases during the first 4 years of storage and then declines over the next 5 years. Bewley and Black (2) attribute this increase in response to an applied GA_3 to a probable consequence of after ripening, in which aleurone layers become more responsive to GA_3 and/or capable of α-amylase synthesis. Aging grains lose one or both of these capacities.

The loss of major food reserves is not an important consequence of seed deterioration, but the capacity to utilize these reserves might be. The status of reserve-degrading enzymes and their synthetic machinery is immaterial in a nonviable seed. In some viable seeds, enzyme activity and synthesis do decline, but this does not seem to affect establishment of seedling vigor. There are changes in hydrolytic enzyme activity in different cereal grains and seeds during storage, affecting both seedling vigor and final crop yield (2).

Along with starchy substances, storage fungi also degrade lipids to their constituents (27), although such events might also occur in the absence of contaminating microflora (19). The loss of seed viability may be attributed partially to the accumulation of free fatty acids within seeds. Baddeley and Hanson (47) noted that long-chain unsaturated fatty acids caused swelling of isolated mitochondria and impaired their normal functions. In some seeds, deterioration is accompanied by a decrease in total unsaturated fatty acids.

The cultivars of rice were subjected to accelerated aging to study changes in the food reserve (starch and protein) contents (47a). A significant reduction in starch and protein was observed with increasing aging period from 0 to 20 days. Genetic variability was noted in the composition of starch and protein of the rice seeds.

Disruption of membrane systems is closely associated with the diverse metabolic changes and failures resulting from the deterioration and loss of viability of seeds. According to Bewley and Black (2), imbibition of viable seeds is accompanied by a rapid but transient efflux of inorganic and organic substances through the plasmalemma/tonoplast membranes. The integrity of the mitochondrial membrane, which is incomplete for several hours after imbibition, becomes stable and organized, perhaps because of some unknown enzymatic mechanism. In lower nonviable seeds, such enzymatic repair mechanisms are inefficient or totally absent, or else these membranes are damaged permanently, making the repair mechanism ineffective.

The loss of integrity of the plasmalemma and tonoplast in deteriorating seeds has been implied from metabolites leaking into the imbibition medium from such seeds to a great extent than from viable seeds (48,49). Abdul-Baki and Anderson (48) also observed that correlation between uptake into and loss of materials from accelerated aged sorghum grains, and their vigor was poor. Correlation between sugar leakage and viability loss was also poor for 10-year-old perennial rye grass grains, as well as between electrolyte leakage and viability (50).

Leakage of cellular metabolites indicates the possibility of damaged and perhaps irrepairable membrane systems in the deteriorating seeds. Other implications of observed leakages include the loss of respirable substrates (sugars). The amount of glucose leaked from deteriorating barley embryos was as high as 60–70% of its total pool, and sucrose loss was 20–30% of the total (48). It is not known whether the higher substrate leakage from nonviable seeds contributes significantly to their loss of germinability. According to Bewley and Black (2), loss of substrate is a consequence rather than the cause of cellular disorders, and extensive leakage of metabolites is one manifestation of several changes seeds undergo during the aging process. Increased leakage of organic metabolites from deteriorating seeds indirectly hastens their destruction by encouraging the growth of contaminating microorganisms.

The mechanism of degradation of membranes and macromolecules in aging seeds during storage has not been fully investigated. Aging seeds at a higher moisture content hydrolyze their important cell components and membranes that cannot be repaired owing to lack of an appropriate synthetic apparatus. At both high and low water contents (the latter below a critical level to allow for enzyme activity), certain adverse reactions occur, resulting in the formation of free radicals. Biological oxidations, both enzymatic and spontaneous, which generate cytotoxic free superoxide radicals (O_2^-) have been described. The latter can react with H_2O_2 to produce single oxygen and hydroxyl ions (OH^-). Highly potent oxidants like singlet oxygen are capable of damaging large polymers and membrane lipids (2). In hydrated tissues, the free radicals produced are controlled by absorbents like tocopherols or scavenging enzymes such as superoxide dismutase (SOD), which converts free O_2^- to H_2O_2, which is in turn removed by catalases. In aging tissues, the balance between free radical–producing and scavenging reactions might be upset in favor of the former. Thus, formation of free radicals in stored aging seeds can result in the progressive inactivation of synthetic enzymes, denaturation of other proteins, and disruption of the integrity of DNA and RNA molecules (2). Membranes could become more permeable and de novo synthesis of enzymes impaired, preventing cell division and elongation. Such fundamental changes in membranes and macromolecules can cause slow or abnormal growth or cessation of growth depending upon the magnitude of the damage.

The evidence for the free radical hypothesis of aging seeds is rather limited, and studies on several species of seeds of different ages using the electron spin resonance technique have failed to find free radical concentrations at sufficient levels to account for associated genetic damage. However, as the proponents of the free radical hypothesis point out, the buildup of free radicals in aging seeds is not expected, since aging is a slow process and free radicals formed may be removed by immediate reaction without allowing them to accumulate to detectable levels.

Fatty acids with two or more unsaturated bonds are thought to be more labile to oxidation and prone to form free radicals than highly saturated fatty acids. It has been argued that if fatty acid oxidation and free radical formation occur during aging, then the content of highly unsaturated fatty acids of seeds should decrease with seed deterioration, with the level of saturated fatty acids remaining constant. Harman and Mattick (51) observed that in pea seeds and axe the level of palmitic, stearic, and oleic (saturated) acids did not change during aging and neither did that of the unsaturated linoleic acid in the axis, although it did decline in the cotyledons. Another unsaturated fatty acid, linolenic acid, however, declined significantly in the axis, as did the seed vigor. This study indicated that free radicals are formed during aging, reflected in the loss of one of the two major unsaturated fatty acids. Stewart and Bewley (52) stored soybean seeds at elevated temperatures (40°C) and found that imbibitional leakage was increased. During this accelerated aging treatment, the viability of seeds decreased for 2 days, followed by further deterioration of axes occurring as postaging deterioration. The peroxidation of unsaturated fatty acids occurred during aging, with linolenic acid declining appreciably on the first day and linoleic declining by the end of the second day (Table 4). The levels of palmitic, stearic, and oleic acids did not change significantly during aging, but declined only slightly; as a proportion of total fatty acid content they increased. Lipid peroxidation was evident in the formation of malondialdehyde, a product of the oxidation of linolenic acid. By the second day of aging, the level of malondialdehyde increased upon imbibition of the axes. When the soybean axes were stored under conditions of high temperature (45°C) in an atmosphere of low RH, viability was retained, there was no increase in imbibitional leakage, and the fatty acid content of the phospholipid fraction remained unchanged (Table 4), suggesting that lipid peroxidation, especially that of phospholipids (membrane constituents), occurs during accelerated aging under unfavorable conditions. Since the changes associated with aging are probably many and also variable, the seed deterioration occurring during aging cannot be attributed solely to lipid peroxidation and the resultant disruption of membrane phospholipids. Bewley and Black (2) concluded that fatty acid peroxidation can occur during seed aging but is probably

Table 4 Content of Fatty Acids in the Polar Lipid Fraction from Soybean Axes During Aging at High Temperature and High or Low Humidity

Aging treatment	Fatty acid content (% of total)				
	Palmitic	Stearic	Oleic	Linoleic	Linolenic
Unaged seeds	16.6	3.1	3.8	57.3	19.3
Days of high-humidity aging:					
1	20.6	4.2	3.5	56.4	15.4
2	22.6	5.9	5.0	52.4	14.0
3	32.1	7.1	9.6	40.5	10.7
4	42.1	8.1	16.6	28.8	4.4
6	39.6	13.0	7.0	33.9	6.5
Days of low-humidity aging:					
2	19.6	3.1	2.1	60.2	14.9
4	19.4	3.0	2.4	58.1	17.0
6	21.1	3.0	2.2	58.8	14.9

Source: Ref. 52.

not an essential component of its viability loss. Studies carried out on maize (53) and on *Ouercus robur* (54,55) employing desiccation-sensitive material that was dehydrated provide convincing evidence for the involvement of free radical mechanisms in dehydration-related loss of viability.

Considerable information has accumulated concerning orthodox longevity and the factors influencing the retention of viability (2,4,17,56,57). Smith and Berjak (6) described deteriorative changes associated with the loss of viability of stored desiccation-tolerant (orthodox) and desiccation-sensitive (recalcitrant) seeds. Free radical mechanisms are important in the loss of viability of stored orthodox seeds. Most of the damage probably occurs on imbibition, when the presence of water permits the aqueous-based deleterious reactions to occur. In wet-stored recalcitrant propagules, damage occurs because of water stress imposed by the withholding of necessary water. Evidence available indicates that the underlying biochemical processes leading to viability loss are similar in orthodox and in recalcitrant seeds. The difference in the two types of seeds lies in the rate at which these events proceed. The anhydrous nature of stored orthodox seeds limits the rate of damage. Wet-stored recalcitrant seeds, although water stressed are sufficiently hydrated to allow rapid damage (2).

REFERENCES

1. C. R. Quick, How long can seed remain alive? *Seeds: The Yearbook of Agriculture*, U.S. Department of Agriculture, Washington, DC, 1961, p. 94.
2. J. O. Bewley and M. Black. *Physiology and Biochemistry of Seeds in Relation to Germination*. Vol. 2, *Viability Dormancy and Environmental Control*, Springer-Verlag, Berlin, 1982.
3. J. F. Harrington, Seed storage and longevity, *Seed Biology* (T. T. Kozlowski, ed.), Academic Press, New York, 1972, p. 145.
4. O. L. Justice and L. N. Bass, *Principles and Practices of Seed Storage*, U.S. Department of Agriculture Handbook No. 506. 1978.
5. P. Berjak. J. M. Farrant, N. W. Pammenter, C. W. Vertucci and J. Wesley-Smith, Current understanding of desiccation-sensitive (recalcitrant) seeds: Development, state of water and responses to dehydration and freezing, *Fourth International Workshop on Seeds: Basic and Applied Aspects of Seed Biology* (D. Come and F. Corbineau, eds.), AFSIS, Paris, 1993, p. 705.
6. M. T. Smith and P. Berjak, Deteriorative changes associated with the loss of viability of stored desiccation-tolerant and desiccation-sensitive seeds, *Seed Development and Germination* (J. Kigel and G. Galili, eds.). Marcel Dekker, New York. 1995, p. 701.
7. H. F. Chin, F. B. Krishnapillay, and P. C. Stanwood, Seed moisture: recalcitrant vs. orthodox seeds, *Physiology of Seed Deterioration* (M. B. McDonald and C. J. Nelson, eds.), Crop Science Society of America, Madison, WI, 1989, p. 15.
8. L. O. Copeland and M. B. McDonald, *Principles of Seed Science and Technology*, 3rd ed., Chapman & Hill, New York, 1995.
9. P. Barjak, J. M. Farrant, D. J. Maycock, and N. N. Pammenter, Recalcitrant (homoiohydrous) seeds: the enigma of their desiccation sensitivity. *Seed Sci. Technol.* 18:297 (1990).
10. H. F. Chin and E. H. Roberts. *Recalcitrant Crop Seeds*, Tropical Press, Kuala Lumpur, Malaysia, 1980.
11. R. H. Ellis, The longevity of seeds. *Hortic. Sci. 26*: 1119 (1991).
12. J. M: Ferrant, N. W. Pammenter, and P. Berjak, Recalcitrance-a current assessment, *Seed Sci. Technol.* 16:155 (1988).

12a.   J.S.Merlin and V.Palanisamy, Seed viability and storability of jackfruit (*Artocarpus heterophyllus* L.), *Seed Res. 28*: 166 (2000).

12b.   A.C.S. Andrade, The effect of moisture content and temperature on the longevity of heart of palm seeds, *Seed Sci. Technol. 29*: 171(2001).

12c.   N.B.Machado-Neto, C.C.Custodio and M.Takaki, Evaluation of naturally and artificially aged seeds of *Phaseolus vulgaris* L., *Seed Sci. Technol. 29*: 137(2001).

12d.   U.Dhar, Y.P.S.Pangtey and A.Tiwari, Seed deterioration studies in Indian butter tree (*Aisandra butyracea* (Roxb.) Baehni), *Seed Sci. Technol. 27*:963(1999).

12e.   S.Kumar, N.C.Singhal and S.Prakash, Intervarietal variability for seed longevity in pea (*Pisum sativum*), *Indian J. Genet. Plant Breed. 57*: 204 (1997).

12f.   J.A.Oladiran and P.M.Mumford, The longevity of large and small seeds of *Amaranthus gangeticus* and *A. hybridus*, *Seed Sci. Technol. 18*:499(1990).

12g.   N.Choudhuri and R.N.Basu, Maintenance of seed vigor and viability of onion (*Allium cepa* L.), *Seed Sci. Technol. 16*: 51(1988).

13.   J. F. Harrington, *Proceedings of 1959 Mississippi Short Course for Seedsmen*, 1960, p. 89.

14.   J. F. Harrington. Problems of seed storage, *Seed Ecology*, (W. Heydecker, ed.), Butterworths, London, 1973, p. 251.

15.   C. M. Christensen. Microflora and seed deterioration. *Viability of Seeds* (E. H. Roberts, ed.) Chapman & Hall, London, 1972, p. 59.

16.   P. K. Agrawal, Seed deterioration during storage, *Proceed. Int. Congress of Plant Physiol.*, Vol. 2, New Delhi, 151/N18 Feb. 1988, p. 15.

17.   E. H. Roberts, Storage environment and the control of viability. *Viability of Seeds* (E. H. Roberts. ed.). Chapman & Hall, London, 1972, p. 14.

18.   E. H. Roberts, Predicting storage life of seeds, *Seed Sci. Technol. 1*:499 (1973).

19.   E. H. Roberts, Problem of long-term storage of seeds and pollen for genetic resources conservation, *Crop Genetic Resources for Today and Tomorrow*, (O. H. Frankel and J. G. Hawkes, eds.), Cambridge University Press, I.B.P., Cambridge, UK, 1975.

20.   L. N. Bass, Controlled atmosphere and seed storage. *Seed Sci. Technol. 1*:463 (1973).

20a.   B.Varghese and S.C.Naithani, Desiccation induced loss of vigor and viability during storage in neem (*Azarrachta indica* A.Juss) seeds, *Seed Sci. Technol. 28*: 485 (2000).

21.   R. P. Moore, Effects of mechanical injuries on viability, *Viability of Seeds* (E. H. Roberts. ed.) Chapman & Hall. London. 1972. p. 94.

22.   M. H. Dickson and M. A. Boettger, Factors associated with resistance to mechanical damage in snap beans, (*Phaseolus vulgaris* L.), *J. Am. Soc. Hort. Sci. 101*:541 (1976).

23.   E. H. Roberts and F. H. Abdullah, The influence of temperature, moisture and oxygen on period of seed viability in barley, broad bean and peas, *Ann. Bot.* (Lond.) *32*:97 (1968).

24.   R. H. Ellis and E. H. Roberts, Improved equations for prediction of seed longevity, *Ann. Bot.* (Lond.) *45*:13 (1980).

25.   R. H. Ellis and E. H. Roberts. Influence of temperature and moisture on seed viability period in barley (*Hordeum distichum* L.). *Ann. Bot.* (Lond.) *45*:31 (1980).

25a.   R.F.Vieia, M.N. de-Faria, J.A-de-O-Ramos, C. Vieira, S.M.L.Donzeles and R.F.T. de Freitas, Seed germination of six grain legumes during storage at room conditions in Vicosa, Minas Gerais State, Brazil, *Seed Sci. Technol. 26*: 489 (1998).

26.   C. M. Christensen, Moisture and seed decay, *Water Deficits and Plants*. Vol. IV (T. T. Kozlowski. ed.), Academic Press, New York, 1978, p. 199.

27.   C. M. Christensen and H. H. Kaufmann, *Grain Storage: The role of fungi in quality loss*, University of Minnesota Press, Minneapolis, 1969.

28.   J. B. Sinclair, The Seed: A microcosm of microbes, *J. Seed Technol. 4*(2): 68 (1979).

28a.   K.Tylkowska and H.Dorna, Onion (*Allium cepa*) seed and plant health with special reference to Botrytis allii, *Phytopathologia-Polonica 21*:55 (2001).

28b. K.Tylkowska and H.Dorna, Effects of cinnamon, garlic, greater celandine, ginger and chosen fungicides on the growth of pathogenic fungi isolated from onion, cabbage and carrot seeds, *Phytopathologia-Polonica, 21*: 25 (2001).

28c. A.M.Abdelmonem, Status of seed pathology and seed health testing in Egypt, *Seed Sci. Technol. 28*: 533(2000).

29. R. W. Hoover, Insects attacking seeds during storage, *Seed Biology,* Vol. III (T. T. Kozlowski, ed.), Academic Press, New York, 1972, p. 247.

29a. K.Bhanu Prakash, H.S.Yogeesha and L.B.Naik, Physiological and biochemical changes during onion seed aging, XI National Seminar on Quality Seed to Enhance Agricultural Productivity, UAS, Dharwar, Jan 181/N20, 2002, *Seed Tech News* 32: (A.Gaur, A.K.Vari, J.L.Varshney, and K.Kant, eds), Indian Society Seed Technology, Division of Seed Science and Technology, IARI, New Delhi, March 2002, p. 151

30. T. M. Ching, Adenosine triphosphate content and seed vigor, *Plant Physiol. 51*:400 (1973).

31. J. D. Anderson, Adenylate metabolism of embryonic axes from deteriorated soybean seeds, *Plant Physiol. 59*:610 (1977).

32. Association of Official Seed Analysis, Tetrazolium testing handbook for agricultural seeds. *AOSA Handbook No. 29*, (D. F. Grabe, ed.), 1970.

33. Association of Official Seed Analysts, Rules for testing seeds, *Proc. Assoc. Off. Seed Ana. 60*(2): 39 (1978).

34. Association of Official Seed Analysts, Radiographic analysis of agricultural and forest tree seeds, *AOSA Handbook No.* 31 (E. Belcher and I. Bozzo, eds.), 1979.

35. I. Gadd, Biochemical tests for seed germination, *Proc. Int. Seed Testing Assoc. 16*:235 (1950).

36. R. C. French, J. A. Thompson, and B. C. H. Kingsolver, Indoxyl acetate as an indicator of cracked seed coats of white beans and other light colored legume seeds. *Proc. Am. Soc. Hort. Sci. 80*:377 (1962).

37. L. W. Woodstock, Physiological and biochemical tests for seed vigor, *Seed Sci. Technol. 1*: 127 (1973).

38. R. G. McDaniel, Relationships of seed weight, seedling vigor and mitochondrial metabolism in barley, *Crop Sci. 9*:823 (1969).

39. R. C. Styer, D. J. Cantliffe, and C. B. Hall, The relation of ATP concentration to germination and seedling vigor of vegetable seeds stored under various conditions, *J. Am. Soc. Hort. Sci. 105*:298 (1980).

40. B. E. Roberts and D. J. Osborne, Protein synthesis and loss of viability in rye embryos: the viability of transferase enzymes during senescence, *Biochem. J. 135*:405 (1973).

41. S. Sen and D. J. Osborne, Decline in ribonucleic acid and protein synthesis with loss of viability during the early hours of imbibition of rye (*Secale cereale* L.) embryos, *Biochem. J. 166*:33 (1977).

42. A. A. Abdul-Baki, Biochemical differences between embryonic axes from green and sun bleached lima bean seeds: synthesis of carbohydrates, proteins and lipids, *J. Am. Soc. Hort. Sci. 96*:226 (1971).

43. M. Edwards, Metabolism as a function of water potential in air dry seeds of charlock (*Sinapsis arvensis* L.), *Plant Physiol. 58*:237 (1976).

44. P. Linko and M. Milner, Enzyme activation in wheat grains in relation to water content, glutamic acid, alanine transaminase and glutamic acid decarboxylase, *Plant Physiol. 34*:392 (1959).

45. D. Aspinall and L. G. Paleg, The deterioration of wheat embryo and endosperm function with age, *J. Exp. Bot. 22*:925 (1971).

46. J. D. Anderson, Metabolic changes in partially dormant wheat seeds during storage, *Plant Physiol. 46*:605 (1970).

47. M. S. Baddeley and J. B. Hansen, Uncoupling of energy linked functions of corn

mitochondria by linoleic acid and monomethyldeceenyl succinic acid, *Plant Physiol.* *42*: 1702 (1967).

47a. C.B.Singh, V.P.Kanaujia and C.P.Sachan, Effect of accelerated aging on biochemical composition of rice (*Oriza sativa* L.), XI National Seminar on Quality Seed to Enhance Agricultural Productivity, UAS, Dharwar, Jan 181/N20, 2002, *Seed Tech News* 32: (A.Gaur, A.K.Vari, J.L.Varshney, and K.Kant, eds), Indian Society Seed Technology, Division of Seed Science and Technology, IARI, New Delhi, March 2002, p.187.

48. A. A. Abdul-Baki and J. D. Anderson, Viability and leaching of sugars from germinating barley, *Crop Sci.* *10*:31 (1970).

49. S. Mathews and W. T. Bradnock, Relationship between seed exudation and field emergence in peas and French beans, *Hortic. Res.* 8:89 (1968).

50. T. M. Ching and I. Schoolcraft, Physiological and chemical differences in aged seeds, *Crop Sci.* 8:407 (1968).

51. G. E. Harman and L. R. Mattick, Association of lipid oxidation with seed ageing and death, *Nature* *260*:323 (1976).

52. R. ·R. C. Stewart and J. D. Bewley, Lipid peroxidation associated with accelerated ageing of soybean axes, *Plant Physiol.* *65*:245 (1980).

53. O. Leprince, R. Dettour, P. C. Thorpe, N. M. Atherton, and G. A. F. Hendry, The role of free radicals and radical processing systems in loss of desiccatio!1 tolerance in germinating maize (*Zea mays* L.), *New Phytol.* *116*:573 (1990).

54. G. A. F. Hendry, W. E. Finch-Savage, P. C. Thorpe, V. M. Atherton, S. M. Buckland, K. A. Nilsson, and W. E. Seel, Free radical processes and loss of seed viability during desiccation in the recalcitrant species *Quercus robur* L., *New Phytol.* *122*: 1(1992).

55. W. E. Finch-Savage, R. I. Grange, G. A. F. Hendry, and N. M. Atherton, Embryo water status and loss of viability during desiccation in the recalcitrant species *Quercus robur* L., Fourth International Workshop on Seeds: Basic and Applied Aspects of Seed Biology, AFSIS, Paris, 1993, p. 723.

56. M. D. McDonald and C. J. Nelson, *Physiology of seed deterioration*, Crop Science Society of America, Special Publication No. II, Madison, WI, 1986.

57. D. A. Priestley, *Seed Aging*, Comstock, Ithaca, NY, 1986.

6
Basic Principles of Seed Production

I. INTRODUCTION

The production of high-quality seed with high analytical quality, species and cultivar purity, high germination capacity, vigor, and uniformly large size, free from weeds and seedborne diseases and low moisture content is a formidable task requiring technical skill and knowledge of cytogenetics and plant breeding. Maintaining the genetic purity of the seed is of utmost importance and will enable growers to exploit the full benefits of introducing improved varieties of seeds. Thus, seed production needs to be carried out under skilled supervision and standardized, well-organized conditions. Agrawal (1) divided the general principles of seed production into two categories: (a) genetic principles and (b) agronomic principles.

II. GENETIC PRINCIPLES

A. Terminology

Crop plants show a wide range of variation in form and habital requirement. This variation can be obvious (e.g., between two species) or imperceptible to the human eye (e.g., between two plants of the same variety). Such differences are either genetic or induced by the growing environment. The genetic make-up of the plant is referred to as the genome, whereas the outward or perceptible appearance of plant is called the phenotype, which is considered to be the product of genotype and environmental conditions.

The scientific basis for the inheritance of genetic characteristics of plants was first established by the Austrian botanist Gregor J. Mendel, who demonstrated how parental traits are transmitted to offspring. The systematic study of the mechanism of inheritance of genetic traits (heredity) is known as genetics, and Mendel's findings are known as Mendel's law of independent assortment and the law of segregation.

The overall activity of the cell is controlled by the information provided by the genes (units of heredity) located on the chromosomes. The alternate form of a particular gene is called an allele. An allele can be homozygous (containing identical genes) or heterozygous (containing dissimilar genes). Biochemically, each gene is made up of deoxyribonucleic acid (DNA), and one or more genes can control one character. The effect of the gene is manifested through the synthesis of different biochemical products (enzymes, proteins, and alkaloids). Production of anthocyanin pigments in flowers is an example of gene-mediated effects in plants.

Gene effects can be dominant (e.g., heterozygous plants are similar to homozygous lines for that character) or recessive (e.g., gene effect is suppressed). In some cases,

111

the effects of genes may be incomplete or partial. Genes can undergo spontaneous changes, known as mutations.

In breeding, the generations are designated by a specific terminology. Plants involved in the first crossing are known as the parental generation, which are usually homozygous genotypes that vary from each other for a particular character. The progeny from the parental generation is referred to as first filial (F_1) generation, and subsequent generations are designated as F_2, F_3, F_4, and so on. The plants, although heterozygous, are nonsegregating in nature. Hybrid vigor, or heterosis, is usually expressed in the first filial generation. Subsequent generations consist of segregating populations. The cross of F_1 with a parent is known as a backcross. Backcrossing helps to concentrate the desirable qualities of the best parent in the hybrid.

A particular characteristic may be a monogenic-controlled by one gene; for example, flower color or multi(poly)genic in nature (controlled by two or more genes). In the case of multiple genes, one gene pair would arrange independently with no effect on the other pair. The genetically controlled characteristics are either quantitative (continuous) or qualitative (noncontinuous). Quantitative characteristics are generally controlled by several genes, each of which contributes an incremental effect to modify the expression of character (e.g., yield). Individual effects are very difficult to ascertain in quantitative characteristics (2).

Several genes controlling different characteristics and located on a single chromosome may be inherited together. This condition is referred to as *linkage,* which may occur in genes controlling qualitative and quantitative characters. In such cases, large populations are needed to isolate the most desired forms (e.g., high yield and disease susceptibility).

An increase in the vigor of a hybrid over either of the parents is known as heterosis, or hybrid vigor, which may take the form of an increase in yield, size, or number of plant parts, disease resistance, and so forth. The hybrid is a plant type resulting from fusion of dissimilar gametes or those having heterozygous gene pair for a particular character. The expression of hybrid vigor is usually more highly pronounced in cross-pollinated crops than the self-pollinated ones.

B. Breeding Methods

Genetic variability is a prerequisite for any plant improvement. Each plant originated in one of eight centers of origin: Abyssinia, Asia Minor, Central Asia, Central America, China, Mediterranean, South Asia, and South America. This range of origin provides genetic diversity for different plant species that can be used by plant breeders to effect improvement.

Collection and assessment of available germplasm for identifying genetic lines having desired qualities is the first step in plant breeding. In most countries, there is a highly specialized organization for introduction and assessment of germplasm of different crop species. Plant breeders can obtain breeding material from such organizations.

The present plant-breeding programs are limited by available genetic variability. Natural or spontaneous mutations (sports) add to the variation, but these are relatively rare. Chimeras resulting from the mutation of somatic cells may also contribute to variation in a small way, but when propagated they are not stable. It is possible, however, artificially to induce genetic variability by exposing plant tissues to chemical mutagens or ionizing radiation. Seeds, meristem (actively growing points), or other tissues are used for this purpose. Changes resulting from such treatments are generally uncontrolled and require large populations to obtain the desired plant type. In recent years, cell, embryo, and tissue culture techniques are being used successfully to induce variability in crop plants (see Chapters 22–24).

C. Deterioration of Genetic Purity

The genetic purity of a variety or trueness to its type deteriorates because of several factors during the production cycles. Kadam (3) listed the following important factors responsible for deterioration of varieties:

1. Developmental variations
2. Mechanical mixtures
3. Mutations
4. Natural crossing
5. Minor genetic variations
6. Selected influence of diseases
7. Technique of the plant breeder

Of these factors, mechanical mixtures, natural crossing, and selective influence of diseases are the most important reasons for the genetic deterioration of varieties during seed production, followed by raising the seed crops in areas removed from their natural adaptation sites, which may cause developmental variations and genetic shifts in varieties (1).

When seed crops are grown under environments with differing soil, fertility, climate, photoperiods, or at different elevations for several consecutive generations, developmental variations may set in as differential growth responses. It is, therefore, preferred to grow the varieties of crops in the areas of their natural adaptation to minimize developmental shifts.

Mechanical mixtures, the most important reason for varietal deterioration, often take place at the time of sowing if more than one variety is sown with the same seed drill, through volunteer plants of the same crop in the seed field, or through different varieties grown in adjacent fields. Two varieties growing next to each other in a field are usually mixed during harvesting and threshing operations. The threshing equipment is often contaminated with seeds of other varieties. Similarly, the gunny bags, seed bins, and elevators are also often contaminated, adding to the mechanical mixtures of varieties. Roguing the seed fields critically and using utmost care during seed production and processing are necessary to avoid such mechanical contamination.

Mutations do not seriously deteriorate varieties. It is often difficult to identify or detect minor mutations occurring naturally. Mutants such as "fatuoids" in oats or "rabbit ear" in peas may be removed by roguing from seed plots to purify the seeds.

Natural crossing can be an important source of varietal deterioration in sexually propagated crops. The extent of contamination depends upon the magnitude of natural cross fertilization. The deterioration sets in due to natural crossing with undesirable types, diseased plants, or off-types. In self-fertilized crops, natural crossing is not a serious source of contamination unless the variety is male sterile and is grown in close proximity with other varieties. The natural crossing, however, can be a major source of deterioration in cross-fertilized crops. The main factors deciding the extent of contamination due to natural crossing are the breeding system of the species, isolation distance, varietal mass, and pollinating agent (4). The isolation of seed crops is the most important factor in avoiding contamination of the cross-fertilized crops. The direction of prevailing winds, the number of insects present and their activity, and the mass of varieties are also important considerations in contamination by natural crossing.

Minor genetic variations can occur even in varieties appearing phenotypically uniform and homogeneous when released. These variations may be lost during later production cycles owing to selective elimination by the nature. The yield trials of lines propagated from plants of breeder's seed to maintain the purity of self-fertilized crop varieties can overcome these minor genetic variations (5). Due care during the maintenance of nucleus and breeder's seed of cross-fertilized varieties of crop is necessary.

New crop varieties often are susceptible to newer races of diseases and pests caused by obligate parasites and thus selectively influence varietal deterioration. The vegetatively propagated stock also can deteriorate quickly if infected by viruses, fungi, or bacteria. Seed production under strict disease-free conditions is, therefore, essential.

Serious instabilities may occur in varieties owing to cytogenetic irregularities in the form of improper assessments in the release of new varieties. Premature release of varieties, still segregating for resistance and susceptibility to diseases or other factors, can cause significant deterioration of varieties. This failure can be attributed to the variety testing programs.

In addition to these factors, other heritable variations due to recombinations and polyploidization may also take place in varieties during seed production, which can be avoided by periodical selection during maintenance of the seed stocks.

D. Maintaining the Genetic Purity of Seed

Agrawal (1) described the following steps to maintain the genetic purity of varieties during seed production:

1. Use of only approved seed in seed multiplications
2. Inspection and approval of seed plots prior to planting
3. Field inspection and approval of growing crops at all critical growth phases for verification of genetic purity, detection of admixtures, weeds, and for freedom from noxious weeds and seedborne diseases
4. Sampling and sealing of cleaned lots
5. Growing of samples of potentially approved stocks for comparison with authentic stocks

Hartmann and Kester (6) recommended the following steps for maintaining the genetic purity of cultivars:

1. Provision of adequate isolation to prevent contamination by natural crossing or mechanical mixtures
2. Roguing of seed plots prior to growth phases at which seed crop gets contaminated
3. Periodic testing of varieties for genetic purity

4. Growing crops only in areas of their adaptation to avoid genetic shifts
5. Certification of seed crops to maintain genetic purity and quality of seed
6. Adoption of generation system.

In the generation system, seed production is restricted to four generations; that is, starting from the breeder's seed, the seed can be multiplied up to three more generations, namely, foundation, registered, and certified seed.

The following measures have been suggested to safeguard the genetic purity of seed during seed production (1):

1. Use of seed of an appropriate class and from an approved source to raise the seed crop. The four generally recognized classes of seeds are breeder's seed, foundation seed, registered seed, and certified seed. The Association of Official Seed Certifying Agencies (AOSCA) has defined these seed classes as follows:
   a. *Breeder's seed:* The seed or vegetatively propagated material directly controlled by the originating or the sponsoring breeder or institution, providing for the initial and recurring increase of foundation seed.
   b. *Foundation seed:* The seed stock handled to maintain specific identity and genetic purity, which may be designated or distributed and produced under careful supervision of an agricultural experimental station. This seed is the source of all other certified seed classes either directly or through registered seed.
   c. *Registered seed:* The progeny of the foundation seed so handled as to maintain its genetic identity and purity and approved and certified by a certifying agency. It should be of quality suitable to produce certified seed.
   d. *Certified seed:* The progeny of foundation, registered, or certified seed that is handled so as to maintain satisfactorily genetic identity and purity and that has been approved and certified by the certifying agency.
2. Preceding crop requirements as have been fixed to avoid contamination due to volunteer plants.
3. Isolation of seed crops from various sources of contamination by natural crossing with other varieties grown alongside and off-types present in the field by wind and insects.
4. Avoiding contamination due to mechanical mixtures at the time of sowing, harvesting, threshing, processing, and handling of seeds, and contamination due to seedborne diseases.
5. Roguing of off-types differing in characteristics from those of the seed variety. The off-types may arise because of the presence of some recessive genes in heterozygous conditions arising from mutations. The heterozygous plants segregate for the characters affected by the particular gene(s) in later production cycles to give rise to off-types. Volunteer plants may arise from accidentally planted seed or from seed produced by earlier crops contributing to off-types.
6. The genetic purity of seed under commercial seed production can be maintained through a system of seed certification to distribute crop seeds, tubers,

or bulbs of true-to-type and high genetic quality. Qualified and experienced personnel of a seed certification agency should inspect seed crops at all appropriate stages of growth and verify seed lots or purity and quality.

7. Periodic testing of genetic purity of seeds by growing the crop to ensure maintenance of quality.

The varieties of self-fertilized species of crops, although theoretically completely homogeneous, may show considerable variation in practice, especially in the newly released varieties. It is therefore necessary to purify these varieties during the maintenance of nucleus/breeder's seed. Harrington (7) outlined the procedure for the maintenance of nucleus seed of prerelease or newly released varieties of crops. The sampling for nucleus seed may be restricted to 15 new varieties of any crop, obtaining about 200 plants from the central 3 m of border rows of replicates. Discarding poor plants and those with few tillers, the plants should be pulled 4–5 days before full maturity to avoid shattering. The plants are then threshed separately and the seed examined in piles on the table, discarding off-types or diseased seeds. The seed is now ready to be sown in a purification nursery called a *nucleus*. The nucleus seed is grown on clean, fertile land at an experimental station having grown no crop of the same kind in the previous year. The 200 progenies of nucleus seed should be sown in a block of 200 double-row plots in four series of 50 double-row plots each, with a plot-to-plot distance of at least 45 cm. The plot must be isolated properly to prevent contamination by natural crossing and spread of diseases. The nucleus plot should be examined critically for removal of off-types throughout the growth phases. At least 180 of 200 original plants should be harvested individually and threshed, avoiding mixing of one nucleus with any other. The seed is then cleaned and examined for uniformity (180 or more piles), with off-types being discarded. Uniform seed can be massed together and treated with fungicide and insecticide, bagged, labeled, and stored as "breeder's stock seed" for use the next year.

The breeder's stock seed from the nucleus is sown on clean, fertile land, having grown no crop of the same kind the previous year. The area required for planting breeder's stock is about 1.2 ha for wheat and 3 ha in the case of transplanted rice. The field should be properly isolated. Adopting standard agronomic practices of sowing, interculture, and harvesting, the seed should be produced at the experimental station in the area where the new variety has been bred. Row spacing should be sufficient to permit examination of plants in rows for possible mixtures or off-types. Plants not typical of the variety should be pulled and removed (roguing) before flowering, as was done for nucleus/breeder's stock seed. Where plants are removed after flowering and pollen has escaped, all surrounding plants within 1 m should be discarded. The breeder's stock is harvested, threshed with clean equipment, and bagged and labeled. The seed should be about 99.9% pure to its variety. This breeder's seed is now ready for multiplication as foundation seed. The breeder's stock should be continued each year to furnish a fresh stock of seeds to the growers of foundation stock until the newer ones replace the variety.

The genetic purity of the breeder's seed of the established varieties of crops can be maintained by growing the crop in isolation and by rigorous roguing during different phases of crop growth. The purity can be further enhanced by bulk selection, wherein 2000–A2500 plants typical of the variety are selected, harvested, and threshed separately. The seeds are critically examined, discarding off-types, if any. The uniform seeds are bulked to constitute breeder's seed. This process may be continued until deterioration sets in, changing plant characteristics of economic importance. Breeder's seed, therefore, needs to be included in yield tests. The breeder must carry over at least enough seed to safeguard against the loss of variety due to a complete failure of crop during the multiplication of foundation seed. The production of breeder's seed is an expensive process, with the associated risks of contamination by repeated multiplication and loss due to adverse growing conditions. Such risks can be minimized and the continuity of the seed program better assured by producing sufficient breeder's seed at one time to meet the requirements of two to three productions of foundation seed (carryover of breeder's seed). The carryover seed must be stored under optimum conditions in order to maintain its vigor and viability (1).

Maintenance of the genetic purity of hybrid seeds is complicated, requiring elaborate procedures. The nucleus seed of inbred lines can be maintained by self-pollination, sib-pollination, or a combination of the two procedures (hand-pollination). Some breeders prefer "sibbing," because it maintains vigor. "Selfing" is used to stabilize inbred lines if a change in breeding behavior is noticed. Some parental material is preferably maintained by alternate selfing and sibbing from one generation to other. Individually selfed or sibbed ears should be examined critically, discarding off-types or inferior characteristics (texture, color, seed size, chaff color, and shape of ear). The uniform ears are then threshed separately and planted ear to row to easily detect and discard off-types from individual ears if any. Alternatively, all of the ears from an individual inbred line may be composited for bulk planting in the next season.

The hand-pollinated seed is sown on clean, fertile soil having no previous crop of the same kind or variety during the previous year (barring maize). It is rather important to ensure that the crop is well isolated, with the requirement varying from crop to crop and depending upon the nature of the material to be protected by isolation, the nature of the contaminant, and the direction of the prevailing wind. The isolation can be achieved either by distance or by time (maize). The inbred lines need good growing conditions to exhibit genetic potential.

Maintenance of genetic purity of inbred lines through hand-pollination and adequate isolation alone is not enough to achieve perfection. The isolated fields must also be critically rogued for off-types and other impure types prior to the shading of pollen. The nucleus seed crop is harvested after physiological maturity if artificial drying facilities exist. Ear to harvest lines are harvested separately and piled; these are again critically examined for ear characteristics, sorting out all off-colored, diseased, or otherwise undesirable ears. If the overall percentage of off-types exceeds 0.1%, hand-pollination should be repeated to produce the second year's breeder's seed. The uniform ears are bulked, dried in a clean dry bin at temperatures not exceeding 43°C, shelled, cleaned, treated with pesticides, and stored under ideal storage conditions as breeder's stock seed. This seed may be increased the following season by paying adequate attention to isolation, roguing, and so forth, to maintain the high genetic purity of the seed.

To maintain the genetic purity of the nucleus seed of noninbred lines, the number of plants for hand-pollination should be large enough to preserve the genetic make-up of the variety, narrowing the genetic base by sibbing only a few plants (about 5000 plants or more). The sibbed ears are examined critically, discarding off-color, textured, or diseased ones. Uniform ears are bulked, dried, shelled, cleaned, treated, and stored as usual. Other practices of seeding sibbed nucleus seed are similar to those described earlier for inbred lines. Roguing, however, needs to be observed more critically by individuals with good knowledge of the material. The breeder's stock seed thus produced from the nucleus seed can be utilized to increase the breeder's stock of non-inbred lines, paying adequate attention to land requirements, isolation, roguing, harvesting, and handling of seed to achieve maximum genetic purity.

Raising breeder's seed crop in isolation and roguing the crop thoroughly at various stages can maintain the breeder's seed of the established varieties of cross-fertilized crops. It is often purified by mass selection. The crop is grown in isolation and rogued carefully as described earlier. At maturity, about 20,000–25,000 true-to-type plants are selected, harvested separately, and bulked after careful examination. This constitutes the breeder's stock seed. The seed may be carried over to ensure against possible failures or unforeseen shortages.

All varieties of apomictic species should be genetically pure, since they reproduce asexually. However, in many species, a certain degree of sexual seed formation is noticed. Since the plants in these varieties are normally highly heterozygous, each plant derived from a seed formed sexually will be different from the typical plant of the variety. In addition to mutations producing aberrant types, natural admixtures from other varieties cause further deterioration. To maintain the varietal purity of such species, a large number of plants are grown thinly enough for critical and separate observation. All inferior plants are then removed before flowering, and the seed from the remaining uniform types is bulked and used as new breeder's seed.

The maintenance of varieties of some artificial polyploids is made more difficult by admixtures of diploid plants. Owing to the differences in the seed-setting capacity of tetraploids and diploids, the latter would tend to increase rapidly from one generation to another, with eventual disappearance of tetraploid material (e.g., red clover). Every seed lot intended for further multiplication, therefore, needs to be examined thoroughly for its content of diploids, which can be removed by mechanical sieving, owing to differences in the seed size (1).

E. Newer Biotechnologies for Plant Breeding

Using a newer biotechnological technique, restriction fragment length polymorphism (RFLP) mapping, in which DNA fragments are used as the genetic markers to follow chromosomal segments through segregating generations, scientists at the International Crops Research Institute for the Semi-Arid Tropics (ICRISAT), India, and elsewhere are now able to produce improved crop genotypes (drought-tolerant or disease-resistant) much faster than was possible using traditional breeding methods. In 1994, ICRISAT published the first genetic linkage map for pearl millet based on RFLP. Because this map is about one-fifth the size of the rice map, fewer markers are required to obtain an accurate picture of the arrangement of genes on the chromosome. Smaller RFLP maps are available for only a few crosses. To locate genes in crosses for which no maps are available, a skeleton map from a previously mapped cross is transferred with small maps and fewer markers are required to transfer the skeleton map to the new cross. This skeleton map provides a rough framework of the gene linkages in the new cross; the details are filled in later by focusing on specific areas of interest using more markers. The small maps are associated with high "linkage drag" during backcrossing. When a desirable gene from one genotype is transferred to another by backcrossing, other genes also tag along with the desirable gene. This difficulty can be overcome by further generations of backcrossing and/or random mating (8).

Application of this map to pearl millet improvement is already in progress at the ICRISAT Asia Center (IAC) in collaboration with the National Agricultural Research Schemes (NARS) in India, in Niger, and at laboratories in the United Kingdom. Genes conferring strain-specific resistance to populations of pearl millet downy mildew from India, Niger, Nigeria, and Senegal have been tagged. Exploratory marker-assisted backcrossing will transfer some of these genes into elite hybrid seed parents of commercial importance in India, Columbia, and Niger. Other target traits for ongoing mapping studies include seedling heat tolerance and terminal drought tolerance.

A homozygous line of a crop with a "fixed" trait for a particular characteristic (e.g., seed size, tillering ability) is used in plant breeding for crop improvement. Using conventional breeding methods, it can take several years to develop a homozygous line, especially in a highly heterozygous, cross-pollinated crop like pearl millet. Starting with haploids (plants having half the normal number of chromosomes), the latter can be converted to dihaploids (which are homozygous) by doubling the number of chromosomes. ICRISAT biologists use a two-step method to produce haploid plants. First the anthers are cultured in a suitable tissue culture medium to form a callus. The latter is then transferred to another medium to allow the cells to differentiate and eventually form embryoids, from which haploid plants can be regenerated. Scientists at the IAC have successfully managed the first step, inducing callus development in cultured pearl millet microspores, and are now working on step two—regeneration of whole plants from calluses.

Temperature, light, and chemical composition of the culture medium influence androgenesis—the development of a haploid from a microspore. The androgenic response has been found to improve if the millet panicles are pretreated at low temperature (14°C) for several days before being placed in the medium. Also, incubation in red light is preferable to keeping anthers in the dark. The sugars (especially sucrose) in the culture medium have important effects on androgenesis. The effects of various media, hormones, and gelling agents are presently being investigated. These studies are expected to result in a protocol for the production of haploids from microspores (8).

Riera-Lizarazu et al. (9) attempted direct hybridization between wheat (*Triticum aestivum*) and barley *(Hordeum vulgare)* for transferring desirable traits, which proved to be ineffectual because of genetic dissimilarities between diverse gene pools. Applying growth regulators prior to and after pollination and using crossable lines have achieved successful hybridization. Detached spikes were used to evaluate the effects of 2,4-dichloroacetic acid and gibberellic acid on crossing success. Seed was set when Fukuho wheat was the maternal parent, and more embryos were obtained by adding higher concentrations of 2,4-dichloroacetic acid to the detached spike culture medium (up to 225 μM/kg). The 2,4-dichloroacetic acid had no effect on embryo formation when barley (Luther) was used as the female parent. The gibberellic acid (8.3–29.8 μM/kg) did not affect embryo formation regardless of cross direction. The 2,4-D–induced increase in embryogenesis was attributed to enhanced embryo survivability (9).

In Japan, biotechnologies such as embryo, ovule and anther culture, cell selection, cell fusion, and genetic engineering are being employed increasingly in breeding new varieties of crops. Cell and tissue culture methods are being used to propagate and produce virus-free seedlings (10,11) (see Chapter 22).

A broad outline of the work being done on "plant genetic systems" in the field of plant biotechnology has been published. This includes genetically engineered crop plants resistant to insects and herbicides (12,13). Montagu (14) has given a broad overview of genetic engineering in crop plants with particular reference to genetic transformation, identification of new genes, the advantages of biotechnology in providing environmentally safe crop protection, engineering plants to produce pharmaceutical products, and hybrid seed production via an anther-specific promoter. (The biotechnology of the production of artificial or synthetic seeds is discussed in Chapter 23.)

F. Production of Synthetic/Artificial Cultivars

Breeding of cereal and other agronomic crops has contributed significantly to the growth of agribusiness worldwide. The invention and development of hybrid maize (corn) during the late nineteenth and early twentieth centuries in the United States signaled a revolution, not only in crop production, but in the way research was carried out by the land-grant universities and agribusiness (8).

In normally self-fertilized crops, new variability may be created by hybridization, followed by the selection of desired cultivars in which desirable characteristics from two or more parents are combined. The type of hybrid cultivar obtained will depend upon the genetic background of the chosen parents as well as on the method of selection used. A similar situation arises when new variability is artificially induced through mutations.

In "pure-line" theory of classic plant breeding, a pure line is defined as all the descendants of a single homozygous individual by continued self-fertilization, resulting in a homogeneous cultivar. Hybridization, however, results in significant heterogeneity. The multiplication of such heterogeneous progeny in bulk to select homozygous individuals would be a gigantic task. Most modern hybrid cultivars are, therefore, selected at an early stage (F_2) as subsequent lines and probably released at the F_8 and F_{12} generations. These are obviously not as homogeneous as a pure line.

Cultivars can also be selected by producing multilines. Whereas normal line selection seeks to produce a new cultivar on the basis of one line or a few lines that are very similar, multiline cultivars are essentially different from each other in their characteristics, such as resistance to pests and diseases or environmental stresses. Thus, by incorporating different sources of resistance, the newly synthesized cultivar is buffered against changes brought about by virulent pathogens. These cultivars are, however, not very stable compared to those produced by the conventional methods of selection. A change in the prevalence of a virulent pathogen may eliminate certain lines from the cultivar. It is, therefore, necessary to return the cultivar to the plant breeder for its reconstitution. This may be advantageous, because it enables plant breeders to substitute new sources of resistance in the material (2).

Alternatively, the plant breeder can create a composite cross by bulking the F_2 generations of several crosses. The composite is allowed to develop for several generations during which natural selection may occur. If the composite is grown at more than one location, a locally adapted cultivar may be developed in time. The composite constitutes a gene pool from which the plant breeder can select a cultivar with desirable characteristics for further multiplication.

An alternative to the composite is the synthetic or artificial method of plant breeding in which a number of lines are put together by the plant breeder in predetermined proportions. A synthetic line generally has a limited life, because the proportions of the constituent lines are likely to change over a number of generations. The plant breeder must plan for seed production on a limited generation basis. This system can be extended by using mixtures of cultivars—claimed to be advantageous in some species—over a single cultivar, especially if different resistance genes are present in each cultivar. This method adds to the cost of mixing, which can be reduced by growing a seed crop for one or two generations after mixing before using it for crop production.

A hybrid cultivar results from a controlled cross between a male and female parent, the seed being harvested from the female parent only and used for crop production.

In self-fertilized species, it is easy to produce hybrid cultivars if male sterile lines are available that can be used as female parents. Kelly (2) discussed developments in the agrochemical industry, producing certain materials which, when applied to wheat, can prevent self-pollination. These substances may act as gametocides, destroying the pollen of the desired female parent, or as inhibitors that prevent pollen produced by the female parent from effecting fertilization. The advantage of the synthetic hybrid cultivar lies in heterosis. Special expensive measures are required to produce seed that is harvested from the female parent only. The resultant heterosis, therefore, must have a profitable effect to compensate for the cost of the production of synthetic hybrid cultivars in the self-pollinating crop species.

In the cross-pollinated crop species, plant breeders look for parent plants that have good combining ability. These plants, when allowed to multiply together, produce a desirable combination of characteristics. Cross-fertilization results in greater heterozygosity in these plants than in the self-fertilized plants and therefore less homogeneity. Each generation of an open-pollinated cultivar is thus a mixture of hybrids. The open-pollinated cultivars are generally grown for a limited number of generations and returned to the plant breeder's maintenance material after each cycle of seed production to produce commercial quantities of seeds.

Putting together a large number of parent plants and allowing random pollination to occur can create composites. A composite in a cross-fertilized species is generally the product of the first generation of such random pollination.

Production of synthetic cultivars begins with a limited number of specific parents, which are permitted to interpollinate. The number of generations of multiplication is strictly limited so as to recreate the synthetic/artificial cultivar at the end of each multiplication cycle.

As with the self-fertilized species, synthetic hybrid cultivars of cross-fertilized species are created by controlling pollination to ensure that seed is produced from a desired crossing. This can be achieved by the following methods (2):

1. By emasculating the female parent, as is done in monoecious plants like maize, by removing the male flowers before the release of pollens.
2. By using male sterility in the female line, so as to avoid the physical removal of male flowers.

3. By using self-incompatibility. In this system, the seed crop is harvested as a whole, since all plants are contributing and receiving pollen (e.g., in brassicas). The self-incompatibility, however, is not always complete, and there may be production of some inbred plants. With the excessive production of such plants, the advantage of heterosis in the subsequent crop is diminished.

The advantage of the synthetic hybrid cultivar in cross-pollinated species is not restricted only to heterosis. Most hybrids are based upon inbred lines. Normally, cross-fertilized plants require inbreeding for several generations to reduce heterozygosity and to include desirable genes in synthetic cultivars. A controlled cross between two such inbreds produces heterosis and a desirable combination of genes in the form of a synthetic cultivar.

The major disadvantage of the production of synthetic cultivars is the higher cost of plant breeding and seed production, requiring considerable time-consuming work to produce desirable inbreds, which alone can be used to synthesize new artificial hybrids. The final seed crop is not fully productive when male sterility or emasculation is used, because only the female parent is harvested for seed.

Therefore, various other hybrids have been produced. The hybrid resulting from the cross of two inbred lines is a "single cross," whereas the F_1 resulting from the cross of two single-cross hybrids as parents is known as a "double cross." In a "three-way cross," an inbred is mated with an F_1 hybrid. A "top cross" is the F_1 resulting from a cross between an inbred or a single-cross and an open-pollinated cultivar. All of the forms of hybrid cultivars require a particular cycle of seed production to produce the seed used in crop production (2).

III. AGRONOMIC PRINCIPLES

Besides the more important basic principles of genetics for the maintenance of seed purity and quality, standardized seed production technology involves agronomic principles used to preserve the high quality of increased seed yields, including agroclimate and location of the seed plot, previous cropping, variety, sowing, seed rate, artificial pollination, roguing, weed control, irrigation, nutrition, plant protection, harvesting, threshing, drying, and storage (1,2).

A. Agroclimate and Location

The variety of crop to be grown for seed production must have a suitable agroclimate adapted to the photoperiodic and temperature conditions prevailing in that location. Specific selected locations would be needed economically to grow crop varieties sensitive to photoperiodism (short days vs long days) and temperatures. According to Agrawal (1), regions with moderate rainfall and humidity are much more suitable for seed production than locations with high rainfall, humidity, and extreme temperatures. Most agronomic crops require a dry sunny period and moderate temperatures for flowering and pollination. Excessive dew and rains affect normal pollination, resulting in poor seed set. Extreme temperatures may cause desiccation of pollen and poor seed set. Very hot and dry weather conditions adversely affect flowering of several crops, especially vegetables, legumes, and fruit crops, which fail to set seed. These crops invariably require cooler climates with low atmospheric humidity to flower and pollinate normally. Oil seed crops may tolerate hot weather during flowering, but very high temperatures can result in premature flowering and the production of poor-quality seeds. Extreme cold temperatures also damage seed quality in the early phases of seed maturation. Thus, locations with an extreme agroclimate (summer heat and cold winters) are generally not suitable for seed production. Excessive rainfall conditions normally result in a higher incidence of diseases and pests, making the harvesting and other operations of seed production extremely difficult. They may also cause delayed maturity and pregermination of seed in many standing crops. A mature seed crop becomes increasingly susceptible to shattering, strong winds, and heavy rainfall. Ample sunshine, moderate rainfall and climate, and the absence of strong winds are ideal for the production of high-quality seed.

Agrawal (1) lists the following desirable characteristics for land selected for seed crops:

1. The seed plot should have soil texture (light, well drained) and fertility characteristics as required by the crop.
2. The plot should be free from volunteer plants and seeds of weeds and other crop plants.
3. The soil of the selected plot should be comparatively free from soilborne diseases and pests.
4. The same crop or a variety thereof should not have been cultivated during the previous season on the plot selected for seed production.
5. The plot must be leveled and feasible for isolation as per the requirements of certification standards.

B. Isolation, Variety, and Sowing

The seed crop must be sufficiently isolated from nearby fields of the same or other contaminating crops as per the requirements of certification standards. The seed crop should be isolated by providing enough distance between seed plots and contaminating fields. In the case of hybrid maize seed production, time isolation can be followed if distance isolation is not feasible. On a small scale in nucleus/breeder's seed production, isolation may be achieved by enclosing individual flowers or by removing male flower parts and employing artificial pollination. Even after the seed crop is harvested, effective isolation of seed from different varieties is essential to avoid mechanical contamination. Bags and other equipment must be thoroughly clean to maintain seed purity.

After the land is prepared for improving germination, including freedom from weeds and uniform irrigation, the selected crop variety is carefully planted. The variety selected should suit the prevailing agroclimatic conditions, be high yielding, and possess desirable attributes such as disease resistance, earliness, and grain quality. Similarly, the seed should be of known purity, appropriate class, and obtained from an authorized official agency. The seed may require treatment before sowing, if not treated already. Seed treatment may be with appropriate fungicides or involve bacterial inoculation for legumes or for breaking dormancy. Seeds having hard seed coats may require soaking in water overnight to facilitate germination. The seed must be planted at its normal planting time in soil having an adequate moisture content for germination. Lower than usual seed rates of commercial crops will facilitate the roguing and inspection of seed crops.

The seed crop is generally sown in rows by mechanical drillers, which allow the desired quantity of seeds to be planted at a uniform depth. The sowing equipment must be thoroughly clean to avoid any contamination. Sowing in rows facilitates effective plant-protection measures, roguing operations, and field inspection. Adequate spacing within rows and the distance between rows are given as per the requirements of the crop to allow air and sunlight to reach developing inflorescences at the plant bases. For hybrids, male and female parent lines are planted in 4:2 or 6:2 proportions to ensure that the seeds of the male and female parent lines are not mixed while planting. Small seed is generally sown shallow and large seed a little deeper to secure good planting. Seed emergence is better from greater depths in sandy soils than in clayey soils and as well as from warmer soil.

C. Roguing, Pollination, and Weed Control

Adequate and timely roguing constitutes the single most important operation in seed production. Rogues differing from normal (weak or sickly plants, bolters, and off-types) are pulled out and discarded at the earliest possible phases, before flowering, especially in cross-pollinated crops, to avoid genetic contamination. Plants obviously differing in height, color of vegetation, leaf size, shape, and orientation, or any other characteristic as well as malformed and diseased plants should be removed completely. In some crops, roguing at the early vegetative stage may be necessary to remove virus-affected plants. Undesirable plants not distinguishable earlier should be removed soon after emergence of the ear head/tassel. In hybrids, where male sterility is employed, special care is required to remove pollen shedders. While removing the plants with ear heads infested by seedborne diseases or loose smut, precautions must be taken to ensure that spores are not spread to healthy plants. Roguing at maturity is also necessary to remove off-plants not distinguishable earlier and contaminants affecting the physical purity of seed. Roguing and sorting of harvested ear heads may be necessary in some crops to remove off-textured, off-colored, diseased, or malformed ear heads. In root and vegetable crops, roguing at harvest time may be needed to confirm the fruit, tuber, or root quality of the crop.

Supplementary pollination provided by honey bees in hives in close proximity to seed crops that are cross-pollinated by insects may be necessary to ensure good seed set and thereby increase seed yields (2).

Production of high-quality seed requires thorough control of weeds on the seed plot. In addition to reduction in seed yields, weeds are often a source of contamination by way of mixing at the time of harvest. Weeds in the seed plot or nearby areas may also harbor a number of pests and diseases. Effective control of weeds at all the phases of crop growth is essential, and they must not be allowed to flower or set seed in any case. Planting seed crops on clean, fallow land or following crop rotations is generally recommended to keep weeds at a minimum. Hand weeding, interculture operations, or chemical weed control may be necessary.

D. Irrigation and Plant Nutrition

Because drier climates are more suitable for producing high-quality, disease-free seeds, irrigation is essential to obtain good seed yields. Irrigation may be required before planting and at suitable intervals up to flowering. One or two irrigations may be desirable for many seed crops. The frequency of irrigation and amount of water supplied depend upon the physical texture of the soil, rainfall, and crop requirement. Maximum benefits from irrigation can be derived only with adequate crop nutrition in the form of organic manures and fertilization, especially readily available sources of nitrogen and phosphorus.

Seed crops are rather sensitive to moisture stress at the vegetative, flowering, and maturity stages. Adequate soil moisture is also necessary for uniform seed germination necessary to further crop stand and good seed yields. Both excessive moisture conditions and prolonged drought will adversely affect germination, growth, and development of the seed crop. Water may be applied by surface irrigation, sprinkler, drip, or overhead irrigation or subsurface irrigation. All of these systems have merits and drawbacks, with the choice of method depending upon circumstances. Sprinkler irrigation has several advantages, but it tends to favor foliage diseases; hence it should be used with discrimination. Subsurface irrigation may be used to overcome certain troublesome diseases of seed crops (1). The irrigation should be stopped 2–3 weeks before seed maturity to ensure the drier conditions needed for harvesting.

Adequate amounts of nitrogen, phosphorus, potassium, and other essential minerals are crucial for the proper growth and development of the seed crop. It is, therefore, necessary to know the nutritional requirements of any individual seed crop and to ensure proper nutrition at all the stages of crop growth. Split applications of nitrogen are generally advocated to avoid lodging of a crop due to excessive vegetative growth. Application of nitrogen at the time of flowering leads to an increase in yield and quality of the seed of most crops. In some early crops, nitrogen dressings at flowering may tend to delay ripening. Whereas most grasses and peas are benefited by early applications of nitrogen, lettuce crops respond well to nitrogen application at the time of flowering (1).

Phosphorus and potassium favor root growth, increased strength of straw, fruiting, and seed development. They also hasten plant maturity and increase disease resistance. Potassium improves the photosynthetic efficiency of plants and favors both protein and lipid metabolism in oil seeds. Deficiencies of other essential secondary and micronutrients also need to be monitored carefully, using soil test measures.

E. Plant Protection

Effective control of all pests, including diseases and insects, is essential to produce a healthy seed crop. In addition to heavy reductions in seed yields, diseases and pests damage the quality of the produce. Systemic seedborne diseases such as loose smut in wheat, which are passed on to following crop generations, must be controlled scrupulously. Nonsystemic diseases may also leave their spores on seed coats and may be carried to following seasons.

Planting seed chemically treated with the appropriate fungicides effectively checks the seedling and many of the seedborne diseases. Applying the appropriate fungicides and insecticides in proper quantities and at the right time can effectively control most seed crop pests. Adoption of appropriate schedules of plant protection and roguing of diseased plants and ear heads from time to time will further check the spread of disease and insects.

F. Harvesting, Drying, and Storage

After the completion of essential cultural operations and approval of seed fields for certification, the crop is ready for harvest. The appropriate time of harvest to ensure maximum seed yield and quality is of great significance. Fully mature seed is easily harvested and cleaned with minimal harvest losses. Although early harvests may make combining difficult, with increased losses in threshing and cleaning, harvesting at later stages may result in increased losses due to weather, lodging, and seed shattering (2).

Seed moisture content is a good indication of the optimum time of harvest. Combines do not normally operate well above 15% seed moisture. Whereas soybeans may be harvested best at seed moisture content of 13%, for wheat the best moisture content varies from 15 to 17% (1). Harvesting of seed crops at seed moisture contents of less than 20% minimizes mechanical damage to seed. If maize ears are picked and dried, they may be harvested at 30–35% seed moisture content.

A seed crop may be harvested manually or mechanically, taking care to avoid mechanical injury to seeds during harvesting and threshing operations. Care must also be taken to avoid any chance of mechanical mixing of seeds and to maintain lot identity. Cemented threshing floors or use of tarpaulins is preferred to maintain the quality of seeds. A crop may be harvested by directly combining in the field using mechanical combines. Sun drying of seeds on threshing floors, spreading the seed in thin layers, may be necessary to reduce its moisture content and improve the storage quality. Drying of a seed crop to its safe moisture limit to preserve its viability and vigor must be carried out rather quickly. If the seed is to be dehydrated mechanically, it should be taken to the processing plant soon after harvesting. Care must be taken at all stages to avoid mechanical mixing and to maintain the identity of seed lots.

Seed may be stored in sacks or bags for short periods. Bags may need to be disinfected with DDT solutions, dried, and cleaned before use. They should be labeled properly and stacked on wooden pallets. Storage facilities should be dry, cool, and clean, disinfected with malathion, and fumigated if necessary.

REFERENCES

1. R. L. Agrawal, *Seed Technology*, Oxford & IBH, Delhi, 1980.
2. A. F. Kelly, *Seed Production of Agricultural Crops*, Longman, New York, 1988.
3. B. S. Kadam, Deterioration of varieties of crops and the task of the plant breeder, Indian J. Genet. *Plant Breeding*, 2: 159, (1942).
4. A. J. Bateman, Contamination of seed crops: I. Insect pollination. J. Genet., 48:257, (1947).
5. H. D. Hann, Maintaining varieties of self-fertilized crop plants, Euphytica 9(e): 37, (1953).
6. H. T. Hartmann and D. E. Kester, *Plant Propagation: Principles and Practices*, Prentice-Hall, New Delhi, 1968.
7. J. B. Harrington, Cereal Breeding Procedure, Food and Agriculture Organization Agricultural Development, Paper No. 28, 1952.
8. International Crop Research Institute for the Semi-Arid Tropics. Report 1994, Patancheru, India, 1995.
9. O. Riera-Lizarazu, W. G. Dewey and J. G. Carman, Gibberellic acid and 2-4-D treatments for wheat X barley hybridization using detached spikes, *Crop Sci.*, 32(1):108 (1992).
10. D. Fitzgerald. *The Business of Breeding Hybrid Corn in Illinois*, 1890–1940, Cornell University Press, Ithaca, NY, 1990.
11. Y. Kobayashi, Biotechnology at the prefront, *Prophyta, No. 5*, 1991, pp. 24, 28.
12. K. Glimelius, Utilization of biotechnology for breeding in agriculture and forestry. I. Agriculture, *Seed Abstr.* 15(12): 4051, (1992.
13. Plant genetic systems, a world leader in plant gene technology, *Biotech. Forum*, 7(3): 202, 204, (1990).
14. M. Van Montagu, New plants, Future in agriculture. *Agro-Ind. Hi-Tech.* 1(1): 8, (1990).

7
Cereal Crops

I. INTRODUCTION

Cereal crops constitute a major food source for humankind. In the developing countries, cereals provide two-thirds or more of the dietary calories. Rice in Asia, maize (corn) in South America, sorghum in Africa, and wheat in the Middle East have been recognized as staple foods upon whose yields famine or feast depends. Even in the developed countries, like Russia and Japan, cereals still provide more than half of the dietary calories. Cereals also constitute an indirect dietary source for people in highly developed countries, like the United States and Canada, where grain is fed to livestock and the population depends on animal food (1).

The production of cereals over the past third to fourth decades has increased more rapidly worldwide than has the population, with an increase in the yield per unit area having contributed much more than an increase in the area planted (2). The world average yield of cereal grain is about 2.5 metric tons per hectare; two to three times that of legume and oilseed crops. Partly because of their higher yielding ability and greater economic returns, especially under subsistence farming, cereals are displacing pulses and legumes in many less-developed countries, although they complement one another both agronomically and nutritionally. The rate of increase in yield on a worldwide scale is also much greater for the major cereals than for the legumes; consequently, cereals are becoming a progressively more predominant component of the total world food supply.

In addition to improved agronomy, such as better weed control and more timely and effective use of agricultural chemicals (fertilizers, pesticides), improved seed has significantly contributed to the recent increases in cereal yields. Plant breeding has played a major role in three ways (3):

1. The selection of disease- and pest-resistant cultivars
2. The development of shorter-statured varieties that do not lodge at high levels of fertilizer applications
3. The selection of cultivars with greater yield potential that can respond to higher inputs

All three plant-breeding approaches are essential and must be used together.

II. SEED-PRODUCTION TECHNOLOGY

McDonald and Copeland (4), describing the principles and practices of seed production, categorized the latter as follows:

1.   Seed production in area of use
2.   Seed production outside the area of use

The seeds of most agronomic crops are produced best in the area where they are planted for commercial production. When adequate quantities of high-quality seed can be produced in the area of use, it is best to do so, because crop varieties normally produce higher seed yields in the area to which they are climatically adapted. In addition, other important issues include transportation and marketing of cereals, soybean, and certain other crops involving larger seed volumes and greater planting areas than small-seeded grasses and legumes. Other than small-seeded grasses and legumes and certain specialty crops, the seed of most field crops are preferably grown in the area of their principal adaptation and use.

A large quantity of the seed produced in the more advanced countries like the United States and Canada is also produced outside its major area of use, including seeds of most grasses, legumes, vegetables, and flower crops. Certain regions in North America provide a unique combination of climate, soil, and cultural factors for the specialized production of high-quality seed. Dry harvesting weather and the availability of water, either through assured rainfall or irrigation, are the most important factors for seed production outside the area of use (4).

A. Wheat, Oats, and Barley

1. Introduction

Botany and Origin.   Wheat (*Triticum* spp.) belongs to the tribe Triticeae. The wild diploid progenitor of wheat occurs throughout the Fertile Crescent of the Middle East, where it was first domesticated about 10,000 years ago along with barley and several pulses (5) by selecting nonshattering, larger seeded forms. Tetraploid wheats were also developed in this area at about the same time. The final step in the evolution of wheat was the hybridization of the tetraploids with *Aegilops squarrosa* to give the hexaploid bread wheat *T. aestivum*. Since *A. squarrosa* occupies a wider range of environments than do other wheat progenitors, it confers on wheat the protein characteristics required for bread making and has a greater adaptive range (6). Wheat is thus adapted to both subhumid and semiarid steppes as well as to more acidic soils (7). Wheat was subsequently spread throughout central Europe to higher altitudes and more humid regions.

Oats (*Avena* spp.) are of uncertain origin but are believed to be native to Asia. Oats may have first appeared as weeds in wheat and barley fields in the Middle East. They became a crop of secondary importance as the temperate cereals spread to higher latitudes and cooler wetter climates (6). Oats are derived from a polyploid series, as is wheat.

The origin of barley (*Hordeum* spp.) dates back to 8000 BC (8). It was known to the Greeks and Romans and was also cultivated in ancient China, after which it was introduced to Japan in about 100 BC. Barley thus appears to have been domesticated at the same time and place as wheat, and may have been even more important than wheat in its early stages. An entirely diploid crop developed similar to that of wheat, although it is not as well adapted to extreme cold. The bulk of the barley produced today is utilized for brewing.

Species and Cultivars. Briggle and Reitz (9) classified the wheat species belonging to the tribe Triticeae according to the genomes or sets of chromosomes their somatic cells contain (Table 1). Einkorn wheat is known from the Stone Age and was developed from wild wheat, *T boeoticum*, which still grows wild in Asia Minor and southeastern Europe. It carries two A↓ genomes. The tetraploid wheats were derived from wild "emmer" wheat, *T dicoccum*, still found in Syria and Palestine. They contain the two A-genomes of the einkorn parent plus the two B-genomes derived from wild grass, *Aegilops speltoides*, thus having the AABB genotype. Hexaploid wheats were then developed from a further crossing when the cultivated tetraploid wheat came in contact with the weed *A. squarrosa*. The hexaploid wheat contains two genomes contributed by the weed, so its genotype is AABBDD. The tetraploid wheat *T. durum* grows best in warmer regions and is an important source of semolina flour used to make pasta products. Hexaploid wheat has spread throughout the world, mainly in the form of *T. aestivum vulgare*, which is the wheat of choice for bread making (3).

Table 1 Genomic Classification of Cultivated Wheat Species

Einkorn group: 7 pairs of chromosomes, diploid wheats
    Wild form, fragile rachis, kernel in hull:
        Triticum boeoticum
    Cultivated form, fragile rachis, kernel in hull:
        T. monococcum Einkorn
Emmer group: 14 parts of chromosomes, tetraploid wheats
    Wild form, fragile rachis, kernel in hull:
        T. dicoccoides
    Cultivated form, partly fragile rachis, kernel in hull:
        T. dicoccum Emmer (emmer wheat)
        T. timopheevi (Timopheevi wheat)
    Cultivated form, tough rachis, free kernel:
        T. durum (durum wheat)
        T. turgidum (poulard or rivet wheat)
        T. polonicum (Polish wheat)
        T. carthlicum (persicum) (Persian wheat)
        T. turanicum (orientale)
Vulgare group: 21 pairs of chromosomes, hexaploid wheat
    Wild form, none; synthetic type only
    Cultivated form, partly fragile rachis, kernel in hull:
        T. aestovi, subsp. *spelta* (Spelt wheat)
        T. aestivum subsp. *vavilovi* (Vavilov wheat)
        T. aestivum subsp. *macha* (Macha wheat)
    Cultivated form, tough rachis, free kernel:
        T. aestivum subsp. *vulgare* (common or bread wheat)
        T. aestivum subsp. *compactum* (club wheat)
        T. aestivum subsp. *sphaerococcum* (short wheat)

Source: Ref. 9.

The most common cultivated oat species, the hexaploid *Avena sativa*, was derived from wild oat, *A. fatua*, whereas the cultivated red oat, *A. byzantina*, is believed to have descended from the wild red oat, *A. sterilis*, which is probably the progenitor of all other oat species. Seeds of cultivated and wild oat species are shown in Figures 1 and 2, respectively.

H. distichum, the most common barley species, is a two-rowed type having only the central spikelet fertile and awned. In *H. vulgare* (= *hexasticum*), a six-rowed barley, all three spikelets are fertile and awned. Cultivated barley varieties are probably derived from the wild two-rowed barley, *H. spontaneum*, of Southwest Asia.

2. Methods of Breeding

Most cultivated species of cereal crop plants including wheat have been modified by breeding procedures to increase their usefulness as food, feed, and industrial products. The improvement of wheat has followed the sequence of introduction, selection, and hybridization. Wheat was first brought to the north Atlantic coast area of the United States by the several groups of colonists, who brought varieties characteristic of their points of emigration. Diverse types of wheat thus were available. The ones best adapted to the new country persisted. Additional ones made by successive waves of

Figure 1   Seeds of cultivated oats (*Avena sativa*). (From Ref. 36.)

Figure 2 Seeds of wild oats (*Avena fatua*). (From Ref. 36.)

immigrants supplemented the first introductions. Some types of wheat came to have major importance: Red Lammas in Virginia, Purple Straw in the Southeast, White Australian on the West Coast, Turkey in Kansas, Iowa, and Nebraska, and the durums in Minnesota, North Dakota, and South Dakota. Most older wheat varieties were gradually replaced by newer combinations, which in turn were replaced by still newer improved higher yielding hybrids better adapted to the environment of the area of their culture (10).

The major improvement in wheat has been in its resistance to various production hazards, including greater resistance to hessian fly, green bug, and other insects, a greater winter hardiness, stiffer straw, and a higher resistance to smuts and rusts. Covered smut, a soilborne parasite, was particularly troublesome in the Pacific Northwest. Sources of resistance are now available to all the physiological races of importance. Achieving resistance to stem rust was also problematic. As resistant types were developed and became established commercially, new forms of rusts arose by mutation and hybridization. New sources of resistance had to be found and incorporated into the commercially accepted wheat types.

The use of newly discovered sources of resistance and the newer breeding refinements afforded by chromosomal substitution techniques offered promise of still further improvements. Studies on the clarification of host-parasite relationships and establishment of genetic and chemical bases for physiological specialization hold significant promise for improvement. Dwarf wheats developed in the Pacific Northwest had higher yield potential than varieties in commercial use in 1960. Plant breeders then began to transfer the dwarf characteristics to wheats adapted to other areas.

The breeding method now generally used to develop hybrids is to obtain homozygous lines by inbreeding and then to make all possible crosses between them to test their combining ability. Desirable inbred lines thus obtained are kept isolated to maintain their genetic purity. The hybrid seed is produced by making appropriate cross-pollinations (10).

Male sterility is now being successfully used in breeding. Plants that are male sterile do not need to be emasculated. Cytoplasmic male sterility has been found to occur in wheat, as it does in other crops like onion, sugar beet, carrot, corn, millet, sorghum, pepper, petunia, and tobacco.

The self-fertilized species of crops like wheat, oats, and barley are inbreeding types, and varieties resulting from a breeding program are required to be homozygous and true breeding. In cereals such as wheat, oats, and barley, once a variety has satisfactorily undergone tests and trials for registration, the breeder maintains the variety by means of ears that are threshed separately and grown out in the field as ear-row progenies. These are inspected periodically, and any row that does not conform to the varietal characteristics is discarded. From each row, a single ear may be harvested and grown similarly in the following year. The remainder of the rows are harvested individually and sown in the following year as ear-row progenies either as single long rows or as small plots. The individual ear-row progenies, if satisfactory, are then bulked and labeled as "breeder's seed." This seed may be multiplied further before entering certification as uncertified prebasic seed. Although self-pollination is normal in wheat, barley, and oats, some out-pollination can occur. A barrier area of the same variety commonly surrounds the ear-row progenies. This can be particularly important with some six-row winter barley varieties, where the lateral florets in the ear are known to be open flowering and thus vulnerable to adventitious pollen (11).

Kofoid (12) initiated a recurrent selection program using gridded mass selection to select for increased seed set on individual plants that were treated with a chemical hybridizing agent (CHA). A significant rate X entry interaction was found only among the inbred lines for grain yield, test weight, grain protein concentration, grain weight, and grains per spike. The families from the mass-selected populations had a 20% greater seed set than those from the original population and a 73% greater seed set than the inbred lines when treated with CHA. The inbreds developed from the population had a 45% greater seed set than the conventional inbreds when treated. The increased hybrid seed production was associated with longer spikes and an increase in the number of spikelets per spike. This study indicated that out-crossing ability can be improved in wheat either directly through selection or indirectly through the use of random-mating populations (12).

3. Cultivation

Agronomic practices for the cultivation of cereal seed crops vary not only from one geographical area to another, but often from one location to another within the same geographical area. They are a function of several variables, differing significantly with variety, cultivar, and genotype. Good farming practices necessary for a successful cereal crop are the same whether the crop is grown for seed or for food/feed purposes.

Culture.

ISOLATION REQUIREMENTS AND LAND PREPARATION. Wheat, oats, and barley seed crops should not be grown in fields used for the same crops in the previous season unless the varieties were the same and approved by a seed certification agency for purity.

In addition, fields should be well drained and free of weeds and the soil should be neutral; that is, pH of around 7.0. Fairly long rotation intervals between crops of wheat, barley, and oats are desirable to decrease contamination with seedborne diseases such as *kamal bunt* (13).

The land should be prepared by bringing the soil to a fine tilth by deep plowing and harrowing before the presowing irrigation, which is necessary for uniform germination. A light shallow plowing or discing after presowing irrigation may also be needed. Leveling is an important aspect of seed-bed preparation. Application of 10% Benzenehexachloride (BHC) or some other suitable insecticide to prevent attack by white ants and *Gujhia* (25 kg/ha) before the last harrowing or plowing is recommended. Oats achieve good growth with less soil preparation than do wheat and barley.

Wheat is normally a self-pollinated crop, but natural cross-pollination to the extent of 1–4% may occur. It is usually sufficient to isolate seed fields with a 3-m strip all around, which is planted with a noncereal crop or left uncropped. If, however, a wheat variety is susceptible to diseases like loose smut caused by *Ustilago* spp., an isolation distance of 180 m between seed fields and other wheat fields is necessary (14). According to the Indian Seed Certification Standards, a minimum of 150-m isolation is needed from other wheat fields wherever an infection from loose smut is expected to increase more than 0.1% in the case of foundation seed production and 0.5% in the case of certified seed. Similar standards are applicable to barley and oats.

PLANTING AND CULTIVATION. Long-duration wheat varieties like Kalyan Sona are sown in India during the first 2 weeks of November, whereas short- to medium-duration cultivars (e.g., Sonalika, HD 1982) are sown during the second 2 weeks. Barley and oats may be sown from mid-October to the first week of November. Nucleus, breeder's, or foundation seed should be obtained from an approved certification agency.

Seed crops of wheat, barley, and oats are sown in rows with a seed drill or behind a plow in furrows. The depth of sowing can vary from 5.0 to 7.5 cm, with a row spacing of 20.0–22.5 cm to facilitate rouging and inspection work. The seed rate for wheat, barley, and oat seed crops varies from 75 to 85 kg/ha. A low seed rate and wider spacing between rows can be used to obtain the highest possible multiplication rate, but the yield per unit area may be less. Seed rates as low as 30 kg/ha have been used successfully in fertile soil conditions (15).

Carbonell et al. (15a) evaluated the effects of a stationary magnetic field (MF) on the initial stages of growth of wheat plants. The wheat seeds were subjected to a 125-mT MF for several times of exposure, with nonexposed seeds forming the control. Wheat seeds were placed in rolled filter papers, moistened, and subjected to magnetic treatment. Stem length and total length were measured on 3rd, 7th, and 10th days after sowing (DAS) as well as stem weight and total weight on the 10th DAS. The magnetic treatments generally stimulated early plant development. Chronic exposure to 125-mT MF showed statistically significant differences to the control for all the parameters studied.

B. Irrigation, Fertilization, and Weed Control

Based on soil test values, wheat seed crops maybe fertilized with 80–120 kg nitrogen (N), 50–60 kg phosphorus (P), and 40 kg potassium (K) per hectare, whereas barley and oats require about 50–60 kg N, 30 kg P, and 20 kg K per hectare. Half of the dose of N and all of the P and K fertilizer doses should be applied at the time of sowing, and the remaining quantity of nitrogenous fertilizer should be applied at the time of first irrigation. The fertilizer should be placed 5 cm away from the seed. A band placement of fertilizer gives better results. The seed crops of oats should not be given too much nitrogen to avoid lodging.

Depending upon the soil texture, four to six irrigations may be needed for wheat and three to four irrigations for barley and oats. The field is irrigated first at the crown root initiation state, that is, about 30 days after sowing, followed by others at the late tillering, late joining, flowering, milk, and, dough stages. Barley and oat crops may be irrigated as and when required. Excessive irrigation decreased the nitrogen content of wheat, barley, and oat seeds in plants grown in Utah (Table 2). However, the phosphorus, potassium, calcium, and magnesium contents of these seeds increased. It is not known whether the primary effect of excess moisture is on mineral absorption by the roots or on the rate of grain fill with carbohydrates and the concomitant dilution of basic cell constituents (17). Plants exposed to low soil moisture or drought conditions have been reported to produce wheat seeds with increased protein content (18). Das et al. (61) investigated the effects of tillage and fertilizer on the seed character and seedling vigor of wheat after transplanted rice. Preparatory tillage with either two of four plowings in wheat after transplanted rice showed higher length and width of seed, higher seed yields, higher seed germination percentage, and seedling vigor index of wheat over zero tillage. Increasing levels of N, P, and K also enhanced the seedling vigor index, seed size, seed germination, and seed yields (Table 3). Das et al. (61) recommended a minimum of tow or the conventional four plowings along with 120 kg N, 60 kg P_2O_5, and 60 kg K_2O/ha in wheat grown after transplanted wet rice.

Timely weeding and interculture through periodical hoeing and weeding operations should be ensured to keep the fields of seed crops free of weeds. A foliar spray of 2,4-dichloro-propionic acid at 0.4 kg a.i./ha, dissolved in about 800 L of water, may be

Table 2 Effect of Excessive Irrigation on the Mineral Content of Wheat, Barley, and Oat Seeds

Mineral	Increased or decrease over controls (%)		
	Wheat	Barley	Oats
Nitrogen	−21	−19	−40
Phosphorus	+55	+30	+30
Potassium	+35	+14	+31
Calcium	+155	+41	+22
Magnesium	+32	+9	+65

Source: Ref. 16.

Table 3 Effects of Tillage and Fertilizer Level on Seed Yield, Seed Character, and Germination of Dwarf Wheat Grown After Transplanted Wet Rice

Treatments	Seed yield (q/ha)	Seed length (mm)	Seed width (mm)	Seed germination (%)
a. Number of tillage operations (plowings)				
0	23.65	5.96	1.93	81.25
2	29.43	6.62	2.13	87.50
4	31.09	7.17	2.47	90.00
Mean	28.09	6.58	2.18	86.25
SEM (+)	1.26	0.06	0.03	3.46
CD (P = .05)	4.31	0.22	0.11	NS
b. Fertilizer level (kg/ha) $N + P_2O_5 + K_2O$				
0 + 0 + 0	13.30	5.82	1.71	80.00
30 + 15 + 15	22.93	6.47	2.09	80.00
60 + 30 + 30	33.78	6.89	2.40	90.00
120 + 60 + 60	42.22	7.14	2.50	95.00
Mean	28.07	6.58	2.18	86.25
SEM (+)	1.95	0.09	0.08	4.20
CD (P = .05)	5.65	0.26	0.23	12.19

SEM = standard error of means; CD = critical difference at +5%; NS = not significant.

given 25–30 days after sowing to control broad-leafed weeds. Agrawal (13) recommended a preemergence application of tribuil at 1.5 a.i./ha in 750 L of water to control *Phalaris minor* (wild oats) in wheat seed crop. The latter can also be controlled by incorporating 2.5 L of avadex dissolved in 600 L of water applied to 1 ha just before sowing.

Control of Pests and Diseases. Pest and disease control practice is essential to ensure healthy seed crops with well-filled seeds. A schedule of pesticide application to control the insects and diseases of wheat, barley, and oats (Table 4) should be scrupulously followed.

Roguing. Two of three roguings may be needed to bring the seed plots of wheat, barley, and oats to seed certification standards. The first roguing may be carried out just before or during flowering to remove any obvious off-type plants. This will ensure prevention of natural crossing of rogues with normal plants and variation in the following year. Obvious rogues at this stage include plants differing in color, plants susceptible to different diseases, tall plants in a dwarf variety, dense heads or other head variations, smutted plants, and early heading plants. A gang of workers with gunny bags and paper bags can easily rogue out the smutted plants. The ear heads of smutted plants should be covered with the paper bag and then the whole plant uprooted without allowing its spores to fall onto other plants. Burning or burying may destroy the rogued plants.

A second roguing may follow just after flowering is completed and before the crop starts to turn color. Rogues at this time include those mentioned in the first

Table 4 Schedule of Pesticide Treatments to Control Insect Pests and Diseases of Wheat, Barley, and Oats

Pest	Control measures
I. Insects	
Termites	5% aldrin or 10% BHC dust at 25 kg/ha in soil at the time of last plowing. Termites in standing crop can be controlled by applying aldrin 30 emulsifiable concentration at 5 L/ha along with irrigation water
Brown white mites	Dimecron 100 at 250 mL/ha or metasystox 25 EC at 650 mL/ha on first appearance of pest.
Hairy caterpillars and army worms	Dusting of 10% BHC at 25 kg/ha
II. Nematodes	
Molya disease	DBCP, 60% at 300 mL/ha applied through irrigation water before planting
Ear, cockle, and tundu disease	Use clean seed free of nematode balls
III. Other diseases	
Rusts	Diathane Z-78 at 2.5 kg/ha in 1000 L water sprayed 4–5 times at interval of 10–15 days. 100 mL of sandovit may be added to solution
Loose smut	Seed treatment with Vitavax at 2.5 g/kg seed. Seeds may also be subjected to solar treatment
Alternaria and *Helminthosporium*	Seed treatment with organomecurial fungicides
Karnal bunt	Spray of Plantavax, 75% wettable powder at boot stage at 400 mL/ha
Powdery mildew	Two sprays with benomyl at 0.5 kg/ha at interval of 10–15 days

Source: Ref. 13.

roguing plus tall varieties that are late in heading, thus escaping the first roguing. Agrawal (13) stressed the importance of the first two roguings, since lodging of plants in highly fertile spots may make a final roguing difficult or impossible. A third roguing may be carried out after the ear heads turn color and begin to mature. In addition to the types previously mentioned, rogues may now be identified on the basis of differences in the color of heads and awns and variations in ear head types. In addition to off-types, objectionable weed plants such as *Convolvulus arvensis* (hiran khuri), *Lathyrus* spp. (chatri-matri), and *Vicia* spp. as well other crops such as barley and oats should be removed.

1. Harvesting and Threshing

Soon after maturity, seed crops of wheat, barley, and oats should be harvested to avoid seed losses due to shattering and uncertain weather conditions. Delay of harvesting in rainy and stormy weather may result in the sprouting of seeds in the ear and rejection of the crop for seed. In the developing countries, manual harvesting with sickles is very common. The crop is cut near the ground, allowed to dry in the field, and transported to the threshing yards. In the developed countries, harvesting of large fields by combines is a common practice. Splinter (19) described losses of seeds in combines. Harvesting of wheat after the seed was dried to more than 30% moisture content resulted in a field loss of 12 lb/acre for each day of delay in harvesting. The cutter bar losses varied from 0.5% at 26% seed moisture content to 1% at 13% seed moisture content (20). Modern oat cultivars can be harvested satisfactorily by combine, although the seeds in an oat panicle tend to ripen less uniformly than those of wheat or barley. The oat seed is hard and difficult to mark with the thumbnail, and the straw tends to retain more green color when the oat seed is ripe than does the straw of other cereals (15).

The traditional methods of threshing include trampling the harvested wheat plants under the feet of bullocks or with wooden sticks in open threshing yards. Considerable losses occur because of birds, rodents, spillage, and incomplete threshing. Threshing is followed by hand winnowing to remove the finely cut stem from the seed. An inefficient thresher may cause significant seed breakage, leading to infestation by insects and mold damage. Care must be exercised to ensure that the harvested seed is not mixed with other seeds. Threshing floors must be cleaned thoroughly to prevent any possible mixing. Combines should be used only when the seed moisture content is 16% or less. Combining should be done in the late morning hours after the dew has dried up. Combines must be cleaned thoroughly to avoid mixing of varieties.

2. Postharvest Operations

Drying. Seeds of wheat, barley, and oats are usually dry (10–12% moisture content) at harvest time. If the grain moisture content is high at harvest, the seeds must be dried to a safe storage moisture level (below 13.5%) to avoid biodeterioration (21). In the tropical countries, sun drying is a common practice for small quantities of produce. However, sun drying may result in seed losses due to incomplete drying, birds, rodents, and spillage.

Mechanical drying is essential in large-scale production to avoid seed losses due to birds, rodents, and unexpected rains and to ensure uniform drying of seeds. The type of drying system used depends on the relative humidity (RH) of the atmosphere. Natural air drying may be employed effectively at RH below 50% when the equilibrium moisture content is below 14%. At RH above 60%, heated air drying may be used to reduce seed moisture to a safe level. Muckle and Stirling (22) recommended a maximum of 49°C for drying seed-purpose wheat. Splinter (19) described various methods of mechanical drying for wheat. The seed moisture content for oat seeds should be below 16% for short-term storage and 14% or less when the seed is to be stored for some months. For long-term storage, the moisture content should be below 10%, and samples should be reduced to 8% before being placed in moisture-proof containers. The dry air temperature should not exceed 44°C when the initial moisture content is high, but this can be increased to 49°C as drying proceeds (15).

Cleaning, Treatment, and Bagging. The seeds of wheat, barley, and oat crops must be cleaned, treated, and bagged immediately after threshing and drying to maintain their high quality. Seed may be cleaned efficiently after threshing and drying to maintain their high quality. Seed may be cleaned efficiently by using a air/screen cleaner or indented cylinder and a gravity separator. In northern latitudes, very little time is usually available between harvest and sowing, and greater seed-cleaning capacity may be needed to deal with peak loads.

Gosteva et al. (23) described a large floor-mounted apparatus for separating weed seeds. Different seed fractions are recovered as the material is fed through a system consisting of a pneumatic separating channel, cylindrical seed sorters, and sieves with varying mesh dimensions and a vibrating surface. Kachru and Sahay (24) developed a medium-capacity pedal- or power-operated air screen grain cleaner for wheat, soybean, and chickpea. For pedal operation, the maximum purity of separated grain was 99.9% for chickpea, with a screen effectiveness of 80.6%, and a minimum at 99.5% purity for wheat, with a screen effectiveness of 71.3%, whereas for power operation, a maximum purity of 99.9% for soybean and a minimum of 99.6% for wheat were obtained.

Chen et al. (25) used an image analysis system with the capacity to acquire and combine three-dimensional laser range data and two-dimensional camera contour-extracted images to discriminate between cereal grains and wheat seeds and between soft white and club wheats, two- and six-row barleys, and rye and triticate grains. Discrimination between wheat, soybeans, oats, wild oats, dent maize, flint maize, wild buckwheat, and sorghum was almost 100% accurate.

Seeds with more than 16% moisture should not be treated chemically. The treated seed must be labeled properly to show the treatment applied. It usually has a limited storage life, and it is desirable to treat only as much seed as is likely to be required. Wheat seeds are treated as follows for the following seedborne diseases:

Disease	Treatment
1. Brown foot rot and ear blight (*Fusarium* spp.)	Organomercury, benomyl, guazatine, thiophanatemethyl imazalil
2. Bunt or stinking smut (*Tilletia caries*)	Organomercury, benomyl-thiram carboxin-thiram, guazatine triadimenol/fuberidazole/imazalil
3. Glume blotch (*Septoria nodorum*)	Guazatine, triadimenol/imazalil, fuberidazole
4. Loose smut (*Ustilago nuda*)	Carboxin, benomyl, triadimenol/fuberidazole/imazilil

Loose smut may also be treated by a hot-water soak (32°C for 4 hr) and immersion in water (54°C for 10 min) followed by cooling and drying (15). Other wheat pests which are not seedborne can also be controlled by seed treatment:

Pest	Treatment
Wheat bulb fly (*Leptohylemgia coarctata*) Wire-worm (*Agriote* spp.)	For early sowing: carbophenothion, chlorfenvinphos For later sowing: γ-BHC (HCH) γ-hexachlorocyclohexane

When slugs or snails are excessive in a field, wheat seed may be mixed with methiocarb pellets, but the seed should be used within 3 months. Organomercurials are cheaper and effective, but widespread use of mercury is now discouraged and its use has been restricted or banned in many countries (15).

Oat seeds are treated as follows:

Disease	Treatment
Brown foot rot and ear blight (*Fusarium* spp.)	Organomercury, guazatine, benomyl, thiophanate-methyl triadimenol/fuberidazole/imazalil
Covered smut (*Ustilago hordei*)	Organomercury, carboxin/thiram-guazatine + imazalil
Leaf (stripe) spot (*Pyrenophora avenae*)	Organomercury, guazitine + imazalil
Loose smut (*Ustilago avenae*)	Organomercury, guazatine + imazalil carboxin/thiram

The main seedborne diseases of barley are leaf stripe (*Pyrenophora graminae*), net blotch. (*P. teres*), covered smut (*U. hordei*), and loose smut (*U. nuda*) The first three are controlled by treating seeds with organomercurials or dithiocarbamates. Loose smut can be controlled by hot water treatment, but this is cumbersome. Chemicals such as a fenfuram can also be used to control loose smut. Organomercurials give good protection against many soilborne fungi. Fungal mildew (*Erysiph graminis*) of seedlings can be effectively treated with a formulation based on benomyl or triforine dimethylformamide.

Storage. Bagged wheat imbibes more moisture and suffers greater losses due to infestation as compared to bulk lots. Sarid et al. (26) have recommended bulk storage in bins at the farm level and storage in silos at the commercial level. Wheat stored in concrete bins picks up moisture from the atmosphere during the rainy season. Seeds are more often spoiled in concrete bins because of heat damage, development of acidity, and loss viability than seeds stored in aluminum bins. Salunkhe et al. (27) described various biological factors of seed deterioration during storage and their control measures.

Seed Yields. Average seed yields of 40–60 quintals/hectare of wheat and 30–40 quintals/hectare of barley and oats in India have been reported (13).

C. Maize and Grain Sorghum

1. Introduction

Botany and Origin. Maize (corn) (*Zea mays*) is the only cultivated cereal to originate on the American continent, where it has been cultivated for at least 4000 years. Maize is native to tropical Central America. The wild plant most closely related to maize is *Z. mexicana* (= *Euchlaena*), or teosinte, which is, however, not the true ancestor of cultivated maize. Both plants have a common ancestor, now extinct, which by crossing with another grass, *Tripsacum*, produced both maize and teosinte. Since maize is intolerant to both shade and drought, it probably originated in areas with alternate wet and dry seasons where day length controls the life cycle of plants. (6). Most tropical races of maize are short-day plants, whereas modern temperate zone cultivars are almost totally indifferent to day length (28).

Maize has unisexual spikelets, having separate male and female inflorescences on the same individual plant. The staminate flowers from the terminal panicle (tassel and the pistillate flowers in spikelets form the spadix, or cob, from which the grain develops. The cob arises laterally in the axil of the foliage leaf in which it is ensheathed.

Sorghum (*Sorghum vulgare, S. bicolor*) belongs to the tribe Andropogoneae. It was known as a cereal in ancient Egypt by 220 BC. Sorghum was probably domesticated in Africa about 5000 years ago in the savanna belt (29). From there, sorghum spread through Africa and India to China. Many tropical sorghums are strict short-day plants in which the local adaptation of day-length response is very important (6). Selection of early-maturing varieties and hybridization has made possible the spread of sorghum from the southern United States to higher latitudes (30). Sorghum is not as well adapted to cooler climates as is maize, but it is more drought resistant.

Species and Cultivars.   *Z. mays* var. *Saccharata* is the common sweet corn eaten as "corn on the cob" in the United States. *Z. mays* var. *Everta* is a special popcorn cultivar. Its small, hard seeds have a hard, glossy outer endosperm. The kernels burst easily at high temperatures, everting the soft palatable endosperm. Other varieties include dent corn, *Z. mays* var. *Americana*, characterized by a depression on the top of the grain, which is caused by shrinkage of the soft endosperm. Flint corn, *Z. mays* var. *Praecox* (= Indurata), with a hard endosperm and no identation (depression), is mostly cultivated in Europe. Flour, or soft, corn, *Z. mays* var. *Amylacea*, is without the horny endosperm and is cultivated almost exclusively by native Americans for their own use. Finally, waxy corn, although not regarded as a distinct variety, is valuable, because its starch consists entirely of amylopectin (3).

The most common species of sorghum is *S. vulgare*. The major cultivated varieties include *S. vulgare* var. *Durra* (durra sorghum), var. *Caffrorum* (kaffir sorghum of Africa), var. *Rexburgi* (Indian sorghum, shallu), and var. *Nervosum* (Chinese sorghum Kaoliang). Sorghum grows well in warmer climates of the tropics and is an important food for people in India, China, and Africa. Figure 3 show a well-growing sorghum crop at the International Crop Research Institute for Semi-Arid Tropics (ICRISAT) Center, Hyderabad, India.

2. Methods of Breeding

Hybrid maize has been widely adopted in the United States since the 1940s and 1950s. The development of hybrids of grain sorghum followed the discovery of male sterile characteristics in 1935 and later. Grain sorghum hybrids were accepted by farmers more quickly than were maize hybrids. The use of hybrid seed requires the production of new seed each year owing to a reduction in hybrid vigor from inbreeding.

The production of hybrid seed of maize and sorghum involves the following three steps:

1. Maintenance of parental lines (inbred lines).
2. Production of a single cross.
3. Production of a commercial hybrid—three-way cross, double-cross hybrid, or a double top-cross hybrid. The first two, inbred line and single cross, are

Figure 3 Grain sorghum crop grown at ICRISAT Center, India. (From Ref. 31.)

referred to as foundation seed, whereas the latter (double cross, double top cross, or three-way cross) are known as certified seed. The open-pollinated cultivars do not require different seed-production management that are applied in the production of a feed crop. For hybrid cultivar production, the female parent must be emasculated before the silks become receptive to ensure that desired crossing takes place. When cytoplasmic male sterility is used, it is necessary to ensure at regular intervals that none of the female parent plants have produced tassels that may shed viable pollen (15).

Utilizing cytoplasmic male sterility produces the hybrid sorghum seed. The steps involved in the production of hybrid sorghum are as follows:

1. Maintenance of parental lines, namely, the male sterile line (Line A) carrying cytoplasmic genetic male sterility, the maintainer line (sister strain of Line A Line B), which is male fertile nonpollen restoring), and the restorer line (Line R) to be used as a male parent for the purpose of producing hybrid seed, male fertile, pollen-restoring line.
2. Production of hybrid seed, which involves crossing of male sterile line (Line A) with restorer line (Line R) as in an isolated field.

The first stage of increase, that is, maintenance of parental lines, is referred to as foundation seed production, and the production of hybrid seed is known as certified seed production. Agrawal (13) has described the details of production of hybrid corn and sorghum seeds.

Great strides have been made in breeding tropically adapted sorghum with high productivity, especially for grain. However, the narrow genetic base of the elite gene pool causes concern about genetic vulnerability and limits the opportunities for future gains. According to a recent ICRISAT report (31), there is limited resistance to major biotic and abiotic stresses and under representation of duel-purpose and forage types in the elite gene pool. Sorghum breeding at the ICRISAT Asia Center endeavors to widen the genetic base in sorghums adapted to the semiarid tropics. The breeding approaches being followed include population improvement, development of parental lines for resistance to the major pests, diseases, and striga (Fig. 4), development of alternate cytoplasmic seed parents, and development of materials and strategies for superior adaptation to post–rainy season environments (32–35).

According to Redenbaugh and associates (59), there are strong technological and commercial reasons to explore maize multiplication by the production of synthetic/artificial seeds through somatic embryogenesis.

3. Cultivation

Culture.

ISOLATION REQUIREMENTS AND LAND PREPARATION. The cultural practices for growing maize and sorghum seeds are similar in many respects. The fields selected

Figure 4 A sorghum strain resistant to striga developed at ICRISAT Center, India. (From Ref. 31.)

should have fertile soil with favorable temperature and moisture conditions. In the United States, some corn seed is produced in irrigated areas, but much is produced in parts of the Corn Belt where normal rainfall is adequate. Grain sorghum is regularly grown in irrigated areas of the western Corn Belt.

Before the availability of hybrids, seed was usually grown in areas to which the variety was adapted. Hybrids are crosses of specific and uniform inbred lines, which change little when they are grown in geographical areas with different climates. The location of seed fields is decided upon by taking into consideration factors such as potential yield in seed fields, weather risks (such as drought, high temperatures, hot winds, and hail), length of season, maturity of the hybrid parents, time of freezing temperatures, isolation from undesirable varieties, freight costs to the planned markets, and other economic factors (36).

The selected seed fields should be free of volunteer maize and sorghum plants, well drained, and well aerated. Since maize is normally cross-pollinated by wind, fields should be isolated from other maize fields by at least about 400 m for foundation seed and 200 m for certified seed to prevent contamination by foreign pollen.

Sorghum is generally a self-pollinated crop, but cross-pollination to the extent of 5–6% may occur; this varies from 0.6 to 50.0% for different places and varieties and occurs more in loose panicles than in compact ones. Seed fields, therefore, must be isolated from possible contaminants by a distance of about 200 m for foundation seed and 100 m for certified seed from fields containing other varieties of grain, dual-purpose sorghum, and the same variety not conforming to varietal purity requirements for certification. Sorghum fields should be about 400 m away from fields containing Johnson grass (*S. halepense*) and forage sorghum (13).

After presowing irrigation, the land is prepared by plowing to a good tilth. One plowing and two to three harrowings followed by leveling is normally sufficient.

Planting and Cultivation. Maize is usually sown about 2 weeks prior to the onset of monsoon in India; that is, the second week of June to mid-July. In the United States, early planting is preferred to achieve early maturity in the fall and to have the maize at safe moisture levels in the event of an early freeze. Early planting in the Corn Belt, however, increases the need to use chemicals to control corn borers. Sorghum in kharif season is sown in the fourth week of June, which may last up to the first week of July, whereas in the southern states of India rabi sorghum is sown in mid-September. The nucleus/breeder's or foundation seeds of maize and sorghum should be obtained from a source approved by a seed-certification agency.

Maize is sown in rows with the help of a maize planter or is dibbled by hand in furrows of 5- to 6-cm depth and spaced at 60–75 cm between rows and 20.0–22.5 cm between plants. Sorghum is also sown in rows at 3- to 4-cm depth and with a spacing of 45 cm between rows and 15 cm between plants. The seed rates for maize and sorghum vary from 16 to 18 kg/ha and from 12 to 15 kg/ha, respectively. Sorghum seeds hardened with 2% KH_2PO_4 showed highest field emergence (68%) and excelled others in maximizing the seed yield (1170 kg/ha) (36a).

4. Irrigation, Fertilization, and Weed Control

Based on soil test values, seed crops of maize may be fertilized with 120–150 kg N, 50–60 kg P, and 40–50 kg K per hectare. An application of 25 kg/ha of zinc sulfate may be needed in zinc-deficient soil. The first dose of 40–50 kg N and the entire quantity of P,

K, and Zn fertilizers should be applied as a basal dose before or at the time of planting by placing fertilizer in bands to the side or below the seed. A second dose of 40–50 kg N should be applied as a side dressing about 30 days after sowing when the plants attain a height of 45–60 cm. The last dose of N should be applied just before flowering.

In rain-fed areas, sorghum is fertilized with 40 kg N, 40 kg P, and 35 kg of K per hectare based on soil test values. For irrigated sorghum, a first dose consisting of 60 kg N, 50 kg P, and 40 kg K per hectare is applied by placing fertilizer 5 cm below and 5 cm to the side of the seed at the time of sowing. A second dose of 40–60 kg N for rain-fed sorghum and 60–70 kg N per hectare for irrigated sorghum is applied between the rows 30 days after planting. Gagare et al. (36b) reported that application of 50 kg/ha of ZnSO$_4$ supplemented with two foliar sprays (0.5%) enhanced pollen production in the sorghum restorer line (CS-3541) as evidenced by higher seed yield.

Maize is more sensitive to both excess water and drought conditions than is sorghum. Irrigations should be scheduled to ensure adequate moisture in the soil. Drainage of the maize field is as important as irrigation. The sorghum crop is irrigated as and when needed to avoid leaf wilting at any stage. An adequate moisture content of the soil must be ensured throughout crop growth, especially at the time of flowering. Proper drainage should also be ensured to avoid water logging. Kharif sorghum needs one presowing irrigation and one or two more irrigations depending upon rainfall, whereas rabi sorghum will need three to four irrigations.

Fields must be kept free from weeds by timely weeding and interculture operations. Intercultivation of maize should not be more than 3–5 cm deep to avoid root damage. The last interculture should include earthing up of the crop to avoid lodging.

Preemergence soil application of tafazine (50% simazine) or atrataf (50% atrazine) at 1 kg/ha in 1000 L of water has been recommended to control weeds in maize fields. An application of atrazine (atrataf) at 1 kg chemical (50% wettable powder) in 1000 L of water controls weeds in sorghum fields.

5. Control of Pests and Diseases

Agrawal (13) described a plant-protection schedule for maize and sorghum seed crops (Table 5).

6. Roguing

The seed crop inspection of maize should particularly ensure that isolation is satisfactory and that, for hybrid seed production, the female parent is not producing pollen. In any case, it is, however, necessary to check that cultivar identity and purity are maintained satisfactorily. When the seed crop follows a maize crop in the preceding season, an additional check must be made to keep the field free from volunteer plants. The crops producing hybrid seeds of cultivars are rejected if at the final inspection there are more than 0.1% off-plants in the female parent. For open-pollinated cultivars, the maximum permitted off-type plants is 0.5% in basic seed and 1% in certified seed (37).

Inbred lines of maize are relatively true breeding strains, which necessitates rigorous roguing for off-types before pollen shedding commences. Abnormally tall and vigorous plants at the knee-high stage should be removed first. At preflowering, off-types identified on the basis of leaf shape, size, and color should be removed. Roguing is continued during the flowering stage to remove plants differing in tassel or silk characteristics. It is important to remove all off-types before pollen shedding

Table 5 Plant-Protection Schedule for Maize and Sorghum Seed Crops

Pests and Diseases	Control measures
Maize	
Insects	
Stem borer	1. Spray 15 days after sowing endrin 20 emulsifiable concentration 0.03% solution, 1.5 mL/L water or Thiodan 35 EC, 0.1% solution (3 mL/L water) or Sevin, 50% wettable powder, 0.2% solution (4 g/L water)
	2. After 20 days apply lindane granules, 1 or 3% Sevin granules in the whorls at 12.5–15 kg/ha
	3. If infestation occurs after ear formation, spray 0.2% Sevin or 0.2% BHC in 1000 L water/ha
Hairy caterpillars, armyworms, hoppers, maize beetles	Application of 5% BHC at 15 kg/ha at the time of sowing or incorporation in the soil
Shoot bugs and aphids	Spray metasystox, 0.25% (1 mL/L water solution at 600 L/ha 1 week after sowing and subsequently at 10-day intervals if needed
Diseases	
Seedling blight	Use treated seed
Leaf blight	Spray Zineb at 2.5 kg/ha 3–4 times depending upon the intensity of disease at interval of about 10 days
Downy mildew	Spray 1.5 kg dithane M-45 in 500 L water/ha
Brown stripe	Spray 3 kg dithane M-45 in 1000 L water/ha 4–6 times as soon as symptoms appear
Bacterial stalk rot	Apply bleaching powder at 3.3 g in 10 L water at plant base (800 L/ha)
Pythium stalk rot	Provide adequate field drainage spraying of 2.5 kg captan or thiram in 100 L water/ha on stems 30–35 days after sowing
Cephalosporium stalk rot	Rogue affected plants
Charcoal rot	Rogue affected plants
Leaf rust	Spray dithane Z-78 at 2–3 kg/1000 L water/ha, 3–4 times
Head smut	1. Use treated seed.
	2. Apply Phomasan at 40 mg/m^2 or the liter formation in 10 gal/m^3
Helminthosporium leaf blight	Spray zineb at 2.5 kg/ha in 1000 L water at interval of 7–10 days
Sorghum	
Insects	
Shoot fly	1. Treat seeds with 5 parts a.i. of carbofuran/100 parts of seed
	2. Apply 3% carbofuran granules at 3 g/m$_2$ of row or 10% phorate granules at 1.5 g/m$_2$ row or disulfoton, 5% at 3 g/m row in seed furrows at sowing time
	3. Spray mixture of 120 mL 20% endrin and 125 g 50% carbaryl in 450 L water/ha 5 days after germination
Stem borer	1. Uproot and burn stubbles to reduce borer population
	2. Apply one of the following insecticides to plant whorls, 2–3 times at 10-day intervals starting 20 days after seedling emergence at 8, 10, and 12 kg/ha, respectively. a) Endosulfan 4%

Table 5 Continued

Pests and Diseases	Control measures
	b) Carbaryl 4%
	c) Lindane 1%
	d) Malathion 10% or spray
	Apply these insecticides at the same dose of a.i. in 500–600 L water/ha
Midge fly	1. Burn panicle residues and threshing chaff to destroy larvae
	2. Spray earheads before flowering with endosulphan, 35% EC at 1.5 L or carbaryl, 50% wettable powder at 3 kg, or lindane, 20% EC at 1.25 L/500 L water/ha, at 4- to 5-day intervals or dust a.i. at the same rate
Earhead bugs	Treat earheads with 10% carbaryl, dust with 18:2 sulfur or 1.3% dust of 20 kg sulfur/ha
Mites	Dust crop with 25 kg sulfur/ha or spray with solution of wettable sulfur at 5 g/L water (290 L/ha)
Aphids	Spray crop with 1080 L/ha of malathion, mixing 50 mL 50% malathion in 45 L water
Red hairy caterpillars, army worms	Dust at 2% parathion or 5% malathion at 18 kg/ha
Rodents	Poison-bait with zinc phosphide. Mix 25 g zinc phosphide and 25 g edible oil and molasses with 750 g cereal grains/ha
Diseases	
Sugary disease	Spray at boot leaf stage with 0.2% Ziram twice at 5-day intervals. Mix Sevin (WP) during second and third sprays at 0.1%. Rogue out all ears showing honey dew at the time of spray.
Ear molds	Spray with aureofungin 200 ppm or 0.02% or captan 0.2% at the grain-setting stage
Leaf spot disease	Spray dithane Z-78 0.2% twice at 15-day intervals starting 45 days after sowing

Source: Ref. 13.

commences. Stalk rot–affected plants should also be removed. Sorghum leaves infected by anthracnose are shown in Figure 5.

Off-type plants and volunteers of sorghum should be removed before pollen shedding to avoid possible crossing. Rogued plants must be pulled out along with the roots to prevent their regrowth. Off-types of sorghum can be easily distinguished on the basis of plant characteristics. Related plants such as Johnson grass, Sudan grass, forage plants, as well as diseased plants (kernel or grain smut, head smut) should be removed from time to time. Sorghum is largely self-pollinated, but may be out-pollinated to the extent of 5–10%. Hence, for the purpose of practical seed growing, sorghum should be regarded as cross-fertilizing, and suitable isolation must be provided from seed crops.

7. Detasseling

A single cross of maize is the result of crossing two inbred lines. To enable two parents to cross, one of them, the seed parent, has to be detasseled so that only its silk will be

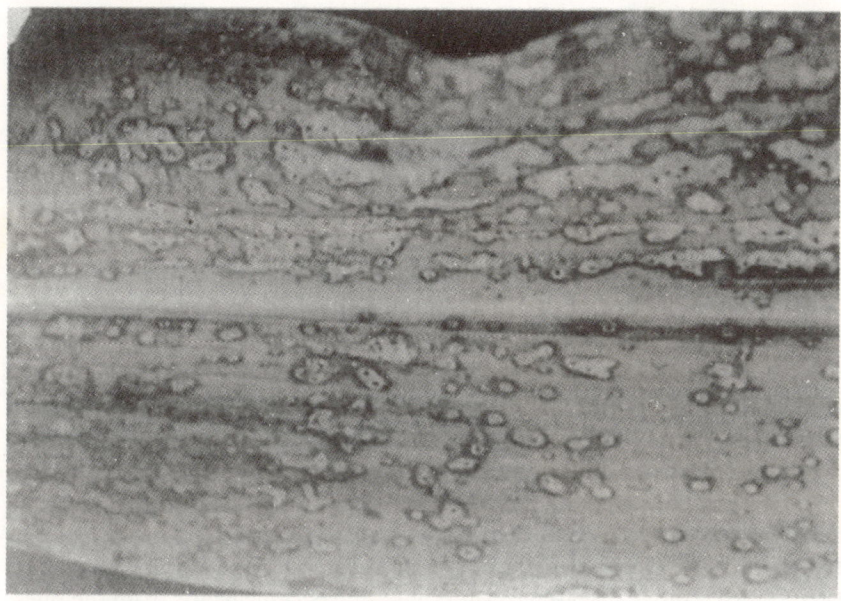

Figure 5   A disease sorghum leaf. (From Ref. 31.)

fertilized by the pollen from the other plant. The pollen-furnishing parent is called the male parent, and the one that is detasseled is known as the female parent, or seed parent. All of the tassels must be removed from the female rows before they shed any pollen. A close check is essential during the flowering period to ensure the genetic purity of the crop.

Detasseling may be begun when the tassel is well out of the leaf sheath but before the anthers begin to shed pollen, which generally occurs 1 or 2 days after the tassels become visible. In certain strains, tassels shed pollen even before their emergence. In such cases, it would be necessary to open the leaf whorl to remove the tassel. Detasseling may need to be carried out every day until it is completed.

The stalk is held with the left hand a little below the tassel, and the tassel is removed by pulling upward steadily, taking the entire tassel in the right hand. A loose or imperfect grasp of the tassel may cause a portion of it to be left on the stalk, which will later produce contaminating pollen. The leaves should not be broken or removed, which may reduce the yield and quality of the seed produced. Immature detasseling may cause a few spikelets to remain, which may emerge and shed pollen. Detasseling should be performed uniformly from the same side of the field every day, especially in large fields. All male rows may be marked at both ends. The removed tassels should be dropped on the ground, since carrying them in hand could contaminate receptive silks.

8.   Harvesting and Threshing

Maize ears can be harvested at a relatively high seed moisture content (30–35%) if facilities for artificially heated air drying are available; otherwise harvesting should be delayed until the seed moisture has decreased to about 15%. Early harvest prevents seed losses in the field due to birds, rodents, stalk breakage, and ear rots. In the case of hybrid seed production, male rows are harvested first and kept aside for commercial use so as to avoid mixing of male ears with female ears.

The mechanical harvesting of maize is more suited to larger operations and is commonly practiced in developed countries like the United States. For mechanical harvesting, grain moisture content, lodging, and type of harvester are important factors. A revamped picker-sheller or combine is preferred. The grain moisture content should be about 20–22% to avoid ear shedding and losses due to dryness, brittleness of stocks, and shellage. In excessively lodged crop, losses tend to be significant during mechanical harvesting. According to Rossman (38), maize should be harvested when cobs can be safely cribbed or dried to prevent spoilage.

Sorghum seed is physiologically mature when its moisture content falls to 30% (39), although according to Chopra (40), it should be 25%. At this stage, however, seed is still soft and prone to damage if handled roughly. In many cases, it may be preferable to wait for the seed to harden and for the moisture content to fall below 15% before harvesting (15). Sorghum is prone to sprouting in wet weather, so it is preferable to harvest at the first opportunity in bad weather.

The seed crop of sorghum must be fully ripe before harvesting. The harvested heads are sorted out to remove diseased or otherwise undesirable heads from the piles. Under hybrid seed production, male rows are harvested first and their heads kept separately to avoid mixing. Female rows are then harvested.

Hand shelling of maize cobs is the most efficient method to avoid losses (27), but is applicable only on a limited scale. The beating of cobs in the sheller results in considerable breakage. Mechanical shellers adjusted to operate at grain moisture levels below 13% experience minimum losses and could be used for large-scale operations. The wooden hand shellers developed by the Tropical Products Institute, London, are best suited to increase the efficiency of manual shelling.

Sorghum ear heads are often threshed manually using a wooden stick or trampled under the feet of cattle to separate grains. Stonerollers, tractor wheels, or metal-disk threshers are also employed. Winnowing against the wind separates the bhusa or glumes and earhead chaff. Losses result from spillage and incomplete separation of grains from the chaff and bhusa.

Samokhvalov et al. (41) described a method of harvesting sorghum using a combine mechanism. The grain is separated from the ears without threshing. Unlike the thresher, the combining mechanism has no concave, and the grain heap can be removed quickly from the equipment, which reduces the extent of grain damage. The results of field tests on a combine gave a throughput of 1.4–2.0 ha/h, seed collection of at least 95%, with grain damage of less than 0.5%.

9.   Postharvest Operations

Drying, Cleaning, and Seed Treatment.   Maize can be dried either on the cob or as shelled grains in an improved crib developed by the African Rural Storage Center, Ibadan, Nigeria (42). In western Nigeria, a drying rate of 1% in 10 days has been achieved in locally built storage cribs with main cobs having an initial moisture content of 21% (43). The optimum dimensions for both round and rectangular cribs have been described to give maximum airflow. Forced air ventilation using unheated air can be used where the initial moisture content of seed is 30–40%. In humid areas, air may be heated to reduce the relative humidity to 70%. In shelled maize, drying must be rapid to avoid fungal growth on the damaged kernels. Salunkhe et al. (27) have described the use of mechanical driers such as in-bin, tunnel, or batch driers for shelled corn, with maximum drying temperatures of 44°C for seed-purpose maize. Arora et al. (44) reported that a temperature of 60°C could be used for seed-purpose maize.

Maize ears for seed are usually dried in cribs constructed with wire mesh to allow free flow of air. Cribs should be smaller in more humid regions. Alternatively, ears may be dried in the open or in an airy building with protection from rain and intense direct sunlight. If heated air is used, its temperature should not exceed 42°C.

Dried maize ears (13% moisture) are easily shelled by either hand or machine. Special shellers have been designed to minimize damage. Shelled seed with 14% moisture content may be stored for a short time: For longer durations, seed moisture content must be reduced to 10–12% or even to 8–10% for long-term storage of seed stocks.

Seed is precleaned to remove debris such as broken cobs, husks, and dirt. Sub - sequent cleaning may be performed by an efficient air screen cleaner. Screens are also used to separate the small, medium, and large-sized seeds: This can also be achieved through gravity separators.

Several fungi, such as *Pythium*, *Fusarium*, *Drechslera*, *Diplodia*, and *Helminthosporium*, infect maize seeds. Treatment of seeds with organomercurials or thiram, quinone, captan, or benzimidazole controls the fungal infection. Captan also controls the bacterial wilt caused by *Erwinia stewartii*.

After hand harvesting, sorghum seed must be dried to 12% moisture or less before threshing. Seed heads can be sun dried on a prepared floor by spreading them not more than 20 cm deep and with frequent turning. Threshed seed can be dried using forced air at a temperature not exceeding 40°C (15).

Sorghum seed is very vulnerable to storage pests and molds and must not be stored in an environment where the moisture content is above 10–11%. All stores must be kept scrupulously clean, and the same applies to drying and threshing floors. Threshed seed can be cleaned either by winnowing at the threshing floor or in the cleaning plant to remove the broken straw and other debris. Final cleaning can be achieved on an efficient air screen cleaner: A gravity separator may also be used.

Seeds may be treated with organomercurials for the control of smuts (*Sphacelotheca* spp.). Other chemicals like sulfur, chloranil, thiram, and carboxin may be used alternatively. Reddy et al. (60) treated sorghum seed (cv. CSH 5) with cow dung ash and wood ash at 200 and 400 g/kg, attapulgite dust and Pulsafe at 5 and 10 g/kg, and malathion 5% dust at 3 g/kg and stored them in cloth bags under ambient conditions. The prestorage treatments with cow dung ash, attapulgite dust, and Pulsafe were effective in maintaining 75% germination up to 9 months and in checking the grain damage by stored grain pests up to 12 months (Table 6).

Storage. In Asia, both bag and bulk storage of shelled maize are common. Bins of concrete, iron, steel, or aluminum are rarely used. For large-scale produce, silos of one or more bins constructed from metal, steel, or concrete are recommended. Cobs are more susceptible to insect infestation than is shelled corn (45); hence, grain storage should be adopted. Of the different storage structures, the metal or mud bins with plastic linings (Pusa bins) or welded-wire mesh bins are better suited for on-farm storage of shelled maize or cobs (27).

Table 6 Effects of Inert Materials on Germination and Incidence of Insect Pests of Sorghum Seed at 6- and 12-Months' Storage

Treatment (g/kg seed)	Moisture (%)		Germination (%)		Grain damage by *Rhizopertha* spp.	
	6 months	12 months	6 months	12 months	6 months	12 months
Cow dung ash (200)	10.1	13.1	81	19	0.0	0.0
Cow dung ash (400)	9.6	12.9	83	19	0.0	0.0
Wood ash (200)	9.9	12.9	81	23	8.7	35.7
Wood ash (400)	9.7	12.0	76	17	11.3	49.5
Inert clay, ABCD (5)	9.8	12.2	85	15	0.0	0.0
Inert clay, ABCD (10)	9.9	12.3	80	22	0.0	7.0
Activated clay (5)	10.2	9.9	83	16	0.0	7.0
Activated clay (10)	10.2	13.1	82	15	0.0	0.5
Malathion, 5% dust (3)	10.1	12.9	84	17	0.0	0.0
Control	10.4	12.4	82	9	11.0	42.3
CD	—	—	4.66	N.S.	7.99	3.83
CV (%)	—	—	9.03	23.38	45.62	18.46

CD = Critical difference at + 5%; CV = coefficient of variation; NS = not significant.
Source: Ref. 60.

Sorghum is traditionally stored in bags in storage structures that are inefficient and susceptible to losses. Iron, concrete, metal, or plastic-lined mud bins (Pusa bins) are the most efficient storage structures for sorghum. Iron-bin storage is best suited to sorghum (46).

D. Rice, Rye, and Triticale

1. Introduction

Botany and Origin. Rice (paddy), *Oryza* spp., belonging to the tribe Oryzeae, originated in the warmer tropical climates. It is a cereal used to a large extent by nations of the Far East and Southeast Asia. Rice has been cultivated in China for over 5000 years, and it is an ancient cereal crop of India. Alexander the Great took it to Europe, but its cultivation started only when the Moors began to grow rice in Spain in 711 A.D. Italy presently grows rice in large quantities.

Rice has two parallel series of species ranging from wild perennial to cultivated annual. *O. glaberrima* was domesticated in West Africa, whereas *O. sativa* was developed in Asia (6). Rice is predominantly a rain-fed crop well adapted to the flooded fields of the tropics. Like other cereals of tropical origin, the response of rice to day length has been modified considerably as the crop spread to higher latitudes. However, most wild forms and tropical cultivars are short-day plants, exhibiting strong photoperiodism, althouth a few of the traditional upland rice varieties are comparatively insensitive to day length (6).

Rye (*Secale cereale*), one of the most recently developed cereals, was known to the ancient Greeks and Romans. It is thought to have originated in Afghanistan and Turkey, where its wild ancestor, *S. montanum*, is still found (47). Another form of rye, *S. anatolicum*, is also found in Syria and Iraq. Similar to oats, rye may have appeared as a weed in the ancient wheat and barley fields of the Middle East (3). Like barley, rye is a diploid with notable winter hardiness and a capacity to grow on light and acid soils (16).

Triticale is the first man-made cereal and is a product of a cross between the genera *Tritucum* and *Secale*. A. S. Wilson, in 1876, first reported the production of two sterile plants by crossing hexaploid wheat (*T. aestivum*) and diploid rye (*S. cereale*) (48). The F_1 hybrids were very vigorous but sterile. Many years later in prerevolutionary Russia, spontaneous chromosomal doubling occurred in some wheat $\times$ rye F_1 hybrids, resulting in the first true breeding diploid triticale.

Species and Cultivars. *O. sativa* is the most common rice species, and its varieties are subdivided into three subspecies: *japonica, javanica,* and *indica*. The *japonica* types are short grained, the *javanica* are intermediate types, and the *indica* rites are long grained. The *japonica* rices have generally adapted to cooler climates and longer days better than the *indica* rices in China and Japan. At higher latitudes, the cultivars become even less sensitive. There are more than 2400 cultivated varieties (49,50).

Rice grows uniquely in flooded fields, which normally drain a few weeks before harvest, creating favorable conditions for grain development. Some varieties of rice grow like other cereals in soils that are not flooded (upland or hill rices). Although these are the oldest cultivated forms, they are today of no economic importance. The bulk of rice consumed is grown in flooded fields (wet, aquatic, or lowland rice). The seedlings planted in the flooded fields are produced in nurseries that are also either wet or dry (3).

There are only a few cultivated varieties of rye, and more than 90% of the rye produced worldwide comes from Europe, where rye bread is preferred (e.g., Germany, Austria, Czech Republic, and Poland). The former Soviet Union is the largest producer of rye, since the crop is well adapted to its cooler climate and short summers.

Self-fertility prevails in rye crops, and it is possible to produce inbred lines for the production of F_1 hybrid cultivars. Some hybrids have not reached commercial significance. Synthetic cultivars produced by the interpollination of several inbred lines for a limited number of generations have also been tried, with a limited seed production experience using these types (15).

Triticale is primarily produced in developed countries that are noted for their high levels of small grain production. Spring triticale is commonly grown in Australia, Argentina, and Canada, whereas winter triticale is grown largely in the former Soviet Union, the United States, France, and China.

2. Methods of Breeding

Hybrids of rice based on cytoplasmic male sterility (CMS) have been developed in China and grown extensively since 1975, occupying about 22% of the Chinese rice crop area (15). Only one CMS is currently in wide use, which may render the hybrids vulnerable to disease epidemics. The difficulty in the production of hybrid cultivars is that pollen dispersal is naturally inadequate and has to be assisted to achieve a reasonable degree of cross-fertilization. Cultural practices are designed to obtain a high panicle population by adjusting plant spacing, fertilizer, and irrigation. Fertilization occurs best at air temperatures between 24 and 28°C. In addition to this, pollen dispersal is assisted by removing the flag leaves to expose the panicles and shaking the plants by beating with sticks or dragging a rope across the field. This requires much manual labor, since shaking must be repeated two or three times each morning during the period of pollen shedding. It is also important to synchronize flowering between male and female parents. Employing three fraternal lines can extend the flowering period of the male parent. For producing the desired hybrid rice, seed crops need an isolation distance of at least 50–100 m. The traditional techniques presently yield about 3 tons of hybrid rice per hectare. In the United States, suitable technology is being developed to mechanize various operations. Hybrid cultivars have been reported to yield about 20–30% more than the usual varieties (50).

Maruyama et al. (51) described mechanical production of F_1 hybrid seeds of rice by mixed planting. Traditionally, a male sterile line and a pollinator line are planted in alternate rows, requiring much manpower. By mixing these seeds, hybrid seed is produced in a mechanized cultivation. The mechanized method ensures a higher ratio of seed set on the male sterile plants, because the average distance between the pollinators and male sterile flowers is closer than that under the alternate-row method. The seeds are harvested in a mixed form of pollinator seeds and F_1 seeds, which are then separated on the basis of their color, size, and other traits. If a female sterile pollinator is made available, F_1 seeds only are harvested. Separation also can be achieved by incorporating a herbicide-sensitive gene into a pollinator, which is subjected to a spray of the specific herbicide just after pollination, to harvest only F_1 seeds.

3. Cultivation

Culture.

ISOLATION REQUIREMENTS AND LAND PREPARATION. Fields selected for rice crop seed production should be such that paddy was not grown in the previous season unless the variety grown was the same. Rice crops grow well on neutral or slightly acidic soils (pH 6.5), having a clayloam texture. Paddy is mainly a self-pollinated crop, although cross-pollination occurs to the extent of about 0.1–4.0%. For the production of genetically pure paddy, fields must be isolated by at least 3 m from other paddy crops (13). For upland rice, a distance of 2 m is required when there is no physical barrier at the edge of the crop. For rice grown in shallow or deep water, seed crops will normally occupy a whole enclosure or bund. When a rice cultivar is changed on land used continuously for growing paddy or when weed rice (*O. rufipogon*) becomes too prevalent, an interval of 2 years free of rice is needed to guard against volunteer plants.

Although triticale is largely self-fertilized, some cross-fertilization does occur in certain cultivars. Depending upon the cultivar, isolation is provided by a physical barrier or a gap of about 2 m. If out-pollination is a possibility, a distance of 250 m should be kept from other triticale crops, which may be increased to 300 m for earlier generations.

An interval of 2 years free from cereals, especially triticale, wheat, rye, and barley, should be allowed before growing a triticale crop; an exception can be made when the preceding crop is triticale of the same cultivar grown using an authentic seed. Weeds causing problems in wheat or rye crops may also cause difficulty in triticale. Crop management, layout, and inspection are similar to those described for wheat.

PLANTING AND SOWING. Seed crops of rice, rye, and triticale do not differ in their cultural and management requirements from crops grown for food production. Paddy is normally grown more successfully by transplanting than by direct sowing. For the production of seed, it is desirable to grow paddy under puddle and transplanting systems. Transplanting gives a much higher multiplication rate and is particularly useful in the early stages of a seed-multiplication cycle. Several innovative technologies have been introduced in Japan to improve rice production (52): direct sowing instead of transplanting seedlings, automatic seed coating, broadcasting, ferrite-coated seeds with electromagnetic acceleration, and harvesting with a combine husker, which has a husking function for high-moisture content rice. An image-processing system is used to study the quality of the rice harvested in this way.

NURSERY. The soil of the nursery bed should be well pulverized by repeated plowings. The plot is then flooded, puddled, and left for 2 days to set with a thin layer of water. Raised seed beds (6 × 15 m) with 0.5-m wide channels are prepared and are leveled after applying fertilizer. The bed should contain standing water (3–5 cm). About 50–60 beds (6 × 1.5 m) are needed to raise enough seedlings to transplant to an area of 1 ha. Each bed is manured with 450 g of superphosphate at the time of preparation. A spray of 5.0 kg zinc sulfate and 2.5 kg lime dissolved in 1000 L of water per hectare may be given 10 days after sowing if khaira disease is prevalent in the area.

Nucleus/breeder's/foundation seed should be procured from a source approved by the certification agency to raise in the nursery. About 500–600 g seed per bed (9 m²) of a coarse variety (30–35 kg/ha) and 400–500 g a fine variety (25–30 kg/ha) is adequate to raise the seed bed. Seed may be broadcast as such or after sprouting in the puddled beds. Paddy seeds may be sprouted by loosely packing in gunny bags and soaking in water for 16–20 hr. Seeds are allowed to germinate by covering the seed bag with wet gunnies. The soaked seeds sprout within 16–20 hr of soaking. Nursery beds should be provided with adequate protection from birds and rodents. The beds are kept moist, draining out excess water due to heavy rains. Nurseries must also be kept free of weeds.

Seedlings are ready for transplanting 3–4 weeks after sowing depending upon the variety. Seedlings should be uprooted gently, discarding those that are weak, diseased, or differing in any way from the usual varietal character.

TRANSPLANTING. Land is prepared for transplanting by plowing it two to three times to a fine tilth and soft soil with fairly impervious subsoil for the quick establishing of the transplanted seedlings: The plowed field is flooded for a week before transplanting. Two to three seedlings per hill are transplanted 2–3 cm deep using seedlings of proper age. The spacing for long-duration varieties 20 × 20 or 20 × 30 cm. Transplanting in lines makes later operations such as roguing easier.

Fertilization, Irrigation, and Weed Control. Paddy crops require about 100–120 kg N, 50–60 kg P, and 50–60 kg K per hectare depending upon the soil-test values. One-fourth of N and all of the P and K fertilizers are applied just before the final puddling, The remaining quantity of N is applied in two stages: one-half at the time of midtillering and the other half at the panicle-initiation stage. A foliar spray of 2% urea solution at the active tillering stage helps to boost paddy yields (13). An application of 15 kg calcium hydroxide in 1000 L of water per hectare is recommended in zinc-deficient soils.

Cultivars of subspecies *indica* do not tolerate excess nitrogen and produce excessive vegetative growth, but subspecies *japonica* can utilize higher application rates. It is important to ensure adequate supplies of phosphorus (15).

Within a day of transplanting, 2.5–5.0 cm of water is allowed to stand in the field, which should be maintained until the hard dough stage or harvesting, using irrigation if necessary.

Plots are kept free of weeds by two to three weeding operations before heading. Spraying 1 kg a.i. of MCPA (4-chloro-2-methylphenoxy acetic acid) or 2,4-D in 150–200 L of water per hectare about 20–25 days after transplanting may also control broad-leafed weeds. For controlling other grasses, a spray of 3 kg a.i. of propanil (Stam-F-34) in 300–400 L of water per hectare about 10–12 days after transplanting has been recommended.

Control of Pests and Diseases.   The following measures have been recommended to control the insect pests and diseases of the paddy (13):
Roguing.   Roguing of off-types should be performed once before flowering and then again at the time of flowering and maturity. Scrupulous roguing immediately after flowering and near maturity are recommended to rogue out wild rice plants, plants

Pests	Control measures
Insects	
Stem borer	Spray endrin 25 EC at 125 L/ha. Mix diazinon 5% granules or gamma-BHC granules at 20 kg/ha twice in soil at 20-day intervals while maintaining 5-cm water level
Brown plant hopper	Spray 0.05% sevin or 25 kg 5% sevin dust/ha, maintaining 3-cm level water in field
Gundhi bug	Dust 25 kg of BHC/ha
Diseases	
Blast	Spray benlate 150 g/ha in 250 L water or hinosan 625 mL/ha in 625 L water once in nursery (third week of sowing) and two to three sprays after 25, 35, and 45 days of transplanting, one at the time of panicle emergence and a final spray after 5 days of panicle emergence
Bacterial leaf blight	Spray 75 g of agromycin 100 plus 500 g of copper oxychloride in 500 L water/ha three to four times at 10- to 12-day intervals
Bacterial leaf streak	Spray 12 g streptocyclin or 75 g agromycin 100 in 500 L water/ha once or twice at 10- to 12-day intervals
Brown spot	Spray 0.25% dithane M-45 or Zineb 4–6 weeks after transplanting three to four times at 10- to 12-day intervals
Seedling blight and foot rot	Treat seeds with benlate at 2 g/kg seed
Khaira disease	Spray zinc sulfate at 5 kg $ZnSO_4$ plus 2.5 kg calcium hydroxide in 100 L water/ha

infested by stem borers, and diseased plants, especially those affected by tungro virus and false smut.

The most difficult weed is *O. rufipogon*, which can cross-fertilize with *O. sativa* and is virtually impossible to remove (53). Hand roguing can only be undertaken after grain formation. Seed growers, therefore, should ensure that the seed crop is free from wild rice to avoid contamination. The seeds of wild rice in small quantities of paddy seeds can be hand picked at the early stages of cultivar maintenance. Purseglove (53) recommended soaking the seeds in water to enhance the red color before hand picking in addition to the following methods:

1. Avoid ratooning or drop seed crop to be taken
2. Plow early after the harvest to encourage dropped seed to germinate, and then graze
3. Plow deep to discourage germination of red rice seed
4. Plow early after harvest and flood for 2–3 weeks to rot out the dropped seed
5. Rest the field from rice at least once every 4 years
6. Clean thoroughly all equipment, particularly the harvesting machinery
7. Introduce markers into the cultivars to be grown so that foliage can be distinguished from that of wild rice to allow early roguing (e.g., anthocyanin pigment in foliage)

Apart from wild rice, other weeds for which standards exist in seed-certification schemes include *Echinochloa* spp., *Panicum* spp., *Sorghum halepense*, and *Cyperus rotundus* (15).

Harvesting and Threshing. The mature paddy seed crop is harvested when the seed moisture content reaches 17–25%. As the seed begins to ripen, water is drained from the field to make it dry and ready for harvest. The seed is usually ready for harvest 4–6 weeks after flowering. The seed should be hard (18% moisture content) but should not exceed a 25% moisture content when the crop is cut. Small areas can be cut by hand with a sickle, the panicles being gathered in bunches and stacked in the field or stacked on racks. The panicles with no straw should be removed to a drying floor. Combines can be used for larger areas and are quite suitable for upland rice. When used for lowland crops, combines are usually fitted with "flotation" tires to prevent the machine from sinking into the soft soil. Seed at a moisture content higher than 25% may suffer injury, which can impair germination, and at a moisture content below 15%, there is increasing danger that an unacceptably high proportion of rice seed will be dehulled.

Rye is harvested in a manner similar to that used for wheat or barley. The seed is ready to harvest when it is hard and difficult to mark with the thumbnail and the straw loses most of its green color, as in the case of triticale. The harvested crop is threshed when the seed moisture content is less than 20% to avoid damage to the seed germ. Rye straw is generally rather tough and may tend to cause blockages if the crop is not dried sufficiently. Excessively dried grain may suffer mechanical injury during harvesting and postharvest operations, affecting seed germination. Using an air screen cleaner or indented cylinders and gravity separators cleans rye seed. Rye is rather susceptible to damage and requires careful handling to maintain its high seed quality.

Triticale seed should be hard and difficult to mark with the thumbnail, and the straw will lose its color when the crop is ready to harvest. The seed does not shed easily, but in some cultivars, the ears may break off at the neck if the crop is allowed to remain too long in the field before harvest. Shriveled or poorly developed seed may be a problem in some cultivars.

Lee et al. (54) reported the effects of combine harvesting and drying methods on grain quality in two rice cultivars—Milyang 23 and Chucheong: Grain yield losses were 5.17 and 2.0%, respectively, with manual cutting and threshing and 6.15 and 2.85%, respectively, with combine harvesting and threshing. The harvested hulled rice weighed 6.50 and 6.73 t/ha, respectively, with manual harvesting and 7.16 and 6.82 t/ha, respectively, with combine harvesting. Drying to 14% grain moisture content took 10 days; for manually harvested and sun-dried grain; 5–9 days for combine-harvested, sun-dried grain 2–3 days for combine, harvested, natural air in-bin–dried grain; and 15 hr for combine-harvested hot air–dried grain. Hot air drying increased the occurrence of cracked and broken grain, whereas combine harvesting increased the incomplete grain ratio in both cultivars.

Seed crops are dried and threshed similar to wheat crops.

4. Postharvest Operations

Small quantities of crop can be threshed using a hand flail or a wooden stick, beating the panicles on a prepared clean floor, or employing a small thresher.

Cleaning. Seeds are winnowed after threshing to separate chaff, dust, empty husks, and light grains. Essential seed-cleaning equipment includes the air screen cleaner coupled with indented cylinders. A gravity separator may also be used.

Seed Treatment. The main seedborne diseases of paddy are blast (*Pinicularia oryzae*), brown spot (*Cochliobolus miyabeanus*), and Bakanae disease (*Gibberella fujikoroi*), which can be controlled by treating the cleaned seed with organomercurials or alternatively with benomyl, thiabendazole, aureofungin, and kasugamycin (15).

Pasha et al. (55) reported that rice seed treatment with 0.2% thiram prevented fungal infection during storage for 150 days. Thiram treatment as a dusting or slurry was superior to the sleeping method. These two methods of treatment preserved germination and seedling vigor. CGA 173506, a new phenyl pyrrole fungicide developed by Ciba-Geigy, Ltd., Agro Division, Switzerland, has been reported to be highly active as a seed treatment against pathogens at low rates of use. It is a very effective seed treatment against *Gibberella* in rice. In cereals, 5 g a.i./100 kg seed gave control of *Gerlachia nivalis, Fusarium culmorum, Tilletia caries* (*T. tritici*), and seedborne *Septoria nodorum* equivalent to that given by the best commercial products. It also controls a broad spectrum of fungi (*Ascomycetes, Basidiomycetes*, and *Deuteromycetes*) on noncereal crops (56).

Triticale is relatively free from seedborne diseases except loose smut (*Ustilago nuda*), which may be treated similar to that in wheat seed. Ergot (*Claviceps purpurea*) is a major problem with both triticale and rye, and seed stocks must be maintained free of ergot, since there is no effective chemical treatment. A small quantity of seed may be stored in a 20% solution of common salt to float the ergot out. The salt is washed from the seed and dried immediately after treatment. Alternatively, tritcale seed may be stored for 1 year before use, as ergots are normally short lived. This is an added reason to avoid planting a seed crop following a triticale crop. Field hygiene is essential, and grasses growing at the edges of fields should be removed before anthesis to avoid ergot development (15).

Drying and Storage. After winnowing or combining, paddy seed must be dried to a 13% moisture content before storage. The seed should not be stored for any length of time if its moisture content exceeds 12% in the more humid climates or 14% in drier areas. For longer term storage, it is essential to dry the seed below 8%.

Feng et al. (57) packaged husked seeds of *japonica* and *indica* rice and treated them with 40% CO_2. After storage for 14 months, the CO_2-treated seeds had a higher protein content and vigor than the controls. It was concluded that 40% CO_2 was a reliable method of storing rice seeds.

E. Millets

1. Introduction

Compared to other cereals, the total production of millet in the world is small. Among the millets, pearl millet is widely cultivated in Asia, Africa, and the United States. Millets exhibit superior performance to other cereals under moisture stress and low soil fertility. They can withstand unpredictable fluctuations in ecological conditions and have a phenomenal capacity to respond to improved agricultural input and technology. Hence, in areas of rain-fed agriculture, the millets constitute important cereal crops. They are a staple food for a large segment of the population in Asia and Africa; areas that contribute most of the world millet production (27).

Botany and Origin. Pearl millet, or bulrush millet (*Pennisetum glaucum*) (= *americanum*) (= *typhoideum*) (a diploid), has been known in Asia and Europe since prehistoric times and seems to have originated in tropical Africa (3). It is cultivated mainly in India and Africa, where it is ground into flour and made into bread or cooked as porridge.

Finger millet (*Eleusina coracana*) is a tetraploid and an important food crop in parts of Africa and India. This is the only variety of millet belonging to the tribe Chlorideae; all others belong to the tribe Paniceae. In various parts of the world, finger millet is also known as ragi, nagli, telabun, marua, Korakan, bird's food millet, or African millet. It probably originated in India and is widely cultivated in India, Malaya, China, and parts of Central Africa.

Foxtail millet (*Setaria italica*) is probably of Asiatic origin and was already cultivated in German, as a Hungarian, or Siberian millet. In Europe, it was generally grown by the lake dwellers; however, today it is cultivated only for fodder.

Japanese barnyard millet (*Echinochloa crusgelli* var *frumentgacea*), also called sanwa millet, is cultivated in Japan and Korea as a human food and used mostly in porridge. In the United States, it is also grown as a forage crop.

Proso millet (*Panicum miliaceum*), also known as hog or broom millet, is the true millet of the ancient Romans, who called it milium. It is thought to have originated in Egypt or Arabia, after which it spread to the Soviet Union, India, China, and Japan, where it is mainly cultivated today.

Species and Cultivars. Cultivars of pearl millet are classified on the basis of breeding system (open pollinated, synthetic, inbred lines, and hybrids); maturity (early, medium, and late); and photoperiod sensitivity (insensitive, intermediate, and sensitive). The cultivars can be distinguished from the characteristics of their culm thickness (thin, thick); inflorescence shape (cylindrical, conical, spindle, club, dumbbell, lanceolate, oblanciolate, and globose); bristle length (shorter than seed, longer than seed); seed exposure (seed exposed in spikelet at maturity, not exposed); seed shape (obovate, lanceolate, elliptical, hexagonal, and globular); and seed color (ivory, cream, yellow, gray, brown, purple, and black) (15).

The cultivars of finger and other millets are less well defined and described. Finger millet cultivars are classified into two groups (53).

1. African highland types with long spikelets, long glumes, long lemmas, and with grains enclosed within the florets
2. Afro-Asiatic types with short spikelets, short glumes, short lemmas, and with mature grains exposed out of the florets

Finger millet cultivars can be distinguished according to their plant height (dwarf, tall); color of vegetative organs (green, purple); tillering (little, profused); type of inflorescence (spikes straight and open, spikes incurved and closed, and spikes branched resembling a cockscomb); length of spikes (short and long); number of spikelets per spike (few, many); length of spikelet (short, long); tightness of grain packing (dense, lax); length of glumes (short, long); and seed color (white, orange, red, deep brown, purple, or black) (53).

2. Methods of Breeding

Hybrid pearl millet seed is produced by utilizing CMS in a manner similar to that of sorghum, involving the following steps:

1. Maintenance of parental lines; namely, the male sterile line, maintainer line, and restorer line
2. Production of hybrid seed

The maintenance of parental lines is referred to as foundation seed production, whereas the production of hybrid seed is known as certified seed production.

The CMS (Line A) used for pearl millet seed production carries male sterility due to cytoplasmic-genetic factors. It is maintained by crossing with a male fertile, non–pollen-restoring strain (Line B), which is a sister strain of Line A in an isolated plot. Line B is essentially similar to Line A except that it is pollen fertile. In a crossing field, the usual planting ratio of Line A to Line B is 4:2. In addition, eight border rows are planted with the seed of Line B all around the field. The seed harvested from Line A is male sterile and is used for hybrid seed production. The seed harvested from Line B can be used in future increases of Line A (13).

Seed of hybrid pearl millet is produced by crossing male sterile (Line A) with a specified restorer line (Line R) in an isolated field. This is the hybrid seed that is sold to farmers. The seed of the restorer line (Line R) is produced in an isolated field similar to CMS line except that in the production of Line R, only one parent is present instead of the two parents required for CMS.

The genetic enhancement research on pearl millet at ICRISAT (31) has moved further into strategic areas. A recent report illustrates the extent of diversity among pollinator collection entries and characteristics that make major contributions to this diversity. A new source of cytoplasmic nuclear male sterility that could he immensely useful for breeding forage hybrids has been reported. Population-improvement research evaluated the potential of selection indices in improving grain and fodder yields and assessed the utility of the ICRISAT Asia Center (IAC) as a selection site to breed open-pollinated varieties for different locations. A molecular mapping population for seedling heat tolerance has also been developed, and identification of quantitative trait loci (QTL) is underway. This population has been used to identify QTL for downy mildew resistance, flowering time, and 1000-grain mass (58). Pearl millet affected by downy mildew is shown in Figure 6.

Finger millet is grown in a similar manner as pearl millet. There are no hybrid cultivars. The crop is largely self-fertilized, but cross-pollination may occur to the extent of about 1%. Hence, some isolation should be provided. In areas where finger millet is grown, wild species are common and may cause problems either by releasing pollen or as mechanical mixtures in the seed crop.

3. Cultivation

Culture.

ISOLATION REQUIREMENTS AND LAND PREPARATION. Pearl millet has two types of flowers. The hermaphrodite flowers mature first and are followed by a second burst on anthesis from male flowers, which have only stamens. The plant is thus mainly cross-pollinated and needs an adequate isolation distance for seed crops. Kelly (15) recommended a minimum isolation distance of 200 m; however, for earlier gener-

Figure 6 A pearl millet plant affected by downy mildew. (From Ref. 31.)

ations, 1000 m is considered to be a safe distance. Between strongly contrasting cultivars, for example, those grown for grain and fodder purposes, an isolation distance of 200 m would be required. Isolation time is not recommended because of the extended period of pollen release of pearl millet.

Millets do not require highly fertile soils and can perform well even in shallow and light-textured soils. For the production of hybrid pearl millet seed, a field is selected in which no type of pearl millet crop was grown the previous season. The selected field should be well drained and aerated. The land is prepared to fine tilth by deep plowing and two to three harrowings followed by leveling.

PLANTING AND CULTIVATION.   A nucleus/breeder's/foundation seed obtained from a source of an approved certification agency is sown in the second half of July for a kharif crop and from mid-October to mid-December for the rabi crop in the southern states of India.

The male and female lines are planted in rows at a ratio of 2:4, which is achieved mechanically by planting the seed rows and tracking the tractor back over the male rows without planting. All of the rows are marked with flags, tags, pegs, or marker plants (sun hemp or dhaincha). Eight border rows are provided on all sides of the field as per the requirements of the certification standards. Mate et al. (57a), in a 4-year study, found that a planting ratio of 2:16 (male:female) was optimum for obtaining the highest seed yield in commercial hybrid seed production. A seed drill calibrated to use for pearl millet may be used for sowing, or it may be dribbled manually not deeper than 2.5–3.5 cm. Transplanting can be adopted to economize on seed. About 625 g of pearl millet seed is sufficient to transplant a 1-ha area, with a spacing of 90.0×22.5 cm. The transplanting may be done from 10 to 25 days after sowing. Early transplanting encourages tillering under favorable growing conditions. The plant height of the transplanted seedlings should not exceed 10–15 cm. Transplants should be removed gently from the nursery, kept under a moist clothe, and planted in a moist furrow and carefully covered.

A seed rate of 1.5 kg (Line A) and 0.75 kg (Line B) would be required with a direct seeding method, whereas this could be reduced to 400–625 and 200–300 g, respectively, with the transplant system.

Fertilization, Irrigation, and Weed Control.   In addition to about 2 tons of farmyard manure/ha applied to the soil at the time of the last plowing, pearl millet requires 50 kg N, 60 kg P, and 40 kg K per hectare as a basal dose at the time of sowing. The fertilizer is placed 5 cm away from and below the seed. A second dose of 50–60 kg N/ha should be applied as top-dressing 25–30 days after sowing or transplanting. In case of very light soils or areas of high rainfall, the second dose of N may be split into two equal quantities, the first 25–30 days and the second 40 days after sowing or transplanting.

One or two irrigations may be necessary if the rainfall is inadequate to ensure appropriate soil moisture at the time of tillering and flowering. The leaves should not be allowed to wilt. Hybrid pearl millet is susceptible to water logging in the earlier stage of growth.

The field should be kept free of weeds with one or two weedings in the early crop stages. A shallow interculture at a 3.5-cm depth may also be given to destroy weeds and promote soil aeration.

Control of Pests and Diseases.   The following plant-protection schedule is recommended to control the pests and diseases of pearl millet seed crops (13).

INSECT PESTS. An application of 10% benzene hexachloride, 5% aldrin, 5% chlordane, or 6% heptachlor in the form of dust maybe applied at 20–25 kg/ha to the soil if it is infested with white ants or other insects.

White grubs and shoot flies are serious pests in some parts of Rajasthan and Gujrat in India. An application of 10% phorate granules or 5% disulfoton (thiodemeton) at 25 and 50 kg/ha can control these pests. Alternatively, a spray of 2 kg of carbaryl (50% WP or 1 L endrin (20% EC) or a mixture of endrin (375 mL) and methyldemeton 250 mL (25% EC) in 500 L of water per hectare may be applied twice: 3–4 days after germination and 7 days after the first spray. For later infestation, a spray of carbaryl (50% WP) at 3 kg in 600 L of water per hectare is recommended.

Red hairy caterpillars are controlled in the early stages of growth by applying 10% BHC dust at 15–20 kg or 1.25 L thiodan (35% EC) in 600–800 L of water per hectare. Leaf-eating insects like leaf rollers, grasshoppers, and armyworms can be controlled with 10% BHC at 20 kg/ha. Sucking insects such as aphids and jassids are controlled using dimethoate, 30% EC at 250 mL in 500 L of water/ha.

Midges usually attack pearl millet earheads and blister beetles, which may be controlled with a spray of carbaryl, 50% WP at 3 kg in 600 L of water/ha or 2% malathion dust at 15 kg/ha.

DISEASES. Pearl millet seeds are normally treated with thiram to protect germinating seeds from fungal disease. The seeds may also be treated with agrosan GN in a ratio of 1:400 at the time of planting to achieve protection from smut. The rust of pearl millet can be checked by spraying zineb, 50% WP, at 1 kg/ha in 800–900 L of water.

To prevent secondary infection of pearl millet with green ear disease, the crop should be sprayed with zineb at 1200 g/ha in 800–900 L of water. The spray may be repeated if cloudy weather continues after 10 days. This treatment also controls leaf spot disease. All infected plants are cut at the harvest and burned after removing healthy ears.

Pearl millet ergot can be avoided by adopting the following practices:

1. Deep plowing soon after the harvest of previous pearl millet crop inverts the soil to bury topsoil containing ergot bodies to deeper layers.
2. Ears of pearl millet are sprayed with ziram (cuman), 0.15%, or a mixture of copper oxychloride (fytolan) and zineb (Dithane-X-78) at a 1:2 ratio. An application of 0.15% mixture at the boot leaf stage in 500–600 L of water per hectare may be repeated two or three times at 5-day intervals.
3. The first ear observed with honeydew should be removed and destroyed promptly to prevent further infection.
4. All heaps of pearl millet ears left after threshing grain, especially from disease-infected fields, should be burned.
5. Sowing time may be adjusted to avoid coincidence of flower opening with continued cloudy and wet weather.

Roguing. Roguing is an essential aspect of the production of high-quality seed. Roguing can be performed thoroughly and efficiently by observing the following guidelines (13):

1. Start roguing before flowering to avoid any contamination from foreign pollen in the field.

2. Cut all rogues and volunteers from the ground level or pull them out with their roots to prevent regrowth.

3. Remove off-types both from seed parent and pollinator rows. The sterile types have only the stigma or a few shriveled anthers exerted. Normal fertile plants will have plump anthers, which are full of pollen, growing out of the tips of both lobes. On shedding, these rupture longitudinally along one side and discharge pollen. The seed parent should be rogued at least once a day while the heads are emerging and are becoming receptive to pollen. This is the time of appearance of fertile heads. The whole plant should be removed by pulling or cutting below the ground surface. One person can examine only two rows at a time. Each head should be examined.

4. Remove all plants our of place; that is, plants in between lines or male plants in female rows and vice versa. Give special attention to the ends of rows.

5. In addition to removing off-types and volunteers from within the isolation distance, eliminate all sources of undesirable pollen before pollen shedding begins.

6. Rogue the field thoroughly before harvesting along with the diseased plants with affected green ears, ergot, and grain smut. The true plant and seed characteristics are usually apparent only after the seed has matured.

4. Harvesting and Threshing

As the seed matures, harvest the male rows first and keep their heads separate to avoid mixing. Female rows should then be harvested and both lines examined for any undesirable heads that could be rejected.

Pearl millet seed crop can be harvested manually when the moisture content reaches around 25%, but it is advisable to wait until moisture fall to 15%. Pearl millet is prone to sprouting in the ear, so harvest should not be delayed in areas of unsettled weather conditions. Combines usually harvest larger areas.

5. Postharvest Operations

Drying.   Seed heads must be dried to 12% moisture before threshing, which may be achieved by sun drying either in the field or on a prepared drying floor. In the latter case, seed heads should not be spread more than 20 cm deep. Under humid conditions, ear heads may be dried using forced air or heated air with temperature not exceeding 40°C (15).

Cleaning.   Scrupulous cleaning of the grains required to avoid storage pests, although pearl millet is less vulnerable in storage than is sorghum. Precleaning of threshed seed is desirable and is carried out either by winnowing or on a precleaner at the cleaning plant. Subsequent cleaning can be completed on an efficient air screen cleaner. Occasionally, a gravity separator can also be used.

Seed Treatments.   Pearl millet is prone to soilborne or airborne diseases. Pearl millet normally does not require treatment to protect against storage pests.

Storage.   Both pearl millet and finger millet store well and are less prone to storage pests than sorghum.

REFERENCES

1. S. S. Deshpande and S. Damodaran, Food legumes: Chemistry and technology. *Adv Cereal Sci. Technol* 10:147 (1990).
2. *Production Yearbook.* Food and Agriculture Organization, Rome, 1988.
3. S. S. Deshpande. B. Singh, and U. Singh. Cereals: *Foods of Plant Origin-Production Technology and Human Nutrition* (D. K. Salunkhe and S. S. Deshpande. eds.). Van Nostrand Reinhold, New York, 1991, p. 6.
4. M. B. McDonald and L. O. Copeland. *Principles and Practices of Seed Production,* Chapman & Hall. New York, 1995.
5. J. R. Harlan and D. Zohary, Distribution of wild wheats and barleys, *Science* 153:1074 (1966).
6. L. T. Evans and I. F. Wardlaw, Aspects of the comparative physiology of grain yield in cereals, *Adv. Agron.* 28:301 (1976).
7. L. A. J. Slootmaker, Tolerance to high soil acid in wheat related species, rye and triticale, *Euphytica* 25:505 (1974).
8. R. H. M. Langer and G. D. Hill, *Agricultural Plants,* Cambridge University Press, Cambridge, UK, 1982.
9. L. W. Briggle and L. P. Reitz, Classification of *Triticum* species and of wheat varieties grown in the United States. USDA Tech. Bull. No. 1278, Washington, DC, 1963.
10. K. J. Rogers and K. A. Lucken, Hybrid wheat seed production in North Dakota, North Dakota Agricultural Experiment Station Report No. 806, *Farm Res.* 30(6): 4 (1973).
11. J. D. Bowring, A. W. Evans, and J. L. Sneddon, Objects and methods of seed production, *Seed Production* (P. D. Hebblethwaite, ed.), Butterworths, London, 1980, p. 3.
12. K. D. Kofoid, Selection of seed set in a wheat population treated with a chemical hybridizing agent, *Crop Sci.* 31(2): 277 (1991).
13. R. L. Agrawal, *Seed Technology*, Oxford & IBH Publishing, New Delhi, 1980.
14. Agricultural and horticultural seeds: their production, control and distribution, *F.A.O. Agricultural Studies* No. 55, Rome, 1961.
15. A. F. Kelly, *Seed Production of Agricultural Crops*, Longman, New York, 1988.
15a. M.V. Carbonell, E. Martinez, A. Raya, and J.M. Amaya, Effects of 125 mT stationary magnetic field in the initial stages of growth of wheat (*Triticum durum* L.), Crop Harvesting and Storage—Quality, energy, environment, Proc. of Int. Conf., Raudonvaris, Lithuania, 14–15 Sept, 2002, *Seed Abstr.* 25 (6): 251 (2002).
16. J. E. Greaves, and E. G. Carter, The influence of irrigation water on the composition of grains and the relationship to nutrition, *J. Biol. Chem.* 58:531 (1923).
17. L. O. Copeland and M. B. McDonald, *Principles of Seed Science and Technology*, 3rd ed., Chapman & Hall, New York, 1995.
18. A. D. Karathanasis, V. A. Johnson, G. A. Peterson, D. H. Sander, and R. A. Olsen, Relation of soil properties and other environmental factors to grain yield and quality of winter wheat grown at international sites, *Agron. J.* 72:329 (1980).
19. W. E. Splinter, *Harvesting, Handling* and *Storage, Wheat: Production and Utilization* (G.E. Inglett, ed.), AVI Publishing, Westport, CT, 1974, p. 52.
20. W. H. Johnson, Efficiency in combining wheat, *Agric. Eng.* 40: 1 (1959).
21. S. M. Illyas and B. R. Birewar, Drying technique for food grains, *Bull. Grain Technol.* 15:206 (1977).
22. T. N. Muckle and H. G. Stirling, Review of the drying of cereals and legumes in the tropics, *Trop. Stored Prod. Res.* 22:11(1971).
23. M. I. Gosteva, G. F. Koharova, E. V. Tereshkova, and E. A. Sokolov, Grain analyser for the quarantine service, *Seed Abst:* 14(11/12): 4190 (1991).
24. R. P. Kachru and K. M. Sahay, Developing and testing of pedal-cum-power operated air screen grain cleaner, *Agric. Mechan.* Asia, *Africa Latin America* 21(4):29 (1990).

25. C. Chen, Y. P. Chiang, and Y. Pomeranz, Image analysis and characterization of cereal grains with a laser range finder and camera contour extractor, *Cereal Chem.* 66(6): 466 (1989).

26. J. N. L. Sarid, K. Krishnamurthy, and S. V. Pingale, Studies on the large-scale storage of food grains in India, Part II: Studies on the relative suitability of cement and aluminum bins for storing wheat, *Bull. Grain Technol.* 3:135 (1965).

27. D. K. Salunkhe, J. K. Chavan, and S. S. Kadam, *Postharvest Biotechnology of Cereals*, CRC Press, Boca Raton, FL, 1985.

28. J. C. Stevenson and M. M. Goodman, Ecology of exotic races of maize. Leaf number and tillering of 16 races under four temperatures and two photoperiods. *Crop Sci.* 12:864 (1972).

29. J. R. Harlan, Agricultural origins: Centers and non-centers, *Science* 174: 468 (1971).

30. W. M. Ross and J. D. Eastin, Grain sorghum in the USA, *Field Crops Abstr:* 25:169 (1972).

31. Asia Region, Annual Report, 1994, International Crops Research Institute for Semi-Arid Tropics, Patancheru, India, 1995.

32. S. Pande, R. P. Thakur, R. I. Karunakar, R. Bandojopadhyay, and B. V. S. Reddy, Development of screening methods and identification of stable resistance to anthracnose in sorghum, *Field Crops Res.* 38:157 (1994).

33. H. C. Sharma, A. B. L. Agrawal, C. V. Abraham, P. Vidyasagar, K. F. Nwanze and J. N. Stenhouse, Registration of nine sorghum lines with resistance to sorghum midge: ICSV 692, 729, 730, 731, 736,739,744; 745 and 748. *Crop Sci.* 34:1425 (1994).

34. G. Tenegne, R. Bandopadyay, T. Mulatu, and Y. Kebede, Screening for ergot resistance in sorghum, *Plant Dis.* 78:873 (1994).

35. N. Seetharam, C. W. Magill: and F. R. Miller, Molecular markers for cold tolerance in sorghum, *Use of Molecular Markers m Sorghum and Pearl Millet Breeding for Developing Countries* (J. R. Witcombe and R. R. Duncan, eds.), Overseas Development Administration, London, 1994, p.432.

36. J. M. Airy, L. A. Tatum, and J. N. Sorenson, Jr., Producing seed of hybrid corn and grain sorghum, *Seeds: The Yearbook of Agriculture*, U.S. Department of Agriculture, Washington, DC, 1961, p. 145.

36a. K. Parameswari and K. Vanangamudi, Hardening and pelleting in maximizing the productivity of rain fed sorghum, cv. CO26, XI National Seminar on Quality Seed to Enhance Agricultural Productivity, UAS, Dharwar, Jan, 18–20, 2002, *Seed Tech News* 32: (A. Gaur, A.K. Vari, J.L. Varshney and K. Kant, eds), Indian Society Seed Technology, Division of Seed Science and Technology, IARI, New Delhi, March 2002, p. 124.

36b. K.C. Gagare, R.B. Patil, and Y.B. Suryawanshi, Response of $ZnSO_4$ on pollen production and viability in male parent of sorghum hybrid-CSH-9, XI National Seminar on Quality Seed to Enhance Agricultural Productivity, UAS, Dharwar, Jan, 18–20, 2002, *Seed Tech News* 32: (A.Gaur, A.K.Vari, J.L.Varshney, and K.Kant, eds), Indian Society Seed Technology, Division of Seed Science and Technology, IARI, New Delhi, March 2002, p.21.

37. Maize seed scheme, *OECD Schemes for the Varietal Certification of Seed Moving in International Trade*, Organization for Economic Cooperation and Development, Paris, 1977.

38. E. C. Rossman, Picker performance, *Proc. 10th Corn Conf. American Seed Trade Association*, 1955, p. 88.

39. L. R: House, *A Guide to Sorghum Breeding*, 2nd ed., International Crops Research Institute for the Semi-Arid Tropics, Patancheru, India, 1985.

40. K. R. Chopra, *Technical Guidelines for Sorghum and Millet Seed Production*, Food and Agriculture Organization, Rome, 1982.

41. A. I. Samokhvalov, V. I. Kiryukhin, and V. I. Loktev, Device for combining sorghum, *Seed Abst.* 14 (1): 236 (1991).

42. *Postharvest Food Losses in Developing Countries*, National Academy of Sciences, Washington, DC, 1978.

43. M. A. Comes and J. Riley, An investigation of drying rates and insect control in a maize crib with improved ventilation, West African Stored Products Research Unit, Annual Technical Report No. 12, 1962, p. 72.

44. B. K. Arora, A. P. Bhatnagar, and A. S. Bakshi, Critical temperature for drying maize seeds, *J. Agric. Eng.* 10:14 (1973).

45. P. S. Hindmarsh and T. A. MacDonald, Field trials to control insect pests of farm stored maize in Zambia, *J. Stored Prod. Res.* 16:9 (1980).

46. S. Vogel and M. E. Graham, *Sorghum and Millet: Production and Use*, International Development Research Center, Ottawa, 1979.

47. K. Lorenz, Rye: Utilization and processing, *Handbook of Processing and Utilization in Agriculture* (A. Wolff, ed.), Vol. 2, *Part 1: Plant Products*, CRC Press, Boca Raton, FL, 1982, p. 243.

48. B. Skovmand, P. N. Fox, and R. L. Villareal, Triticale in commercial agriculture: progress and promise, *Adv. Agron.* 37:1 (1984).

49. R. H. M. Langer and G. D. Hill, *Agricultural Plants*, Cambridge University Press, Cambridge, UK, 1982.

50. Y. Pomeranz, *Modern Cereal Science and Technology*, VCH Publication, New York, 1987.

51. K. Maruyama, H. Kato, and H. Araki, Mechanized production of F1 seed in rice by mixed planting, *Jpn. Agric. Res. Quart.* 24(4): 243 (1991).

52. N. Ito, Rice production mechanization system, Proc. Int. Agric. Eng. Conf. Exh., Bangkok, Thailand, Dec. 3–6, 1990 (V.H. Salokhe and S.G. Ilangantilekel, eds.), Asian Institute of Technology, Bangkok, Thailand, p. 355.

53. J. W. Purseglove, *Tropical crops: Monocotyledons*, 5th ed., Longman, London, 1985.

54. H. J. Lee, J. H. Seo, and U. W. Lee, Effects of combine harvesting and drying methods on grain quality in rice cultivars, *Korean J. Crop Sci.* 35(3): 282 (1990).

55. M. M. Pasha, K. A. A. Appaiah, and N. G. K. Karanth, Efficacy of thiram as a seed dressing fungicide and its effect on seedling vigor of paddy, *Trop. Sci.* 31(2):123 (1991).

56. A. J. Leadbeater, D. J. Nevill, B. Steck, and D. Nordmeyer, CGA 173506: A novel fungicide for seed treatment, *Bright on Crop Protection Conf. Pests and Diseases*, Vol. 2, British Crop Protection Council, Thornton, Heath, UK, 1990, p. 825.

57. R. Y. Feng, Y. Z. Gao, and B. Si, Several simple grain storage methods and their effects on grain quality, *Plant Physiol. Commun.* (4): 36 (1989).

57a. S.N. Mate, R.B. Patil, Y.B. Suryawanshi, K.C. Gagare, M.R. Mantare and R.W. Bharud, Planting ratio studies for commercial seed production in pearl millet, XI National Seminar on Quality Seed to Enhance Agricultural Productivity, UAS, Dharwar, Jan, 18–20, 2002, *Seed Tech News* 32: (A. Gaur, A.K. Vari, J.L. Varshney, and K. Kant, eds), Indian Society Seed Technology, Division of Seed Science and Technology, IARI, New Delhi, March 2002, p. 22.

58. E. S. Jones, J. R. Witcombe, C. T. Hash, S. D. Singh, M. D. Gate, D. S. Shaw, and C. J. Li, Mapping QTLS controlling resistance to downey mildew in pearl millet and their applications in plant breeding programmes, *Use of Molecular Markers in Sorghum and Pearl Millet Breeding for Developing Countries*, Proc. of ODA Plant Sci. Res. Program Conf., 29 Mar–1 Apr., 1993, Norwich, UK (J. R. Witcombe and R. R. Dunean, eds.), Overseas Development Administration, London, 1994, p. 76.

59. K. Redenbaugh, D. Slade, P. Viss, and J. A. Fujii, Encapsulation of somatic embryos in synthetic seed coats, *HortScience* 22(5): 803 (1987).

60. S. V. Reddy, T. Ramesh Babu, and S. H. Hussaini, Effect of inert material on germination and grain damage by stored grain pests of sorghum, *Seed Research*, Vol. 1 (S. P. Sharma, ed.), Indian Society of Seed Technology, New Delhi, 1993, p. 283.

61. N. R. Das, N. Gosh, and N. N. Mukherjee, Seed character and seedling vigour of wheat (*Triticum aestivum*) after transplanted rice (*Oryza sativa*) as affected by tillage and fertilizer, *Seed Research*, Vol. 2, (S. P. Sharma, ed.), Indian Society of Seed Technology, New Delhi, 1993, p. 756.

8
Pulse Crops

I. INTRODUCTION

Leguminosae (or Fabaceae) is the third largest family of flowering plants (after Compositae and Orchidaceae) and is second only to grasses (Gramineae) in economic importance. It contains a currently estimated 16,000–19,000 species belonging to about 750 genera. The domesticated legumes, or pulses, are members of a subfamily Papilionoidae, which is further classified in 10 botanical tribes, 3 of which contain the major vegetable and grain legumes (Fig. 1).

Pulse crops are legumes that are harvested mainly for their dry seeds to provide protein in the human diet. Some pulses (beans and peas) are harvested for their green pod vegetables and fodder. Some legumes like groundnuts and soybeans are harvested as dry seeds mainly for their oil content (oilseeds), whereas some are rich sources of protein in addition to their oil content. Other legumes are used as fodder crops as companions for grasses, which are dealt with elsewhere in this volume (see Chapter 13).

Some pulse crops, such as alfalfa and red clover, are perennial in nature, but all are treated as annuals for the purpose of seed production. Some require cooler climates to induce flowering (e.g., autumn-sown cultivors of *Vicia faba*). The inflorescences are racemose and seeds develop in a pod (legume). In some pulse crop species, flowering takes place over a long period of time so that seed ripening is not synchronized, which makes harvesting difficult (1).

Leguminous seeds, grain legumes, and pulses are second only to cereals as a source of human food and animal feed. Nutritionally they are two to three times richer in protein than cereal grains. In addition to seeds, legumes offer a variety of other edible products. Immature pods of several food legumes are edible 2 or 3 weeks before the fibers lignify and harden. Although the green and succulent pod vegetables have less protein than the mature seeds, they are rich in vitamins, minerals, and soluble carbohydrates. The relative proportion of essential amino acids in pulses is not as well balanced for human dietary requirements as it is in meat, milk, or fish. Most legume proteins are deficient in methionine and tryptophan. Leguminous seeds also contain a variety of toxic constituents, such as flavonoids, alkaloids, and uncommon proteins, often found in leaves, pods, and seeds. Many of these toxins may, however, be eliminated by simply soaking seeds in water or by thorough cooking.

Of the cultivated legumes, only about 20 are commercially important (Table 1). Of the various legumes, soybeans, groundnuts, dry beans, peas, broad beans, chickpeas, and lentils are the major food legumes grown all over the world. The major producers of food legumes are China, India, the former Soviet Union, the United States, Brazil, Mexico, and Ethiopa (2).

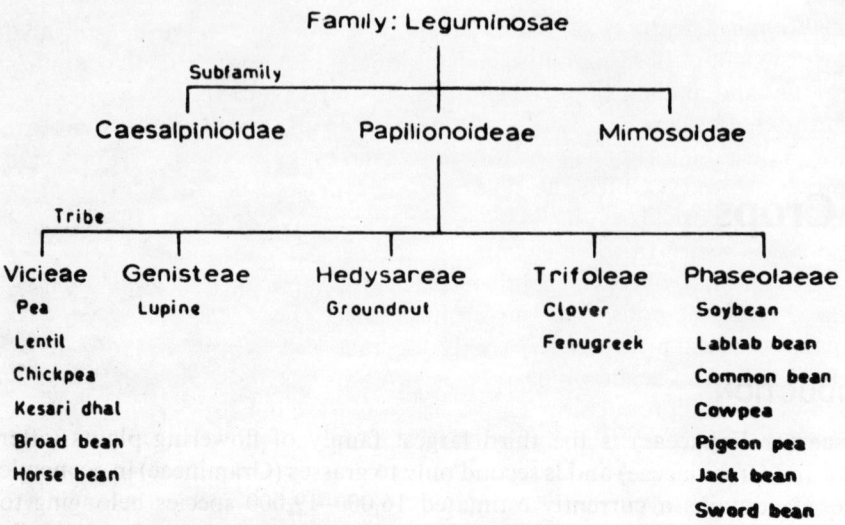

Figure 1 Botanical classification of food legumes.

Table 1 The Most Common Food Legumes

Scientific name	Common name
Arachis hypogaea	Groundnut, peanut
Cajanus cajan (L.) Millsp.	Pigeon pea, red gram, Congo pea, arhar, tur
Cicer arietinum L.	Chick pea, Bengal gram, gram
Glycine max (L.) Merr.	Soybean, soya
Lablab purpureus (L.) Sweet	Hyacinth bean, Egyptian bean, Val
Lathyrus sativus L.	Kesari dha, chickling vetch, grasspea
Lens culinaris Medik	Lentil, masur
Lupinus albus L.	White lupine
Lupinus angustifolius L.	Blue lupine, New Zealand blue lupine
Lupinus luteus L.	European yellow lupine
Macrotyloma uniflorum (Lam.) Verdc.	Horse gram, Madras gram, kulthi
Phaseolus lunatus L.	Lima bean, butter bean
Phaseolus vulgaris L.	Bean, common bean, French bean field bean, pinto bean, navy bean, dry bean
Pisum sativum L.	Common or garden pea, dry peas
Psophocarpus tetragonolobus (L.) DC	Winged bean, Goa bean, four-angled bean, Manila bean, princess pea
Vicia faba L.	Broad bean, faba bean, horse bean
Vigna aconitifolia (Jacq.) Marechal	Moth bean, mat bean
Vigna mungo (L.) Hepper	Black gram, urd
Vigna radiata (L.) Wilczek	Green gram, golden gram, mung bean
Vigna umbellata (Tunb.) Ohwi and Ohashi	Rice bean, mambi bean
Vigna unguiculata (L.) Walp.	Cowpea, black-eyed pea
Voandzeia subterranea (L.) Thouars	Bambarra groundnut

Source: Ref. 2.

Several botanical features of pulses are both a bane and boon in seed production. The growth of pulses generally is indeterminate; that is, the growth continues from the terminal and axillary buds while flowering and seed formation are both in progress. Mature seed, therefore, is ready to harvest on the lower part of the plant while new flowers are still forming at the top, which makes the decision regarding time of harvest rather arbitrary. Cutting too late permits ripe seeds to fall to the ground and be lost, whereas early harvest may gather an excessive amount of immature, green, and shriveled seeds. In some leguminous species, the pods open and the seed drops as soon it is ripe, whereas, in others, special equipment may be needed to remove the seed from the pods. A large variation in the size of the pulse crop seed influences many cultural practices, especially the rate and depth of planting and method of harvest. Most leguminous species spread vigorously and are hard to maintain in rows for seed production. Variability of pulses applies to genera, species, and varieties. Cultural practices for the production of pulse seeds vary for each species (3).

II. METHODS OF BREEDING

Room exists for considerable improvement in most pulse crops using the traditional plant-breeding techniques of selection with or without hybridization. Both mass selection and pedigree selection can be usefully employed. Consideration of space, time, labor, and finance together determine the practices adopted and the modifications to the standard pedigree selection method. Smartt (6) described the bulk, bulk pedigree, complex bulk population, multiple crosses, and backcross methods used for pulse improvement. The protein content and protein quality in terms of its amino acid composition and content of essential amino acids like tryptophan and methionine as well as elimination of certain undesirable antinutritional factors like cyanogenetic glycoside, protease inhibitors, and hemagglutinins are the most important considerations in pulse improvement by breeding, including chickpea and pigeon pea. According to Smartt (6), there is enormous room for improvement both using traditional plant-breeding methods and developing newer techniques to increase the output and improve the quality of pulse crops.

International Crop Research Institute for Semi-Arid Tropics, Hyderabad, India (ICRISAT) (7) produced the world's first pigeon pea hybrid in 1992, using genetic male sterility, in which sterility is controlled by genes in the cell nucleus. Although the hybrid exhibited the classic performance advantages over open-pollinated varieties, it was costly to produce, because in a genetic male sterility system, only half of the plants are male sterile and the remaining plants must be manually uprooted to ensure that they do not pollinate the plants producing hybrid seed. The development of cytoplasmic male sterility (CMS) in which sterility is controlled by the cytoplasmic genes can solve this problem. In this system, every plant is male sterile and pure-breeding male-sterile lines (and from them, hybrids) can be produced cheaply and reliably. The ICRISAT scientists are trying to develop a CMS system for pigeon pea and metods to use molecular markers to classify CMS lines. At the request of the Indian Institute of Pulse Research, ICRISAT has convened a working group on CMS systems involving Indian National Agricultural Research Schemes and seed companies. In 1994, two advanced-generation male-sterile progenies based on a cross between cultivated pigeon pea and *Cajanus sericeus* (a wild relative) were found to be promising in greenhouse tests at the ICRISAT Asia Center (IAC). Experimental tests showed that about 60–70% of plants were male sterile under field conditions compared to the ideal 100% in stable CMS.

Cajanus platycarpus, a distant relative of cultivated pigeon pea (C. cajan), has many desirable characteristics such as extra and early flowering and maturity, annuality, photoperiod insensitivity, high harvest index, rapid seedling growth, salinity tolerance, and resistance to phytophthora blight and pod borers. Until recently it was not possible to transfer these characteristics to cultivated genotypes. But by fine tuning existing techniques and developing embryo-rescue techniques for pigeon pea ICRISAT has succeeded in producing the first ever hybrids between C. platycarpus and C. cajan. Many fertile hybrid plants have now been produced and are being used to raise the next generation of hybrids. This population shows much variation, but some plants flower very early and are insensitive to photoperiod. According to a ICRISAT report (7), efforts have been made to develop a stable hybrid population that can be effectively used in pigeon pea–improvement programs.

Despite decades of research, ascochyta blight continues to ravage chickpea fields in India and other countries. Resistance to this disease has been found in Cicer pinnatifidum, a wild species collected in Turkey and conserved in the ICRISAT gene bank. By developing workable embryo-rescue and tissue-culture techniques, hitherto not available for chickpea, IAC scientists have produced hybrids between cultivated chickpea (C. arietinum) and C. pinnatifidum. These hybrid plants are being grown and maintained in vitro. Once it is ensured that these hybrids can survive in the soil, they will be crossed with cultivated varieties to develop genotypes that farmers can use (8). After screening more than 1500 germplasm accessions and more than 150 advanced breeding lines, ICRISAT scientists have identified a few drought-resistant chickpea genotypes, most notably ICC 4958, which has unusually long roots (9).

An investigation on seed and seedling growth parameters was carried out on 19 pigeon pea (C. cajan) genotypes. The results showed substantial variability for the parameters studied. Correlation coefficients suggested positive and significant associations among the traits. Bold seeds produced vigorous seedlings, and the observations indicated the possibility of selecting a genotype based on a seed or seedling growth parameter for adverse agroclimatic conditions (9a).

McDonald and Drake (9b) described a new commercial approach to cultivar identification by electrophoresis of Phaseolus vulgaris seed proteins using the Pharmacia Phast electrophoresis system. The system employs miniature polyacrylamide gels and a fully automated electrophoretic process. A typical electrophoretic run with associated gel staining and destaining is completed in 90 min and costs US$0.60/seed. P. vulgaris isolines could be differentiated using this system. Advantages of the Phast system include the availability of commercially precast polyacrylamide gels, completion of electrophoresis in 90 min, the capability of evaluating 24 seeds in one run on two gels, and computerization of the electrophoretic run. Disadvantages include problems in the uniformity of sample protein application, limitations in the total number of seeds evaluated, difficulties in the interpretation of electrophoretic patterns due to the small gel size, the inability to evaluate multiple enzyme-staining patterns from one gel, and a lack of information on standardization of the system and interpretation of results. It is concluded that the Phast electrophoresis system presents seed analysts with a more automated and less expensive approach to varietal identification than current techniques.

III. SEED-PRODUCTION TECHNOLOGY

A. Chickpea and Pigeon Pea

1. Introduction

Botany and Origin. Chickpeas (*Cicer arietinum*), belonging to the tribe Vicieae, are commonly grown in the subtropics and during cooler seasons in the tropics. They are widely cultivated in India and in several Near East and Mediterranean countries. They are moderately resistant to drought and high temperatures. Chickpeas are herbaceous annuals that branch close to the ground. Pigeon pea (*C. cajan*) belongs to the Fabaceae family of the tribe Phaseolaceae. It is commonly grown in several tropical and subtropical countries of Asia and Africa. It is one of the most prized pulse crops of India. The pigeon pea is a short-lived, 1-to 4-m high, perennial shrub, but is usually grown as an annual crop, often for green manure or for cover in plantations. Its deep and penetrating root systems make it especially valuable as a renovating and contour-hedge crop for checking soil erosion. Mathur (9c) described the role of a genetic marker in the production of chickpea, using light-dependent purple pigmentation (LPD); two genes complement the inheritance of this marker: a, *p* gene, which produces purple coloration on the light-exposed plant surface, and a booster gene, *B*, which intensifies the pigmentation. The genes can be transferred in the agronomically superior cultivars and is expressed in the F_1 generation. The LDP markers with the quantity traits of cultivar can be an added advantage in the future use of varietal characterization.

Species and Cultivars. Two closely related annual species of chickpea, *C. reticulatum* and *C. echinospermum*, occur in Turkey. Wild *C. reticulatum* is interfertile with the cultivated pulse and closely resembles morphologically cultivated *C. arietinum* (4). Cream-seeded kabuli chickpeas (of Meditarrean and Middle Eastern origin) have the largest seeds and grow well under irrigation. Local Indian cultivars have smaller seeds. Hybrids between kabuli and indigenous (desi) cultivars have produced strains of medium-sized seeds with fair yields. An improved chickpea cultivar developed by ICRISAT scientists is shown in Figure 2.

According to the International Rules of Botany Classification, *C. cajan* (L.) Millsp. has been adopted for the species known earlier as *C. flavus* and *C. bicolor*. These two species were formerly described under *C. indicus* Spreng (5).

Kelly (1) has described the main distinguishing characteristics of the cultivars of chickpea and pigeon pea on the basis of the morphological characteristics of their leaves, stems, flowers, pods, and seeds, as well as their growth and flowering patterns.

2. Cultivation

Culture.

ISOLATION REQUIREMENTS AND LAND PREPARATION.   Fields selected for the seed production of chickpea and pigeon pea should not have been used for the same crops the previous year unless the same variety was grown using a certified seed. The land selected should be light in texture, well drained, and neutral in reaction, having a pH of around 7.0.

Chickpea flowers are normally self-pollinated before they open; hence, an isolation distance of about 20 m for the foundation seed and 10 m for certified seed

Figure 2 An improved chickpea cultivar by the ICRISAT Scientist. (From Ref. 7.)

would be sufficient to maintain the genetic purity of the seed. According to Kelly (1), except for the very early stages of multiplication, chickpea cultivars may be classified as self-pollinating, requiring separation from other crops only by a clear demarcation. Pigeon pea is partially self-pollinated and partially cross-pollinated. Although the anthers burst before the flowers open, considerable cross-fertilization may take place via bees and other insect pollinators. Natural crossing to the extent of about 65% has been reported (10). According to Purseglove (11), however, cross-fertilization in pigeon pea varies from 5 to 40% and is normally about 20%. The seed crops of foundation and certification seed classes should therefore be isolated at least by 200 and 100 m, respectively, from other fields growing the same crop.

For chickpeas, land need not be plowed to a fine tilth—it is enough to turn the soil to allow some aeration before planting. For pigeon peas, a medium tillage followed by two harrowings is recommended.

PLANTING AND CULTIVATION.   High-quality seed (nucleus/breeder's/foundation) should be obtained from a source approved by a seed-certification agency. Chickpeas are best sown in the third and fourth weeks of October in India. Delayed sowings may result in decreased yields of seed crops. Pigeon peas are sown in the first week of June to obtain a good crop with higher seed yields. The seed crop of chickpeas is sown in rows by drilling 3–4 cm deep, spacing 45–60 cm between rows. About 45–50 kg of seed is needed to sow an area of 1 ha depending upon seed size.

Pigeon peas may be sown either with a seed drill or by plow in furrows 5 cm deep. Spacing between rows can vary from 60 to 75 cm and between plants from 20 to 30 cm. About 10–12 kg of seed is required to sow an area of 1 ha. Joshi et al. (11a) found that a row female to male ratio of 1:4 was optimum for hybrid seed production in pigeon pea under Punjab conditions in India.

Fertilization, Irrigation, and Weed Control. In addition to about 1 ton of farmyard manure and seed treatment with rhizobium culture, chickpea seed crops require about 25 kg of N and 50 kg P per hectare as a basal fertilizer dose. A foliar application of 0.2–0.3% diammonium phosphate once after 30 days of sowing and again at flowering has been found to increase chickpea seed yield.

The fertilization requirements of pigeon peas are the same as for chickpea crops. The fertilizer may be drilled at the time of sowing, placing it 10–15 cm deep and to the side of the seed. Band application of fertilizers yields better results than the traditional broadcasting method. Parameswari and Vanangmudi (11b) reported that pigeon pea seeds inoculated with *Rhizobium* and applied with 25:50:25 kg of NPK per hectare produced 26% higher seed yield over the control.

One presowing irrigation may be given for chickpea seed crops if the soil moisture is not adequate followed by one or two light irrigations at 45–75 days of crop growth. Pigeon peas may require one or two light irrigations prior to the onset of monsoon. During irregular rainfall situations and prolonged dry weather conditions, one irrigation at the time of flowering and another at the seed development stage are necessary for pigeon peas.

Seed crop fields should be kept free of weeds by one or two weedings in the early stages of crop growth. Pigeon pea seed plots may be hoed when crops are 30 and 60 days old if needed.

Control of Pests and Diseases. *Helicoverpa annigera* (Hubmer) is the major insect pest of chickpea followed by a sernilooper, *Autographa nigrisigna* (Walker), which damages chickpea pods. In stores, chickpeas can be severely damaged by bruchids (*Callosobruchus* spp.). *Ascochyta* blight, *Botrytis* gray mold, and *Alternaria* blight, rust, and *Stemphylium* blight are important diseases of chickpea.

Hussaini and Ramesh Babu (12) reported that prestorage treatment using inert clay (5 g/kg) was effective against pulse beetle (*Callosobruchus chinensis* L.) damage up to 12 months of storage due to total mortality of the released adults without adversely affecting germination of pigeon pea (*C. cajan* L.), *mung* bean (*Phaseolus radiatus* L.), and *urd* bean (*P. mungo* L.) seeds (Table 2). Prestorage seed treatments with *karanj* oil and castor oil (10 mL/kg seed) effectively controlled the pest without adverse effects on

Table 2 Effects of Inert Clay on Adult Mortality, Oviposition, and Adult Emergence of Pulse Beetle (*Callosobruchus chinensis* L.) Under Inoculated Conditions

Treatment	Adult mortality (%)		Oviposition (%)		Adult emergence (%)	
	6 months	12 months	6 months	12 months	6 months	12 months
Mung bean (T)	100	100				
Mung bean (C)			100	100	95	94
Pigeon pea (T)	100	100				
Pigeon pea (C)			100	100	96	96
Urd bean (T)	100	100				
Urd bean (C)			100	100	96	95

T = treated with inert clay (5 g/kg seed); C = untreated control.
Source: Ref. 12.

seed germination. The oils of groundnut, neem, and mustard, although effective against pulse beetles, reduced germination of mung bean slightly (Table 3). Siddiqui and Agrawal (13) also reported that a newly developed nonconventional dust, atta-pulgite-based clay dust (ABCD), was effective as a prestorage treatment (at 0.5% w/w) for about 1 year in controlling *C. chinensis* of pulse seeds (green gram, red gram, and black gram) as measured by adult mortality, oviposition, and adult emergence at regular intervals up to 12 months (Table 4). In seed lots treated with ABCD, oviposition was nil, and hence adult emergence was 0%, whereas in the untreated control, it was 100%. Prestorage treatment with ABCD was found to be effective even 1 year later. Adult insects released in treated seeds recorded 100% mortality, whereas the untreated ones emerged totally (13).

Cutworms of chickpea can be controlled by applying 10 kg aldrin dust or 10–20 kg of 5% sevin dust per hectare, incorporating the pesticide in the soil. If the damage reappears, the soil may be drenched with 250 mL parathion plus 1 kg DDT in 1000 L of water per hectare. To control the chickpea caterpillar, the crop should be dusted with 10% sevin at 10 kg/ha. Alternatively, a spray of thiodan at 450 mL in 250 L of water or nuvacron at 250 mL in 250 L of water per hectare may be applied.

Pod flies of pigeon pea can be controlled with a spray of thiodan at 1.25 L/ha or nuvacron at 625 mL/ha in 250 L of water. Pod bugs, plume moths, caterpillars, and cow bugs can be effectively controlled with a spray of 625 mL of nuvacron or 25 kg of 10% sevin dust per hectare. The seed crop may be dusted with 10% folidol at 25 kg/ha to control tussock caterpillars and leaf rollers.

Roguing. Off-types and diseased chickpea plants affected by blight and wilt are removed from time to time to enhance seed purity. The seed crop of pigeon pea should be kept pure by removing off-types and diseased plants infested with wilt, leaf spot, stem canker, yellow mosaic virus, and sterility virus as and when required during the crop growth.

Table 3 Effects of Seed Treatment on the Germination of Mung Bean Seed During Storage

Treatment	Dose	Germination (%)	
		6 months	12 months
Inert clay	5 g/kg	89	82
Thiram	3 g/kg	89	82
Delsan-30	3 mL/kg	84	81
Carbendazim	2 mL/kg	87	d
Neem oil	10 mL/kg	82	83
Karanja oil	10 mL/kg	86	89
Mustard oil	10 mL/kg	80	86
Groundnut oil	10 mL/kg	79	86
Castor oil	10 mL/kg	95	90

d, seed completely damaged by *C. chinensis*.
Source: Ref. 12.

Table 4 Effects of ABCD on Adult Mortality, Oviposition, and Adult Emergence of *Callosobruchus Chinensis* Under Inoculated Conditions and Viability (Germination) of Pulse Seeds[a]

Treatment	Seed Viability (% germination)					
	2 months	4 months	6 months	8 months	10 months	12 months
Green gram (T)	93	92	92	89	89	85
Green gram (C)	94	92	91	91	90	87
Red gram (T)	93	92	92	90	90	–
Red gram (C)	94	94	93	92	91	–
Black gram (T)	96	94	93	90	89	–
Black gram (C)	96	95	95	94	93	–
	Adult mortality (%)					
Green gram (T)	100	100	100	100	100	100
Green gram (C)	0	0	0	0	0	0
Red gram (T)	100	100	100	100	100	100
Red gram (C)	0	0	0	0	0	0
Black gram (T)	100	100	100	100	100	100
Black gram (C)	0	0	0	0	0	0
	Oviposition (%)					
Green gram (T)	0	0	0	0	0	0
Green gram (C)	100	100	100	100	100	100
Red gram (T)	0	0	0	0	0	0
Red gram (C)	100	100	100	100	100	100
Black gram (T)	0	0	0	0	0	0
Black gram (C)	100	100	100	100	100	100
	Adult emergency (%)					
Green gram (T)	0	0	0	0	0	0
Green gram (C)	94	96	95	98	96	94
Red gram (T)	0	0	0	0	0	0
Red gram (C)	96	95.	96	91	94	96
Black gram (T)	0	0	0	0	0	0
Black gram (C)	97	95	96	94	96	95

T = treated with ABCD at 0.5% W/W; C = untreated control.
[a] A sample of 100 seeds replicated four times was drawn from the treated stock and five pairs of adult bruchids were released to study the mortality, oviposition, and emergence of adults.
Source: Ref. 13.

3. Harvesting and Threshing

Chickpea seed matures when the leaves turn reddish brown. The crop is generally harvested either by pulling the plants by hand or by cutting them with sickles. The cut plants are stacked in small heaps and allowed to dry in the sun. The harvested crop is then transported to clean threshing floors. The seeds are threshed by trampling, beating with wooden flails, or under the wheels of a tractor.

Pigeon pea seed crops are harvested when the seed is mature by cutting the plants with sickles. The plants are left in the field to dry for about 1 week. The dried produce is taken to a clean threshing yard and threshed manually with wooden flails or by trampling. Mechanical harvesters may be employed to harvest larger areas.

4. Postharvest Operations

Cleaning and Drying.   Winnowing in natural wind generally performs cleaning of the threshed produce. The cleaned seed is dried to 8–10% moisture before storage. Necessary precautions must be taken to avoid mechanical admixtures during harvesting and all postharvest operations.

Seed Treatment and Storage.   The seedborne diseases of pigeon pea and chickpea are not considered to be important enough to warrant seed treatments. Seeds are stored at the farmer, trader, and government levels in various types of storage structures, as described by Salunkhe et al. (2).

Seed Yields.   Seed yields of chickpea and pigeon pea in various parts of the world can vary from 400 to 1800 and from 500 to 1100 kg/ha, respectively (10). According to Agrawal (10), the seed yields of chickpea and pigeon pea in India range from 15 to 20 and 20 to 25 quintals/hectare, respectively.

B. Green Gram, Black Gram, and Cowpea (*Vigna* Spp.)

1. Introduction

Botany and Origin.   Green gram (*Vigna radiata* L., syn. *Phaseolus aureus*) and black gram (*Vigna mungo* L., syn. *Phaseolus mungo*) are botanically very similar pulses. Black gram can be distinguished from green gram by its much shorter, stout, very hairy pods and longer oblong seeds, which vary in color from blackish to olive. Both crops are commonly grown in several tropical Asian countries. They thrive best under a hot dry climate. Green gram is comparatively less tolerant to drought and humidity than the black gram. Among all the pulses grown in India, black gram is highly prized for its nutritious value and food uses.

Cowpea or black-eyed pea (*V. unguiculata* L. Waif) is a dwarf (determinate and semideterminate), climbing pulse crop that probably originated in India but is now widely cultivated in different parts of the world. It is grown in semiarid to subhumid tropics and is tolerant to drought and high temperatures, but has specific day-length requirements. Cowpea is a vigorously growing annual legume with a strong taproot, bearing numerous horizontally spreading lateral roots.

Sixty-five cultivars of *V. mungo* were grown in the kharif (monsoon) season of 1985 and in summer 1986, and screened after harvest for percentage of hard seeds. When grown in the monsoon season, 14 cultivars produced 30–76%, 8 cultivars 20–30%, and 43 cultivars <20% hard seeds. In the summer, the highest proportion of hard seeds was 15% in cv. UH84-4, and 40 cultivars produced <2% hard seeds. In a laboratory experiment, the effects of scarification using sandpaper, hot-water treatment at 80°C for 2 min, or treatment with concentrated H_2SO_4 for 30, 60, and 120 s in increasing germination percentage were tested in 11 cultivars with the highest proportions of hard seeds. Treatment with H_2SO_4 for 60 and 120 s and scarification with sandpaper gave germination percentages of 86.46, 86.0, and 80.78, respectively, compared with 28.76% in controls. In a further experiment, the percentage of hard seeds declined from 50% immediately after harvest to 7% after 12 weeks storage under ambient conditions (13a).

Species and Cultivars.   About 2000 strains and botanical varieties of green gram are known, which are grouped as follows:

1. *V. radiata* var. *sublobata* (Roxb.) Verdcourt (*syn. P. sublobatum* Roxb.); *P. trinervis* wight and Am., *V. opisotricha*, A. Rich; *V. brachycarpa* (Kurtz.)
2. *V. radiata* var. glabra (Roxb.) Verdcourt (syn. *P. glaber* Roxb.); *P. mungo* var. *glaber* (Roxb.) Bak in Hookf; *P. calcaratus* var. glaber (Roxb.) Prain

About 100 varieties of green gram are known in China and other Asiatic countries, which differ in growth habit, plant size, period of maturity, pod color, and size and color of seeds. The main golden cultivars grown in the United States are *P. aureus*, and the green ones are *P. typica* or *radiata*. In India, the black-seeded types are known as grandis, and the brown-seeded ones as bruneus (2).

Many cultivars of black gram are recognized in India; namely, early-and late-maturing and small-seeded with black or olive-green seed coat. *V. radiata* and *V. mungo* are both derived from *V. radiata* var. *sublobata* (Roxb.) Verdc. (*P. sublobatus* Roxb., *P. trinervis* Wright and Am), which grows wild in India and Indonesia (14).

2. Cultivation

Culture.

ISOLATION REQUIREMENTS AND LAND PREPARATION.   A light-textured and well-drained field should be selected for the seed crops of green gram, black gram, and cowpea. The same kind of crop should not have been grown in the previous two seasons unless the previous crop was of the same variety and the seed was certified.

According to Purseglove (15), green gram is fully self-fertile and almost entirely self-pollinated. Therefore, isolation sufficient to prevent mixture at harvest (about 3 m) would be sufficient. Agrawal (10), however, recommended an isolation of 20 m for foundation seed crop and 10 m for certified seed crop to avoid mechanical admixtures. Since pollen shedding takes place long before the petals open, self-pollination is the rule in these pulses.

The degree of cross-pollination in cowpea varies considerably between different areas and is greater in more humid regions (15). About 20% out-crossing has been reported in cowpea. Some isolation is therefore desirable in the early generation of seed production up to and including crops producing basic/foundation seed (16). In areas of abundant insect activity, cowpea varieties should be isolated by a distance of at least 130 m (17).

Fields should be prepared well by plowing and one or two harrowings followed by leveling. If the seed crop is being raised after wheat or paddy (in the south), no land preparation should be necessary, and the crop can be seeded after one presowing irrigation.

PLANTING AND CULTIVATION.   Green gram and black gram are usually taken as kharif crops. They can, however, be sown either in the second week of February or in April soon after the wheat harvest. In the southern parts of India, these two pulses are also grown as rabi (winter) crop in lighter soils with moisture-retentive characteristics. They can also be grown in rice fallows after paddy. The best time to sow a cowpea seed crop is June or July.

Nucleus/breeder's/foundation seed obtained from an approved certification agency should be used for sowing. The recommended spacing for green and black gram seed crops grown in different seasons are as follows (10):

Kharif and rabi crops	Row to row, 30–45 cm
	Plant to plant, 7–10 cm
Spring and summer crops	Row to row, 20–25 cm
	Plant to plant, 7–10 cm

Cowpea seed crops may be sown in line 2–3 cm deep with row to row spacing of 25–30 cm and plant to plant distance of 1–2 cm.

Fertilization, Irrigation, and Weed Control.   In addition to 5 tons of manure, green gram and black gram crops require about 20 kg N and 35–40 kg P per hectare during kharif season. On medium- to low-fertile soils, cowpea seed crops should be provided with 20–40 kg N and 60–75 kg P per hectare based on soil test values. In zinc-deficient soils, a foliar spray of 0.5% zinc sulfate plus 0.5% lime solution has been recommended (10).

Spring and summer seed crops of green gram and black gram will require frequent irrigations to maintain adequate soil moisture during crop growth. Kharif crops may not require irrigation in areas of assured rainfall. One or two irrigations may be required if there is a prolonged dry period. Cowpea seed crops may also require one or two irrigations during the entire growth period.

The fields of seed crops must be kept clean and free of weeds by one or two weeding or hoeing operations in the early stages of crop growth. A spray of 1 kg treflan, a.i., in 1000 L of water per hectare at the time of final field preparation is recommended to check weeds in the seed plots.

Control of Pests and Diseases.   The aphids and white flies of green gram and black gram may be controlled with a spray of 625 mL metasystox in 625 L of water per hectare, which may be repeated at 15- to 20- day intervals if necessary. Dusting with 10% sevin or folidol at 25 kg/ha controls caterpillars and pod borers.

Anthracnose and cercospora diseases can be checked with a spray of copper oxychloride at 1.25–2.50 kg in 625 L of water per hectare as soon as initial symptoms of the disease appear. For dry root rot, soil is drenched with copper oxychloride at 5 kg/ha in 250 L of water. Powdery mildew can be controlled with 25 kg sulfur dust/ha or a spray of 25 kg wettable sulfur or benlate at 150 g/ha. Rusts can be controlled with a spray of wettable sulfur at 2.5 kg/ha or 2.5 kg fine sulfur dust/ha at 15-day intervals (10). Spraying with 0.1% lindane, 0.04% monocrotophos, or 0.07% endosulfan can effectively control stem-boring maggots, aphids, jassids, and white flies of the cowpea seed crops.

Colletotrichum dematium has been reported to be seedborne in most legumes, and Smith et al. (17a) have shown that *C. dematium* is also seedborne in cowpea (*V. unguiculata*) in Southern Africa. Eight fungicides, that is, benomyl, captab, fludioxonil, imazalil/iprodione, iprodione, mancozeb, thiram and triadimenol, and a commercial *Bacillus subtilis* isolate were evaluated for their efficacy in reducing in vitro mycelial growth of *C. dematium*. Seeds were artificially inoculated with *C. dematium*, treated with captab [captan], benomyl, fludioxonil, thiram, imazalil/iprodione, and mancozeb at the recommended rates and hot water, and the percentage germination and infection determined in vivo. The percentage of emergence and disease incidence were also determined in greenhouse experiments for these seeds. Only imazalil/

iprodione (half the recommended rate) and thiram significantly reduced the percentage of germination. All treatments significantly decreased the percentage of infection of artificially inoculated cowpea seeds. All treatments, except benomyl and the hotwater treatment, significantly increased the percentage of emergence and decreased the percentage of disease incidence in greenhouse experiments when compared to the control (17a).

Roguing. Off-types and severely diseased plants of green gram, black gram, and cowpea (blight, mosaic, and anthracnose-affected plants) should be rogued from time to time if necessary.

3. Harvesting and Threshing

Summer and spring seed crops of green gram and black gram are ready to harvest when the pods turn black. The kharif crop is harvested when most of the pods have turned black. Harvesting practices vary in different regions. In green gram, the fruiting process extends over a long period of time. Hand-picking of pods, therefore, is a common practice. In certain varieties, 75% of the crop can be harvested at first picking and the remainder about 10 days later (14). The plants along with the roots are pulled and stacked for a week in the sun on a threshing floor. After the pods are thoroughly dried, they are threshed by beating with sticks or trampling by oxen.

Saini et al. (18) reported the effects of sowing time and stage of harvest on seed yield and quality of mung beans. The harvesting of seed crops after 2 and 3 weeks of pod initiation produced higher seed yield compared to harvesting after 4 weeks (Table 5). Seed yield and standard germination were significantly higher when the seed crop was sown in July as compared to August sowing, although seed weight was higher in the latter.

Small cowpea seed crops can be harvested manually. Picking is essential for climbing cultivars grown on supports. If the pods mature over a wide interval, more than one picking may be beneficial. The pods are ready for picking when they turn yellowish brown and seeds are firm. For larger areas, the crop may be windrowed and the pods allowed to dry out before gathering with a combine or carting to the thresher.

Srimathi et al. (18a) harvested cowpeas cv. Co. 2 at 10- (T1), 20- (T2), 30- (T3), and 60- (T4) day intervals after 75 days after sowing. The highest seed yield (932.2 kg/ha) was recorded in plants harvested at T2, followed by T1 (916.4 kg/ha). Seeds collected at these treatments also had higher germination compared to T3 and T4. Hard seed content was highest (10.3%) at T4 and lowest (1.8%) at T3. The root and shoot lengths were highest at T4 (29.2 cm) and T2 (32.0 cm). The dry weight and vigor index of seedlings were highest at T2 and lowest at T4.

4. Postharvest Operations

Drying and Cleaning. Small quantities of seed are normally dried by spreading on a drying floor, which if covered should be well ventilated. In humid regions, artificial dryers can be used to dry seed; the air temperature in these cases should not exceed 35°C (1). Seed-cleaning requirements are similar to those described for other pulses.

Seed Treatment and Storage. Cowpea seeds can be treated with captan or thiram to protect tem against anthracnose (*Collectotrichum lindomuthianum*), wilt (*Fusarium*

Table 5 Effects of Sowing Time and Stage of Harvesting on Seed Yield and Quality of Mung Bean (cv. K 851)

Parameters	Sowing date			C.D. at 5%	Weeks after pod initiation			CD at 5%
	July 20	July 26	Aug. 3		2	3	4	
Seed yield (g/ha)								
1984	11.04	10.31	9.26	NS	11.35	10.35	8.74	1.55
1985	12.86	12.22	10.83	1.45	12.91	11.73	10.12	1.45
Mean	11.95	11.27	10.05	–	12.12	10.04	9.43	–
1000 Seed weight (g)								
1984	33.0	32.4	35.3	1.49	31.8	34.0	34.8	1.49
1985	33.5	33.1	35.9	1.83	32.4	34.5	34.9	1.83
Mean	32.2	32.9	35.6	–	32.1	34.3	34.6	–
Seed moisture (%)								
1984	12.9	11.4	13.0	0.29	14.0	12.3	11.0	0.29
1985	13.1	11.8	13.3	0.31	14.5	12.9	11.4	0.31
Mean	13.0	11.6	13.1	–	14.2	12.6	11.2	–
Germination (%)								
1984	90.0	91.5	83.8	3.24	85.9	88.1	91.3	3.24
1985	91.1	91.4	81.3	4.12	83.8	87.7	91.2	4.2
Mean	90.5	91.5	82.6	–	84.8	87.5	91.3	–
Root length (cm)								
1984	17.6	17.3	17.9	NS	18.6	17.2	17.0	NS
1985	16.3	16.5	16.6	NS	16.9	16.3	16.1	NS
Mean	16.9	16.9	17.3	–	17.8	16.8	16.6	–
Shoot length (cm)								
1984	22.8	23.1	23.1	N.S.	23.5	22.6	22.9	NS
1985	29.3	30.1	30.3	N.S.	30.8	29.1	29.5	NS
Mean	26.1	26.6	26.7	–	27.1	25.9	26.2	–

CD = critical difference at $P = .05$; NS = not significant.
Source: Ref. 18.

oxysporum), and damping-off disease (*Pythium* spp.). Strepnomycin sulfate controls bacterial blight (*Xanthomonas vignicola*). According to Kelly (1), the best way to combat seedborne diseases of pulse crops is to use only healthy seed from a clean crop. Studies conducted in India (19) showed that cowpea (*V. unguiculata*) and horse gram (*Dolichos biflorns*) seeds treated with malathion and stored in polyethylene bags retained their viability over a period of 8 months. The powdered leaves of *Cymbopogon flexuosus* (2–4% by weight) gave adequate protection to cowpea seeds against bruchids (*Callosobruchus maculatus*). Salunkhe et al. (2) have described various improved storage structures to store pulse crop seeds without significant losses in quantity or quality.

C. Pea, French Dry Bean, and Broad Bean

1. Introduction

Botany and Origin. Peas (*Pisum sativum* L.) are widely cultivated in cool, moist, temperate climates or in cool seasons at higher altitudes in the tropics and subtropics.

They are susceptible to high temperatures and drought. The pea plant is a short-lived herbaceous annual, glaucous and climbing by means of leaflet tendrils. Peas are completely cross-fertile.

French beans (dry beans or common beans) (*Phaseolus vulgaris*), native to the American continents, are widely grown throughout the world in warm temperatures or subtropical climates. They are also grown in the cool seasons or at higher altitudes in the tropics. These crops are susceptible to high temperatures, frost, and drought.

Broad beans (faba beans) (*Vicia faba*) are widely grown in temperate and subtropical countries with good rainfall in the Mediterranean region. During cool seasons, they are also cultivated at higher altitudes in the tropics. The crop is highly susceptible to drought and high temperatures.

Species and Cultivars. More than 1000 varities of peas have been identified, which vary in height, maturity period, pod size and type, and seed characteristics. In the species *P. sativum*, the following six subspecies have been identified (20): *hortense, syriacum, absyssinicum, jomardi, elatius,* and *arvense.*

There are two types of garden pea: smooth seeded and wrinkled seeded. The latter is sweeter than the former. The three main types of peas used for human consumption are (12):

1. Vining peas used for canning fresh (garden peas), quick freezing, or artificial drying
2. Threshed dry peas, which are sold dry or canned,
3. Pulling peas sold as fresh peas in pods (12).

More than 14,000 cultivars of French beans have been recorded in the species *P. vulgaris*, grouped into following four types:

1. Red kidney beans—1.5 cm or longer, used in Latin America,
2. Medium field beans—1–2 cm long, pinkish with brown spots, grown in United States as pinto beans
3. Marrow beans—1.0–1.5 cm long, grown as yellow eye bean
4. Navy beans—8 mm or less, grown in California

The species of broad bean (faba bean) (*Vicia faba* L.) have been subdivided into subspecies or botanical varieties depending mainly upon seed size. However, modern cultivars do not fit well into such groupings because of alterations in seed size. According to Kelly (1), faba beans are grown for the following main purposes:

1. Broad beans are harvested when the seeds are tender for use as a vegetable either fresh or after freezing or canning. The crop in some areas may be treated as a pulse crop; that is, harvested when mature and the dried seed eaten later.
2. Field beans are harvested when mature, and the dried seed is used as a stock food. Field beans are sometimes subdivided into tic beans, which have small seeds and horse beans, which have larger seeds.
3. Faba beans may also be harvested as whole plants for stock food, either hay or silage. Field beans are generally used for this purpose.

Cultivars are classified on the basis of a melanin spot on the wing petal (absent, present) and the color of the testa of ripe seed (beige, green, red, violet, or black). Faba bean cultivars can be distinguished on the basis of plant-height, pod length, 100-seed weight, and time of 50% flowering (1).

2. Cultivation

Culture.

ISOLATION REQUIREMENTS AND LAND PREPARATION.   A field having good drainage-neutral soil on which the same crop was not grown in the previous year (unless the variety of the crop was the same and certified) should be selected as a seed plot.

Peas are mainly self-pollinated crops with little natural crossing. Adequate isolation provided in the form of a physical barrier or a distance of 3 m to avoid physical mixing at harvest should be sufficient. French bean is normally regarded as a self-pollinating crop, but some out (cross)–pollination can occur. Although isolation of about 3 m is usually enough, some more highly specialized cultivars used as vegetables for freezing or canning may isolated by about 100 m for crops intended for further multiplication and 50 m for other crops. These distances may be doubled when the crops are planted on areas of 2 ha or less.

V. faba is partly self- and partly cross-fertilized. The flowers are adapted for insect pollination, and bees mainly effect cross-fertilization. Smaller fields of 2 ha or less should therefore be isolated by a distance of 200 m when the seed to be produced is for further multiplication and by 100 m for other crops. For fields greater than 2 ha, the corresponding distances are 100 and 50 m. These distances may not be enough when broad bean cultivars are being grown as seed crops; these require 500- or 300-m distances.

Land is prepared to the desired tilth by one deep plowing and three to four harrowings followed by leveling. The nucleus/breeder's/foundation seed should be obtained from a source approved by the certification agency.

PLANTING AND CULTIVATION.   Peas for seed crop are sown from mid-October to November in the plains and from February to March in the hills. Seed should be sown in lines with a seed drill or behind a plow 3–4 cm deep, with a spacing of 45–60 cm between rows and 5 cm between plants. About 60–75 kg pea seed is required to cover an area of 1 ha.

French beans are sown from January to February or from August to September in the plains, whereas in the hills, they may be sown from March to May. Sowing may be performed with a drill or seed may be dibbled manually in lines 60–75 cm apart at 10–15 cm distance and 2.0–2.5 cm deep. Adequate soil moisture should be ensured for optimum germination. A seed rate of about 25–30 kg (pole varieties) and 85–90 kg (bush varieties) are needed for 1 ha. Seed may be inoculated with appropriate nitrogen-fixing bacteria for quick nodulation.

Faba beans may be sown from January to February or from August to September (on the plains) and from March to May (in the hills). The recommended spacing for an autumn-sown crop is generally wider (30–35 cm) than for spring-grown crops (20 cm). Kelly (1) recommended 60 × 12 cm spacing for better results. Wider spacing may also be advantageous in drier areas, and where interrow cropping is practiced. Seed rates of faba bean vary considerably from 175 to 200 kg/ha for the small-seed cultivars to as high as 250 kg/ha for larger seeded varieties. Lower rates are generally used for free-tillering autumn-sown cultivars than for those sown in spring, which tiller less. Faba beans should be sown deeper than most crops (8 cm deep). Broadcasting the seeds followed by plowing to put the seed deep in soil or putting the seed box on the plow are more satisfactory methods than sowing the seed with drillers. Deep sowing prevents bird damage.

Suzuki and Khan (20a) investigated the effects of temperatures and duration of seed humidification on seed and seedling performance in snap bean (*P. vulgaris* L.) to determine the optimal treatment. Seed vigor was improved by humidification as measured by enhanced germination, ACC (1-aminocyclopropane-1-carboxylic acid)–derived ethylene production, and seedling emergence and growth. The invigoration effects of seed humidification were obtained at treatment temperatures of 15–40°C, although the effective treatment duration was relatively short at 40°C. When seeds were humidified at 25°C, treatment durations of 6–8 days were optimal for germination and ethylene production. However, seedling emergence and growth in a soil media were similarly improved by seed humidification for 1–8 days. The different responses of germination and seedling growth to humidification duration suggested involvement of multiple seed humidification factors. These results showed a relatively wide range of effective temperatures and duration for maximum benefits of seed humidification in snap bean.

Demir et al. (20b) investigated the effects of seed moisture (8, 10, 12, 14, 16, and 18%) on field emergence (three sowing dates in 2 years in Turkey), seedling dry weight, and electrical conductivity in two different colored cultivars (red: 4F-89; white: Yalova-5) of *P. vulgaris*. Increasing seed moisture up to 14–16% improved emergence, whereas seedling dry weight continued to increase until 18%. Both cultivars gave a similar response to moisture content, although the white-coated cultivar performed worse than the red one in all criteria. At 8% seed moisture content, hardseededness contributed to the failure of emergence in the red cultivar. Soil temperatures above 13.5°C did not affect emergence and seedling growth.

Fertilization, Irrigation, and Weed Control. Pea seed crops require about 25 kg N, which is top-dressed in two doses during the early growth period and flowering time, and 50–70 kg P and 50 kg K in addition to about 5 tons of manure added at the time of land preparation.

For French beans, manure requirements are greater: about 25 tons of manure and 15 kg ammonium sulfate, 500 kg superphosphate, and 125 kg potassium sulfate at the time of solving. A second dose of 125 kg ammonium sulfate should be applied as top-dressing 25 days after germination.

Two to three irrigations may be needed at intervals of 15–20 days according to weather and soil conditions before the flowering and fruiting of pea seed crops. French bean and road bean seed crops should be irrigated as and when required to maintain adequate soil moisture conditions at the time of flowering and pod development.

Fields should be kept free of weeds by two or three hoeings or weedings, especially during the early stages of crop growth.

Control of Pests and Diseases. Pea seed crops should be sprayed with 0.25–0.5% DDT emulsion to control aphids, jassids, and pod borers at the beginning of flowering. Mildew may be prevented by dusting the top with fine sulfur once or twice in its later stages of growth.

Agrawal (10) recommended the following schedule of plant protection to control insect pests and diseases of French bean:

	Control measures
Insects	
Bean beetles	Spray with 4% rotenone at 1.5 kg/100 gallons water/ha or cryolite at 1.5 kg/50 gallons water/ha
Bean aphids, thrips	Spray with parathion at 625 mL/625–750 L water/ha at 15-day intervals
Diseases	
Anthracnose	Use seed treated with ceresan; spray crop with copper fungicides
Powdery mildew	Dust crop with sulfur or lime sulfur, beginning at first appearance of symptoms and repeated at 10-to 15-day intervals
Rust	Dust finely ground sulfur at 25–30 kg/ha before infection
Root diseases and nematodes	Fumigate soil with nematicides like DDT or nemagon

The plant-protection measures for faba bean are similar to those of French bean or pea. Great care, however, needs to be exercised while applying insecticides close to or during flowering time to ensure that bees and other pollinators are not harmed. In addition to pollination, bees "trip" flowers to enable fertilization to occur. Cross-pollination results in the production of more vigorous plants than those derived by self-fertilization (1).

The increased cultivation of field beans and peas in Austria since 1984 has generated an increase in infestation rates of the bean beetle (*Bruchus rufimanus*) and the pea beetle (*B. pisorum*). Austrian seed-certification regulations forbid the presence of living seed beetles in leguminous seeds. Customary investigation methods do not truly reflect actual infestation levels. Girsch et al. (20b) presented a new method for the evaluation of seed beetle infestation in leguminous seeds. Beetle-proof polystyrol boxes were inlaid with filter paper saturated with 100 mL (beans) or 80 mL (peas) of a solution of 0.1% NaOCl and 1% chinosol. Samples of 4×100 g field beans or peas were deposited in the boxes and incubated at 30°C in complete darkness for 7 days. A comparison with the traditional Rohloff method demonstrated the superiority of the newly developed method, which reliably determined true infestation levels of seed samples and prevented interfering fungal attack without reducing beetle viability. Tests showed that the total population of live beetles could be determined after 6 days. Furthermore, beetle emergence from seeds was shown to decrease linearly from harvest to next June. The new method has been in use in Austria for 6 years and has given best results for approximately 2500 samples (20c).

Roguing. The off-types and the diseased plants affected by pea mosaic, root rot, and blight should be rogued out from the field from time to time. French bean seed crops require careful roguing to maintain their genetic purity. Plant characteristics such as foliage color, plant type, flower, and pod characteristics should be used to remove off-types from the field at preflowering, flowering, and maturity stages. The plants affected by diseases like bacterial blight, anthracnose, ascochyta blight, and bean mosaic should also invariably be removed from time to time as required. The roguing requirements of faba bean are similar to those of French bean and pea seed crops.

3. Harvesting and Threshing

The harvesting of pea seed crops may begin when 90% of the pods turn brown. The plants are uprooted and stacked in small heaps and allowed to dry in the field for about a week.

French beans and faba beans should be harvested when a large percentage of pods are fully ripe and most of the remaining pods have turned yellow. Harvesting should be in before the lower pods become dry enough to shatter. As with peas, small areas are harvested by hand. The longer bean stalks can be bound into sheaves and stacked in the field to dry to allow the seeds to mature further. Larger areas are usually combined. Windrowing is possible provided wet weather conditions are not prevailing in the field. Faba beans tend to ripen irregularly on the plant, because flowering is not well synchronized, the plants being indeterminate. Therefore, the appropriate time of harvest is difficult to judge. The maturing plant loses its leaves, after which the stem and pods begin to turn black. For windrowing, at least 25% of the pods should have changed color, and for direct combining, it is advisable to wait until up to 90% have changed. The seed moisture content at this stage will be around 30% or less; for direct combining, it should fall below 20%.

Threshing is carried out after drying by flailing with wooden sticks or trampling by bullocks. Mechanical threshing easily damages the seed. Careful setting of drum clearance and seep is therefore essential. Much seed can be lost by shattering at the cutter bar, requiring an appropriate reel setting. Very ripe crops combine best when slightly damp with dew to reduce shattering.

4. Postharvest Operations

Cleaning and Sorting. Seed is mainly cleaned by air screen cleaners. For stained peas, an optical sorter an be used. Cracked or damaged pea seeds can be separated on a pin drum. However, when weed seeds lodge in the holes, they are difficult to remove, resulting in higher cleaning losses. Pea seeds should be handled carefully during all postharvest operations, including threshing and cleaning, as they are liable to crack or split. Care should also be taken to ensure that seed is not damaged during the long drops of seed from hoppers or elevators. The cleaning equipment should not be overloaded, which could cause friction between the seed and the moving parts.

Seed Treatments. Foot rot and leaf and pod spot are the main seedborne diseases of peas caused by *Ascochyta pisi*, for which the use of disease-free seed is advocated. A small quantity of seed can be treated with thiram in warm water (37°C) by soaking seeds for 12–24 hr followed by immediate drying. The seed should be handled with extreme care. Slurry treatments with a fungicide like benomyl are more effective than powders (1).

D. Soybean, Lentil, and Hyacinth Bean or Horse Bean

1. Introduction

Botany and Origin.   Soybeans (*Glycine max*) originated in China and have been extensively cultivated here since prehistoric times. They are now widely grown in the United States, Brazil, East Asia, and to some extent in Africa and the Near East. Soybeans thrive in subtropical climates with good rainfall and warm sun and require irrigation in semi-arid regions. Soybeans are very sensitive to photoperiod and are inherently short-day plants.

Lentils (*Lens esculenta*) are cool season crops of the Mediterranean region and the ear East and are now cultivated in various parts of the world. They prefer a warm, dry climate and, although unsuited to cultivation in the wet tropics, can be raised in cool, dry seasons at higher altitudes in the tropics under irrigation. The crop is quite susceptible to winter frost. Lentils are a herbaceous annual pulse characterized by higher dehiscent pods.

Hyacinth beans (*Dolichos lablab* or *Lablab purpureus*) and horse gram (*D. biflorus*) are grown extensively in the southern parts of India with moderate rainfall. They are drought resistant and hardy and tolerate a wide range of soil types. Lablab bean is used both as a pulse and as a vegetable.

Species and Cultivars.   Three soybean species have been identified: *Glycine ussuriensis* (wild), *Glycine* max (cultivated), and *Glycine gracilis* (intermediate). *G. max* is commonly grown throughout the world.

In the United States and Canada, soybean cultivars have been divided into 12 maturity groups designated as 00, 0, and 1 to X. Group 00 is adapted to the longer days of higher latitudes, and Group X is the latest to mature. The cultivars of soybean adapted to higher latitudes flower earlier if they are planted in short-day conditions, so that in the tropics most cultivars mature at about the same time. Soybeans are sensitive to temperature and moisture conditions at flowering time, so that maturity groups in North America do not necessarily apply to other areas at similar latitudes. Five groups of soybean identified by the International Board for Plant Genetic Resources (IBPGR) (21) only correspond to certain U.S./Canadian groups; the remaining nine IBPGR groups do not relate to the U.S./Canadian groups (22).

IBPGR group	U.S./Canadian groups
1	00, 0
3	I, II
5	III, IV
7	V, VI, VII
9	VII, IX, X

Soybean cultivars are also classified on the basis of hair color (gray, tawny), flower color (white, violet), and hilum color (gray, yellow, brown, dark brown, or black). Kelly (1) has described the main distinguishing characteristics of soybean cultivars.

The cultivars of lentil are distinguished on the basis of flower color and shape and color and mottling of seeds. There are two groups of subspecific rank:

1. *Lens culinaris* spp. *macrosperma* (Baumb.) Baroulina—large white or blue flowers grown in the Mediterranean Africa and Asia
2. *Lens culinaris* ssp. *microsperma* (Baumb.) Baroulina—small-seeded cultivars with convex pods, grown principally in Asia and Africa

Lens orientalis (Boiss) Hand-Mazz is morphologically and genetically closer to *L. culinaris* and is grown as a cold season crop throughout the tropics, particularly in India, North Africa, and Central and South America (23).

The cultivars of *L. purpureus* are recognized based on variability of size, shape, and color of the pods. Duke et al. (24) described the following three subspecies of the hyacinth bean or kidney bean:

1. *L. purpureus* ssp. *uncinatus* Verdcourt—has more slender inflorescence and smaller pods (1.5 × 4 cm). Synonyms: *L. uncinatus* A. Rich; *D. lablab; Forma uncinatus Penzig; D. uncinatus* (Schweinf.); *D. lablab* var. *uncinatus* (Schweinf.) Chiov; *L. niger* var. *uncinatus* (A. Rich.) *Curf., L. niger* var. *Crenatifructus* Curf.
2. *L. purpureus* ssp. *bengalensis* (Jacq.) Verdcourt—has pods similar to lima beans but quite dissimilar to those of other species, with free racial interbreeding. Synonyms: *D. bengalensis* Jacq. *D. lablab* ssp. *bengalensis* (Jacq.) Rivals; *L. niger* ssp. *bengalensis* (Jacq.) Curf.
3. *L. purpureus* var. *rhomboideus* (Schinz) Verdcourt—has more glabrous leaves, pods with broader bracteoles, and pods that are minutely puberulous rather than pubescent. Synonyms: *D. lablab* var. *rhomboideus*; *D. pearsonii* Hutch.

2. Breeding

Specht and Graef (25) described an alternative breeding method which attempts to combine the best features of the diallele selective-mating (DSM) system and the conventional breeding (CB) system into a male-sterile–facilitated cyclic breeding scheme (MSFCB). In the new method, the 10 annually chosen elite parents are not directly mated to each other (as in the CB system), but are placed instead in an isolation nursery containing male-sterile plants. Insects normally transfer pollen from the 10 elite parents to the male-sterile plants. At least one F_1 seed is harvested from each of the available male-sterile plants and, when sown in a winter nursery, provides the F_1 plants that will be bulk threshed to generate F_2 seed. Most of the seed proceeds through the breeding program, but a portion of the F_2 seed is cycled to the next year's isolation nursery, thus providing the male-sterile plants that will be mated to the next set of 10 elite parents. The F_2 plants in the isolation nurseries segregate three male-fertile plants to one male-sterile plans. The male-fertile plants are rogued as soon as they produce an examinable flower (the shrunken anthers bear no pollen). Unlike the DSM system, the MSFCB system allows the annual inflow of 10 new parents. Such breeding methods are termed "recurrent introgression" rather than "recurrent selection." The MSFCB mating design thus provides a "recombinational link or connection" between the intermating phases that commence each successive annual breeding cycle. The MSFCB allows empirical testing of the use of single-seed descent (SSD), beginning with F_1 seed rather than the traditional F_2 seed. This breeding scheme optimizes genetic recombination and thus also genetic variability. The obvious advantages of this system are (22):

1. Convenient production of many F_1 individuals from many crosses,
2. A more rapid "recombinational link" between successive cycles

Ranch (26) described the production and potential of soybean somatic embryos from primary explant tissue (cotyledons), including the large-scale somatic embryos, scale-up of proliferation, and plant performance.

3. Cultivation

Culture.

ISOLATION REQUIREMENTS AND LAND PREPARATION.   As in the case of other seed crops, the land selected for growing soybeans, lentils, and hyacinth beans should not have been used for the same crop in the previous season unless the seed crop was of the same variety. The selected field should be neutral in reaction (soil pH, 7.0) and well drained.

Soybean anthers dehisce in the bud itself before the flowers open, hence self-pollination normally occurs, with less than 1% crossing due to insects. An isolation distance of about 3 m from other soybean fields is therefore sufficient to maintain genetic purity and to prevent physical mixing during harvesting.

Although lentils are mainly self-fertilized, some out-pollination can occur. Early generations in a maintenance-breeding program should therefore be safeguarded with an isolation distance of 20 m for foundation seed class and 10 m for certified seed class from other lentil fields. For later generations, a distance of 5–10 m is considered to be adequate (1,10).

The selection of land and isolation requirements for the hyacinth bean is similar to those of the French bean. Seed fields are prepared to a fine tilth by deep plowing and two to three harrowings followed by leveling.

PLANTING AND CULTIVATION.   Soybean seed crops are best sown in the first half of July, using seed (nucleus/breeder's/foundation) obtained from a source approved by the seed-certification agency. It is essential to inoculate the seed with the rhizobium culture, especially when the crop is grown newly on a soil. Agrawal (10) described two types of soybean culture: peat culture and soil culture. For inoculation with peat culture, 100 g of sugar is dissolved in 1 L of water, boiled for 15 min, and cooled to ambient temperature. This solution is sprinkled over the soybean seeds. The peat culture is then dusted on the seeds and mixed well. One-half kilogram of peat culture mixed with the soybean seed is sufficient to cover an area of 1 ha. For inoculation with soil culture, slurry is prepared by mixing the soil culture with a sugar solution (prepared as in peat culture) and mixed thoroughly with the seed. The seed should be inoculated in shade. The amount of culture should be tripled if the crop is to be raised on soils new to the crop.

Soybean seed is planted in rows with a seed drill 2–3 cm deep in soil having moisture content or 4 cm in lighter soils. The recommended spacing between rows varies from 40 to 60 cm and from 4 to 5 cm between plants. About 65–75 kg of seed is sufficient an area of 1 ha.

Lentils may be sown around mid-October, with seed sown in lines 2–3 cm deep with a row to row spacing of 25–30 cm and plant to plant spacing of 1–2 cm. The seed rate varies from 25 to 30 kg/ha for small-seeded cultivars and from 35 to 40 kg/ha for the large-seeded varieties.

Hyacinth beans are best dibbled at a distance of 100 × 30 cm, using 12–15 kg seed/ha. Other requirements are similar to those for French beans. This crop, however, needs staking of vines for their proper spread.

Fertilization, Irrigation, and Weed Control.   Soybeans require about 20–25 kg N, 80–100 kg P, and 30–40 kg K per hectare at the time of planting depending upon the soil fertility status. In zinc-deficient soils, a spray of 0.5% zinc sulfate plus 0.25% lime solution may be beneficial.

For lentils, 20–40 kg N, 60–75 kg P, and 30–40 kg K per hectare are applied depending on soil tests. Zinc deficiency can be overcome by spraying the crop with a solution of 0.5% zinc sulfate and 0.25% lime in a mixture. Hyacinth beans or dolichos beans are normally fed with 90 kg ammonium sulfate and 40 kg superphosphate per hectare.

Soybean crops should be irrigated as and when required to maintain adequate soil moisture throughout the crop growth, particularly during flowering, seed development, and maturation. Lentil and hyacinth bean seed crops may need one or two irrigations depending on soil moisture conditions. It is important to ensure optimum soil moisture during flowering and pod development.

Fields of the seed crops must be kept free of weeds to obtain good yields. Effective weed control from the beginning of plant growth is essential. This may be achieved two to three hand weedings or by applying weedicides like lasso (4 kg/ha), senecor (0.5 kg/ha), or both together.

Control of Pests and Diseases.   Yellow mosaic disease of soybean can be controlled by spraying with a mixture of 0.1% malathion and 0.1% metasystox on the 20th, 30th, 40th, and 50th days after sowing. Malathion may be replaced by thiodan to control hairy caterpillars. The spread of rust may be prevented by spraying soybean crops with 2.5 kg dithane Z-78 per hectare on the 60th day of sowing, repeating once 10 days later if necessary. No serious pests affect lentil seed crops. The hairy caterpillars and pod borers of this crop can be controlled by spraying 0.04% monocrotophos solution (1000 L/ha) at the appropriate time.

The plant-protection measures for the hyacinth bean seed crop are similar to those for French bean.

Roguing.   Soybean plants infected with yellow mosaic virus should be rogued out as soon as they appear in the field (during the first 2–3 weeks) to prevent their further spread. This operation should be continued up to the last phases of crop growth and maturation. At the time of flowering, off-types may be removed on the basis of flower color and other plant characteristics. Final roguing may be carried out at crop maturity to rogue out off-types on the basis of pod characteristics.

Off-types and diseased plants of lentil, which are affected by blight as well as weed plants (*Lathyrus* and *Vicia* spp.) should be rogued out from the seed field from time to time as required. The roguing operation for hyacinth bean is similar to that described for the French bean.

4. Harvesting and Threshing

Soybean seed crops should be harvested as soon as the seed matures and has reached a moisture content of about 13–14%. Early harvests (second week of October) of soybean carried out at Mississippi State University were reported to retain higher seed germinative quality compared to the seed crops harvested later (second week of December) (10).

The crop is harvested manually or by combine when it is completely dried. In the developed countries, harvesting by combine is very common. Since all seeds on a soybean plant essentially mature at the same time, the leaves drop rapidly and the stems dry (27). The combine needs to be frequently adjusted during the day to reduce harvesting losses, which may amount to 10–20% (28). Special care should be taken to prevent mechanical damage to the seed when harvesting by combine. As seed moisture drops below 12%, the germination capacity of seed may be affected as the mechanical injury increases. Combines with appropriate cylinder seeds thresh the crop properly without cracking the seed. In soybean varieties that tend to shatter, harvesting should be carried out when the pod color starts changing from yellow to brown and leaves begin to shed. A delay of even 2 days may result in pods bursting and eventual seed shedding.

Saini et al. (29) examined three threshing methods for soybean: hand threshing, stick beating, and a soybean thresher at 300, 400, and 500 rpm. Hand shelling maintained maximum viability and vigor of the seed during 2 years of storage. The crop threshed at 500 rpm had the highest mechanical damage owing to a lower percentage of healthy seeds. The seeds obtained by threshing at 400 and 500 rpm showed relatively greater loss of viability and vigor in storage during both years, as well as under an accelerated aging test.

A small amount of produce can be threshed by gently beating the harvested plants on a plank on a tarpaulin sheet or prepared floor.

The harvesting of lentil crops may be carried out when the plants become yellow and seeds resist pressure when pressed between the fingers. The lower pods turn brown when the crop is ready for harvest. The plants are pulled by hand and placed in small bundles to dry in the field for about a week. Flailing with sticks carries out threshing.

Hyacinth bean species are harvested by cutting the crop to the ground or by pulling by hand when the pods are dry and yellow. Plants are stacked in the field to dry in the sun. Ripe pods may also be picked from the standing crop manually. After drying, the crop is threshed by flailing with sticks or trampling by bullocks.

5. Postharvest Operations

Cleaning and Drying.   Soybean seeds are particularly vulnerable to damage by rough handling after harvest and will deteriorate rapidly under poor storage conditions. It is essential that seed moisture is 4% or below during storage. A moisture content of 12% is also recommended to maintain seed viability. Soybean seed does not store well, and even good, healthy seed will begin to deteriorate after 6 months if the temperature and humidity are not controlled.

The threshed soybean produce is best cleaned on an air screen cleaner or by using a gravity separation. Winnowing against natural wind to separate plant debris and dust from the seed may clean a small amount of produce.

Cleaned soybean seed may be dried in the sun or using heated air if the initial moisture content is high (above 20%). The drying temperature should not exceed 38–40°C (1,2), and drying should be performed gradually.

Storage and Seed Treatment.   Soybean seeds packaged at 9% moisture in either multiwall paper or polyethylene bags maintained germination through 40 months in a cool but un–air-conditioned warehouse (30). Seed vigor, however, was substantially reduced after 24 months (Fig. 3). For long-term storage (5–8 years) of valualbe seed germplasm, a highly sophisticated storeroom with minimal transmission of moisture vapor through the walls is needed. Such rooms are generally well insulated and equipped with a refrigeration system capable of maintaining a storage environment of about 10°C and 50% relative humidity (RH) (30). Byrd and Delouche (31) compared the efficiency of several widely used vigor tests with the germination test to evaluate potential during storage (Fig. 4). The germination percentage of soybean seeds stored at 30°C and 50% RH did not significantly decrease until after 7 months. Accelerated aging and cold test responses, however, significantly decreased after 1–4 months of storage, as did the field emergence percentage (31). Typical bulk storage bins used for storage and aeration of soybean seeds are shown in Figure 5.

Treatment with a fungicide such as thiram immediately after cleaning can prevent fungal infection spreading during storage. The seed, however, should not be treated if it is intended subsequently to inoculate with the rhizobium. Inoculation with rhizobium is usually necessary when soybeans have not been grown for some time on a field. When the seed is inoculated, other seed treatment is not advisable; however, the treated seed will not affect the rhizobium already present in soil.

Treating the seeds with thiram, captan, or carboxin can control the fungal seedborne diseases of soybean, lentil, and hyacinth bean.

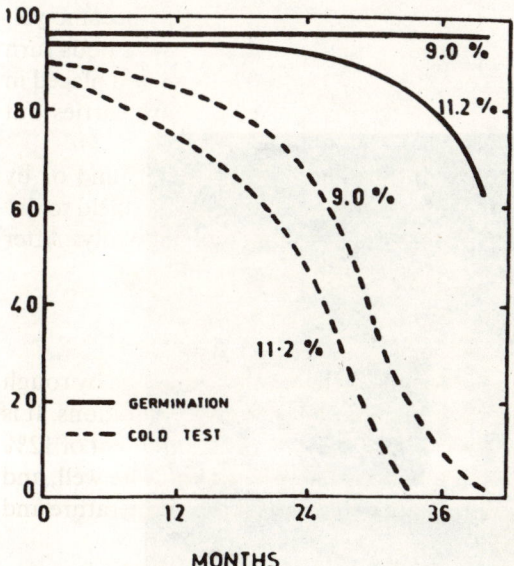

Figure 3 Germination and cold test percentage of soybean seed (cv. Clark) packaged in three-ply multipaper bags during un–air-conditioned storage for 40 months. (From Ref. 30.)

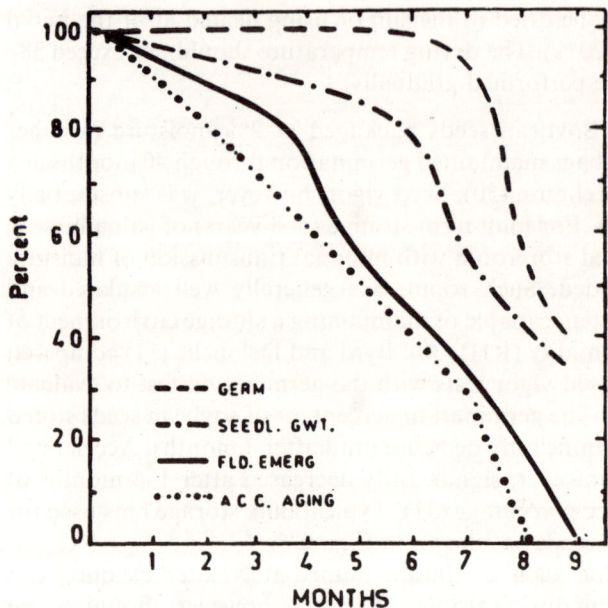

Figure 4 Response reactions of soybean seeds (cv. Lee 68) during storage at 30°C and 50% RH, relative to levels of responses at the beginning of storage, i.e., 0 months = 100%. (From Ref. 31.)

Figure 5 Typical bulk storage bins used for storage and aeration of soybean seeds. (From Ref. 30.)

Seed Yields. The average seed yields of soybean and lentil vary from 20 to 25 Q/ha, whereas that of hyacinth bean range from 300 to 600 kg/ha.

E. Other Minor Pulses

1. Scarlet Runner Bean

Runner beans (*Phaseolus coccineus* L.) are utilized in a similar manner as French beans, with most of its cultivars being climbers requiring support. There are some dwarf cultivars. Runner beans are long-day plants sensitive to temperatures. Seed set and pod development is best at 20–25°C. Bees and other insects need to trip the flowers for seed fertilization. Although runner beans can self-fertilize, out-pollination does occur up to 40% (32). The seed crop, therefore, would require an isolation distance of at least 100 m from other fields. The other seed-growing requirements and production technology are similar to those described for French beans.

2. Lima Bean or Sieva Bean

Lima beans (*Phaseollus lunatus* L.) are native to tropical America, but are widely grown throughout the humid and subhumid tropics and subtropics. They are fairly tolerant to high temperatures, poor soil conditions, and drought but are more susceptible to frost. Lima beans are perennials but are often cultivated as annuals or biennials.

Lima beans can be successfully grown on all types of soils, but the best results are obtained when raised on moderately fertile, well-drained soils. Highly fertile soils are conducive to greater vegetative growth and low seed production. Heavy soils prone to water logging are detrimental to this crop. Seed rates for lima beans vary according to the seed size of the variety used, ranging from 50 to 150 kg/ha. Lower seed rates are required when lima beans are grown as space-planted pole beans. The crop takes 7–8 months to mature fully, and yields of dry seeds range from 10 to 30 Q/ha (34).

3. Rice Bean

Rice beans (*Phaseolus calearatus* L.) are mainly grown in the hills of India up to an altitude of 2000 m in the western Himalayas. It is usually grown as a kharif crop and sown by broadcasting at 60–80 kg/ha. Rice bean grows rather vigorously, giving a thick, dense mass of foliage, and therefore is often used as a cover crop to check soil erosion on the slopes. The cultivation and seed-production technology of rice bean is similar to that of green and black grams. The seed yields of rice bean vary from 250 to 300 kg/ha. Among the pulses, rice beans are a rich source of minerals, especially calcium (34).

4. Moth Bean

Moth beans (*Phaseolus aconitifolius* L.) are, like rice beans, grown mainly on the Indian subcontinent. It is an herbaceous annual with a spreading, prostrate habit. Among the kharif pulses of India (those sown before monsoon to ripen in autumn), moth bean is considered to be the most drought resistant. It is usually grown as a dry crop on light sandy soils and requires minimal land preparation. The plant is normally sown by broadcasting the seed with the onset of monsoon rains, using 3–4 kg/ha seed rates. Moth bean matures within 3 or 4 months of sowing, and yields of seeds obtained range from 1500 to 2000 kg/ha. The cultivation, harvesting, threshing, and drying practices are similar to those used for other dry land pulses of India.

5. Cluster Bean, or Guar

Guar, or cluster bean (*Cyamopsis tetragonoloba*), has been grown in India since ancient times for vegetable and forage purposes. The crop was introduced to the United States from India in 1905 (33). Recently, the industrial applications of guar have been recognized, and guar gum is now widely used as a thickening agent in various food products and for sizing textiles and paper products (34). Guar is a hardy, bushy annual and resists drought conditions. It grows well in deep alluvial and sandy loams, but is quite susceptible to water logging and excessive wetness. It is sown by broadcasting when raised for green manure or forage, but it is usually planted in rows (45–60 × 15–30 cm), with seed rates varying from 10 to 25 kg/ha. Pods begin to set about 45–55 days after sowing, and seeds mature 110–160 days after sowing. Pods are resistant to shattering losses. The crop should be harvested soon after its maturity when the pods turn brown and dry with a seed moisture content of less than 14%. The pods are harvested by hand and threshed manually. Seed yields of about 900 kg/ha can be harvested when the crop is grown under dry land conditions, although yields can double with irrigation.

6. Kesari Dhal or Chickling Pea

Kesari dhal, or chickling pea (*Lathyrus sativus* L.), is mainly grown in India. It is a hardy, drought-resistant annual crop and thrives well even when grown on poor soils. In central parts of India, it is cultivated after cereals have failed because of drought (famine crops). *Lathyrus* is a winter crop, requiring cooler climates. However, it is quite resistant to water logging as ell as drought conditions during the growing season. It is usually grown on clay loams, which retain moisture for longer periods of time. The crop requires little tillage when grown in rice fields after their harvests. The seed is sown by broadcasting on lands submerged in water until late October to November, after which it sprouts and grows in spite of subsequent hardening of the soil. The seed rates range from 35 to 45 kg/ha when broadcasted. The crop needs little interculture and aftercare. It matures within 4–6 months, and is harvested soon after leaves begin to turn yellow, before the pods become fully ripe, when they dehisce and seeds shatter easily. The harvested crop is dried in heaps in the fields for about a week, threshed on a threshing floor, and winnowed to separate seeds. Seed yields of 10–15 Q/ha are common.

7. Lupine

Lupines (*Lupinus* spp.) were used as human food by the ancient civilization surrounding the Mediterranean and Andean highlands. The most commonly used food lupines include *L. albus*, *L. luteus*, and *L. mutabilis*. The first two species are native to southern Europe, and the third is native to the Andean regions (34).

Lupines have been important green fodder and silage crops. Their use in human nutrition has been limited owing to the presence of bitter-testing alkaloids, although modern varieties have been bred for low alkaloid content (sweet lupines). Blue-yellow and white flowering types of lupines are known. Blue lupines are widely grown in western Australia and thrive well on deep, fertile soils but are susceptible to *Fusarium* wilt and ripen very late. Yellow lupines are grown for seed in northeastern Europe, but are more useful for green manuring and reclaiming very acidic, sandy soils. White lupines are the most promising for grain production. They have high-yielding potential (up to 50 Q/ha) and are early ripening (120–140 days) and resistant to diseases. They are also resistant to seed shattering by dehiscence. White lupines are normally grown on acid sandy soils, but perform better when raised on deep, well-limed and free-draining loams. The crop requires a thoroughly prepared seedbed and seed rates vary from 50 to 60 kg/ha.

Lupines ripen rather unevenly, causing harvesting problems. Also, large, fleshy pods need prolonged drying periods and shatter far too easily. The crop, if combined, needs care to prevent harvesting losses. The threshed seeds need further drying to about a 13–14% moisture content before storage.

8. Winged Bean or Goa Bean

Winged bean (*Psophocarpus tetragonolobus* L.) probably originated in India and is grown in parts of Southeast Asia, the Pacific region, Africa, and the Caribbean. It is mainly grown for its immature edible pods and cooked as a vegetable. Its leaves, young sprouts, flowers, and fruits are also used as vegetables. The dried seeds of winged beans are a rich source of protein and oil. It is a perennial vine, twining and glabrous, but is usually grown as an annual crop. The main lateral roots run horizontally at a shallow depth, later becoming thick, tuberous, and nodulous. The crop can be grown on a range of soils provided they are well drained. The winged bean is sensitive to drought, frost, salinity, and water logging (34).

Winged bean is normally space planted (about 10,000–12,000 plants/ha), sowing seed 5–6 cm deep with a spacing of 30–60 cm apart. The crop requires stakes or trellises for support. The flowers are usually plucked to increase root yields. After fruiting, the top dies, but the plant is perennial. The stored food in tubers allows resumption of growth the following year when rains begin.

The green pods are ready to harvest 6–8 weeks after sowing (or 2 weeks after pollination). Pods become tough after 3 weeks, and seeds mature after 3 more weeks. The plant may bear pods indefinitely, with declines in production. For crops grown for seed purposes, production technology including cultivation, harvesting, and drying practices is similar to that for other seed legumes. Dry seed yields of winged bean vary from 500 to 2000 kg/ha and of edible tubers from 20 to 100 Q/ha.

REFERENCES

1. A. F. Kelly, *Seed Production of Agricultural Crops*, Longman, New York, 1988.
2. D. K. Salunkhe, S. S. Kadam, and J. K. Chavan, *Postharvest Biotechnology of Food Legumes*, CRC Press, Boca Raton, FL, 1985.
3. M. W. Pedersen, L. C. Jones, and T. H. Rogers, Producing seeds of legumes, *Seeds: The Yearbook of Agriculture*, U.S. Department of Agriculture Washington, DC, 1961, p. 171.
4. S. Chandra, J. A. Duke, H. Pollard, C. F. Reed, L. J. G. Van der Maasen, and D. Zohary, *Cicer arietinum. Handbook of Legumes of World Economic Importance* (J. A. Duke ed.), Plenum Press, New York, 1981, p. 53.
5. G. N. Pathak, Red gram. *Pulse Crops of India*, Indian Council of Agricultural Research, New Delhi, 1970, p. 14.
6. J. Smartt, *Tropical Pulses*: Tropical Agriculture Series, Longman, London, 1976, p. 261.

7. ICRISAT, Report 1994. International Crops Research Institute for Semi-Arid Tropics, Patancheru, India, 1995, p. 9.

8. H.A. Rheeman and M. P. Haware, Mode of inheritance of resistance to ascophyta blight *Ascophyta rabiei* (Pass) Labr. in chickpea (*Cicer arietinum* L.) and its consequences for resistance breeding, *Int. J. Pest Manage. 40* (2): 1666 (1994).

9. K. B. Singh. R. S. Malhotra, M. H. Halila, E. J. Knights, and M. M. Verma, Current status and future strategy in breeding chickpea for resistance to biotic and abiotic stresses, *Expanding the Production and Use of Cool Season Food Legumes* (F. J. Muehlbauer and W. J. Kaiser, eds.), Kluwer, Ac, Dordrecht, Netherlands, 1994, p. 572.

9a. N.K. Biradarpatil, M. Shekhargouda and D.P. Biradar, Analysis of seed and seedling growth parameters in *Cajanus cajan* L., Millsp. *Adv. Plant Sci. Res. (India) 9*: 31(1999).

9b. M.B. McDonald Jr. and D.M. Drake, An evaluation of a rapid and automated electrophoresis system for varietal identification of seeds, *Seed Sci. Technol. 18*: 89(1990).

9c. D.S. Mathur, Role of genetic marker in seed production of chickpea, XI National Seminar on Quality Seed to Enhance Agricultural Productivity, UAS, Dharwar, Jan, 18–20, 2002, *Seed Tech News* 32: (A. Gaur, A.K. Vari, J.L. Varshney, and K.Kant, eds), Indian Society Seed Technology, Division of Seed Science and Technology, IARI, New Delhi, March 2002, p. 2.

10. R. L. Agrawal, *Seed Technology*, Oxford and IBH Publishing, New Delhi, 1980.

11. J. W. Purseglove, *Tropical Crops: Monocotyledons*, 5th ed., Longman, London, 1985.

11a. D.P. Joshi, P.S. Sidhu, S.S. Gill, G. Bassi, N. Batra, S.R. Sharma and Amandeep, Optimum row ratio of parental lines used for hybrid seed production of pigeon pea in Punjab, *Int. Chickpea & Pigeon Pea News Lett. 5*:31(1998).

11b. K. Parameswari and K. Vanangmudi, Influence of biofertilizer inoculation and chemical fertilizer nutrition on seed yield and quality of pigeon pea hybrid COPH-2, XI National Seminar on Quality Seed to Enhance Agricultural Productivity, UAS, Dharwar, Jan, 18–20, 2002, *Seed Tech News* 32: (A. Gaur, A.K. Vari, J.L. Varshney, and K. Kant, eds), Indian Society Seed Technology, Division of Seed Science and Technology, IARI, New Delhi, March 2002.

12. S. H. Hussaini and T. Ramesh Babu, Effect of inert clay, fungicides, plant oils and insect growth regulators on the development of *Callosobruchus chinensis* and viability of pulse seed, *Seed Research*, Special Vol. 1 (S. P. Sharma, ed.), Indian Society of Seed Technology, New Delhi, 1993, p. 306.

13. M. K. H. Siddiqui and P. K. Agrawal, Attapulgite-based clay dust (ABCD) for seed treatment and storage of food grains, *Seed Research*, Special Vol. 1 (S. P. Sharma, ed), Indian Society of Seed Technology, New Delhi, 1993, p. 568.

13a. R.P.S.Tomer and P.Kumari, Hard seed studies in black gram (*Vigna mungo* L.), *Seed Sci. Technol. 19*:51 (1991).

14. J. A. Duke, B. P. Pandya, C. F. Reed, and J. K. P. Weder, *Vigna mungo* (L.) Hepper, *Handbook of Legumes of World Economic Importance* (J. A. Duke, ed.), Plenum Press, New York, 1981, p. 291.

15. J. W. Purseglove, *Tropical Crops: Dicotyledons*, Longman, London, 1984.

16. N. W. Simmonds, ed., *Evolution of Crop Plants*, Longman, London, 1984.

17. *Agricultural and Horticultural Seeds: Their production Control, and Distribution*, Food and Agriculture Organization, Agricultural Studies No. 55, Rome, 1961.

17a. J.E. Smith, L. Korsten and T.A.S. Aveling, Evaluation of seed treatments for reducing *Colletotrichum dematium* on cowpea seed, *Seed Sci. Technol. 27*: (1999).

18. S. L. Saini, G. Kaur, and R. K. Chowdhury, Effect of seed maturation and sowing time on productivity and quality of seed in mung bean, *Seed Research*, Special Vol. 2 (S. P. Sharma, ed.), Indian Society of Seed Technology, New Delhi, 1993, p. 956.

18a. P. Srimathi, K. Malarkodi and V. Palanisamy, Influence of harvesting intervals on the seed yield and quality in cowpea, *Madras Agric. J. 86*: 501(1999).

19. H. Usha, S. Javare Gowda, and H. Ramaiah, Influence of containers, chemical and bioproducts treatment on storability of cowpea (*Vigna unquiculata*) and horse gram (*Dolichos biflorus*) seeds, *Legume Res.* 13(1): 13 (1990).

20. H. S. Gentry, Pisum resources: A preliminary survey, *Plant Genet. Res. News Lett.* 25:3 (1971).

20a. H. Suzuki and A.A. Khan, Effective temperatures and duration for seed humidification in snap bean (*Phaseolus vulgaris* L.), *Seed Sci. Technol.* 28: 381 (2000).

20b. I. Demir, A. Gunay and Y. Ceylan, Seed moisturization as an enhancement treatment for emergence and seedling growth in bean (*Phaseolus vulgaris* L), *Seed Sci. Technol.* 26: 261(1998).

20c. L. Girsch, P.C. Cate and M. Weinhappel, A new method for determining the infestation of field beans (*Vicia faba*) and peas (*Pisum sativum*) with bean beetle (*Bruchus rufimanus*) and pea beetle (*Bruchus pisorum*), respectively, *Seed Sci. Technol.* 27: 377(1999).

21. *Descriptors for Sorghum, Soy Bean*, International Board for Plant Genetic Resources, Food and Agriculture Organization, Rome, 1984.

22. Soya bean (*Glycine max* (L.) Merrill; sunflower *Helianthus annuis* L. and *H. debilis* Nut), *Guidelines for the Conduct of Tests for Distinctness, Homogeneity and Stability*, International Union for the Protection of New Varieties of Plants, Geneva, 1983.

23. D. Zohary. The wild progenitor and the place of origin of the cultivated lentil; *Lens culinaris, Econ. Bot.* 26:326 (1973).

24. J. A. Duke, A. R. Kretchmer, Jr., C. F. Reed, and J. K. D. Weder, *Lablab purpureus* (L.) Sweet, *Handbook of Legumes of World Economic Importance* (J. A. Duke, ed.), Plenum Press, New York, 1981, p. 102.

25. J. E. Specht and G. L. Graef, Breeding methodologies for chickpea; New avenues to greater productivity, *Chickpea in the Nineties*, Proc. Second Int. Workshop on Chickpea Improvement, 4–8 Dec. 1989, ICRISAT, Center, India, International Crops Research Institute for the Semi-Arid Tropics, Patancheru (ICRISAT), India, and International Center for Agricultural Research in the Dry Areas (ICARDA), Aleppo, Syria, 1990, p. 217.

26. J. P. Ranch, The potential for synthetic soybean seed, *Synseeds: Applications of Synthetic Seeds to Crop Improvement* (K. Redenbaugh, ed.), CRC Press, Boca Raton, FL, 1993, p. 329.

27. B. E. Cadwell, R. W. Howell, R. W. Judd, and H. W. Johnson, *Soybeans: Improvement, Production and Uses*, Agronomy Series No. 16, American Society of Agronomy, Madison, WI, 1973, p. 681.

28. T. E. Devine, J. A. Duke, C. F. Reed, and R. J. Summerfield, *Glycine max* (L.) Merr, *Handbook of Legumes of World Economic Importance* (J. A. Duke, ed.), Plenum Press, New York, 1981, p. 83.

29. S. K. Saini, J. N. Singh, and P. C. Gupta, Effect of threshing method of seed quality of soybean. *Bull. Grain Technol.* 18:105 (1950).

30. J. C. Delouche, Seed quality and storage of soybeans, *Soybean: Production, Protection and Utilization*, Proc. Conf. INTSOY Ser. No. 6 (D. K. Whigham, ed.), College of Agriculture, University of Illinois, Urbana, 1975, p. 86.

31. H.W. Byrd and J. C. Delouche, Deterioration of soybean seed in storage, *Proc. Assoc. Off. Seed Anal.* 61:41 (1971).

32. R. A. T. George, *Vegetable Seed Production*, Longman, London, 1985.

33. F. J. Poats, Guar, a summer row crop for the South West. *Econ. Bot.* 14:241 (1960).

34. U. S. Deshpande and S. S. Deshpande, Legumes, *Foods of Plant Origin: Production, Technology, and Human Nutrition* (D. K. Salunkhe and S. S. Deshpande, eds.), Van Nostrand Reinhold, New York, 1991, p. 137.

9
Oilseed Crops

I. INTRODUCTION

Oilseed crops are grown primarily for the extraction of edible oils for human consumption as well as nonedible oils for industrial use. In addition, these crops provide a valuable source of protein, which is present in the residue after the extraction of oil. The specific analytical requirements of both the oil and the residue after extraction make it necessary to preserve carefully the genetic purity of the oilseed. Even a small contamination of the oilseeds with other genetic material may render the oil of inferior quality or contaminate the protein meal with toxic substances. The seed grower should, therefore, be careful to avoid mixture or contamination from undesirable pollen during seed production.

Groundnut (peanut), sunflower, cottonseed, and soybean are the major oilseed crops of the world. Oilseeds have been treated as tropical crops.

Several improvements in oil quality have been made by way of introducing new hybrid varieties, particularly in rapeseed. New rapeseed varieties have significantly low contents of erucic acid glucosinolate. Plant breeders have succeeded in altering the fiber content of the oilseed, which has necessitated a change in the worm assembly of the prepress machine. The change in fiber content and the ever-increasing demand for better quality oil, both for human consumption and industrial purposes, have resulted in major alterations in cooking procedures and worm assemblies of prepress machinery (1).

According to Zwartz and Hautvast (2), breeding of oil-bearing seeds, fruits, and nuts should have two objectives: improvements in the ratio of polyunsaturated fatty acids to saturated fatty acids and in the energy density of diets. Some vegetable oils, like rapeseed, sunflower, soybean, and cottonseed, contain toxic substances, which can be eliminated through a well-designed breeding program. Although breeders have succeeded in removing many of these substances in the past, adequate oilseed yields have been difficult to achieve. Cotton, for example, is more susceptible to pest attack when the toxic substance gossypol is removed by breeding. Groundnut varieties with a low aflatoxin load have been identified and used in a breeding program to minimize this antinutritional factor in the groundnut.

A number of serious oilseed diseases and pests are carried via the seed. Some control measures can be attained by producing seed that is resistant to these pests and diseases. Considerable success has been attained in breeding indehiscent varieties of oilseeds whose seeds do not shatter at maturity. Improved varieties of oilseeds may be developed that have both acceptable yield and oil quality and can be grown by complete mechanization (3).

A. Groundnut and Sunflower

1. Introduction

Botany and Origin. Groundnut or peanut (*Arachis hypogaea* L.) belongs to the division Papilionaceae of the family Leguminosae. Groundnuts probably originated in the region of eastern South America where a large number of species are found growing wild; for example, A. *monticola* (K.&R.). On the basis of genome donation, *A. hypogaea* is thought to have originated from a hybrid between *A. cardenasii* and *A. batizocoi* (K.&G.), with both parents occurring in reasonable proximity in Bolivia (5).

Groundnuts were cultivated in precolumbian native societies of Peru between 2000 and 3000 BC, well to the northwest, from where it is believed to have spread to other parts of the world. During the time of Columbus, groundnuts were widely distributed in South and Central America and the Caribbean region. They were probably brought to West Africa from Brazil in the sixteenth century and from there to the east coast of Africa and to India (6).

The taproot system of groundnut enables the plant to explore a large volume of soil up to a depth of 2 m for moisture and nutrients, with abundant nodules on main and lateral roots. The leaves are alternate and pinnate, with three to four leaflets carried on long petioles. Leaves are obovate and softly hairy and about 3–5 cm long. Yellow flowers carried on axillary branches are small; about 12 mm long. Groundnut is a self-pollinating crop, although out-crossing to the extent of about 2% takes place in the areas of high bee activity (7). About 7–10 days after fertilization, the receptacle thicknens, elongates, and forces the ovary downward into the ground. The long carpophore bearing the fertilized ovule is known as the "peg" and the action of burying of the immature pod is called "pegging." Mature pods (8.0 × 1.0–2.5 cm) are fibrous, containing two to four spherical or ovoid kernels. The kernels are without endosperm and consist of 40–45% oil and 25–30% protein. The oil content varies significantly with variety, seed size, season, and cultural practices.

Sunflower (*Helianthus annuus* L.) belongs to a large and successfully cultivated family of flowering plants, Compositae. It probably originated in the southwest United States–Mexico region, where it grows profusely in its natural state. As a cultivated plant, sunflower was introduced to Europe during the sixteenth century, and from there it was reintroduced to North America in the late nineteenth century. It almost certainly reached Europe from Mexico via Spain. Sunflower became very popular as an ornamental plant and was soon established as an oilseed crop in eastern Europe. It was imported into Russia from Holland in the eighteenth century, with the first commercial production occurring between 1830 and 1840. At one time, the former Soviet Union had more than 6 million hectares planted with this crop. The discovery of cytoplasmic male sterility and fertility restoration made possible the efficient production of high–oil-content hybrid sunflowers, making it one of the most important oilseed crops in the world.

The cultivated sunflower is a tall, erect, unbranched, coarse annual with a distinctly large, golden head with oil-bearing seeds. The crop has five distinctly recognizable stages: first true leaves, bud emergence, first anthesis, late anthesis, and physiological maturity. The sunflower is characterized by a shallow but substantial root system, with its taproot growing up to 3 m long. The main stem and branches bear terminally the disc-shaped head, capitulum, about 10–30 cm in diameter, depending upon cultivar, season, soil fertility, and other factors. The number of sound seeds per head and per unit area are more important in determining seed yield than is the diameter of the head.

Flowering in most hybrid sunflowers is remarkably uniform, with 80–90% of the heads opening within 3–4 days of the first to do so. Short days favor floral initiation. Flowers are cross-fertilized, pollination being influenced by insects walking on their surface. Honeybees are the main and almost exclusive pollinators. Inadequate pollination can be a major cause of poor seed set in the commercially grown crop.

The sunflower "fruit" or seed varies in color from black to white, but brown, stripped, or mottled seed can also occur. There seems to be a relationship between a dark hull and high average seed oil; however, an increasing number of hybrids have light-colored seeds with high oil content.

Species and Cultivars. The cultivated groundnut, *A. hypogaea*, is an allotetraploid (2n = 40) with the basic chromosome number of the genus *Arachis* being 10 and most wild species being diploids (2n = 20). Gibbons et al. (8) and Hammons (9) classified groundnut species as follows:

Arachis hypogaea		No floral axis on main axis
subsp. *hypogaea*		Alternating pairs of vegetative and floral axes along lateral branches
var. *hypogaea*	Virginia	Less hairy, short branches
var. *hirsuta*	Peruvian runner	More hairy, long branches
subsp. *fastigiata*		Floral axes on the main axis, continuous runs of multifloral axes along lateral branches
var. *fastigiata*	Valencia	Less branched
var. *vulgaris*	Spanish	More branched

Groundnut varieties destined for edible and industrial markets have markedly different characteristics. For edible purposes, a relatively low–oil-content nut with high protein and sugar combined in a bold and clean seed is favored. Industrial peanuts should have a high oil content, high shelling percentage, long dormancy, and good storability. Early maturity, flower fertility, seed dormancy, yield, and resistance to pests and diseases are other important considerations in breeding groundnut varieties.

Seshadri (10) classified groundnut cultivars grown in India on the basis of duration and habit of growth and characteristics of stem, leaf, flowers, pods, and kernels. The four main varieties under cultivation in India are Coromandel, Bold, Peanuts, and Red Netal, marketed under different names. The first two varieties are spreading or creeping types, whereas the latter two belong to the bunch or erect type (3). Some promising varieties of groundnut in respect of their yield and high oil content (about 50–55%) cultivated in India are M-13, Jyoti, TG-1, DH-3-30, Chandra Kadiri-3, M.H.-2, TMV-10, Phule Pragati (JL-24), Co-I, M-37, BP-1, and Bp.2.

Of the 67 known species of sunflower, about 17 species are cultivated; mainly for ornamental purposes. The varieties commercially grown for seed are considered to be *H. annuus* var. *macrocarpus* (DC) CK II, with *H. annuus* subsp. *lenticularis* being the nearest wild relative. *H. argophyllus* (native to Texas) is considered to be the most closely related species of *H. annuus*.

The cultivated sunflower is a tall, erect, unbranched, coarse annual with a distinctly large, golden head with oil-bearing seeds. Three main subspecies of the common sunflower, *H. annuus* L., are considered to be valid: *H. annuus*, subsp. *lenticularis* Dougl., *H. annuus* subsp. *annuus*, and *H. annuus* var. *macrocarpus* Dogl. In addition to these, Russian scientists recognize *H. petiolaris* and divide *H. annuus* subsp. *annuus* further into *annuus*, *pustovojtii*, *armeniaca*, and *australis*. Disease-resistant genes were introduced from *H. tuberosus* (Jerusalem artichoke) in an intergeneric cross in the Russian. discovery of cytoplasmic male sterility in crosses between *H. petiolari* and *H. annuus* (11), combined with fertility restorer genes (12), allowed breeders to produce hybrids and cultivars of sunflower with short stems suitable for mechanical production. Male and female sterility in sunflowers can be induced chemically (13,14). Inbred hybrids have shown the biggest increase in yield and oil content. Synthetic varieties (combination of inbred lines) have also produced high seed and oil yields. The cultivars suitable for mechanical harvesting are Peredovik, Mennonite, and Sunrise, which have been introduced to Europe and the United States from Russia. Improved cultivars of sunflower grown in India are Co-I, EC 68414 (Tamil Nadu), EC 68415 and Surya (Karnataka, Andhra, Maharashtra and U.P.), PKV SUF 72–37 (Maharashtra), BSH-1 (Karnataka), and Morden (India). The early varieties grown in Britain are South Cross, Pole Star, Mars, Sunrise, Jupiter, and Saturn (3,4).

2. Breeding

Diploid members of *Arachis* can be crossed with *A. hypogaea*, although rapid pathways for introgression are lacking. Groundnut may also be vegetatively propagated. Since artificial crossing is both difficult and tedious, the procedure can save much time. This method allows several plants to be grown from each seed (6).

Increased yield and oil content have been the main objects of breeding. The work has concentrated on bunch types grown commercially on a large scale. For edible purposes, a relatively low oil content nut with high protein and sugar combined in a bold, clean seed is preferred. These factors appear to be related to lower seed yields. Industrially used nuts should have a high oil content, high shelling percentage, long dormancy, and good storability. Other important factors are early maturity, flower fertility, and seed dormancy.

The introduction of pest and disease resistance to bunch types has had limited success. The selection for disease and insect resistance should concentrate on diploid-level hybrids rather than individual species. Their progeny could then be directly crossed with *A. hypogaea* (15). Where resistance has been obtained, it is often related to undesirable characteristics such as low yield or longer maturity. The runner types offer a much greater opportunity for improvement, and simple selection methods can achieve rapid results (16). Breeding and selection programs require seed of known percentage and quality, the supply of which is often restricted. Field multiplication techniques should ensure that no seed is wasted and that harvested seed is of good quality. Groundnut crops have considerable breeding potential (6) owing to their wide genetic variability and the unexploited wild species available.

International Crop Research Institute for Semi-Arid Tropics, Hyderabad, India (ICRISAT) scientists are attempting to resolve one of the worst problems of groundnut; namely, early leaf spot disease. Very few germplasm lines are resistant to this disease. Some wild species do have resistance but are incompatible with cultivated groundnut (*A. hypogaea*). Two wild species from South America, *A. appressipila* and *A. paraguanensis*, have been used to develop hybrids using embryo-rescue and tissue culture techniques at the ICRISAT Asia Center (IAC). Some of these pollen-sterile hybrids which carry resistance genes are being used to develop adapted cultivars. A number of partially rust (*Puccinia arachis*)–resistant groundnut genotypes have also been developed wherein the disease develops much more slowly and the crop is harvested before rust damage becomes too severe. Rust resistance is measured in terms of five components: infection frequency, incubation period, lesion diameter, percentage of leaf area damaged, and sporulation index. These components have been studied in 143 rust-resistant genotypes, and it was found that all five components were significantly correlated with each other and with mean field rust scores. This allows breeders to select for any one component while breeding for resistance. Seventeen of the screened genotypes had low sporulation indices and long inoculation periods, indicating that they could be used for resistance breeding. An *Arachis* wild species, *A. hoehnoi*, has been identified as having a high level of resistance to aphids. Only immature pods were obtained when *A. hypogaea* was crossed with *A. hoehnoi* (pollination to pod efficiency of 18%). *A. hoehnoi* has been found to be resistant to leaf minor, and both mature and immature pods were obtained when it was crossed with *A. hypogaea*, with a pollination to pod formation efficiency of 21% (mature pods) and 9% (immature pods). Breeding work for abiotic stresses indicated that groundnut cultivars ICGV 86699 and ICGV 86743 are drought resistant (17). The ICRISAT gene bank's groundnut collection displays a wide range of variation in seed color (Fig. 1).

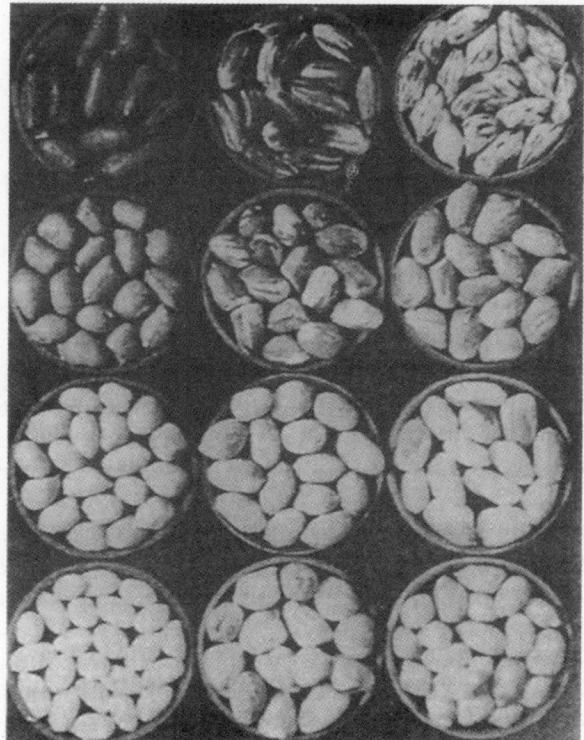

Figure 1 The ICRISAT gene bank's collection of groundnut (*Arachis hypogaea*). (From Ref. 17.)

Sunflower breeding aims mainly to increase seed oil content, reduce height, and introduce pest and disease resistance. The first objectives have proved to be easier to achieve than the last. Strong positive correlations exist between seed yield and plant height, head diameter, and seed weight, whereas oil content is negatively correlated to hull thickness. Breeding (or selection) of varieties whose seed has a space between the hull and kernel without loss of oil content would assist seed processing. Also, cultivars suitable for saline soils would be of great value.

The finding of cytoplasmic male sterility in crosses between *H. petiolaris* and *H. annuus*, combined with fertility restorer genes, has enabled breeders to produce hybrids with short stems suitable for mechanical production. Some success in introducing resistance against major pathogens of sunflower has also been achieved, especially mildew and rust. Hybrids of inbred lines have given the biggest increase in yield and oil content. Synthetic varieties, a combination of inbred lines, are also available (6). Hybrid sunflowers have an advantage over the open-pollinated varieties—the former produce very high seed and oil yields—flower over a short period, and have short, sturdy stems with a single head. Many hybrids are also drought resistant or resistant to some insect pests and diseases. Hybrids are also self-compatible (self-pollinating) and thus do not wholly depend on insects (bees) for pollination, facilitating chemical control of insect pests and diseases. Weiss (6) has described a method of producing hybrid sunflower seed.

Changes in seed vigor attributes and protein content in sunflower (*H. annuus*) hybrid APSH11 were compared with its parental lines (7-1A [female] and RHA271 [male] under accelerated aging conditions of 80% relative humidity and 30°C. The hybrid was superior to its parents, or on par with the better storer (7-1A) parent with respect to germination, membrane leakiness, dehydrogenase (oxidoreductase) activity, and protein content. The seed protein profiles were significantly altered after 20 days of aging in all genotypes, with the most prominent changes being in RHA271 (17a).

3. Cultivation

Culture.

ISOLATION REQUIREMENTS AND LAND PREPARATION. As for all seed crops, the fields selected should be such that groundnut and sunflower crops were not raised during the previous one or two seasons unless the crop variety was the same and was grown from certified seed. In areas where bacterial wilt is a serious problem, groundnut seed crops should not be rotated either with groundnuts or with solanaceous crops like tomato, potato, and brinjal. Groundnut prefers well-drained sandy loam soils with high organic matter. The sunflower field selected should also be well drained, having deep, fertile soils with neutral properties.

Groundnut is a completely self-fertilized crop, with little natural crossing. Cross-pollination does not take place, because the stigma remains closed in the keel even in fully opened flowers. An isolation distance of about 3 m from other fields is, therefore, sufficient for pure seed production.

Sunflower is both a self- and cross-pollinated crop, the extent of crossing varying from 17 to 62% depending upon insect activity (18). Fields of foundation (basic) and certified seeds of sunflower, therefore, need to be isolated by at least 1000 and 500 m from other sunflower fields and wild sunflowers, respectively.

Land is prepared by one deep plowing and two to three harrowings followed by leveling to bring it to the desired tilth for planting.

PLANTING AND CULTIVATION. Groundnut crops in India are sown from mid-June to the first week of July, using nucleus/breeders/foundation seed obtained from a source approved by a seed-certification agency. The seed should be treated with mercurial fungicides before planting. Groundnut is sown in lines either behind the plow in 5- to 8-m deep furrows or by seed planter 5–8 cm deep depending upon soil type and moisture conditions. The row to row spacing for errect, bunchy types should be about 30 cm, whereas for spreading types, it can vary from 45 to 60 cm, with 10–15 cm between plants for both types. Seed rates of groundnut for bunchy and spreading types range from 80 to 100 and 60 to 80 kg/ha depending upon seed size.

The sunflower seed crop, unlike most other crops, is not season dependent. Barring extreme freezing, sowing can be adjusted according to the availability of land for planting, avoiding the coincidences of maturity periods of the crop with rains, which adversely affect seed quality. A seed obtained from an approved source is sown in rows 2–4 cm deep with a spacing of 80×20 cm, using 8–10 kg seed/ha.

Gopal Singh and Sudhakar (19) reported the response of sunflower hybrids to hydration-dehydration under field conditions. Fresh accelerated aged seeds of sunflower hybrids (MSFH- 8 and APSH-II) were hydrated by soaking in water and a solution of p-hydroxybenzoic acid (10^{-5} M) and sodium chloride (10^{-3} M) for 2 hr and dried again to their original 8% moisture content just before sowing. This treatment produced better field performance of both the hybrids as measured in terms of germination, emergence index, plant stand, leaf area index, and seed yield (Table 1). The response of the deteriorated (accelerated aged) seed to the water-soaking

Table 1 Response of Sunflower Hybrids to Hydration-Dehydration Under Field Conditions

Treatment	Germination (%)	Emergence index	Plant stand (per ha)	Leaf area index 30 days	Leaf area index 60 days	Seed yield (Q/ha)
Fresh seed						
Control	92	20.43	50,929	0.596	1.89	25.63
Water	95	21.75	52,643	0.637	1.94	27.44
Sodium chloride	85	17.69	45,500	0.426	1.68	21.53
p-Hydroxybenzoic acid	87	19.30	48,413	0.457	1.78	23.51
Naturally aged seed						
Control	70	16.21	42,325	0.489	1.68	18.28
Water	78	21.39	50,659	0.540	1.74	25.35
Sodium chloride	75	16.50	47,484	0.503	1.69	21.29
p-Hydroxybenzoic acid	78	19.32	49,738	0.533	1.73	22.35
Accelerated aged seed						
Control	50	2.30	8,460	0.314	1.21	1.70
Water	59	11.09	28,698	0.390	1.32	12.07
Sodium chloride	51	4.64	16,270	0.409	1.55	5.19
p-Hydroxybenzoic acid	53	6.33	21,489	0.433	1.56	7.30
CD at $P = .05$	4.83	3.31	9,087	0.128	0.267	5.51

CD = critical difference.
Source: Ref. 19.

chemicals appeared to be much better than that of the fresh seed (19). Rao and Singh (20) also reported that hydration-dehydration treatments with 2×10^{-5} M tannic acid and 10^{-5} M benzoic acid or 10^{-3} M NaCl increased the germination percentage and seedling vigor index in sunflower, safflower, black gram (*Vigna mungo*), and green gram (*V. radiata*) seeds compared with water-soaked and dry seeds. In another experiment, seeds of two sunflower cultivars (Morden and MSFH- 8) were soaked in distilled water or treated with 500 ppm KNO_3 or $CoNO_3$, 100 ppm $CaCl_2$, or 200 ppm ascorbic acid. All of the seed treatments were found to increase the germination percentage and seedling vigor compared with the untreated control. Seed treatment with KNO_3 produced the highest germination and the most vigorous seedlings (21).

Fertilization, Irrigation, and Weed Control.   Groundnut crops require about 20 kg N, 50–80 kg P, and 30–40 kg K per hectare depending on soil test values. Soils not rich in organic matter will need a basal dose of about 5–10 tons of farmyard manure or compost. Adequate soil moisture must be ensured at the time of flowering, seed development, and maturation, applying one or two irrigations if needed. For sunflower seed crops, presowing irrigation is necessary if the crop is sown in spring or summer, and is desirable for rabi-sowing for uniform germination and better stand. Sunflowers are more drought tolerant than other oilseed crops. Depending upon rainfall, two to four irrigations may be required to maintain adequate soil moisture during the flowering and grain-filling stages.

Seed crops of groundnut should be kept free of weeds; weeding can be done when the crop is 2–3 weeks old, at flowering, and at the time of peg formation. Weeds can be controlled by applying herbicides like prometrin or lasso at 1–2 kg a.i. dissolved in 500–600 L of water per hectare immediately after sowing.

Sunflower crops need one or two weedings during the first 6 weeks after germination. Earthing-up once before and, if needed, after irrigation, around 45–50 days after sowing, will prevent lodging of sunflowers owing to their large and heavy heads.

Control of Pests and Diseases.   The hairy caterpillars of groundnut can be controlled with a spray of thiodan, 35 EC, at 0.5% solution in 800 L of water per hectare. Infestation by aphids should be checked by spraying metasystox, 25 EC, 1 L dissolved in 100 L of water per hectare, and that of grubs can be controlled by applying carbofuran, 10% granules at 12 kg/ha at the time of sowing or dusting with 1% lindane at 40–60 kg/ha.

Spraying the crop with 0.25% solution of dithane M-45 as soon as the symptoms of the disease are noticed may control tikka disease of groundnut; three to four sprayings at an interval of 15 days may be required to control the disease.

Although no serious pests infest sunflower crops, one should watch for attack by cutworms during the seedling stage and for head-borer damage and jassid attack at the bloom stage. These insect pests can be controlled by applying 5% heptachlor dust to the soil at 15 kg/ha and one or two sprayings of 0.025% metasystox (25 BC), respectively.

Birds may cause extensive damage to a sole maturing field of sunflower when no other seasonal crop is present in the green stage. Seed crops, therefore, need protection from birds.

Roguing.   The off-types of groundnut can easily be distinguished on the basis of plant size and color of leaflets and flowers. Diseased plants affected by rosette, mosaic, and root rot should be removed and destroyed from time to time.

Sunflower crops require a minimum of two roguings: one at preflowering and a second at crop maturity. Before flowering, off-types such as tall, very early or very late, branched as well as weak, and wild and diseased plants should be rogued out. At about 75% maturity, wild, ornamental, diseased (wilt, blight, or charcoal rot), or otherwise damaged plants as well as those not conforming to the botanical characteristics of the variety under seed production should be removed by scrupulous examination.

4. Harvesting and Threshing

Since groundnuts develop underground, it is difficult to judge crop maturity correctly. Soil moisture conditions are important in reducing groundnut losses during harvesting. Flowering is a progressive phenomenon in groundnut, taking place over 2–3 months, requiring about 2 months to attain full development. Thus, it is not unusual to find nuts in different stages of maturity as harvest time approaches (10). To obtain a maximum yield of best-quality produce, the proper time for harvesting aims at the recovery of a high percentage of well-formed and good-quality nuts. The crop is preferably harvested during periods of bright sunshine for satisfactory drying and handling operations. Improper soil moisture conditions and delayed harvests can cause heavy losses of groundnut due to sprouting in the field (3,4). Weete et al. (22) used the Arginine Maturity Index (AMI) to assess the maturity of groundnut pods. The use of the AMI generally gave higher yields than other methods used to determine harvest dates. The most accurately predicted dates of groundnut maturity were obtained with samples taken 3–4 weeks before harvest. The groundnut crop is ready to harvest when leaves turn yellow and begin to fall, pods become reticulated, and seed gets separated from the pod shell. Plants are uprooted by hand or dug out and left in the field 2–3 days for drying. Long sun-drying may result in splitting of seed cotyledons. Pods are either picked by hand or threshed using suitable machines. Picking should be done when pods separate readily from stalks and the seed rattles in the pods.

Lifting of groundnut pods from the soil and separation can be carried out mechanically on a large scale. A tractor-mounted peanut shaker is used to lift, shape, and windrow two peanut rows loosened by tractor blades. Equipment for picking peanuts from the windrow or an attached picker is also used. White and Roy (23) described a once-over harvester for groundnut, which gave 50% higher total harvested yield than conventional digging or combining methods.

Sunflower seed crops can be harvested when the top leaves dry and flowers become shriveled. Tall and short varieties of sunflower are normally harvested approximately 120–160 and 90–120 days after planting, respectively. Like most other crops, sunflower heads ripen before the leaves start to wither. The central florets loosen, followed by weakening of the neck and bending over of the head, which simultaneously changes in color from green to yellow. Leaves then start to fade from the bottom up, indicating the time of harvest. The heads may be allowed to dry out in the field before cutting. Under favorable conditions, the heads dry out to a very low seed moisture content on the standing plant, and neither "stocking" nor "fencing" may be required. Seed is ready to harvest when the back of the flower heads turn yellow or brown, dropping the seed moisture content to 10–12%.

Heads can be threshed directly by combine or thrown into a harvester. The use of chemical defoliants like diquat, magnesium chlorate, or dipyridil phosphate accelerates drying of the standing crop and assists combine operations (24).

5. Postharvest Operations

Cleaning and Drying.   The threshed pods are cleaned and further dried for 3–4 days to reduce the seed moisture contents to about 10%. Overdried pods become brittle and are easily prone to damage, so that care must be exercised not to dry too fast or to a moisture content much below 10%. The drying air temperature should be about 35°C and should not exceed 38°C. Cultivars may differ in their drying characters and requirements. It is, therefore, necessary to check carefully during the drying process to ensure that no damage occurs. The shelled seed can be stored provided it is dry and has not suffered any physical damage. Threshed sunflower seed should be cleaned as soon as possible using air screen cleaners and dried to a less than 10% moisture content to protect seed and oil quality.

Kachru and Sahay (25) modified a pedal-cum-power-operated air screen cleaner used for separating cereals and pulses for the efficient cleaning and grading of sunflower, safflower, linseed, and mustard seeds. The machine and operating parameters for optimum separation were identified based on the determination of some physical properties of oilseeds. The adapted cleaner has been reported to be highly suitable for seed processing.

In hot tropical countries, sacks are allowed to stand with the necks open, preferably under shade, to dry the seed quickly. In humid and temperate regions, artificial drying may be necessary. Cold air is quite effective in most cases; however, if heated air is used, its temperature should not exceed 50°C as a general rule (26).

Storage and Seed Treatment.   Groundnuts are normally stored in the form of unshelled nuts. Groundnuts used for the purpose of seed must be stored 7–8 months. Groundnuts used for sowing in India are generally stored in earthen pots, mud bins, bamboo baskets, or other types of wicker-work receptacles, usually plastered with mud and cow dung. Smaller quantity of groundnuts may also be stored in gunny sacks; however, these are prone to damage by dampness, rodents, and other storage pests.

Sunflower seeds are best stored at a moisture content not exceeding 9.5%. Freshly threshed seed crops may be heated rapidly during storage owing to enzyme and bacterial action at a higher moisture content of seeds (3,4).

Inoculation with rhizobium is usually not required, although groundnut is a leguminous crop. However, if groundnut seed is inoculated with rhizobium, it should not be treated chemically. The treatment of seed with organomercurials or thiram provides protection against fungal diseases during seed germination.

Sunflower seeds may carry various fungi, which may cause damage during storage as well as after sowing. Kelly (28) advised dressing of sunflower seeds with captan or similar fungicides a few months before the seed is utilized. The seed also needs protection against certain insect pests during storage.

B. Rapeseed and Mustard

1. Introduction

Botany and Origin.   Commercial rapeseed and mustard are obtained from the genus *Brassica*, belonging to the family Cruciferae. The name rape is derived from the Latin *rapum*, meaning turnip. Brassicas are widespread and constitute the main oilseed crop in the temperate climates of Europe and North America and parts of the nonhumid regions of Asia and Africa. The two most widely grown oilseed rapes, namely, *Brassica napus* var. *oleifera* and *B. rapa* (which includes *B. campestris*), are botanically separate species but are treated alike for most practical farming and cultivation purposes (27). In parts of the Indian subcontinent, *B. juncea* is grown for oil. In more temperate areas, it is grown as a mustard, and together with *Sinapsis alba*, it is grown for seed in the same manner as oilseed rapes.

Oilseed rapes have a lengthy taproot system, which makes the crop relatively drought resistant (6). Leaves are dark green, glaucous, lyrate, pinnatifid, and stalked on the lower sessile, clasping the stem to some extent. The inflorescence is an elongated raceme, borne terminally on the main stem as well as on the branches, carrying bright flowers varying in color from orange to pale yellow. According to Free and Williams (28), about 68% of flowers become pods, with vernalization (low temperature in the early growth stage) being a major factor influencing flower bud development and seed yield. The flowers of *B. rapa* (*B. compestris*) are cross- and self-pollinated (with the exception of some Indian forms), whereas those of *B. napus* are basically self-fertile, so that the amount of cross-pollination is greater in the former than in the latter. The fruit is a narrow pod, siliqua, 5–10 cm long, and consists of two carpels separated by a false septum, which shatters after maturity and is a varietal characteristic not desirable in an oilseed strain. Both *B. campestris* and *B. napus* produce dark brown to black seeds, with 1000 grain weight varying from 4 to 6 g depending upon variety and season.

Species and Cultivars. The genus *Brassica* contains about 160 annual and biannual species. In India, three distinct types of *B. campestris* are known: brown sarson, yellow sarson, and loria. Another species, *B. juncea*, called rai, is also known as a minor oilseed crop. A number of related species are cultivated for their seeds grouped under the general name mustard, including *B. juncea* (brown mustard), *B. nigra* (black mustard), *B. carinata* (Ethiopian mustard), and *Sinapsis alba* (white mustard).

A wide range of rapeseed cultivars varying in degree of drought resistance is available, with *B. napus* being more drought tolerant than *B. campestris*. The origin and seed chemical characteristics of a wide range of rapeseed cultivars have been described (3,29). The Indian Council of Agricultural Research, New Delhi, recommended several improved varieties of loria, brown sarson, and mustard (30).

2. Breeding

Hybrid cultivars of rapeseed and mustard have not been successful, but some synthetic cultivars have shown promising results. The basic concept is to mix a number of selected lines in given proportions and to multiply this mixture for a restricted number of generations. Except for the limitation on the number of generations, the growing of seed is similar to that for conventional rapeseed cultivars (26).

Of the four diploids (*B. nigra*, *B. oleracea*, *B. campestris*, and *B. tournefortii*) and three allopolyploids (*B. napus*, *B. juncea*, and *B. carinata*), *B. napus* crosses with *B. campestris* but is known to cross with *B. oleracea* only with extreme difficulty. Low erucic varieties are hybrids produced from high- and low-acid types; the F_1 generation often contains less than 0.5% erucic acid in its oil. These low–erucic acid, "Lear" varieties have yields comparable with high-yielding "Hear" varieties. An increase in the linoleic acid level with a decrease in linolenic acid (which causes rancidity during storage) is also desirable.

Low yield, late maturity, shattering, and lack of pest and disease resistance are the main problems in Indian rapeseeds. Pure line selection has helped to overcome these problems to some extent. Nuclear magnetic resonance spectroscopy to determine seed oil content without impairing its viability has aided the breeding program greatly in addition to the degree of out-crossing that exists or can be expected. Wide variation exists in the percentage of cross-fertilization for individual plants of *B. napus* (10–100%). Thus, the isolation distance depends on the strains or lines used in a breeding program. Utilization of cytoplasmic male sterility (CMS) in *B. napus* is being investigated. It has been found to be easier to raise the oil content of seed than to greatly improve seed yields. Rapeseed is tolerant to both salinity and moisture stress. Efforts are being made to enhance this property in newer varieties. The feeding value of meal residue has been raised considerably by discovering varieties of rapeseeds with minimum glucosinolate contents. However, there has been little success in introducing pest resistance in rapeseed, although resistance to some fungal diseases has been obtained.

Tay et al. (31) reported that germination of encapsulated rape (cv. Primor) embryoids was 100% when alginate (polyanion) formed the inner matrix and chitosan (polycation) formed the outer layer. However, when the matrix make-up was reversed, embryoids did not germinate. The artificial seeds produced by this complex coacervation were hardened in dilute alkaline solutions of NaOH and $Ca(OH)_2$.

The research carried out regarding transferring disease-resistance genes from wild allied species into cultivated *Brassica* species, extending the range of genetic variability for economic traits among allopolyploid cultivated *Brassica* species and producing alloplasmic male sterile lines for hybrid seed production, has been summarized (32). Wide hybridization was achieved by protoplast fusion, and hybrids were characterized by morphological, cytological, and molecular analyses.

3. Cultivation

Culture.

LAND PREPARATION AND ISOLATION REQUIREMENTS.   Fields for the production of rape and mustard seed should not have been used for species of *Brassica* in the previous season unless the variety of the crop grown was the same and the seed used was certified. Brassica seeds are very long-lived in the soil, and great care is needed to ensure that volunteer plants do not occur in a seed crop. Such plants may be a source of mixture in the harvested seed and may cause further damage by cross-pollinating some of the crop plants. Kelly (26) recommends a gap of 5 years for earlier generations of multiplication and at least 3 years for the final generation. The commercial seed crops of rape and mustard seed grown for oil should also be excluded from the field for the required interval to avoid seed shedding. Indeed, it is advisable to exclude all *Brassica* and closely related species grown for forage and vegetables to avoid risk of building diseases.

B. napus is usually self-fertilized, but honeybees have been known to increase seed yields. *B. juncea* is also largely self-pollinated. On the other hand, *B. rapa* and *S. alba* are cross-pollinated to a large extent. Good isolation must be provided to prevent contamination between rapeseed cultivars with different characteristics. For the multiplication of seed, isolation distances of 400 m (for foundation seed) and 200 m (for certified seed) have been recommended (18,26). It is beneficial to place honeybee hives in a seed crop at the time of flowering.

Land is prepared by plowing and three to four harrowings followed by leveling to bring the seed plot to a desired tilth.

PLANTING AND CULTIVATION. In India toria is generally sown from the second to fourth weeks of September sarson from mid-October to mid-November, and rai from the end of September to the first week of October, using nucleus/breeder's/foundation seed from a source approved by a seed-certification agency. The seed crop is sown in lines at a depth of 1.25–1.50 cm with a seed drill. The spacing between rows can vary from 30 to 60 cm and between plants from 4 to 5 cm. The seed rate ranges from 5 to 8 kg/ha.

Fertilization, Irrigation, and Weed Control. Rapeseed and mustard require about 40–90 kg N/ha, depending upon variety and region, and 20–25 kg P/ha. The crop is known to respond well to additions of organic manures, which should be well mixed at the time of land preparation. If only fertilizers are used, the higher doses of nitrogen may be split into two, with the first applied at the time of planting as a basal dose and the remainder should be given at the time of first irrigation.

One presowing irrigation and one irrigation at the time of flowering are recommended for obtaining higher seed yields.

Charlock (*S. arvensis*) and wild radish *Raphanus raphanistrum*) are the two most difficult weeds of the rapeseed crop, because the weeds have seeds very similar in size to that of *Brassica* crops. *B. nigra* is also a weed in some areas, which could be difficult be eliminate from a seed crop. These weeds should be removed before their preemergence stage. Herbicides may be used both at preemergence and postemergence stages but require careful handling and expertise. It is advisable to select clean land for a seed crop. Another weed, cleavers (*Galium aparine*), in a *B. napus* crop can be controlled postemergence with carbetamide.

Roguing. All off-type plants, easily distinguishable on the basis of plant characteristics, should be removed before flowering to ensure the genetic purity of the seed crop. Remaining off-types, if any, identified on the basis of siliqua characteristics of the rapeseed fruits should be rogued out before maturity. Satyanashi (*Argemone mexicana*) is a highly objectionable weed in rapeseed and mustard seed crops, which should be removed altogether as and when observed.

4. Harvesting and Threshing

Rapeseed and mustard seed crops should be harvested as soon as plants begin to turn light yellow. At this stage, most of the siliqua become light yellow and seeds turn light brown. Rapeseed sheds easily, and it is important to harvest the crop at the right time—a period often no longer than a week.

The crop is harvested either by windrowing and picking up later (by combine or for threshing), by desiccating and direct combining later, or by direct combining. The first method is used for hand harvesting, wherein the crop is cut into bundles and allowed to dry. When the crop is ready for windrowing, the seed is firm but not hard and can be marked with a thumbnail. The seed moisture content at this stage varies from 20 to 35%.

The crop can be threshed when the seed moisture has decreased to 14%. Desiccation (with diquat) can take place at a somewhat later stage than windrowing, but this technique is not recommended for the seed crop owing to risks to germination of the harvested seed. Direct combining should start when the plants have changed color and seeds are dark and at about 14% or less moisture content. According to Kelly (26), special harvesting equipment is available to minimize seed losses during harvesting. Rapeseed flows very easily and is small; combines should be sealed to prevent seed from escaping through gaps that would not admit larger seeds.

5. Postharvest Operations

Drying and Cleaning.   The harvested rapeseed is very prone to damage during storage if it has a high moisture content. The high oil content of the seed means that it can deteriorate very rapidly during storage if not dry enough. The moisture content of rapeseed should, therefore, be less than that for a cereal seed—usually less than 9% for medium-term storage, less than 7% for longer periods, and less than 5% for long-term storage.

Small seed packs more tightly than larger seed and resists airflow. The depth of the seed spread for on-floor drying should not be more than 1 m. A depth of less than 1 m may also cause pockets of undried seed to form between air inlets (6). Ward et al. (33) advocated a maximum air temperature of 5°C above ambient for on-floor drying and 7°C above ambient in ventilated bins, which should not be overfilled.

If continuous-flow cereal driers are used for rapeseed, care must be taken to prevent small seed from escaping through gaps. The depth of the seed on flatbed driers should be adjusted so that airflow is not restricted and bubbling is prevented. Ward et al. (33) recommended maximum safe seed temperatures in continuous-flow driers as follows:

Moisture content of seed (%)	Maximum safe temperature of seed (°C)
10–17	66
19	60
21	54
23	49
25	43
27	38
29	32

Rapeseed and mustard seed can be effectively cleaned using an airscreen cleaner.

Seed Treatment and Storage.   Phoma, or blockleg, caused by *Leptosphaeria maculans* is a widespread seedborne disease of rapeseed. Resistant cultivars now available provide an effective way to combat this disease. Seed treatment with iprodine gives some protection. *Alternaria* leaf spot can also be controlled by iprodine or by organomercurials. The insecticide lindane is also used in conjunction with the fungicide captan to prevent the development of these pests (26).

Singh et al. (34) investigated the effects of storability of mustard seed under three conditions of controlled environment. Mustard seed lots dried under controlled conditions to 3.18 and a 5.25% moisture content with an initial viability of 100% were sealed in laminated alluminum foil packets and stored for 36 months at 25, 10, and −10°C. The mean germination period was maximum when seeds were stored at 25°C and minimum at −10°C. The number of abnormal seedlings after 26 months of storage differed significantly at the three temperatures. The seedling vigor as measured by germination rate and seedling root and shoot lengths declined prior to the decline in seed viability. Reducing the seed moisture from 5.25 to 3.18% had no significant effect on seed viability and seedling vigor. The seed viability dropped to 50% at 33 months when seeds were stored at 25°C, whereas at 10 and −10°C, the decline was only 1–2% (Table 2).

Seed Yields. Rapeseed yields under good crop management range from 15 to 21 Q/ha.

C. Safflower and Sesame

1. Introduction

Botany and Origin. Safflower (*Carthamus tinctorius* L.) belongs to the family Aste-raceae of the broad group. Compositae. It is a warm-temperature crop cultivated widely in Asia, Africa, Russia, and China. The cultivated safflower probably origi-nated in the region between the eastern Mediterranean and the Persian Gulf (35). About 25 valid species of the genus *Carthamus* are believed to be distributed from Spain and North Africa through west Asia and India.

Safflower is a highly branched, herbaceous, thistlelike annual, generally with yellow flowers. The seeds, resembling smaller sunflower seeds, contain 35–45% oil. The plant has a well-defined taproot with numerous thin laterals. The stem is stiff, and cylindrical and becomes brittle when mature. Plant height, varying from 50 to 200 cm, is influenced by variety, climate, and cultural practices. The leaves are large, deeply serrated on the lower stem, short, stiff, and ovate to obovate at the inflorescence. The lower leaves are generally spineless, but the number of spines on the upper leaves is a

Table 2 Seed Viability and Seedling Vigor of Mustard Seed Lots at Two Moisture and Three Temperature Levels

Storage (months)	Temperature (°C)	Moisture content (%)	Germination (%) Normal	Germination (%) Abnormal	Root length (cm)	Shoot length (cm)
26	25	3.18	71	3	7.18	4.75
	25	5.25	71	3	8.35	4.85
	10	3.18	98	2	10.62	4.95
	10	5.25	98	2	11.77	4.90
	−10	3.18	99	1	11.00	3.25
	−10	5.25	99	1	11.21	3.50
30	25	3.18	64	10	6.32	4.70
	25	5.25	64	12	7.55	4.70
	10	3.18	98	2	7.30	3.50
	10	5.25	98	2	8.88	3.10
	−10	3.18	99	1	8.08	3.19
	−10	5.25	99	1	9.21	3.00
33	25	3.18	51	10	6.01	3.57
	25	5.25	51	16	7.20	4.00
	10	3.18	98	2	7.14	3.00
	10	5.25	97	3	8.70	3.00
	−10	3.18	98	2	6.70	2.38
	−10	5.25	99	1	7.50	2.70
36	25	3.18	45	22	4.98	2.81
	25	5.25	43	22	6.50	3.00
	10	3.18	97	3	6.50	2.64
	10	5.25	97	3	7.30	2.96
	−10	3.18	98	2	6.50	2.30
	−10	5.25	98	2	7.00	2.50

Source: Ref. 34.

varietal characteristic. The inflorescence consists of numerous florets closely collected together on a circular, flattened receptacle. The number of florets varies from 20 to 180. Safflower is a self-pollinated crop, but bees and other insects aid in optimum fertilization for maximum yields. Seed with 4–10 g 100–seed weight varies in size (6–9 mm) (Fig. 2).

Sesame, or sesamum (*Sesamum indicum* L. or *S. orientale* L.), belongs to the order Tubiflorae of the Pedaliaceae family, comprising 16 genera and about 60 species. In addition to *S. indicum*, other species, such as *S. angustifolium*, *S. radiatum*, and *Ceratotheca sesamoides* (false sesame), are cultivated to a limited extent in Africa. Sesame originated in Africa and spread through western Asia to India, China, and Japan and then to other parts of the world. The varieties and strains differ considerably in size, form, growth, flower color, and seed size, color, and composition. Sesame is typically an erect branched annual (occasionally perennial), 0.5–2.0 m in height with a well-developed root system. It is multiflowered, and its fruit is a capsule containing a number of small oleaginous (oily) seeds (Fig. 3).

Sesame leaves may be opposite or alternate in different varieties, and their arrangement influences the number of flowers borne in the axils and thus the yield per plant. An opposite arrangement of leaves encourages multiple flowering. Flowers occur singly on the lower leaf axils with multiple flowers on the upper stem or branches, borne on very short peduncles, having a corolla of five lobes, which vary in color from white or very pale pink to dark purple.

Sesame is normally self-pollinating, but insect pollination is common—without crossing up to 10%, occasionally reaching 50% in some cultivars (36). Low temperature at flowering can result in sterile pollen or premature fall of flowers, and high temperatures ($\geq 40°C$) at flowering can seriously affect fertilization, reducing the number of capsules produced. Mazzami (37), however, reported from Venezuela that the highest sesame seed oil content was obtained at the highest mean temperatures.

Species and Cultivars. Agriculturally important species of safflower are closely related to *C. tinctorius* and *C. oxyacantha* Bieb., and *C. lanatus* (wild safflower). The former is a medium-sized, branching, spiny annual safflower grown in the region from the Caucasus to northern India and Afghanistan. The improved varieties of safflower grown in India are AI, Manjira, Bhima (S4), N7, S144, A300, APRR1, APRR3, JSF2, Annigiri, and Tara. Gila is a well-known variety of safflower grown in California.

In addition to *S. indicum*, other cultivated species of sesame for seed purposes are *S. capanse* (*S. alatum*), *S. schenkii*, *S. laciniatum*, *S. angolense*, *S. prostratum*, *S. occidentale*, and *S. radiatum*. ICAR (30) has recommended several improved cultivars of sesame for cultivation in different Indian states.

2. Breeding

The cultivated safflower *C. tinctorius* has 12 pairs of chromosomes (2n = 24), as have *C. oxyacantha*, *C. palaestinus*, and *C. flavascens*. The first three can readily cross and yield fertile hybrids. The improvements related to seed composition of the reduced hull seed types were important. Development of cultivars with high levels of oleic acid have improved their nutritional quality. A rapid method of determining the degree of resistance to fungal infection in segregating populations based on leaf extract has been developed in India. The resistant varieties exhibit polyphenol oxidase activity, which inhibits spore germination (38).

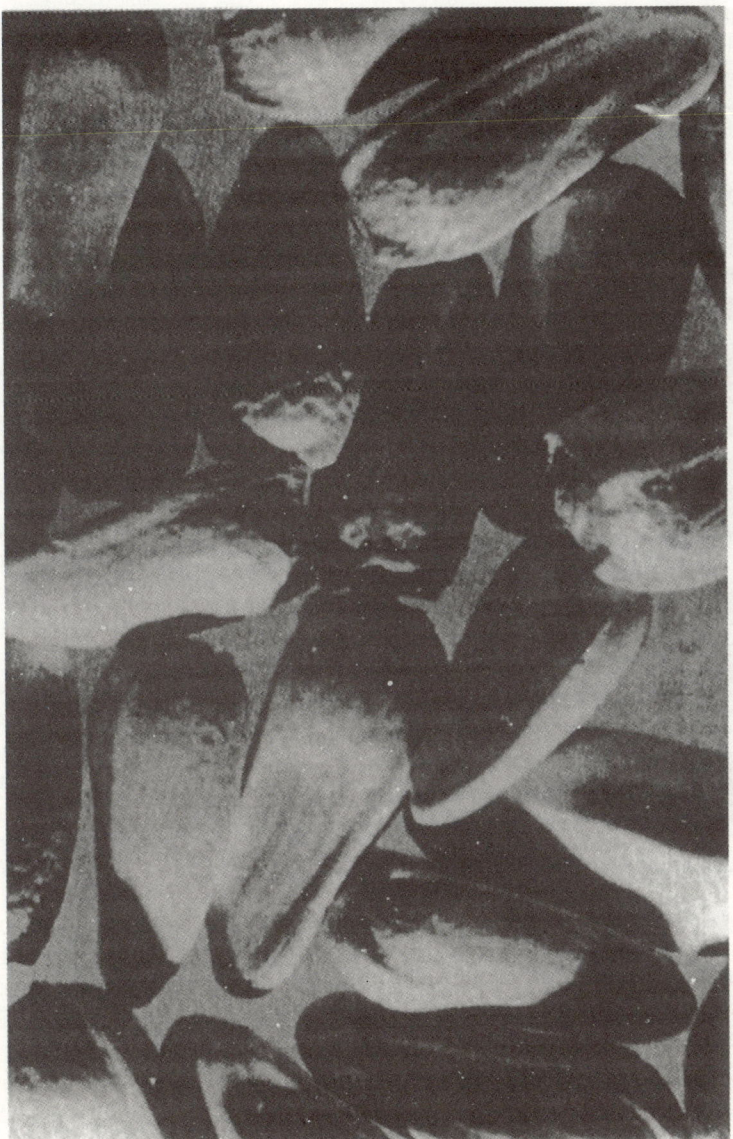

Figure 2 Safflower (*Carthamus tinctorius* L.) seeds. (From Ref. 57.)

A high degree of out-crossing in thin-hulled types and introduction of CMS are greatly assisting breeding work. Vegetative hybridization between safflower and sunflower through grafting was reported from the former U.S.S.R. Heterosis has been widely reported in safflower for yield, oil percentage, plant height, and other characteristics.

Polyploidy in sesame can be induced chemically by colchicine treatment. Major breeding and selection programs undertaken in the United States, India, Venezuela, and other countries seek to introduce indehiscence characteristics for mechanized crop

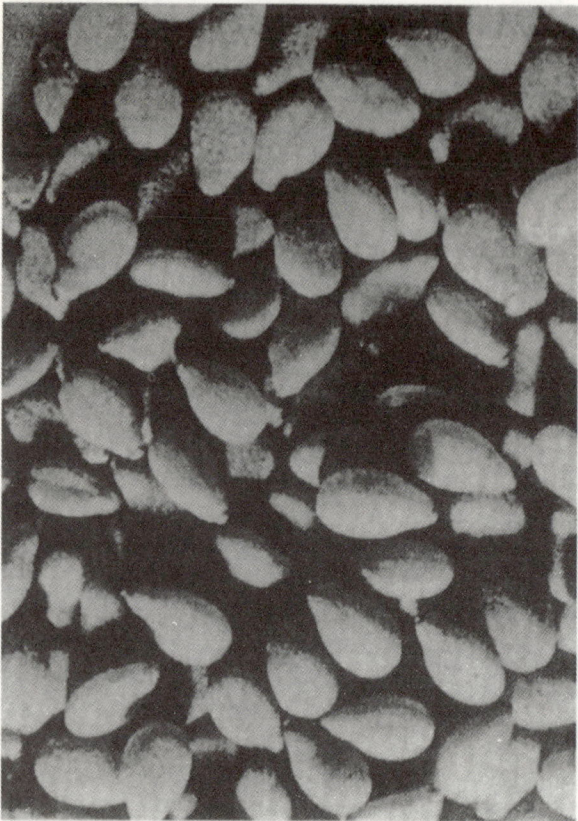

Figure 3   Sesame (*Sesamum indicum*) seeds. (From Ref. 57.)

production and other suitable agronomic characteristics as well as disease resistance, especially against bacterial leaf spot caused by *Psuedomonas sesami*, which is a worldwide and damaging disease. A similar situation exists with *Fusarium* wilt and phyllody. There are presently no sesame cultivars that are completely resistant to any major disease.

3. Cultivation

Culture.

LAND PREPARATION AND ISOLATION REQUIREMENTS.   Fields on which safflower and sesame crops were not grown in the previous season should be selected for seed production unless the crop variety was the same and the seed used was from a certified source. The soil should be rich in organic matter, well drained, aerated, and free of weeds. Land should be prepared by one plowing and two or three harrowings, followed by leveling to bring it to a desired tilth.

Both safflower and sesame are basically self-pollinated crops, although cross-pollination to the extent of 10% (safflower) and 10–50% (sesame) occurs, depending on the cultivar being grown and the conditions under which flowers develop. Because safflower seed dormancy is comparatively short, the shed seeds germinate and seedlings can be destroyed. However, 1 year free from safflower between seed crops is advisable. For seed crops of earlier generations in a multiplication sequence, an isolation distance of 400 m is desirable for both crops. For the final generation, seed crops can be isolated by a definite barrier or a gap of 3 m, although if the cultivar to be grown is prone to out-crossing, a greater distance may be required.

PLANTING AND CULTIVATION. Safflower is sown in June–July by drilling at a depth of 3–5 cm with spacing of 35–60 cm (occasionally 90 cm) between rows using seed rates varying from 30 to 60 kg/ha for a large commercial crop.

Sesame is sown from the first week of June to the first week of July for kharif crop. In southern India, it is grown as a rabi crop, and is sown from mid-October to mid-November. Seed germinates well when average temperatures are between 25 and 27°C. Nuclcus/breeder's/ foundation seed obtained from a source approved by a seed-certification agency is sown in lines with a seed drill 2.5–3.5 cm deep with a spacing of 30–45 × 15–22 cm. Seed rates vary from 2.5 to 5.5 kg/ha.

Fertilization, Irrigation, and Weed Control. Safflower crops require about 20–30 kg N, 20–80 kg P, and 10–20 kg K per hectare depending upon soil test values. Safflower is sensitive to excess moisture and high humidity. Thus, crops may require one or two irrigations when grown in dry areas. Mechanical weeding of young safflower is difficult; hence, weed populations need to be controlled prior to planting. However, depending upon the weed population, crops may require two or three weedings during the growth period.

Sesame responds well to organic manure. If the soil is low in fertility, sesame may require about 35 kg N, 20 kg P, and 35 kg K per hectare depending upon soil test values. Highly fertile soils will need little fertilization. Irrigation would be necessary during prolonged dry weather conditions to maintain adequate soil moisture during flowering and seed-maturation stages.

Two to three hoeings and weedings may be needed to keep the field free from weeds. The first hoeing may be done prior to irrigation when plants are 15–20 cm high. Broad-leafed weeds may be controlled by spraying sesone at 1.4–2.8 kg/ha. Safflower is very susceptible to weed competition at the rosette stage, and clean fields are necessary to obtain higher yields.

Grass weeds such as *Sorghum halepense* and other *Sorghum* and *Digitaria* species have seeds similar in size and shape to those of sesame. It is not possible to control these difficult weeds in a seed crop; it is therefore necessary to choose a clean seed plot (26).

Control of Pests and Diseases. *Uroteucon compositae* (Theo) (*Doctynotus compositae*) causes heavy losses in safflower yields. This pest can be controlled by spraying the crop with 0.05% phosphomidon as soon as it appears. Although safflower aphids can be removed mechanically, they can also be controlled by 0.34% malathion and 0.04% trithion. Aphids can also be controlled by spraying 0.05% dimethoate or 0.05% monocrotophos or by one or two sprays of 1.5% quinolphos.

Alternaria leaf spot can be controlled by spraying the crop with dithane-M-45 (75% WP) at 2.5 g/L. One or two sprays of 0.05% Calixin or 0.25% diathane M-45 at 15-day intervals effectively controls the safflower rust. The leaf spot caused by *Rumularia* can be controlled by 0.25% diathane M-45 or 0.1 % Bavistin (50% wettable powder) one to three times at 16-day intervals (30).

The caterpillars, gallfly, and leaf rollers of sesame crops can be controlled by dusting 5% BHC at 20 kg/ha. ICAR recommended three sprayings of 0.07% endosulfan or carbaryl 0.25% WP at 750 L/ha at 15-day intervals or carbaryl 5% dust at 20 kg/ha to control leaf roller and capsule borer. The gallfly can be controlled with a spray of 0.03% dimethoate or 0.07% endosulphan at 750 L/ha. The attacks of hawk moth and hairy caterpillar can be checked by dusting crops with 5% carbaryl and spraying with 0.07% endosulfan or 0.03% monocrotophos, respectively (3).

Roguing. The off-types and diseased seedlings of safflowers with *Alternaria* leaf spot should be removed in the early stages of crop growth. Heads infested with head rot, head blight, or gray mold (*Botrytis cinerea*) should be rogued at appropriate stages. Head rot-affected plants (*Ramularia carthami* Z.) should also be removed.

The off-types of sesame plants and diseased plants affected by rosette and leaf spot blight should be rogued out of the seed crop from time to time as required.

4. Harvesting and Threshing

Safflower is ready to harvest about 35–40 days after maximum flowering when the plant is quite dry but not brittle, the bracts on the head turn brown, and seeds have a moisture content of less than 8%. Safflower seed should be harvested at as low a moisture content as possible; preferably between 5 and 8%. The crop does not normally lodge, and seed shedding is unusual. Therefore, it pays to wait for the moisture content to decrease. The seed can be tested for maturity by squeezing several heads between the fingers. If the seed separates easily from the head, it is ready to harvest.

Grain combines are suitable for harvesting large safflower crops, although a threshing operation is not fast enough. Mechanical harvesting damages the seed considerably if the seed moisture content is not appropriate. Combining can start when the seed moisture content is about 10–12%. In some cases, the crop can be windrowed when the seed moisture content is 20–25% and picked when it is reduced to below 8%. The seed will need to be dried immediately to a safe moisture content. In addition to visible damage, there can be hidden damage to the embryo even when the hulls appear unbroken.

Direct harvesting costs less and usually produces higher yield and better-quality seed. Cutting and windrowing prior to combining should, therefore, be restricted to commercial crops only.

Sesame seed crops are ready to harvest when plants turn yellow with their capsules still green—about 80—150 days after sowing. The crop is cut manually by hand and stacked in bundles, which are piled in a circle on the threshing floor with the root end outward. For about a week, the stacks are opened out and bundles spread each day in the morning and collected and stacked again in the evening. After the plants are dried completely, they are threshed by shaking or beating with sticks.

Dehiscent types should be invariably allowed to dry off to avoid seed losses. With the introduction of nonshattering varieties, it has been possible to harvest sesame crops mechanically using either a reaper-binder or combine harvester. However, mechanical harvesting is better adapted to commercial crops than to seed crops because of damage to the seed. Sesame seed is easily damaged; even microscopic cracks in the seed coat adversely affect both the viability and oil quality. Slow working with optimum speed and accurate setting of the concave and cylinder are important in mechanical harvesting and threshing (3).

5. Postharvest Operations

Drying and Cleaning. Safflower seed is best dried on the plant in the field, but if harvested at a moisture content above 8%, immediate drying is essential. Bin drying is normally preferred, but continuous-flow drying with air temperature not exceeding 40°C can also be employed. For safe storage, seed moisture should be less than 8%.

Drying of the sesame crop is usually not necessary provided the seed is harvested at about 8% moisture. The seed should also be free from green matter after threshing. Seed can be cleaned satisfactorily on an air/screen cleaner.

Storage. Safflower seed is best stored under refrigeration. Grain bins are quite suitable for bulk storage of safflower seed provided the moisture content is below 5%. Smaller quantities of safflower seeds can be stored at 8% grain moisture content with little loss of viability, but for safe, long-term storage, more stringent environmental control is necessary. Bass (39) noted that safflower seeds with a 4–7% moisture content had about 80% germination after 8 years of storage at −10 to −12°C. Kole and Gupta (40) reported that the percentage germination of safflower seeds with the moisture content increased to 15% and stored for 0,7, 14, or 21 days decreased from 96 to 79% during the last week of storage.

Clean and dry sesame seed is stored in bulk. Seed will become heated if contaminated with extraneous matter, producing discolored and rancid oil.

Seed Treatment. Safflower seed is prone to several seedborne fungi, which can be effectively controlled with organomercurial seed dressing. *Alternaria* leaf spot can be prevented by treating the seed with thiram at 3 g/kg of seed.

According to Neargaard (41), seedborne disease of sesame caused by *Pseudomonas sesami* can be prevented by treating seed with 0.1% streptomycin and disease caused by *Xanthomonas sesami* by Abavit B. ICAR (30) recommended steeping sesame seed in 100 ppm Agrimycin or 0.05% streptocycline suspension to prevent seedborne infection of bacterial blight. *Alternaria* leaf spot and *Rhizoctonia* stem rot of sesame can also be checked by treating seed with 0.3% thiram.

Seed Yields. Seed yields of safflower and sesame range from 8 to 10 and 2 to 6 Q/ha, respectively.

D. Castor, Linseed, and Niger

1. Introduction

Botany and Origin. Castor (*Ricinus communis* L.) belongs to the Euphorbiaceae, or spurge, family, containing a vast number of plants native to the tropics. It is a warm-season plant indigenous to eastern Africa and probably originated in Ethiopia. There is a great range of wild and semicultivated types of castor in Ethiopian East Africa, which spread down to the Nile and across the Red Sea (6).

Castor is usually a fairly tall, many-branched perennial, but when cultivated commercially, it is short-lived, erect, little branched, and treated as an annual. Castor plants vary greatly in growth habit, color of foliage, stems, and seed size, color, and oil content.

There are basically two types of castor plants: tall and short, commonly known as a giant and dwarf castors, with maturity periods varying from 140 to 160 days.

Leaves are alternate and borne on long, stout petioles. The inflorescence, forming a pyrimidal raceme, is borne terminally on the main and lateral branches. The lower portion of the receme bears male flowers, and the upper bears female flowers; the ratio between male and female flowers is a varietal characteristic influenced by climate. High temperature favors maleness. The flowers are produced over an extended period. The fruit is a globular capsule, spiny to some degree, becoming hard and brittle when ripe and occasionally shattering at maturity. Castor is essentially a long-day plant, but can adapt with some loss of yield to a fairly wide range of day length. Castor beans are oval in shape with black and white stripes (Fig. 4).

Linseed (*Linum usitatissimum* L.) belongs to the family Linaceae. The genus *Linum* includes most herbs and shrubs found in the temperate and subtropical regions bordering the Mediterranean Sea.

Linseed is an erect annual about 60–120 cm high and is cultivated on the plains as well as up to an altitude of 1800 m in India. The small flowers are blue, bluish violet, or white, growing in terminal panicles. Fruits are capsular with five cells, each containing two seeds, which are yellowish or blackish brown, small, flattened, and oval with a smooth shining coat. Linseed is cultivated for both oil and fiber. The seeds yield a quick-drying oil, and the stalks yield flax (a textile fiber).

Figure 4 Castor bean (*Ricinus communis*) seeds. (From Ref. 57.)

Niger (*Guizotia abyssinica* Cass., syn. *G. oleifera* DC) is a member of the Compositae (sunflower) family, belonging to the tribe Heliantheae. The genus *Guizotia* is thought to have originated in the Ethiopian highlands.

Niger is a stout, erect, moderately branched annual herb, about 1 m in height, with attractive yellow flowers. The leaves are opposing (sometimes alternate on the upper stem), sessile, lanceolate, and softly hairy on both surfaces. The flowers develop from leaf axils in clusters of two to five. The heads flower over an extended period of 15–30 days, resulting in uneven ripening and high shattering losses. The period from emergence to flowering is about 90 days. Niger is basically cross-pollinated, with some self-fertilization. The fruit is an achene, typical of the Compositae, small (1.5 × 3–5 mm) and lanceolate in shape. The fruit contains about 15–30 mature seeds, with a 1000-seed weight ranging from 3 to 5 g. Unlike safflower, mature niger seeds are easily dislodged and shattered. Niger is a short-day plant of the temperate region but is adapted to a semitropical environment (3,4,6).

Species and Cultivars. Zukovsky (42) described three species of castor: *R. communis*, *R. macrocarpus*, and R. microcarpus. The most common subspecies are *R. communis persicus* (Persian subsp.), *R. communis chenensis* (Chinese subsp.), *R. communis zanzibarensis* (Zanzibar subsp.), *R. communis sanquineus* (Crimson subsp.), and *R. communis gibsoni* and *R. communis cambogenesis* (purple subsp.). In addition to these, Popova (43) recognized *R. communis africanus* and *R. communis mexicanus*. Castors are sometimes classified as tall, intermediate, and dwarf types based on node number. Salunkhe and Desai (3) have listed improved cultivars of castor grown in India.

Three species of *Linum*—*L. usitatissimum*, *L. bienne* (syn. *L. angustifolium*), and *L. grandiflorum*—have been recorded in India. *L. usitatissimum* is widely cultivated for its oilseed, whereas others are grown as ornamentals.

Fiber-type linseed varieties are generally slender, tall-growing, nontillering, and sparingly branched, whereas the seed-purpose cultivars are usually dwarf in habit, more branched, and profusely tillering. Two ecologically distinct types of linseed grown in India are (a) the gangetic types of North India, and (b) penninsular types grown in the south of Ganges and Jamuna. Both types are variants of *L. usitatissimum* introduced from central Asia. They also may have been derived from crosses of *L. usitatissimum* with *L. strictum* in the north India or with *L. perenne* and *L. mysorense* in south India. Several improved varieties of linseed are grown in India (30).

Three species of the genus *Guizotia* are native to tropical Africa, and *G. abyssinica* grows wild as a weed in Ethiopia. It is also grown in India and Pakistan and is an important local oilseed crop.

2. Breeding

For the production of single-cross hybrid castor seed, lines giving a 1:1 ratio of pistillate and heterozygous monoecious plants are used. In the crossing plot, the latter are rogued out 1–5 days before flowering commences. The female plants are then cross-pollinated by a selected male pollinator line planted in every sixth or eighth row. The crop may be required to be rogued about five or six times to keep self-pollination to a minimum. The hybrid seed can also be produced using 90–100% pistillate lines, which eliminates roguing, but the seed produced is not entirely uniform (18).

Castor varieties suitable for mechanized cultivation and reduction in height have been widely sought. The use of male or female sterile lines has aided breeding work.

Introduction of a gene for short internodes in the United States has drastically reduced plant height. Efforts are being made to increase the oil content of these cultivars. The varying thickness and strength of capsule walls and resistance to pests and diseases are other important factors in castor breeding. The first complicates mechanical hulling, which needs accurate adjustment of machinery to produce undamaged seed sample. Introduction of thornless varieties and salinity tolerance are other desired characteristics in a breeding program.

The most important objectives in the breeding program of niger is to increase seed yield, which is the major limiting factor in its profitability. This has partially been achieved in India through selection of new strains producing 1 ton/ha compared with the usual 300–400 kg/ha. Seed yield positively and significantly correlates with the number of branches and capitula per plant, with both these characteristics correlating with node number. Strains more suitable for semimechanized agriculture, with a more uniform flowering/nonbranching habit and increased seed and oil yields, are now being developed through breeding and selection. Also, strains with a high seed number per head with reduced shattering losses are desirable. The range of phenotypes available suggests that these objectives will eventually be achieved.

3. Cultivation

Culture.

LAND PREPARATION AND ISOLATION REQUIREMENTS. Fields used for the seed crops of castor, linseed, and niger should not have been used for these crops in the previous season unless the crop grown was of the same variety using certified seed from an approved source. The fields selected should also be well drained, aerated, and rich in organic matter. Land should be prepared by deep plowing, two to three harrowings, followed by leveling to bring it to the desired tilth.

Castor is an open-pollinated crop; cross-pollination occurs by wind, varying from 5 to 35% depending upon the prevailing climatic conditions. For pure seed production, the seed field of castor must be isolated from other fields of castor by at least 300 m for foundation seed and 150 m for certified seed.

Linseed is mainly a self-pollinated crop, but some cross-pollination does occur through insects to the extent of about 3%. The seed crop should, therefore, be isolated from other linseed fields by about 300 m for foundation seed and 100 m for certified seed.

Niger is a basically a cross-fertilized crop with little self-fertilization. The crop should be separated from other niger fields by at least 1000 m for foundation seed and 500 m for certified seed.

PLANTING AND CULTIVATION. Kharif castor is generally sown in June and July or in September–October (as a rabi crop). In other parts of India, it may be sown from August to September (Gujrat) or in April (Karnataka). Seed obtained from a certified agency is sown in lines with a seed drill or using a plow in furrows 7.5–10.0 cm deep with 90 cm between lines; the distance between plants may vary from 45 to 90 cm (18). The seed rate of castor crops varies from 11 to 18 kg/ha depending upon spacing, seed size, and method of sowing. Rainfed crops may require seed rates up to 33–44 kg/ha.

Sesame is planted from the first week of June to the first week of July (kharif crop). In southern parts of India, it is grown as rabi a cold-season crop, when it is sown from mid-October to mid-November. Seed germinates well at temperatures around 25–27°C. Nucleus/breeder's/foundation seed obtained from a source approved by a seed-certification agency is sown in rows with a seed drill, about 2.5–3.5 cm deep, with row to row spacing of 30–45 cm and plant to plant distance of 15–22 cm, using 2.5–5.5 kg seed/ha.

Niger is usually sown by broadcasting and is mixed with sand to assist in its even distribution. It is sown in July or August (south India) after heavy rains about 1–3 cm deep. In the northern states, it is drilled with spacing of 60 × 10 cm; seed rates vary from 6 to 12 kg/ha.

Fertilization, Irrigation, and Weed Control. Castor seed crops respond well to the application of organic manure. Fertilizer requirements vary considerably in different areas, ranging from 20 to 80 kg N, 10 to 40 kg P, and 10 to 40 kg K per hectare depending upon soil fertility. Two to three irrigations are required, depending upon rainfall, to ensure adequate moisture at the time of flowering and seed maturation. Castor crops should be kept free of weeds, especially for the first 60 days after planting, by two or three hoeings or weedings. Annual dicot weeds may be controlled by applications of 3–4 kg 2,4-dichloropropionic acid or triflurelin/ha.

Linseed seed crops require about 50–60 kg N, 40 kg P, and 40 kg K per hectare. Nitrogen may be given in two splits, with half being applied at the time of sowing and the remaining half at the time of first irrigation. Irrigation during the early growth period and flowering greatly enhances seed yields. One weeding during the early stage of crop growth and one after the first irrigation are necessary to control weeds.

Control of Pests and Diseases. Semilooper of castor crop is controlled by spraying 0.35% thiodan or 0.03% dimecron. The jassids of castor can be controlled by dusting the crop with 3% DDT. Castor pod borer and caterpillars are controlled by spraying thiodan (1.2 L/ha) and 5% BHC dust or spray of 0.04% malathion, respectively. The *Phytophthora* blight and *Cercospora* diseases of castor can be checked with a spray of Bordeaux mixture (4:4:50) at 15-day intervals or 0.25% diathane M-45 two to three times at 15-day intervals.

For controlling laphigma, semilooper, and linseed fly, crops should be dusted with 5% gammexene at 20 kg/ha. The greasy cutworm can be checked with 2% gamma BHC dust at 20 kg/ha. For linseed rust, crops should be sprayed with 2 kg zineb dissolved in 100 L of water per hectare. Powdery mildew can be controlled by spraying crops with 0.2% thiovit solution or dusting with sulfur powder at 20 kg/ha.

Caterpillar, bollworm, fruit fly, shield bug, hawk moth, weevil, aphid, and leafworm are the important insect pests affecting niger crops, which can be controlled as described for safflower crop. The diseases of *Cercospora* and *Alternaria* (leaf spot) and bacterial blight (*Pseudomonas* spp.), root rot (*Phytophthora* spp.), and powdery mildew (*Sphaerothera* spp.) can be controlled using measures as described for castor and linseed crops.

Roguing. All off-types of castor seed crops must be removed before flowering. Similarly, diseased plants affected by *Phytophthora* blight and *Cercospora* leaf spot should be rogued out as soon as they are noticed. These plants should be destroyed to prevent any further spread of disease.

4. Harvesting and Threshing/Hulling

Harvesting and hulling are the most difficult and time-consuming operations involved in growing castor, although suitable machines and varieties for large-scale and mechanical harvesting are available. In India, castor harvest is generally spread over a period of 5–10 weeks or more, as all capsules or fruits on the branches or spikes do not mature at the same time. The first-formed fruits ripen and begin to dry about 120–150 days after planting depending upon soil and variety. Immature kernels are light in weight and low in oil content compared to fully matured ones. Delaying harvest will result in heavy loss due to dehiscence and shedding of seeds.

If mechanical harvesting is employed, harvesting of seed crops requires great care and special skill in operating combines. Hand or manual harvesting may be preferred for smaller plots. Hulling can be mechanized, provided working speeds are kept low, feeding dry capsules regularly. Hulled seeds should be sorted immediately to remove damaged ones. The harvested fruits (light yellow in color) are placed in piles to dry in the sun until they blacken; later they are threshed by beating with sticks.

Linseed crops are harvested soon after the plants turn brown and brittle and seeds rattle in the bolls. Crops are cut with sickles, or plants may be pulled by hand and left in the field for 1 or 2 days to dry. They are then transported to the threshing yards, where they are threshed by flailing the siliquas with sticks. Stationary threshers can also be used, or the crop can be directly or indirectly combined.

Niger is ready to harvest about 120–150 days after emergence; early-maturing strains are ready to harvest within 90–120 days. Mechanical harvesting of niger is not feasible because of nonuniform crop maturity over a period of 15–20 days. Harvesting niger is a problem due to easy dislodging and shattering of seed; about 25% seed can be lost in this way.

In India, niger plants are manually cut with a sickle, dried in the field for 2–3 days, bundled, and carried to the threshing floor, where they are threshed by hand, usually with sticks.

5. Postharvest Operations

Cleaning.   The threshed produce of castor, linseed, and niger is cleaned by winnowing in the wind. Special concrete threshing floors or a heavy tarpaulin or plastic sheet are employed to obtain clean seed. Damaged seed and broken pieces of capsules must be removed at once to avoid damage to seed in store. Air/screen cleaners are normally used to clean the seed effectively.

Storage.   Castor seeds are large and occupy considerable space in the storehouse in relation to their weight. Unlike bagged groundnuts, castor seed cannot be stored in the open except for short periods, as both heat and sunlight will reduce seed quality. Castor seed must be bagged carefully and handled as little as possible. Wooden scoops or shovels and rubber conveyor belts are generally employed. Low-temperature storage (-5 to $-7°C$) significantly decreases the viability (from 93 to 3%) and germination capacity of castor seed (44); however, if castor is stored at a very low temperature of $-196°C$, its high-germination percentage can be retained (45).

Clean, whole seeds of linseed and niger can be stored easily in well-protected containers. Because of its small size, niger seed occupies relatively little space. The seeds are prone to attack by insects and other storage pests if not well protected. A small quantity of seed can be stored in 200-L oil drums fitted with lids after the seeds are treated with appropriate insecticides.

Seed Treatments. Castor, linseed, and niger seeds can be effectively protected from seedborne fungal diseases caused by *Botrytis, Colletotrichum,* and *Phoma* by treating the seed with thiram. *Alternaria* and *Phoma* can be prevented by seed treatment with organomercurials. The *Fusarium* wilt of linseed caused by *F. oxysporum* f. sp. *lini* can be prevented by treating the seed with thiram or Topsin at 2.5 g/kg seed.

E. Other Oilseed Crops

1. Coconut: Botany and Cultivation

Coconut (*Cocos nucifera* L.) is a member of Palmae family of the monocotyledons. All palms are characterized by their unbranched trunks, crowned by fan-shaped, pinnate or binnate leaves. The inflorescence and fruit are usually produced in the leaf axis, or in some species at the end of the mature trunk. Coconut is included in the tribe Cocoideae, consisting of about 20 genera. Genus *Cocos* is considered to be monotypic, containing only *C. nucifera,* but has considerable genetic variability.

Coconut is monoecious, both male and female flowers being on the same inflorescence. The spadix develops within two stout sheaths or spathes. Of the 15–20 female flowers produced on each spadix, only four to five sets (buttons) grow rapidly to become pointed, cornlike nutlets. Each spadix bears 250–300 male flowers, which are smaller in size than female flowers. The receptive period lasts 4–6 days, ensuing cross-fertilization from a spadix opening later.

Coconut fruit is a drupe, with a thin epicarp overlying a thick, fibrous mesocarp; inside the mesocarp is the hard shell-like endocarp, enclosing the endosperm. The latter provides, as the fruit ripens, the white meat, or copra, which is rich in oil, and the watery milk of the central cavity. The coconut palm begins to bear fruit in about the sixth year, comes to full bearing in about the twelfth year, producing 80–150 nuts per year. The embryo, which develops within the nut, fills the central cavity and emerges through one of the three pores (eyes) in the shell (3).

The dwarf coconut cultivars can be clearly distinguished from the tall ones. The former show a degree of self-pollination and form stable populations. The tall palms are predominantly cross-pollinated and vary greatly in their characteristics. Hybrids between tall and dwarf varieties occur in nature and have also been produced by artificial cross-pollination; they are vigorous with broad stems and a large number of leaves. About 63% of the hybrid progenies flower in less than 4.5 years and develop nuts that are large and heavy (46). Mathai (47) compared the performance of hybrid coconuts and reported that Tall × Tall hybrids were the shortest. In a comparative trial with several Indian cultivars, the East Coast Tall and Laccadiv Ordinary were the most promising cultivars (48). Two distinct forms of dwarf coconuts occur in the west coast of India; namely, Dwarf Orange (DO) and Dwarf Green (DG). About 95% of the natural progenies of DG and 80% of DO breed true to type. The off-type progenies of dwarfs are thought to be the result of the outcross with Talls. The open-pollinated progenies of DO were superior to all other cultivars.

Fully ripe coconuts are harvested to obtain high-quality copra. For ball copra, the nuts should be allowed to mature fully and fall naturally. Coconuts are picked by human climbers or cut using knives attached to the end of long bamboo poles. Copra is dried to about a 6–7% moisture content and stored in a well-ventilated dry storage room at uniform temperature conditions.

2. Oil Palm: Botany and Cultivation

The oil palm (*Elaeis quineensis* Jacq.) belongs the family Palmae, a distinct group of monocotyledons. It is grouped with *Cocos, Corozo*, and other genera of the order Palmates, under the tribe Cocoineae. *Elaeis* is derived from the Greek word *elaion*, meaning oil, and the specific name *quineensis* indicates its origin on the Guinea Coast. Oil palm is a large-feather palm having a solitary columnar stem with short internodes. It has short spines on the leaf base and within the fruit bunch. It gets its characteristic appearance from the irregular set of the leaflets on the leaf. Like coconut, oil palm is normally monoecious with male or female flowers, but sometimes is hermaphroditic, with inflorescences developing in the axis of the leaves. The fruit is a drupe borne on a large compact bunch. The fruit pulp, which provides oil, surrounds a nut, the shell of which encloses the palm kernel (49).

According to Redenbaugh et al. (50) and Redenbaugh (51), both oil palm and cotton have strong technological and commercial bases and a great potential for propagation through the production of synthetic/artificial seeds by somatic embryogenesis.

A fully domesticated oil palm is one that is taken from its natural habitat and consciously reproduced for successive generations under man-made conditions; the oil palm has the ability to reproduce itself under these conditions. The plant favors growth under conditions of uniform rainfall (2000 mm), distributed evenly throughout . the year, temperature variation of 23–30°C, and constant sunshine for at least 5 hr a day all year and 7 hr a day for certain months.

Two fruit forms of palm, namely, *dura* and *tenera*, have been distinguished by shell thickness. Janssens (52) divided both the common fruit type *nigrescens* and the green-fruited *virescens* into three forms: *dura, tenera*, and *pisifera*. The white-fruited *albescens* was also recognized. Smith (53) recognized both mantled and unmantled *nigrescens* and *virescens* fruit and called them types, dividing all four into thick-shelled and thin-shelled forms. This procedure has thus established the use of fruit-type and fruit-form classification of oil palms, eliminating the need for the term *variety* for material that might be heterozygous in many of its character: The term *variety* was inappropriately used for the *tenera* form.

Palm oil yields largely depend on rate of growth and development and maintenance of the leaf surface, number and size of leaves produced, and the proportion of the leaves providing female inflorescence. Maintenance of the leaf surface is also an important factor in determining the sex ratio and the proportion of female flowers to the total number of flowers (inflorescences). Palm is susceptible to adverse climatic conditions at germination and early growth stages. The most remarkable growth feature of the oil palm is the leaf area duration which is the integral of the leaf area index over the growth period and a measure of its photosynthetic potential. Initiation of anthesis takes about 33 months, and fruits ripen 38–39 months after planting (49).

Oil palm fruits are harvested when they turn reddish orange in color, which is about 5.5–6.0 months after pollination and 3.5 years after planting. Whereas underripe fruit contains less oil than ripe fruit, overripe fruit provides oil with a higher free fatty acid content. Ideally, fruits should be collected when the fully ripe fruits drop from the bunches, examined daily, and then harvested when the majority of the fruit are fully ripe but none is overripe or damaged. Harvesting oil palm is a compromise between yield and quality. The harvested bunches are transported to the place of oil extraction for processing.

Palm oil seed is stored after shade drying on a large scale in 44-gallon drums covered with wooden lids. The safe moisture content of the seed for storage is 14%. At a high seed moisture content, seed may require lower temperature for storage.

3. Crambe: Botany and Cultivation

Crambe (*Crambe abyssinica* Hochst) is a member of the Cruciferae, tribe Brassicae, and is closely related to the rapes and mustard. The genus *Crambe* contains about 30 species, mostly perennial herbs, although a few are shrubs or annuals, distributed mainly in the Mediterranean, Euro-Siberia, and the Turko-Iranian region (6). The only cultivated member is *C. abyssinica*, also known as Abyssinian kale. It is an herbaceous annual, about 1 m in height, naturally much branched and bears long recemes of yellow flowers, producing a large number of small, brownish seeds with an oil content of about 40%. The plant has a substantial root system typical of the Cruciferae. The stem is stiff and branches close to the ground to form about 30 or more secondary branches, which again branch to form tertiary branches. The extent of branching, which depends on cultural methods, is considered to be a disadvantage for mechanical harvesting, since each terminates in a raceme. Although branching may increase individual plant yield, it may also increase loss from frost and pests or the number of immature seeds harvested by extending the flowering period.

The two species *C. abyssinica* and *C. hispanica* are distinguished on the basis of the shape of the basal leaves (54). The two species cross readily, and this characteristic is easily transferred (55).

The fruit is a capsule (siliqua), initially pale green, but turning yellow with maturity, when it becomes covered with a well-defined network of small ridges. Each capsule contains a single spherical seed, greenish brown or brown in color. Seed size varies considerably in diameter (0.8–2.6 mm), being influenced by the number of seeds per plant, soil fertility, and rainfall. The 100–seed weight thus varies from 25 to 50 g. The oil content of hulled seed ranges from 35 to 60% and protein from 20 to 40%. Crambe oil contains erucic acid ranging from 35 to 65%.

Crambe prefers moderate rainfall and a warm climate but can be adapted to colder or drier areas. Naturally restricted to Mediterranean-type climates, crambe is grown commercially over a much wider climatic range from sea level to an altitude of 2000 m. A temperature range of 15–25°C is required during its vegetative period, and it is susceptible to frost. It is not dependent on rainfall, being able draw on subsoil moisture, and is a comparatively drought-resistant crop. However, as a commercial crop, it is not suited to arid areas with low water availability.

Crambe requires well-drained sandy loam soils neutral to slightly alkaline in reaction; below pH 6.0, its growth and yield are usually adversely affected (6). It requires about 15–20 kg N and 50 kg P per hectare depending upon soil test values. The seed is sown 1.5–2.5 cm deep using a seed driller from mid-June to mid-July (Kenya) or in April–May (Poland) with row to row spacing of 20–30 cm. The soil moisture and occurrence of frost determine the time of planting; a period of adequate moisture and 100 frost-free days should exist from planting to harvest. Crambe is a vigorously growing plant and, once past the seedling stage, competes with any but the most rapidly growing weeds. Seed plots should, therefore, be keep free of weeds in the initial stages. Depending upon rainfall, the seed crop may need one or two irrigation. Crambe is attacked by several insect pests damaging other cruciferous crops, which can be controlled by the pest-control measures described for these crops. The *Alternaria* leaf spot and *Fusarium* wilt can be prevented by seed dressing with organomercurials or thiram.

Crambe is ready to harvest about 90–100 days after planting when the majority of leaves fall, the upper stem is yellow, and about 75% of the capsules turn yellow. The crop is cut, windrowed, and threshed with a combine or direct combine harvested. The harvesting and postharvest operations are similar to those for rapeseed. Crambe may be windrowed with a seed moisture content of 25–35%, but should not be threshed until it has dropped to 12–15% moisture. Below this level, hulls are damaged, affecting seed and oil quality during storage. Seed yields of about 15–20 Q/ha are usually achieved. Crambe seed should be cleaned and dried to a 10% moisture content for storage. Jablonski (56) reported that seed with a low moisture content withstands low temperatures without a major loss of viability. Seeds were found to lose viability when stored for 2 years at 21°C and 71% RH. However, when placed in controlled storage at 10°C and 50% RH or at −1°C and 40% RH, crambe seed remained viable for 8 years.

4. Jojoba: Botany and Cultivation

Jojoba (*Simmondsia chinensis* [Link.] Schneider [syn. *S. californica* Nutt. and *S. pabulosa Kellogg*]) is the only species of the genus Simmondsia, one of the six genera in the Buxaceae, family Euphorbiacae. It grows wild in southwestern areas of the United States and northwest Mexico.

Jojoba is a multistemmed, woody shrub varying in height from 0.5 to 6.0 m, usually around 2.0–2.5 m. It is almost prostrate, evergreen, drought resistant, dioecious, with nutlike seeds producing a liquid wax possessing unique qualities for a vegetable oil.

Jojoba produces male and female apetalous flowers on separate plants, although hermaphroditic plants are known to exist. The fruit is a green, elongated oval capsule, enclosing one to three seeds, although usually only one seed develops, which resembles a peanut or acorn in shape (12–18 × 6–12 mm). The 100–seed weight varies from 40 to 80 g (or 1600 seeds/kg).

Jojoba requires well-drained sandy clays or alluvial soils and needs little land preparation. Seedlings grown in nurseries respond to the application of nitrogen and phosphorus.

Jojoba seed is planted either by direct seeding in the field or by nursery production of seedlings for late transplanting. Seed bed temperatures of 25–35°C are optimum. The seed is sown 3–5 cm deep depending on soil moisture availability, using a seeder mounted on a toolbar. Sowing seed in nurseries and then transplanting to prepared fields speeds planting establishment and saves seed. The optimum distance between plants depends on the local environment. In Arizona, 5 m between rows is considered to be minimum to allow interculture operations, with seedlings planted every 45 cm in rows or 15–30 cm if direct seeded. Irrigated seedlings and young plants grow faster than unirrigated plants. Weeds must be controlled from time to time. Jojoba is comparatively free from major insect pests and diseases (3).

Jojoba seed is usually harvested manually. The mechanical or manual pickers invariably harvest a large proportion of seed with a high moisture content together with twigs, leaves, and so forth; which will need both cleaning and drying. An average seed yield of 3–4 kg per tree annually is required for profitable production.

The harvested seed is sun dried to about 9–10% moisture. The seed stores well if protected from pests and vermin.

REFERENCES

1. J. Davis and L. Vincent, Extraction of vegetable oils and fats, *Fats and Oils: Chemistry and Technology* (R. J. Hamilton and A. Bhati, eds.), Applied Science, London, 1980, p. 123.
2. J. A. Zwartz and J. G. A. J. Hautvast, Food supplies, nutrition and plant breeding, *Plant Breeding Perspectives* (J. Sneep and A. J. T. Hendriksen, eds.), Holbek, O. Coed, Centre of Agricultural Publication and Documentation, Wageningen, Netherlands, 1979.
3. D. K. Salunkhe and B. B. Desai, *Postharvest Biotechnology of Oilseeds*, CRC Press, Boca Raton, FL, 1986.
4. D. K. Salunkhe, J. K. Chavan, R. N. Adsule, and S. S. Kadam, *World Oilseeds: Chemistry, Technology and Utilization*, Van Nostrand Reinhold, New York, 1992.
5. J. J. Smartt, W. C. Gregory, and M. P. Gregory, The genomes of *Arachis hypogaea*, *Euphytica* 27:329 (1978).
6. A. Weiss, *Oilseed Crops*, Tropical Agriculture Series, Longman, London, 1983.
7. R. W. Gibbons and J. R. Tattersfield, Out-crossing trials with groundnuts, *Rhod. J. Agric. Res.* 17(1):71 (1969).
8. R. W. Gibbons, A. H. Bunting, and J. Smartt, The classification of varieties of groundnut, *Euphytica* 21:78 (1971).
9. R. O. Hammons, Genetics of *Arachis hypogaea*, *Peanuts: Culture and Uses*, Oklahoma State University, Stillwater, 1973, p. 135.
10. C. R. Seshadri, *Groundnut*, Indian Central Oilseeds Committee Himayatnagar, Hyderabad, 1962.
11. P. Leclerq, Une sterile male cytoplasmique chez Ie trounesol, *Amim. Ameliot: Plantes* 19:99 (1969).
12. M. L. Kinman, Key to hybrid sunflower found, Crops Soils 23:21 (1970).
13. F. F. Campos, Use of ethrel in the induction of male sterility, Central Luzon State University, Munoz, Philippines, 1978.
14. G. L. Torres, J. D. Gimenez, and J. F. Martinez, Male and female sterility induced in sunflower with GA, *An. Inst. Nac. Invest. Agron. Prod. Veg.* (Spain) 9: 147 (1979).
15. H. T. Stalker and J. C. Wynne, Interspecific hybrids in the genus *Arachis* section *Arachis*, *Agron. Abstr*: (63) (1978).
16. J. J. Smartt, *Groundnut Production in Zambia*, Government Printer, Lusaka, Zambia, 1967.
17. ICRISAT, Asia Region, Annual Report, 1994, International Crops Research Institute for Semi-Arid Tropics, Patancheru, Hyderabad, India, 1995, p. 71.
17a. M. Dadlani, R. Mathur, D. Choudhury and A. Varier, Manifestation of seed vigor in sunflower hybrid under accelerated aging, *Plant Physiol. Bichem.* 22: 17 (1995).
18. R. L. Agrawal, *Seed Technology*, Oxford & IBH Publishing, New Delhi, 1980.
19. B. Gopal Singh and P. Sudhakar, Response of sunflower (*Helianthus annuus*) hybrids to dehydration under field conditions, *Seed Research*, Sp. Vol. No. 2 (S. P. Sharma, ed.), Indian Society of Seed Technology, New Delhi, 1993, p. 776.
20. G. R. Rao and B. G. Singh, Effect of hydration-dehydration on germination and seedling growth of certain oil seeds and pulse crops, *Ann. Agric. Res.* 13(1):67 (1992).
21. B. G. Singh and G. R. Rao, Effect of chemical soaking of sunflower (*Helianthus annuus*) seed on vigour index, *Indian J. Agric. Sci.* 63(4):232 (1993).

22. J. O. Weete, W. O. Branch, and T. A. McArdle, Determining peanut harvest dates in Alabama by the Arginine Maturity Index (AMI), *Bull. Agric. Exp. Sta. Auburn Univ.* 516:31 (1979).

23. P. M. White and R. C. Roy, A once-over peanut harvester, *Proc. Am. Peanut Res. Educ. Soc.* 14(1):116 (1982).

24. R. I. Smirnova, Effect of desiccation on sowing qualities and epiphytic microflora of sunflower seed, *Tr: Novosib. Skh. Inst.* 84:203 (1975).

25. R. P. Kachru and K. M. Sahay, Separation of oilseeds by pedal-cum power operated air screen cleaner, *J. Oilseed Res.* 5(2):132 (1988).

26. A. F. Kelly, *Seed Production of Agricultural Crops*, Longman, New York, 1988.

27. F. R. Harper and B. Berkenkamp, Revised growth-key for *Brassica campestris* and *B. napus*, *Can. J. Plant* Sci. 55:657 (1975).

28. J. B. Free and I. H. William, The insect pests of oilseeds rape, Proc. 5th Int. Rapeseed Conf., Swedish Seed Association, Svalov, Sweden, 1978, p. 310.

29. C. S. Hoveland, J. W. Odom, R. L. Haaland, and M. W. Alison, Jr., Rapeseed in Alabama, Alabama Agric. Exp. Sta. Bull. No. 532, Auburn University, Auburn, 1981.

30. ICAR, Rapeseed, mustard; Taramira: Package of practices for increasing production, Technologies for better crops, No. 2, Indian Council of Agriculture Research, New Delhi, 1981.

31. L. F. Tay, L. K. Khoh, C. S. Loh, and E. Khor, Alginate-chitosan coacervation in production of artificial seeds, *Biotech. Bioeng.* 42(4):449 (1993).

32. V. L. Chopra, P. B. Kirti, S. B. Narasimhulu, S. Prakash, K. K. Abduraman, and B. Dominic, Somatic hybridization for improvement of crop brassica, *Biotechnology in Agriculture*, Proc. 1st Asia Pacific Conf. on Agricultural Biotechnology, Beijing, China, Aug. 20–24, 1992 (C. B. You, Z. L. Chen and Y. Ding, eds.), Kluwer, Dordrecht, Netherlands, 1993, p. 18.

33. J. T. Ward, W. D. Basford, J. H. Hawkins, and H. M. Holliday, *Oilseed Rape*, Farming Press, Ipswich, 1985.

34. N. Singh, P. P. Khanna, and N. K. Chaudhary, Storability of mustard (*Brassica juncea*) in three different controlled environments, *Seed Research*, Special Vol. 1 (S. P. Sharma, ed.), Indian Society of Seed Technology, New Delhi, 1993, p. 317.

35. P. F. Knowles, Centres of plant diversity and conservation of crop germplasm: safflower, *Econ. Bot.* 23(4):324 (1969).

36. H. A. Van Rheenen, Aspects of cross-fertilization in sesame. *Trop. Agric.* 57(1): 53 (1980).

37. B. Mazzami, Variation in the oil content of local cultivars of sesame in Venezuela, *Agron. Trop.*/(Maracay, Venez.) 9(1): 3 (1959).

38. A. O. Karve and A. K. Deshmukh, Leaf extract assay for *Alternaria* resistance screening in safflower, *Indian J. Gene. Plant Breed.* 37(1): 154 (1977).

39. L. D. Bass, Data supplied by plant physiologist, National Seed Storage Laboratory, Fort Collins, CO, 1968.

40. S. N. Kole and K. Gupta, Biochemical changes in safflower (*C. tinctorius*) seeds under accelerated aging, *Seed Sci. Technol.* 19(1):47 (1982).

41. P. Neargaard, *Seed Pathology*, Vols. 1 and 2, Macmillan, London, 1977.

42. P. M. Zukovsky, Cultivated plants and three wild relatives, Sovetskaja Nauka Mosco, 1950, an abridged version published by CAB, 1962.

43. E. V. Popova, Castor oil plant, Lenin Academy of Agriculture, p. 63, Scientific and Industrial Application, Botany Men. Culture, Leningrad, 1939.

44. A. P. Blagdyr and L. B. Sevastyanova, Changes in sowing quality of sunflower and castor seed stored under low temperatures, *Sel. Semenoved* (Moscow) 1:59 (1975).

45. V. A. Fedosenki, The use of super-low temperatures for long-term seed storage: Methods

and techniques, Byull. *Vses. Ordena Lenina Ordena Druzhby Narodov Nauchnolssled. Inst. imeni N.I Vavilova* 77:53 (1978) (Russian).

46. R. J. A. W. Lever, Pests of Coconut Palm, Agriculture Studies No. 7, Food and Agriculture Organization, Rome, 1969.

47. G. Mathai, Adult performance of hybrids involving three varieties of coconut (*C. nucifera* L.), *Indian - Coconut J.* 9(11): 1 (1979).

48. N. N. Potty, B. J. Naik, L. Rajamony, and P. K. R. Nambiar, Comparative performance of eight coconut varieties in red loam soil, *Indian Coconut J.* 11(5):1 (1980).

49. C. W. S. Hartley, The *Oil Palm*, Tropical Agriculture Series, Longman, London, 1967.

50. K. Redenbaugh, D. Slade, P. Viss, and J. A. Fujii, Encapsulation of somatic embryos in synthetic seed coats, *HortScience* 22(5):803 (1987).

51. K. Redenbaugh (ed.), *Synseeds: Application of Synthetic Seeds to Crop Improvement*, CRC Press, Boca Raton, FL, 1993.

52. P. Janssens, Le palmier a huile au Congo Portugais et dans l'enclave de Cabinda, Descriptions des principales varieties de palmier (*Elaeis guineensis*), *Bull. Agric. Congo Belge* 18 (29 and 59), 1927.

53. E. H. G. Smith, A note on recent research on empire products *Bull. Imp. Inst. London* 33(3):371 (1935).

54. G. A. White, Distinguishing characteristics of *Crambe abyssinica* and C. *hispanica, Crop Sci.* 15(1):91 (1975).

55. L. C. Beck, K. J. Lessman, and R. J. Buker, Inheritance of pubescene and its use in outcrossing measurements between a C. *hispanica* and C. *abyssinica, Crop Sci.* 15(2):221 (1975).

56. M. Jablonski, Contribution to the physiology of germination and determination of the value of Crambe, *Crambe abyssinica, Alb. Thaer: Arch.* 6(9): 649 (1962).

57. *Seeds: The Yearbook of Agriculture*, U.S. Department of Agriculture, Washington, DC, 1961.

10
Vegetable Crops

I. INTRODUCTION

Successful vegetable production depends upon a supply of high-quality seeds. At the present time, the seed industry plays an important role in both the production and the distribution of vegetable seeds. One important reason that the demand for vegetables has increased in many parts of the world is the development of improved and more diverse methods of vegetable preservation, such as canning, freezing, and dehydration. Such technology has led to a larger range of vegetable types and demands by processors for increased quantities. There have also been rapid advances in the technology of vegetable production on a large scale in Europe and the United States, where the rapid adoption of herbicides suitable for individual crops has led to crop establishment in a relatively weed-free environment and the subsequent use of precision drilling. This increasing interest in vegetables, no matter the scale of production, has led to local, national, and international activity to improve the supply and quality of vegetable seed.

Vegetable production areas range from large-scale farm enterprises and market gardens growing for profit to private gardens or homesteads, where vegetables are essential to families' efforts to supplement their diets or income. Vegetables are also cultivated in some communities as physical recreation or as a pastime or hobby. With the further extension and development of urban communities, the commercial producer has continued to play an increasingly important role in meeting the vegetable requirements of the population. Commercial production has extended considerably during the last few decades in many parts of the world as large-scale enterprises endeavor to provide continuity of supply for the fresh market, processors, and export.

Vegetables can be classified according to the part of the plant eaten (Table 1). When vegetables are classified according to their season of growth or climatic area, the diversity of species is apparent and there is some indication of each species' environmental requirements. Also, this can indicate the type of climate where seed production mayor may not be successful (Table 2) (1).

II. TOMATO, PEPPERS, AND EGGPLANT

Tomato, peppers, and eggplant (or brinjal) are three important "fruit" vegetables belonging to the Solanaceae, or nightshade, family. Also included in this important family are potato, petunia, tobacco, and belladonna. Warmer tropical climates favor most members of this group. A flower with a five-lobed calyx and a gamopetalous

Table 1 Vegetables Classified According to Part of Plant Used as Vegetable

Part consumed	Scientific name	Common name
Seedling	*Glycine max* (L.) Merr	Soy bean
Shoot	*Asparagus officinalis* L.	Asparagus
Leaf	*Amaranthus cruentus* L.	African spinach
Bud	*Brassica oleracea* L.	Brussels sprout
Root	*Daucus carota* L.	Carrot
Bulb	*Allium cepa* L.	Onion
Flower	*Cynara scolymus* L.	Globe artichoke
Fruit	*Lycopersicon esculentum* Mill.	Tomato
Seed	*Phaseolus vulgaris* L.	Bean

Source: Ref. 1.

corolla characterizes them. The ovary is generally superior and fruit is usually a berry or a capsule.

A. Tomato

1. Introduction

Botany and Origin.   Tomato (*Lycopersicon esculentum*) is a warm-season vegetable crop of great commercial interest. It requires warm weather and plenty of sunshine for its best development. Tomato is intolerant of frost and often unfruitful when temperatures are too high. Although it is grown under a wide range of climatic conditions, its growth may be checked or the fruiting abilities permanently impaired by prolonged cold, cloudy weather or repeated low temperatures without actual freezing.

Although in tropical South America, where it is native, tomato is a perennial crop, it is usually grown as an annual for both seed as well as fruit purposes. The plant is large and heavily branched with alternate, pinnate compound leaves. Leaves and young stems are covered with fine hairs. The initially erect plant soon becomes decumbent when the branches grow more than about I m long, adventitious roots being developed readily from the stems. The stems are round, soft, brittle, and hairy

Table 2 Examples of Classification of Vegetables According to Season

Scientific name	Common name
Warm-season crops	
Hibiscus esculentus L.	Okra
Zea mays L.	Sweet corn
Cool-season crops	
Apium graveolens L.	Celery
Beta vulgaris L.	Beet root, red beet

Source: Ref. 1.

when young, but become angular, hard, and almost woody when old. Tomato plants develop a rapidly growing, strong taproot if allowed to grow naturally, but in the transplanted crop, the taproot is broken and the plant develops a dense fibrous root system. All roots branch profusely, spreading over an area of about 1 m^3.

The tomato inflorescence, or flower cluster, is borne laterally, terminating the growth of lateral branches. The plant axis is made up of a succession of lateral axes. In some varieties, flower clusters occur every third intermode of a main axis. Such varieties have an "intermediate" growth habit, with plant height continuing as long as the environment favors growth. In others, inflorescence occurs more frequently, sometimes at every node, until the formation of a terminal cluster, which ceases the elongation of that branch. This type of growth habit is called "determinate" or "self-pruning." The latter plant types produce a greater number of inflorescences within a given vine length. Also, these types tend to mature more quickly than the indeterminate types. The number of flowers per cluster in most cultivars varies from four to five and sometimes more. Small-fruited types usually have a greater number of flowers. In most commonly grown field varieties of tomato, about two to four flowers set fruits within each cluster.

The tomato flower has a 5- to 10-parted calyx, which persists until the fruit matures. The yellow petals are united in a short tube with five or more lobes, which are often recurved. The five stamens are attached to the base of the corolla tube. The long anthers of the stamens are partly united in the form of a cane surrounding the pistil. The latter consists of a multicelled ovary and a long slender style reaching the tip or projecting from the staminal cone as much as 2 mm, with a capitate, single, narow, or sometimes bulbous stigma. The buds, flowers, and fruits develop progressively within an individual cluster. There is no definite flowering peak in tomatoes. Anthesis appears to be correlated with temperature and soil moisture.

Tomato is normally a self-pollinated crop, self-fertilization being favored by the position of the receptive stigma within the cone of the anthers and the normal pendant position of the flower. Although the stigma is receptive at the time of anthesis, anthers do not dehisce until about 24–48 hrs later. Cross-fertilization of tomato flowers to the extent of about 5% may occur through insects such as bees. Pollen tube growth is relatively slow even at optimum temperatures. The tomato fruit is a two- to many-celled berry with a fleshy placenta of variable shape, size, and color depending upon the variety.

The enveloping mucilaginous sheath around the seeds must be removed before they are dried, when they obtain a tan to light brown color. Seeds are flattened in an obovate fruit, conspicuously pubescent around the periphery, distinguishing them from the seeds of other solanaceous vegetables. Like fruit size, the number of seeds per fruit also varies greatly, ranging from about 150 to 300 or more in some commercial varieties. The 1000–seed weight of tomato is approximately 2.4 g.

Species and Cultivars. The genus *Lycopersicon* consists of a few annual or short-lived perennial species. Bailey (2) classified tomatoes into two species: *L. pimpinellijolium* and *L. esculentum*. The latter is the parent of several commercial tomatoes, and practically all belong to the variety Commune, the common tomato. Others included in this class are grandifolium (a large-leafed or potato-leafed type), Validum (an upright or dwarf type with dense, dark green foliage), Cerasiformae (a cherrylike type with two-celled, globular fruits and "standard" foliage), and Pyriformae (a pearlike or oblong-type fruit with "standard" foliage). In addition to the two species mentioned above, four more species of tomato have been recognized: *L. cheesmani*, *L. peruvianum*, *L. hirsutum*, and *L. glandulosum*.

Earliana was probably grown more extensively in the United States than all other early cultivars combined because of its excellent characteristics such as extreme-earliness, bright red resistance to fusarium wilt, and good processing quality. Bonny Best Marglobe and Pink are other important cultivars. Tigchelaar (3) included several new introductions in the New Vegetable Varieties List XXI. Other cultivars of tomato include Pearmech, Early Cascade, Full House, Baron, Florida MH-1, Kewalo, La Fayette, Vermillion, Royal Red Cherry, Auburn 76, Bonner Patriot, Ohio 736, 7663 and 7681, Sandpoint, Mountain Pride, Cherokee, and Oregon 5-4, a new parthenocarpic line. Berry and Gould (4) reported the release of Ohio 7870, which is an early-maturing, main-season cultivar suitable for mechanical harvesting and canning.

The Indian Agricultural Research Institute, New Delhi, has released the following improved strains of tomato: Sioux, Pusa Red, Plum, Pusa Ruby, Pusa Early Dwarf, Co-I, Keck-Ruth Ageti, Keck-Ruth, Punjab Tropic, Punjab Chhuhara, Krishnasagar-S-20, and Tomato Sweet-72. Other important cultivars of tomato grown in India are Bonny Best, Large Red, Golden Queen, Best of All, Ponderosa, Kalianpur Type-I, Fireball and Early Leth-bride Rome, Italian Red Pear, and Pusa Ruby.

2. Breeding

Earlier work on tomato breeding was restricted to the improvement of fruit shape, size, color, and quality. During the 1950s and 1960s, many fine tomato varieties were found to be susceptible to several diseases and pests and to lack adaptability to certain regions, U.S. Department of Agriculture (USDA) investigators developed Marglobe, a fusarium wilt-resistant variety of tomato, which became one of the most important commercial varieties in the world. Marglobe belongs to the *L. esculentum*. Soon an immune line was discovered in *L. pimpinellifolium*, which encouraged the trend toward the development of immune varieties (e.g., Pan America and Southland).

In addition to the efforts to develop tomato varieties resistant to fusarium wilt, varieties resistant to other diseases such as nailhead rust, leaf spot, leaf mold, verticillium wilt, mosaic, and curly top were sought. Breeders in this area achieved considerable success. Varieties adaptable to the short growing season of North Dakota and those producing fruit under high-temperature conditions in the South were also developed; for example, Sumerset, Summer Prolific, and Allseason. Studies on the inheritance of specific characteristics have been of great practical value in developing newer tomato varieties. Weaver (5) reported that pollen viability has a major role in determining the fruit set of tomatoes at high temperatures and can serve as a basis of screening for tomato plants that should yield more fruit in high-temperature environments.

Hedrick and Booth (6) first observed the phenomenon of hybrid vigor in tomatoes, with the intervarietal hybrids producing about 21% more than the more productive parent. This indicated good commercial possibilities for first-generation hybrids (F_1). Crane (7) first reported male sterility in tomatoes, which was followed by other reports. The arrangement of the male-sterile plants in the field influenced the hybrid seed production of tomato under field conditions. Setting out fertile and male-sterile plants in pairs to intermingle branches was found to be desirable. Although male-sterile plants provided an easier way to produce hybrid tomato seed, few hybrids had good combining ability. The production of hybrid tomato seeds was pointless unless they exhibited definite hybrid vigor.

A further development in tomato breeding has been the use of F_2, or second-generation, seed from a controlled cross. Such seed may have greater commercial significance than using male-sterile mutants in the production of F_1 seed. The F_2 seed is not true hybrid seed as is the F_1. Any F_1 variety or strain must be thoroughly tested against the original parents as well as other improved strains before releasing it for commercial use. Falavigna et al. (8) proposed a new technique of production of F_1 seeds based on hand pollination of intact flowers from a male-fertile seed parent lines, genetically marked by the bs trait. The new technique was compared with those involving emasculation and hand pollination using a male-fertile line homozygous for bs and c traits. The two commonly used techniques for the production of commercial F_1 tomato seeds, hand-emasculation and pollination of male-fertile seed parent lines and (2) hand pollination of male-sterile lines possessing a seedling marker tightly linked to the male-sterile characters, are handicapped by the high cost of hand operation requiring a great deal of time and skilled labor. Exploitation of natural cross-pollination (NCP) could reduce the cost of these hand operations, including that of pollen collection. Schemes based on the opticomechanical separation of hybrid seeds obtained through natural out-crossing of two male-fertile seed parent lines genetically marked by different alleles have been proposed. The benefits of hand pollination of unemasculated tomato flowers as compared with that involving both emasculation and hand pollination are evident from the data reported in Table 3.

Breeding work in Bulgaria has led to the development of the orange-fruited cultivar Karobeta from crosses involving *L. pimpinellifolium* f. galapagos as source of the β-carotene gene. Karobeta is determinate, vigorous, early ripening, and resistant to tobacco mosaic virus. The firm multilocular fruits ripen uniformly, are resistant to cracking, transport well, and have an agreeable sweet flavor. Sokhi et al. (9) evaluated 167 lines of tomato for their reaction to *Meloidogyne* spp., 20 of which showed resistance. Six resistant lines were crossed with susceptible Punjab Chhuhara. The F_2 segregation indicated that a single dominant gene controlled resistance. In Hawaii 5229, the resistance to *M. incognita*, *M. acruta*, *M. arenaria*, and *M. javanica* was

Table 3 The Mean Fruit Set, Fruit Weight, Seeds per Fruit, Contamination (% Brown Seeds), and Amount of F_1 Seeds of Tomato Produced per Hour Under Hand-Pollination of Unemasculated Flowers[a] Compared with Both Emasculation and Hand-Pollination

Parameter	With emasculation	Without emasculation
Mean fruit set	69.9[b]	87.1[b]
Fruit weight (g)	105.4	116.1
Number of seeds/fruit	85.6	88.8
Contamination (% brown seeds)	1.1[b]	6.2[b]
F_1 seeds produced per hour	10.7[c]	26.1[c]

[a] 5415-Mo (bs,c) male-fertile parent line.
[b] Significant at $P = .05$.
[c] Significant at $P = .01$.
Source: Ref. 8.

monofactorial and completely dominant. In Nematex, Small Fry, and Cold Set, a single gene for resistane was identified, being dominant in Nematex and Small Fry but recessive in Cold Set.

A common breeding method for resistance to rough blossom-end scarring in large-fruited fresh market tomatoes is to select genotypes with pointed blossom end morphology. Unfortunately, in many breeding lines, pointedness is associated with adaxial leaf curl (10,11), which may result in increased foliar disease problems (12). Pointedness is usually most pronounced in young immature fruit and gradually disappears as the fruit matures. Barten and Scott (13) studied the characterization of blossom-end morphology genes in tomato and their usefulness in breeding for smooth blossom-end scars and reported that various blossom-end morphology genes may be backcrossed into desirable breeding lines and complementing parents may be intercrossed to obtain optimal smoothness in the hybrid without undesirable pointed mature hybrid fruit.

A number of hybrid combinations give good yields. A combination of Pusa Ruby and Best of All is preferred as the F_1. It not only yields 50% more than Pusa Ruby, but its fruit quality is also very attractive. Even the F_2 generation of this combination gives a 15–20% higher yield and the segregating lines are all marketable. The F_1 not only gives a higher yield of attractive fruits, but is also resistant to root-knot nematodes.

For hybrid seed production, hand pollination is carried out with or without emasculation. Where emasculation is used, buds are first emasculated and then enclosed in grease-proof bags fastened with pins or threads the day before flowering. The flowers are hand pollinated with forceps or a camel hairbrush and then rebagged. Four of 5 days later, the bags are removed when fertilization is assured. Bud pollination is not used. The time taken for hand pollinating each flower is about 50–60 secs. In the United States, the use of stored pollen at 27°C is preferred to fresh pollen at this temperature. Fresh or stored pollen is applied with a blackened matchstick to male-sterile flowers at temperature below 29.°C. If functional male sterility is used, it is unnecessary to emasculate flowers, as pollen cannot reach the stigma without artificial aid.

Alcohol dehydrogenase and esterase isoenzyme genes were used to estimate hybridity in 1594 tomato seeds. Parental lines and hybrid cultivars taken from different lots and experimental stations in Bulgaria were studied. The banding patterns were obtained by means of vertical electrophoresis in polyacrylamide gels. As genetic markers of hybridity, isoenzymes of the Adh-1 and Est-1 *loci* were used. It was established that quantitative variation of isoenzyme No. 1 at locus Est-1 is as being indicative of hybridity as isoenzyme expression at Adh-1 (13a).

3.   Seed-Production Technology

Culture.   The culture of tomato for seed is more or less identical with that used for crops grown for open market or for processing. Seedlings are raised on seed beds, preferably in cold frames or hot beds to overcome the danger of frost and transplanted to the seed production field. In parts of the United States such as the Midwest, tomatoes grown for seed production are directly seeded in the field.

LAND PREPARATION.   Tomatoes grow best in neutral (pH 6.0–7.0) loam soils with good drainage, although the crop can be grown on a variety of soils, ranging from sandy loam to heavy clay. Liming is advocated in acid (pH < 5.0) soils. Lighter textured soils are preferred where earliness is of great importance. Plowing, harrowing, discing, and leveling prepare land. Tomatoes require well-drained soil. The soil surface should be fairly smooth and free from clods.

PLANTING. Tomato seedlings are raised in nursery beds. About 500 g of seed provides enough seedlings to cover 1 ha of land. Seeds are sown in shallow rows by broadcasting on a well-prepared seedbed and lightly covering with soil. Seedbeds should be irrigated immediately after sowing. Seedlings must be protected from direct sun, heavy rains, and frost.

Seedlings are ready to transplant within about 3–4 weeks. In the plains, tomatoes can be grown from June to November, the first crop being sown in June, the second in August, and the third in October–November. If frosts commonly occur in winter, tomatoes should be transplanted in the spring. Spacing of tomato crops varies from 30 to 120 cm within the row and 1 to 2 m between the rows. Both ridges and furrows and flatbed systems are used to plant tomatoes, but the ridge and furrow method provides good support for plants.

Young seedlings need regular watering and protection from cold and heavy rains. Support or staking may be necessary in certain cases. In the northern European countries, tomatoes are commonly staked. Dipping the seeds in 2,4-dichloropropionic acid at 5 ppm for 24 hrs before sowing enhances the germination of seed (14). Finch-Savage and McQuistan (15) studied the effects of abscisic acid (ABA) on the germination of tomato seeds. The seeds were treated with an aerated 10^{-4} M ABA solution for 15 days at 15°C. This treatment reduced both the time to emergence and the spread of emergence times and increased the percentage emergence compared with untreated seeds. The minimum, optimum, and maximum temperatures for seed germination were reported to be 10, 25, and 30°C, respectively (16). Seed germination has been found to decrease with increasing fruit storage time (17). Low-temperature presowing treatment of tomato seed increases the rate of germination (18). Fierro et al. (19) reported that enrichment with CO_2 (900 µL/L, 8 hr/day) and supplementary lighting for about 3 weeks before transplanting increased accumulation of dry matter in shoots by about 50% compared with the control, whereas root dry weight increased 49% for tomatoes and 62% for peppers. Early yields increased by about 15% and 11% for tomatoes and peppers, respectively.

Variation in seed quality during maturation was determined in tomato (*L. esculentum* cv. Rio Grande) in two subsequent years (1998 and 1999 in Bursa, Kenya, and Turkey, respectively) to relate to changes in seed dry weight and fruit maturity. Seed quality was assessed by germination at low (15 and 18°C) and high (35°C) temperatures, germination at low water potential (−0.5 MPa), emergence percentage, and seed quality constant (K_i). Seed dry weight did not change between 60 and 90 and 50 and 80 days after anthesis (DAA) in 1998 and 1999, respectively. However, seeds harvested 70 DAA attained not only maximum germination under stress conditions but also maximum emergence and K_i values in both years. Delayed harvests (80 and 90 DAA) caused a decline in seed quality. Results showed that the occurrence of maximum tomato seed quality is related to changes in fruit color but not in seed dry weight (19a).

Yogeesha et al. (19b) conducted studies on stigma receptivity, pollen viability, and effect of fruit picking on seed yield and quality using parental lines of two popular tomato hybrids, Pusa hybrid 1 and Pusa hybrid 2 during the 1996–1997 post–rainy season. The emasculated female flowers of both hybrids became fully receptive after 2 days of emasculation and remained so for another 2 days. The pollen of both male parents stored under room conditions (mean maximum temperature 26 °C and relative humidity 47%) retained high viability even after 5 days of storage, whereas under a refrigerated condition (constant temperature of 9–10 °C), they showed no significant drop in viability even after 7 and 5 days of storage in Pusa hybrid 2 and Pusa hybrid 1, respectively. The number of seeds and seed weight per fruit obtained from pollination with these pollen treatments did not differ significantly. Seven fruit pickings performed at 3-day intervals showed no significant drop in the number of seeds, seed weight per fruit, and seed quality attributes except the last picking. Considering the pollination period of 30 days, it is likely that the last 4–5 days of pollination resulted in poor-quality fruits with poor see yield and quality. Therefore, the pollination period may be restricted to 25 days only.

Bio-osmoprinting is a combination of osmoconditioning and biopriming procedures that simultaneously hydrates seeds and applies a bacterial coating in a single treatment. The bacterium applied is *Pseudomonas aureofaciens* (*P. chororaphis*) AB254. Processing tomato (cv. OH 8245) subsamples (5 g each) from a single seed lot received one of four treatments; untreated, osmoprimed, AB254 bioprimed, and bio-osmoprimed. Osmoprimed treatments soaked the seeds in aerated −0.8 MPa $NaNO_3$ for 7 days and then dried them back to their original moisture content (approximately 14%). Seeds were bio-osmoprimed by soaking in aerated −0.8 MPa $NaNO_3$ for 4 days at which time a mixture of nutrient broth, polyalkylene glycol, and bacterial stocks were added. The seeds were then hydrated for an additional 3 days. In the absence of pathogen pressure, no differences in germination were observed between bio-osmopriming and conventional osmopriming. Bio-osmoprimed treatments consistently contained 105 bacterial colony-forming units (cfu)/seed compared to 108 cfu for AB254-coated treatments. When seeds were planted in soil-less media infected with *Pythium ultimum*, AB254 coatings protected tomato seeds from infection equally as well as the fungicide metalaxyl. Bio-osmopriming also provided protection, but at a slightly lower rate. This technique improves the chances of the seed lot establishing a healthy stand (19c).

IRRIGATION, FERTILIZATION, AND WEED CONTROL. Tomato crops are irrigated to maintain adequate soil moisture, but water logging should be avoided at all times during crop growth. Winter crops may need irrigation every 2–3 weeks, whereas summer crops may need it once a week. Irrigating tomato crops during periods of frost helps to keep temperatures above freezing. Heavy irrigation after a long dry spell may result in fruit cracking. Similarly, irrigation late in the season may result in watery fruits of poor quality.

A crop yielding about 40 tons of tomato fruit requires as much as 93 kg N, 20 kg P, and 126 kg K per hectare. A fertilizer mixture containing 40–60 kg N, 60–80 kg P_2O_5, and 100–120 kg K_2O can be applied prior to transplanting. The following doses of fertilizers have been recommended (20):

Early (rainy)-season crop—Five tons of farm yard manure (FYM) and 500, 312, and 85 kg/ha of ammonium sulfate, superphosphate, and muriate of potash, respectively

Winter crop—Five tons of FYM and 400, 250, and 68 kg/ha of ammonium sulfate, superphosphate, and muriate of potash, respectively

Spring-season crop—Five tons of FYM and 300, 188, and 50 kg/ha of ammonium sulfate, superphosphate, and muriate of potash, respectively

The higher levels of phosphorus applied to mother plants increases the total seed yield, and the combination of higher nitrogen and phosphorus has been found to increase the germination and seedling emergence rates of the progeny in greenhouses (21).

A deficiency in boron may lead to cracking of the skin with pitted and corky areas, deformed shape, and uneven ripening of tomato fruit. An increased boron supply improves fruit shape and reduces hollowness in the fruits (22). Govindan (23) reported that with increasing availability of zinc in the growth medium, the ascorbic acid content of tomato fruits increased but the carbohydrate content decreased. Fields should be kept free of weeds through frequent intertillage operations, which should be shallow as the plants grow so as not to injure roots growing deeper than the 5-cm surface soil.

TRAINING AND PRUNING. In some Western countries, branches of tomato plants are trained vertically with the help of wires or ropes. This practice is claimed to result in early ripening, less disease incidence, easier interculture and harvesting, clean and healthy fruits, and higher yields of better-quality fruits. Pruning of side shoots to produce single-stemmed plants and staking have been claimed to have the following advantages: (a) higher yield per hectare (average 6 kg/plant), (b) higher cluster and early crop, (c) more efficient interculture, spraying, harvesting, and so forth, and (d) cleaner, more uniform, and larger fruit. Vine-ripened fruits were found to be more nutritious and better tasting.

Roguing. Little roguing is required in the production of market seed if stock seed plantings are rogued carefully. Commercial production of market seed over large areas makes roguing inefficient and expensive. However, careful roguing is essential for stock seed production in which rogues are eliminated on a plant basis. Off-type plants should be removed to avoid any possibility of cross-pollination. At the maturity stage, fruits and plants should be examined for overall characteristics and type. If a large proportion of the fruit on a plant fails to meet the required external and internal characteristics (e.g., shape, color, size, nutritional value), the plant should be removed.

Isolation of seed fields is highly desirable to prevent even the limited cross-pollination that may take place. An isolation distance of about 16 m between two different varieties and even a greater distance for the production of stock seed have been recommended.

Control of Pests and Diseases. Tobacco caterpillars (*Prodenia litura* F), jassids (*Empoasea devastants*), epilachnabeetles, and fruit worms (*Heliothis armigera*) are the major insect pests of tomato. Spraying 0.1 % DDT or endrine at 2-week intervals can control most of these. For jassids, 0.2% endrine or 0.02% parathion is recommended. A mixture of 2.02% benzenehexachloride (BHC) and 0.1% dichlorodiphenyltrichloroethane (DDT) or 0.02% endrine or 0.02% parathion, repeated two to three times during the infestation, is more effective.

The major diseases of tomato in India are damping off (*Pithium* spp., *Phytophthora* spp., and *Rhizoctonia* spp.), defoliation (*Stemphylium solani* [Weber] and *Alternaria solani* [Ellis and Martin]), early blight (*A. solani*), fruit rot (*Phytophthora* spp.), fusarium wilt (*Fusarium lycopersici*), powdery mildew (*Leveillula taurica*), and viral diseases such as tomato mosaic and leaf curl. Damping off of tomato can be controlled by adopting the following control measures: (a) select nursery beds with good drainage, (b) drench seed beds with 2:2:50 Bordeaux mixture or other copper fungicide 2–3 days before sowing, and (c) spray the seedlings with 2:2:50 Bordeaux mixture or copper fungicides (Agrosan or Cerasan, Cuprovit, Pernox) at 3-day intervals for 1 week. Defoliation of tomato leaves, flowers, buds, and fruits can be checked by 4:4:50 or 4:2:50 Bordeaux mixture, 0.25% Copramat, or 0.2% Diathane Z-78. Weekly spraying with DDT or endrine in the early stages of crop reduces insect vectors of viral diseases of tomato.

Harvesting. Tomatoes grown for seed are harvested in much the same way that they are picked for market or for processing. Although cracked or bruised fruits can be used for seed, care should be taken to avoid fruit damage. Fruit should not be left on the vine to overripen. Ripe tomatoes should be picked leaving the stalk on the vine if possible to avoid fruit injury. The fruit, while grasped in the hand, should be dislodged from the vine by twisting, with the thumb kept pressed against the vine. Fruits may be harvested mechanically with either stationary or mobile harvesters. Mechanical harvesting has made bulk handling of tomatoes possible, with harvest rates reaching as high as 30 tons per hour. Dev and Sharma (23a) found that for quality seed production in tomato (cv. Roma), fruits harvested from first picking proved to be superior.

Basave-Gouda et al. (23b) harvested tomato cv. L-15 plants raised during the kharif season of 1996 at three different stages—fully matured with uniform yellow skin, fully matured with yellowish red skin, and fully matured with uniform red skin—at 5-day intervals. Seeds from the fruits of 11 pickings were extracted separately to assess seed qualities (1000–seed weight, germination, dry matter production, and vigor index). The 1000–seed weight, germination, and vigor index remained almost the same up to the sixth picking; thereafter the values decreased. The germination potential of the seeds decreased with advanced stages of fruit maturity. The fruits harvested at the fully mature and ripe stage with red skin had the highest 1000–seed weight (2.09 g), germination (86%), dry matter production (16.62 g), and vigor index (1437).

Valdes and Gray (23c) recorded the percentage seed germination, percentage normal seedlings, mean germination time, seed water, and dry matter contents throughout fruit maturation in several crops of tomato (grown in greenhouses, in polytunnels, or in the field in Chile) in order to identify the optimum harvest date for seed quality as defined by the fruit-maturity stage. Maximum percentage seed germination and percentage germination of normal seedlings were obtained as early as the breaker stage of the fruit, and thereafter they did not change as the fruit ripened. Germination was most rapid from seeds at the red stage of fruit maturity, and, in general, further delay in harvest led to deterioration of the seed.

4. Postharvest Operations

Seed-Extraction Methods. Tomato seed can be collected using juice-extracting equipment or equipment similar to that employed for vine seeds followed by fermentation or acid treatment. The juice and the pumice (pulp, skin, and seeds) are extracted using canning equipment such as pulpers or juice extracters. Fruits are scaled to loosen the skin. The crushed material is usually passed into a revolving cylindrical screen, which allows the seeds and juice to pass through the mesh, whereas the fruit debris passes through the cylindrical screen to drop in the field. Alternately, the debris is collected separately for later disposal if the operation is stationary. The juice and seed mixtures are collected in separate containers. Field operation of this type of machine is shown in Figure 1. This method is rapid and leaves the seed practically free of the gelatinous tissue that surrounds it in the fruit. Large quantities of seed extracted from field crops are usually washed in long water troughs (Fig. 2).

Machines similar to that used for vine crops lack heavy knives. The cut and crushed fruit is passed between corrugated rollers before falling into the revolving wire mesh cylinder. The material and juice passing through the screen is then poured into large tubs or vats and seed is extracted by fermentation or acid treatment.

FERMENTATION. The masses of fruit pulp and juice are allowed to ferment without adding water. Fermentation disintegrates the mucilaginous matter adhering to the seed, allowing the seed to sink to the bottom. The undecomposed pulp floats to the top, leaving a layer of clear liquid in between. The contents of the vat should be stirred frequently to release the seeds entrapped in the floating pulp and to prevent a fungal attack starting at the surface of the mass, which might discolor or even injure the seed. Satisfactory separation of seed and pulp may be achieved in 2 days if the process continues at 23.9–26.7°C. With long-term fermentation, the germination percentage declined severely at ≥ 25°C (Fig. 3). At 15°C, more than 50% of the seeds were viable even after 25 days of fermentation. Fewer of the seeds lost their viability and good vigor at 15°C even after 15 days of fermentation (Fig. 4) (24).

EXTRACTION WITH ACID. Hydrochloric acid is added to pulp at the rate of about 85 g/11.34 kg of material. After a thorough mixing of acid and pulp, seed may be washed free within about 15–30 mins. Soaking tomato seeds in 0.6 M HCl or 0.05% 2-phenylphenol for 15 mins was highly effective in reducing both the number of infected seeds in the treated samples and primary canker incidence in field plots to zero or near-zero levels (25).

The chemical extraction procedure has the following advantages: (a) seed can be extracted and dried the same day, (b) the problem of low and high temperatures are

Figure 1 Tomato fruit crusher and seed extractor. (From Ref. 1.)

Figure 2 Washing extracted tomato seeds in water troughs. (From Ref. 1.)

avoided, and (c) discolored seed resulting from fermentation is entirely eliminated. A distinct disadvantage of the acid extraction method is that, if the bacterium causing canker is present, the acids used in this process do not kill it. If canker is suspected of being present, the acid-extracted seed should be further treated with acetic acid.

Washing and Drying. The extracted seed may be washed using either a shaker-washer or a sluiceway. The former consists of two screens, one suspended above the other in the same frame, which are constantly shaken back and forth by a mechanical device. Each screen consists of a tray with wooden sides (about 2.5 × 10–15 cm). Pulp and seed are periodically placed on the upper screen, which prevents the passage of the coarse pulp and skin, allowing the seed and smaller particles to reach the lower screen. Additional water in the lower screen floats off the remaining remnants of pulp and rind as well as some light-weight seeds. After sufficient (5–10 kg) seed is accumulated on the lower screen, it is removed and allowed to drain. The seed is then dried.

      Another method involving the use of a current box, washing flume, or sluiceway is very effective when large quantities of seed need to be washed. The flume is three-sided and about 3 m long and 25–35 cm wide. Cross pipes or riffles about 75 cm high are placed every 30–35 cm. The seed, along with a gentle stream of water, enters the

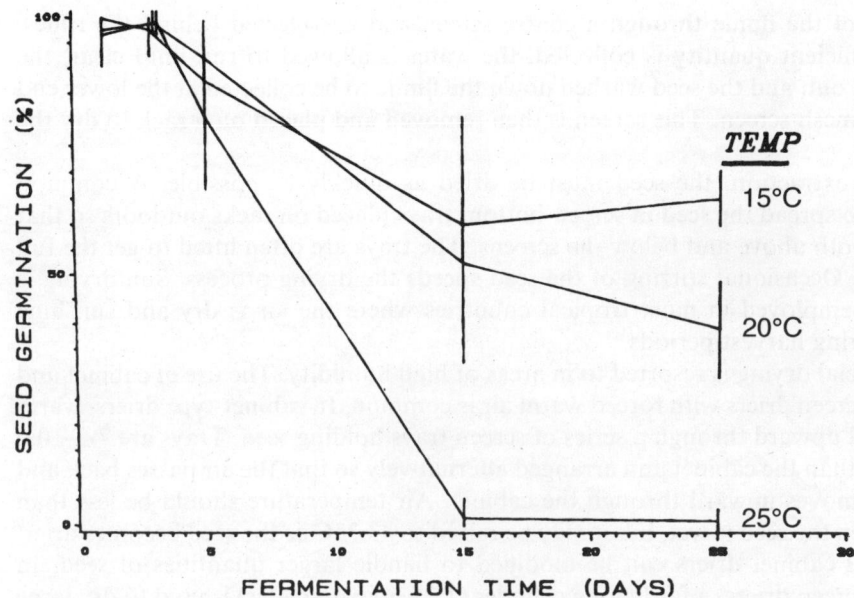

Figure 3 Long fermentation times reduce seed germination. The percentage of viable seeds after fermentation for 1, 3, 5, 15, or 25 days at 15, 20, or 25°C. The results are reported as means with standard errors and are from four groups of 100 seeds each. (From Ref. 24.)

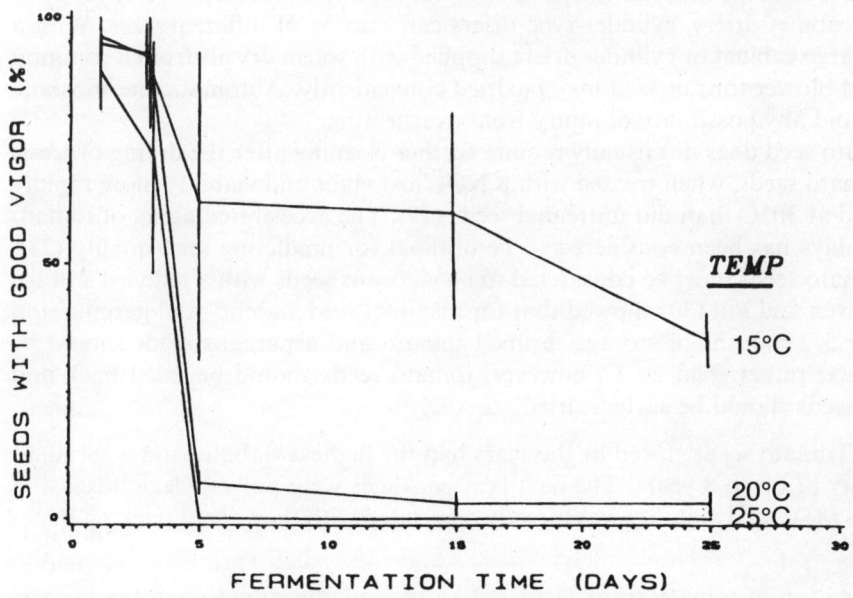

Figure 4 Long fermentation times reduce seed vigor. Percentage of seeds with good vigor (hypocotyls $\geq$ 16 mm after 94 hr) each fermentation for 1, 3, 5, 15, or 25 days at one of the following temperatures: 15, 20, or 25°C. The results are the means of four groups of 100 seeds each. Standard errors of the means are also given. (From Ref. 24.)

upper end of the flume through a coarse screen and is collected behind the riffles. When a sufficient quantity is collected, the water is allowed to run until clear, the riffles taken out, and the seed washed down the flume to be collected at the lower end on a small-mesh screen. This screen is then removed and placed on a rack to dry the seed.

After extraction, the seed must be dried as quickly as possible. A common method is to spread the seed in screen-bottom trays placed on racks outdoors so that air passes both above and below the screens. The trays are often lifted to get the full solar effect. Occasional stirring of the seed speeds the drying process. Sun drying is commonly employed in most tropical countries where the air is dry and sunshine prevails during harvest periods.

Artificial drying is resorted to in areas of high humidity. The use of cabinet and revolving-screen driers with forced warm air is common. In cabinet-type driers, warm air is forced upward through a series of screen trays holding seed. Trays are 7.5–10.0 cm shorter than the cabinet and arranged alternatively so that the air passes back and forth as it moves upward through the cabinet. Air temperature should be less than 37.8°C when the seed is wet, but it may be raised to 43.3°C as the seed becomes drier. These small cabinet driers can be modified to handle larger quantities of seed. In revolving-screen driers, a fairly long cylinder (1.25 m in diameter) is used to dry large quantities of seed rather rapidly. As the cylinder slowly turns and tumbles the seed within it, a stream of warm, dry air is blown through from one end to other. The screen mesh is maintained such that the seed will not fall through. A door in the jacketed cylinder allows loading and unloading of seed. Seed may be dried to a moisture content of about 12%. At higher drying temperatures of 45, 50, and 55°C, the percentage germination decreased from an initial 95% to 81, 70, and 64%, respectively (26).

Like cabinet driers, cylinder-type driers can also be of different sizes. With a battery of large cabinet or cylinder driers supplied with warm dry air from a common furnace and blower tons of seed may be dried conveniently. Automatic thermostatic controls avoid any possibility of injury from overheating.

Tomato seed does not usually require further cleaning after the drying process. Primed tomato seeds, when treated with KNO_3, lost vigor and viability more rapidly when stored at 30°C than did untreated seeds (27). The accelerated aging of tomato seed for 8 days has been considered to be optimal for predicting seed quality (28). Primed tomato seeds must be considered to be vigorous seeds with a reduced storage life (29). Owen and Pill (30) showed that for maximal seed viability and germination rate after 1.5–3 months of storage, primed tomato and asparagus seeds should be stored at 4°C rather than 20°C; however, tomato seeds should be dried back and asparagus seeds should be surface dried.

Packing.   Tomato seeds stored in glass jars had the highest viability and a germination capacity of up to 8 years. The next best packages were polyethylene boxes and plastic bags (31). Seed viability was directly correlated with the type of packing material used.

Yields.   One ton of tomato fruits yield 3–7 kg of seed depending upon the variety. Under California conditions, average seed yields of about 283 kg/ha are obtained. Yields per unit area, however, vary considerably with the variety, crop stand, and proportion of the crop saved for seed. In India, Singh et al. (32) recorded an average yield of 145 kg/ha of tomato seed for 3 normal years. The quantity of fruit required to produce 1 kg of tomato seed varied from 160 to 210 kg according to variety. Whereas the big-fruited, mild-season varieties of Marglobe and Rutgers required 190 and 210, respectively, kg fruit to produce 1 kg of tomato seed, the medium-sized, midseason variety Best of All required only 160 kg.

B. Pepper (*Capsicum* spp.)

1. Introduction

The chili, or pepper plant (*Capsicum* spp.), is native to tropical South America (Brazil). It is grown as an annual crop in the temperate regions. Warm and humid climates of tropics favor the growth of capsicum. Peppers are essentially a hot weather crop and susceptible to frosts. High light intensities increase yield but reduce the content of capsalcin, the pungent principle in the fruit.

Botany. Bailey (2) divided peppers (*C. frutescens*) into five groups based on fruit shape:

1. Cerasiforme—the cherry pepper, a pungent variety
2. Conoides—the cone pepper, also pungent, with conical or oblong-cyndrical fruits
3. Fasciculatum—red cluster with effect fascicled fruit, about 7.5 cm long and 0.6 cm thick, red in color, and extremely pungent
4. Longum—long pepper with dropping elongated, pungent fruit, 7.5–30.0 cm long and tapering at the apex
5. Grossium—bell or sweet pepper, having large, puffy red or yellow fruit with a depression at the base and usually furrowed sides and a mild flavor

The pepper plant branches dichotomously, so that the first division (branching) has no main or control stem. The younger herbaceous stem and branches turn woody and brittle with age. Leaves are simple, flat, glabrous, and vary in shape from ovate to elongate. The leaf size varies considerably from fairly large in sweet peppers to small in the pungent varieties. Roots spread to a depth of more than 1 m in the form of a dense network. Capsicum flowers are normally solitary, but are occasionally borne in small cymes of leaf axils. The calyx is five lobed and the corolla is five-parted and white (or occasionally purple) in color. The five stamens attached to the base of the corolla are separated. The bluish anthers dehisce by splitting longitudinally. The single style is usually longer than the stamens and the stigma is club shaped. The ovary generally has three locules, but the number may vary from two to four or more. Peppers (cv. Ruby King) tend to blossom and set fruit earlier under short-day conditions, blossoming being greatly delayed when the natural daylight period is lengthened by incandescent light (33).

Capsicum fruit is a podlike berry with a short, thick peduncle. The shape of the fruit varies from a flattened oblate to long, slender, and tapering. The size also ranges from very small (e.g., Tobasco, a pungent type) to large fruits or sweet peppers (e.g., California Wonder, World Beater). The fruit pericarp may be thin or thick according to the variety. Fruit set is highest in the early stages of anthesis, but subsequently declines as the majority of flowers are aborted (34).

Seeds within the fruit mature as the fruit ripens, the fruit color usually changing from green to red or yellow. The fruit base is often divided into two to four or more locules. The seed is borne in a compact formation on the placenta, usually at the basal end of the fruit. The varied fruit characteristics and their combinations constitute a fairly large number of pepper cultivars. Capsicum seed is slightly larger than that of tomato, with the 1000–seed weight of pepper being approximately 6 g.

Cultivars.   Tigchelaar (3) described the following pepper cultivars: Bellringer, Big Bertha, Early Jalupeno, Gypsy, Hybelle, Lady Bell, Mild California, Petile Yellow Sweet, Pip, Registant, Giant-4, Starr, and Sweet Belle. Garden Sunshine is a new mosaic-resistant, bell-type sweet pepper with lemon yellow immature fruits that ripen to a bright red color.

2. Breeding

Mass selection was the common method for varietal improvement in the early part of twentieth century (e.g., selection of Chilli No. 9 from the native Mexican types), The Connecticut Station developed a new sweet pepper, Windsor A, a cross between California Wonder and Bountiful. This was followed by a release of Pennwonder, a cross between Harris, Earliest, and California Wonder.

Joshi and Khalatkar (35) irradiated genetically pure *C. annum* seeds (2–20 kR gamma-radiation ^{60}CO) and studied various quality parameters of the population raised from these seeds. Low-radiation doses enhanced the number of seeds per capsule as well as fruit size and number. Higher doses reduced seed and capsule production. The lower doses also induced mitotic and meiotic abnormalities as well as pollen sterility, although the frequencies for these parameters were much greater with higher doses. The stimulation in growth and capsule and seed production appeared to be physiological rather than genetic, since it was not carried forward in the M_2 generation.

Attempts at mutation breeding in *C. annum* have resulted in the induction of a number of mutants for growth habit, capsule, and seed characteristics (36). Sodium azide, a very potent chemical mutagen, induced large variability in the number of capsules per plant, number of seeds per capsule, seed size, and weight of 100 seeds. The number of seeds per capsule varied from 18 to 84, and the 100–seed weight ranged from 0.5 to 1.3 g. These mutants also exhibited a higher germination percentage and vigor than the parental variety.

Restriction fragment length polymorphism (RFLP) patterns were detected between two parental lines of the commercial hybrid cultivar Maccabi. A suitable probe was chosen from a *Capsicum* genomic library and the HindIII digest of the pepper DNA was selected for the assay. The observed polymorphism was inherited as a simple codominant allelic marker, enabling the distinction between the parents themselves and between parent and progenies (36a).

3. Seed-Production Technology

Culture.   Peppers grown for seed are cultivated in practically the same way as those grown for market or processing. Seedlings are raised protected by glass or cloth or more usually on seed beds. In certain southwestern parts of the United States, seeds are occasionally sown directly in the field.

LAND PREPARATION.   Peppers grow best in well-drained loam soils rich in organic matter. Sandy or sandy loam soils are preferred where the growing season is short. Acid soils and water-logging conditions are not suitable for peppers. The land should be plowed and planked smooth to prepare a seed bed of suitable size. Farmyard manure should be mixed during plowing.

PLANTING.   In regions with mild winters and moderate monsoons, peppers can be sown year round. Although crops can be grown at any time of the year, April to June is the best time both for the hills and the plains (37). In India, large-fruited chili varieties are grown from August to September in the plains and from April to May in the hills. About 1.0–1.5 kg of seed sown in a 230 m^2 area can provide sufficient seedlings for planting an area of 1 ha. For uniform sowing in nursery bed, seeds are mixed with sand before broadcasting. About 6- to 8-week-old seedlings (15–20 cm tall) are considered to be ideal for transplanting. Seedlings are transplanted in the field with 60–100 cm between rows and 30–60 cm between plants B1F −2 (tribenzimidazolido phosphate) applied at 10-8% concentration as a seed treatment and sprayed at flowering has been reported to produce higher yields and improve fruit quality (38). Macit (39) demonstrated that potassium salt, sodium hypochlorite, gibberellic acid (GA), indoleacetic acid (IAA), and kinetin ruptured the seed coats of hot peppers, increasing their germinability.

IRRIGATION, FERTILIZATION, AND WEED CONTROL.   In addition to about 25 tons/ha of FYM, peppers require 45 kg N, 80 kg P_2O_5, and 40 kg K_2O/ha for irrigated crops and 30 kg N, 50 kg P_2O_5, and 25 kg K_2O/ha for unirrigated crops. Increasing N application from 125 to 200 kg/ha significantly increased fruit and seed yields (40).

The seed crop is first irrigated after transplanting followed by subsequent irrigations at 5- to 7-day intervals depending upon the season and moisture-retention capacity of the soil. During rainy and winter seasons, the interval between two irrigations may be 10–15 days or longer. Two to three shallow hoeings during the early stages of crop growth will keep weeds under control.

Osman and George (41) reported that in a greenhouse trial with sweet peppers (cv. World Beater), application of 1900 kg N, 640 kg P, and 2800 kg K per hectare resulted in the highest fruit and seed yields and germination percentage. Seeds from fruits lower on the plants had the highest mean seed weight and germination percentage and required the shortest time for germination and seedling emergence.

Roguing.   When a selected stock seed is grown well isolated, little roguing is necessary. Plants are rogued based on the plant and fruit characteristics as a whole rather than the individual characteristics. Off-types should be removed as soon as they are observed. If a plant with small leaves that is likely to be a pungent variety is crossed with an adjacent sweet pepper, rogues with pungent fruit will appear in the next generation. Plants with undesirable fruit types can be rogued out as the first fruits are nearing edible maturity. This eliminates off-type seed from the harvest, and more importantly prevents further cross-pollination with normal adjacent plants. Occasional off-color plants can be removed at the fruit-ripening stage.

Pepper is predominantly a self-pollinated crop, but some cross-pollination may occur if plants are placed closed together. A minimum distance of about 400 m between varieties is desirable, and a greater distance is considered safer between distinct types (e.g., pungent and sweet peppers). The distance between varieties should be even greater for the production of stock seed.

Control of Pests and Diseases.   Pepper is susceptible to a number of diseases caused by bacteria, fungi, and viruses. In India, chili thrips and tiny sucking insects that feed on leaves are very common. A spray of 0.025% nicotine sulfate or 5% benzenehexachloride (BHC) dust at 20 kg/ha effectively controls the pest. Pepper weevils (*Anthonomus eugenii*) and pepper maggots (*Spilographa electa*) are serious insect pests. Application of 5% dichlorodiphenyltrichloroethane (DDT) dust every 10 days beginning at setting and dusting with talc to keep the fruits covered can control pepper weevils and maggots, respectively.

The important diseases of pepper are damping off (*Pythium aphanidermatum*, *Phytophthora* spp., and *Pellicularia filamentosa*), anthracnose (*Collectotrichum capsici*), die-back (*Vermicularia capsici* Syed), leaf curl or mosaic, blight or wilt, blossomend rot, and root knot. Seed treatment with Cerasan or a 1% spray of 5:5:50 Bordeaux mixture is recommended to control damping off and anthracnose. Die-back, blight, and wilts also can be checked with a 5:5:50 Bordeaux mixture or a copper fungicide at 2 kg in 150 L of water. The incidence of die-back was least after seed treatment with carbendazim and spraying with mancozeb. Fruit not was least after seed treatment with thiram and spraying with mancozeb (42). Seed treatment with emisan gives effective control of *Aspergillus*, *Colletotricum*, and *Rhizopus* spp. (43). *Colletotrichum gloesporioides* was found to colonize 41% of the seed coats and 36% of the endosperm layers in red pepper (44).

Harvesting.   Ripe fruit is picked, cut, and macerated mechanically to separate the seed. Roller-type machines discharge the seed into a large rotary screen in which it is washed free of pulp and skins. Fermentation is generally not necessary. The best time to harvest is 50–60 days after planting when the fruits are bright to deep red. Early harvest of immature fruits affects germination of seeds, which may be less than about 10%. Doijode (45) showed that seed harvested at the ripe fruit stage showed high germination capacity and seedling vigor. However, fruits could be harvested at the breaker stage for an early seed crop without affecting their quality. The seed viability was highest (>40% after 30 months) in seeds extracted from ripe fruits (46).

4. Postharvest Operations

Drying.   After washing, the seed is dried in thin layers on screen-bottomed trays exposed to the sun and air or by artificial dehydration, as described for the tomato.

Milling.   Dried seed may require milling to remove fragments of dried fruit present in the seed. Thorough washing can eliminate the need for milling. Small seeds may be separated and rejected using appropriate screens.

Packing.   The packing material used to store pepper seed significantly influences its viability and germination. Seeds preserved in glass jars followed by polyethylene boxes and plastic bags had higher germination capacity than those stored in cloth bags or tin containers for approximately 8 years. In all types of packing, seed viability diminished gradually but not evenly, and was directly correlated with the type of packing material used (31).

Yields.   The average yields of pepper seed vary from 40 to 60 kg/ha or more.

C. Eggplant

1. Introduction

Eggplant, or brinjal, probably originated in India and China. Eggplant is a very tender plant and requires a long, warm growing season for successful production. Although cold nights and short summers do not favor satisfactory yields, it can be grown in the warmer sections of northern areas.

Botany. Eggplant (*Solanum melongena*), belonging to the Solanaceae, or nightshade, family, is a bushy plant that grows to a height of about 0.5–1.5 m. The leaves are large and alternate on the stems and are more or less oval in shape. Eggplant has a strong taproot, penetrating to a depth of about 1 m within 6–7 weeks. The mature plant has an extensive root system. The flowers are large, violet-colored, and solitary or form clusters of two or more. On most varieties, perfect flowers are borne singly and opposite of leaves. The large fleshy calyx is usually five lobed. The expanded purple corolla is about 5 cm in diameter. The anthers are arranged in a cone around the style and discharge the pollen through terminal pores. The stigma often projects beyond the anthers.

The eggplant fruit is a fleshy berry and forms in a pendant position. It is held by the calyx, which, after the corolla has withered, enlarges considerably, enclosing the entire basal portion of the mature fruit. For market, the fruits are harvested by cutting the peduncle, but for seed, the fruit is allowed to mature and ripen on the plant until an abscission layer develops between the fruit and the calyx, causing the fruit to fall to the ground.

Cultivars. The popular cultivars of eggplant in India are Black Beauty, Long Black, Muktakeshi, Round Purple, Banaras Giant, Pusa Kranti, and several hybrid lines such as S-1, S-4, S-5, 5-8, and 5-16. H-4 has been developed at Hissar and A-16 at Pantnagar (47). The high-yielding cultivars grown in Maharashtra (India) are Manjari Gota, Surti Gota, and American Purple. The Indian Agricultural Research Institute, New Delhi, has developed a new hybrid between Pusa Purple Long and Hyderpur. It is a promising cultivar with attractive, long to oblong, purple fruit. Virgin Long is a new variety of eggplant welladaptable to alkaline soils; it is high yielding and resistant to *Phomopsis* rot (48).

2. Breeding

The first hybrid variety of eggplant, New Hampshire Hybrid, was released in 1937. It is an extremely productive early season type. Badger State, developed by Wisconsin Station, has plants and fruits somewhat larger than the New Hampshire Hybrid. The Central Experimental Farm Ottawa in Canada developed a small-fruited but a noticeably prolific variety, Blockie et al. (49) investigated hybrid vigor and combining ability in several promising American varieties. They crossed seven eggplant varieties in all possible combinations. The resulting hybrids showed a consistent increase in both early and total yields as compared with the mean of both parents. The most promising crosses with respect to earliness and total yield were Early Long Purple, New Hampshire Hybrid, Early Long Purple × Badger State, and New Hampshire Hybrid × Florida High Bush. Fruits of some varieties contain more than 2500 seeds. According to Chauhan (37), the first-generation eggplant hybrids in India have given early and 80–100% higher yields than the better parent. The cost of hybrid seed production can be reduced by the use of male-sterile lines.

Petrov et al. (50) noted that the stage of fruit maturity invariably influenced the seed quality. Seeds from the fruits harvested at the time of complete botanical maturity had the highest sowing quality. Postharvest storage of fruits for 3–5 days favorably influenced the seed quality and productivity of plants obtained from them. Best results with heterotic seed production were obtained when the flowers were pollinated with abundant amounts of pollen. Two- or threefold pollination was recommended, for which pollen should be collected from more plants. In the case of pollen storage, pollination should be carried out no later than the fourth day of storage and 2–3 days following emasculation.

Hybrid Seed Production. Hybrid eggplant seed can be produced with reasonable expenditure by the emasculation and pollination technique, as its floral morphology favors rapid emasculation and pollination. Also, many seeds are formed in a single fruit (800–1000 seeds in long brinjal and 1000–1500 in round brinjal). Therefore, very few fruits produce a sizable number of seeds.

In producing hybrid seed, the variety setting a large number of seeds in a single fruit should be treated as the female parent so that a large amount of seed can be obtained in a single attempt. The flower buds that are expected to open the next day are selected on the female parent. With the help of forceps, the flower buds are opened and the stamens, the number of which varies from five to seven, are removed one by one. This process of removing the stamens is called "emasculation." These emasculated buds are then bagged in muslin bags to prevent pollination with undesirable pollens. While emasculating the flower, care should be exercised to ensure that no anther is ruptured or crushed. If this happens, those flowers should be rejected and the forceps sterilized with alcohol. The flower buds of the male parent should also bagged to avoid contamination.

The next morning, the flowers that are bagged for the pollen grains are plucked and collected in a Petri dish. The female buds are then uncovered. The anther from the male flower is removed and held between the arms of the forceps. As the pollen grains in the anthers of brinjal are released through apical pores, the anther is held perpendicular to the stigma surface, keeping the apical pores of the anther opposite the stigma surface. The forceps are tapped and the yellow-colored pollen powder is dusted onto the stigma. This process of dusting the pollen grains onto the stigma is known as "pollination." The pollinated buds are again bagged to prevent cross-pollination.

Emasculation and pollination can be performed simultaneously. The success of fruit setting when this method is followed is marginally reduced, but the labor and time required for bagging the emasculated buds and unbagging them for pollination is effectively saved. The emasculated and unpollinated buds and male buds are bagged with bags of different colors so that each set of buds can be distinguished.

Somatic Embryogenesis. Seeds of *S. melongena* cv. Giulietta were cultured on MS medium in the dark before transferring 20-day-old seedlings to MS medium supplemented with 10 mg NAA/L for 1 month. Calluses were graded according to size and transferred to half-strength MS medium. Bipolar somatic embryos developed in 10 days and were regenerated on solid MS medium. Embryos of 3–5 mm were mixed in 2% Na-alginate solution and dropped individually into calcium chloride. Gibberellic acid (GA_3) at 1, 2, and 8 mg/L or sucrose at 20 g/L was added to the Na-alginate solutions to evaluate their effect on germination. The highest regeneration rate was seen in somatic embryos derived from calluses measuring 120–150 μm. The best germination rates (67%) of artificial seeds were obtained using a Na-alginate solution of pH 6.0. GA_3 and sucrose had negative effects on synthetic seed germination (51).

3. Seed-Production Technology

Culture. Although eggplant cultivation for seed is more or less identical with that for the fresh market and the crop is handled in much the same way as are tomatoes and peppers, its requirements, especially in the seed bed, are more exacting. Since a prolonged warm season is necessary for higher yields, in temperate conditions, eggplant is usually started under glass. Similarly, a greater care is essential while transplanting eggplant to avoid root injury. Young plants will wilt easily if exposed to direct sunlight.

LAND PREPARATION. Deep, fine- or loamy-textured soil with good drainage is most favorable for the growth of eggplant. Soil pH should be neutral or slightly acidic. The land should be plowed four or five times and smoothed before transplanting of seedlings. Manure or compost fertilizer (5–10 tons/ha) should be incorporated into the soil during the first plowing. Beds are prepared after the land is well prepared and leveled.

PLANTING. In India, seed may be sown in November to yield fruit in April–May or in April for the July–August crop. The weather conditions and availability of irrigation determine the time of sowing and transplanting. About 0.5–0.75 kg seed is sufficient to produce seedlings to cover a 1-ha area. Spacing varies from 45 × 60 cm to 60 × 75 cm for long and round varieties and 75 × 120 cm for high-yielding varieties.

IRRIGATION, FERTILIZATION, AND WEED CONTROL. Eggplant responds well to fertile soil conditions and needs a good water supply. The Indian Council of Agricultural Research, New Delhi, has recommended 40–60 kg N, 60–80 kg P_2O_5, and 100–120 kg K_2O per hectare. The field should be irrigated every third or fourth day during the summer and after 12–15 days during the winter. Timely irrigation is very important to obtain high yields of eggplant. Interculture or hoeings should be fairly deep and close when plants are small and shallower as the plants develop.

Roguing. Off-type eggplants may be rogued in the early stages before blossoming as well as after the fruit sets. It is possible to remove off-type plants while the first fruit is still only partially developed. Roguing can also be performed based on fruit size, shape, color, and other overall plant characteristics. Two or three systematic roguings are essential for the production of stock seed.

Self-pollination is more common than cross-pollination in eggplant. The cone-like shape of anthers favors self-pollination, but since the stigma ultimately projects beyond the anthers, there is ample opportunity for cross-pollination. The extent of cross-pollination in eggplant may vary from 0.2 to 46.8%; averaging about 6.75%. Because of significant cross-pollination, a minimum distance of about 400 m is desirable between two varieties of eggplant. For the production of seed stock, the distance should be considerably greater.

Control of Pests and Diseases. In a tropical climate, eggplant is attacked by several insect pests such as fruit and shoot borers (*Leucinodes orbonalis* S), stem borers (*Euzophera perticella* R), bud worms (*Gnorimoschema blapsigona* M), jassids (*Emposca* spp.), sucking insects (*Vrentious sentis*), and *Epilachna* beetles. The fruit, shoot, and stem borers and bud worms can be controlled by spraying 0.1 % lindane, 0.02% endrine, or 0.1% DDT five times at 2-week intervals. A spray of 0.04% parathion also can check stem borers and bud worms. Other insects can be controlled by spraying 0.1% DDT or BHC or 0.02% endrine.

The important diseases of eggplant are damping off (*Pythium* spp. *Phytophthora* spp.), leaky fruit rot (*Pythium* spp.), blight (*Phomopsis vexans*), little leaf (virus), and root knot nematodes. Treatment of seeds with hot water (52°C) for 30 min, Cerasan, or dipping of seedlings with Cuprovit or any other copper fungicide has been recommended to control damping off disease of eggplant. *Phomopsis* blight can be checked by treating the seeds with Cerasan or Agrosan and a copper fungicide spray like Bordeaux mixture (4:4:50) in the nursery. Little leaf, caused by a virus, is spread by a leaf hopper (*Hisbimonas* spp.). Parathion sprays will keep this insect under control. Following the appropriate rotations, the use of resistant varieties and soil fumigants like Nemagon effectively control root knot.

Harvesting. Eggplant fruits are allowed to mature beyond the edible stage when harvested for seed. Seeds in fruits nearing edible maturity continue to mature slightly for several days if not removed from the fruits. Seeds of eggplant are extracted by cutting, crushing, or macerating the well-matured or ripe fruits with equipment similar to that used for cucumbers. Since simple crushing and squeezing is not sufficient for eggplant, special machines are used to cut, macerate, and squeeze the pulp through a circular screen to force the seed from its placenta. For high-quality seeds, eggplant fruit should be harvested about 80 days after anthesis. Fruits harvested at the completely yellow stage had the highest seed yield (102.55 kg/ha), and harvesting fruits prior to or later than the full yellow stage resulted in lower seed yield and quality (52).

4. Postharvest Operations

Extraction of Seed. The seeds of eggplant are extracted by treating the pulped fruits with concentrated HCl at 30 mL/kg of seeds with constant stirring for 20 min. The seeds are conditioned to a uniform 8% moisture content and graded by a flotation technique into floating and sinking batches. The endosperm weight of the seeds that sink is double that of the seeds that float. The seeds that sink give better results in all respects (53).

Milling and Grading. After thorough washing of eggplant seed and pulp portions by a method similar to those employed for tomato seed, they are dried immediately either in thin layers or screen-bottomed trays exposed to sun or wind or by the artificial drying methods. The dry eggplant seed may require some milling to remove the fragments of dried fruit tissue depending upon the seed condition. Grading also may be necessary to separate very small seeds from the lot. Seed germination and vigor index were acceptable up to 18 months of storage, but declined rapidly thereafter. They were highest in seeds treated with 1 g thiram plus 1 g bavistin per kilogram of seeds and stored in paper–aluminum foil–polyethylene–lined pouches (54). Rudrapal and Nakamura (55) reported that halogen vapor treatment of eggplant and radish seeds employing chlorine, bromine, and iodine for 8–48 hr at 25°C significantly slowed down the deterioration of seeds under accelerated as well as natural aging conditions. The halogen vapor stabilizes the unsaturated fatty acid components of the lipoprotein membranes of cells, which might render them less susceptible to peroxidative changes during aging. The hydration-dehydration treatments of eggplant and radish seed extend the viability of these seeds by reducing free radical damage to cellular components (56).

    Halogen vapor treatment of harvest-fresh aubergine and radish seeds employing chlorine, bromine, or iodine for 8–48 hr at 25°C significantly slowed down the deterioration of seeds under accelerated as well as natural aging conditions. Beneficial effects of halogen treatments were associated with greater membrane integrity as indicated by lower electrical conductance of seed leachates and reduced leakage of sugars from the treated seeds. Higher activity of dehydrogenase enzymes in the germinating seedlings and reduced lipid peroxidation were noted after treatment. It has been suggested that the halogen vapor stabilizes the unsaturated fatty acid components of the lipoprotein membranes of the cells, possibly rendering them less susceptible to peroxidative changes during aging (56a).

III. ONIONS AND RELATED VEGETABLES

The cultivated bulb vegetables belonging to the family Alliaceae of the order Amaryllidaceae are onion, garlic, Welsh onion, shallot, and chive. All belong to the genus *Allium* and are propagated by seed, except garlic, which is grown from its cloves. Onion and garlic are commercially important bulb vegetables grown in India, China, Japan, Russia, the United States, and several European and Far Eastern countries. Asparagus belongs to the family Liliaceae, in which all the bulb crops were formerly included.

A. Onion

1. Introduction

The onion probably originated in Central Asia and the regions around the Mediterranean Sea. Although it is grown widely in most temperature countries, it is chiefly limited to regions of low-atmospheric humidity. It requires cool weather during its early development and during the early growth of the seed stalk. Later a moderately high temperature and dry atmosphere favor bulb maturity as well as the production of seed during the second year.

Botany. The common onion bulb (*Allium cepa* L.) is generally a biennial plant, but it is grown as an annual. The genus *Allium* contains about 300 widely distributed species. Some of the wild species produce bulbils instead of seed in the flower cluster, as does the tree onion.

Onions form bulbs in the first year and seed in the second year. The leaves arise from a shortened crown stem (stem plate). The sheaths of the older or outermost leaves enclose the younger ones. The basal portions of the leaves encircle the stem and thicken to form the bulb. The stem elongates during the second year, forming the flower stalk.

The onion has a fibrous root system extending to a depth of more than 1 m, but most are about 0.5 m. Flowers are borne in simple umbels at the apex of a floral stem, hollow and round in cross section, and somewhat swollen at the middle or near the base. Most onion varieties produce seeds stalks more than 1 m in height. The number of seed stems per plant may vary from 1 to 20 or more depending on the variety, size of the mother bulb, and time of planting. Before expanding, the umbel is enclosed within a papery spathe consisting of two or three bracts, which are split open by the pressure of the developing flower buds. The number of flowers per umbel varies considerably from 50 to over 2000. Temperatures of around 21.1°C favor vegetative growth, whereas lower temperatures, around 12.8°C, are conducive to seed stalk formation. Short-day conditions are also favorable to seed production.

The white or bluish flowers of the onion have an outer and an inner whorl of three stamens each. The anthers of the inner stamens dehisce first. The pistil has a three-celled ovary, ech with two ovules. The style is about 1 mm in length, while the flower opens first. It is not receptive until it elongates to a length of about 5 mm, 1 or 2 days after all of the anthers have dehisced. Opening of flowers usually continues for a period of 2 weeks or more, and onion plants may bloom for over 30 days. The fruit is a three-lobed, three-celled capsule, each containing one or two black seeds at maturity. The embryo, consisting mostly of the cotyledon, is spirally twisted, cylindrical, and embedded in the endosperm, which forms the major portion of the seed.

Cultivars.   Onion cultivars continue to change; every year newly developed hybrid cultivars are added to the old ones. The hybrid cultivars are higher yielding and more uniform in bulb size than the old cultivars. The American, or pungent type, and the foreign, or mild type, are the two general types of onion grown in the United States. Red, white, and yellow are the most common bulb colors observed. The American cultivars produce bulbs of smaller size, denser texture, stronger flavor, and better keeping quality, varying in shape from oblate to globular. The important yellow cultivars grown in the United States are Brigham Yellow Globe, Yellow Globe Danvers, Early Yellow Globe, Mountain Danvers, and Ebegezer. The best-known red cultivars are Red Whether Field, Southport Red Globe, and California Early Red. The most popular white onion varieties for storage, dehydration, and processing are Southport White Globe and White Creole. These have an excellent white color and high flavor or pungency. White Portugal, or Silverskin, is a green onion used for pickling. The Bermuda onion is the most popular foreign cultivar and is grown extensively in Texas. Yellow Bermuda and its various strains have flattened bulbs with a mild flavor. Crystal Wax is a white onion cultivar of the Bermuda type. Early Grano and its strains are early maturing, globular, yellow onions grown in Texas, the Southeast, and California. Many strains of Yellow Sweet Spanish onions are grown in the western parts of the United States. White Grano or White Balbosa, White Sweet Spanish, Southport White Globe, and Barletta are important white cultivars.

The new onion cultivars released in the late 1970s and early 1980s include (3) Apache Chief, Autumn Glo, Autumn Surprise, Brahma, Bumper-C, Capable, Carmen, Copper Lustre, Durango, D 5551, Early Pak, Enterprise, Excel G, Festival, Golden Cascade, Granex 429, Harvestmore, Inca, Magnum, Matador, Nutmeg, Ontario M, Prime Beauty, Red Harvest, Redron, Rielto, Ringmaker, Sunglow, Taurus, Valdez, and White Spear Bunching. Nutmeg, Sunglow, and Taurus have excellent storage properties.

The Indian Agricultural Research Institute (IARI), New Delhi, classified onion cultivars into two groups: large-sized (5–10 cm in diameter) bulbs and small-sized (1.2–2.5 cm) bulbs. The large-sized deshi (local) cultivars are Red Globe, White Globe, Yellow Globe, White Patna, Large Red, Poona Red, Nasik Red, and Yadgiri or Beauty Red. The IARI has released Early Garno (yellow bulb), Pusa Red, and Pusa Ratnar, which have large-sized bulbs. The small-sized bulb cultivars of onion grown in India are Prize Taker, Market Gardner, Yellow Globe, Queen of Paris, and Silverskinned.

2. Breeding

The early breeding work concentrated on the elimination of off-type plants to obtain varieties with uniform and desirable plant characteristics. This was accomplished by mass selection, inbreeding, and thorough testing followed by massing of desirable inbred lines.

Although some selfing does occur, onion is essentially a cross-pollinated crop. There is thus a need to prevent intervarietal crossing, which can be accomplished by using paper bags, cheesecloth, or wire- or plastic-screen cages. The use of flies as pollinators has been recommended to achieve good seed setting. Honeybees can also be used as pollinators when large cages made of plastic screen are employed for selfing.

In a study carried out in Poland, bee colonies were placed at one edge of the fields, as well as cages, with no insects inside to serve the control plot. The results of this study allowed Woyke (57) to come to following conclusions:

1. The presence of honeybees and other pollinating insects increase fruit set, the number and mass of seed, and seed yield of onion.
2. At the time of first flowering and under bad weather conditions, honeybees work on blooming onion more intensively the closer their hives are situated to the field. When the flowering near the honeybee colonies is ended, they move their activity to further parts of the field.
3. Bee foraging on the open pollinated onion cultivars in Poland is satisfactory. Spraying of certain attractants like Citral, Geraniol, or Anise appear to be essential only when more attractive competitive flowering plants are present in the neighborhood of the onion field or if there are few bee colonies surrounding it.

Globerson et al. (58) studied the nature of flowering, seed maturation, and seed quality of onion to determine the proper time for the mechanical harvesting of seeds. The number of flowers per umbel ranged from 200 to 1000 and varied according to cultivar, date of planting, bulb size, and growth conditions. Within an umbel, 25–31 days elapsed from the opening of the first flower to that of the last. In the field, there was a time lag of 20–30 days between umbels at the beginning of flowering. These variations depended mainly on the temperature at flowering time. Seeds from individually marked flowers, which were collected 45–50 days after anthesis, were able to germinate, and younger seeds had poor germination capacity.

Measurement of seed yields and crossing levels in cage studies conducted in the United Kingdom showed that there was little overall difference in the efficiency of blowflies and honeybees as pollinators of synthetic and open-pollinated onions. However, substantial variation occurred from one season or part of a season to another in seed yield per plant and in levels of crossing between cultivars, which was attributed to the probable differences in the relative number of umbels of the cultivars or in the synchronization of flowering. This variation was greater than the differences caused by the pollinator used (59).

Ahmed and Abdalla (60) reported that flower abortion was markedly reduced when honeybees were used to pollinate onions under cages. Seed yield per seed bed was significantly higher as a result of the increase in flowers that set seed. Seed yield/plot also increased compared with open pollination. The number of seeds per capsule, however, remained unaffected. The amount of dry matter or total soluble solids (TSS) of the onion bulb is an important factor, correlating to its keeping quality. Two outer scales can be removed for testing bulbs for breeding purposes. In addition to the dry matter, other characteristics such as bulb color, bulb diameter, neck thickness, and disease resistance have been employed in the breeding program.

The F_1 hybrids of onion showed increased vigor, but it was too expensive to produce such seeds, since it required emasculation. This problem was overcome by using male-sterile lines, which do not produce fertile pollen. The male-sterile condition results from an interaction between a recessive nuclear factor and a nonnuclear or cotyplasmic factor. All plants having normal cytoplasm (N) produce viable pollen, whereas all male-sterile plants have the sterile type cytoplasm (S). In addition to the S cytoplasm, male-sterile plants also have the recessive nuclear factor ms. This gene has no effect on the pollen in the presence of normal (N) cytoplasm. All male-sterile plants thus have the genetic constitution Smsms. A dominant Ms factor produces viable pollen whether it is in a homozygous or a heterozygous condition. Pollen-producing plants, therefore, have any one of the following genotypes: Nmsms, NMsMs, SMmMs, or SMsms. The nonnuclear cytoplasmic factor is inherited only through the female parent, so a SMsms × Nmsms cross will produce male-sterile plants only. Similarly, the Smsms × NMsms cross will produce plants in the ratio of one male-sterile to one male-fertile and all plants of the Smsms × NMsMs cross will be male-fertile.

The original male-sterile line, Italian Red 13–53, has been used in the development of male-sterile lines in most of the widely grown onion varieties, which was made possible by finding plants in these varieties with the genotype Nmsms. If such a genotype cannot be found in a commercial variety, then it is necessary to develop such a line. This can be achieved by crossing the male-sterile line Smsms with the fertile line having the genetic constitution NMsMs. The F_1 plant, SMsms, may be used as the male parent and backcrossed to NMsMs. Such a cross would result in plants with a genotype of NMsMs or NMsms. The former genotype will produce only plants homozygous for MS, whereas the latter (NMsms) when selfed will produce one-fourth NMsMs, one-half NMsms, and one-fourth Nmsms. The genetic constitution of the plants can be determined by backcrossing to the male-sterile line. After developing the desired male-sterile line, it can be maintained by continually backcrossing to a male-fertile line with a genotype NMsms. Male-sterile lines may also be maintained by a vegetative method.

The developed male-sterile lines are required to be tested for combining ability with other lines of the same variety or with lines of different varieties. It is usually desirable to have many male-sterile lines in order to be able to discard all but those that give the best results. Similarly, a large number of inbred lines of the pollen parent used to produce hybrid seed enables vigorous selection.

The development of male-sterile onion varieties has revolutionized onion breeding. It is now possible to obtain hybrids incorporating desirable characteristics from two varieties in commercial quantities. In addition, hybrid varieties are expected to give increased yield, greater uniformity, and adaptation to a specific purpose such as, for example, earliness and resistance to salinity.

Breeding of onion for disease resistance to downy mildew caused by *Peronospora destructor* includes the use of the male-sterile line 13–53, which has immune seed stalks and highly resistant leaves. This line was one of the parents of Calred, which is a highly resistant variety to downy mildew. This led to further breeding of onion varieties resistant to pink rot, purple blotch, and yellow dwarf. The USDA has released an onion highly resistant to pink rot and smut as well as to yellow drawf, namely, Bettsville Bunching, which is a cross between the common onion (*A. cepa*) and Welsh onion (*A. fistulosum*). GA_3 (2000 ppm) and $GA_4 + GA_7$ (5000 ppm) sprayed 10 times at 2- to 3-day intervals starting from the bolting stage were most effective as gametocidal agents for common onion for breeding purposes. This is advisable when a limited amount of hybrid seeds is needed for breeding purposes (61). GA_3 appears to hasten flower bud formation by activating meristematic tissue (62).

3. Seed-Production Technology

The bulb-to-seed and seed-to-seed methods are two methods used to produce onion seed. The former is most commonly employed, and involves first the production of onion bulbs as for fresh market and then replanting them to produce seed. The bulbs produced from 1 ha are usually sufficient to plant 2–5 ha for seed.

Bulb-to-Seed Method.

CULTURE. Onion bulbs for seed are grown in more or less the same way as they are produced for market. Since medium-sized onion bulbs have been found to be satisfactory, many seed growers use higher seed rates (about 3.5–7.0 kg/ha) than commonly used for the production of market crop (2.2–4.5 kg/ha). The 30 × 45 cm spacing is considered best for onion seed crops (63). Onions can be grown successfully on different types of soils except low-lying, marshy, and heavy clay. Sandy loam, silty loam, and deep friable soils retaining adequate amounts of moisture are highly preferred for onion cultivation. Land is plowed shallow, three to four times to a fine tilth, and leveled.

Planting time varies according to the location and the variety. In regions with mild winters, seed is sown in the fall; in colder areas, it is sown in early spring. In the northern parts of India, seed is sown in October and transplanted in November. On the hills, seed is sown in late February–May and transplanted in March–June. In some parts of India in addition to the two usual crops (June–October and October–January), a third summer crop is raised from January to June. Krishnaveni et al. (64) found that planting in the second and third weeks of November gave the highest bulb and seed yields and the best seed quality. No seed set was observed after a third week in January planting (Table 4).

The Indian Council of Agricultural Research, New Delhi, has recommended 50 kg N, 50 kg P_2O_5, and 100 kg K_2O per hectare. A side dressing of 50 kg N/ha may be given 1 month after planting. An onion yield of 300 Q/ha removes about 75 kg N, 36 kg P, and 68 kg K. In addition to the application of inorganic fertilizers, about 25 tons/ha of manure is also recommended. Application of 40 and 80 kg N/ha significantly increased plant height, number of leaves, sprouts and scapes per plant, diameter of umbel, and seed yield compared with no control (65). The highest seed yield, 6.24 Q/ha, was obtained from plants spaced at 45 × 30 cm and receiving 80 kg of N/ha (66).

A light irrigation is applied immediately after planting; the subsequent irrigation may be given at 7- to 10-day intervals. Fields should kept free·of weeds to produce good onion crops. Two or three hoeings can control the weeds effectively. Cultivation should be no deeper than is necessary to control weeds. Chemical weed control may also be applied.

Field trials conducted over two seasons showed that N application increased plant height, flower stalk thickness, and seed yield. The increase in seed yield was mainly due to an increase in the number of florets produced per seed head. The application of P had no effect on seed yield in the absence of N, but P plus N led to a highly significant increase in onion seed yield (67). Thiourea at 200 ppm gave maxi-

Table 4 Effects of Dates of Planting on Quality Parameters of Onion

Treatment	1000–seed weight (g)	Germination (%)	Shoot length (cm)	Root length (cm)	Dry matter production (g)	Vigor index
November, 2nd week	3.86	90.0	9.5	6.9	0.973	88.0
November, 3rd week	3.40	86.0	9.0	6.7	0.971	83.3
November, 4th week	2.96	85.0	9.2	6.7	0.969	82.3
December, 1st week	2.93	82.0	8.8	6.3	0.964	79.0
December, 2nd week	2.80	81.0	8.8	6.5	0.963	78.3
December, 3rd week	2.66	80.0	8.8	6.4	0.961	77.6
December, 4th week	2.66	78.0	8.9	6.2	0.961	75.6
January, 1st week	2.78	78.0	8.7	5.8	0.846	74.4
January, 2nd week	2.81	77.0	8.6	5.9	0.845	74.0
January, 3rd week	2.60	76.0	8.5	5.9	0.844	72.0
January, 4th week	—	—	—	—	—	—
February, 1st week	—	—	—	—	—	—
February, 2nd week	—	—	—	—	—	—
February, 3rd week	—	—	—	—	—	—
February, 4th week	—	—	—	—	—	—
CD (p = 0.05)	0.26	1.901	0.183	0.202	0.0021	5.505

Source: Ref. 64.

mum plant height and fresh weight of leaves and roots. GA at 100 ppm significantly increased both the fresh weight (FW) and dry weight (DW) of bulbs, and leaf DW and the percentage of bulb yield were also highest with this treatment (68). Onion plants spaced at 75×20 cm produced higher seed yield than those with 75×5 cm spacing (69).

ROGUING.   Roguing should be carried out before the bulbs are harvested based on foliage color or other plant characteristics. Roguing at this stage is especially important in the production of stock seed, for it is easier to remove late-maturing bulbs before the crrop is harvested. The harvested bulbs also should be carefully rogued. Further roguing may be necessary at the end of the storage period to remove sprouts and rots. Roguing has become more complicated with the introduction of hybrid varieties, because it is necessary to rogue each of the inbreds used as parents. The characteristics of these parents may differ from each other as well as from the resulting hybrids.

HARVESTING, STORAGE, AND REPLANTING. Onion bulbs harvested in late spring or early summer are stored and replanted in the fall. Those harvested in the fall may be stored for a few weeks and planted again in the spring. Cooler temperature, relatively dry weather with low humidity, and ample ventilation minimize storage losses of onion bulbs. Szalay (70) reported that between August 8 and 15 was the most appropriate time for harvesting onion for seed production, as the seeds harvested at this time showed optimum germination ability, highest 1000–seed weight, and minimum seed losses due to shattering during mechanical harvesting. Treatment with CCC (Chloromequat) and Camposan M (ethephon), when applied in combination, reduced plant height and increased the resistance of the peduncles to bending, thereby facilitating mechanical seed harvesting (71).

The optimum time for planting bulbs for seed production in India is the second half October. Post–monsoon rains may damage early-planted crops, and if planting is delayed, the reduced vegetative growth decreases seed yield. Late blooms and the incidence of thrips are other disadvantages of later plantings. Nehra et al. (72) noted that seed yield declined as planting was delayed until November 15.

Bulbs may be planted 7.5 cm deep in flat beds or along one side of ridges 45 cm wide with 30 cm between plants. The selected mother bulbs for planting should be cut across the middle. The lower half with the disclike stem and roots is utilized for planting, as it hastens sprouting. About 11250–13710 bulbs are needed to plant an area of 1 ha. Mechanical planting may be adopted in areas of labor shortage. Bulb size greatly influences the seed yield of onion, with a higher yield being obtained with an increase in the bulb size. Gill and Singh (73) obtained the highest number of bolting stems per plant (12.54) and seed yield (9.64 Q/ha) with grade A bulbs. The highest number of days to 50% bolting (104.38) was observed with grade C bulbs. Wheeler and Ellis (74) found that soil moisture was the main determinant of the rate of seedling emergence from all but the driest field seed beds. The highest seed quantity was produced by plants raised from bulbs planted on September 25 (14.75 Q/ha) and October 10 (14.95 Q/ha), whereas plants raised from seedlings gave the lowest seed yields irrespective of the planting date (75).

Seed-to-Seed Method.

CULTURE. Although less seed by volume is produced by the seed-to-seed method than by the bulb-to-seed method, the former enables farmers to grow seed from varieties that are difficult to store; for example, White Sweet Spanish. The seed yield using the seed-to-seed method is often higher than that obtained from a bulb-to-seed planting owing to the greater number of plants and seed heads per unit area.

Seed is sown sufficiently early so that the plants will bolt the following spring. The sowing time depends mainly on the variety and the region. Planting should be earlier in areas with shorter growing seasons. Leveling of seedbeds and preirrigation are advisable. Shallow cultivation is performed as soon as weeds show up. Soil should have an adequate moisture content during seed germination, especially when the seed is sown in midsummer. Land may be furrowed prior to the time of drilling. The distance between rows is 0.75–1.0 m. Approximately 2.5 kg of seed may be required to cover 1 ha of land. Garniely et al. (76) showed that the yields of Granex 33 and Behairy onions were closely correlated with the weight of the seed used to establish the stand.

Elemental content was consistently higher in heavier seeds, but elemental concentrations in the seeds were generally negatively related to seed weight, onion growth, and yield.

ROGUING. Thorough roguing is not possible with the seed-to-seed method. It is, therefore, recommended that only the highest quality stock seed produced by the bulb-to-seed method should be sown, which requires little roguing if any. It may be possible to remove off-color bulbs in certain white varieties, which may be carried out in the spring.

CONTROL OF PESTS AND DISEASES. Heavy losses of onion crop may be caused by thrips (*Thrips tabaci* and *Helipthrips indicus*). This pest is more serious during dry weather. Spraying with 0.5% BHC can control thrips. Dusting with 0.65% lindane at 22.4–28.0 kg/ha or 6.5%. Lindane spray is also satisfactory.

Singh et al. (77) screened 44 onion cultivars for preference by *Heliothis armigera*, a gram caterpillar that damages the seed crop by feeding on developing umbels. The six cultivars least preferred were Sel 102-1, Large Red, S-76, Verma Giant, S-207, and S-243; 16 cultivars were highly preferred.

Onion blight (*Alternaria porri* and *A. palandui*) is a common disease occurring in several onion-growing countries. *A. porri* also causes purple blotch. Blight is a serious disease on seed crops in certain localities. Seed treatment with 0.2% thiram is recommended to prevent this disease. Onion smut (*Urocystis cepulae*) is also a serious disease in several countries. Seed treatments with fungicides like hexachlorobenzene (HCB) are recommended. Dipping onion seedlings for 4 hr in propolis alcoholic extract and Sumisclex at 40 g/L as dipping treatment reduced the white rot incidence under greenhouse conditions (78). Four sprays of Dithane M-45 at 12-day interval commencing from disease appearance controls downy mildew of onion caused by *Peronospora destructor* (78a). In addition to downy mildew of onion, other diseases of minor importance are pink rot (*Phoma terrestris*), fusarium rot (*Fusarium* spp.), neck rot (*Botrytis* spp.) and yellow dwarf and aster yellow viruses.

HARVESTING. Onion is harvested in late June–July in California and during August or early September in the northern sections of the United States. The heads should be picked when the fruit opens and exposes the black seed. At this stage, practically all of the seed is mature enough to give a good germination. The seeds turn black during the milk stage and take a bit longer to mature. It may be desirable to pick the field over twice, but for the average grower, once-over harvests are usually cheaper and more practical.

The best time to harvest onion seeds mechanically is at 60–70% dry matter content, which occurs 45–60 days after the beginning of flowering in the field. Another criterion to judge harvest maturity is when 1–3% of the umbels in the field have mature seed; that is, when the capsules are open and black seeds are visible, usually about 10–12 days prior to the traditional hand harvesting of the umbels (58). Data reported in (Figs. 5 and 6) indicate that harvesting of onion in mountain regions can start when the dry matter content of seeds reaches 65%, which occurs approximately 50–60 days after the field starts to flower. In other areas, harvesting can be started earlier when the dry matter reaches 20–30% (Fig. 6). Seed heads may be cut or snapped off with a quick twist of the hand, leaving a short piece of stem attached.

4. Postharvest Handling

Drying. The seed heads are dried on canvas or specially built drying trays—a tray 60 × 150 × 10 cm is preferred. (Fig. 7). Such trays are convenient to handle and can

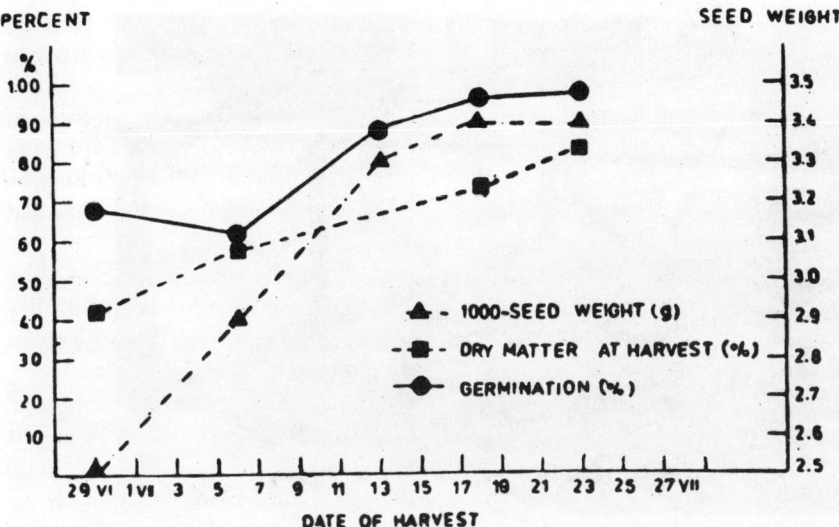

Figure 5 Effects of harvesting time on 1000–seed weight(g), dry matter and harvest (%), and germination (%) of onion seeds (cv. Ori). (From Ref. 58.)

also be used to store onion bulbs. The trays should not be filled with seed heads to a depth of more than about 15 cm to facilitate efficient curing. Seed heads may be stirred occasionally during the first few days after picking to bring about uniform drying and to prevent molding and overheating of the seed. The onion seed may also be dried in burlap bags by emptying one fruit-picking bucket into each bag and hanging the bags on fences to dry. Bags should be shaken every day to prevent heating.

Under humid conditions, seeds may be dried in sheds with ample air circulation. Frequent stirring may be needed while the seed is dried. Since natural seed drying often

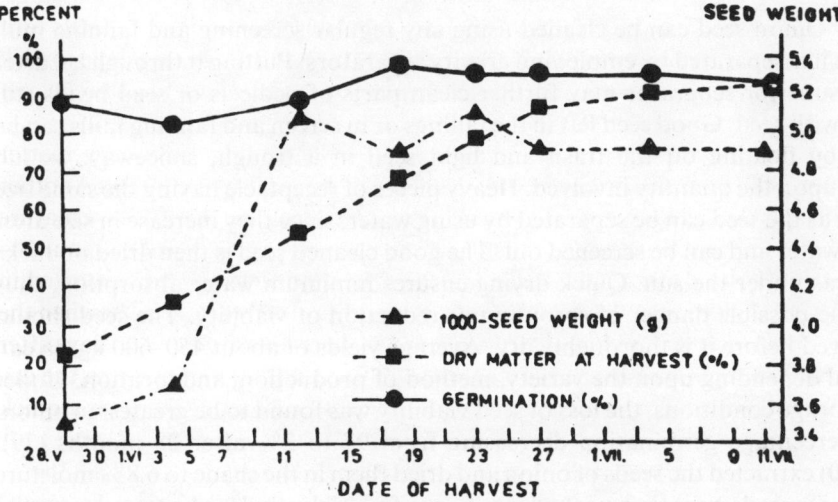

Figure 6 Effects of harvest time on 1000–seed weight (g), dry matter and harvest (%), and germination (%) of onion seeds (cv. Ori.) (From Ref. 58.)

Figure 7 Drying onion seed using a bin drier. (From Ref. 58.)

requires 2–3 weeks, some growers prefer to dry seed quickly using artificial dryers or dehydrators. An onion seed–drying bin is shown in Figure 7.

Seed Extraction/Threshing. After adequate drying, seed is easily separated from the heads. A combine harvester or specially designed onion thresher is used to extract onion seed from the heads. Some seed is threshed out on canvas with a large roller. Flailing or using a rubbing board may thresh small quantities. Care should be taken to avoid chipping of seed in the machine, which may be examined periodically with the help of hand lenses. Shattered seed at the bottom of the tray or on the canvas should be combined with the threshed seed for recleaning. The tailings may need to be returned to the thresher to recapture the escaping seed in the first turn.

Cleaning. Onion seed can be cleaned using any regular screening and fanning mill, after which it is separated by employing gravity separators. Putting it through a Carter Disk or a superior separator may further clean parts of pedicels or seed heads still remaining with seed. Good seed left in the tailings or in screen and fanning mills can be recovered by floating off the trash and light seed in a trough, sluiceway, or tub depending upon the quantity involved. Heavy pieces of receptacle having the same size and weight as the seed can be separated by using water, since they increase in size after absorbing water and can be screened out. The good cleaned seed is then dried on racks or on canvas under the sun. Quick drying ensures minimum water absorption, thus avoiding the possible danger of sprouting or reduction of viability. The seed should not be bagged before it is thoroughly dry. Average yields of about 450–600 kg/ha can be obtained depending upon the variety, method of production, and location. Under ambient storage conditions, the loss of seed viability was found to be greatest in onion, with the percentage germination decreasing from 97 to 3% after 28 months (79). Doijode (80) extracted the seeds of onion and dried them in the shade to 6.8% moisture content. These seeds were then packed into paper-foil–polyethylene bags under partial vacuum or in paper bags (control). All of the bags were stored in ambient conditions

(16–35°C and 25–90% RH) for up to 30 months. No germination was recorded after 30 months, and after 18 months the vacuum-packed and control onion seeds gave 54 and 25% germination, respectively. Choudhuri and Basu (81) reported that the improved storability of hydrated-dehydrated onion seeds was associated with greater dehydrogenase activity and appreciably lower lipid peroxide formation in cells. Taylor et al. (82) studied the amino acid leakage from onion, leek, cabbage, tomato, and pepper seeds aged at 45°C and 90% RH. They observed that as seeds aged, the amino acid content in dry seeds increased slightly for all species, whereas a large increase in the percentage of amino acids from imbibed seeds leaked into the soak water from onion, leek, and cabbage only. The predominant amino acids found in the nongerminable seed leachate were alaine (onion, leek, and cabbage), glutamic acid (leek and onion), and arginine (leek and cabbage). Tomato and pepper seeds had relatively impermeable seed costs, whereas leek and onion were intermediate with respect to amino acid leakage and tetrazolium chloride (TTC) uptake (83). The semipermeable layer in seed coats of leek and onion was composed of cutin, whereas in tomato and pepper, the layer was composed of suberin (84).

B. Leeks

1. Botany

The leek (*Allium porrum* L.) is a native of Europe and western Asia. It is a tall hardy biennial with white, narrowly ovoid bulbs and broad leaves. Like the onion, it does not form large bulbs. The leek has a fibrous, widespread root system. During the first year, the stem is platelike, like that of the onion, but of greater diameter. Leaves are solid with bread-keeled blades, each half being flat. In the second year, the stem elongates, forming a single stalk, about a meter or more in height, terminated by a single umbel containing several thousand flowers. The spathe protecting the undeveloped flower buds is much longer than that of the onion. Bees mainly pollinate the pinkish flowers.

2. Seed-Production Technology

Leek seeds are sown in the spring in rows 75–90 cm apart with 7.5–10.0 cm between plants. Plants are then allowed to overwinter. Leek plants are occasionally transplanted. The cultural requirements regarding irrigation, fertilizers, interculture, and plant protection are in general similar to those for onion. The fertilizer requirements for leek are slightly higher than those for onion because of the heavier foliage growth of the plants.

Leek seeds are harvested in a manner similar to that described for onion. The seeds require more time for drying; the use of an artificial drier is, therefore, recommended. In humid regions, seed heads should be threshed immediately after drying, as the seed is more difficult to remove if the capsules absorb moisture. Leek seed is more difficult to thresh than onion and may need to be run through the machine several times to extract it from the capsules. Yields of leek seeds average about 550 kg/ha.

C. Chives

1. Botany

Chives (*Allium schoenoprasum* L.) are native to Europe and adjoining parts of Asia. They are grown in India as a garden crop. The chive plant is a hardy perennial growing in dense clumps with greenish or purple-tinted leaves and clusters of narrow, weakly developed bulbs at the base of the leaves. The onion-flavored leaves and bulbs are used for seasoning salads. The leaves of chive are circular in cross section and about one-fourth the size of onion leaves. Chives may be propagated either vegetatively or by seed.

2. Seed-Production Technology

Seed stocks developing early in the spring are short—about 30 cm high. The small inflorescence contains about 25–100 flowers. The chive seed may be sown directly or seedlings may be transplanted. The seed is harvested as soon as capsules begin to open, showing a black seed to avoid seed loss due to shattering. Chive is considered to be a shy seeder. The postharvest handling (curing, threshing, cleaning) of seed is the same as that for onion seed except that screen sizes are slightly smaller to suit the smaller chive seed.

D. Shallots

Shallots (*Allium ascolonicum*) are perennial vegetables producing a compound bulb made up of several bulblets enclosed in a common scale. The leaves are narrow, round in cross section, and hollow. Although the crop is usually grown for its bulblets or corns, it produces seed if it is planted in the fall. The umbels are smaller than those of onion, having about 200–250 flowers. Harvesting, postharvest handling, and plant-protection methods are similar to those for onion.

E. Asparagus

1. Introduction

Botany. Asparagus (*Asparagus officinal*) is a dioecious perennial leafy green vegetable with an erect branching stem. It forms a woody rhizome or underground stem, which gives rise to large fleshy roots. The aboveground stem arises from the bud in the rhizome. The scalelike leaves are found at the nodes subtending the main branches and the small leaflike stem. The male flowers have an outer and an inner whorl of three stamens each and a rudimentary pistil. The female flowers have a pistil and rudimentary stamens. Asparagus crops started from seed have an approximately one-to-one ratio of male-to-female plants. Asparagus is indigenous to temperate Europe and Asia, although the crop can stand both hot and cold conditions.

Cultivars. Few varieties of asparagus are grown in India. Mary Washington and Martha Washington are best adapted to Indian conditions. Both varieties are resistant to rust.

2. Seed-Production Technology

The culture of asparagus grown for seed is more or less the same as that of the market crop. In seed production, one male plant is used for every five to six female plants. Only large crowns should be selected for planting. None of the spears in a crop grown for seed should be cut for fresh use. The entire planting should be used for seed production.

Land Preparation. Deep, loose, well-drained, fertile, sandy loams are considered best for the growth of asparagus. Asparagus favors neutral soils and tolerates saline conditions to a certain extent. Land should be prepared by deep plowing and harrowing.

Planting. Seeds are soaked in warm water for 1 day before sowing to facilitate quick germination. The seed is sown during winter and autumn, beginning in nursery beds. About 6- to 8-week-old crowns or vigorously growing seedlings are then planted in the field with 6.0 cm between rows and 7.5–10.0 cm between plants in December–January. About 8–10 kg seed/ha is required.

Irrigation and Fertilization. During warm weather, the crop should be irrigated at intervals of about 7–10 days. Irrigation may not be required from July to January. Application of 25 tons of manure and 100 kg each of muriate of potash and superphosphate per hectare have been recommended.

Control of Pests and Diseases. The important pests and diseases of asparagus are beetle (*Crioceris asparagi* and *C. duodecimpunctata*), rust (*Puccinia asparagi*), and fusarium wilt (*Fusarium* spp.).

Harvesting. Plants are cut in the fall before frost and allowed to dry in the field. A pick-up harvester may be used to thresh the crop. During this process, much of the seed is knocked out of the berries, but a fairly large amount is still held in the sticky mass coming from the harvester. This mixture should be washed immediately after threshing to remove sugars and to prevent fermentation, which would injure seed. Special machines have been designed for washing asparagus seed. Small quantities of seed can be threshed out of the berries from the dried plants with a flail. The berries are put into a sack and crushed. The crushed material is then put in large vats or tubs where the seed can be separated from the mixture by washing with water. The pulp and light seed are floated off, leaving the good seed at the bottom. The washed seed is dried immediately by spreading it thinly on screen-bottomed drying trays. The seed may also be dried artificially at about 32–38°C. The dried seed does not require any additional milling.

IV. CARROTS, CELERY, PARSLEY, AND PARSNIPS

Carrots, celery, parsley, and parsnips all belong to the family Umbelliferae, the members of which are characterized by their typical inflorescence, or umbel. Some minor vegetables also included in this family are celeriac, dill, fennel, chervil, caraway, anise, and coriander.

A. Carrots

1. Introduction

Botany. Carrots (*Daucus carota* var. *Savita*) are native to Europe, Asia. Northern Africa, and possibly North and South America. The carrot is an annual, but it is normally cultivated as a biennial vegetable when grown for seed. During the first year, it produces a rosette of leaves and a fleshy taproot; the seed stalk develops in the second year, extending to a height of 1–2 m or more. The leaves are pinnately compound with long petioles.

The edible and salable portion of the root is actually an enlarged fleshy taproot. It consists of the cortex (or phloem) and the core (or xylem). High-quality carrots should have a large cortex and a minimum of core. The color of carrot roots of commercial interest may be yellow or orange-red. The fibrous root system is fairly extensive and widely branched reaching a depth of about 2 m, although the mass of roots remains in the upper 45–60 cm of soil.

The inflorescence is a compound umbel, usually called a head, which is a single, central, or primary head. The umbels terminating the branches are called umbels of the secondary order. There may be similar third-order and fourth-order branches and heads. The primary head may be 12.5–15.0 cm or more in diameter. The flowers and later the seeds occur in a dense formation within the head. The surface of the head is more or less flat but may become cupped as the plant matures. The secondary and other heads are smaller in size and bear fewer seeds. Flowers first open on the periphery of the central head followed by those on secondary and other heads. Flowering on a single plant may continue for about 4 weeks. The entire umbel may take about 7–10 days to complete flowering depending upon its size and environment.

The flowers are perfect, having usually white, small petals; petals may sometimes be greenish white. The calyx is whole with five petals and five stamens. Both the petal tips and stamens tend to turn inward when young, although those at the circumference are enlarged and turn outward. The ovary is inferior, consisting of two locules, each with a single ovule. Cross-pollination is common, usually brought about by bees and other insects.

Cultivars. The most important bunching cultivars of carrot grown in the United States are Imperator, Gold Spike, and Gold Pak, all having long slender carrots with a smooth exterior. Red Cored Chantenay, Royal Chantenay, and Autumn King are popular processing cultivars. The highly flavored French, or Chantenay, cultivars are not suitable for processing. Royal Chantenay grows about 2.5 cm longer and is more cylindrical than Red Cored Chantenay. Nantes is a good home garden carrot cultivar. Other popular cultivars are Rubicon, Perfection, Bunching, Hutchinson, Bagley, Touchon, and Coreless.

Dainello and Heineman (85) reported that the hybrids Woodland, 13 C × 24, and Fanci Pak were the most promising carrot cultivars grown in southwest Texas. A new carrot inbred, Florida 524, has been released, which is valued for its usefulness as a male parent in experimental hybrids (86). Tigchelaar (3) mentioned the following cultivars of carrot: Bonanza, Cutlass, Grenadier, Imperator406, Manipak, Packmaster, Target, Trophy, and Woodland. The Indian Agricultural Research Institute, New Delhi, released the following carrot varieties: Nantes, Chantenay, and Pusa Kesar, which is a selection from a cross between Local Red and Nantes (47).

2. Breeding

Before the 1950s there was little controlled breeding of carrots; seed growers mainly depended on root selection followed by mass selection for improvement. The chief characteristics used for such selection were root shape, exterior and interior color, cortex width, size and color of the core, and size and toughness of the neck.

For selfing either an entire plant or several heads in different stages of development are required to be enclosed in a cage with flies. To make a cross, an individual umbel of the female parent is caged, and after another dehiscence is complete, an umbel of the male parent is introduced before its anthers dehisce together with some flies. The stalk of this umbel should be placed in jar of water adjacent to the plant. Any seed setting on the umbel of the female parent is the result of cross-pollination.

Since carrots cross-pollinate easily, varieties should be sufficiently isolated from each other. A distance of about 1.6 km is advisable between carrots of different colors (white, yellow, or orange) grown for seed production. This distance should be increased when the crop is grown for stock seed.

Synthetic varities are produced by randomly mating several parents so that all possible matings between parents have an equal possibility of occurring. The parents are collectively called the Syn 0 and their offspring the Syn 1. The amount of seed of a synthetic variety is multiplied by successive generatiõns of random mating without selection; these generations are called Syn 2, Syn 3, and so on. Thus, a synthetic variety is an expanding population consisting of few individuals in Syn 0 and many individuals in advanced generations. Because there are few individuals in Syn 0, relatives will mate in advanced generations, resulting in inbreeding. Synthetic seed will consist of either a quiescent or nonquiescent somatic embryo with or without a protective encapsulation. The exact form of synthetic seed required will depend upon its specific application (87).

Somatic embryogenesis is recognized as a technique for producing large quantities of a clone and is consequently of great interest in development of artificial seeds (88). Somatic embryogenesis in liquid culture, particularly in a bioreactor, is thought to be an economical approach to future commercialization of artificial seeds (89–91). In establishing a bioreactor culture system, certain factors are important, including cell density, bioreactor design, foaming, preculture of embryogenic cells, and dissolved O2 concentrations. Teng et al. (92) reported that during somatic embryogenesis, dissolved O2 concentrations decreased to 33% of saturation and then increased up to 80% when embryo development approached maturity and mature embryos germinated. Bioreactor-cultured embryos germinated with relatively short cotyledons and long roots, whereas flask-cultured embryos germinated with relatively long cotyledons and short roots.

Kitto and Janick (93) studied the effect of hardening treatments on the survival of synthetic-coated asexual embryos of carrot and showed that survival was increased by hardening embryos with 10^{-6} M ABA during embryo initiation and development. The use of pharmaceutical-type capsules has been developed as a coating system for the production of synthetic seeds. The capsule body was covered on its inner surface by a watertight film composed of polyvinyl chloride (PVC), polyvinyl acetate (PVA), and bentone as a thickener to control the nutrient supply and the subsequent development of somatic embryos. The germination medium and the embryo were then placed within this improved capsule. Carrot somatic embryos were able to sustain conversion rates of 90% in such coatings (94). The effects of encapsulation or biochemical treatments on the physiological behavior of carrot somatic embryos were studied by Bazinet et al. (94a); they reported that treatment with a high concentration of $CaCl_2$, mannitol, or sucrose at 4°C was required in addition to encapsulation in order to decrease the growth rate or to slow down metabolic activity and to preserve the integrity of somatic embryos during long-term storage. The germination rate, respiration, and ethylene emission decreased with increasing alginate concentration. This inhibitory effect of encapsulation resulted principally from a limitation of oxygen diffusion to the somatic embryo, since germination and radicle growth of naked somatic embryos were inhibited in atmospheres containing less than 15% O_2. Radicle growth was almost totally inhibited in 5% O_2 (94b).

3. Seed-Production Technology

Root-to-seed and seed-to-seed are the two distinct methods of growing carrot seed. The root-to-seed method is the commonly employed procedure in the seed trade.

Root-to-Seed Method.

CULTURE. In the first season, the method of culture is much like that followed in the production of carrot for the fresh market. The main difference is the planting time. In parts of Idaho and Utah, the seed crop is planted from June 1 to 15, whereas in California, it is planted in July; carrots are best sown from August to October in Arizona. Rows in steckling seedbeds are 30–45 cm apart or may be in pairs on slightly raised beds. During the first few weeks of planting, frequent irrigation may be necessary, which may be reduced at later stages.

All seed growers do not necessarily grow stecklings. Some seed companies grow and distribute them to farmers to grow a seed crop. This affords seed companies greater control over the stecklings used. The stecklings from 1 ha are sufficient to plant 10–15 ha of seed plants. On the basis of weight, about 3.5–5.0 tons of sorted healthy stecklings of medium size would be required to plant 1 ha of land. Stecklings are lifted from field, topped, and placed in cellar storage in late October (Idaho and Utah), November–December (California), or January (Arizona). The stecklings can be ideally stored at 0°C and 90–95% RH. Tops may be cut or removed prior to lifting the roots. Digole and Shinde (95) reported the highest seed yield (15.4 Q/ha) with stecklings where shoots and roots were cut back by one-quarter and one-half, respectively.

In India, the seed is sown in the first week of September and the roots are ready for transplanting in January. The roots grown on 1 ha provide enough seed material to transplant to 5–8 ha. Average seed yields of about 7–10 Q/ha are obtained (47). Malik et al. (96) obtained the highest seed yields of carrot (10.9–11.1 Q/ha) from the closely spaced plants (45 × 20 cm) with all umbels left on the seed. Quality was not affected by either spacings or number of umbels allowed to develop on the plant. Seed yield of carrot increased with increasing N rates from 10.5 Q/ha at 50 kg N/ha to 11.49 Q/ha at 150 kg N/ha as compared with the control yield of 9.6 Q/ha. Spacing, however, had little effect on seed yield (97).

ROGUING. Roguing begins before replanting by removing the small, diseased, cracked, or injured and off-type stecklings. Characteristics such as shape and exterior or interior color aid in roguing off-type stecklings. When the crop is grown to raise stock seed, cutting to examine the interior color and core size should be followed.

REPLANTING. Stecklings stored for a considerable length of time should be regarded before replanting to remove decayed or badly shriveled roots. The soil should have an adequate amount of moisture, through irrigation if necessary, at the time of planting. If irrigation is not available, a small amount of water poured around each steckling at the time of planting helps to increase crop stand and seed yield. Setting out stecklings as soon as possible after removal from storage rather than exposing them to open air for several days may increase seed yields.

In the United States, stecklings are commonly set out in rows 90 cm apart, whereas in India, they are planted with 60 cm between rows and 45 cm between plants. Various machines are available for large-scale plantings.

Seed-to-Seed Method.

CULTURE. In the seed-to-seed method, the seed for the steckling crop is sown, the roots grown, and the seed harvested in one location. This is not a common practice for raising carrot seed. This method necessitates a prolonged period of cool weather for seed stalk formation.

Jacobsohn and Globerson (98) conducted laboratory experiments in growth chambers and commercial fields at temperatures ranging from 15 to 35°C; in most cases, with a Nantes variety of carrot. Larger seeds germinated and performed better under more adverse conditions; namely, deep planting and elevated temperatures. Results of emergence and establishment tests conducted under commercial field conditions during the summer indicated no difference between seedling emergence and establishment from large and ungraded seeds based on the number of viable planted seeds. Large seeds resulted in a higher root yield in spring and summer, but these differences were diminished in autumn-planted carrots, having a longer growing season. Planting graded seeds did not improve carrot root uniformity at harvest time.

Hermansen et al. (98a) studied the effects of hot-water treatments of carrot seeds on seedborne fungi, germination, emergence, and yield. Seeds infected with *Alternaria dauci* were hot-water treated at temperatures ranging from 44 to 59°C at intervals of 5°C for periods of 5–40 mins. Different grades of healthy carrot seeds were hot water treated at 50–55°C. In some experiments, seed treatments with *Trichoderma harzianum*, *Streptomyces griseovirides*, or fungicides were included. Seed germination and survival of fungi were investigated in the laboratory and emergence was determined in a growth chamber test. Emergence and yield measurements under field conditions and postharvest evaluation of carrots were made in some experiments. Hot-water treatment of carrot seeds at 44, 49, and 54°C generally improved germination of infected seeds and reduced the incidence of *A. dauci*. Treatments at 54°C for 20 mins eradicated *A. dauci* without adversely affecting germination, emergence, or yield. A significant correlation ($r = 0.97$) between the incidence of *A. dauci* and the occurrence of abnormal seedlings was found. Small seeds were more sensitive to hot water than larger seeds. In two field experiments, seed treatment with iprodione gave higher emergence and seedling weight than hot-water treatment, possibly because iprodione may have reduced attack by soilborne fungi, whereas hot water-treated seeds would not have been protected against such organisms. Hot-water treatment of seeds and seed treatments with the biological control agents had no effect on carrot yield and storage quality but reduced the incidence of the saprophyte *Ulocladium atrum* on the seeds. The germination percentages found in the laboratory correlated well with the emergence percentages in the growth chamber. The hot-water treatment was found to be an alternative to fungicides to eradicate seedborne pathogens in carrots in organic farming systems.

Seeds from primary, secondary, and tertiary umbels were harvested and threshed separately. Primary umbel seeds germinated better than seeds of equal size and weight from secondary and tertiary umbels. The ratio of primary umbel to seed increased with higher densities (98).

PLANTING.   Seed is sown in time to allow the plants to enter the winter period when most of the roots have attained their widest minimum diameter of about 0.6–1.2 cm. In the southern parts of United States, late July or early August is an appropriate time for planting. About 0.56–0.86 kg of seed is sufficient to plant 1 ha of land with a spacing of 90 cm between two rows. Schoneveld et al. (99) noted that seed germination was highest when seeds were sown at a depth of 1.5–2.5 cm, which is achieved with oblique pressure. Plantings carried out in dry summer weather require frequent irrigations until seedling emergence in about 10 days. The soil should have an adequate moisture content at the time of sowing and throughout seedling emergence. A complete loss of stand may occur if the soil dries even for a period of few hours. Too deep planting and failure to maintain proper soil moisture are the most frequent causes of a poor crop stand from summer plantings. The seed should be sown in soil in which no carrots have been grown during the last 4 years, and it should be isolated as far as possible from other carrot seed fields. Such precautions should be carried through to the second year in a root-to-seed crop.

Gray (100) found the responses of seed yield and quality in carrots to result in a range of plant densities from 110,000 to 2560,000 plants/ha (seed-to-seed method) and 100,000 to 800,000 plants/ha (root-to-seed method). Seed yields from the root-to-seed method increased from about 1100 to 1500 kg/ha as the plant density increased from 100,000 to 800,000 plants/ha. For the seed-to-seed method, yield increased from 700 to 2400 kg/ha over the range of plant densities studied. At the highest densities, about 60% of the seed yield was contributed by the primary-order umbels as compared with less than 20% at the lowest plant density (Table 5). Plant density, however, did not significantly influence plant height, time of flowering, or crop maturity. The effect of plant density on 1000–seed weight was not consistent (Table 6).

Table 5   Effects of Plant Density on Total Yield of Cleaned Carrot Seed Contributed by Various Umbel Orders

| Umbel order | Yield (% by weight) | |
	100,000 plants/ha	800,000 plants/ha
Primary	19.1	58.9
Secondary (S_1)	10.4	9.1
S_2	31.1	16.1
S_3	9.9	7.3
S_4	8.7	6.8
S_5	8.3	0.6
S_6	6.7	0.7
S_7	6.0	0.1
S_8	4.5	0.0
S_9	2.8	0.1
S_{10}	2.3	—
S_{11}	1.1	—
S_{12}	0.5	—
All tertiaries	6.4	—

Source: Ref. 100.

Table 6   Effects of Plant Density on 1000–Seed Weight of Carrot

Root-to-seed method		Seed-to-seed method	
Plants/ha (thousands)	1000–seed weight (g)	Plants/ha (thousands)	1000–seed weight (g)
100	0.958	110	1.755
200	0.943	320	1.729
400	0.942	720	1.772
800	1.000	1800	1.816
SED	±0.0136		±0.0234

SED = Standard error of difference between two means.
Source: Ref. 100.

Only stock seed of the highest quality should be sown for seed-to-seed plantings. A stock seed developed from single plant selections and by modern plant breeding methods gives better results. Such stocks are not likely to produce a great number of off-type rogues. Seed produced by the seed-to-seed method should not be used to produce another seed-to-seed crop.

Seed yields of about 750–1150 kg/ha can be obtained from an efficiently handled crop. The seed-to-seed method of seed production saves time and labor otherwise used in lifting the steckling, replanting, and storage. Normally, higher seed yields are obtained from this method than from the root-to-seed method. However, in the seed-to-seed method, the type of root cannot be accurately determined and systematic roguing is not practical.

IRRIGATION, FERTILIZATION, AND WEED CONTROL.   About 20–25 tons/ha of well-rotted farmyard manure should be applied during land preparation. In addition, 200 kg of muriate of potash and 150 kg of superphosphate should be supplied per hectare before sowing. The Indian Council of Agricultural Research, New Delhi, recommended 40–50 kg N, 40–50 kg P, and 80–100 kg per hectare in addition to basal manuring (37). The highest average seed yield of 17.19 Q/ha was obtained with the highest N rate (120 kg/ha). Phosphorus had a less pronounced effect on seed yield (101).

The first irrigation should be applied immediately after sowing followed by a second after 4–6 days. Fields should be irrigated sparingly in winter and after 2 weeks of dry weather. Heavy irrigation may result in excessive foliage growth and poor-quality roots, delaying crop maturity. Gypsum block may be used to monitor soil moisture more accurately. Seed quality, as measured by seedling root length, has been found to improve with increasing water application (102).

The field should be kept free of weeds, especially during the early phases of crop growth. The crop should be thinned, leaving space around each plant. Herbicides may be used to control weeds. In trials with carrots grown for seed production, an application of Stomp (pendimethalin) at 1.5 kg preemergence plus 1 kg postemergence/ha or linuron at 2 plus 1 kg/ha effectively controlled weeds, markedly increased seed yield, and improved the sowing quality of the carrot seeds. The application of Tetral (chlorthal-dimethyl) and Brometryne was also found to be effective against weeds and for increasing yields (103).

CONTROL OF PESTS AND DISEASES. Insect pests are a comparatively minor problem in the production of carrot seed. The most important pest is the lugus bug (*Lugus campestris*, *L. oblmeatus*, *L. hesperus*, and *L. elisus*). Another important insect pest is the carrot rust fly (*Psila rosae* Feb).

Watery soft rot (*Sclerotmia sclerotiorum*) gray mold (*Botrytis cinera*), black rot (*Alternaria radicma*), soft rot (*Erwmia carotovera*), and root-knot (*Heterodera maroni*) are the important diseases of carrot, affecting roots stored for the winter. They may occasionally also affect crops in the field.

Bacterial blight (*Xanthomonas carotae*) is a potentially serious disease of the above ground portion of carrot, becoming very conspicuous at the time of flowering and seed setting. Seed should be treated with hot water (52.2–53.0°C for 10 min) before planting to prevent this disease. Aster yellows and green dwarf (virus) diseases have also been reported in carrot. Prolonged heat treatment and higher temperatures were particularly effective in reducing populations of seedborne *Alternaria dauci* (104). The treatment of carrot plants with potassium metabisulfite or ascorbic acid has been found to reduce the incidence of white rot caused by *Sclerotinia sclerotiorum*, increasing the carrot seed yields (105). According to Sharma et al. (78a), steckling rot of carrot can cause 100% losses during favorable climatic conditions in heavy soils of northeast areas of the Punjab in India resulting in poor seed recovery. Steckling dip treatment in Emisan (0.1%) for 24 hr before transplanting exhibited a significant control of the disease.

HARVESTING. Carrot seed is generally harvested when the secondary heads ripen completely and third-order heads begin to turn brown. Shattering is not a very serious problem in carrot. In the United States and Europe, carrot seed is usually harvested in September. The crop is either pulled out or moved. Pulling is much easier and quicker. With mowing machines, some branches are likely to be left on the ground during the threshing operation. Pulled plants are placed in windrows in small piles of three to five plants and are allowed tocure under the sun for 4–5 days. Under humid conditions, the curing process may require 10–15 days or longer. The crop is ready to thresh when the stems become brittle. A combined harvester and thresher may be used.

4. Postharvest Operations

The spines must be removed from the threshed carrot seed. Since only the rubbed seed can be handled satisfactorily by seeding machines, all carrot seed sold in the United States today is exclusively a rubbed seed. The carrot seed is first put through a despining machine, where the spine is broken through some sort of rubbing action. The barley debearder is also frequently employed to despine carrot seed. The chaff and dust are then removed in a screen-fanning mill with specific screen sizes. The final cleaning is performed on a gravity separator.

Tylkowska and Bulk (105a) studied the effects of priming on seed infestation with fungi and germination of carrot (*D. carota*) seed lots naturally contaminated with seedborne *Alternaria* spp. Three seed lots of two cultivars, Karita and Nantes Topscore F_1, varying in the initial incidence of *Alternaria* spp., were primed in polyethylene glycol (PEG) 6000 or hydroprimed under laboratory conditions. The lots differed in their response to priming, both in terms of germination and the presence of *Alternaria* spp. Priming resulted in a faster germination for all three seed lots. The uniformity of germination was improved in two lots. The overall percentage germination did not increase. Both osmotically and hydroprimed seeds gave rise to a decreased number of normal seedlings compared with untreated seeds. This decrease resulted from an increase in the number of seedlings with fungi and/or decay and dead seeds, which was related to a significant increase in the number of seeds contaminated with fungi after priming. In the seed lots with intermediate- or high-incidence levels of *A. dauci*, *A. radicina*, or *A. alternata*, priming resulted in significantly higher levels of contamination. Moreover, fungi invaded the inner parts of the seeds, especially after PEG priming. These findings indicate that the level of fungal contamination of seeds should be evaluated prior to the selection of seed lots for priming.

The seed yields of carrot vary from about 560 to 675 kg/ha with 90 × 90 cm spacing and from 900 to 1125 kg/ha with 90 × 30 cm spacing. In the seed-to-seed method, the yields are slightly higher than in the root-to-seed method. Also, smaller rooted varieties such as Nantes usually yield about 50 kg less than Chantenay under the same conditions.

B. Celery

1. Introduction

Following carrot and parsley, celery is the third most important umbelliferous vegetable grown in the United States, Canada, and the Netherlands. As a salad crop, it ranks second only to lettuce.

Botany. Celery (Apium Graveolens) is a Biennial Crop. The leaves are pinnately compound, and blades may be light or dark green according to the variety. The important varietal characteristics include color, size, length, and cross section of the leaf and the shape of the of the petiole. Celery develops a well-formed taproot, which normally is broken at the time of transplanting. A large mass of roots develops within a limited area. The inflorescence consists of small white flowers in compound umbels. The umbels (or heads) of celery are smaller and less compact than those of carrot. The flowers, although potentially self-fertile, are normally cross-pollinated by insects. The seed (mericarp) results from the splitting of the schizocarp (fruits) and is also ribbed and much smaller than carrot seed.

Cultivars. Tigchelaar (3) described the following cultivars of celery with improved characteristics:

> Clean Cut—Open-pollinated, duration 125 days from transplanting, excellent shipping quality, large heavy petioles of good length, relatively few side shoots, similar to Utah 52-70
>
> Florigreen—Attractive, uniform, vigorous widely adapted green stalked cultivar developed to provide a distinct petiole type for fancy grade pack, tolerant to brown spot, wide adaptability, similar to Florida 683
>
> Transgreen—Wide thick petioles and excellent yield potential, similar to Florida 683

Thompson and Kelley (106) grouped celery cultivars into two classes: green cultivars with dark green foliage and yellow or self-blanching cultivars with yellowish green foliage. The green cultivars may be of an early type, which are easy to blanch, or of a later type and slower to blanch. Golden Self Blanching is the most extensively grown yellow cultivar. Other important yellow cultivars are Michigan Improved Golden, Cornell 19, Supreme Golden, Golden Plume, Wonderful, Golden No. 15, and

Cornell 619. The Indian Agricultural Research Institute, New Delhi, has recommended Standard Bearer and Wright Grieve Grant as improved cultivars of celery in India.

2. Breeding

The important characteristics of celery in a breeding program are length of petiole, crispness. tenderness, lack of pithiness, lack of ribbing, shape of cross section, and flavor. Premature bolting, observed in certain varieties, has also interested breeders. Varieties resistant to fusarium yellows of celery have also been developed.

Somatic embryogenesis in celery was first reported by William and Collin (107) and became a model system for the study of this phenomenon. Somatic embryos were induced from the primary callus of celery arising from leaf blade explants placed on Murashige and Skoog (MS) salts and vitamin medium supplemented with 9µM 2,4-D–free MS medium. Embryos proceeded in a standard developmental pattern through the globular-heart and torpedo-shaped stages. Secondary somatic embryos occurred on the cotyledons and hypocotyl of primary embryos (108).

A cell culture system for the production of somatic embryos from the F_1 celery line 1026-2 was developed and scaled up in a bioreactor system, which increased somatic embryo productivity and promoted uniformity of development in the presence of ABA (109). Dehydrated high-quality somatic embryos were coated with an alginate gel (1.5–20.0%) and artificial endosperm using encapsulation machines. The 10,000 plants derived from these artificial seeds were tested in the field and in hydroponic culture. Some morphological variation in leaf shape was observed. Cho and Yang (110) studied the effect of PEG and ABA on storage starch accumulation and the survival of desiccated somatic embryos of celery, and reported that embryo maturation in the presence of PEG plus ABA increased the percentage survival after 3 days of desiccation. An increase in starch accumulation occurred at the same time as a decrease in amylase activity in somatic embryos treated with PEG plus ABA, which might account for the desiccation tolerance.

3. Seed-Production Technology

Culture. In California, seed beds are sown in July or sometimes in December–January. The partly grown seedlings are then transplanted to the field. Plants with off-color foliage or poor growth habits may be rogued out at this stage. In other parts of the United States with more severe winters, crops are started in the early spring by sowing in greenhouses or hotbeds, and seedlings are dug in the fall, rogued for type, and cold-stored until planting time the following spring. The tops are stored at 0°C and 90–95% RH, maintaining moisture and good ventilation around the roots. The roots are placed in moist soil. The withered and decayed leaves should be removed when plants are transplanted with about 90 cm between rows and about the same space between plants. The closer spacing results in higher seed yields.

For the fresh market, celery is successfully grown on muck or peat soils or on loams with a high organic content. Celery needs high soil fertility, usually maintained by the application of balanced commercial fertilizers. Urea used in combination with low N improved the percentage of shoot dry matter and improved leaf area, shoot and root dry weight, and root-to-shoot ratio of the seedlings. Urea proved to be beneficial in improving transplant yield potential under high N fertilization (111). The crop also requires an adequate and constant supply of moisture, since droughts, even if short, seriously impair crop quality.

The balance of endogenous gibberellins, cytokinins, and inhibitors like abscisic acid regulates celery seed dormancy, and the seed hormone status largely determines its performance when sown in any environment. Within and between cultivars the extent of dormancy of celery seeds can be measured quantitatively as the germination response to a mixture of gibberellins (Ga_4, GA_9, and GA_4 + GA_7) in the dark at a temperature inhibitory to germination. It has been demonstrated that seeds that have a low dormancy and require only short exposure to red light in order to break dormancy respond to GA_4 + GA_7 treatment alone, whereas seeds that are very dormant and require prolonged red light exposure germinate only when other growth regulators such as cytokinins, daminozide, or ethephon are supplied in addition to GA_4 + GA_7. Thomas et al. (112) used this information to develop seed-soak treatments to improve the germination of natural or pelleted celery seeds (Fig. 8). The dormancy-breaking effect of red light for celery seeds incubated at 2°C was stimulated by exogenous application of GA_4 + GA_7 together with permissive compounds including cytokinins, ethephon, daminozide, and fusicoccin (113). The effect of light, temperature, and seed priming on the germination of celery seeds has been studied by Perez-Garcia et al. (114), who reported that at relatively high temperatures (25°C), light exposure increased germination, although the magnitude of this increase was variable and depended on cultivar. Similarly, priming treatments did not affect final germination percentage but were effective in reducing the time to 50% of final germination percentage (T_{50}).

Thomas et al. (112) showed that seeds from different umbel positions on parent celery plants had quite different dormancy and germination characteristics (Table 7).

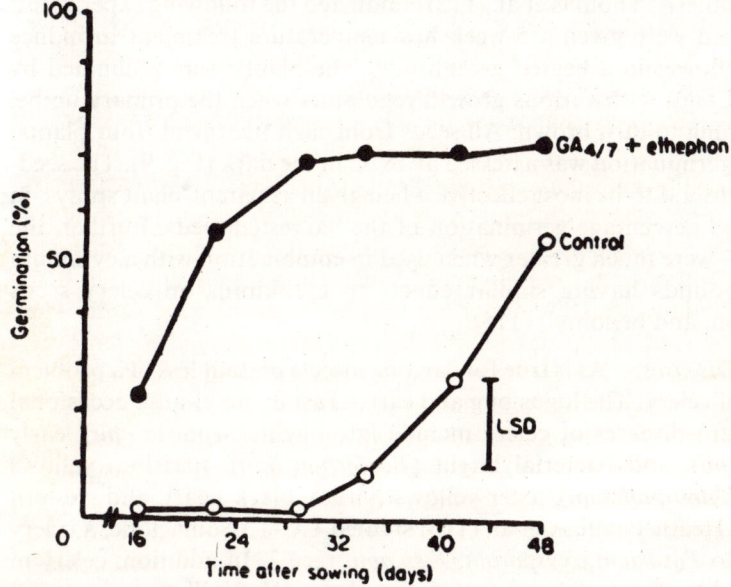

Figure 8 Effects of prepelleting growth-regulator soaks on emergence of celery (cv. Lathom Blanching) seedlings in a greenhouse at 15–20°C maximum temperature. (From Ref. 112.)

Table 7 Characteristics of Seed from Different Umbels of Celery Plants (cv. Lathom Blanching)

Umbel position	Mean seed weight (mg)	Germination at 18°C in light (%)	Germination 18°C in dark (%)		Upper temperature united for dark germination (°C)
			$GA_4 + GA_7$	$GA_4 + GA_7 + BA$	
Primary	0.474	50	38	92	17.6
Secondary	0.438	72	49	95	17.7
Tertiary	0.380	94	30	96	16.7
Quarternary	0.348	82	10	98	15.3
LSD at 1%	0.069	9.2	9.2	9.2	1.8

GA = gibberellic acid; BA = benzyl adenine; LSD = least significant difference.
Source: Ref. 112.

Seeds from primary and secondary umbels were larger, less dormant, and less viable than those from tertiary and quaternary umbels. The response of primary and secondary seeds to $GA_4 + GA_7$ alone at 2×10^{-4} M in the dark at 18°C was less than that of tertiary and quaternary seeds, but seeds from all umbels showed a similar maximum response to $GA_4 + GA_7$ when used in combination with benzyl adenine (BA) (Table 7).

It is adduced that later-formed (tertiary and quaternary) seeds are either deficient in cytokinin-type compounds or contain other compounds which interfere with gibberellin and cytokinin utilization. The application to parent plants of cytokinins or compounds giving the same effect (e.g., daminozide and ethephon) decreases the dormancy of tertiary and quaternary seeds, improving the rate and uniformity of germination.

To test this hypothesis, Thomas et al. (112) conducted the following experiment: plants grown from seed were given a 5-week low-temperature treatment to induce flowering and left to flower in a heated greenhouse. The plants were pollinated by blowflies and sprayed once with various growth regulators when the primary umbel seeds were just beginning to turn brown. All seeds from each treatment (four plants) were bulked, and the germination was assessed at 18°C in the dark (Fig. 9). The seed-soak treatments were found to be most effective when given as parent-plant sprays, in increasing the rate and percentage germination of the harvested seeds. Further, the effects of $GA_4 + GA_7$ were much greater when used in combination with a cytokinin (BA) or other compounds having similar effects to cytokinins on celery seeds (daminozide, ethephon, and benomyl) (115).

Control of Pests and Diseases.   As is true for carrots, insects present less of a problem than do the diseases of celery. The lugus bug and carrot rust fly may cause occasional damage. The important diseases of celery include late blight (*Septaria apii*), early blight (*Cercospora apii*), and bacterial blight (*Bacterium apii*), fusarium yellows (*Fusarium apii* and *F. apii pallidum*), aster yellows (virus), black heart, and western celery mosaic (virus). Heath-Pagliuso et al. (116) studied UC-T3 Somaclone, a celery germplasm resistant to *Fusarium oxysporum*, *F.sp.apii*, race 2. In addition, celery in storage is also infected by *Sclerotinia sclerotiorum* and several *Bacillus* spp., causing soft rot, pink rot, and bacterial soft rot. A storage temperature of 0°C and good ventilation to prevent water-saturated air will help to control these diseases.

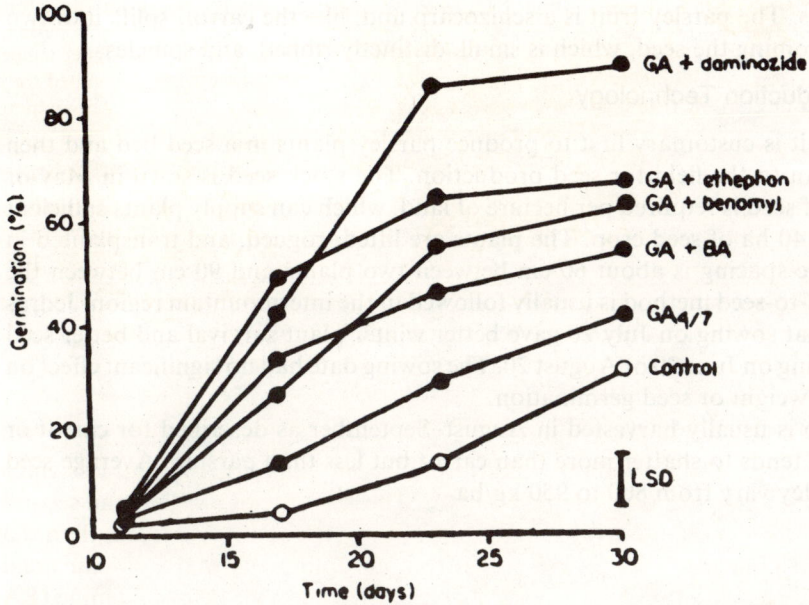

Figure 9 Effects of growth-regulator sprays applied to celery (cv. Lathom Blanching) during seed maturation or germination of seeds at 18°C in dark. (From Ref. 115.)

Harvesting. Celery seed is usually ready to harvest during August or early September. As in the case of carrots, plants are pulled, windrowed, and cured in the sun before threshing. Appropriate harvest maturity is important to avoid shattering. With small plantings, harvested plants may be placed on canvas to reduce seed loss due to shattering. The use of combine harvesters is very common in the United States and other developed countries. Plant bunches are tossed onto moving canvases and fed to the machine directly from the window. Seed can be cleaned easily with a screening mill, followed by a gravity separator. Ramin and Atherton (117) reported that seed yield per umbel was reduced by chilling during seed development.

C. Parsley

1. Botany

Parsley (*Petroselinum crispum*) is normally a biennial vegetable crop. During the first year, a dense rosette of leaves is formed and the white taproot becomes thickened. The leaves are largely segmented. They may be flat, curled, or finely divided depending upon the variety. The Hamburg, or turnip-rooted, variety has plain leaves. In the second year, a seed stem develops with a number of second- and third-order branches, reaching a height of about 1.5–2.0 m depending upon variety, soil fertility, spacing, irrigation, and other conditions. The taproot of parsley may reach a depth of over 1 m, but the laterals rarely grow more than a meter.

The inflorescence is characteristic of the umbelliferous type, and as for carrot, it is compound. The flowers are small, greenish yellow, and potentially self-fertile. The umbels are less dense than those of carrots. Significant cross-pollination occurs through insects. The parsley fruit is a schizocarp and, like the carrot, splits into two mericarps, becoming the seed, which is small, distinctly ribbed, and spineless.

2. Seed-Production Technology

In California, it is customary first to produce parsley plants in a seed bed and then transplant them to the field for seed production. The stock seed is sown in May or June; 2–3 kg of seed is required per hectare of land, which can supply plants sufficient to raise about 40 ha of seed crop. The plants are lifted, rogued, and transplanted in December. The spacing is about 60 cm between two plants and 90 cm between the rows. The seed-to-seed method is usually followed in the intermountain region. Jedras (118) found that sowing on July 20 gave better winter plant survival and better seed yield than sowing on June 20 or August 20. The sowing date had no significant effect on the 1000–seed weight or seed germination.

The crop is usually harvested in August–September as described for carrot or celery. Parsley tends to shatter more than carrot but less than parsnip. Average seed yields for parsley vary from 800 to 950 kg/ha.

D. Parsnip

1. Botany

Parsnip (*Pastinaca sativa*), like most other umbelliferous crops, is a biennial. It develops a rosette of leaves and a fleshy taproot in the first year, as in carrot. The pinnately compound leaves and their segments are much larger than those of carrot. In the second year, the stem elongates, growing to height of 1–2 m or longer. The basal leaves are petiolate and the upper stem leaves are sessile. Of the umbelliferous vegetables, parsnip grows most slowly.

Compound umbels characterize the parsnip inflorescence, which are less compact than those of carrot and similar to those of celery. The yellow-green flowers arise from the summit of the ovary; that is, they are epigynous. Stigmas are not receptive until about 5 days after the anthers in the stamen split into two mericarps, forming the seed of the trade. Parsnip seed is distinguished from that of celery or parsley by two lateral ribs that are expanded into wide, flat wings. The seed retains its viability for 1 or 2 years if stored well.

2. Seed-Production Technology

In California, seed is sown in late July at a rate of about 3.4–4.5 kg/ha. Soaking of seeds in GA3 at 500 ppm for 24 hr accelerates seedling emergence by 5–6 days (119). Seedbeds are prepared as described for carrot. The roots are dug out, rogued, and transplanted in January–February. If the tops are not already frozen back, they are cut off about 7.5 cm above the crown before the roots are lifted. A spacing of 90 × 90 cm is generally recommended. The steckling field is started earlier in the spring in Idaho than in California, and the stecklings are lifted in late October or early November and stored until spring. The root-to-seed method of producing seed is more common than the seed-to-seed method. In the eastern United States, parsnip roots are allowed to overwinter in the ground, after which the stecklings are lifted, graded, and reset in the field for the seed crop (root-to-seed method).

Harvesting and postharvest operations are similar to those described for carrot. Because shattering is serious problem, proper crop maturity for harvest is important.

E. Other Minor Umbellifers

Seeds of following minor vegetable crops are also produced to a limited extent: celeriac (*Apium graveolens repaceum*), dill (*Anetheum graveolens*), caraway (*Carum carvi*), coriander (*Coriandrum sativum*), fennel (*Foeniculum dulce*), chervil (*Anthriscus cerefolium*), and anise (*Pimpinella anaisum*).

V. LETTUCE, ENDIVE, SALSIFY, AND CHICORY

Lettuce, endive, salsify, and chicory belong to the family Compositae, with lettuce being the most important vegetable crop grown for both market and seed production. Two other minor vegetable crops belonging to this family are dandelion and globe artichoke.

A. Lettuce

1. Introduction

Botany. Lettuce (*Lactuca sativa*) is a warm-season vegetable grown as an annual. It is botanically related to wild lettuce. There are two classes of head lettuce: crisp head and butter head. Cool weather with a mean monthly temperature of 12.8–15.6°C favors the growth of lettuce. It is grown mainly in areas having cool summers and mild winters.

Lettuce forms a shortened fleshy crown stem from which basal leaves arise. Leaves may vary in color from light to dark green, and some may be reddish brown completely masking the green color. They may be circular in shape or long and spatulate. The degree of their compactness also varies considerably. In the crisp-headed varieties, the extreme compactness of heads may interfere with the emergence of the seed stalk. Lettuce has a strong, fleshy taproot, elongating to a length of about 2 m or more.

The seed stem with its numerous branches forms a bushy inflorescence, growing to a height of about 60–120 cm or more. There is considerable variation in the compactness of the inflorescence depending on the variety. The branches, commonly bearing clasping leaves, are terminated by a flower head. The inflorescence is a panicle consisting of a cluster of heads, each containing 15–25 or more flowers. The head terminating the inflorescence is the oldest; the secondary or lateral heads arise later in the axil of the leaves. The flowers are perfect and the yellow corolla is sympetalous. The ovary is one celled, the style is single, but the stigma is two lobed. The five stamens are each attached separately at the base of the corolla tube, but the anthers are united forming a cylinder around the style. The lettuce seed is actually a one-seeded fruit developed from the one-celled ovary. The seed matures about 2 weeks after anthesis.

Although the mechanism of pollination generally favors self-pollination, a certain amount of crossing may take place in the field. The bracts subtending the flower head begin to open as a result of the development of flower heads, which elongate quickly during the 24 hr before anthesis. The style and stigma of the pistil are covered with small brush hairs. The anthers dehisce prior to the elongation of the pistil. As the pistil elongates, the hairs on the side of the pistil brush the pollen out of the pollen sacs of the dehisced anthers. The lobes of the stigma separate, allowing the pollen to fall on the stigmatic papillae located on the inside surface of the stigma. The flower remains open only for a short time. Wild bees and other insects frequently visit the flowers.

Cultivars. The most important commercial cultivars are in the class crisp head, having a cabbage-heading habit and crisp or brittle foliage. New York and Imperial are two of the most popular crisp-head cultivars grown in the United States, and many strains have been developed to adapt these cultivars to different climates. Imperial 44, New York 515, and New York 12 are best adapted to the northeastern states. Great Lakes and Pennlake are other important cultivars of the crisp-head type.

Several strains of Imperial and Great Lake are known. White Boston, Big Boston, May King, Salamander, and Wayahead are the best-known cultivars of the butter-head type. Cos or Roman lettuces are distinguished by their upright habit, long heads, and spatulate leaves.

Ryder (120) reported that Sea Green, a new lettuce cultivar, resists big vein disease caused by viruslike organisms. Great Lake (head type) and Chinese Yellow (loose-leaf type) cultivars are commonly grown in India. The Indian Agricultural Research Institute, New Delhi, has released the following improved lettuce varieties: Great Lakes, Chinese Yellow, Imperial 859, and Slow Bolt.

2. Breeding

Because of the lack of environmental adaptability of head lettuce, it was necessary to develop varieties specific for specific regions (e.g., summer and winter production). It was also necessary to develop lettuce varieties resistant to pests and diseases such as brown blight, downy mildew, and tip burn. Both selection and crossbreeding followed by inbreeding and testing have been employed for these purposes.

Inbreeding of lettuce is relatively simple. The plant is enclosed in an insect-proof cloth bag at the start of flowering, ensuring that no head has bloomed before bagging. The bag is secured to a stake, which should be at least as high as the seed stalk. The mouth of the bag should be securely fastened to the plant stem. When the seed is mature, the stem should be cut just below the bag, leaving it tied to the seed stalk until the seed is threshed.

For crossing, the small delicate florets are grouped together into a tight head, making mechanical emasculation impossible. Since self-pollination occurs as soon as the stigmatic lobes begin to curl, the pollen should be removed from the stigma before that time, which can be accomplished by careful washing of the flowers with a fine stream of water. The washed flowers are dried with a few vigorous air puffs from the mouth or with a jet of air from a pressure sprayer or syringe. The clean stigmas are then gently rubbed, with the stigmas carrying pollen from the male parent. Clipping of the ray petals from the head of the pollen parent aids in the pollinating procedure. It is advisable to leave only crossed heads on each branch.

With this technique, some selfing may occur if all the pollen grains are not removed by washing. Since it is often difficult to distinguish hybrids from their parents, the backcross method is of little value in lettuce breeding. Sowing of hybrid seed between alternate rows of the maternal parent has been found to be useful in determining hybrid plants. Hybrids are more vigorous and thus can be detected. F_1 plants may be easily detected using dominant marker genes in some cases. A practice of inbreeding following single crossing has been found to be the best method of breeding lettuce varieties because of the difficulties in emasculation and hybrid detection.

Complete hybridization (100%) in lettuce was accomplished consistently using the clip-and-wash method of emasculation, which is a combination of washing and clipping for pollen removal.

Wash methods and clip methods produced 98 and 95% hybridization, respectively. The wash-and-clip method is quick and easy and eliminates inadvertent self-pollination (121).

Lettuce breeding in the past has resulted in outstanding success in the development of varieties resistant to brown blight, downy mildew, tip burn, and premature bolting as well as those with improved quality. The first brown blight–resistant Imperial varieties were developed in the United States by selecting the plants of some varieties grown in the New York region. Because of the large number of biological races of *Bremia lactucae*, which causes downy mildew in lettuce, developing lettuce varieties resistant to this disease has been more difficult. It has not been possible to develop lettuce varieties resistant to all the prevailing races of downy mildew.

Grand Rapids and Great Lakes were developed as varieties resistant to tip burn. The latter is also a slow-bolting type with excellent adaptation to different regions of the United States. Further selections, such as variety No. 456, were more resistant to tip burn on muck soils of New York than the Great Lakes variety. The variety Progress was developed by crossing Imperial 44 with a hybrid whose parentage included New York, Hanson, Old Iceberg, and White Paris Cos. This variety was similar to Imperial 44 in its adaptability but had the advantage of less tip burn. Wurr (122) discussed various aspects of outdoor crisp-head lettuce production, including new cultivars, seedling vigor, seedling raising, and temperature responses. He described a new seedling-production technique—the techniculture system. It involves sowing seed in a tray with 400 cylindrical holes, each containing a small plug of peat stabilized with a binding compound so that the seedlings can be handled by an automatic transplanter.

Seed dormancy causes a problem for lettuce growing. Eenink (123) reported that a constant temperature of 20°C induced a clear distinction for degree of dormancy between lettuce cultivars. Diallele crosses were carried out between four lettuce cultivars—Magiola and Portato (dormant) and Deci-Minor and Valore (nondormant)—to study the inheritance of dormancy. On the basis of the behavior of the F_1, F_2, and F_3 seeds, Eenink (123) concluded that the dormant cultivars (Magiola and Portato) and nondormant cultivars (Deci-Minor and Valore) differ from each other with respect to one major gene for dormancy, possibly accompanied by modifier genes.

Most lettuce varieties can be vernalized; that is, they can be made to extend stems and flower more rapidly following a period at near freezing temperatures. This effect is inductive because no visible changes occur during the cold treatment in contrast to the effect of long photoperiods and raised temperatures, which act by hastening floral initiation. Vernalization in lettuce has a great potential to shorten the generation time in breeding programs and to hasten flowering for commercial seed production, especially in late-flowering varieties. However, the vernalization response in cultivated lettuce varieties is relatively small compared with that in closely related lettuce species.

Prince (124) described the technique of vernalization as a means of hastening seed production in *L. sativa* varieties dealing with the physiology of vernalization, responsive stages, duration of cold treatment, devernalization, interaction with photoperiod and temperature, and the range of the vernalization response. The standard procedure used in the laboratory is to place the seeds on two Whatman No. 1 filter papers in a Petri dish containing 4 mL of distilled water. Individual seeds are separated to minimize the spread of seedborne pathogens. The dishes are wrapped in a poly-ethylene bag containing wet absorbent paper to reduce evaporation, and these are held at 2–4°C for 28 days, preferably in the light. After this treatment, seeds are allowed to germinate in the light at about 15°C up to the cotyledon stage and transferred to soil.

Suitable genotypes for breeding for bolting resistance could be found in areas having long, warm but not excessively dry summers or from cool areas with a short growing season where a strong vernalization response ensures that nonvernalized plants do not bolt. The latter type has the advantage that artificial vernalization would allow seed production even in a cool climate. *L. serriola* offers a greater range of variation than exists in *L. sativa* for either strategy, and although the same may be true with *L. virosa* and *L. saligna*, hybridization is easier with *L. serriola* (124).

3. Seed-Production Technology

Culture.   Up to the harvesting stage, there is little difference in the method of growing lettuce for seed or for the fresh market.

LAND PREPARATION.   Land should be prepared by plowing thoroughly, al-though it is not necessary to pulverize the seed bed to a fine texture, since a granular surface results in better seed germination than a powdery one.

PLANTING.   Depending upon the climate of the region, lettuce grown for seed is planted from October to January (California) or in March–April (areas with severe winters). Under irrigated conditions, crops are grown either on raised beds (50–55 cm wide) or in rows spaced equally across the field. The center-to-distance between beds is about 90–100 cm. The seed is sown by drilling, adjusting the drills to achieve about 1–2 cm depth of sowing. The seed should be barely covered when it is drilled in dry soil following by irrigation. Where this method is practiced, the beds should be kept moist until the seed is germinated. About 0.5 kg of seed is sufficient to cover an area of 1 ha. Good seed may be mixed with dead (heated) seed to obtain a thin planting. Thinning should be carried out to maintain about 20–30 cm or more spacing in the row. Saini et al. (125) reported that the participation by endogenous ethylene was essential for the light-induced relief of thermoinhibition of lettuce seed germination. However, light did not act exclusively via ethylene, since exogenous ethylene alone in darkness did not promote germination. Growth regulators have been reported to increase the germi-nation percentage of salt-stressed seeds and reduce germination delay; kinetin was more effective than GA_3 (126). The minimum germination temperature in lettuce cultivars ranged from $-2.7°C$ (in cv. Salina) to $0.0°C$ (incv. Hilro) (127). Measure-ments of the germination potential showed that embryos from seeds produced in cool conditions were less able to cope with high temperatures than those from warmer conditions (128). The two cultivars showed similar patterns of response to the mean weight of individual seeds, which fell with increasing the temperature of environment, and the mean weight of seeds yielded per plant, which was lowest in cool conditions and highest in warm conditions; this was more pronounced with cv. Saladin (Table 8).

IRRIGATION, FERTILIZATION, AND WEED CONTROL.   Lettuce responds to fertiliza-tion in soils with low fertility levels, especially to the application of phosphorus. About 25–50 kg N and 100–150 kg P_2O_5 per hectare are recommended depending upon the soil fertility. Under semiarid conditions, the crop should be irrigated at the time of thinning and again after deheading followed by another irrigation at the time of flowering. Weeds can be kept down if the soil is not kept too moist. Intercultivation should be minimum to maintain the field free of weeds.

Table 8 Effects of Temperature of Seed Crop Environment on Characteristics of Seeds of Two Lettuce Cultivars

Regimen	Temperature (°C)	Sabine			Saladin		
		Mean yield per plant		Mean seed weight from 50-seed samples (mg/seed)	Mean yield per plant		Mean seed weight from 50-seed samples (mg/seed)
		Number of seed (approx.)	Weight of seeds (g)		Number of seeds (approx.)	Weight of seeds (g)	
Cool	20/10	1000	1.5	1.498	7900	13.8	1.714
Warm	25/15	3450	3.7	1.005	30100	30.4	1.008
Hot	30/20	4250	3.5	0.821	26100	24.3	0.910

Source: Ref. 128.

Roguing.   Off-type plants should be removed as soon as they are detected. Trueness to type is especially important in the heading varieties of lettuce. Plants are rogued based on uniformity of maturity, head size, leaf cover, foliage color, and leaf type. Rogued plants should be destroyed completely by cutting at least 2.5 cm below the ground surface, since shoots may develop on plants cut off just below the lower leaf, causing contamination if they produced seed. The practice of harvesting a seed crop after removing a crop of market lettuce, followed by some growers, is not desirable. If plants are not carefully rogued before the heads are harvested for market, the percentage of off-type plants producing seed may be very large.

Deheading.   With certain crisp-head varieties, it is usually necessary to remove or treat the heads in some way to achieve normal seed stalk development, because the stalk tends to curl inside the head. It may eventually break through, but it may cause head rotting or delay in crop maturity and reduction in seed yield.

Among various treatment methods, one involving quartering the upper portion of the head with a tool fashioned with knives set at right angles and fastened to the lower end of a long handle has been found to be fairly satisfactory. In this method, the operator plunges the knives down into the head without causing injury to the growing point of the stem. A more common method is to peel back the leaves on each plant by hand to expose the growing point. This method does not cause injury to the stem, but it is laborious and costly.

The deheading operation must be timed properly. It should be carried out as soon as the head has reached its full size but before the stem has elongated. The elongating stem may be injured or broken off in the deheading process. The deheading process thus must be completed quickly to obtain the best results. When the head is cut off, leaving a few basal leaves, several seed stems develop instead of just a central one. These are weaker, forming smaller inflorescences and fewer seeds.

The effects of photoperiod and application of growth regulators to parent lettuce plants on seed yield, weight, and performance have been studied. Thomas and O'Toole (115) reported that treatment of parent plants grown under long-day (17 hr) conditions with soil-drench applications of 2000 mg/L Chlormequat chloride reduced the length of the flowering stalks and delayed flowering and seed ripening, but considerably increased the percentage germination of harvested seeds. Seed yield from treated plants was considerably lower than that from control plants (Table 9), indicating growth-retarding effects of the chemical in the progeny seedlings.

Chemicals other than growth regulators also have been found to influence the quality of seed produced from treated plants. In a greenhouse experiment, lettuce plants (cv. Great Lakes) were given soil-drench treatments with a range of sublethal concentrations of simazine to determine their effects on seed yield and progency development. The results shown in Table 10 indicate the beneficial effects of simazine on seedling growth. The total seed yield and individual seed weights were little affected by the treatment, but the progeny growth from such seeds was enhanced considerably. The application of GA_3 (15 ppm) promoted seed stalk formation in lettuce, but did not significantly increase seed yields (129).

Control of Pests and Diseases.   Aphids are the most serious pests of lettuce, reducing the seed yield. Spraying with 3–4% nicotine dust controls the pest most effectively. Organophosphorus insecticides like parathion or malathion can also be used.

Table 9 Effects of Chlormequat Chloride Spray on Flowering and Seed Production of Lettuce (cv. Proftuin's Blackpool) Grown Under Long-Day (17 hr) Photoperiod

Treatment	Flower stalk height (cm)	Time from sowing to 50% seed ripening (days)	Seed weight per plant (g)	100-seed weight (g)	Time to 50% germination (days)	Final germination (%)
Control	18	187	1.73	191	8	58
Chlormequat chloride (2000 ppm)	7	205	0.28	195	4	84
LSD at 5%	3.2	7	0.32	32	2	9

LSD = least significant difference.
Source: Ref. 115.

Table 10 Effects of Simazine Pot-Drench on Seed Production and Progency Performance of Lettuce (cv. Great Lakes)

Rate of simazine application (kg/ha)	Seed weight per plant (g)	100–seed weight (mg)	Germination at 22°C (%)	Seedling growth after 42 days	
				Height (mm)	Fresh weight (g)
—	2.175	146	86	18.9	0.24
0.075	1.939	158	90	19.1	0.30
0.150	2.578	137	75	19.2	0.48
0.30	2.683	136	91	22.7	0.39
0.60	3.610	129	85	19.1	0.42
0.90	0.305	102	65	17.3	0.25
LSD at 5%	0.56	24	8	2.3	0.13

LSD = least significant difference.
Source: Ref. 115.

Important diseases seriously limiting the seed yield of lettuce are down mildew (*Bremia lactucae*), slimy soft rot (*Botrytis* spp.), spotted wilt, aster yellow, brown blight, and mosaic. Plants affected with downy mildew show light green or yellowish spots on the upper surface of the leaves. A downy white growth appears on the underside of the diseased parts of the leaf. Lettuce varieties resistant to this disease should be used. Soft rot causes considerable damage to plants of the heading varieties of lettuce if they are not deheaded early. Timely removal of heads and maintenance of uniform soil moisture conditions have been found to be advantageous. Spotted wilt, aster yellow, and mosaic are the viral diseases of lettuce. Spotted wilt is spread by thrips. Mosaic is a seedborne disease.

Harvesting. Lettuce does not bloom at any one time but rather there are peaks of blooming that are reflected in peaks of harvesting. It takes approximately 12 days from the opening of blossoms to seed maturity. Lettuce is generally harvested when about 50% of plants feather out, showing the white pappus. Better yields are sometimes obtained when the crops are allowed to reach the full-feather stage. However, the longer the plants are left after beginning to feather out, the greater the chance of seed loss through shattering caused by wind or rain. In case of severe shattering, it is advisable to leave the plants for 2–3 weeks to allow as many of the later flowers as possible to mature.

Lettuce may be harvested by cutting the plants by hand, which if carried out carefully keeps shattering losses of seed to a minimum. Stems are cut several inches above the ground with long-handled shears. The plants are placed in small piles or windrowed and later picked up by a combine. Crops are usually harvested early in the morning when plants are still moist with dew, which helps keep shattering to a minimum. The cut plants may be dried on a canvas.

In mechanical harvesting, the machine cuts the plants and pushes them off the tray or pan into piles on the ground. With such machines, considerable shattering may occur, since the crop is roughly handled. Also, the machine piles the lettuce instead of windrowing it continuously. Weeds, if present, are cut along with the lettuce, making the cleaning process difficult. The seed should be threshed as soon as the plants are dried sufficiently, which may require 3–4 days in hot dry weather. A combine using a pick-up attachment usually carries out threshing. The small piles of lettuce are forked directly into the machine. The best-quality seed was produced by harvesting without cutting off the stem followed by pneumatic grading and sorption drying. GA treatment of 0.002% or precooling at 10°C for 24 hr and germination at 20°C for 7 days have been found to be the most suitable testing methods for fresh seeds (130).

Machines are also employed to combine the standing crop directly from the field. This method works best if the crop has a good stand. When the plants are thin, many will fall to the ground when cut instead of falling on the draper. Combining, however, mixes a certain amount of green foliage and branches with the seed. It is necessary to dry this material quickly after it has been threshed to prevent discoloration and reduction in the percentage of germination. This can be performed either with artificial driers or by spreading the seed out thinly on a canvas in the sun. Occasional raking will hasten the process of seed drying.

In general, leaf-type lettuce produces much more seed than head type (e.g., Great Lakes). The yields of Imperial varieties and strains range from about 225 to 450 kg/ha, whereas leaf varieties like Prize Head or Grand Rapids may yield about 550 kg/ha or more.

4. Postharvest Operations

After threshing, lettuce seed usually requires cleaning. Normally, the seed first goes through a fanning mill followed by separator. Indent separators also may be used. The milling process generally reduces the weight of the threshed seed about one-third to one-half. Air-dried seeds of *L. sativa* L. (3% moisture content) were subjected to gamma-irradiation from a ^{60}CO source with doses ranging from 3.4 to 23.8 kGy, and root dip tissues were prepared for electron microscopy following imbibition. At the lowest radiation dose, germination was retarded but not viability (131).

B. Endive

1. Introduction

Botany. Endive (*Cichorium endivia*) is a biennial vegetable belonging to the Compositae family. It probably originated in eastern parts of India. It is eaten mainly as a salad green and is sometimes consumed as a potherb. The cut, curled, and frilled leaves are decorative and frequently used for garnishing and for flavoring soups.

Like lettuce, endive forms a large taproot and a rosette of leaves before producing a seed stalk. The stem elongates in early summer of the second year. The branches of stem are much coarser and fewer in number than in lettuce. The endive flowers occur in heads and are perfect with five-notched, ligulate corollas. The heads are much larger than those of lettuce. The flowers are pale blue in color, varying from 18–20 per head. The flowers are mostly self-pollinated. The endive fruit (seed) is an achene with a short scalelike pappus.

Cultivars. Endives are classified into two general groups: the curled or fringe-leaved cultivars and the broad-leaved cultivars. The former are highly ornamental and grown largely as popular salad vegetables. The Green Curled Ruffic, Deep Heart Fringed, Green Curled Pancalier, and White Curled are the curled or fringe-leaved types. The varieties representing the broad-leaved class are Broad-Leaved Batavian, Full

Heart Batavian or Escarole, and Florida Deep Heart. These are used mainly in stews and soups.

2. Seed-Production Technology

Culture.   Endive seed is sown in November–December or from August 15 to November 15, depending upon the climate and region. It is drilled at a rate of about 1.0–1.5 kg/ha in rows about 75 cm apart. Plants are thinned in the spring, keeping about 20–25 cm distance between plants. The requirements of endive for soil, fertilizer, irrigation, and interculture are about the same as those for lettuce.

Harvesting.   Seed heads do not mature uniformly, so some shattering losses occur during harvesting. The plants are mowed and either windrowed or piled to dry. Cut plants may be dried on canvas for several days. Curing requires about 10 days or more depending upon the climate. The crop is then threshed with a combine. Average yields are about 675 kg/ha for smooth varieties, with curled varieties normally yielding about 30% less.

C. Salsify

1. Botany

Salsify (*Tragopogon porrifolius*) is a hardy biennial vegetable. In the first year, the plant develops a long, fleshy taproot and a rosette of leaves from a very short crown or stem. In the second year, the stem elongates and branches may grow to 1 m or longer when in full bloom. The flowers are perfect, large, and purple in color and are borne in single heads. Like those of endive, the flowers open early in the morning and close before noon. The flowers are self-pollinated. The fruit is an achene, long, beaked, and containing a single seed.

2. Seed-Production Technology

Culture.   The cultural requirements of salsify for soil and climate are similar to those of lettuce. The crop is grown for seed in July–August, and stecklings are transplanted in late fall or early winter with spacing of 30 cm between plants and 90–120 cm between rows. Off-type plants should be rogued.

Harvesting.   Salsify seed is usually harvested by hand picking the individual heads, which is often carried out in the early morning to minimize shattering losses. The harvesting operation may continue for several weeks if the plants are maintained in good growing condition. Seed yields of 1900 kg/ha may be obtained with a good crop.

D. Chicory

1. Botany

Chicory (*Cichorium intybus*) is a perennial vegetable crop. The plant forms an enlarged root and a spatulate-shaped rosette of leaves. The seed stalk is produced in the second and following years. Both the seed stalks and flowers are similar to those of endive. The chicory seed is actually a fruit called an achene. Two types of *C. intybus* are known. The types used as vegetables are known as Witloof, Witloof Chicory, and French endive.

2. Seed-Production Technology

Chicory seed is sown in the early spring at the rate of about 1.0–1.5 kg/ha in rows approximately 75 cm apart. Plants are thinned to maintain about 20–25 cm between plants. Crops do not require a high fertility level. The seed matures in June–August depending upon the region and climate. The method of harvesting and cleaning of seed is similar to that for endive. Average seed yields are about 675–800 kg/ha.

E. Minor Compositae

Dandelion (*Taraxacum vulgare*) and globe artichoke (*Cynara scolymus*) also belong to the Compositae family. The latter is usually propagated vegetatively.

VI. CUCURBITS

Cucumber, melons, gourds, squashes, and pumpkins belong to the family Cucurbitaceae, comprising a number of important vegetable crops such as cucumber, muskmelon, watermelon, summer and winter squashes, and pumpkin. Also included in this group are certain minor vegetables like citron, gherkin, and chayote.

A. Cucumber

1. Introduction

Botany. Cucumber (*Cucumis sativus*) is an annual trailing vine vegetable with a main stem and branches. Branch length may vary from 0.5 to 1.5 m when fully grown; a well-developed plant measures about 1.9–2.4 m across. It is frost sensitive but can grow in cold weather conditions. The plant starts flowering early, producing marketable fruits within about 2–3 months depending upon cultivar, region, soil, and climate. It takes several more weeks of growth before the fruits mature and ripen sufficiently to harvest the crop for seed. Most cucumber varieties develop seed within about 100 days.

Cucumber plants develop a strong taproot soon after seed germination, growing to a depth of about 90 cm. Most cucumber varieties, being monoecious, bear staminate and pistillate flowers. Pistillate flowers are mostly borne singly in the first half-axils of the secondary (or fruiting) branches. Occasionally there may be two or more pistillate flowers in one axil. The staminate flowers generally occur in clusters of five. Staminate flowers are more abundant than pistillate ones, their ratio being influenced by the length of day. Long days tend to produce more staminate than pistillate flowers.

In pistillate flowers, the ovary is located below the calyx and corolla. The calyx tube has five slender lobes. The yellow corolla also has five lobes, which are somewhat recurved at their tips. The stamens are rudimentary in pistillate flowers, but the pistil consists of a well-developed three-loculed ovary with a thick short style; ovaries with four to five locules may also occur occasionally. Each locule contains several parallel rows of ovules. The staminate flower is characterized by the absence of the inferior ovary. The single whorl of three stamens is inserted at the base of the corolla; two of these have two anthers each and the third has only one. Pollination usually occurs through honeybees. Although significant cross-pollination occurs in the presence of such insects, selfing is not unusual. About 30–35% natural self-pollination has been reported under field conditions.

Cucumber fruit varies from about 8 to 30 cm in length depending upon the variety. Fruits are usually rounded triangular in cross section. There is no cavity in the locules; the space is entirely filled with placental tissue and the adhering seeds. Seeds are still soft and immature when the fruit has almost reached its full size. At this stage, it is picked for market as a slicing fruit.

The fruit must be allowed to mature fully for seed production, when the white-spined (slicing) varieties become yellowish white, whereas the steckling types with black spines become golden-yellow, orange, or brown. These mature colors of cucumber fruits enable the seed growers to spot certain off-type plants that might have been overlooked previously. Although pollination is not always necessary for fruit production, it is essential for seed production. Some cucumber varieties can produce fruits parthenocarpically without pollination. In such fruits, seed never matures. These fruits have partially developed ovules having the appearance of small embryonic seed. The fully matured seed is flattened and ovate. The average-sized fruit contains about 400–600 seeds.

Cultivars. Cucumber cultivars are classified as slicers, or table stock, and picklers. The fruits of slicers are shapely and round ended, about 16–20 cm long, and of a dark persistent green color. The fruits of picklers are shorter, very prolific, and used mainly for small-sized pickles. Tigchelaar (3) listed the following important cultivars of cucumber. Addis, Calypso, Charger, Campass, Cypress, Dasher, Earlipik 14, Green Spear 14, Greenpack, Greenstar, Hyslice, Liberator, Liberty Bel, Lucky, Strike, Mariner, Multipik, Pacer, Panorama, Peppi, Perfecto Verde 14, Peto Triplemech, Philly Pic-of-the Pickle, Pik Master, Pot Luck, Sampson, Slice Master, Slicerite, Spear-It, Super Slice, Tally, Tasty Green, Triple Grown, and Trispear.

The cultivar Wisconsin 2757 released by the USDA and the University of Wisconsin has a unique combination of disease resistance and horticultural characteristics (132). Nath and Dutta (133) classified a large number of indigenous cultivars of cucumber available in India into four groups: (a) Balam Khira type, (b) long-spined green type, (c) spineless long green, and (4) Sikkim cucumber. Balam Khira and Poona Khira are the most popular cucumber cultivars of India. The Indian Agricultural Research Institute, New Delhi, has released the following three improved cultivars of cucumber: Japanese Long Green (early maturing with long green fruit), Straight Eight (early fruit, medium to long in size, thick, straight, and cylindrical with round ends and green skin), and China (medium late, very hard, and prolific with slender fruit about 50 cm long and deep green–spined or white-spined skin).

2. Breeding

Seed growers mainly developed improved varieties of cucumber prior to the 1980s through breeding, selection, and introduction. Characteristics such as fruit length, color at the time of edible maturity, size of seed cavity, and texture and flavor of flesh have received prime consideration in the breeding of cucumber. Crossing American varieties of cucumber with English types developed early varieties such as Deltus, Irondequoit, and Davis Perfect. The ability of cucumber (*C. sativus* L.) cotyledons to develop flowers in vitro was reported by Msikita etal. (134). The majority of flowers in culture were staminate (up to 48%); a few pistillate flowers (up to 12%) were obtained through media manipulation. Gy4 and Gy5 are multiple disease-resistant gynoecious cucumber inbreeds. They have a high level of resistance to anthracnose (*Colletotri-*

chum orbiculare) and angular leaf spot (*Pseudomonas syrigae* pv. *lachrymans*) (135, 136).

Milotay (137) investigated the possibility of field seed production by gynoecious cucumbers. In a field trial with cv. Kecskemeti Keseredesmentes Konzerv, silver nitrate or silver thiosulfate at 600–800 ppm applied twice at the first true leaf stage induced development of staminate flowers and adequate pollination and seed production. Silver thiosulfate was more efficient, as it had greater stability and less sensitivity to pollution and water quality of the treatment solution. Earlier research also showed that silver compounds are better than gibberellins for inducing staminate flowers on all-female cucumbers. Application of ethephon (400 ppm) significantly increases the number of female flowers in cucumber (138).

Hybrid Seed Production. Hybrid cucumber seed can be produced by hand-pollination. In the United States, gynomonoecious lines are being used for hybrid seed production. These lines produce two kinds of plants: gynoecious (in which all flowers are female) and monoecious (in which male and female flowers occur separately on the same plant). To produce hybrid seed, the gynomonoecious line is used as a female parent and planted adjacent to a selected monoecious variety. At about the 10-node stage, all monoecious and intermediate plants are removed from the gynomonoecious line, leaving the gynoecious plants to bear hybrid seed. Zhang et al. (139) showed that male sterility (MS) is of practical importance in cucumber breeding, because it can facilitate F_1 hybrid seed production without hand-pollination. But the use of pollen sterility (PS) in cucumber F_1 hybrid seed production is not practical.

Somatic Embryogenesis. Cotyledons of cucumber (*C. sativus* L.) produce organogenic callus, but plant regeneration has been proven to be difficult. Somatic embryos have been regenerated from cotyledon protoplast (140) and from anther cultures: (141), but the incidence of embryogenesis was low. Callus was induced on immature (15 days postpollination) zygotic embryos placed on semisolid MS salts and vitamin medium containing 9 μM 2,4-D and 87.6 mM sucrose. This medium was solidified with 3 g Gelrite/L, and pH was adjusted to 5.8 before autoclaving (142). Ladyman and Girard (143) studied the effect of cucumber somatic embryo development in various geling agents and carbohydrate sources, and found that when sucrose was compared with fructose in the presence of either agar or Gelrite as geling agent, the subsequent germination of the somatic embryos was higher when the tissue was multiplied on Gelrite in the presence of 3% sucrose. Maltose (3%) severely inhibited tissue multiplication. Lou and Kako (144) reported that the highest yield of somatic embryos in cucumber occurred in cultures initiated with high sucrose levels (9 or 12%), although 12% sucrose inhibited callus formation and growth.

3. Seed-Production Technology

Culture. The cultivation of cucumbers for seed is much the same as growing the crop for the fresh market except that it is allowed to reach full maturity, requiring an additional growth period of about 1 month.

LAND PREPARATION. Cucumber grows rapidly in sandy or sandy loam soils, but silt or clay loam soils are preferred for higher yields. Hard soils without organic matter are not suitable. The land should be prepared by plowing two to three times and leveled by planking. Seedbeds of suitable size are then prepared.

PLANTING.   Soil temperature should be higher than 10°C to obtain good germination. The germination percentage increases with increase in temperature up to 29.4°C. Seeds are sown in prepared beds either in rows or in pits 120–150 cm apart. May–June and February–March are suitable sowing times for rainy-season and summer varieties of cucumber, respectively.

The highest fruit yields (899 g/plant) and seed yields (13.76 g/plant) were obtained by sowing cucumber on January 30 with a foliar application of 200-ppm ethephon. The germination percentage of the seeds different significantly owing to sowing date and was generally higher with December sowing (145). Presowing laser irradiation of seeds (cv. Sandra), stimulated embryonal root growth, photosynthesis, and peroxidase activity but reduced the plastid pigment content in the leaves (146). Treating seed with ethephon (800 mg/L) and oxygen increased the yield (147). Jennings and Saltveit (148) reported that a longer chilling treatment is required to reduce growth and cause irreversible damage in 16-hr-old germinated cucumber seeds than in seeds geminated for 28 or 32 hrs. These results suggest that some factor that confers chilling injury tolerance is gradually lost during the early stages of germination following imbibition. They also studied the effect of imbibition temperatures on fresh weight changes during germination of cucumber seed and showed that the rate and extent of imbibition at 25°C was much greater than at 2.5°C (Fig. 10). Turkan (149) showed that on fumigation with exhaust gas for 3 days at 10 min/day, germination of cucumber seed was reduced from 94% (untreated control) to 81% and to 60% at 3 hr/day.

Brushing (40 strokes/1.5 min twice daily) or moisture stress conditioning (MSC) (daily nonlethal dry-down cycles) reduced seedling growth of two cucumber (*C. sativus* L.) cultivars and three squash (*Cucurbita pepo* L.) cultivars. Brushing increased the rate of water loss from detached leaves of cucumber, squash, and watermelon, whereas MSC decreased water loss from leaves of cucumber and squash (150).

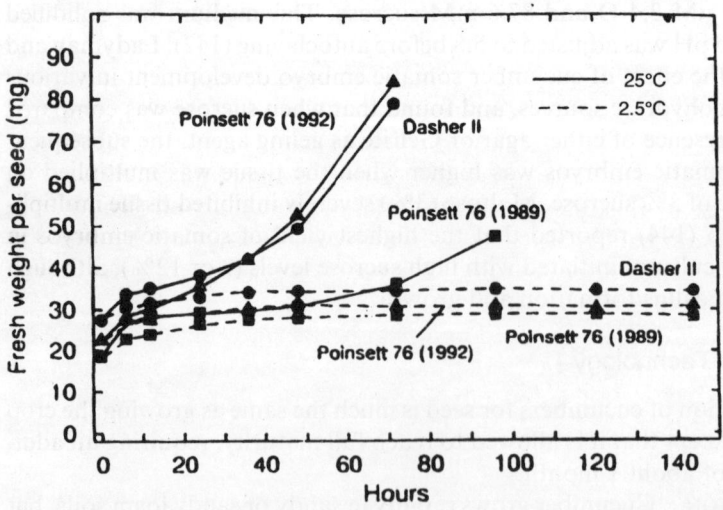

Figure 10   Effect of imbibition temperatures of 2.5 and 25°C on fresh weight changes during germination of Dasher II and Poinsett 76 cucumber seed from 1992 and Poinsett 76 seed from 1989. (From Ref. 148.)

FERTILIZATION, IRRIGATION, AND WEED CONTROL. The rainy-season crop requires little irrigation, but the hot-season crop may require irrigation occasionally. About 35–50 tons of manure per hectare should be mixed well with soil during plowing. Applying additional quantities of N, P, and K fertilizers should ensure adequate soil fertility. Yawalkar (47) recommended the addition of about 1.5–2.0 Q/ha of ammonium sulfate in two split doses: the first when vines have started spreading and the second when fruiting begins. Weeds should be controlled by adequate intercultural operations.

Roguing. Most off-type cucumber plants can be removed until the time of market maturity. Roguing should be carried out before flowering and natural cross-pollination occurs, but it is usually difficult to determine rogues prior to fruit development. Plants with off-type fruits should be removed from the field as soon as they are detected to prevent possible contamination. The spine color of the fruit (white or black) aids in roguing. An off-type spine color of fully matured fruits is easily distinguished according to variety. Removal of off-type plants at this stage helps to prevent mixing with the rest of the seed. Cross-pollination, however, may have already occurred, which will result in off-type plants in succeeding generations. Removal of the off-type fruit without removing the entire plant should be avoided.

Because bees mostly pollinate cucumbers, varieties must be sufficiently isolated to prevent crossing. When a crop is being grown for the stock seed, a distance of about 1.5 km between cultivars is desirable. The cucumber, however, does not need to be isolated from muskmelons, watermelons, squashes, or pumpkins, because cross-fertilization between any of these and cucumber does not occur. The adjacent planting of different species of cucurbits may result in reduced seed yields, because the pollen of one species may occasionally stimulate the production of parthenogenic fruit in another. Schultheis et al. (151) evaluated the effectiveness of two commercial bee attractants—Bee-Scent and Beeline—for enhancing pollination of cucumber (*C. sativus* L.) and watermelon (*Citrullus lanatus* Thunb) by counting the number of bee visitations to blossoms of cucumber and watermelon and their effects on fruit quality, yield, and crop profitability. The compounds did not improve the number of visitations for either pickling cucumbers or watermelons. There was no significant improvement in cucumber or watermelon yield or in monetary returns.

Control of Pests and Diseases. Striped and the spotted cucumber beetles cause severe damage to cucumber crops. The insects chew the leaves and tender shoots throughout the summer. The larvae of this insect may also injure cucumber roots. The striped beetle is known to carry the bacteria causing disease as well as to transmit mosaic virus from affected plants to healthy ones. A 5% DDT or 0.75% rotenone spray can be used effectively to control this insect.

The fruit fly (*Daucus cucurbitae*), red pumpkin beetle (*Raphidopala foveicollis*), and epilachna beetle (*Epilachna emplicata*) are the major insect pests of cucurbits in India (47). Spraying of 0.02% endrine effectively controls fruit flies. The red pumpkin beetle can be checked by dusting plants with 5% BHC or 5% DDT or 1% lindane at 7–10 kg/ha. Dusting 4% carbatyl or endosulfan has also been recommended to control the red pumpkin beetle. Using any of the following pesticides can control the epilachna beetle: 0.04% malathion spray, 0.5% lindane dust, or 0.05% endosulfan spray.

The minor insect pests of cucurbits include stinkbugs, aphids, pumpkin caterpillar, plume moth, leaf-eating caterpillar, blister beetle, flea beetle, gallfly, and mites.

Downy mildew (*Pseudoperonospora cuhensis*), powdery mildew (*Erysiphe cichoracearum*), leaf spot (*Cercospora* spp.), bacterial wilt (*Erwinia tracheiphila*), angular leaf spot (*Phytomonas lachrymans*), anthracnose (*Colletotrichum lagenarium*), scab (*Claadosporum cucumerinum*), mosaic (virus), and root-knot (nematodes) are the important diseases of cucurbits, including the cucumber.

Harvesting. The cucumber seed is ready to harvest when the fruits have turned pale yellow (white-spined type) or golden (black-spined type). A vine harvester can be used for harvesting and extraction of cucurbit seeds (Fig. 11). The fruits are picked up and tossed directly into a mobile seed extractor. The adhering pulp portion surrounding the seed can be removed by fermentation or by mechanical and chemical methods.

FERMENTATION. The pulp is poured into wooden barrels and allowed to ferment for several days. The fermenting material should be stirred occasionally to prevent any discoloration or blackening of seed from that may form on the material floating at the surface. The separated seed sinks to the bottom. At 15.5–21.1°C, the fermentation process takes 3–6 days to complete. Very slow fermentation rates may reduce seed viability. Fruits of four cucurbit crops (cucumber, melon, watermelon, and squash) were harvested 25, 35, and 45 days past anthesis (dpa) and their seeds were extracted immediately or extracted after 10 or 20 days of preextraction storage (pes). Upon extraction, the seeds were subjected or not subjected to fermentation, washing, and drying. Cucumber, melon, and watermelon reached full germinability by 35 dpa, but squash required a longer period. Fermentation and drying were important for improving the germinability of immature seeds of cucumber, melon, and watermelon. Fermentation had a deleterious effect on immature seeds, but drying and washing improved germinability of squash seeds (152).

Figure 11 Vine harvester for harvesting and extraction of cucurbit seeds. (From Ref. 1.)

MECHANICAL AND CHEMICAL METHODS.   Special machines may be used to separate seeds of cucurbits from their fruits, in which the fruits are cut and macerated and pulp is squeezed through a circular screen to separate the seed forcibly. Machines using centrifugal force are also employed. A considerable amount of water is required to wash the seed toward the end of the operation. The washed seed must be dried quickly to a safe moisture content.

Acid or alkali may be used to extract the seed from its pulp. The 25% technical-grade ammonia at 12 parts per 1000 of the material can be used. Hydrochloric acid is used toward the end of the process to restore the natural color of the seed. This method has been claimed to be rapid, simple, and cheap. The seed separated by such treatment has better germination than that obtained by the fermentation process.

4.   Postharvest Operations

Washing and Drying.   A shaker-washer or a sluiceway can be used to wash cucumber seed. Washing of cucumber, melon, and watermelon seeds increases the rate of germination but not the percentage. The extracted seed must be dried as quickly as possible. The seed is usually spread on screen-bottomed trays, which are placed on rocks in the open so that air passes from both above and below the screen. Occasional stirring speeds the drying process. Artificial dehydrators may be used if there is not ample sunlight. Large rotary driers (Fig. 12) are used by watermelon seed producers, especially for a preliminary drying period.

Milling.   The dried cucumber seed may still need further cleaning with some type of screening and fanning mill. This process removes pieces of dried and shriveled pulp and rind. Milling also eliminates any lighter seeds that still remain.

Figure 12   Rotary drier for preliminary drying of large-seeded cucurbits. (From Ref. 1.)

Storage.   Seed storage is important for germplasm preservation, and most cucumber germplasm is stored as seed. Seeds may be stored for decades depending upon the storage conditions. Nawab Ali et al. (153) stored cucumber seeds at 3°C and 38% RH for up to 26 years. They showed that seeds older than 13 years did not germinate. Cultivars stored 10 years gave 80% germination.

Seed Yields.   About 0.45 kg of seed is produced from 45 fruits of good size of a slicing cucumber variety, and yields of 325–450 kg/ha are common, but yields as high as 900 kg/ha have been reported (47).

B. Muskmelon

1. Introduction

Muskmelon is probably a native of northwest India from which it was spread to China and southern Europe. Its growth is favored by hot, dry weather. Environment significantly influences the flavor and sweetness of muskmelon.

Botany.   Muskmelon (*Cucumis melo*) is an annual with vine growth similar to that of cucumber. The crop requires slightly higher temperatures for optimum growth than does the cucumber.

Fruit maturity is achieved within about 85–120 days depending upon the variety, region, soil, and climate.

Muskmelon has a root system similar to that of the cucumber. Many European varieties of muskmelon are monoecious, but most American varieties are andromonoecious. In both groups, staminate flowers are borne in clusters of three to five in al leaf axils except those occupied by the pistillate or perfect flowers. The latter are borne singly in the first and second axils of the branches. A lengthening branch ceases to bear pistillate flowers after the second axil. In perfect flowers, there is a whorl of stamens with anthers containing viable pollen. Honeybees are the chief pollinating agent, cross-pollination being of common occurrence. Even in self-fertile perfect flowers, self-pollination is meager unless aided by bees or by hand.

Muskmelon fruit is fleshy, varying considerably in shape, size, and appearance according to variety. Most varieties have 10 longitudinal sutures and ribs; a few varieties lack such ribbing. When the fruit of most varieties reach edible maturity, a slight crack appears at the point of their attachment to the stem. When this crack completely encircles the attachment, the fruit has reached its edible maturity and is also ready for seed harvest. A single muskmelon fruit may contain 400–600 seeds, which resemble cucumber seeds in size and appearance.

Cultivars.   Some investigators have classified muskmelons into two groups: cantaloupes (*C. melo*, type *reticulatus*) and winter melons (*C. melo*, type *inodorous*). Tigchelaar (3) described cultivars of cantaloupe released in the 1970s: Alaska, Ambrosia, Cameo, Chando, Chieftain, Dixie Jumbo, Don Juan, Earli-Dew, Earlisweet, Early Dawn, Mainstream, Roadside, Ship Master, Sprint, Summet, Top Score, Topset, and Zenith. The important cultivars of muskmelon grown in India are Arka Rajhans, Arka Jeet, Pusa Sharbati, Hara Madhu, Durgapura Madhu, Sharda, Safeda Gola, Hara Gola, Motia, Batia, Amritsari, Lucknow Sweet, and Kutana Selection-A (47).

2. Breeding

The early development of muskmelon varieties depended mainly on selections in populations resulting from natural crossing; controlled cross-pollination was hardly if ever practiced. Characteristics such as fruit shape, size, color, thickness, flesh flavor, resistance to disease and pests, as well as shipping quality were the prime considerations in the breeding program. The problem of emasculation in andromonoecious varieties was a greater handicap in the search for good muskmelon F_1 hybrids. Spraying muskmelon and watermelon plants with 3.36 mM/dm^3 (2-chloroethyl) phosphonic acid (ethephon) and $0.305 \text{ mM/dm}^3 Zn^{2+}$ (as an organic salt) led to the absence of male flowers for 40 days. The application of zinc also increased the level of ethylene in the leaves (154). Abbott (155) studied the fungicidal inhibition of pollen germination and germ-tube elongation in muskmelon and showed that cupric hydroxide, mancozeb, and chlorothalonil reduced the percentage of pollen that germinated and the rate and length of germ-tube elongation regardless of cultivar. Benomyl had very little overall effect on pollen germination or germ-tube elongation.

For many years, considerable attention has been given to developing muskmelon varieties resistant to pests and diseases. The joint breeding program established by the USDA and the California Agricultural Experiment Station has helped to develop several new resistant varieties suitable for both shipping and fresh market. Boyhan and Norton (156) identified breeding line AC-82-37-2 as having resistance to alternaria leaf blight caused by *Alternaria cucumesina* Elliot in muskmelon. AC-70-154, a gummy stem blight–resistant muskmelon breeding line, was developed by Norton and Cosper (157).

3. Seed-Production Technology

Culture. The cultivation of muskmelon seed is more or less the same as when it is grown for the fresh market.

LAND PREPARATION. The requirements of soil and land preparation are the same as for cucumber. Heavy clay soils should be avoided. Seedbeds should be prepared after one deep and several shallow plowings.

PLANTING. In India, seeds are sown from January to the middle of March and from the middle of February (plains) to April–May (hills). The seed rate varies from 3.3 to 4.6 kg/ha. Early crops may be sown in ridges and furrows (150 cm wide with a distance of 60 cm between furrows) on both sides of the ridges with 90 cm between plants. Four to six seeds are sown at each hill, 1.5 cm deep, followed by a light irrigation to facilitate seed germination. Alvarez (158) reported that the highest ethephon concentration (8000 mg/L) decreased the percentage germination, but ethephon at 2000 and 4000 mg/L increased the speed of germination. Muskmelon (*C. melo* L.) seed typically does not germinate in soils near or below 15°C, and seedlings exposed to less than 18°C show little grow and are susceptible to chilling injury (159).

In the United States, muskmelons are usually drilled in rows about 180×240 cm apart. In the West, beds are slightly raised between which are furrows for irrigation. The vines and fruits may then be trained on the beds to keep them away from the water. The distance between two plants varies from 45 to 60 cm. Close spacing produces higher seed yields. The muskmelon responds well to high soil fertility and an abundant supply of soil moisture.

IRRIGATION, FERTILIZATION, AND WEED CONTROL.   A second irrigation may be performed after a week. Crops should be irrigated frequently during dry weather, and irrigation should be stopped before fruits begin to ripen. Melons can be grown using saline water throughout the growing period (160). For each hill, 5 kg of farmyard manure, 50 g of superphosphate, and 100 g of castor cake have been recommended. The Indian Council of Agricultural Research, New Delhi, as recommended 340 kg ammonium sulfate, 540 kg superphosphate, and 136 kg sulfate of potash per hectare. The addition of 20 g of lime per square meter is also useful for melons. The field should be maintained free of weeds by two to three shallow hoeings.

Roguing.   Off-type muskmelon fruits are easily spotted earlier than are cucumber off-types. Plants bearing off-type fruits should be destroyed as soon as they are detected to prevent further crossing. The entire plant rather than the individual fruit should be considered to be the basic unit in the roguing operation. The need for desirable internal quality (e.g., flavor and sweetness) further complicates roguing in muskmelons, because only cutting the fruit and making an actual examination can evaluate such quality. Such examination of the fruit interior is an important operation in the production of stock seed.

When grown for seed, two muskmelon varieties should be at least 200 m apart from each other as well as from any other member of *C. melo*. The distance must be considerably greater when seed stock is being grown. Muskmelon can be safely grown adjacent to cucumber, pumpkin, squash, or watermelon, since it will not cross with any of these crops.

Control of Pests and Diseases.   The insect pests of most cucurbits are the same and have been dealt with under cucumber. Like cucumber, muskmelon is also subject to a number of diseases, the most common ones being bacterial wilt (*Erwinia tracheiphila*), verticillium wilt (*Verticillium alboatrum*), fusarium wilt. (*Fusarium bulbigenum* var. *niveum*), anthracnose (*Colletotrichum lagenarium*), downy mildew (*Pseudopernoospora cuhensis*), powdery mildew (*Erysiphe cichoracearum*), fruit rot (*Phytophthora capsici*), root rot, root-knot, curly top, and mosaic (virus). Gaikwad and Sen (161) found that Bavistin (carbendazim) and Cercobin (thiophanate) were the most effective systemic fungicides for inhibiting spore germination and growth of *F. oxysporum* in vitro and wilt disease in vivo. Proper crop rotation, use of disease-resistant varieties, and control of insect vectors spreading diseases have been recommended to control other diseases.

Harvesting.   Muskmelon fruit reaches a stage known as "full-slip" when a slight crack appearing at the point of attachment to the stem completely encircles the stem. Although the seed has practically matured at this stage, seed growers usually wait until there are enough ripe melons to justify a general harvesting operation. Fruits may be placed in piles or windrows from which they are cut individually to extract seed or tossed into a mobile seed extractor. The punctured fruits of muskmelon can be harvested for seed extraction without any adverse effect on seed germinability. Singh et al. (162) noted that quality was best when the seeds were harvested at the full-slip stage. Dhillon (163) reported that planting six seedlings per hill and roguing male fertiles for 10 days resulted in cost-effective F_1 hybrid seed yield (109 kg/ha) (Table 11).

Table 11 Muskmelon Hybrid Seed Yield (kg/ha) of Four
Female-Parent Seedling Densities with Three Roguing Periods
(means of 2 years)

	Roguing periods (days)		
Seedlings per hill	10	15	20
3	56.4	53.7	54.0
4	69.3	66.6	64.6
5	83.7	87.0	84.8
6	109.0	114.8	116.7
LSD at 1%		8.4	

LSD = least significant difference.
Source: Ref. 163.

4. Postharvest Operations

Seed is separated from pulp by any one of the methods described for cucumber. The
washing and drying operations are also the same. Dried seed must be recleaned by a
milling operation using a screening and fanning mill with a suitable screen size. Seed
yields of muskmelon vary with variety, soil, climate, and incidence of pest and diseases.
Average yields of about 125–340 kg/ha are generally encountered. Hand-pollination
methods have been used to increase fruit set in muskmelons.

C. Watermelon and Citron

1. Introduction

Botany. Watermelon (*Citrullus vulgaris*) and citron (*C. vulgaris citroides*) require
relatively long periods of hot weather for fruit development. Watermelon is an annual
crop with stems that are angular in cross section, and its leaves are distinctly and
pinnately lobed, and may be further lobed and toothed. Watermelon plants are much
larger than either cucumber or muskmelon plants—the individual branches may
extend up to 6 m. It takes 80–110 days for fruit maturity depending upon variety,
region, soil, and climate. The edible maturity of the fruit coincides with the harvest
maturity of the crop for seed.

The watermelon root system is similar to that of muskmelon but more extensive.
Watermelons are generally monoecious with certain exceptions, like Angelinoi,
Chilean, Baby Delight, and Winter Queen, all of which are andromonoecious. Flowers
are mostly borne singly; female flowers may occur at every 3d, 4th, 9th, or 10th node.
Staminate flowers are borne on nodes at which there are no female flowers. Flower
types are essentially like those of muskmelon, but are slightly smaller and have a
greenish yellow tubular corolla. Honeybees are the chief pollinating agents, and considerable cross-pollination occurs: Even the hermaphroditic flowers are rarely self-
pollinated, because they require insects for pollination. There is significant natural
crossing in the field.

Fruit weight varies from 2 to 20 kg depending upon the variety and growing
conditions. In some varieties, fruit may be even larger (35–40 kg). Fruit shape may be
globular, oval, or oblong, and exterior color may be grayish white, green, or striped

and mottled. The flesh of immature fruit in all varieties is usually white, but changes to pink, red, or yellow with maturity. Unlike muskmelon, watermelon does not develop an abscission layer for its detachment from the vine. A thumping technique is often used to judge harvest maturity; a dull, dead sound usually indicates a ripe melon, whereas a sharper, more metallic ring indicates an unripe fruit.

Cultivars. Watermelon cultivars with good shipping quality are Dixie Queen, Florida Giant, Tome Watson, Congo (anthracnose resistant), Leesburg (fusarium resistant), Blacklee (fusarium resistant), Klondite Strains, Stone Mountain, Flairfax, and Charleston Gray. The last two are resistant to anthracnose and wilt.

Alena is an open-pollinated tetraploid variety with watermelon fruits of excellent horticultural quality (flesh is dark red and very firm, with up to 12.5% sugar) (164). Other recently released watermelon varieties include Prince Charles, Sugar Bush, Sugar Doll, Sun Shade, Sweet Meat II WR, Yellow Doll, Blue Belle, and Mirage (3). New Hampshire Midget, Asahi Yamato, Sugar Baby, Pusa Badana (seedless hybrid between Textra-2 of the United States and Pusa Rassal of India), and Arka Jyoti are the important cultivars of watermelon grown in India.

2. Breeding

The objectives of watermelon breeding include characteristics such as earliness, tough rind (for shipping), fruit shape, flesh color and flavor, eating quality (texture), and freedom from white heart. California Agricultural Experiment Station scientists, working mainly on uniformity, flesh color, and eating quality, have developed improved cultivars such as California Klondite, Striped Klondite, and Long Mountain. An extremely early but small variety, Northern Sweet, originated from the Minnesota Agricultural Experiment Station. New Hampshire Midget, an extremely early and small (1.2–2.2 kg) variety, was developed at the New Hampshire Agricultural Experiment Station.

Conqueror is a wilt-resistant watermelon variety released and developed by the USDA. A number of other wilt-resistant varieties were developed at the Iowa, Florida, Missouri, Tennessee, and California Agricultural Experiment Stations. The USDA also released an anthracnose-resistant variety, Congo, in 1949, followed by Flairfax, a variety resistant to both anthracnose and wilt.

Triploid, seedless watermelons have been achieved by crossing tetraploid watermelons with regular diploid ones. Tetraploids are first produced from diploids using colchicine; once produced, they breed true to type. Pollinating a tetraploid with a diploid then produces a seedless triploid watermelon. Polyploid watermelons have few or no seeds, and thus for the consumer have a great advantage over diploids. Tetraploids produce about 100 seeds per fruit, many fewer than diploids. Tetraploid seeds are produced by open-pollination in isolation, but the relatively low yield of seeds per fruit results in high seed costs. Triploid seeds are even more expensive, as they can be produced only if the tetraploid acts as the seed parent.

Polyploid watermelons are grown in some regions, but not as commonly as one might expect. One important reason is the low germination tendency of the expensive polyploid seeds. Poor germination is especially common in early spring, when suboptimal temperatures (<25°C) prevail (165). Parthenocarpic haploid watermelon embryos were obtained after pollination with gamma-irridiated (200 or 300 Gy) pollen (166). Adventitious shoot regeneration has been reported from cotyledons of tetraploid watermelon; however, the percentage of explants that produced shoots was low (167). High-frequency shoot regeneration (>75%) has been obtained from diploid watermelon using cotyledon ex plants from 5-day-old in vitro germinated seedlings (168). Chompton and Gray (169) demonstrated that the most important factors for adventitious shoot regeneration in diploid watermelon were plant genotype and seedling age at the time of explant preparation.

Capillary electrophoresis (CE) in conjunction with sodium dodecylsulfate–polyacrylamide gel electrophoresis (SDS-PAGE) was used to identify nine cultivars of melons. The identification was based on qualitative differences shown by seed storage proteins (globulins) separated by the two techniques. Although CE showed good reproducibility, it failed to identify all of the cultivars examined. The use of CE in conjunction with SDS-PAGE increased the possibility of cultivar identification. It has been suggested that the inability of these techniques to discriminate between all of the nine cultivars examined might be due to the rather conservative nature of the globulins extracted from the same members of the Cucurbitaceae family (169a).

3. Seed-Production Technology

Culture. The culture of watermelon for either the fresh market or seed is practically identical. Watermelon, being a cross-pollinated crop, needs adequate isolation and thorough roguing. About 1.5 kg of seed is usually sufficient to cover 1 ha of land. The drilling method requires slightly more seed per unit area. However, the plants must be thinned to maintain a distance of about 1 m in the row.

LAND PREPARATION. Deep, well-drained, sandy loams favor watermelon growth. Land should be prepared by plowing two to three times and planking to prepare seed beds 2.4 m wide with 60-cm furrows between them.

PLANTING. In northern parts of India, seed is sown in February and in central India from February to June. In southern India, it is sown from mid-January to the end of March. Growers use slightly higher seed rates (5.6–7.0 kg/ha) when crops are raised for the market. Seeds are sown in flat or raised beds on both sides or in pits (three to four seeds per pit). Tetraploid Alena and the diploid Sugar Baby seeds were subjected to one of several treatments prior to germination at 17,21, and 25°C, including internal splitting, soaking in H_2O (aerated or nonaerated), $GA_4 + GA_7$, BA for 24 hr or in KNO_3 for 5 days followed by drying. These treatments were successful in increasing the germination percentage of Alena embryo length. Alena seeds had larger seed counts and absorbed more water than those of their diploid counterpart, Sugar Baby (170). Yang and Sung (171) studied the effect of suboptimal temperature on the germination of triploid watermelon seeds of different weights and showed that significant differences existed between seed weight classes and temperature treatments for final germination.

Demir and Venter (171a) subjected watermelon seeds to osmoconditioning (2% KNO_3, 20°C, 6 days) or hydropriming (30°C, 18 hr) and incubated at 15, 25, and 38°C. Mean time to germination was decreased and germination increased by both priming treatments at 15°C with osmoconditioning being superior to hydropriming. However, neither osmoconditioning nor hydropriming treatment affected germination significantly at 25 and 38°C. The effect of priming on root growth and on emergence from deep sowings was also assessed after synchronization of radicle emergence between treatments to exclude the confounding effect of germination rate differences. Root growth was not significantly improved by priming, but emergence at 15°C was enhanced. Improved emergence after priming was not due to the beneficial effect on radicle emergence only but also to improved hypocotyl growth. Osmoconditioning increased germination under osmotic stress.

percentage and mean germination time (Table 12). The observed delay in germination at suboptimal temperature (15°C) might be linked to the slower developmental patterns of glyoxylate cycle enzymes.

IRRIGATION, FERTILIZATION, AND WEED CONTROL. About 25 tons of manure should be applied during land preparation, which can be incorporated during plowing. As a top dressing, the addition of 680 kg of ammonium sulfate, 60 kg of superphosphate, and 272 kg of potassium sulfate has been recommended. The ICAR's recommendations for these fertilizers are 340, 540, and 136 kg/ha, respectively. An application of 20 g of lime/m^2 has been found to be beneficial. The crop should be lightly irrigated after sowing. If the field is deficient in moisture, the crop should be irrigated sparingly once every 7–10 days. After the fruit sets, vines should be trained to cover the beds and avoid contact with water. Irrigation may be stopped when fruits begin to ripen. The field should be kept free of weeds by two or three hoeings.

Roguing. If the drilling method is followed, individual off-type plants can be easily detected. Fruit shape and exterior color will aid in roguing. Prompt roguing is necessary to prevent the contamination of surrounding plants by off-type pollen. As in the case of other cucurbits, the plant should be considered the unit in roguing operations rather than individual off-type fruits. The large number of varieties and strains has complicated roguing in watermelon. A minimum distance of 400 m should be maintained between varieties when watermelon or citron is being grown for seed; for stock seed the distance should be considerably greater.

Control of Pests and Diseases. Watermelon is susceptible to all of the common insect pests and diseases of the cucurbits. In the United States, the important diseases of watermelon are fusarium wilt (*Fusarium niveum*), anthracnose, and root-knot.

Harvesting. When watermelon fruits reach edible maturity, they are ready for seed harvest. Harvested melons may be placed in windrows or piles. They may then be individually cut to scrape out the seed. The soft flesh of half of the melon may be pulled using the hand and fingers, allowing the seeds and pulp to fall onto a screen. Seed-extracting machines used for cucumbers and muskmelons may also be employed to separate watermelon seeds. The fruits are tossed into a hopper at one end of the ma-

Table 12 The Effect of Temperature on Germination Parameters of Triploid Watermelon (cv. Phong Sen No. 1) Seeds

| | Seed weight | | | | |
| | heavy | | light | | |
Particulars	25°C	15°C	25°C	15°C	LSD 0.05
Germination (%)	72.3	59.8	63.3	50.5	4.1
Mean germination time (days)	2.1	6.0	2.6	9.2	0.4
Seed leakage (μmho/cm/q fresh wt.)	360.5	275.8	595.5	363.1	28.4

LSD = least significant difference.
Source: Ref. 171.

chine, or they may be carried to the machine from conveyor belts extending on each side. This arrangement eliminates the necessity of windrowing the fruits.

4. Postharvest Operations

The separated seed is washed, dried, and cleaned in much the same way as described for cucumber and muskmelon. The seed may be recleaned using a screening mill after it is dried. Yields of about 225–280 kg/ha are obtained from fields grown exclusively for watermelon seed. The yield varies considerably according to variety, region, soil, and climate.

Citron fruit in general resembles a small watermelon (4–7 kg). It is usually globe shaped, and the exterior color is normally green with light striping. The flesh is white and rather firm in texture. There are about 200–250 seeds in an average-sized fruit. The color of mature seeds may be tan, red, brown, black, or mottled or patterned with these colors. They are usually flat with a smooth surface, except in mottled ones, which may be slightly roughened. Citron seed is usually either a clear shiny green or red depending upon variety.

D. Pumpkin and Squash

1. Introduction

The terms *squash* and *pumpkin* have been applied rather loosely to various varieties without regard to the species of *Cucurbita* to which they belong. The common varieties have been classified into three botanical species: *C. pepo*, *C. moschata*, and *C. maxima*. It is generally agreed that all types of *C. pepo* (zucchini) and *C. moschta* are called pumpkins, and all types and cultivars of *C. maxima* are called squash. Both pumpkins and squashes thrive well in cool, moist weather conditions, but they are susceptible to frost.

Botany.   Pumpkins and squashes are annuals. *C. maxima*, which includes most of the winter squashes, has long vinelike branches characteristic of other cucurbits. It is the most vigorously growing vine of all the cultivated cucurbits. The main branches may be about 3–7 m or more in length, with a total plant spread of about 10 m or more. Leaves are large and kidney shaped. The plant matures within about 100–120 days. *C. moschata*, which includes some pumpkins and some winter squashes, has spreading vines resembling *C. maxima* in size. The stem is round or smoothly five angled. The leaves are usually large and blades are three to five-lobed. Plants require 90–120 or more days to mature. *C. pepo*, to which many of the common pumpkins and summer squashes belong, may have bushy or long trailing vines, with a spread of 6–10 m. *C. pepo* requires about 90–120 days to reach seed maturity.

The root systems of both *C. maxima* and the spreading type of *C. pepo* are similar and resemble those of muskmelon and watermelon. Squashes and pumpkins, irrespective of species, are monoecious, having both the staminate and pistillate flowers borne singly. The staminate and pistillate flower counts of squash fluctuate according to environmental conditions, and maintaining production over a range of planting dates will depend on careful cultivar selection (172). Flowers are yellow and similar to but much larger than those of other cucurbits. The male and female flowers are borne in a certain sequence, and temperature and photoperiod influence the duration of the developmental phases of the vine. High temperature and long days tend to maintain the plant in the male phase with more staminate flowers. Low temperature and short-day conditions, on the contrary, tend to increase the frequency of pistillate flowers at nodes.

Honeybees are the main pollinators. In both squashes and pumpkins, there is considerable cross-pollination within species. Fruit type varies considerably depending upon variety. Fruit weight ranges from 0.5 to 50.0 kg or more. Shape also varies greatly: there are small and large, globular and oblate types, as well as oval, oblong, and cylindrical types. The fruit surface may be soft or hard, smooth or warty, or even creased and ridged. The exterior color may be white, yellow, golden, salmon, light or dark green, gray, or various intermediate colors. The peduncle of *C. maxima* is very large, thick, and round in cross section with a soft, corky surface. In *C. moschata*, it is smoothly five angled without a corky surface. *C. pepo* has mature penduncles, which are also sharply five angled. In most varieties, regardless of the size and shape of the fruit, there is a seed cavity at maturity. The flesh and placental tissues are drier than in cucumber and melons. Fruits of all varieties of squash and pumpkin are allowed to mature fully for seed harvest.

Cultivars.   Tigchelaar (3) described the origin and characteristics of the following cultivars of squash released in the 1970s: Black Eagle, Black Magic, Burpee Golden Zucchini, Butterbush, Cracker, Daytona, Early Butternut, Eldorado, Genie, Gold Rush, Gold Slice, Golden Eagle, Golden Girl, Golden Swam, Greenzine, Ingot, Lemondrop, Moneymaker, Napolini, Patty Green Tint, Right Royal, Scallopini, Seneca Gourmet, Slenderella Summer Sun, Sundance, Table Ace, and Zish. The ICAR has released three cultivars of summer squash: Early Yellow Prolific, Australian Green, and Pusa Alankar. The last one is a hybrid that is early, high yielding, with dark green fruits and tender, delicious flesh. The improved cultivars of pumpkin grown in India are Arka Suryamukh, Arka Chandan, IHR-83-1-1-1, Co. 1, and Co. 2. Except for Arka Suryamukh, which belongs to *C. maxima*, the remaining are of *C. moschata* types.

2. Breeding

Many early varieties of pumpkin and squash developed in the mid-1930s were associated with Native American tribes. Buttercup, developed by the North Dakota Station, is a variety of *C. maxima*, followed by Kitchenette and New Brighton by the Minnesota Station. The California Agricultural Experiment Station introduced Grey Zucchini No. 1, and the Connecticut Experiment Station introduced Connecticut Straight Neck in 1936, followed by Caserta, an early type of Cocozelle. All three varieties belong to *C. pepo*. The large amount of natural crossing between varieties of squash and pumpkin has resulted in a great deal of variation within stocks but has not led to the development of many new cultivars.

Male-sterile lines would be a great asset in the development of first-generation hybrids, since they would eliminate the need for emasculation in the seed parent. Matlob and Basher (173) reported that an application of indole-3-acetic acid (IAA) (50–200 ppm) to *C. pepo* L. (cv. Mullah Ahmed) suppressed staminate flowers, enhanced pistillate flower development, and increased total yield by increasing the ratio of pistillate to staminate flowers. An application of ethephon (300 ppm) to four field-grown cultivars of bush and trailing forms of *C. maxima* and *C. pepo* caused leaf epinasty, earlier female flower production, and late production of male flowers. The delay in male flower production was more marked in *C. pepo* cultivars. Both one- and two-spray treatments resulted in more female, fewer male, and fewer total flowers per plant than in controls after 60 days. All the male main buds already formed at the nodes at the two true leaf stage aborted, and all the secondary buds developed into functional female flowers (174). Plant regeneration from tissue cultures of summer squash has been observed (175).

3. Seed-Production Technology

Culture.   The culture of squash or pumpkin for seed is more or less similar to that of cucumber Bush-type pumpkins may be spaced at 1 × 1 m, but larger vining types should be placed 3 × 4 m apart. In India, seed of *C. pepo* is generally sown from January to April in the plains and October–November for forced production to catch the early market. Seed of *C. maxima* is usually sown from January to June in the plains. Both are basically warm-season vegetables. There are about 400 and 160 seeds per ounce of *C. pepo* and *C. maxima*, and the seed rate varies from 8 to 10 and 5 to 7 kg/ha, respectively. The seeds of both crops remain viable for about 4 years under good storage conditions. Land preparation, irrigation, fertilization, and interculture operations are similar to that of cucumber. An application of 25–35 tons of farmyard manure per hectare before preparing the seed bed and an additional 60 kg N, 30 kg P, and 30 kg K per hectare should be given as a top dressing for *C. pepo*. For *C. maxima*, in addition to 25–35 tons of manure per hectare, about 103 kg N, 106 kg P, and 40 kg K per hectare should be applied as top dressing. The field should be irrigated after sowing and again about 7–10 days thereafter. In dry weather, crops should be irrigated every 4–5 days. Fields should be kept free of weeds.

Roguing.   Off-type plants are usually detectable fairly early. Bush-type plants in a normal runner-type variety can be easily detected. The early shape of fruit or shape of the ovary at flowering may aid in the roguing operation. Off-type plants should be immediately rogued to avoid further damage by cross-pollination, considering the entire plant as a roguing unit. Varieties of the same species should be planted at a minimum distance of about 400 m, and for stock seed even further apart.

Control of Pests and Diseases.   The important insect pests of squashes and pumpkins are the squash bug (*Anasa tristis*) and the squash vine borer (*Melittia satyriniformis*). The latter bores tunnel into the stem, destroying plant tissue and causing death of the plant. Only varieties of *C. maxima* and *C. pepo* are susceptible to the borer. Squashes and pumpkins are less susceptible to the common diseases of cucurbits. They are, however, subject to attack by downy mildew, bacterial wilt, and anthracnose.

Harvesting.   In the United States, seed harvest of squash and pumpkin usually follows the first frost in autumn. After frost has killed the foliage, the fruits are easily seen, making the harvesting operation more efficient. Fruits may be windrowed or placed in piles. Vine-seed threshers similar to those used to extract cucumber and watermelon seed are commonly employed on commercial seed farms. Most winter squash fruits (*C. maxima*) are threshed by cutting them in half and scooping the seed out by hand. Dematte (176) showed that higher fruit (*C. moschata*) yields per hectare were due to a larger number of fruits and not larger fruits. The average fruit weight decreased as the number of fruits per hectare increased. The heaviest fruit yields, the heaviest seeds, and high yields of below-average fruits were correlated with low seed yields, but these seeds had the highest germination percentage. Arora et al. (177) reported that ethrel at 100 and 250 ppm significantly increased the number of fruits per vine and yield per hectare because of a reduction in the sex ratio resulting in male fruits per plant. GA at 10 and 25 ppm improved vine growth and seed content.

4. Postharvest Operations

The seed and pulp of *C. maxima* are usually not fermented owing to the danger of discoloration and damage. With pumpkin and summer squash (*C. pepo*), fermentation should not be allowed to proceed beyond the point of separation of the placenta from seed. For seeds for which fermentation is not advisable, the pulp may be gradually loosened from the seed by rolling and raking. If placed in wate and vigorously stirred, the seeds tend to float and may be skimmed from the surface.

Seeds should be washed with water using a shaker washer. The shaker washers is particularly effective with both *C. maxima* and *C. pepo*, especially when they are not fermented. The washed seeds should be dried quickly, and may be recleaned by milling with some type of screening and fanning mill. Screen sizes may vary with species and their varieties. The average seed yields of summer squash in the United States range from 675 to 900 kg/ha, and those of winter squash range from 450 to 560 kg/ha.

VII. CRUCIFERS (COLE CROPS)

Several vegetables belonging to the family Cruciferae are of considerable economic importance. Cruciferous vegetables grown for their aboveground parts include cabbage, cauliflower, heading and sprouting broccoli, kale, collard, mustard, Chinese cabbage, kohlrabi, and Brussels sprouts. Some minor cruciferous vegetables usually included under the class salads are cress, sea kale, upland cress, and watercress. Although cruciferous vegetables look different, they have similar flowers, which are always regular, perfect, and cruciform. All cole crops are hardy and thrive best in cool weather. They are grown in hot, tropical, and subtropical southern climates mainly during the winter. Most cole crops of the Cruciferae or mustard family belong to the same genus, *Brassica*, and to the same species.

A. Cabbage

1. Introduction

Cabbage, one of the most important vegetable crops worldwide, is grown in many regions including the United States, China, Japan, India, and several European countries.

Botany. Cabbage (*Brassica oleracea*, type Capitata Linn) has been found to grow wild in England, Denmark, and France; wild cabbage is herbaceous and usually a perennial. Cultivated cabbage is biennial, although it is grown as an annual for market crop. The cultivated cabbages differ greatly in size, shape, and color of leaves and in size, shape, color, and texture of the head. Cabbage thrives best in a relatively cool and moist climate.

The cabbage has an extensive, fibrous, and finely branched root system. At the time of head maturity, the taproot and many laterals penetrate the soil obliquely up to a depth of about 2 m. When cabbage plants are uprooted and set out again, new roots begin to grow. Cabbage flowers are borne in terminal racemes, which develop on the main stem and all its branches. The bright yellow hypogenous flowers borne on slender pedicels are perfect, regular, with four sepals, four petals, six stamens, and a two-celled ovary containing many ovules per cell. The spreading terminal portion of the petals forms a cross, which is the chief diagnostic characteristic of the Cruciferae.

Pollination occurs mainly through bees, for example, the cuckoobee (Nomadidae), the leaf-cutting bee (Megachilidge), and the mining bee (Andrenidae), which are probably more important than honeybees in cabbage seed production, because they work at temperatures below 15.5°C. Under natural uncontrolled conditions, cross-pollination is more prevalent than self-pollination in *Brassica*. The finding that the stigma of *Brassica* is receptive for about 5 days before and 4 days after anthesis led to the development of a bud-pollination technique, which is commonly used by cabbage breeders. In this method, flower buds are artificially opened and pollen is then applied to the exposed stigmas. Fertilization occurs about 5 days after the pollen has been applied to the stigma.

Brassica seed is small, globular, and smooth. The seed is mostly embryo, with the endosperm being absorbed in the formation of the cotyledons, which are conduplicate. The cabbage pod contains about 12–20 seeds. A large, open-pollinated plant may produce about 225 g of seed with a 1000–seed weight of about 3.2 g.

Cultivars. Cabbage cultivars are classified into the following eight groups:

1. Wakefield and Winning Stadt group—pointed or conical, early and intermediate (e.g., Early Jersey Wakefield, Charbeston Wakefield, and Winning Stadt)
2. Copenhagen Market Group—globular, early and intermediate (e.g., Golden Acre, Copenhagen Market, Large Late Copenhagen, Midseason Market, Globe, Bonanza, and Enkhuizen Glory)
3. Flat Dutch or Drumhead Group—flattened heads, early or late (e.g., Early Round Dutch, All Head Early, Succession, All Seasons, Sure Crop, Premium Late, and Flat Dutch)
4. Savoy Group—flat or round, early, intermediate, or late (e.g., Drumhead Savoy, Improved American Savoy, and Chieftain)
5. Danish Ballhead Group—globular to slightly flattened, late and long-keeping (e.g., Danish Ballhead, Hollander, and Wisconsin Ballhead)
6. Red Cabbage Group—various forms, intermediate or late, with deep purplish red color (e.g., Mammoth Rock Red, Large Red, and Red Danish)
7. Albha Group
8. Volga Group

The last two groups are considered to be obsolete.

Tigchelaar (3) described several varieties of cabbage released in the 1980s: Blue Heaven, Blue Pak, Cole Cash, Defender, Excel, Grand Slam, Guardian, Jackpot, Regal, Satellite, Shamrock, Sunup, and Venus. He classified the commonly grown Indian cultivars of cabbage into the following four groups:

1. Round head or Ballhead types (e.g., Golden Acre, Pride of India, Copenhagen Market, Mammoth Rock Red, and Express)
2. Flat head or Drumhead type (e.g., Pusa Drumhead)
3. Conical head type (e.g., Jersey Wakefield)
4. Savoy type (e.g., Chieftain)

Of these cultivars, Golden Acre, Pusa Drumhead, and Pride of India are grown most extensively in India.

2. Breeding

All species of *Brassica* cross with each other. The characteristics of suitable varieties vary greatly depending upon the locality of culture and purpose for which the crop is grown; for example, fresh market, shipping, or processing. Odland and Noll (178) developed an entirely new resistant variety, Penn Valley, by crossing Penn State Ballhead with the resistant Wisconsin Ballhead.

A fairly rapid and accurate method of determining solidity of cabbage heads consists of weighing each head in water after it has been tightly covered with a rubber cap to prevent the penetration of water. Inbreeding followed by selection is often a desirable procedure to develop newer strains. Inbreeding of cabbage for more than two germinations loses vigor rapidly. Also, the greater uniformity exhibited by inbreds as compared to open-pollinated lines is often a disadvantage in connection with certain characteristics of the cabbage inflorescence. Individual plants of inbred lines of cabbage or related crops may not only be incompatible with their own pollen but also with that of other plants in the same line.

In Japan, a commercial F_1 hybrid cabbage seed was produced by setting out cross-compatible (but self-incompatible) plants of two lines in such a way that they alternated in the row. This encouraged crossing, and it was possible to obtain about 98% crossed seed. If this seed is used to produce a transplanted crop, the few selfed will be weaker than the hybrids, and they can be eliminated at the time of transplanting.

Odland and Noll (179) outlined a breeding program describing the steps involved in the production of F_1 hybrid cabbage seed that eliminated many of the difficulties inherent in earlier methods. Two inbreds are developed from one variety or strain of cabbage; these lines must be able to cross reciprocally. Similarly, another two inbred lines are developed from another variety or strain, and these too must also be able to cross reciprocally. The resulting hybrids of each of these two pairs of line then must be able to cross reciprocally with each other. If these conditions are met and the final hybrid is commercially acceptable, all four hybrid lines are increased to a point at which a relatively large isolated field or plot may be planted to alternate rows of lines 1 and 2 and a similarly isolated field planted to alternate lines 3 and 4. If the hybrid from lines 1 and 2 is designated as X and that from lines 3 and 4 as Y, then the final commercial F_1 seed is produced by growing X and Y in alternate rows in another isolated field. Seed is harvested in bulk from all the plants, and the seed from any field is almost genetically identical. This method still requires the slow, tedious initial building up of the inbred lines, involving considerable bud pollination. Asexual propagation may, however, be used to increase 400-fold vegetative growth in a single season. Cuttings are first made and rooted from the stored heads in midwinter, which are then set in the field in spring. Further cuttings are made and rooted when the plants reach the loose head stage. Full-sized heads are formed on these plants by fall. Heel cuttings (a leaf with axillary but attached) are most successful and practical.

Plants from cuttings taken from old or mature head leaves as well as from somewhat young leaves and from roots were obtained. This method is useful to increase breeding stock and to carry it over seasons when production by seed is not readily applicable. North (180) made stem and leaf cuttings of Brussels sprouts and heading broccoli. Sections of heading broccoli curd, each with a portion of leaf attached, were treated with a hormone, and the cuttings were then set in pots under glass at 15.5°C. Adventitious buds arose after 2 months at the base of the cuttings. These were used as shoots, each producing a new plant.

The degree of self-incompatibility of a number of inbred lines has been investigated. Cabbage plants showed 90–100% crossing when mixed pollination were made. These plants when selfed produced only plants that were highly incompatible in self and sib matings. Temperatures varying from 12.8 to 21.1°C have been found to be favorable for pollination and ultimate seed testing. However, there is a great tendency toward successful crossing in a temperaure range of 10.0–14.4°C than 14.4–21.1°C. Determination of incompatibility and testing may, therefore, be carried out under temperatures as high as or slightly higher than those anticipated In the field when the cabbage is in bloom.

3. Seed-Production Technology

Cabbage seed can be produced either by the headed plant-to-seed method or by seed-to-seed method. The former method is usually employed to produce stock seed, since proper inspection and roguing are possible. The latter method is often used to produce market seed.

Seed Production from Headed Plants.

CULTURE.   During the first year, the culture of cabbage for seed is much the same as that for market or for processing. When cabbage is grown for seed, the time of planting is adjusted to obtain full head maturity just prior to winter if the heads are to be stored in a cellar. The earlier matured heads will not have satisfactory storage conditions unless artificial cooling is employed. Larger seed yields of cabbage can be obtained by planting crops in August or early September. Tur (Chlormequat) applied at 1.5 and 2.5% during the full-rosette stage has been reported to give the best seed yields, namely, 1.42 and 1.49 tons/ha, respectively, compared with 1.34 tons/ha for controls (181).

ROGUING.   Off-type plants are more effectively rogued when the head crop is nearing maturity. Plants are removed when they do not conform to the accepted standard. After stripping the basal and outer loose leaves, the exposed may be rogued for shape during its storage.

STORAGE.   Seed cabbage is stored at 0°C and 90–95% RH. At low temperature, cabbages remain dormant and the growth of pathogens is inhibited. High humidity, although favorable to certain decay-causing organisms, prevents dehydration of the cabbage plants. When cabbage is stored in cellars, it is desirable to place the entire plants with their roots on wire mesh or slatted shelves arranged in tiers as close together as is practical for operation.

REPLANTING.   The mature cabbage plants are dug out and immediately reset in the field in October–November. Prior to planting, the top of each head is crosscut. Planting is carried out in furrows so that when a second furrow is made adjacent, soil is thrown over the heads, covering them to a depth of about 5 cm. The cabbage plants are thus stored in the location where they will grow and produce seed the following year.

The plants stored in the cellar should be set out early in the spring as soon as the soil is dry enough to work. Cabbage can tolerate moderate frosts and snow after planting. The spacing varies from 45 to 90 cm within rows and 100 to 200 cm between rows depending upon the variety and region. To facilitate normal development of the seed stalk, two cuts about 2.5–5.0 cm deep are usually made at right angles across each head at or soon after planting. Repeated cutting at later stages may be necessary; the cutting, however, should not injure the growing point. At the time of planting, plants are set at uniform distances along the furrows and plowed in. Root growth is favored when the soil has adequate moisture and good tilth.

Seed-to-Seed Method. In this method, the cabbage crop is allowed to remain in the same location where it is set out in the seedling stage. Plants are either allowed to form partial heads, or they may enter the winter season while still in an advanced rosette stage. The seed-to-seed method is more common than the root-to-seed method, with nearly all marked and some stock seed in the United States being produced by this method. This method, however, necessitates the use of a high-quality stock seed, since critical roguing is not feasible under large-scale commercial productions.

CULTURE. Seed is sown in seed beds from May 15 (late varieties) to July 15 (early varieties), so that plants reach the same stage of development at the time of transplanting. The transplanting is carried out in late July or early August at distances of 40–60 cm within rows and 120–180 cm between rows. Mechanical planting may be carried out in large-scale operations. For successful seed production, the cabbage plant requires a low temperature (3–4°C) over 2 months followed by a summer with small amounts of well-distributed rainfall (182).

ROGUING. Off-type plants should be removed as much as the stage of development of plants will allow in November or December. It is not possible to detect off-types accurately until the head is fully formed with the outer rosette leaves not stripped off (the shape and color of the leaves help to detect certain off-type plants).

WINTER PROTECTION. Unlike some root crops, cabbage needs protection from freezing in the winter when it is grown by the seed-to-seed method.

Cultural Practices Common to Both Methods.

STAKING. The developing flower stalks need support. Stakes about 2 m tall are driven in the ground at 25- to 30-cm intervals along the row. Heavy twine is strung between the stakes to keep the stalks in an upright position.

FERTILIZATION, IRRIGATION, AND WEED CONTROL. Cabbage, being a heavy feeder, responds well to fertilizer application. Nitrogen application is especially important, and under irrigated conditions up to 170 kg of N per hectare has been recommended depending upon the soil fertility level. About 80–100 kg N/ha may be used under unirrigated conditions. Phosphorus and potash may also be applied if the soil is not well supplied with these nutrients.

If cabbage is grown in regions where precipitation is low, additional water supplied through irrigation may be highly desirable. Timely control of weeds is essential in cabbage seed production. Weed killers such as isopropyl phenyl carbamate give excellent control of grass weeds. Wild mustard should not be allowed to grow with the cabbage crop.

ISOLATION. Isolation is one of the most difficult and complicated problems in the production of cabbage seed, because cabbage varieties cross easily not only with each other but also with the subspecies of *B. oleracea*. For the purpose of isolation in seed production, cruciferous vegetables are generally grouped into two classes: (1) those belonging to *B. oleracea*—Brussels sprouts, cabbage, cauliflower, collard, kale, and kolhrabi, and (2) others—radish, turnip, rutabaga, mustard, Chinese cabbage, and the various cresses.

The honeybee is a common pollinator of cabbage—it usually does not fly more than 8–10 km in search of nectar. A minimum distance of about 400 m is needed between plants that will naturally cross, but for production of seed stock, a distance of at least 1.6 km is necessary.

CONTROL OF PESTS AND DISEASES. The important insect pests of cabbage are aphid (*Brevicoyne brassicae*), cabbage worm (*Pieris rape*), diamond black moth (*Plutella maculipennis*), looper (*Autographa brassicae*), seedpod weevil (*Ceutorhynchus assimilis*), and cabbage maggot or larva. Application of 5% chlordane dust at 40–55 kg/ha gives good control of cabbage maggot and seedpod weevil. Nicotine dust or tetraethyl pyrophosphate is often effective against aphids. For nicotine, an atmospheric temperature of 18.3°C or more is necessary. The sweet potato whitefly has recently expanded its host range and increasingly damaged *B. oleraceae*. In California, damage has been particularly severe in broccoli and cauliflower (183). Elsey and Famham (184) studied the response of *B. oleracea* L. to *Bemisia tabaci* (Gennadius) and reported that cabbage and broccoli were less infested than other crops in screen cage tests, with kale, collard, and Brussels sprouts experiencing relatively high and kolhrabi intermediate infestations.

Diseases that can be a serious hazard to cabbage seed crops include mosaic (virus), ring spot (*Mycosphaerella brassicicola*), downy mildew (*Pernospora parasitina*), stalk rot (*Sclerotinia sclerotiorum*), black leaf spot (*Alternaria brassicicola* and *A. brassicae*), clubrot (*Plasmodiophora brassicae*), and yellow disease (*Fusarium oxysporum, F. conglutinans*).

The virus causing mosaic is transmitted by the cabbage aphid and the green peach aphid. Isolation of plant beds from the infected fields and control of aphid vectors are recommended to control mosaic. Wettable sulfur (1–2 kg/50 L) and Fermate (0.75 kg/50 L) can control downy mildew. Treating the seed at 50°C for 30 mins can prevent alternaria black spot. Benomyl and thiabendazole can effectively control *Mycospherella brassicicola*. Benomyl, NK 483, a mixture of phenolic compounds and sodium tetraborate, has significant activity against clubrot in transplanted cabbage (185).

Alternaria brassicae, A. brassicicola, and *A. japonica* cause black spot disease in crucifers. Iacomi-Vagilescu et al. (185a) developed a polymerase chain reaction (PCR)–based diagnostic assay to detect pathogenic *Alternaria* in some species of cruciferous seeds. The assay successfully revealed the presence of *A. brassicicola* and *A. japonica* of crucifers after two incubations, even at levels of infection as low as 10%.

HARVESTING. Cabbage seedpods, which are borne in a raceme, ripen in more or less the same sequence with which the flowers open. Harvesting should be performed before the older pods become dry to prevent shattering losses. An appropriate stage of maturity at which the seed will not crush or split when rubbed between the hands should be selected for harvesting. This occurs when a noticeable proportion of the pods turn yellow. Individual plants may be cut with a machine or long-handled pruning shears and placed on stakes laid across the intervening furrows between low ridges. The cabbage plants may also be piled on large canvases to collect the shattered seed.

The dry cabbage seed threshes easily because of the natural tendency for the pods to dehisce. Threshing may be carried out either in a regular grain thresher or in combine harvesters. It is advisable to operate the cylinder at the lowest speed to achieve satisfactory threshing. Vigorous threshing methods may crack or hull the cabbage seed.

Cabbage seed yields range between 450 and 1000 kg/ha or slightly higher. Seed yield is significantly influenced by premature bolting caused by changes in cultural and environmental factors like early planting, depth of planting, soil fertility, growth checks, rest period, and time of transplanting. Periods of low temperature extending over 30–60 days result in bolting. Seeding can be effectively hastened by storing mature plants for 2 months at about 4.4°C and then growing them at an average temperature of 21.1°C. Temperature thus exerts considerable influence on seed stalk formation. In addition to environmental factors, crop heredity also influences premature bolting.

4. Postharvest Operations

Cabbage seed is easily cleaned using any standard mill utilizing screens and air blasts or suction. The small, smooth, and globular cabbage seed is separated after cleaning. Arya et al. (186) obtained seed yields varying from 110.3 to 1203.8 kg/ha with an average of 568 kg/ha in India. Ellis et al. (187) studied the long-term storage of crucifer seeds at very low moisture contents (1.2–6.1%), and reported that seed storage at very low moisture content is an acceptable procedure for the long-term maintenance of seed accessions in gene banks.

B. Cauliflower and Heading Broccoli

1. Introduction

The seeds of cauliflower and heading broccoli (sometimes called cauliflower-broccoli) have high value, since they are relatively difficult to produce. The cultural and environmental requirements of these two vegetables, although similar to those of cabbage, are more exacting, especially in respect to temperature. Cauliflower is much more susceptible to freezing injury than is cabbage, yet a uniformly cooler temperature must prevail for the normal development of the curd. Both crops require fairly high humidity.

Botany. Cauliflower and heading broccoli (*B. oleracea* var. *Botrytis*) are biennial vegetable crops. The curd, for which the market crop of cauliflower is grown, forms in the first year in a spring-sown crop. Later, the floral parts form, developing an inflorescence in the second year that is smaller and more umbrella shaped than that of cabbage. There is no central main stem above the point where branching begins. The inflorescence rarely exceeds 60–75 cm in height. The inflorescence of heading broccoli, with its 5000–8000 flowers, opens about 10–14 days after the initiation of flowering; cabbage usually requires longer. The leaves of cauliflower and heading broccoli are usually longer and narrower than those of cabbage. The inner leaves are smaller and tend to curve inward and protect the curd. The two vegetables do not differ from cabbage in respect to pollination.

Cultivars. Snowball and Erfart, available in several types, depending upon earliness, plant size, head size, and length of leaves, are the most widely grown cultivars of cauliflower. Super Snowball is one of earliest cultivars to mature; others of a similar type include Early Snowball A, Super Junior, and Snowcap. Erfart and Snowball types are mostly later maturing and larger than Super Snowball types, with curds being more rounded, thicker, and heavier. This group includes Early Snowball, Snowdrift, White Mountain, Snowball × Snowball Imperial, and Improved Holland Erfact. Cossa and April are late-maturity cultivars (in the spring); Early Purple Head has purple heads and does not require blanching. The cultivar K1 is a valuable addition to the Snowball cauliflower group. Italian Green Sprouting and Calabrese are the most popular cultivars of broccoli. Emperior, Futura, Moran 143 A, and Packer are new broccoli cultivars.

2. Breeding

According to Tigchelaar (3), Snowball 42 and Snowball 76 were introduced as new cultivars of cauliflower. The former is an open-pollinated cultivar, having more concentrated maturity than Snowball M, and has a deep white curd, which holds well. Snowball 76 is also an open-pollinated cultivar, having a more concentrated maturity than Snowball x. Several new hybrid lines of broccoli have been developed, which include G 1117 A, G 1102A, and G 1106A.

3. Seed-Production Technology

Culture. The cultural requirements of cauliflower and heading broccoli are more specific than those of cabbage. Cauliflower heads cannot be grown and stored like cabbage. In British Columbia, cauliflower seed is sown in outdoor seedbeds in early September or in a cool greenhouse in early October; in late October or early November, the seedlings are usually large enough to transplant. The spacing varies according to variety and locality 75 × 100 cm for early varieties and 99 × 120 cm for later, large varieties). Dry cauliflower seeds show an immediate rapid phase of imbibition upon the addition of water. The intact testa of cauliflower seed is capable of acting as a barrier to water influx and a high–water-uptake rate is damaging to the embryo (188). Cauliflower requires a constant but not excessive supply of moisture. Sometimes a crosscut is made in the curd at maturity to allow the rainwater to drain away and to give the developing branches better aeration. A Bordeaux mixture may be sprayed after cutting the curd to prevent decay.

Roguing. Careful roguing is essential, especially when the crop is grown for seed stock. Cauliflower varieties vary in their morphological characteristics, especially at the time of maturity. Off-season plants thus may be rogued out to achieve uniformity in the variety. The head characteristics such as size, depth, color, compactness, and uniformity should be considered when roguing. The cauliflower seed producer must ensure sufficient isolation of the crop from other cauliflower fields as well as from any other brassicas.

Control of Pests and Diseases. The insect pests of cauliflower and broccoli are the same as those that attack cabbage, with aphids being the most serious. Cauliflower clubrot can be effectively controlled by Captafol applied as a root drench at 0.1 g/200 mL for each plant (189). All diseases that affect cabbage also affect cauliflower and broccoli. In addition, cauliflower is also attacked by whiptail, a result of a deficiency of molybdenum and bacterial soft rot caused by *Bacillus carotovorus*. The latter attacks cauliflower when the curd begins to develop into the inflorescence. Spraying with Bordeaux mixture or using a copper-lime dust during crosscutting prevents the disease. Endosulfan at 0.05%, methyl-demeton at 0.02%, and malathion at 0.05% were found effectively to check population build-up of *Brevicoryne brassicae* infesting a cauliflower seed crop in India (190).

Harvesting. Like cabbage seed, cauliflower seed must be harvested at the appropriate maturity to minimize shattering losses. After cutting, the crop may be stacked on large canvases to collect the shattered seed. Harvesting may be carried out in September– October, which is later than for most cruciferous crops. The plants may be cut individually with long-handled pruning shears while they are still yellow and with the pods being mostly brown and undehisced. The plants are placed in an inverted position on wires stretched across a Dutch barn and cured. A canvas spread on the floor collects seed shattered during the curing process. The threshing, milling, and cleaning of the seed are carried out in the same manner as described for cabbage. Average seed yield of cauliflower and broccoli vary from about 110 to 170 kg/ha. Higher yields (400 kg/ha) may be obtained under exceptional conditions. Seed yield increases with curd size, but curd size has no effect on germination (191). Application of 0.2% boron as a seedling root dip with 1 kg boron/ha applied to the soil resulted in the highest seed yield (2.05 q/ha) and a 97% increase in cauliflower diameter as compared with the control. An application of molybdenum and lime had no effect on cauliflower size. Liming (250 kg/ha soil application in the form of paper mill sludge) has been found to increase seed yield, whereas molybdenum has no effect (192).

The effects of different curd-cutting methods (scooping, curd pruning, and half-curd removal) on the seed yield and yield attributes of five cauliflower cultivas (Pusa Deepali, Pusa Early Synthetic, Improved Japanese, Pant Subhra, and Pusa Himjyoti) were investigated in New Delhi, India. All cutting treatments were effective in increasing seed yield. However, the highest values for siliques per primary branch (195.6), siliques per plant (1260.0), seeds per silique (13.1), and seed yield (67.4 g/plant) were found in the scooping method. Among the cultivars, Pusa Deepali and Pusa Early Synthetic had the highest values for siliques per primary branch (185.0 and 22.8), siliques per plant (1437.4 and 1604.2), seeds per silique (14.5 and 13.1), and seed yield (58.9 and 65.7 g/plant) (192a).

The cauliflower cultivar Kibo Giant was sown in 1994–1995 and 1995–1996 at Udhagamandalam, Tamil Nadu, India, to determine the influence of the stage of harvesting on seed yield and quality. A comparison of six stages of harvesting revealed that maximum seed yield and quality parameters were obtained 283 and 290 days after sowing, respectively. The high temperature and relative humidity along with frequent and considerable rainfall the period of seed development and heavy rainfall just prior to harvesting reduced seed viability, vigor, cell membrane integrity, protein, and oil contents. The harsh weather conditions had also caused lodging of plants and shattering of siliques. The results suggested that Kibo Giant could be safely harvested from 283 to 290 days after sowing without significant reduction in seed yield and quality (192b).

C. Sprouting Broccoli

1. Introduction

Sprouting broccoli is better adapted to warmer and less humid conditions than those preferred for cabbage. Although the low-temperature requirement is not as essential for broccoli seed production as it is for cabbage, broccoli must be grown during the cooler months of the year in southern regions.

Sprouting broccoli (*B. oleracea* var. *Italica*) differs from heading broccoli and cauliflower in plant habit and in having heads that are green rather than white. The central stem is about 60 cm or longer, depending upon variety and cultural conditions, bearing at its top a fairly large central head. The latter is, however, smaller than the cauliflower curd. Numerous side heads also develop over a period of many weeks, and are comparatively smaller than the central head. Unlike the cauliflower curd, the head of sprouting broccoli when ready for market is a dense cluster of flower buds instead of prefloral primordial tissue. Other botanical characteristics such as roots, pod, and seed are identical with those of cabbage.

2. Breeding

Very little systematic breeding work has been carried out to improve broccoli strains. One variety, Texas 107, released by the Texas Agricultural Experiment Station, was the result of inbreeding and selection to obtain greater uniformity, early formation of the central head, and vigorous production of side shoots during the following 6–7 weeks. Some seed companies in the United States have also developed varieties suitable for a specific season and maturity, as well as for freezing.

3. Seed-Production Technology

Culture. The seed of sprouting broccoli may be sown directly in the field where the plants ultimately mature to produce seed, or seedlings may be raised in seedbeds for later transplantation. About 1.7 kg of seed is sufficient to cover an area of 1 ha. Broccoli grown for seed is sown in rows spaced about 75–90 cm apart, and plants are thinned or transplanted 35–60 cm apart. In Arizona, broccoli blooms by February; seed harvesting begins in early May and is completed by early June. Like cabbage, sprouting broccoli responds to fertilizer and an adequate supply of moisture.

Roguing. Off-type plants must be removed both at the time the central head is ready for market and later as the side shoots form. Uniformity in both appearance and performance in a strain is desirable. Tightly formed heads that lack leaves with buds of similar type and color are preferred. Head production on side shoots is also an important characteristic. A profusion of well-formed heads on side shoots is desirable in broccoli strains selected for the frozen food industry. Also, the foliage of individual plants should be similar in type, size, and color.

Control of Pests and Diseases. The insect pests and diseases that attack the seed crop of sprouting broccoli are primarily the same as those attacking cabbage and cauliflower. Broccoli, however, has been reported to have a high natural resistance to yellow disease, and sprouting broccoli does not show symptoms of boron deficiency in the field as early as cauliflower.

Harvesting. Broccoli grown for seed is harvested similar to that of cabbage. The seed stalks are cut by hand, windrowed, and cured. Under warmer regions, curing will be completed within 3–5 days. A harvester with a pickup attachment may be used. Threshing, milling, and cleaning processes are essentially the same as for cabbage and other cruciferous crops. Seed yields of about 900–1100 kg/ha have been reported.

D. Minor Brassica Vegetables

1. Introduction

Kale and collard (*B. oleracea* var. *Acephala*), Brussels sprouts (*B. oleraceae* var. *Gemmifera*), and kohlrabi (*B. oleraceae* var. *Caulorapa*) are minor biennial vegetable crops belonging to the *Brassica* group.

The minor biennial *Brassica* vegetables differ from each other in general appearance, with all having the blue-green foliage typical of the cabbage group. Kale has an open type of growth; lacking entirely any head or curd, it resembles the original wild type more than most cultivated brassicas. Varieties and strains of kale differ in size and height as well as in form and color of leaves.

Collard varieties are less variable. Collard has a thick, stiff main stem, which is terminated by a loose head of cabbagelike leaves. The plant reaches a height of about 1 m. Collard thus has an upright plant habit compared to the smaller varieties of kale. Brussels sprouts have a single, erect stem extending to a height of 60–75 cm. The small edible buds or sprouts are borne in the axils of the many leaves, which may hide the main stem completely from view at the time of edible maturity. Kohlrabi (knol khol) is a small plant grown for its edible stem, which enlarges just above the ground. The leaves are cabbagelike, but smaller and sparse. The bulblike stem is prominent, and the root system is extensive.

The inflorescence of all four minor *Brassica* vegetables is essentially the same as that of cabbage, and all crops require cooler climates for flower initiation. The minor *Brassica* vegetables are similar to cabbage in respect to their pollination.

2.   Seed-Production Technology

Culture.   With the exception of Brussels sprouts, minor *Brassica* vegetables are usually grown by the seed-to-seed method. In southern regions, the seed is sown in the late summer or at the beginning of winter so as to allow the crops to complete their vegetative growth during the cooler months and flower the following spring. Occasionally Brussels sprouts are also grown by direct seeding, allowing the crop to overwinter in the field. Like cabbage, Brussels sprouts can be stored in shallow trenches. Collards and larger kale varieties are usually sown in rows at least about 90 cm apart. Kohlrabi and smaller varieties of kale may be spaced at 75 cm. The distance between plants varies from 45 to 60 cm.

Roguing.   Even minor vegetable crops and their varieties should be true to type and uniform in size, general appearance, and time of maturity of the edible plant part. Leaf character is important in kale and collards, whereas in Brussels sprouts, it is essential to have buds of good size that are uniform and firm in texture. Similarly, kohlrabi should have an attractive edible stem of uniform size, shape, and color. Crops should be rogued at the time of their prime market maturity.

Control of Pests and Diseases.   Minor *Brassica* vegetables are susceptible to the same common pests and diseases that attack cabbage, cauliflower, and other crucifers. Brussels sprouts are highly resistant to yellow disease, but kohlrabi and kale are susceptible to it. Although Brussels sprouts, kale, and kohlrabi are affected by boron deficiency, the symptoms are slightly different from those in other crucifers.

Harvesting.   Harvesting of all minor *Brassica* vegetables is similar to that for cabbage. The individual plants are cut when the pods are yellow but not dry. During the curing process, some pods may split open and dehisce. Seed maturity and time of harvesting vary with the crop variety and region where the crop is grown. For collards that flower in the southern regions in late February and March, harvesting of seed begins in May or early June.

The methods of threshing, milling, and cleaning of seed are essentially the same as described for other cruciferous crops. The seed yields of Brussels sprouts and kale are usually lower than that of cabbage (about 350–700 kg/ha), and kohlrabi normally yields 900–1500 kg/ha. Collard yields are usually somewhat higher than sprouting broccoli.

E. Mustard and Chinese Cabbage

1. Introduction

In addition to *B. oleracea* vegetables, there are several other leafy cruciferous crops belonging to the genus *Brassica* and even to other genera. Mustard and Chinese cabbage are annuals, without the glaucous foliage characteristic of members of *B. oleracea*. Both mustard and Chinese cabbage tend to bolt with the onset of warmer climate, although higher temperatures may not be the exclusive causal factor.

Botany. Although mustard (*B. juncea* and *B. alba*) and Chinese cabbage (*B. pekinensis* and *B. chinensis*) share several common botanical characteristics, they belong to different species of *Brassica* and are cytogenetically distinct from one another. *B. juncea* has 18 pairs of chromosomes, whereas *B. pekinensis* and *B. chinensis* have only 10 pairs of chromosomes each. Black mustard (*B. nigra*) is grown commercially for seed. Although principally an oilseed crop, mustard is commonly grown as a salad and green vegetable in the northern parts of India.

Mustard forms an open rosette of leaves. The crop is hardy, and in the southern regions, it may be sown from September to March. The leaves are usually long and broad. The most commonly grown, *B. pekinensis*, represents two types (varieties)—Chihili and Wong Bok. Both form solid heads, but in the former they are longer, slender, and somewhat pointed at the tip. The heads of Wong Bok are much shorter and blocky. There are a number of outside loose leaves, folding around the head, which are long and greenish in color.

The inflorescence of both mustard and Chinese cabbage resembles that of other brassicas in form. The pods and seeds are somewhat smaller. Pollination occurs mainly through bees; natural cross-pollination and self-fertility are most common in both the crops. The 1000–seed weight of Chinese cabbage is about 1.6 g compared to 3.2 g for cabbage and other forms of *B. oleracea*.

Fast and simple procedures were developed for DNA extraction and RAPD analysis of head Chinese cabbage (*B. pekinensis*) in order to determine its hybrid purity. Reproducible and consistent RAPD results were achieved with DNA isolated by a modified method. One person could analyze about 100 samples in 1 day with the method. From the 20-decamer primers used for the analysis of four inbreeding lines and two hybrids by RAPD markers, five primers could be used to identify some of the parents and F_{1s}. Primer OPH09 was successfully used to determine the genetic purity of Renewed Qingza No. 3 hybrid seed. The results showed that RAPD markers could be used to identify Chinese cabbage F_1 hybrids and parents to determine the genetic purity of hybrid seeds (192c).

Cultivars. The cabbage-leafed mustard (*B. campestris*, type Rugasa Roxb) has radical leaves, which form a loose cabbagelike head. Fordlock Fandy or Ostrich Plume and Southern Giant Curled are popular cultivars of Japanese mustard grown in the United States. In India, the brown or Kali Sarson (*B. campestris* L, type Oichotoma

Watt) and Yellow Sarson (*B. campestris* L, type Glaucazo sp. Roxb) are grown in the north.

Two species of Chinese cabbage are grown: Pe-tsai (*B. pekinensis*) and Pakchoi (*B. chinensis*). The Pe-tsai resembles cos lettuce but produces a much larger head, which is elongated and compact. The Pakchoi cultivars resemble Swiss chard in growth habit, but they do not form solid heads. The most popular cultivars of Chinese cabbage are Chihili and Wong Bok.

2. Seed-Production Technology

Culture. The production of seed in annual mustard and Chinese cabbage is much simpler than that of the biennial brassicas. Both crops are sown directly in the field where the seed is to be produced. In southern warm regions, the seed is sown in September–October, whereas in the cooler northern areas, it is sown in the early spring. Recommended spacings are 45–60 cm between rows for mustard and 50–60 cm for Chinese cabbage. Mustard responds well to fertilizer application, especially nitrogen (up to 150 kg/ha). Fields should be free of wild mustard and other related weeds to avoid probable crosses.

Roguing. In the production of stock seed, careful roguing is highly desirable. Planting may be timed and plants thinned to obtain a better crop stand and to produce seed stalks in the least amount of time. Type and variety must be considered when roguing. The size, shape, and firmness of heads are of prime importance with Chinese cabbage. Mustard and Chinese cabbage belong to one specific group of crucifers (radish, turnip, rutabaga, mustard, Chinese cabbage, and various cresses), and neither will cross naturally with any member of *B. oleracea*.

Control of Pests and Diseases. Some of the diseases and insect pests affecting cabbage are also found on mustard and Chinese cabbage. The important diseases are black rot, black leaf spot, and anthracnose.

Harvesting. Both mustard and Chinese cabbage can be harvested like any other grain crop by cutting and threshing in a combine harvester directly in the field. It is necessary that the field be free of weeds, and the crop should have uniform maturity. The crop may also be harvested by cutting the individual plants, as in the case of other crucifers.

Threshing, milling, and cleaning operations are essentially the same as those employed for other cruciferous seeds except that slight modifications in screen sizes are required to suit smaller sized seeds. Average commercial yields for mustard range from 450 to 560 kg/ha and for Chinese cabbage from 560 to 700 kg/ha.

F. Cresses and Sea Kale

1. Botany

The common cress (*Lepidium sativum*) is an annual upland cress, winter cress (*Barbarea verna*) is a hardy biennial, and watercress (*Roripa nasturtium aquaticum*) and sea kale (*Crambe maritima*) are perennials. Crops sown in early spring flower with the arrival of hot weather and are ready for harvest before the summer is over.

Upland, or winter, cress is a slow-growing biennial. It is hardy enough to overwinter even in northern regions and produces seed the next spring. The leaves are oval, deeply serrate, and resemble those of watercress in shape and flavor. Watercress is a trailing perennial commonly found growing along streams and waterways. Sea kale is commonly grown in Europe and is native to western coasts. The inflorescence usually appears during the second year followed by flower production every year.

2. Seed-Production Technology

Common cress is normally sown in early spring in rows about 15 cm apart and harvested like grain crops in August. It may be cut with binder, stacked, and threshed later. All cresses are insect-pollinated and cannot cross with each other or with other cruciferous crops.

VIII. CHENOPODS

The garden, or table, beet, spinach, Swiss chard, and orach or mountain spinach are commonly grown vegetables included in the family Chenopodiaceae. Both spinach and beets are economically important vegetables. Although garden beets and Swiss chard belong to the same species, *Beta vulgaris*, the former is primarily grown as a root crop, whereas the latter is grown for its foliage. All chenopods are, however, characterized by simple, alternate leaves and small bisexual or unisexual flowers which lack petals and sometimes even the calyx.

A. Garden Beet

1. Introduction

Beet is a cool-season crop grown in areas having high humidity. Sandy loam or somewhat heavier soils with good moisture-retention capacity are well suited for growing beets for seed production.

Garden beet is naturally a biennial vegetable crop producing a rosette of leaves and a fleshy taproot in the first year. In the second year, one or more branched seed stalks develop from the crown, reaching a height of over 1 m, but they are not as erect or as tall as those of carrot or cabbage.

The rosette leaves develop in a close spiral with the oldest ones on the outside, and may be triangular or oblong in shape. Environmental factors such as temperature, time of planting, spacing, and soil moisture influence size, shape, and color of the foliage. Leaf blades may vary in color from light green to dark red. The root system of beet is extensive and varies from oblate through globe to long and tapering according to the variety. The exterior color varies from orange-red to dark purple-red, and the internal root color depends on factors such as temperature, season of growth, soil type, and nutrient deficiencies. Immediately around and below the fleshy root there is an extremely dense mass of fairly short and much rebranched roots. Such an extensive root system makes the beet a fairly drought-resistant crop.

Beet plants bear inflorescence in the second year. The flower stalks are large openly growing panicles. Blooming, followed by seed maturity, begins at the base of the panicles. Flowers occur in clusters of two or three, occasionally singly, arising in the axils of bracts of inflorescence axis or its secondary branches. Flowers, being small and having no corolla but only a five-part green calyx, are rather inconspicuous. They are perfect with five stamens, and there is only one ovule per ovary. Pollination generally occurs by way of wind. Beet seed is botanically a fruit and is actually a seed ball containing several seeds. The beet seed has a shiny reddish brown surface, and is about 3 mm long. The 1000–seed weight is about 17 g.

2.   Breeding

Mass selection has been the principal method of developing improved varieties of garden beet in the past rather than by any planned program of controlled inbreeding and cross-pollination. Self-sterility in beet is a common characteristic of plants in an ordinary population, and it takes considerable work to develop self-fertile lines, especially with desirable characteristics.

Male sterility is rather common in *B. vulgaris*. About 2% of plants in the curly-top–resistant sugar beet variety (U.S. No.1) are male sterile but female fertile. When male sterility is inherited cytoplasmically, it is possible by controlled crossing to produce male-sterile offspring in whatever quantity desired. These male-sterile plants are used easily to produce F_1 hybrids.

3.   Seed-Production Technology

Like most other biennial root crops, there are two general methods of growing beet seed: root-to-seed and seed-to-seed. Although a considerable portion of beet seed is produced by the seed-to-seed method, the root-to-seed method is more useful in the production of stock seed.

Root-to-Seed Method.

CULTURE.   In the first year, the beet crop is grown in much the same way as the market crop. In the second year, the stecklings are set out for the seed crop. Seed is usually sown later in the spring than for the market or home garden crop. Beets, being quick growing, require about 2 weeks less time to reach edible maturity than the earliest varieties of carrot. Seed is normally sown from June to August to obtain stecklings of satisfactory size by the end of the growing season. The seed rate may vary from 8 to 20 kg/ha.

Two rows per bed or a single-row system may be followed. Row spacing also varies from 35 to 75 cm. Enough stecklings can be obtained from 1 ha of land to cover an area of about 10–20 ha depending upon the method. During harvesting, care should be taken not to bruise or injure stecklings to prevent storage rots. Head pulling is therefore advocated. Care should be taken to prevent injury to the growing point when the tops are cut, leaving about 2.5 cm of petiole. Mowing machines are used to cut the tops prior to lifting.

ROGUING.   Beet plants are rogued for both foliage and root types. Good stock seed production requires careful and critical roguing, and when such stock seed is used for market crop, little or no roguing is necessary. Off-type foliage plants should be first removed from the steckling field prior to general harvest. Roots failing to meet the required color standard for the variety or strain are then discarded. The cut surfaces of stecklings should be dried before storage.

STORAGE OF STECKLINGS.   The stecklings are usually stored in a pit with the tops still attached; provided the latter are free of aphids. The tops subsequently wither and decay and the roots remain in satisfactory condition. It is desirable to store stecklings in reasonably small quantities instead of in one great pile or large bin whether pit or cellar storage is used. A new type of pit that is rather shallow and not ventilated is employed in British Columbia. The old deeper pits give good protection from freezing, but often cause overheating in spite of ventilation. In the new type of pit, a narrower and much shallower excavation is made (about 60–90 cm wide and 15–30 cm deep). Stecklings are placed in the pit to the level of the top of the ground but no higher. They are lightly covered in early fall and later again with soil plowed and shoveled over them (about 12.5 cm in areas near the coast and 25 cm in southern interiors). Some growers use pits as narrow as 35–43 cm wide, and roots are removed using a potato digger and hauled directly to the field for planting. At temperatures of 7.6–10.0°C during storage, stecklings are conditioned for flower stalk formation more quickly than at lower temperatures of 0.5–6.7°C. Also, a shorter storage period (2–3 months) rather than a longer one (4–5 months) is more effective.

REPLANTING. Stecklings are set out after harvesting from the steckling bed or from storage much the same way as described for carrots and turnips. The stored stecklings are replanted as early as possible in the spring (March–April). In areas like California where no storage is necessary because of the milder climate, they are planted in November–December almost immediately after pulling. Undue drying of stecklings must be avoided during the planting operation. Sometimes stecklings are allowed to remain in open containers for several days or weeks, which hastens shriveling.

Pushing down in the soft mark of the row sets out small- or medium-sized stecklings. The crown should be just below the soil surface and preferably lightly covered with earth. If a planting machine is used, the planters should be followed up to ensure that the soil is firmed about every root. Intimate contact with a moist soil in good tilth will hasten plant establishment. seed crop of beet is usually planted in rows 90 cm apart with stecklings set 45–60 cm or even further apart within rows. Occasionally, they may be set out at 90 × 90 cm; closer spacing, however, may result in higher seed yields.

Seed-to-Seed Method. As in the case of carrots and turnips, the seed-to-seed method of producing beet seed is much simpler than the root-to-seed method. It involves much less labor and expense and does not require storage. The main limitation of this method, however, is that critical and thorough roguing is not possible. Seed is usually sown slightly earlier than for seed-to-seed turnips. In the Pacific Northwest, it is sown in late August or early September depending upon the variety. Slow-growing, late varieties are planted first. Only stock seed of high genetic quality should be used for seed-to-seed plantings. Trueness to type and freedom from rogues of every sort in the crop are highly desirable and practically necessary to obtain a seed of satisfactory quality.

Cultural Practices Common to Both Methods. A seed crop of garden beet needs a thorough culture in the second (seed) year to obtain higher seed yield. Beets respond to both organic manures and commercial fertilizers, especially on lighter soils with low fertility levels. About 25–30 tons of farmyard manure and 500–1000 kg of 5:10:5 fertilizer mixtures (NPK) are usually recommended. A reasonably moist soil is desirable. Irrigation is necessary, particularly when fertilizers have been applied generously. Weeds should be eliminated by shallow cultivation and hoeing during the early stages of plant growth.

Control of Pests and Diseases. Garden beets are attacked by aphids and lugus bugs more seriously than other insect pests. Lugus bugs in particular may cause considerable damage to beet seed.

Mosaic and downy mildew (*Peronospora schachtii*) are the most serious diseases of garden beet. Other diseases affecting this crop are curly top, a viral disease transmitted by the beet leaf hopper (*Circulifer tenellus*), yellow and other viral diseases, and boron deficiency. Mosaic can be controlled by planting steckling beds in areas well isolated not only from other garden beets but also from closely related crops such as mangels, sugar beets, and Swiss chard. Using disease-free beet seed, avoiding fields where the disease has occurred previously, and eliminating infected stecklings can prevent downy mildew. A spray of 4-4-40 Bordeaux mixture of 20–80 copper lime also generally controls this disease. An application of 25–50 kg boraxflla is recommended to overcome boron deficiency on highly calcareous soils. Higher doses may be injurious on acid soils.

Harvesting.   Beet crop is ready to harvest when the seed at the base of the branches is mature and noticeably brown. The plants may be pulled from the soil by hand when the seed at the tip of the branches is still immature. Pulling the plants in the early morning may largely prevent the shattering of ripe seed. Mowing the beet crop is usually difficult and unsatisfactory because of the sprawling tangled nature of the plants. Roots are cut off with a large knife immediately after pulling, and plants are laid in small piles for curing. The field may be harvested in two to three operations, because the crop does not usually mature uniformly. The crop is often stacked on wooden racks, which are left until the crop is threshed.

When the plants and seeds are dry, the crop is threshed using grain threshers, making certain adjustments in the thresher. The air blast should be reduced to a point at which no seed is carried over. It is not practicable to combine beet seed directly as it stands, because beet crop ripens unevenly and an entire crop is rarely ready to harvest at any one time.

4. Postharvest Operations

Beet seeds are first cleaned using a fanning or screening mill, which removes most of the chaff and plant material. Seed balls vary considerably in size and are comparatively larger than most small vegetable seeds. Openings in the mill allow the trash to fall through. Beet seed is finally cleaned on a "canvas draper." Freely moving rounded seed can often be separated from nonrolling material such as small sticks using such a draper. In this equipment, material to be cleaned falls from a hopper, which spreads the seed onto the upper end of an inclined upward-moving continuous canvas belt. By varying the speed at which the belt turns, round seeds can be made to roll or slide down the incline to be caught in the hopper below. Nonround objects like small sticks are prevented by friction from sliding and are carried upward from round seeds like those of garden beet or chard. The incline of the moving canvas and its spread must be adjusted carefully for each lot, since different seed lots vary in trash content.

Seed yields of garden beet vary from about 1600 to 2200 kg/ha. Seed yield is influenced by premature bolting, which is in turn influenced by various factors such as heredity, time of sowing, frost, checks in growth, low temperature, light, transplanting, seed size, planting depth, seed age, seed treatments, and soil moisture.

B. Swiss Chard

1. Botany

Swiss chard (*B. vulgaris* var. *Cicla*) is biennial vegetable grown primarily for its foliage. The commercial chard varieties are distinguished by the color of their foliage, especially the color of the blade and midrib. The rosette of leaves is more effect and taller than that of the beet, and the midribs are much wider and more conspicuous.

Swiss chard has an extensive root system with a strong taproot and many laterals. Taproots may penetrate to a depth of over 2 m in the first year. The inflorescence, pollination, and seed characteristics are similar to those of garden beet. The inflorescence of chard, which may be about 2 m tall, closely resembles that of sugar beet rather than the garden beet in its general vigor, creating a harvesting problem not encountered in garden beet.

2. Breeding

There has not been much systematic breeding work planned in the past for crop improvement. Mass selection and collection of seed from open-pollinated stocks have been the principal methods of developing varieties with improved horticultural characters.

3. Seed-Production Technology

Culture. Seed crop of chard may be grown either by transplanting or by the seed-to-seed method as described for garden beet. Seed is usually sown around mid-June, and the plants are plowed out the following spring and transplanted in much the same way as are garden beets. In the seed-to-seed method, chard is handled much like seed-to-seed sugar beets. The spacing recommended is 90 cm between rows and 30 cm between plants in a row.

Roguing. Plants are rogued depending upon the foliage characteristics at the time of edible maturity or prior to seed stalk formation. Leaf characteristics exhibited before maturity aid in detecting off-type chard plants. As in garden beet, pollination occurs by the wind, and for production of stock seed enough distance (about 5 km) should be maintained between different varieties.

Control of Pests and Diseases. The same pests and diseases attack chard as attack the garden beet.

Harvesting. Chard is more difficult to harvest than garden beets. The sugar beet harvester, a specially built windrower equipped with two cutter bars, is ideal for harvesting chard. The cut plants are cured in the swath and then threshed with a combine having a pickup attachment. Then it is milled and cleaned similar to garden beet. The average seed yields of chard range from about 100 to 200 kg/ha; higher than those of garden beets.

C. Spinach

Spinach is a cool-weather, hardy crop grown as an overwintered crop for the spring market. A cool and humid climate favors the growth of spinach. It can be grown on a variety of soils, but heavy loam soils are preferred for higher seed yields; acid soils are not suitable.

1. Botany

Spinach (*Spinacea oleracea*) is an annual vegetable grown for its foliage. Their foliage characteristics and the general form of the rosette commonly distinguish spinach varieties. Characteristics such as savoying, leaf color, plant size, number of leaves, blade length, and width are influenced by temperature, day length, and length of vegetative growing period. Low night temperatures increase savoying in varieties, and both low temperatures and short day lengths tend to delay bolting.

The root system is extensive with a strong taproot penetrating the soil to a depth of 2 m. Flower stalks develop after a period of vegetative growth. The floral initiation time is influenced by variety, day length, and temperature. Light, temperature, and other environmental factors have complicated interrelationships in the development of spinach.

Spinach is unique among the cultivated vegetables in being tetramorphic; that is, it expresses its sex through four different types of plants:

1. Extreme males—these are usually smaller plants bearing only staminate flowers. The seed stalk is characterized by a lack of foliage or only small leaves barely larger than scales. Extreme males normally bolt ahead of the other three types.
2. Vegetative males—these also bear only staminate flowers with fully developed leaves on the seed stalk.
3. Monoecious plants—these bear varying proportions of staminate and pistillate flowers in the same clusters and are rare in most varieties; leaves are fully developed to the tip of the stalk.
4. Female plants—these bear only pistillate flowers with fully developed leaves to the tip of the stalk.

Extreme male plants are not desirable from any point of view. Their small leaves and early bolting habit detract from their market and home garden value. Since these characteristics are transmitted to seed-bearing plants if the extreme males are present during pollination, they are not desirable in seed plantings. Monoecious plants are most desirable both for market and edible use, and they are capable of producing seed without the presence of male plants. Female plants, although satisfactory for market purposes, set seed only if there are male or monoecious plants present nearby to supply pollen. In most spinach populations, the dioecious condition prevails. In general, there is a 1:1 ratio of male to female plants, and if monoecious plants are present, they usually prevail. The proportion of monoecious plants may vary from about 4 to 9%.

Spinach flowers, like those of beet, are small and lack a corolla. They are borne in clusters of 6–20 in the leaf axils. In addition to staminate and pistillate flowers, there are also perfect or hermaphroditic flowers, although these are rare and usually occur on plants having mostly pistillate flowers. In the staminate flowers, the calyx has four parts, with each segment just below one of the four stamens. Each stamen bears two large anthers, which dehisce by longitudinal splitting. In the pistillate flower, the calyx may have two to four parts enclosing the one-celled ovary. Because spinach pollen is extremely small and lightweight, pollination occurs by way of wind. The pollen is produced abundantly over a comparatively long period of time. Spinach seed of the trade is actually a single-seeded fruit, botanically called "utricle." The 1000–seed weight of spinach is about 10 g.

2. Breeding

There has been no extensive planned breeding program to improve spinach varieties. Since the fine pollen is quickly and easily carried by the wind, certain precautions must be taken, as with beets. Emasculation is easy if both male and female plants are used.

The monoecious spinach plant is often a favorite selection of the vegetable breeders; such a plant is self-pollinated when enclosed in a protective covering and tends to produce a population with a proportion of monoecious plants.

Variety Hollandia was developed by selection from a monoecious plant found in Holland in the early 1920s. At first, this variety was purely hermaphroditic, but male and female plants soon reappeared. Other varieties developed from single monoecious plants are the smooth-seeded varieties like Princess Juliana and Nobel.

3. Seed-Production Technology

Culture. Because spinach is an annual crop, seed production is comparatively easy. Seed is usually sown as early in the spring as possible. In the United States, about 11–83 kg of seed per hectare is sown 50–75 cm apart. Plants may be thinned to achieve 10–15 cm between plants in a row. In the Netherlands, the seed rate is slightly higher—about 13–22 kg/ha when sown in rows 35–45 cm apart. The minimum germination temperature in spinach cultivar Medania was found to be $-3.1°C$ (193). The highest seed yields (7.26–10.08 Q/ha) and total returns (value of seeds and cut leaves) were obtained from plants spaced at 45×30 cm and cut once. More frequent cuttings have been found to reduce the seed yield (194). Crops need cultivation between rows to control weeds. Hoeings should be shallow and repeated only to control weeds. In European countries, more day-to-day attention is given in the form of hand weeding and intercultivation.

Roguing. An initial roguing may be carried out at the time of thinning and weeding operations to remove early off-type plants. In flat-leafed varieties, all plants with savoyed leaves should be removed. Similarly, in savored varieties, all plants with flat leaves should be removed. Thinning combined with appropriate roguing will enhance seed yields in addition to eliminating undesirable extreme male plants. At the initiation of bolting, a second roguing may be essential. Plants not meeting varietal requirements in the prime market stage can be removed at this time. A distance of about 1.6 km is recommended between two varieties to prevent cross-pollination.

Control of Pests and Diseases. Several species of aphids have been known to transmit spinach mosaic. Aphids are difficult to control on spinach because of the plant's habit of growth and the savoyed leaves in some varieties. Downy mildew (*Peronospora effusa*), wilt (*Fusarium oxysporum, F. spinaciae*), damping-off (*Pithium* sp.), boron deficiency, and a group of viral diseases such a mosaic (blight), beet mosaic, aster yellows, curly top, spotted wilt, spinach yellow dwarf, and virus yellows are important. Copper sprays and dusts, although recommended, do not satisfactorily control downy mildew. Mosaic-resistant varieties such as Virginia Savoy and Old Dominion are used to prevent mosaic. An application of borax at a rate of about 10 kg/ha to soil often eliminates boron deficiency.

Harvesting. The seed crop of spinach is normally ready to harvest in July–August. Like garden beet, the crop tends to ripen somewhat unevenly. The entire crop may, however, be harvested as soon as most of the late-ripening plants start turning yellow. The crop may be cut using mowers, reapers, binders, or by hand. The plants are threshed after drying, using a grain thresher or all-purpose combines. Average seed yields vary from 600 to 1100 kg/ha in the United States and from 800 to 2200 kg in the Netherlands.

4. Postharvest Operations

With well-broken seed clusters during threshing, milling and cleaning of seed is comparatively simple using modern equipment. The smooth, round seeds use relatively smaller perforations in the screens to scalp off much of the larger chaff in the first separation. Milling of spiny seed is more difficult. Spinach seed is comparatively heavy; air blast or suction can be used advantageously to clean the seeds.

D. Orach or Mountain Spinach

Orach needs a drier climate than do spinach and beets and can be grown in a wide range of soils from neutral to alkaline. It is a hardy crop like spinach, but grows more vigorously and several feet taller.

1. Botany

Orach (*Artiplex hortensis*) is an annual vegetable crop grown for its foliage. It does not produce a rosette of leaves like spinach does. The intermodes lengthen in the seedling stage, with the main axis growing to a height of about 2–3 m in the drier regions. The leaves are petioled and the blades are cordate, sagitate, or long ovate in shape. Individual leaves are harvested for food rather than the entire plant. Secondary branches arise from the main stem, with the lateral spread extending over 1 m.

Orach is monoecious, and unisexual flowers are borne in clusters. The flowers resemble those of spinach in size and appearance. Two types of seeds are borne on all orach plants—yellow-brown and black. Each of these is again of two types: (a) seed is in a vertical osition and attached to bracts, and (b) seed is in a horizontal position in the perianth and without bracts. Both types of seeds produce plants that look alike, and these in turn will also bear both kinds of seeds. Yellow-brown seeds germinate in 3–7 days; black seeds take more time and have a lower germination capacity.

2. Seed-Production Technology

Culture. In the United States, about 4 kg of bractless seed is sown per hectare in rows about 1.2 m apart. The depth of planting should be less than 6 mm to obtain a better crop stand and higher seed yield. Seed is sown in October–November for overwintering or in April–May. Since blooming and seed development take considerable time, late plantings should be avoided. Plants became well established in the spring, when they are thinned to maintain about 45–60 cm between plants. Fertilizer, irrigation, and weed control are more or less similar to that described for spinach except that irrigation should be held to a minimum. With adequate moisture for seed germination, only one or two irrigations may be necessary to mature a crop.

Roguing. Off-types and other rogues should be removed and destroyed when the plants have made a good vegetative development. Large, dark or grass green leaves characterize Triumph, which is the most widely cultivated variety in the Great Plains region. Gelbe has medium-sized, pale, yellowish green leaves. Lee Giant is the tallest variety, growing over 2–3 m high. Its dark green leaves are the smallest and least fleshy of all the varieties. Deep Blood Red grows less than 2 m high and has dull crimson leaves with petioles and midribs of dark purplish red. A distance of about 400 m or more is necessary between varieties to prevent cross-pollination.

Control of Pests and Diseases.   Several species of aphids and various chewing insects, beetles, and caterpillars attack orach crops seriously. Stomach poisons like arsenate or fluorine compounds can control these insects. The damping off of young seedlings is the most serious disease of orach. It is advocated that orach crops should not be grown in areas where beet diseases are known to be serious.

Harvesting.   Orach plants are cut before the seeds are fully matured and dried. On commercial farms, a combinder is often used to harvest the crop. Close spacing in the row facilitates the mechanical harvesting of orach. The bundles are placed in stacks for drying. Quick and thorough drying may prevent molding and heating of orach seed.

3. Postharvest Operations

How orach is milled depends upon whether the orach seed is to be sold with or without bracts. A seedling machine cannot be used if bracts are to be retained. Both types of seeds are cleaned employing a screening mill, and a gravity separator enables the grower to produce orach seed without bracts. Seed (with bracts still attached) yields range from 3700 to 4500 kg/ha; about 75% of the total seed weight may be lost after the bracts are removed. A further milling process may eliminate 30% of the remaining weight. Yields of recleaned, bractless seeds are about 650–750 kg/ha.

IX. POTATOES

A. Introduction

Both commercial multiplication and maintenance of a potato variety are based on the fact that vegetative reproduction occurs by means of tubers. Consequently, the progeny obtained by planting tubers is identical to the original true seed material from which the variety was derived and identical to all tubers eventually derived from that one seed material apart from mutants that may occasionally occur.

B. Breeding

A new potato variety is usually derived from a true seed material. True seed may be obtained from naturally formed fruits, which contain seeds predominantly resulting from self-pollination. Bumblebees and other insects do visit potato flowers, so cross-pollination is not excluded.

   Growing seedlings until tuber harvest in a greenhouse has the advantage of more efficient control of virus-transmitting aphids. Tubers from such seedlings are usually virus free. The earlier aphids infest the plants, the greater the chance that tubers will become virus infected and, consequently, that many genotypes will be lost. In southern countries with no winter frosts and where potatoes are cultivated all year round, the risk of potato seedlings becoming seriously virus-infected is great.

   One well-established method of maintaining a potato variety is by means of selection of individual plants, preferably from a high or the highest grade or class and subsequent multiplication as a clone for some years, during which a comparison of clones is made. Deviating types and clones that yield less are discarded. In many countries, this work of maintaining potato varieties is guided and controlled by state or semistate inspection and certification services.

C. Seed-Production Technology

Obtaining high yields of seed tubers of the prescribed quality demands techniques appropriate to the environment in which the seed crops are grown. No one environment is likely to be optimal for any particular combination of characteristics required by a grower. The environment will affect seed yield and have a considerable influence on the costs of production: the higher the costs of production, the less successfully will the seed be able to compete with seed grown in more favorable environments.

1. Freedom from Viral Diseases

The first requirement is to start with seed stocks that have been freed from viral infection. The techniques for achieving this are well established. Heat therapy has been used to obtain stocks free from potato leaf roll virus (PLRV) and potato virus Y (PVY), and apical meristem culture has also successfully eliminated viral infections to provide virus-free stocks of important varieties (195).

The maintenance of stocks free from virus is one of the dominant problems in potato seed production; a major source of infection is through aphid vectors, principally Myzus persicae. The most common method of maintaining virus-free stocks is by the selection of locations for seed production where aphids are not favored—usually cool, wet, and windy areas in temperate-oceanic and mountainous climates. Viral spread will occur at a low level and plants with visual symptoms can be removed (rogued).

Insecticides may be used to limit aphid populations, but such measures do not prevent the spread of nonpersistent viruses (e.g., PVY and potato virus A [PVA]), which can be acquired immediately. Aphid populations may be monitored using Moericke traps–shallow pans painted yellow inside and filled with water containing insecticide. Other means of containing viral disease problems include the use of resistant varieties (196).

The multiplication of virus-free plants in vitro from small nuclear seed stocks protected from viral infection by physical/chemical methods may become a practicable proposition following successful laboratory attempts to propagate the potato plant in vitro. The use of shoot tip culture (197) has given rise to 40–100 plants from each potato, and adventitious shoots produced from rachis, petiole, or leaflet blade explants could give rise to large numbers of propagules per parent plant (198,199).

The technique of producing virus-free stock consists of tuber indexing and selecting single-pile units for freedom from viruses and mycoplasms (200).

Detection of Viruses and Mycoplasms. Latent viruses like X and S are detected by serological methods. For detection, a drop of antiserum on a glass slide is mixed with a drop of sap from the leaf of the plant to be tested. The reaction shows agglutination of chloroplast if the virus is present. The viruses Y and A are detected by the A6 test using detached leaves of *Solanum demissum* × *Solanum tuberosum* var. *Aquila*. Detached leaves of A6 are placed in a box containing a moist mass and are dusted with 500-mesh carborundum powder and a few drops of phosphatic buffer solution. Sap from plants is inoculated and the residual carborundum powder is rinsed off with water. In combined tests for viruses Y and A, the leaves are kept for 7–8 days at about 20°C to watch for development of symptoms.

Leaf roll virus (LRV) is detected by the phloroglucinol test. In a modified technique, a piece is taken from near the base of the stem and then transverse sections from the nodal region are put into phloroglucinol solution in concentrated 10–11 N

HCl. After 1 min, the sections are removed and mounted in water. The primary phloem of leaf roll–infected plants are stained orange-red.

Mycoplasms, namely, witch's broom (WB), marginal flavescence (MF), and purple top roll (PTR), are detected visually on the basis of symptom expression. PTR is characterized by rolling and purple-pink pigmentation of the basal parts of the leaflets of the top leaves. MF induces flavascent margins of younger leaves and stunting. WB causes stunting and produces filamentous stems having small leaves and small-sized tubers.

Tuber Indexing. A large number of single-pile units of different varieties raised at the Nucleus Seed Station are inspected for visual symptoms of diseases, varietal purity, and high yields. The healthy plants are harvested individually. Four tubers of each selected plant are numbered individually for greenhouse testing. One eye from each tuber is scooped out and planted in a small pot, and the counterpart is stored. However, the process can be complicated owing to the frequent presence of healthy and infected parts in the same tuber. The main objective is to detect, and eliminate, as much viral/micoplasmal infection as can be done at the time of tuber indexing from the base material so that the infection may be diluted in subsequent stages of nucleus stock multiplication.

Each plantlet from a single leg is serologically tested in the greenhouse simultaneously for PVX and potato virus S PVS using bivalent antiserum. Plants free from S and X viruses are further tested biologically for PVY and A by using detached leaflets of A6 for LF by a histochemical test and visually for WB, MF, and PTR.

Maintenance of Virus-Tested Stocks.

Stage I—The counterparts of the single eyes showing freedom from the above-mentioned viruses is subsequently planted in the field during the low-aphid period in plains or in the hills. Virus-free counterparts are separately planted in rows with a distance of 1 × 1 m so that the foliage does not touch. Each plant is tested serologically for X and S viruses by taking a composite sample of one leaf from each stem. Wherever a positive agglutination reaction is observed, the entire clone is destroyed. The plants are visually examined for other viral and mycloplasmal diseases.

Stage II—All produce of a hill, unit, or clone is separately planted in a plot. For serological testing against viruses X and S, two leaves from each individual plant are taken, with six leaves constituting a sample. Visual inspection is carried out for other diseases. Any positive serological reaction or visual symptom in any of the samples makes it obligatory to rogue out the entire clone.

Stage III—The seed material of each clone selected during stage II is planted in bulk. Viral infection in the third stage is determined by sample testing (serologically for viruses X and S) and visual inspection for other viral and mycoplasmal diseases. The virus-free nucleus stocks are subsequently utilized for production of breeder's seed.

2. Freedom from Fungal and Bacterial Diseases

Production of seed tubers free from infection or contamination with bacterial and fungal diseases has received comparatively less attention than freedom from infection with viruses. Tubers contaminated with disease organisms may not cause disease until after crops are harvested. The conditions at harvesting are important predisposing factors for the expression of disease.

3. Freedom from Nematodes

Ideally, seed tubers should be free from contamination with nematodes, the most important being the golden and white potato cyst nematodes, which are found world-wide. The most effective way of ensuring freedom from contamination is to ensure that seed potatoes are grown on soils free from infestation, which normally requires the use of very wide rotations. Alternative methods of control, such as the use of resistant varieties, has the attendant problems that the multiplication of biotypes to which the variety is not resistant will be favored, whereas chemical control methods are expensive and require special precautions (201).

4. Seed Size

The most important agronomic factor under the control of the farmer is plant density in terms of the number of main stems per unit area. High stem densities may be achieved by using high seed rates of small tubers, which are allowed to develop sprouts toward the end of the storage period, when apical dominance is lost, thus permitting the development of several sprouts per tuber.

5. Physiological Age

Field factors that tend to increase the ultimate physiological age of the planted seed tuber are most likely to operate through their effect on time of breaking of tuber dormancy (202). Early planting, use of sprouted seed, and growing seed in relatively warmer conditions tend to increase the physiological age of seed tubers.

D. Harvesting

The two most important factors related to seed harvesting are time and method of harvesting. Both affect the quality and the yield of salable seed. It is uncommon for seed crops to be harvested when they have senesced naturally. To allow time for the skin to set on the tubers, the tops should be destroyed some 10–14 days before lifting. The tops may be burned off with a flame thrower or destroyed chemically. Pallais et al. (203) reported the optimum harvesting time to be about 10–11 weeks after pollination.

A variety of harvesting techniques are used, ranging from head digging and lifting to completely mechanical harvesting. Whatever the method, the major concern is the avoidance of damage that can lead to infection by pathogenic organisms. Lifting potatoes under very dry conditions should be avoided, as tubers are susceptible to damage when they have lost turgor. At the other extreme, very turgid potatoes are prone to damage during harvesting. Irrigation should be terminated well in advance of harvesting (204). The soil should be reasonably warm, which reduces internal bruising and encourages healing (205). Hide and Lepwood (206) advocated that seed potatoes not be lifted when immature, as they are prone to infection by *Fusarium* species.

Wounds heal rapidly at a temperature range of 2.5–20.0°C, which is essential immediately after harvesting and grading, involving a period of about 2 weeks at 10°C or above (207). After wound healing, the main requirement for seed storage is close temperature control within the range of 2–5°C. The achievement of any particular physiological condition would require the imposition of appropriate temperature regimens, after the break of dormancy.

E. Storage of Seed Potatoes

In most countries, seed potatoes as well as a lare portion of the potatoes to be sold for consumption or processing must be stored for some time. The conditions of storage should be such that a decrease in weight and quality of potatoes is minimal. Depending upon the existing conditions, potatoes can be stored by one of the following methods: clamp storage, compartment storage, refrigerated storage, and storage exposed to light.

1. Clamp Storage

In clamp storage, potatoes are stored in bulk heaps on the ground or in underground holes. This is the simplest and cheapest method, which can be used in areas where temperatures are low enough. Potatoes stored in clamps are protected from rain, cold, and light (to prevent greening). The potatoes may be piled in a heap and covered with layers of straw and soil or plastic. Since the potatoes produce heat and moisture, the heap must not be completely covered to allow some ventilation.

2. Compartment Storage

Small quanties of potatoes can be stored for a relatively short period (2.5–3 months) loose or in bags, crates, or other container in compartments protected from light. If these compartments are not insulated, their mean temperature is the same as the outside mean temperature. Long-term storage may, therefore, not be possible, depending on the prevailing outside temperature. The heap or stack must not be too large so that sufficient ventilation is allowed.

3. Refrigerated Storage

Mechanical refrigeration will be necessary for longer storage periods in high-temperature regions. A common feature of clamp and compartment storage is that there is virtually no ventilation. These simple methods can be used only if the heap or stack is small. However, when larger amounts of potatoes are stored, storage must take place in a building with forced ventilation, fans, and an air-distribution system.

4. Seed Potato Storage with Exposure to Light

A widespread and long-established method of storing seed potatoes is to place tubers in crates or on racks in thin layers and expose them to light, either natural or artificial (e.g., fluorescent tubes). Exposure to light favorably affects the quality of the seed, especially at temperatures of 10–15°C, and potatoes can be stored well for 3–4 months.

Light has several positive effects on the quality of seed potatoes. It retards the physiological aging of tubers by decreasing the number and robustness of sprouts. Light promotes greening of the tuber, thus decreasing its susceptibility to fungal diseases. In general terms, where seed potatoes are to be stored for a relatively long period, it is best to store them in the dark at 2–10°C, whereas they can be best stored in the light at 10–20°C. At temperatures above 20°C, they may be stored in the light but only for short periods of 3–4 months.

F. Seed Plot Technique of Producing Potato Seeds

1. Introduction

The essentials of the seed plot technique for potato seed production in the plains are (a) start with disease-free seed stocks, (b) select suitable locations, and (c) fulfill the requirements for certified seed production, such as roguing and crop inspection.

Land Requirements.   See potatoes should not be produced on land where potatoes are likely to be infected by brown rot, wart, or nematodes, and where potatoes were produced during the previous year. The soil should be well drained, aerated, deep, and slightly acidic (pH 5.2–6.4).

Isolation.   A minimum isolation of 20 m for foundation seed and 5 m for certified seed should be provided around the field to separate it from fields of other varieties.

Culture.   Seed is sown from around September 20 (when rainfall is low) to October 15; delayed planting results in poor yields. About 25–30 Q/ha of potato seed will be sufficient, using normal-sized tubers (4–6 cm). Bhatt et al. (208) reported that 100–seed weight, imbibition, percentage germination, and speed of germination were highest for large-sized seeds. The use of true potato seed is recommended for marginal areas where tuber quantity is more important than quality (209). Sexual seed of potato or true potato seed (TPS) is mostly virus free, and offers a practical alternative to seed tubers (210). The use of larger tubers (55–60 g), application of 40 ppm GA, and pruning to three remaining clusters was found to increase the number of flowers and seed pollination in potato (211). Whole tubers should be used for better results. The tubers are planted 3–4 cm deep in soil having adequate moisture. After October 15, when the temperature goes down, cut tubers can also be used for planting. Each piece of potato tuber used for planting should have at least two to three emerging eyes weighing about 40 g. The planted tubers should not come into direct contact with fertilizers, which should be placed right below the seed. Spacing of 60 × 15–20 cm is recommended. A potato planter may be used, which places fertilizer, insecticide, and seed in proper proportion to each other. If true potato seed is sown at high temperature, genotype is a crucial factor. Sufficient seed storage (more than 18 months) may be essential, and seed priming is more effective than the standard GAI500 treatment (212).

Irrigation and Weed Control.   Potatoes require light but frequent irrigation; the first irrigation should immediately follow emergence. Subsequent irrigations should be applied at regular intervals, which are restricted after the crop has tuberized well. Irrigation is stopped after the third week of December (10–15 days before killing of haulms).
   Seed crops are kept free from weeds. At least one earthing-up is required, which is carried out when plants attain the height of about 15 cm. Exposed tubers should be covered with soil during earthing-up. The haulms must be cut by the end of December or by the first week of January.

Roguing.   Careful roguing is necessary for producing a high-quality potato seed. First roguing is done 25 days after sowing to remove all virus-infected plants and all plants apparently belonging to other varieties, which can be identified from foliage color. A second roguing should be carried out when the crop is fully grown (about 50–60 days after sowing). At this time, tubers are formed, and therefore while roguing not only the upper portion of plant but also all off-type tubers should be removed carefully. Also, at this stage the virus-inffected plan, if any, should be removed. A third and final roguing is done just before cutting the foliage. At this stage, all virus-infected plants and off-type plants, if any, along with their tubers are carefully removed.

Thakur and Upadhya (213) showed that the use of tuberosum as female and andigena as male parents were important features in the development of viable and low-cost technology for hybrid true potato seed production in South Asia. Almekinders and Wiersema (214) reported that application of N during the flowering period did not affect flowering but significantly decreased the numbers of seeds per berry. Pruning lateral stems enhanced flowering of the main stem and decreased the total number of flowers per plant (215).

Harvesting. The crop is ready to harvest 10–15 days after haulm cutting when the skin of the tuber has hardened. Premature harvesting produces poor-quality, soft tubers, which will not withstand transportation and storage. The soil moisture content should be optimum at the time of potato digging to obtain clean tubers. Potatoes are harvested by suitable available equipment, avoiding mechanical damage. The harvested tubers should not be left exposed to the hot sun for a prolonged period; they are immediately lifted and placed in an airy shed in piles for 7–10 days for superficial drying and hardening of the skin. If sheds are not available, piles may be made in the field and tubers covered with dry haulms.

2. Postharvest Operations

The properly cured potatoes are further graded based on shape, color, and depth of eyes, discarding off-types. In addition to the off-types, tubers with cuts, bruises, cracks, or other mechanical damage or showing visible symptoms of such diseases as late blight, dry rot, wet rot, scab, and black scurf are also removed. After sorting and grading, the seed potato is placed in clean labeled hessain bags (50-kg size). The seed potatoes are then moved to the end-use areas or for cold storage. If the ambient temperatures are above 32°C, the seed potatoes should first be placed in precooled chambers or in a cool place for preconditioning, and then stored in cold storage at 2.2–3.3°C and 75–80% RH. Periodic inspection of seed stocks in cold storage is necessary. Turning of bags during the rainy season help to improve aeration. Thakur and Upadhya (216) developed a method for extraction of true potato seed from the berries. Mature berries harvested 45 days after pollination were stored until they were macerated with a hand-operated juice extractor and debris was removed using 3.13-mm mesh netting. Seed and pulp mass were then treated with 10% HCl for 20 min to free the seeds from the remaining pulp and surface-sterilize the seeds. Seeds were dried over a muslin cloth stretched over a wooden frame and then stored in double polyethylene bags kept in desiccators under refrigeration. This method can be used for commercial-scale extraction of true potato seed. The effect of HCl (10%) treatment on germination and shoot growth of TPS-2 open-pollinated true potato seed is given in Table 13.

A satisfactory method of extracting potato seeds from berries has been described (217) in which the macerated berries are placed in a modified funnel similar to the apparatus used by nematologists to extract nematode cysts from soil suspensions (Fig. 13). The apparatus is basically a funnel with a cylinder attached and a rubber bung with coiled copper tubing fitted into the neck of the funnel. The macerated berries

Table 13 Effects of Hydrochloric Acid (10%) Treatment on Germination and Shoot Growth of TPS-2 Open-Pollinated True Seeds

Treatment duration	Seedling germination (%)	Seedling shoot height (cm)
20 min	97.33	2.17
40 min	97.50	2.22
60 min	96.17	2.28
80 min	97.33	2.20
Control	97.17	2.10
LSD (0.05)	0.84	0.23
cv (%)	2.13	8.49

LSD = least significant difference.
Source: Ref. 216.

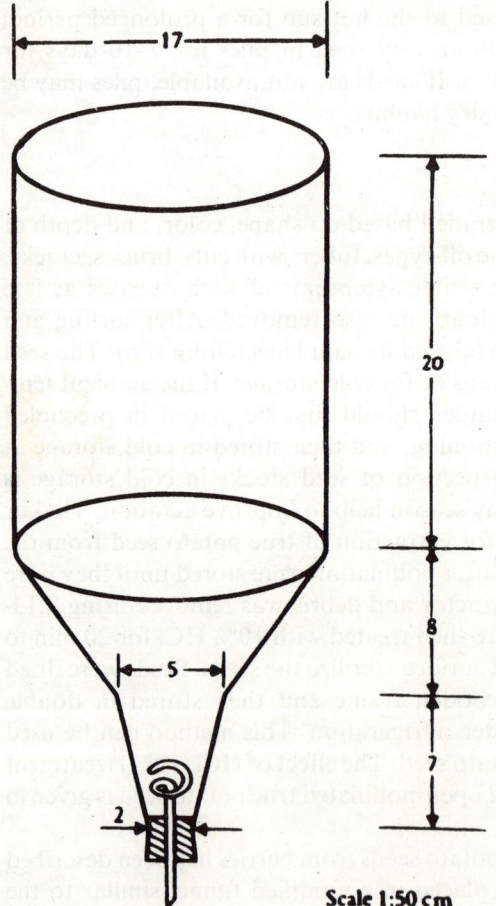

Scale 1:50 cm

Figure 13 Diagram of apparatus to extract true potato seeds from macerated berries. (From Ref. 217.)

are placed in the top, and as water passes through the coiled tubing, it creates a cyclonic agitation. The scum and fruit debris overflow at the top while the seeds are collected at the bottom of the apparatus and are retrieved by removing the bung.

X. MINOR VEGETABLE CROPS

Turnip and radish are the important root crops of the family Cruciferae. Rutabaga, also known as Swedish turnip, is a minor vegetable crop belonging to the same genus as turnip. Although horseradish is normally propagated by root cutting, it has been shown that seed can be produced. Okra is an important vegetable crop in India and other tropical countries. New Zealand spinach and rhubarb are vegetables of minor importance.

A. Turnip

1. Introduction

Botany.   Turnip (*B. rapa*) is a cool-season vegetable grown in deep soils ranging from sandy loams to clay loams in texture. With the exception of certain Asian annual varieties like Shogoin, turnip is a biennial vegetable crop. The fleshy edible root develops in the first year. Leaves arising from a very short stem develop in the form of a rosette. Leaves forming in the first year vary considerably in size and shape depending upon the variety and time of development. In the second year, alternate leaves develop from the stalk and branches of the inflorescence.

The exterior of the fleshy root below the soil surface may be white or yellow; depending upon the flesh color, aboveground it may be red, purple, white, yellow, or green. The structural arrangement of flowers is similar to that of other cruciferous crops. The flowers of the white-fleshed varieties are bright yellow and those of yellow-fleshed varieties are pale orange to yellow in color. Pollination occurs mainly through honeybees. Maintenance of honeybee hives adjacent to the fields ensures pollination of all flowers, which is necessary for good seed setting.

Cultivars.   The most popular cultivars of turnip grown in the 1950s were Purple Top Globe, White Milan, White Flat Dutch, White Egg, Yellow Globe, Golden Ball, and Yellow Aberdeen. The following four cultivars of turnip have been released in recent years:

> Charlestowne is aphid and bolt resistant and has cold-tolerant foliage with a small, white root and dark green leaves, and is similar to Shogoin.
> Roots is open-pollinated with large roots, dark purple tops, and white globes, and is similar to Purple Top White Globe.
> Royal Crown is an F_1 hybrid with a semiflat root. The lower portion of the root is white with an attractive purple top.
> Shortop is an early variety with short tops (25–30 cm).

The Asian or local cultivars of turnip grown in India are Variety 4 (white), Variety 4 (red), Pusasweti, and Pusa Kanchan. The last two have been released by the ICAR, New Delhi (47). The European, or temperate, cultivars of turnip grown in India include Purple Top White Globe, Golden Ball, Snow Ball, and Early Milan. The ICAR-released cultivars of this group are Pusa Chandrima and Pusa Swarnima.

2. Breeding

Successful inbreeding in turnips is difficult because, like all other cruciferous vegetables, turnips are also often self-incompatible. This difficulty may be overcome by bud pollination. It is possible to select lines that are naturally self-fertile.

3. Seed-Production Technology

As with most biennial root crops, turnip seed can be grown by two methods: root-to-seed and seed-to-seed. In the root-to-seed method; the roots are dug and stored during the cooler months of the winter; in seed-to-seed method plants are allowed to overwinter in the field. The former method is commonly used to produce stock seed, and the latter method, which is less expensive, is used to produce market seed.

Root-to-Seed Method.

CULTURE.   In the first year, the root-to-seed method differs little from that of growing turnips for market. Seed is usually sown from June 15. Slightly more seed (3–4 kg/ha) is, however, required than for a market crop, since the roots need not be as large as those grown for sale. Stecklings dug from 1 ha of land can cover an area of 6–10 ha for seed crop the following spring. Stecklings are dug in late fall or early winter, and are stored in cellars or ventilated pits for planting in the following season. A storage temperature of about 0°C or just above and a relative humidity of 90–95% to prevent shriveling have been recommended.

Turnip stecklings are replanted early in the spring. The land is preferably plowed in the fall and well prepared. Roots are set in the field like those of carrots. Mechanical planting may be adopted for a commercial seed crop. Turnip roots should be pushed into the soft soil to a depth of 10–15 cm. The usual spacing recommended for turnip is about 75 cm, but this can vary from 60 to 90 cm between rows and from 20 to 45 cm between plants in a row.

ROGUING.   Off-type plants should be removed in the field prior to harvest. Even during harvesting, small, misshapen, diseased, or any other undesirable roots should be discarded. For stock seed, the stecklings should be examined critically for varietal type based on characters such as shape and external and internal color.

Seed-to-Seed Method.

CULTURE.   The seed is sown directly in the field where the seed crop is to be produced. This method eliminates the work and expense of lifting, storing, and replanting of stecklings. Critical roguing is also not feasible in this method. Practically all of the market seed in the United States is produced by the seed-to-seed method.

The seed is sown in late summer or early fall depending upon the variety and locality. The planting date is adjusted so that the crop will attain sufficient growth before winter to be large enough to withstand any heaving caused by alternate freezing and thawing of the soil. The crop should not be mature enough to have market-size roots, because larger roots are more susceptible to freezing injury. Spacing between rows varies from 45 to 90 cm, with closer spacing being more common and profitable.

The field selected for turnip seed crop should be such that no standing water accumulates following heavy winter rains. Localities having the benefit of snow cover often make good sites for over wintered turnip seed crops. Areas in which wild mustard and wild turnip are abundant should be avoided.

ROGUING. A high-quality, true-to-type stock seed should be used in seed-to-seed plantings, since roguing is usually limited to removal of plants with off-type foliage. Any stock seed having a significant proportion of off-type roots should be avoided as a planting material in the next generation.

Cultural Practices Common to Both Methods.

FERTILIZER, IRRIGATION, AND WEED CONTROL. Application of 200–300 kg of 16-20-0 fertilizer per hectare or its equivalent as a side dressing has been recommended. Higher doses of nitrogen (exceeding 150 kg/ha) may cause lodging of turnip seed crop. Application of balanced fertilization (N, P, and K) is, therefore, important. Chakrabarti (218) reported that, turnip plants responded to nitrogen much more than to phosphorus. Turnip requires moderate, uniform soil moisture for best development. Overirrigation should be avoided. Appropriate cultivation for weed control is necessary. Weeds such as wild mustard and wild turnip should be completely eliminated; if necessary by hand pulling. These weeds, if allowed to mature and harvested along with the turnip seed, will be extremely difficult to separate. With either the seed-to-seed or the root-to-seed method, a minimum isolation distance of about 400 m between any two varieties has been recommended. The turnip seed crop should also be well isolated from mustard, Chinese cabbage, rutabaga, rape, and radish.

CONTROL OF PESTS AND DISEASES. Turnip is susceptible to the same insect pests to which cabbage and other crucifers are susceptible. Aphids can be particularly injurious to turnips in both years. In the second year, inflorescence and developing seedpods may be affected. Turnip is subject to several diseases affecting the leafy cruciferous crops, with club rot disease being the most serious.

HARVESTING AND POSTHARVEST OPERATIONS. Turnip seed crops are usually ready to harvest by midsummer or earlier. The seed shatters as early as that of cabbage, and the method of harvesting is the same as described for cabbage. Crops should be harvested while the bulk of the pods are still yellow but not dry. The seed stalks are cut off near the base with a sickle or mowing machine. The plants are rolled into windrows and threshed as soon as the pods are dry. In Utah and Idaho, all-purpose combines are used to harvest turnips. Since the turnip seed is subject to cracking and hulling injury, cylinder speed should not be higher than is necessary to thresh the seed. Milling and cleaning operations are essentially the same as those for cabbage seed. The average seed yields of turnip range from 450 to 700 kg/ha. Under favorable conditions, higher yields of up to 2200 kg/ha may be obtained.

Lepori and Quagliotti (219) studied the effect of storage conditions on viability of turnip seed stored at 30°C, room temperature (4.5–28.0°C), and 40°C and relative humidity 35,55, and 75%. These tests showed progressive seed deterioration as evidenced by an increase in average germination time, appearance of abnormal seedlings, and reduction in the percentage of seed germinated. A relative humidity of 35% was found to be most suitable for long-time seed storage even at relatively higher temperatures (Figs. 14–17).

Premature bolting significantly influences turnip yields. An annual turnip variety, Shagoin, is particularly prone to premature bolting. Sakr (220) concluded that exposure to low temperature is essential to produce seed under field conditions. Continuous exposure to low temperature (4.4–10.0°C) initiated and developed seed stalks of turnip more effectively than any other method when the temperature was alternatively low and high (15.5–21.1°C). A temperature of 0–4.4.0°C delayed the elongation

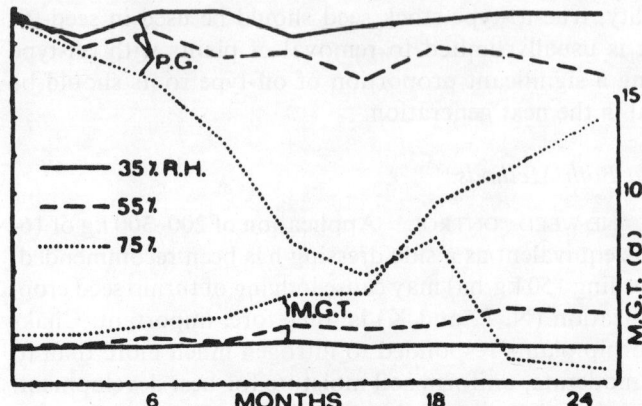

Figure 14 Effects of varying relative humidity on percentage germination and mean germination time (days) of turnip seed stored at room temperature. (From Ref. 219.)

but not the initial differentiation of flower stalk primordia. The roots held at this temperature for 3–4 months were in much better condition than those held at 12.8–18.3°C. Also, roots held at lower temperatures produced larger seed stalks than those held at higher ones. Turnip seed production thus is not likely to be successful in regions where lower temperatures (4.4–10.0°C) do not prevail for a period of several weeks (longer periods being more desirable). Seed yield was higher with larger stecklings and closer spacing (221).

B. Rutabaga

1. Introduction

Rutabaga (*B. napobrassica*), or Swedish turnip, is a more important crop in northern regions. The botany, culture, and seed production of rutabaga closely resemble those

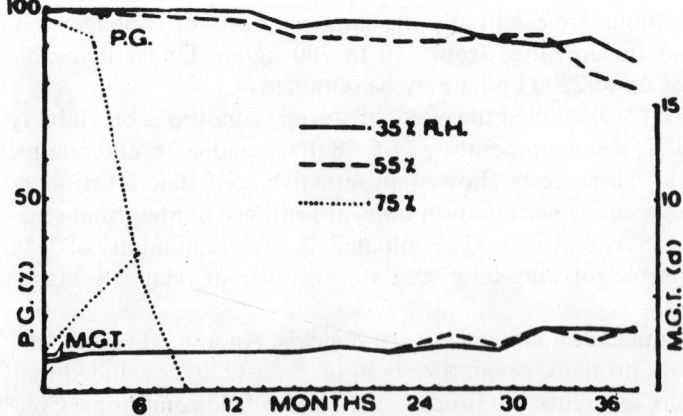

Figure 15 Effects of varying relative humidity on percentage germination and mean germination time (days) of turnip seed stored at 30°C. (From Ref. 219.)

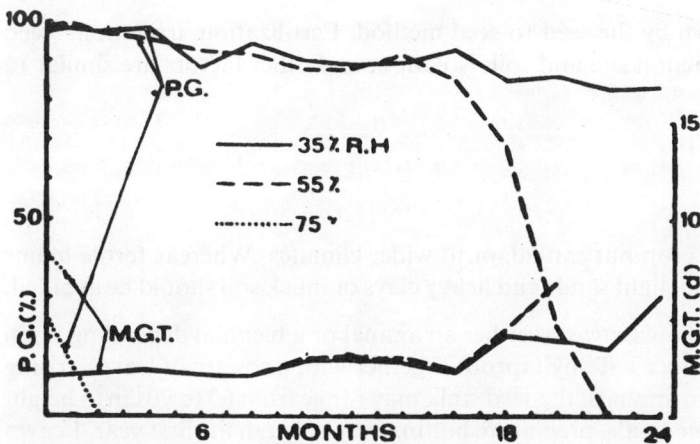

Figure 16 Effects of varying relative humidity on percentage germination and mean germination time (days) of turnip seed stored at 40°C. (From Ref. 219.)

of turnip. Rutabaga is slow growing and produces a larger plant and larger fleshy root than turnip. Turnip is a diploid with 20 chromosomes, whereas rutabaga is an amphidiploid with 38 chromosomes. The exterior color of the fleshy root above the soil surface may be purple, green, or bronze depending upon the variety. Root flesh may be white or buff-yellow with a similar exterior color below the soil. The flower color is correlated with the flesh color. White varieties have bright yellow flowers, and yellow varieties have orange-yellow flowers, similar to those of turnips.

2. Seed-Production Technology

The culture, roguing, and harvesting of rutabaga are the same as those for turnip except that the time of planting is earlier (early June). Rutabaga may also be planted in

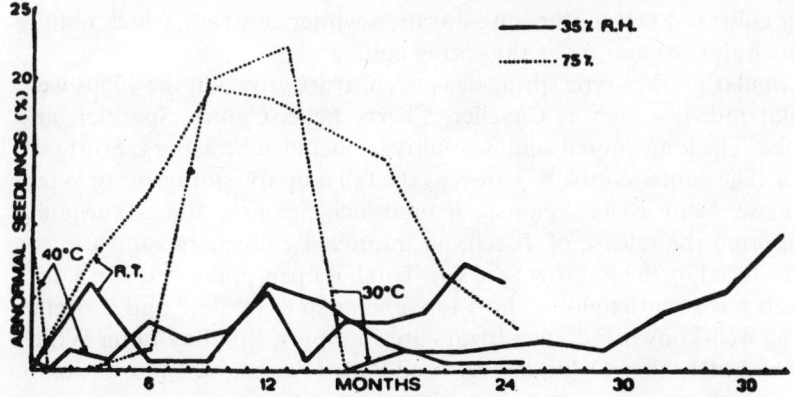

Figure 17 Effects of varying relative humidity and temperature on the percentage of abnormal seedlings at room temperature (4.5–28°C). (From Ref. 219.)

early August when grown by the seed-to-seed method. Fertilization, irrigation, weed control, choice of production site and soil, isolation, and other factors are similar to those for turnip.

C. Radish

1. Introduction

Radish is a cool-season crop but can adapt to wider climates. Whereas fertile loams favor the growth of radish, light sands and heavy clays or muck soil should be avoided.

Botany.   Radish (*Raphanus sativus*) is either an annual or a biennial depending upon the type. Both types produce a fleshy taproot together with a rosette of leaves arising from a shortened stem. In annuals, the seed stalk may range from 60 to 90 cm in height during the first year. In biennials, premature bolting may occur in the first year. Leaves are usually pinnate and partly divided and covered with stiff hairs.

As in the turnip, the fleshy edible portion of the radish root develops from both the primary root and the hypocotyl. Roots vary greatly in size, shape, exterior color, and length of time that they remain edible. The edible portion in annual varieties varies from about 2.5 to 10.0 or 12.5 cm in length. Shape ranges from oblate to long tapering, and exterior color may be white or various shades of scarlet and crimson. The roots o fbiennial varieties may be 20 cm or more in length, and exterior color may be black, purple, or the common colors of the spring varieties. The flesh color of American varieties is white.

The inflorescence is the terminal receme of the crucifers, and the flowers are white, rose, or lilac. Pollination occurs primarily by honeybees. Mature radish seed is first yellowish and turns reddish brown with age. The 1000–seed weight of radish is about 13.3 g as compared to 3.2 g for cabbage, broccoli, Brussels sprouts, and cauliflower.

Cultivars.   There are two types or groups of radish varieties grown for seed purpose: annual types and biennial types, including the less popular Chinese and Japanese or so-called winter radishes. Based on the season of the year in which the radish is grown, radish cultivars can be grouped into three types: (a) the quick-growing spring culti-vars, which include most of the commercially grown cultivars of radish; (b) the late-maturing summer cultivars; and (c) the long-duration winter cultivars, which require about twice as much time to mature as the spring cultivars.

The most popular market-type spring-season cultivars grown in the 1950s were bright red globular radishes such as Cavelier, Cherry Belle, Comet, Sparkler, and Early Scarlet Globe. The long-rooted summer cultivars included Strasburg, Stuttgart, and White Vienna. The winter cultivars grown as the fall crop for storage were White Chinese, China Rose, Long Black Spanish, Round lack Spanish, and Sakurajima. Tigchelaar (3) reported the release of Torch, an improved cultivar of radish in the United States introduced by the Asgrow Seed Co. Torch is open-pollinated 2 to 3 days earlier than Saxa. It has a scarlet globe, short top, and crisp white flesh and is similar to Red Prince. The well-known Indian cultivars are Jaunpuri, Bombay Long White, and Long Red. The IARI, New Delhi, has released Pusa Himani, having white skin, crisp flesh, and sweet flavor. Several Asian or tropical cultivars of radish such as Japanese White, Pusa Desi, Pusa Chetki, which include all the early varieties grown in the United States, and Pusa Reshmi are also grown extensively in India. The Tamil

Nadu Agricultural University, Coimbatore, released Co-1, an improved cultivar of radish to be grown in the plains (222).

2. Breeding

Existing strains of radish with many types, forms, and colors can be further improved by inbreeding and pure line selection. As in case of other crucifers, self-incompatibility is a major problem, which can be overcome by following a system of bud pollination. Selection can be made toward self-compatible lines to eliminate the necessity of bud-pollination in later generations. Pundir et al. (223) found that no pollen germinated following self-pollination within the same flower or within flowers of the same inflorescence of a local cultivar. However, following cross-pollination within the same plants, pollen germination and seed set were 44 and 70%, respectively, and following open-pollination, pollen germination and seed set were 40 and 62%, respectively.

3. Seed-Production Technology

Radish seed can be produced by either the root-to-seed or the seed-to-seed method, with seed being harvested within a year with either method. Full-sized edible roots are formed in both methods. In the United States, radish seed production is mostly limited to that of the annual type, which is relatively simple compared with the production of biennial crops.

Root-to-Seed Method.

CULTURE. This method is used primarily to produce high-quality stock seed. The seed is sown early in the spring in narrow rows or may be broadcasted. The seed rate may be as high as 10 kg/ha depending upon the spacing and method of sowing. The stecklings usually reach edible maturity within 3–4 weeks, when they are dug out, graded for type, and reset in another field about 15 cm apart in rows 1 m apart. Nauriyal and Lal (224) treated radish plants (cv. Japanese White) spaced at 45 × 45, 60 × 60 cm with GA at 50 or 75 ppm, MH at 25 or 50 ppm, or CCC (chlormequat) at 250 or 500 ppm (as foliar sprays) during flowering. Seed yields were highest (10.85 Q/ha) when plants were spaced at 60 × 60 cm and treated with chlormequat; at 500 ppm as compared with the control yield of 7.92 Q/ha. Rawat and Singh (225) obtained the highest seed yield of radish cv. Pusa Rashmi (6.5 Q/ha) when plants were spaced at 45 × 45 cm and received 100 kg N/ha in split doses. GA had no effect on the germination percentage (93%), but morphactin decreased it by 5.5%. Both growth regulators promoted radicle and hypocotyl extension and increased radicle and hypocotyl dry weight (DW) in the first 96 hrs after germination. Radicle DW increased as growth-regulator concentration increased (226).

ROGUING. Off-type radish roots can be rogued critically in the root-to-seed method of producing radish seed. The shape, size, and color of roots are used to rogue out off-type radishes.

Seed-to-Seed Method. Nearly all commercial market seed is grown using this simpler and less time-consuming method. The annual radish crop is sown as early as possible in the spring. Seed yield was found to be highest (12.83 Q/ha) when the crop was sown during the second week of October. Seeds produced from October-sown crops also had the highest germination percentage (227). About 3 kg seed/ha is sown in rows 70–90 cm apart. The seed may be drilled in rows 2.5–5.0 cm apart using a seed drill. It may be necessary to mix some inert material with the seed to avoid too thick a crop stand. Seeds weighing less than 5 mg should not be used for steckling production but could be used for raising a crop for consumption (228). Only seed grown by the root-to-seed method and subject to rigorous roguing should be used to raise seed-to-seed crops.

Cultural Practices Common to Both Methods.

FERTILIZER, IRRIGATION, AND WEED CONTROL.   Irrespective of method of culture, adequate fertilization, uniform and moderate irrigation, and some cultivation for weed control in the seed crop are essential. Seed yield and the 1000–seed weight were statistically higher where N was applied at 50 kg/ha along with P at 25 kg/ha (229). Rastogi et al. (230) showed that the mean seed yield was highest (11.36 Q/ha) when plants were spaced at 60 × 45 cm and received 60 kg N/ha. The highest 1000–seed weight (12.44 g) was obtained with 150 kg N/ha and a plant spacing of 60 × 60 cm (231). In experiments at Ludhiana, India, radish cv. Punjab Safed stecklings comprising an intact root system or 75 or 50% of the total were planted on November 15, December 1 or 15, or January 1. Whole root stecklings planted early (November 15 or December 1 produced plants with luxuriant vegetative growth and a high number of branches (the major seed yield component). Planting in the second half of November was optimum for seed yield, but there was a very strong association of seed yield with steckling size. A significant interaction between planting date and steckling size suggested that increasing the steckling size could make up for a loss in seed yield as a consequence of delayed sowing. However, the planting date had no significant effect on seed quality (231a).

Weeds such as wild mustard, wild turnip, or wild rutabaga and wild radish should be removed carefully from radish seed fields because of possible cross-pollination and cleaning problems. Gambhir et al. (232) reported that an application of 1 kg dalapon, 1.0–1.5 kg fluchloralin, and 2.0–2.5 kg nitrogen per hectare 1 day after transplanting radish stecklings markedly decreased the populations and DW of weeds, including *Cyperus rotundus, Cynodon dacrylon, Poa annua, Chenopodium album, C. murale, Convolvulus arvensis,* and *Meliotus* sp. and resulted in increased seed yields of radish. Fluchloralin was found to be the most effective weedicide, followed by nitrogen and dalapon. Application of 1.5 kg Alachlor per hectare controlled weeds effectively, producing yields similar to weed-free plots with two hoeings. In trials at Ludhiana in 1989–1992, radishes were grown for seed production and given N at 0, 37.5, 75.0, or 112.5 kg/ha and P at 0, 20, or 40 kg P_2O_5/ha. All the P and half the N was applied as stecklings were planted out and the remaining N was applied 1 month later. Seed yields increased as the N rate increased up to 75 kg/ha and as the P rate increased up to 20 kg/ha (232a).

ISOLATION.   A distance of at least 400 m should separate radish varieties from each other. For stock seed, this distance should be a minimum of 1.6 km.

CONTROL OF PESTS AND DISEASES.   Owing to its rapid growth, radish crop escapes many slowly developing pests and diseases. Radish is subject to the widely spread club rot disease, but it is not affected by yellow disease. Radish is particularly susceptible to black rot and alternaria blight. Avoiding sites of production where the trouble is known to exist can control black rot. Blight (*Alternaria raphani*) can be checked by using disease-free seed.

HARVESTING AND POSTHARVEST OPERATIONS.   Annual-type radishes are usually harvested later than turnips during August or September. The plants are allowed to mature fully before harvesting, since there is no natural dehiscence and immature pods are difficult to thresh. Plants are cut when most of the pods are brown with a moving machine and forked into small piles to dry. The drier the pods, the more easily will they break open during the threshing process.

All-purpose combines or machines equipped with rollers, such as bean threshers, are used for threshing. Pods are subjected to a gentle crushing motion releasing the seeds. The threshed seeds should be dried thoroughly to avoid possible damage in the sack. After drying, seed should be properly cleaned to remove broken or sprouted seeds. Average seed yields range from 500 to 1000 kg/ha; higher yields of up to 1300 kg/ha can be obtained under favorable conditions.

The incubation of seeds of radish cv. Early Scarlet Globe with 10 µM aspirin resulted in an increase in the temperature range for germination (232b). The analysis of the percentage germination and germination rates indicated an increase in the optimum temperature from 21.4 to 26.0°C, although at 32.6°C, 80.8% of seeds germinated with aspirin and there was no germination in the control. The analysis of the kinetics of seed germination indicated that aspirin treatment decreased the enthalpy of activation of the process. The aspirin treatment also resulted in the synchronization of seed germination. On the basis of our results, we propose aspirin application in practice to increase the tolerance to high temperature and to synchronize seed germination in radish cv. Early Scarlet Globe.

D. Horseradish

Horseradish (*Armoracia rusticana*) can be induced to produce seed. Weber (233) found some clones of the common horseradish producing ovules in about 5% of the ovaries. Weber pollinated these plants with the pollen from two clones of the Bohemian horseradish having fertile and functional pollen. A horseradish strain producing seed was obtained from this cross.

The common broad-leafed and hollow-petioled horseradish is 100% susceptible to mosaic virus, but the Bohemian horseradish, which is smooth, narrow leaved, and solid petioled, lacking in both root quality and yield, is highly resistant to white rust and mosaic. It is possible to develop a resistant common type of horseradish with higher yield and quality by breeding.

E. Okra

1. Introduction

Botany. Okra, bhindi, or Lady's finger (*Abelmoschus esculentus* L.) is native to tropical or subtropical Africa. It requires a long and warm growing season and grows best in fertile sandy loam soil. It is mainly grown on plains for both market and seed. Okra belongs to the Malvaceae, or mallow, family and forms a fairly heavy taproot. The seed is dark brown in color and round and medium in size. The 1000–seed weight is approximately 52.2 g.

Cultivars. Cultivars with spineless pods range in color from creamy to dark green. Many cultivars produce pods with prominent white ridges, but a few produce smooth pods. Perkins Spineless and Dwarf Long Pod are dwarf cultivars producing green ridged pods about 17.5 cm long. Clemson Spineless is of intermediate height and produces long green pods with moderate ridging. Louisiana Green Velvet is tall with green pods with moderate ridging. Perkings Mammoth Long grows to a height of about 2 m and produces green pods about 20 cm long with prominent ridges. White Velvet produces creamy white pods, and Louisiana Market has smooth pods. Long Green is the most popular cultivar in the United States.

Pusa Makhamali, Pusa Sawani, and Bhindi No. 13 are the most popular okra cultivars grown in India. Perkins Long Green is also grown in northern hilly areas. Both Pusa Sawani and Pusa Makhamali are smooth, bristleless, and green in color. Pusa Sawani is fairly mosaic tolerant and is edible when it grows to about 20–25 cm long. Other cultivars released by the IARI, New Delhi, are Padra Shankar Palli and T1 (Uttar Pradesh). Cultivars suitable for processing are Lucknow Dwarf, Long Green, Velvet Green, Smooth Velvet, Long White, and Red Wonder, all of which are, however, susceptible to mosaic and other diseases (37). Punjab Padmini, released by Punjab Agricultural University, has pods of edible size (10 cm) 4 days after flowering (234).

2. Breeding

Breeding has led to development of spineless okra varieties. Okra is normally self-pollinated, but in the Lady Finger variety, about 8.75% crossing was recorded. Hybrid vigor has been reported in certain crosses of okra.

3. Seed-Production Technology

Culture. Seed is sown in rows 90–120 cm apart, and plants are spaced about 30 cm apart within rows. Soaking seeds in 250-ppm cycocel for 10 hr has been reported to give the highest average fruit yield per plant (235). Planting should be delayed until the danger of frost is over. In the southern United States, seed is sown from April to mid-June. In India, it is sown on ridges 45–60 cm apart depending upon variety, with the dwarf type being more closely spaced. Seed may be soaked for 24 hr to soften the hard seed coat and induce better germination. Maximum seed yield of 1844.12 kg/ha was obtained when the crop was grown at 45×30 cm on June 20 (236).

Fertilizer, Irrigation, and Weed Control. Adequate fertilization, uniform irrigation, and appropriate weed control are essential. The highest mean seed yield (11.84 Q/ha) was obtained from plants spaced at 60×40 cm and receiving 60 kg N/ha (237). Application of 100 kg N and 30 kg P_2O_5 per hectare was found to produce a satisfactory seed yield (760 Q/ha) (238). The first irrigation is given between the furrows immediately after sowing. The subsequent irrigations may be applied at intervals of 5–6 days in hot weather and 10–14 days in cool season. Seed-germination percentage decreased with an increase in the number of rainy days (239). The early crop requiresfour to five weedings and the late crop two to three weedings (47).

Roguing. Off-type plants should be removed well before the flowering stage. Although okra is self-pollinated, a certain amount of cross-pollination occurs because of its attractive flowers.

Control of Pests and Diseases. Okra is susceptible to corn earworm or boll worm (*Heliothis annigera*), which is a pod-borer, southern green stinkbug (*Nezara viridula*), and other plant bugs that extract plant juice. A 10% DDT dust or spray can control these pests. In India, spotted boll worm (*Earias* sp.) and jassids (*Empoasca devastans*) are the serious pests of okra. The use of 0.1% BHC, 0.02–0.03%, endosulfan, or 0.02% parathion can effectively control these pests.

Yellow vein mosaic virus and powdery mildew (*Erysiphe polygoni*) are the important diseases of okra: Spraying of 0.02% oxydematon methyl, 0.05% dimethoate, and 0.05% monocrotophos at 10-day intervals controls mosaic. Dusting with fine sulfur as soon as the disease is noticed can check powdery mildew.

Harvesting and Postharvest Operations. In the United States, the okra crop remains in the field until plants have frozen and become thoroughly dry. The crop is then harvested with a combine. After threshing, the seed is cleaned with ordinary seed-cleaning equipment. The average seed yields of Parking Long Green and Pusa Makhamali in India are about 1045 and 2742 kg/ha, respectively. Hooda et al. (240) treated seeds of Pusa Sawani with Cytozyme (containing chelated micronutrients, enzymes, and growth hormones, mainly the cytokinins) at 5.0, 7.5, or 10% and/or Cytozyme was applied to the foliage at 0.75, 1.0, or 1.25% at 21, 42, and 63 days after sowing. Seed treatment plus foliar application produced the highest yields of the heaviest seeds of okra. Bhatnagar and Sareen (241) obtained the highest yield of okra (cv. Pusa Makhamali) when plants were treated with chlorflurecol at 300 ppm. Increasing rates of chemical treatment reduced plant height and diminished the formation of hermaphroditic flowers but increased the number of pistilate flowers.

F. New Zealand Spinach

1. Introduction

New Zealand spinach (*Tetragonia expansa*) belongs to the family Aizoaceae. The plant is a shrublike decumbent, with leaves that are alternate, fleshy, and triangular with a short petiole. Rowers are small and axillary. The hard, angular, particularly spiny pod (fruit) contains several seeds.

2. Seed-Production Technology

Seed is sown in rows 75–90 cm apart 2.5–5.0 cm deep, and plants are thinned to space them 30–45 cm apart. Flowering, which extends over a long period, starts near the plant base and proceeds toward the tips of the branches as they grow. Seed ripens over a prolonged period of time. The crop is harvested when the seed is mature. Plants are cut at the base and dried on canvas. Seed may be lost due to shattering. The harvested seed requires little cleaning.

G. Rhubarb

1. Botany

Rhubarb (*Rheum rhaponticum*) belongs to the family Polygonaceae. It is a perennial vegetable crop primarily grown in cooler climates. The plant forms a thick, woody rhizome and has a fibrous root system, the crown being just below the soil surface. The inflorescence is paniculate, having several racemelike branches which arise in the axils of the upper leaves. The stigma is not receptive until after the anthers have shed their pollen.

2. Seed-Production Technology

The seed is planted in the spring in fertile and well-prepared soil. The plants are thinned to grow about 7.5–15.0 cm apart. The roots developed may be transplanted early in the fall in milder climates or early in the spring in severe climates. Rhubarb responds well to high soil fertility. The seed stems are cut as soon as the seed is matured and dried thoroughly. The crop is then threshed with a combine, and using an ordinary seed-cleaning device cleans seed.

Rhubarb is susceptible to crown-rot disease (*Phytophtora cactorum*), which can be checked using healthy plants when starting new crops. No other serious pests and diseases cause heavy losses of rhubarb.

REFERENCES

1. R. A. T. George, *Vegetable Seed Production*, Longman, London, 1985, p. 1.
2. L. H. Bailey, *Manual of Cultivated Plants*, 2nd ed., Macmillan, New York, 1949.
3. E. C. Tigchelaar, New Vegetable Varieties List XXI, Garden Seed Trade Association, American Seed Trade Association, *HortScience* 15(5): 565 (1980).
4. S. Z. Berry and M. A. Gould, Ohio 7870 tomato, *HortScience* 17(2): 266 (1982).
5. M. L. Weaver, Screening tomato for high temperature tolerance through pollen viability tests, *HortScience* 24(3): 493 (1989).
6. U. P. Hedrick and N. o. Booth, Mendelian characters in tomato, *Proc. Am. Soc. Hort. Sci.* 5: 19 (1988).
7. M. B. Crane, Heredity off types of inflorescence and fruits in tomato, *J. Genetics* 51:1 (1915).
8. A. Falavigna, M. Badino, G. P. Soressi, and G. Bruzzone, A new technique for hybrid seed production: direct hand pollination of male-fertile bs lines, *Acta Hart.* 111:47 (1981).
9. S. S. Sokhi, G. D. Munshi, and J. S. Jhooty, Tomato lines resistant to diseases, *Indian J. Agric. Sci.* 54 (1): 66 (1984).
10. R. G. Gardner, 'Mountain Delight' tomato; NC 8288 tomato breeding line, *HortScience* 25:989 (1990).
11. R. G. Gardner, 'Mountain Spring' tomato; NC 8276 and NC 84173 tomato breeding lines, *HortScience* 27:1233 (1992).
12. R. G. Gardner and A. F. Nash, Observations on the nipple tip (n) trait and associated characteristics, *Rep. Tomato Genet Coop.* 37:45 (1987).
13. J. H. M. Barten and J. W. Scott, Characterization of blossom-end morphology genes in tomato and their usefulness in breeding for smooth blossom-end scars, *J. Am. Soc. Hort. Sci.* 119(4): 798 (1994).
13a. M. Vodenicharova, T. Stoilova, and M. Markova, Comparative study of esterase and alcohol dehydrogenase loci for tomato hybridity, *Seed Sci. Technol.* 24: 185(1996).
14. A. K. Mehta, R. P. Singh, and G. Lal, Effect of concentrations and methods of application of 2,4- D on yield, fruit quality and seed quality of tomato, *Veg. Sci.* 16(1): 1 (1989).
15. W. E. Finch-Savage and C. I. McQuistan, Abscisic acid: an agent to advance and synchronise germination for tomato (*Lycopersicon esculentum* Mi.) seeds, *Seed Sci. Technol.* 19(3): 537 (1991).
16. S. M. Singer, Seed germination responses to temperatures of several cultivated tomato plants, *Egypt. J. Hort.* 17(1): 77 (1990).
17. N. B. Baldo, V. C. Fono-Hera, D. M. Vallador, and D. E. Panilan, Influence of length and storage temperature on the seed viability as seedling growth of tomato, *CMU J. Sci.* 1(1): 48 (1988).
18. P. Coolbear, R. J. Slater, and J. A. Bryant, Changes in nucleic acid levels associated with improved germination performance of tomato seeds after low temperature pre-sowing treatment, *Ann. Bot.* 65(2):187 (1990).

19. A. Fierro, T. Nicolas, and A. Gosselin, Supplemental carbon dioxide and light improved tomato and pepper seedling growth and yield, *HortSci* 29(3): 152 (1994).

19a. I. Demir and Y. Samit, Seed quality in relation to fruit maturation and seed dry weight during development in tomato, *Seed Sci. Technol.* 29: 453 (2001).

19b. H. S. Yogeesha, A. Nagaraja and S.P. Sharma, Pollination studies in hybrid tomato seed production, *Seed Sci. Technol.* 27: 115(1999).

19c. J. E. Warren and M. A. Bennett, Bio-osmopriming tomato (*Lycopersicon esculentum* Mill.) seeds for improved stand establishment, *Seed Sci. Technol.* 27: 489 (1999).

20. D. K. Salunkhe, B. B. Desai, and N. R. Bhat, *Vegetable and Flower Seed Production*, Agricole Publishing Academy; New Delhi 1987, p. 118.

21. R. A. T. George, R. J. Stephens, and S. Varis, The effect of mineral nutrients on yield and quality of seeds in tomato, *Seed Production* (P. D. Hebblethwaite, ed.), Butterworths, London, 1980, p. 561.

22. O. A. Bevre, Effects of boron on fruit and seed development in the tomato, *Norw. J. Agric. Sci.* 4(3): 233 (1990).

23. P. R. Govindan, Effect of boron on tomato, *Current Sci.* 2(1): 14 (1952).

23a. H. Dev and S. K. Sharma, Influence of picking stages in seed recovery and quality in tomato, XI National Seminar on Quality Seed to Enhance Agricultural Productivity, UAS, Dharwar, Jan 18–20, 2002, *Seed Tech News* 32 (A. Gaur, A,K. Vari, J.L. Varshney, and K.Kant, eds.), Indian Society Seed Technology, Division of Seed Science and Technology, IARI, New Delhi, March 2002, p. 43.

23b. Basave-Gouda, G. H. Ravikumar, P. N. Reddy, Arvind-Kumar, B. Gouda and A. Kumar, Impact of fruit maturity status and picking stage on the seed quality of tomato, cv. L-15, *Karnataka J. Agric. Sci.* 13:33 (2000).

23c. V.M. Valdes and D. Gray, The influence of stage of fruit maturation on seed quality in tomato (*Lycopersicon lycopersicum* L. Karsten), *Seed Sci. Technol.* 26: 309 (1998).

24. A. Liptay, Extraction procedures for optimal tomato seed quality, *Acta Hort.* 253:163 (1989).

25. B. N. Dhanvantari, Effect of seed extraction methods and seed treatments on control of tomato bacterial canker, *Can. J. Plant Pathol.* 11(4): 400 (1989).

26. S. J. Gowda, K. C. Talukdar, and H. Ramaiah, Effect of drying methods on seed quality in tomato (*Lycopersicon lycopersicum*), *Seed Res.* 18(2): 126 (1990).

27. A. D. Alvarado and K. J. Bradford, Priming and storage of tomato seeds; II. Influence of a second treatment after storage on germination and field emergence, *Seed Sci. Technol.* /6(3): 613 (1988).

28. S. D. Doijode, Changes in seed quality on deterioration in tomato, *Prog. Hort.* 20(3–4): 253 (1988):

29. C. A. Argerich, K. J. Bradford, and A. M. Tarquis, The effect of priming and ageing on resistance to deterioration of tomato seeds, J. Exp. Bot. 40(2/4): 593 (1989).

30. P. L. Owen and W. G. Pill, Germination of osmotically primed asparagus and tomato seeds after storage up to three months, *J. Am. Soc. Hort. Sci.* 119(3): 636 (1994).

31. H. P. Popovska, L. T. Miadenovski, and M. Mihajlovski, The influence of packing over germination of pepper and tomato seeds, *Acta Hort.* 911:281 (1981).

32. H. B. Singh, P. M. Bhagchandani, and M. R. Thakur, Vegetable seed production in kuluvalley Vs summer vegetables, *Indian J. Hort.* 21:221 (1964).

33. E. C. Auchter and C. P. Harley, Effect of various lengths of day on development and chemical composition of some horticultural plants, *Proc. Am. Soc. Hort. Sci.* 21: 199 (1924).

34. E. M. Khan and H. C. Passam, Flowering, fruit set and development of the fruit and seed of sweet pepper (*Capsicum annum* L.) cultivated under conditions of high ambient temperature, *J. HortSci.* 67(2): 251 (1992).

35. M. M. Joshi and A. A. Khalatkar, Experimental managenesis in *Capsicum annum* L.:

effect of different doses of gamma radiations on capsule and seed production, *Acta Hort.* 111:55 (1981).

36.   G. V. Umalkar, M. K. Vyawhare, R. M. Kashkar, and S. G. Kashikar, Sodium azide-induced mutations for quality seeds in *Capsicum annum* L., *Acta Hort.* 111:63 (1981).

36a. O. Livneh, Y. Nagler, Y. Tal, S. B. Harush, Y. Gafni, J. S. Beckmann and I. Sela, RFLP analysis of a hybrid cultivar of pepper (*Capsicum annuum*) and its use in distinguishing between parental lines and in hybrid identification, *Seed Sci. Technol.* 18: 209 (1990).

37.   D. V. So Chauhan, *Vegetable Production in India*, 3rd ed., Ram Prasad and Sons, Agra and Bhopal, India, 1981.

38.   V. E. Sovetkina, G. I. Dymova, and G. L. Matevosyan, The effect of growth regulators applied to *Capsicums* grown in protected cultivation, *Agrokhimiya* 7:103 (1988).

39.   F. Macit, Stimulation of pepper seed germination by some chemicals and growth regulators, *Acta Hort.* 111:139 (1981).

40.   K. Vanagamudi, K. S. Subramanian, and M. Baskaran; Influence of irrigation and nitrogen on the yield and quality of chilli fruit and seed, *Seed Res.* 18(2): 114 (1990).

41.   O. A. Osman and R. A. T. George, The effect of mineral nutrition and fruit position on seed yield and quality in sweet pepper, *Acta Hort.* 143:133 (1984).

42.   R. R. Perane and M. B. Joi, Control of fruit rot and dieback of chilli by seed treatment and spray, *J. Mah. Agric. Univ.* 14(3):368 (1989).

43.   T. A. S. Setty, B. C. Uthaiah, K. B. Rao, and K. M. Indiresh, Chemical control of seed microflora of chilli, *Plant Pathol. Newslett.* 6(1–2): 22 (1988).

44.   T. Haenglee and H. Chung, Detection and transmission of seed-borne *Colletotrichum gloeosporioides* in red pepper, *Seed Sci. Technol.* 23:533 (1995).

45.   S. D. Doijode, Studies on vigor and viability of seeds as influenced by maturity in chilli (*Capsicum annum* L.), *Hary J. Hort. Sci.* 17(1–2): 94 (1988).

46.   S. D. Doijode, Seed storability as affected by different stages of fruit development in chilli, *Veg. Sci.* 15(1): 15 (1988).

47.   K. S. Yawalkar, *Vegetable Crops in India*, 2nd ed., Agri-Horticultural Publishing House, Nagpur, India, 1980.

48.   C. Ramcharan and D. S. Padda, 'Virgin Long' eggplant, *HortScience* 17(2): 266 (1982).

49.   M. L. Odland and C. J. Noll, Hybrid vigor and combining ability in eggplants, *Proc. Am. Soc. Hort. Sci.* 51:417 (1948).

50.   H. Petrov, M. Doikova, and D. Popova, Studies on the quality of eggplant seed, *Acta Hort.* 111:273 (1981).

51.   P. Mariani, Eggplant somatic embryogenesis combined with synthetic seed technology, *Capsicum Newslett.* (Special issue): 289 (1992).

52.   M. Jayabarathi, V. Palaniswamy, D. Kalavathi, and P. Balamurugan, Influence of harvesting conditions on the yield and quality of brinjal seeds, *Veg. Sci.* 17(2): 113 (1990).

53.   J. A. Selvaraj and K. R. Ramaswamy, Effect of density grading on seed quality attributes in brinjal, *Seed Res.* 16(1): 117 (1988).

54.   J. A. Selveraj, Studies on storage of brinjal seeds. I. Biocide treatments and containers for storage, *South Indian Hort.* 36(6): 313 (1988).

55.   D. Rudrapal and S. Nakamura, Use of halogens in controlling eggplant and radish seed deterioration, *Seed Sci. Technol.* 16(1): 115 (1988).

56.   D. Rudrapal and S. Nakamura, The effect of hydration-dehydration pretreatments on eggplant and radish seed viability and vigor, *Seed Sci. Technol.* 16(1): 123 (1988).

56a. D. Rudrapal and S. Nakamura, Use of halogens in controlling eggplant and radish seed deterioration, *Seed Sci. Technol.* 16:115 (1988).

57.   H. W. Woyke, Some aspects of the role of the honeybee in onion seed production in Poland, *Acta Hort.* 111:21 (1981).

58. D. Globerson, A. Sharir, and R. Eliasi, The nature of flowering and seed maturation of onions as a basis for mechanical harvesting of the seeds, *Acta Hort.* 111:99 (1981).

59. L. Currah and D. J. Ockendon, Onion pollinated by blowflies and honeybees in large cages, *Ann. Appl. Biol.* 103(3): 497 (1983).

60. J. H. Ahmed and A. A. Abdalla, The role of honeybees as pollinators on onion (*Allium cepa* L.) seed production, *Acta Hort.* 143: 127 (1984).

61. M. Badino, Gametocidal effects of gibberellic acid (GA_3, GA_4 + GA_7) on common onion (*Allium cepa* L.), *Acta Hort.* 111:79 (1981).

62. Y. Shishido and T. Saito, Effects of plant growth regulators on low-temperature production of flower buds in onion plant, *Jpn. Soc. Hort Sci.* 53(1): 45 (1984).

63. D. Singh, H. Singh, S. S. Gill, and M. L. Chadha, Effect of plant density on onion seed yield, *Ann. Biol.* 6 (2): 171 (1990).

64. K. Krishnaveni, K. S. Subramanian, M. Bhaskaran, and K. N. Chinnasami, Effect of time of planting bulbs on the yield and quality of Bellary Onion seed, *South Indian Hort.* 38(5): 258 (1990).

65. B. K. Nehra, M. L. Pandita, and Kirti Singh. Cultural and nutritional studies in relation to seed production in onion. II. Effect of bulb size, spacing and nitrogen on plant growth and seed yield, *Haryana l. Hort. Sci.* 17(1–2): 106 (1988).

66. S. R. Bhonde, Ram Lecchiman, K. J. Srivastava, and U. B. Pandey, A note on effect of spacing and levels of nitrogen on seed yield of onion, *Seeds Farms* 15(1): 21 (1989).

67. I. H. Ahmed and A. A. Abdulla, Nitrogen and phosphorus fertilization in relation to seed production in onion, *Acta Hort.* 143:119 (1984).

68. J. K. Hore, N. C. Paria, and S. K. Sen, Effect of presowing seed treatment on germination, growth and yield of *Allium cepa* L. var. Red Globe, *Haryana l. Hort. Sci.* 17(1–2): 83 (1988).

69. G. Lal, D. K. Singh, and B. Ram, Note on the effect of spacing and time of planting of onion bulbs on seed production, *Prog. Hort.* 14(4): 204 (1982).

70. F. Szalay. The importance of harvesting date in onion seed production, *Hort. Abst:* 54(b): 328 (1984).

71. S. Sarati, I. M. Aleksandrova, I. G. Tarakanov, and T. N. Onushko. The response of onion seed plants to treatment with retardants, *Ref. Zhur.* 50:56 (1987).

72. B. K. Nehra, Y. S. Malik, and A. C. Yadav, Seed production in onions as influenced by time of bulb planting and cut treatments, *Haryana* Agric. *Univ. Res.* 19(3): 225 (1989).

73. S. S. Gill and H. Singh, Effect of bulb size and dates of planting on growth parameters and seed yield of onion, *Seed Res.* 17(1): II (1989).

74. T. R. Wheeler and R. H. Ellis, Seed quality and seedling emergence in onion (*Allium cepa* L.), *HortSci.* 67(3): 319 (1992).

75. B. K. Nehra, M. L. Pandita, and Kirti Singh, Effect of planting material and date of planting on bolting and seed yield of onion, *Veg. Sci.* 17(2): 195 (1990).

76. Sayed Gamiely, D. A. Smittle, and H. A. Mills, Onion seed size, weight and elemental content affect germination and bulb yield, *HortScience* 25(5): 522 (1990).

77. D. Singh, B. B. Deol, S. P. S. Gill, and L. Singh, Field screening of onion cultivars (seed crop) against gram caterpillar, *Punjab Hort.* 23(3–4): 242 (1983).

78. F. G. Fahmy and M. O. M. Omar, Potential use of 'propolis' to control white rot disease of onion, *Assoc. Agric. Sci.* 20(1): 265 (1989).

78a. R. C. Sharma, N. Kohli and S.S, Gill, Prevalence and management of emerging diseases of vegetables, XI National Seminar on 'Quality Seed to Enhance Agricultural Productivity', UAS, Dharwar, Jan 18–20, 2002, *Seed Tech News* 32 (A. Gaur, A.K. Vari, J.L. Varshney and K. Kant, eds.), Indian Society Seed Technology, Division of Seed Science and Technology, IARI, New Delhi, March 2002, p. 65.

79. Varier Anuradha and P. K. Agrawal, Long-term storage of certain vegetable seeds under ambient and reduced moisture conditions, *Seed Res.* 17(2): 153 (1989).

80. S. D. Doijode, Studies on partial vacuum storage of onion and bell pepper seeds, *Veg. Sci.* 15(2): 126 (1988).

81. N. Choudhari, and R. N. Basu; Maintenance of seed vigor and viability of onion (*Allium cepa* L.), *Seed Sci. Technol.* 16(1): 51 (1988).

82. A. G. Taylor, S. S. Lee, M. M. Beresniewicz, and D. H. Paine, Amino acid leakage from aged vegetable seeds, *Seed Sci. Technol.* 23(1):113 (1995).

83. M. M. Beresniewicz, A. G. Taylor, M. C. Goffinet, and B. T. Terhune, Characterization and location of a semipenneable layer in seed coats of leek and onion (Liliaceae), tomato and pepper (Solanaceae), *Seed Sci. Technol.* 23(1): 123 (1995).

84. M. M. Beresniewicz, A. G. Taylor, M. C. Goffinet, and W. D. Koeller, Chemical nature of a semi permeable layer in a seed coats of leek, onion (Liliaceae), tomato and pepper (Solanaceae), *Seed Sci. Technol.* 23(1): 135 (1995).

85. F. J. Dainello and R. R. Heineman, Carrot varieties evaluations in South-West Texas, *Texas Agric. Exp. Sta. Pub. No.* PR-3871, *Hort. Abst.* 55:2288 (1992).

86. M. J. Bassett, J. O. Strandberg, and J. M. White, 'Florida 524' carrot inbred, *Hort Science* 17(2):264 (1982)

87. D. J. Gray and A. Purohit, Somatic embryogenesis and development of synthetic seed technology, *Crit. Rev. Plant Sci.* 10(1): 33 (1991).

88. K. Redenbaugh, P. Viss, D. Slade, and J. A. Fujii, Scale-up: artificial seeds, *Plant Tissue and Cell Culture* (Alan R. Liss, ed.), New York, 1987, p. 473.

89. W. B. Greidziak, B. Diettrich, and M. Luckner, Batch cultures of somatic embryos of *Digitalis lanato* in gas lift fermentors; development and cardenolide accumulation, *Plant. Med.* 56: 175 (1990).

90. S. Takayama, B. Swedlund, and Y. Miwa, Automated propagation of microbulbs of lilies, *Cell Culture and Somatic Cell Genetics of Plants*, Vol. 8 (I. K. Vasil, ed.), Academic Press, New York, 1991, p. 112.

91. W. L. Teng, C. P. Lin, and Y. J. Liu, Regenerating lettuce from suspension culture in a 2-liter bioreactor, *HortScience* 28:669 (1993).

92. W. L. Teng, Y. J. Liu, Y. C. Tsai, and T. S. Soong, Somatic embryogenesis of carrot in bioreactor culture systems, *HortScience* 29(11): 1352 (1994).

93. S. L. Kitto and J. Janick, Hardening treatments increase survival of synthetically-coated asexual embryos of carrot, *J. Am. Soc. Hort. Sci.* 110(2): 283 (1985).

94. J. M. Dupuis, C. Roffat, R. T. DeRose, and F. Molle, Phamlaceutical capsules as a coating system for artificial seeds, *BioTechnology* 12(4): 385 (1994).

94a. C. Bazinet, A. Kersulec, V. Dufrene, F. Corbineau, D. Come, J. N. Barbotin, and D. Thomas, Physiological responses of somatic embryos (*Daucus carota* L.) to mechanical and biochemical stress during encapsulation and storage, *Proceedings of the Fourth International Workshop on Seeds: Basic and Applied Aspects of Seed Biology*, Angers, France, July 20–24, 1992.

94b. A. Kersulec, F. Corbineau, J. F. Hervagault, D. Thomas, and D. Come, Limitations of oxygen diffusion to somatic embryos of carrot (*Daucus carota* L.) by encapsulation; *Proceedings of the Fourth International Workshop on seeds: Basic and Applied Aspects of Seed Biology*, Angers, France, July 20–24, 1992.

95. P. T. Digole and N. N. Shinde, Carrot (*Daucus carota* L.) seed yield as influenced by different shoot and root treatments of variety Pusa Kesar, *Veg. Sci.* 17(I): 20 (1990).

96. Y. S. Malik, K. P. Singh, and P. S. Yadav, Effect of spacing and number of umbels on yield and quality of seed in carrot, *Seed Res.* 11(1): 63 (1983).

97. S. K. Sharma and J. J. Singh, Effect of level of nitrogen and spacing of plants on the yield of carrot seed, *Prog. Hort.* 13(3–4): 97 (1981).

98. R. Jacobsohn and D. Globerson, *Daucus carota* (carrot) seed quality. I. Effects of seed size on: germination, emergence and plant growth under subtropical conditions, and

II. The importance of the primary umbel in carrot seed production, *Seed Production* (P. O. Hebbelthwaite, ed.), Butterworths, London, 1980, p. 637.

98a. A. Hermansen, G. Brodal and G. Balvoll, Hot water treatments of carrot seeds: effects on seed-borne fungi, germination, emergence and yield, *Seed Sci. Technol.* 27 599 (1999).

99. J. A. Schoneveld and W. Maldegem, Research on carrots: Grading is more uniform as a result of better germination, *Vollegrondsgroenten* 1(11): 10 (1991).

100. D. Gray, Are the plant densities currently used for carrot seed production too low? *Acta Hort.* 11:159 (1981).

101. Ahmed Nazeer and M. I. Tanki, Effect of nitrogen and phosphorus on seed production of carrot (*Daucus carota* L.), *Veg. Sci.* 16(2): 107 (1989).

102. J. J. Steiner, R. B. Hutmacher, A. B. Mantel, J. E. Ayars, and S. S. Vail, Response of seed carrot to various water regimes. II. Reproductive development, seed yields and seed quality, *J. Am. Soc. Hort. Sci.* 115(5): 722 (1990).

103. A. A. Iskenderov and A. V. Volvodin, Promising herbicides for seed carrot. *Hort. Abst* 54(5): 249 (1984).

104. J. O. Strandberg and J. M. White, Response of carrot seeds to heat treatments, *J. Am. Soc. Hort. Sci.* 114(5): 766 (1989).

105. V. A. Koltunov, I. O. Ustino, A. F. Ustinova, and E. A. Yatsenko, Effect of biologically active substances on the causal agent of white rot of carrot, *Mshchita Rast.* 1: 718 (1979).

105a. K. Tylkowska and R.W. van den Bulk, Effects of osmo- and hydropriming on fungal infestation levels and germination of carrot (*Daucus carota*) seeds contaminated with *Alternaria* spp., *Seed Sci Technol.* 29:365 (2001).

106. H. C. Thompson and W. C. Kelly, *Vegetable Crops*, 5th ed. McGraw-Hill, New York, 1957.

107. L. Williams and M. A. Collin, Embryogenesis and plantlet formation in tissue cultures of celery, *Ann. Bot.* 40:325 (1976).

108. Yong-Hwan Kim, Origin of somatic embryos in celery tissue culture, *HortScience* 24(4): 671 (1989).

109. M. Hayashi, Practical application of somatic embryogenesis synthetic seed system, *Curr. Plant Sci. Biotechnol. Agric.* 15:305 (1993).

110. H. J. Cho and W. Y. Yang, Effects of PEG and ABA on storage starch accumulation and the survival of dedicated somatic embryos of celery, *Res. Rep. Rural Dev. Admin. BioTechnol.* 33(3): 13 (1991).

111. N. Tremblay and A. Gosselin, Growth, nutrient status and yield of celery seedlings in response to urea fertilization, *HortScience* 24(2): 288 (1989).

112. T. H. Thomas, N. L. Biddington, and D. Palevitoh, Improving the performance of pelleted celery seeds with growth regulator treatments, *Acta Hort.* 83:235 (1978).

113. T. H. Thomas, Hormonal involvement in photoregulation of celery seed dormancy, *Monogr. Br. Soc. Plant Growth Reg.* 20:51 (1990).

114. F. Perez-Garcia, J. M. Pita, M. E. Gonzalez-Benito, and J. M. Iriondo, Effects of light, temperature and seed priming on germination of celery seeds (*Apium graveolens* L.), *Seed Sci. Technol.* 23(2): 377 (1995).

115. T. H. Thomas and D. F. OToole, The effects of environmental and chemical treatments on the production and performance of some vegetable seeds, *Seed Production* (P. O. Hebbelethwaite, ed.), Butterworths, London, 1980, p. 501.

116. S. Heath-Pagliuso, J. Pullman, and L. Rappaport, UC-T3 Somaclone; celery germplasm resistant to *Fusarium oxysporum f.sp.apii*, race 2, *HortScience* 24(4): 711 (1989).

117. A. A. Ramin and J. G. Atherton, Manipulation of bolting and flowering in celery (*Apium graveolens* L. var. duice), I. Effects of chilling during germination and seed development, *J. HortSci.* 66(4): 435 (1991).

118. L. Jedras, The date of sowing for parsley seed production using the seed-to-seed method, *Biul. Warzywniczy* 2:67 (1989).

119. K. Suchorska, Influence of growth regulators upon the seeds germination of parsnip and fennel, *Herba Hung*. 30(1–2): 68 (1991).

120. E. J. Ryder, 'Sea Green' lettuce, *HortScience* 16(4): 571 (1981).

121. R. T. Negata, Clip-and-wash method of emasculation of lettuce, *HortScience* 27(8):907 (1992).

122. D. Wurr, Crisp lettuce: Breeding stronger, attractive varieties, *Grower* 100(23): 17, 19 (1983).

123. A. H. Eenink, Seed dormancy in lettuce: Influence of temperature on induction and expression of seed dormancy and studies on the inheritance, *Acta Hort*. 111:41 (1981).

124. S. D. Prince, Vernalization and seed production in lettuce, *Seed Production* (P. D. Hebbelethwaite, ed.), Butterworths, London, 1980, p. 485.

125. H. S. Saini, E. D. Consolacion, P. K. Bassi, and M. S. Spencer, Control processes in the induction of relief of thermo inhibition of lettuce seed germination, *Plant Physiol*. 90(1): 311 (1989).

126. K. Kabar, Interactions among salt (NaCl), kinetin and gibberellic acid in the germination of lettuce seeds, *Doga Turk-Botanik Derg*. 13(2): 296 (1989).

127. O. Roeggen, Germination at low temperature in lettuce (*Lactuca sativa* L.) and spinach (*Spinacia oleracea* L.), *Seed Sci. Technol*. 17(2): 263 (1989).

128. R. L. K. Drew and P. A. Brocklehurst, Effects of temperature of mother-plant environment on yield and germination of seeds of lettuce (*Lactuca sativa*), *Ann. Bot*. 66(1): 63 (1990).

129. P. A. A. Aguiar, Influence of gibberellic acid on lettuce seed production, *Rev. Brabil Sementes*, 4(1): 89 (1982).

130. S. Stoyanova and S. Neikov, The effect of methods of harvesting treatment and control of after-ripening of lettuce seeds for long term storage, *Rast. Nauki* 26(5): 57 (1989).

131. M. T. Smith, Ultrastructural changes during imbibition in seeds of lettuce (*Lactuca sativa* L.) after gamma irradiation, *Seed Sci. Technol*. 19(2): 385 (1991).

132. C. E. Peterson, P. H. Williams, M. Palmer, and P. Louward, 'Wisconsin 2757' cucumber, *Hort. Sci*. 17(2): 268 (1982).

133. P. Nath and O. P. Dutta, New high yielding varieties of cucurbits, *Indian Hort*. 15:11 (1971).

134. W. Msikita, R. M. Skirvin, J. A. Juvik, W. E. Splittstoesser, and N. Ali, Regeneration and flowering in vitro of Burpless hybrid cucumber cultured from excised seed, *Hort Science* 25:474 (1990).

135. R. L. Lower, Gy4 cucumber inbred and 'Raleigh' hybrid pickling cucumber, *Hort Science* 26(1): 78 (1991).

136. T. C. Wehner, S. F. Jenkins, and R. L. Lower, Gy5 cucumber Gy5 inbred and 'Johnston' hybrid pickling cucumber, *HortScience* 26 (1): 77 (1991).

137. P. Milotay, Possibilities of field seed production by gynoecious cucumbers (Hungary), *Hort. Abst* 54(b): 338 (1984).

138. V. A. Patil, D. B. Bangal, V. R. Kale, and M. B. Jamdagni; Modification of sex expression in cucumber; *J. Mah. Agric. Univ.* 8(3): 283 (1983).

139. Qizhang, A. C. Gabert, and J. R. Bagget, Characterizing a cucumber pollen sterile mutant: inheritance, allelism, and response to chemical and environmental factors, *J. Am. Soc. Hort. Sci.* 119(4): 804 (1994).

140. S. R. Jia, Y. Y: Fu, and Y. Lin, Embryogenesis and plant regeneration from cotyledon protoplasts culture of cucumber, *J. Plant Physiol*. 124:393 (1986).

141. J. F. Lazarte and C. C. Sasser, Asexual embryogenesis and plantlet development in anther culture of *Cucumis sativus* L, *HortScience* 17:88 (1982).

142. Yong-Hwankim and J. Janick, Somatic embryogenesis and organogenesis in cucumber, *HortScience* 24(4): 702 (1989).

143. A. R. Ladyman and B. Girard, Cucumber somatic embryo development in various gelling and carbohydrate sources, *HortScience* 27(2): 164 (1992).

144. H. Lou and S. Kako, Somatic embryogenesis and plant regeneration in cucumber, *HortScience* 29(8): 906 (1994).

145. Sitaram, A. F. Habib, and G. N. Kulkarni, Effect of growth regulators on seed production and quality in hybrid cucumber (*Cucumis sativus* L.), *Seed Res.* 17(1): 6 (1989).

146. N. Shaban and P. Kartalov, Effect of laser irradiation of seeds on some physiological processes in cucumbers, *Rast. Nauki* 25(5): 64 (1988).

147. S. M. Medzhitov, The effect of pre-sowing seed-treatment on the growth, development and productivity of cucumbers, *Ref. Zhur.* 4(55): 385 (1990).

148. P. Jennings and M. E. Saltveit, Temperature effects on imbibition and germination of cucumber (*Cucumis sativus* L.) seeds, *J. Am. Soc. Hort. Sci.* 119(3): 464 (1994).

149. I. Turkan, The effects of exhaust gas on seed germination and seedling growth of cucumber (*Cucumis sativus* L.), *J. Turk. Phytopathol.* 17(2): 81 (1988).

150. J. G. Latimer and R. B. Beverly, Conditioning affect growth and drought tolerance of cucurbit transplants, *J. Am. Soc. Hort. Sci.* 119(5): 943 (1994).

151. J. R. Schultheis, J. T. Ambrose, S. B. Bambara, and W. A. Mangum, Selective bee attractants did not improve cucumber and watermelon yield, *HortScience* 29(3): 155 (1994).

152. H. Nerson, Fruit age and seed extraction procedure affect germinability of cucurbit seeds, *Seed Sci. Technol.* 19(1): 185 (1991).

153. Nawab Ali, R. Skirvin, W. E. Splittstoesser, and W. L. George, Germination and regeneration of plants from old cucumber seed, *HortScience* 26(7): 917 (1991).

154. K. Andrasek, Regulator in the seed production of hybrid F_1 of musk and watermelon, *Acta Hort.* 220:219 (1988).

155. J. D. Abbott, Fungicidal inhibition of pollen germination and germ-tube elongation in muskmelon, *HortScience* 26(5): 529 (1991).

156. G. E. Boyhan and J. D. Norton, Inheritance of resistance to Alternaria leaf blight in muskmelons, *HortScience* 27(10): 1114 (1992).

157. J. D. Norton and R. D. Cosper, AC-70-154, A gummy stem blight resistant muskmelon breeding line, *HortScience* 24(4): 709 (1989).

158. J. Alvarez, Germination of melon (*Cucumis melo* L.) seeds affected by ethephon, *Comun. INIA Prod. Veg.* 20(81): 21 (1989).

159. O. A. Lorenz and D. N. Maynard, *Knotts Handbook for Vegetable Growers*, 2nd ed., Wiley, New York, 1980.

160. S. Mendlinger and D. Pasternak, Effect of time of salinization in flowering, yield and fruit quality factors in melon (*Cucumis melo* L.), *J. HortSci.* 67(4): 529 (1992).

161. S. J. Gaikwad and B. Sen, Chemical control of cucurbit wilt caused by *Fusarium oxysporum* Schlecht, *Veg. Sci.* 14 (1): 83 (1987).

162. HariSingh, L. Tarsem, and R. S. Rana, Effect of puncturing of muskmelons fruits at various stages of maturity and seed quality, *Seed Res.* 16(1): 13 (1988).

163. N. P. S. Dhillon, Flhybrid seed production in muskmelon; Management of the male sterile population, *Seed Sci. Technol.* 22:60 (1994).

164. Z. V. Karchi, A. Govers, and H. Nerson, 'Alena' watermelon, *HortScience* 16(4): 573 (1981).

165. C. F. Andrus, V. S. Seshadri, and P. C. Grimball, Production of seedless watermelons, *USDA Tech. Bul. No.* 1425 (1971).

166. N. Sari, K. Abak, M. Pitrat, J. C. Rode, and R. D. Vaulx, Induction of parthenogenetic haploid embryos after pollination by irradiated pollen in watermelon, *HortScience* 29(10): 1189 (1994).

167. J. W. Adelberg, B. B. Rhodes, H. T. Skorupska, and W. A. Bridges, Explant origin affects the frequency of tetraploid plants from tissue cultures of melon, *HortScience* 29(6): 689 (1994).

168. J. Z. Dong and S. R. Jia, High efficiency plant regeneration from cotyledons of watermelon (*Citrullus vulgaris* Schard.), *Plant Cell Rep.* 9:559 (1991).

169. M. E. Chompton and D. J. Gray, Shoot organogenesis and plant regeneration from cotyledons of diploid, triploid and tetraploid watermelons, *J. Am. Soc. Hort. Sci.* 118:151 (1993).

169a. R .Bonfitto, L. Galleschi, M. Macchia, F. Saviozzi and F. Navari-Izzo, Identification of melon cultivars by gel and capillary electrophoresis, *Seed Sci. Technol.* 27:779(1999).

170. H. Nerson, H. S. Paris, and Z. Karchi, Seed treatments for improved germination of tetraploid watermelon, *HortScience* 20(5): 897 (1985).

171. M. L. Yang and F. J. M. Sung, The effect of suboptimal temperature on germination of triploid watermelon seeds of different weights, *Seed Sci. Technol.* 22(3): 485 (1994).

171a. I. Demir and, H. A. van de Venter, The effect of priming treatments on the performance of watermelon (*Citrullus lanatus* (Thunb.) Matsum. & Nakai) seeds under temperature and osmotic stress, *Seed Sci. Technol.* 27: 871(1999).

172. D. S. Nesmith and G. Hoogenboom, Staminate and pistillate flower production of summer squash in response to planting date, *HortScience* 29(4): 256 (1994).

173. A. N. Matlob and E. A. Basher, Effect of growth regulators on sex-expression and yield of summer squash, *Acta Hort.* 137:361 (1983).

174. R. J. Hume and P. H. Lovell, The control of sex-expression in cucurbits by ethephon, *Ann. Bot.* 52(5): 689 (1983).

175. P. P. Chee, Initiation and maturation of somatic embryos of squash (*Cucurbito pepo*), *HortScience* 27(1): 59 (1992).

176. M. E. Dematte, Yield components of mature fruits and seeds of *Cucurbita moschata*; Duchesne cv. Canhao IAC-3046, *Proc. Tro. Reg. Am. Soc. Hort. Sci.* 25:47 (1982).

177. S. K. Arora, R. N. Vashishta, and P. S. Partap, Effect of plant growth regulators on growth, flowering and yield of pumpkin (*Cucurbita moschata* Poir), *Res. Dev. Rep.* 6(1): 31 (1989).

178. M. L. Odland and C. L. Noll, Penn valley cabbage, *Pa. Agic Exp. Sta. Prog. Rep.*, No. 84 (1952).

179. M. L. Odland and C. L. Noll, The utilization of cross-compatibility and self-compatibility production of F$_1$ hybrid cabbage, *Proc. Am. Soc. Hort. Sci.* 55:391 (1950).

180. C. North, Vegetable propagation of cabbage and applied vegetables, *Imp. J. Exh. Agr.* 20:43 (1952).

181. D. A. Abdurashidov, The effect of the preparation on the growth, development and seed production in white head cabbage, *Ref. Zhur.* 4 (55): 361 (1990).

182. S. Samandamurthi and K. S. Sundaram, A note on seed production in cabbages under Kodaikanal conditions, *South Indian Hort.* 37(3): 183 (1989).

183. T. M. Perring, A. Cooper, D. J. Kazmer, and C. Shields, New strain of sweet potato whitefly invades California vegetables, *Calif. Agr.* 45:10 (1991).

184. K. D. Elsey and M. W. Farnham, Response of *Brassica oleracea* L. to *Bemisia tabaci* (Gennadius), *HortScience* 29(7): 814 (1994).

185. G. R. Dixon and E. Wilson, Evaluation of chemicals for control of club rot in transplanted cabbage, *Crop Protection in Northern Britain*, Scottish Crop Res. Inst., Dandee, UK, 1984, p. 400.

185a. B. Iacomi-Vagilescu, D. Blancard, M. Guenard, V. Molinero-Demilly, E. Laurent, and P. Simoneau, Development of a PCR-based diagnostic assay for detecting pathogenic *Alternaria* species in cruciferous seeds, *Seed Sci. Technol.* 30: 87 (2002).

186. P. S. Arya, B. N. Korla, and P. P. Sharma, Effect of environmental factors on cabbage seed production in Kalpa Valley; *South Indian Hort.* 31(6): 297 (1983).

187. R. H. Ellis, T. D. Hong, M. C. Martin, G. F. Perez, and C. GomezCampa, The long-term storage of seeds of seventeen crucifers at very low moisture contents, *Plant Vat: Seeds* 6(2):75 (1993).

188. A. C. McCormac and P. O. Keefe, Cauliflower (*Brassica oleracea* L.) seed vigor imbibition effects, *J. Exp. Bot.* 41(228): 893 (1990).

189. K. G. Tateand L. H. Cheah, Control of club rot in cauliflower, *N. Z. Commet: Grower* 38(9): 36(1983).

190. D. Raj and B. B. Kanwar, Minimizing insecticide use against cauliflower pests in India, *Trop. Pest Manage.* 36(1): 10 (1990).

191. K. C. Jandial and S. S. Saini, Effect of curd size on seed yield and germination of late cauliflower (*Brassica oleracea* L.) var. botrytis, *Environ. Ecol.* 7(2): 472 (1989).

192. U. C. Panigrahi, N. B. Pattanayak, and C. Das, A note on the effect of micronutrients on yield of cauliflower seeds in the acid red soil of Orissa, *Orissa J. Hort* 8(1–2): 62 (1990).

192a. P. R. Kumar, N. C. Singhal, Ram-Singh and R. Singh, Effect of different curd cutting methods on seed production of cauliflower (*Brassica oleracea* L. var. Botrytis), *Seed Res.* 28: 136 (2000).

192b. C. Gurusamy, Effect of stage of harvesting on seed yield and quality of cauliflower, *Seed Sci. Technol.* 27: 929(1999).

192c. W. H. Zhang and D. S. Wang, A fast procedure for genetic purity determination of head Chinese cabbage hybrid seed based on RAPD markers, *Seed Sci. Technol.* 26: 829(1998).

193. O. Roeggen, Germination at low temperature in lettuce (*Lactuca sativa* L.) and spinach (*Spinacia oleracea* L.), *Seed Sci. Technol.* 17(2): 263 (1989).

194. H. Singh and S. S. Gill, Effect of spacing and leaf cutting on seed yield of spinach (*Beta vulgaris* L.), *J. Res. Punjab Agric. Univ.* 20(3): 261 (1983).

195. B. Kassanis, The use of tissue cultures to produce virus-free clones of some British potato varieties, *Ann. Appl. Biol.* 45:422 (1957).

196. H. W. Howard, The production of new varieties, *The Potato Crop* (P. M. Harris, ed.), Chapman & Hall, London, 1978.

197. P. B. Goodwin, Y. C. Kim, and T. Adisarwanto, Propagation of potatoes by shoot-tip culture; shoot multiplication, *Potato Res.* 23:9 (1980).

198. S. Roest and G. S. Bokelmann, Vegetative propagation of *Solanum tuberosum* L., in vitro, *Potato Res.* 19:173 (1976).

199. S. Roest and G. S. Bokelmann, In vitro adventitious bud techniques for vegetative propagation and mutation in vitro through adventitious shoot formation, *Potato Res.* 23: 167 (1980).

200. G. C. Upreti, *Development of Virus-Tested Nucleus Stocks, Recent Technology in Potato Improvement and Production*, CPRI, Simla, 1977.

201. K. Evans and D. L. Trudgill, Nematode pests of potatoes, *The Potato Crop* (P. M. Harris, ed.), Chapman & Hall, London, 1978, p. 440.

202. D. C. E. Wurr, Seed tuber production in management, *The Potato Crop* (P. M. Harris, ed.), Chapman & Hall, London, 1978, p. 827.

203. N. Pallais, H. Asmat, N. Fong, and J. Santos-Rojas, Factors affecting seedling vigor in potatoes: I. Stage of seed development, *Am. Potato J.* 66(12): 793 (1989).

204. D. Gray and J. C. Hughes, Tuber quality, *The Potato Crop* (P. M. Harris, ed.), Chapman & Hall, London, 1978, p. 504.

205. R. H. Jarvis, Mechanization and crop performance, *The Potato Crop* (P. M. Harris, ed.), Chapman & Hall, London, 1978, p. 355.

206. G. A. Hide and D. H. Lapwood, Disease aspects of potato production, *The Potato Crop* (P. M. Harris, ed.), Chapman & Hall, London, 1978.

207. W. G. Burton, The physics and physiology of storage, *The Potato Crop* (P. M. Harris, ed.), Chapman & Hall, London, 1978, p. 545.

208. A. K. Bhatt, T. C. Bhalla, H. O. Agrawal, M. D. Upadhya, and N. Sharma, Effect of seed size on imbibition and germination of open pollinated true seeds of potato, *Seed Res.* 16(2): 178 (1988).

209. M. A. Contrevas, True potato seed: potential and use, *Simiente* 59(1–2) (1989). 357

210. P. Malagamba. Potato production from true seed in tropical climates. *HortScience* 23:495 (1988).

211. S. Satjadipura. Method of increasing number of flowers and fruit of potato cv. E-1282/ 19, *Bull Penetitian Hort.* 17(4): 44 (1989).

212. N. E. Pallais, N. Y. Espinola, R. M. Falcon, and R. S. Garcia, Improving seedling vigor in sexual seeds of potato under high temperature, *HortScience* 26(3):296 (1991).

213. K. C. Thakur and M. D. Upadhya, Technology for hybrid true potato seed production in India, *Seed Res.* 1:82 (1993).

214. C. J. M. Almekinders and S. G. Wiersema, Flowering and true seed production in potato (*Solanum tuberosum* L.). 1. Effect of inflorescence position, nitrogen treatment and harvest date of berries, *Potato Res.* 34(4): 365 (1991).

215. C. J. M. Almekinders, Flowering and true seed production in potato (*Solanum tuberosum* L.). 2. Effects of stem density and pruning of lateral stems, *Potato Res.* 34(4):379 (1991).

216. K. C. Thakur and M. D. Upadhya, Extraction and processing of true potato seed, *Seed Sci. Technonol.* 18 (3): 589 (1990).

217. S. Sadik, A method for seed extraction in true potato seed, *Letter* 2:3 (1982).

218. A. K. Chakrabarti, Effect of nitrogen and phosphorus on turnip seed crop, *Seed Res.* 11(1): 87 (1983).

219. G. Lepori and L. Quagliotti, Effects of storage on the viability of turnip and broccoli rabs seeds, *Acta Hort.* 111:255 (1981).

220. E. S. M. Sakr, Effects of temperature and photoperiod on seed stalk development in turnip, *Proc Am. Soc. Hort. Sci.* 44:473 (1944).

221. C. B. Singh, M. L. Pandita, and S. C. Khurana. Studies on the effect of root age, size and spacing on seed yield of turnip (*Brassica campestris* var. *rapa* L.), *Veg.Sci.* 16(2):119 (1989).

222. S. Tharnpuraj, K. G. Shanmugavelu, O. A. Pillai, S. Anbu, and C. R. Muthukrishnan, 'Col' radish, a new variety for plains, *South Indian Hort.* 29:152 (1981).

223. N. S. Pundir, R. F. Abbas, and A. A. AI-Attar, Pollen germination and pollen tube growth in *Raphanus sativus* L. following self and cross fertilization, *Phyton* (Argentina) 43(2):127 (1983).

224. M. C. Nauriyal and H. Lal, Effect of spacing and plant regulators on seed production of radish, *Veg. Sci.* 9(2):85 (1982).

225. T. S. Rawat and V. Singh, Effects of spacing and nitrogen application on the performance of seed crop of radish, cv. Pusa Rashmi, *Udyanica* 4:17 (1981).

226. N. R. Singhvi, and H. K. Chaturvedi, Effect of presoaking seed treatment with gibberellic acid and morphactin on various morpho-physiological parameters in *Raphanus sativus* L, *Adv. Plant Sci.* 3(1): 165 (1990).

227. R. Chatterjee, Effect of sowing date on root production and seed yield of radish cv. Improved Chinese Pink, *Hort. J.* 2(1): 55 (1989).

228. T. V. Karivaratharaju, V. Palanisamy, and K. Vanangarnudi, Steckling quality in radish (*Raphanus sativus* L.) as affected by seed weight, *South Indian Hort.* 36(1–2):81 (1988).

229. J. L. Mangal, B. R. Batra, and G. R. Singh. Effect of irrigation and nitrogen and phosphorus interaction on seed production of radish, *Haryana J. Hort. Sci.* 17(1–2):97 (1988).

230. K. B. Rastogi, P. P. Sharma, and B. N. Korla, Effect of different levess of nitrogen and spacing on seed yield of radish (*Raphanus sativus* L.), *Veg. Sci.* 14(2):105 (1987).

231. S. K. Sharma and Lal Gulshan, Effect of certain cultural practices on the test weight of radish seed, *Seed Res.* 18(2): 154 (1990).

231a. S.S. Gill and B.S. Gill, Seed Yield in radish as influenced by the date of transplanting and steckling size, *Seed Res.* 23: 28(1995).

232. O. P. Gambhir, Y. S. Malik, and M. L. Pandita, Chemical weed control in seed crop of radish, *Indian J. Weed Sci.* 15(1): 74 (1983).

232a. S. S. Gill, B. S. Gill, S. P. S. Brar, and B. Singh, Nitrogen and phosphorus requirement of radish seed crop, *Seed Res.* 23: 47 (1995).

232b. M. Takaki and R. E. Rosin, Asprin increases tolerance to high temperature in seeds of *Raphanus sativus* L. cv. Early Scarlet Globe, *Seed Sci. Technol.* 28: 179 (2000).

233. W. W. Weber, Seed production in horseradish. *J. Hered.* 40(8): 223 (1949).

234. M. S. Saimbhi, Studies on pod growth in okra; variety Punjab Padmini, Punjab Veg. Grower 17/18:40 (1983).

235. J. L. Mangal, S. Lal, and S. K. Arora, Studies on the effect of chlorocholine chloride and napthalene acetic acid application on salt resistance and productivity of okra, *Hary. Agric. Univ. J. Res.* 18(3): 191 (1988).

236. K. Singh, D. A. Sarnaik, and C. S. Bisen, Effect of sowing dates and spacings on the yield and quality of okra seed, *Res. Dev. Rep.* 5(1–2): 83(1988).

237. K. B. Rastogi, P. P. Sharma, N. P. Singh, and B. N. Korla, Effect of different levels of nitrogen and plant spacing on seed yield of okra (*Abelmoschus esculentus* L.) Moench, *Veg. Sci.* 14(2): 120 (1987).

238. P. C. Lenka, D. K. Das, and H. N. Mishra, Effect of nitrogen and phosphorus on seed yield of bhindi cv. Parbhani Kranti, *Orrisa J. Agric. Res.* 2(2): 125 (1989).

239. G. Singh and H. Singh, Effect of stimulated rains on seed quality of okra cultivars, *Seed Res.* 16(2): 226 (1988).

240. R. S. Hooda, M. L. Pandita, and A. S. Sidhu, Effect of seed treatment and foliar application of cytozyme on seed yield and yield attributes of okra, *Haryara Hort. Sci.* 12(1): 135 (1983).

241. V. K. Bhatnagar and R. Sareen, Influence of chlorflurenol on growth, sex-expression and yield of okra, *J. Mah. Agric. Univ.* 8(3): 280 (1983).

11
Flowers and Ornamental Crops

I. INTRODUCTION
A. Commercial Flower Seed Production

Commercial flower seed production has undergone unrestricted revolutionary growth in terms of the number of crops and their varieties as well as production. This phenomenal development is the result of technological advancements in the production of flower seeds and developments in the areas of production, packaging, handling, transportation, and marketing of flowers and ornamental plants all over the world.

The world germplasm contains more than 1500 different varieties of flowers and ornamental crops. Each species of flower crop grown for seed has its own specific planting time, culture, problems of pollination and harvesting, and storage requirements. A mild climate with moderate rainfall favors the growth and development of most flowers and ornamental crops.

Hybrid vigor in ornamental plants was exploited for the first time in the 1940s to produce all double F_1 petunia hybrids, which revolutionized the flower seed industry. This technique paved the way for the introduction of numerous outstanding F_1 hybrids of several types of flowers, including petunias, in the United States, the Netherlands, Japan, Denmark, the United Kingdom, and former West Germany. Now the F_1 hybrids are available for almost all flower crops from A (antirrhinum) to Z (zinnia), including ageratum, begonia, calendula, calceolaria, cyclamen, carnation, dianthus, gerbera hollyhock, impatiens, marigold, nicotiana, pelargonium, petunia, portulaca, and stock.

The real breakthrough in hybrid flower production occurred since the 1950s, and there have been tremendous advancements in this area, with numerous creative approaches being employed to produce F_1 hybrids. These include the use of male-sterile lines, synthetic/artificial seed production through somatic embryogenesis (see Chapter 23), the use of the vacuum pump for pollen collection, the use of infrared light for dehiscence of anthers for pollens, long-term storage of pollen grains, and the use of pollination aids to overcome tedious conventional hand-pollination techniques. Cheaper, easier, and quicker methods are being sought to produce hybrid flower seeds. The F_2 seeds developed in recent years have gained popularity, including antirrhinum, petunia, and pansy, and are less expensive than F_1 seeds.

The varied agroclimatic conditions existing in India provide a vast scope for the production of a wide range of flower seeds of both tropical and temperate origin. With the availability of relatively inexpensive human labor, technical expertise, and other materials, it is possible to organize a strong flower seed production program exclusively for export purpose (1). The following four basic factors must be considered in developing an efficient seed program.

1. Assessment of the world seed demand, its consistency or stability, and the approximate pricing structure
2. Identifying best areas for quality seed production and determination of export potential based on the existing global demand for seed and the facilities available
3. Promotion of both internal and international seed trades
4. Development of a basic infrastructure for the production of flower seeds

1. Assessment of World Demand for Flower Seeds

Since flower seeds are in great demand in Western countries, entering into an international flower seed market can be highly remunerative. However, not all seeds are equally in demand, and the demand for seeds of specific flower crops may change rather rapidly. For example, in the United States, the total area planted with aster, petunias, snapdragons, and sweet peas was less in 1970 than in the 1950s, whereas for marigold, zinnia, and stock, it was more. This may be partly due to the introduction of male-sterile lines to overcome the nonavailability of manual labor. The decrease in demand represents either a shifting of production centers or changes in market trends. It is therefore necessary to determine the worldwide demand for varieties of different flower crops by conducting industry surveys in various importing countries. It is also a good idea to develop custom seed production programs with foreign seed companies or plant breeders to assure an advance market.

2. Identification of Production Centers

Each flower crop needs a specific set of soil and climatic conditions to produce quality seeds on a large scale. It is therefore essential to identify the ideal locations for the production of different flower crops. The important agroclimatic considerations are:

1. Soil type, texture, moisture-retentive ability, and soil reaction (pH)
2. Climate (temperature, sunlight, rainfall, and relative humidity)
3. Availability of perennial water source

Trained personnel with knowledge and experience of growing different flower crops for seed is essential. Because flower seed production is labor intensive, inexpensive human labor should be available throughout the year.

3. Promotion of Seed Export

Promotion of an export trade in flower seed requires proper planning and implementation. The seed produced for export must meet quality standards and consumer preferences, and must be adaptable to the soil and climatic conditions of the importing countries.

4. Development of a Basic Infrastructure

The cost and feasibility of undertaking seed production—involving all aspects of production, processing, quality control, packaging, storage, and transportation to the foreign market in an acceptable form—are important considerations. This involves a considerable amount of research on the production and postharvest biotechnologies and investment on sophisticated equipment and basic infrastructure facilities. The problem of seed export can be handled in two ways: custom seed production for a foreign plant breeder, institute, or seed company and production of seeds of indigenous varieties.

B. Types of Flower Seeds Produced by Breeding and Selection

Four types of flower seeds are produced: mixture or open-pollinated seeds, pure inbred lines, F_1 hybrids, and F_2 strains. In addition, some bedding plants, for example, geranium and fuschia, are produced asexually or vegetatively (2).

1. Mixture or Open-Pollinated Seeds

Mixtures refer to a combination of more than one color or plant type in each seed lot. Open-pollination commonly produces natural mixtures in several flower crops. The progenies of such seeds give mixed populations. The general uniformity of plant characteristics is of an important consideration in mixtures. A relatively uniform plant population can be obtained by roguing off-types before flowering. Uncontrolled open-pollination can produce a large number of plants with a particular flower color, which produces more seeds. A proportionate number of plants producing different colored flowers can be maintained to bring about uniformity in population. It is, however, not possible to obtain true-to-type progenies from open-pollinated seeds of flower crops.

Pollination may be controlled to produce uniformity in flower color, and then seeds can be mixed to obtain a balanced color. This technique, known as "formula mixture," involves mixing of inbred lines (F_1 and F_2 strains) in a desired proportion to produce balanced flower color. The major advantage of formula mixture is that the color combination in the progeny can be predicted and reproduced in comparison to natural mixture. The individual genotypes of superquality can be selected for mixing, taking into consideration their growth and flowering habits. The formula mixtures are available for pansy, petunia, and snapdragon.

2. Inbred Lines

Pure inbred lines are produced by repeated selfing to obtain uniform progenies. These lines provide the basic genetic material for F_1 hybrid seed production. In the usual seed-production program, progeny of desirable plant types are raised and only the best plants in the population are selected for selfing or producing inbred seeds. The major characteristics used for selection are growth habit, leaf characteristics, color, type, and size of flower, resistance to pests and diseases, and time and duration of flowering. It may be necessary to continue selfing through several generations to obtain homologous lines.

Inbred lines can be produced two ways:

1. By growing uniform lines in an isolated area to prevent contamination by foreign pollen provided there is no morphological barrier for cross pollination. The seeds thus produced are called open-pollinated seeds and can produce true-to-type plants.
2. By hand pollinating or artificially pollinating self-incompatible types. Costly seeds can be produced under controlled climatic conditions.

In either case, the inbred lines produce uniform progenies, which are superior to mixed seeds, although their performance may not be as good as F_1 hybrids.

3. F_1 Hybrids

F_1 hybrids result from the crossing of two inbred lines. Since the 1950s numerous new and more attractive hybrid varieties of several types of flower crops have become available; for example, geranium, antirrhinum, pansy, marigold, zinnia, ageratum, double dianthus, begonia, impatiens, portulaca, gerbera, and carnation. F_1 hybrids represent a unique combination of vigor and uniformity and have advantages such as dwarfness or compactness, increased basal branching, free-blooming character, early and extended flowering, doubleness, improved flower size and shape, more attractive color, tolerance to adverse agroclimatic conditions, and resistance to pests and diseases. F_1 hybrids are not self-reproducible, thus protecting the plant breeder's interest.

4. F_2 Strains

F_2 strains are produced by selfing F_1 hybrids. Since F_1 hybrids are not self-reproducible, one can expect wide variation in flower color, size, shape, plant habit, and foliage characteristics. This is because diverse genotypes of parents are used to produce F_1 hybrids, and there is no uniform segregation of various characteristics in F_2. The major advantage of F_2 seed is its low cost. By proper selection of inbred lines, it is possible to produce nearly uniform F_2 strains.

Salunkhe et al. (1) described the essentials of the successful seed production of flower crops. Based on type of pollination, flowering annuals can be groupd into:

1. Self-pollinating crops
2. Cross-pollinated crops
3. Intermediate crops

According to Redenbaugh and associates (3,4), ornamental crops like begonia, cyclamen, geranium, gerbera, impatiens, and petunia could benefit commercially from the development of synthetic/artificial seeds through somatic embryogenesis technology, although the technological feasibility of using synthetic seeds of these crops remains to be established.

The value of the accelerated aging test for seed vigor has been limited in small-seeded crops, because moisture uptake is too rapid resulting in fast seed deterioration for some species. Zhang and McDonald (4a) examined a method to retard small seed moisture uptake in an accelerated aging test. High- and low-quality impatiens (*Impatiens wallerana* [*I. walleriana*]) seed lots were exposed to three different saturated salt solutions producing differing relative humidities (KCl 87%, NaCl 76%, and NaBr 55% RH) at 38 and 41°C. All other equipment and procedures were the same as recommended for a standard accelerated aging test. As the relative humidity of the accelerated aging chamber declined, seed moisture content decreased. Seed deterioration of the small-seeded crop increased with increasing relative humidity so that useful accelerated aging results that correlated with seed vigor could be obtained with either KCl or NaCl after 4 days' germination by aging for 72 and 96 hr at 38°C for KCl and NaCl, respectively, or 48 hr at 41°C for both salts. It was concluded that saturated salt solutions modify the relative humidity of an accelerated aging test, making the use of this valuable seed vigor test applicable to impatiens and likely other small-seeded crops (4a).

II. COMMERCIAL FLOWER CROPS

A. Ageratum

1. Introduction

Botany and Origin.   Ageratum (*Ageratum conzoides*), or floss flower, belongs to the Compositae family, and is a popular bedding and potted plant grown primarily for its blue, white, or pink flowers. It is native to Central America. The present-day commercial ageratum is derived from a cross between *A. conzoides* and *A. houstonianum* (*A. mexicana*). It is a small edging or bordering plant attaining a height of about 15–20 cm. The leaves are ovate in shape, dark green in color, and have toothed margins. Flowers are pale lavender to deep mauve in color. White, azure blue, and dark blue are also very common in the trade. The flowers are borne terminally in clusters to give an attractive cover to the foliage, growing out from the bottom of the plant. An F_2 hybrid of *A. houstonianum* showing the typical ageratum tints is pictured in Figure 1.

Cultivars.   Important varieties of ageratum are the F_1 hybrids such as Blue Angel, Blue Blazar, Blue Heaven, Royal Jay, Spindrift, Summer Snow and other Adriatics, Blue Puffs, Blue Surf, and North Sea.

Figure 1 F_2 hybrid of *Ageratum houstonianum* showing segregation of the typical ageratum.

2. Breeding

As in other Compositae members, the inflorescence in ageratum is known as the head, or capitulum, which consists of ray florets with very few, if any, disc florets. It is a cross-pollinated crop and needs proper isolation for seed production, which can be carried out both under field and greenhouse conditions. The open cross-pollinated seeds are also available to the trade.

3. Cultivation

Culture: Land and Isolation Requirements. Ageratum requires well-drained, aerated, light-textured soil with an adequate amount of organic matter for optimum growth. Water logging drastically affects the growth and flowering in this crop. A slightly acidic soil reaction (pH 6.0–6.5) favors growth. A mild climate with scanty rainfall during flowering is needed to achieve a good crop. The optimum temperature is 18–20°C; the crop, however, needs ample indirect sunlight.

Because it is cross pollinated, ageratum must be separated, with an isolation distance of at least 400 m from other crops to obtain high-quality seeds.

Planting and Interculture. Ageratum seedlings should be planted in rows in well-prepared land in the open from October–November in the plains and February–March in the hills. The field should undergo two or three weedings or hoeing operations to control weeds from time to time. Ageratum has no serious pests and diseases needing chemical control.

Roguing. The field should be examined to remove off-types periodically, especially at the time of flowering, to obtain pure seeds.

4. Harvesting and Postharvest Operations

The seed heads are harvested when they are dried and seeds are extracted by threshing. Seeds are then cleaned, graded, and stored in a cool, dry place.

B. Alyssum and Candytuft

1. Introduction

Botany and Origin. Alyssum (*Alyssum maritima* or *Lobularia maritima* L.) and Desy candytuft (*Iberis umbellata*) both belong to the Cruciferaceae family. Alyssum is a short-growing plant (20–30 cm) with narrow light green leaves and white, pink, or yellow inflorescence covering the entire top surface. The plant has a compact but spreading habit; each branch terminates into an inflorescence giving rise to a uniformly spread floral carpet on the soil surface. It is popular for edging, bedding, window boxes, rock gardens, and borders as well as for pot culture. Alyssum is native to Europe and Asia.

Commercial candytuft, also native to Europe (United Kingdom), is a short- to medium-height plant (30–40 cm) with long, narrow leaves and long white, pink, purple, liliac, carmine, or rose inflorescence. The individual flowers are larger than those of alyssum. The plant is less compact than alyssum and is used for borders, rockeries, pots, and window boxes.

Both species are highly cross-pollinated crops, being facilited by insects and honeybees.

Species and Cultivars. Two varieties of *A. saxatile* (*Compactum* and *Sulfureum*) are closely related to present-day commercial types, but are perennials with yellow flowers. Most *Iberis* species (*I. amara grandiflora, I. amara coronaria,* and *I. umbellata alba*) bear white flowers, whereas *I. umbellata carminea, I. umbellata purpurea,* and *I. umbellata liliaciana* have carmine, purple, and liliac flowers, respectively.

The popular commercial varieties of alyssum are Oriental Night (purple), Midnight (purple), New Carpet Snow (white), Royal Carpet (purple), Wonderland (deep pink), and Snow-drift (white).

Garden *Iberis* are divided into two groups: *I. amara* and *I. umbellata.* Another type, *I. odorata.* has sweet-scented flowers. The popular varieties of *Iberis* are Alba (white), Carminea (carmine), Liliacina (liliac), Purpurea (purple) and Red Cardinal and Dwarf Fairy mixed. The hyacinth-flowered strain of *I. amara* is a tall, errect plant bearing long-spiked white flowers resembling hyacinth.

2. Cultivation

Soil and Climate. Alyssum and candytuft require soil and climate similar to that of ageratum and other cool-season annuals. The optimum temperature and relative humidity are 21–24°C and 90–95%, respectively.

Planting. Alyssum seeds are very small (3500–4000 seeds/g) compared to those of candytuft (400–500 seeds/g). The seeds are sown in September and seedlings are planted in October, at 10–15 cm spacing, and using only one healthy seedling.

Isolation, Interculture, and Roguing. Because both alyssum and candytuft are cross pollinated, a minimum isolation distance of 400 m must be maintained from other crops. The field should be kept clean, well aerated, and loose. Off-types must be removed by roguing from time to time to maintain purity.

3. Harvesting and Postharvest Operations

The crop is harvested before shedding of seeds begins. Individual pods of alyssum are removed when they begin to dry, whereas, in case of candytuft, whole plants are harvested when pods turn yellow. Seeds are extracted after the pods are completely dried, which are in turn dried, cleaned, and stored in moisture-proof containers.

C. Aster

1. Introduction

Botany and Origin. China aster (*Callistephus chinensis*), belonging to the Asteraceae family, is native to China, and is one of the important cut flowers grown in both beds and pots in India and used in garden or interior decoration. Both tall and dwarf types are used as flowering or herbaceous border plants.

The name *Callistephus* is derived from a Greek word meaning beautiful crown. China aster is a popular annual flowering plant with erect growth habit, about 20–75 cm tall, and with coarse, hairy branches. The leaves are light to dark green in color and oval in shape with irregular or deeply cut margins and grooved petioles. The daisylike flowers vary in form, size, and color. They may be single or double headed, in curved, anemone, peony, quilled, ruffled, or comet types in form, and small buttonlike pom-pom or large in size. The common colors are white, rose, pink, red, blue-lavender, magenta, yellow, crimson, scarlet, mauve, purple, and primrose. The varieties vary in their keeping qualities.

Species and Cultivars. Aster varieties are classified on the basis of flower type or height. They may also be grouped into garden types (e.g., dwarf green) and florist cut types (e.g., Perfection Strain). The popular dwarf cultivars of aster are Dwarf Queen, Dwarf Triump, Dwarf Kirkwel (wilt resistant), and Pinocchio. The important tall cultivars are American Branching (wilt resistant), American Beauty, Powder Puff (bouquet type is wilt resistant), Grego, Giants of California, Heart of France, Giant Rocket, Super Princes, Pompom Laplata (wilt resistant), Stardust, and Queen of Market.

2. Breeding

The single and semidouble varieties of aster are predominantly cross pollinated, whereas the double-headed ones are generally self-pollinated. An effective aster-breeding program should ensure adequate isolation to safeguard against occasional cross pollination through insects. All modern varieties have been developed from *C. chinensis*. Early improvement work carried out in Europe, especially in France, yielded double flowers like peony flower types. Work done in Germany led to the development of a quilled-flowered–type aster, which became very popular in the United States as the German aster. Later introductions include double types, variegated blue and white types (United Kingdom), dwarf types followed by dwarf-comet types and those with compact long flat rays, cut-flower types with long stems and wide-spreading habit (France), and Semple strains with long stiff stems and large flowers (United States).

Apart from evolving varieties with more desirable flower and plant character-istics, wilt-resistant aster varieties were developed through continuous pure-line selection. The current breeding work concentrates on the production of aster–yellow virus–resistant lines and F_1 hybrids with attractive blooms suitable for both bedding and cut flowers.

3. Cultivation

Culture.

SOIL AND CLIMATE. Loamy or sandy loam soil is best for aster production. Heavy loams with inadequate drainage are not suitable unless the texture is improved by adding sand or riverbed. Aster crops also require an ample supply of organic matter in the soil. The land selected for seed crops should be well drained with a soil pH of around 6.5.

Since asters are very susceptible to frost injury, the time of sowing should be chosen to avoid exposure of the crop to chilling temperatures. The China aster grows best when the night temperature is around 15.4°C and it grows well up to 20°C (night temperature). At higher temperatures, the stems become stretched and bear poor-quality flowers. Asters need ample sunlight for growth, although they can tolerate shaded light.

Seed crops should be isolated from other crops, with a minimum distance of 400 m for single and semidouble strains, which are cross pollinated.

SOWING AND TRANSPLANTATION. In India, aster seeds are sown in August–September earlier in low-rainfall areas) in the plains and in March–April in the hills.

Raised nursery beds prevent water logging. Seeds are sown in shallow rows and lightly covered. Seeds germinate well at 20°C, taking about a week to complete germination. After germination, the temperature can be reduced to 15.4°C. It takes about 3 weeks for seedlings to be ready for transplantation.

The seedlings (with four true leaves) are transplanted to well-prepared land. The spacing varies with the type of soil and cultivar grown (30–40 cm). Dwarf varieties can be planted about 15–20 cm apart.

Irrigation, Fertilization, and Weed Control.   Aster seed crops need frequent irrigation depending upon the soil type and climate. The crop may be irrigated at intervals of 3–4 days during summer and 10–15 days during winter.

The crop is fertilized with 40 kg N, 40 kg P, and 20 kg K per hectare depending upon soil fertility. It is beneficial to apply nitrogen in two or three split doses—the first at the time of transplanting followed by another after 1 month of transplanting; the final dose is applied when flower buds appear. Poddagaudar et al. (4b) showed that GA_3 (200 ppm) spray significantly increased seed germination, root and shoot length, seedling dry weight, and seedling vigor index, which was on a par with 0.1% boron spray.

The plot should be kept free of weeds by two or three weedings or hoeings, which also helps to keep the soil well aerated.

Roguing and Control of Pests and Diseases.   Off-types and diseased plants, if any, should be rogued out from time to time. The important diseases of aster are wilt, damping-off, aster–yellow virus, and mycoplasma, which can be minimized by soil sterilization, clean cultivation, and change of place for growing asters. Aster can become infested with beetles, aphids, leafhoppers, and caterpillars, which can be controlled chemically.

4. Harvesting and Postharvest Operations

The seed heads are ready to harvest when they become fuzzy. The seeds are then dried in the shade, cleaned, and stored in moisture-proof containers in a cool, dry place. Small seeds (450–600 seeds/g) remain viable for about 1 year.

D. Calendula

1. Introduction

Botany and Origin.   Calendula (*Calendula officinalis*), or pot marigold, belonging to the Compositae family, is an important bedding plant grown primarily during the cooler months of winter and early summer in the plains and during the fall and summer in the hills. It can also be grown in shady locations receiving ample indirect sunlight in hot, humid climates.

Calendula is native to the Mediterrnean area of southern Europe, where it has been cultivated since ancient times. The inflorescence, or head, is fully double with ray florets. Flowers range in color from light yellow to deep yellow and orange to dark orange.

Cultivars.   Important commercial varieties of calendula include Coronets, Gypsy Festival Mix, Pacific Beauty Mix, and Mandarin (F_1 hybrid). There are several open-pollinated cultivars, which are suitable as cut flowers.

2. Breeding

Breeding in calendula is similar to that of other Compositae plants.

3. Cultivation

Culture.

SOIL, CLIMATE AND, ISOLATION REQUIREMENTS. Calendulas prefer cool and dry climates with ample indirect sunlight or partial shade. Temperatures beyond 30–35°C are harmful. A well-drained, light-textured soil with adequate organic matter is best suited for the growth of calendula. Heavy soils with poor drainage drastically reduce its growth and flowering. Soils with pH of 6.0–6.5 are optimum for growth.

Seed plots should be separated from those of other Compositae varieties by at least 400 m to avoid contamination.

PLANTING. Calendula is usually propagated through seeds, which are dark brown in color (125–200 seeds/g). They germinate at 20–21°C within about 7–8 days, and seedlings are ready to transplant in about 3 weeks. In the plains, seeds are best sown in September–October, whereas in the hills they should be sown in in February–March or July–August. Seedlings are planted in well-prepared land at 30×30 cm, preferably in the evening.

Roguing and Interculture. Seed crops should be irrigated frequently as and when needed; fields should be weeded or hoed to keep them clean. Periodical roguing should be carried out to remove off-types and diseased plants, if any. Because calendulas are cross pollinated, all atypical plants and cross-compatible species should be rogued as soon as they are noticed.

Harvesting and Postharvest Operations. Crops are harvested when the seeds are partially dried to avoid shattering losses. Harvested seeds must be dried before threshing. The threshed crop is cleaned thoroughly, graded, and stored in airtight containers in a cool, dry place.

E. Dahlia

1. Introduction

Botany and Origin. The dahlia (*Dahlia variabilis*) belongs to the Compositae family and is a native of Mexico. Dahlias are popular both as bedding and as potted plants, exhibiting a remarkable diversity of form, color, and size.

Present-day dahlias are a result of continuous crossing between several wild species and varieties followed by careful selection. It is an herbaceous plant growing to a height of 30–180 cm depending upon the variety and cultural method. The leaves are round, dark green, and opposite with slightly serrated margins. The flowers vary in size, form, and color. Flower sizes range from a few centimeters (miniature) to about 25–30 cm (giant or large varieties).

Species and Cultivars. Important species that have contributed to the development of modern dahlias are *D. coccinea*, *D. imperialis*, *D. merckii*, and *D. jaurezii*. Apart from planned improvement programs, chance mutations occurring in nature have also contributed to the development of some popular cultivars.

Based on the size and shape of their flowers, dahlia varieties can be grouped into 11 distinct types as follows (1):

Type 1, Singles: includes varieties with flowers 10 cm or less in diameter and have one row of ray florets surrounding a distinct central disc. They may either be "show singles," with rounded overlapping petals, or "singles," with more pointed petals. These are very popular as flowerbed plants (e.g., Erances, Kokette, and Liebenswert).

Type 2, Star Dahlia: have small flowers with two or three rows of pointed ray florets, which overlap slightly. The central disc is very prominent (e.g., White Star).

Type 3, Anemone-Flowered Dahlias: have an outer row of ray florets and a central group of disc florets, which are tubular and raised to give a pincushion effect (e.g., Comet).

Type 4, Peony-Flowered Dahlias: produce flowers with two or three rows of flat florets with a prominent disc in the center. The flower size varies from 12.5 to 17.0 cm or more (e.g., Large).

Type 5, Collaratte: blooms of this type usually have flattened outer ray florets with a prominent central disc and a row of narrower florets (collar) surrounding the central discs, which are half the length of the outer petals. The collar may be of the same or different color, and flower size is about 10 cm or more (e.g., Scarlet Queen, Lady Friend, Suntan, and Aureoline).

Type 6, Decorative Dahlias: blooms in this type are fully double, with several whorls of flat ray florets with nearly nonvisible central discs. This class can be subdivided into large (flower size over 20 cm), medium (15–20 cm), small (10–15 cm), and miniature (less than 10 cm) (e.g., Crydon Masterpiece, Liberator, and Peter Ramsey [large]; Peace and House of Orange [medium]; Mary Richards [small]; Arabian Night, Doris Duke, and Newby [miniature]).

Type 7, Pompom: are also fully double, up to 10 cm in diameter, and globular in shape with central florets smaller than outer florets. The florets are blunt, short, and incurved (e.g., Ascog [large], Bonny [medium], Diana [small], and Rosea [small]).

Type 8, Cactus Dahlias: like decorative types, these also have fully double flowers. The petals are straight, in curving, and partially revolute, giving a starlike appearance to the flowers. The ends of the petals in some varieties are split. These are also subdivided into large (over 20 cm), medium (15–20 cm), small (10–15 cm), and miniature (less than 10 cm) (e.g., Arab Queen, Rodeoo, and Colonel [large]; Polar Beauty, Guiding Star, an Eclipse [medium]; Pinnacle, Grace, and Doris Day [small], and Little Mermaid [miniature]).

Types 9–11, Miscellaneous Dahlias: flower types different from those mentioned above (e.g., orchid-flowered, speckled, or stripped-flowered types) are grouped as miscellaneous dahlias (e.g., Giratte or Disneyland). Apart from these, there are other types like show or fancy types and drawf bedding types.

2. Breeding

Breeding in dahlias is very simple, because no emasculation or bagging is necessary. As in other Compositae plants, the ray florets are pistillate flowers with stable stigma, which either open up or become larger and shiny when mature. Transferring pollen from the central disc florets of one flower to another with a fine brush can effect controlled pollination.

3. Cultivation

Culture.

SOIL AND CLIMATE. Deep fertile and well-drained soils with good aeration and a sandy loam texture are best suited to dahlias. The soil should be rich in humus and a good supplier of phosphorus and potassium. Dahlia requires ample sunlight, and low temperatures will restrict the growth.

PLANTING. Dahlias can be multiplied both sexually and vegetatively. The dwarf bedding types are usually sexually propagated, but other varieties, especially the double types, do not breed true-to-type when grown from seeds. Dahlias are usually planted in September and October in the plains or in March–May in the hills. In coastal areas and southern parts of India, seeds can also be sown in May. Seeds germinate in 1 or 2 weeks, and seedlings are ready to transplant in 3–4 weeks.

Dahlias can be grown either in beds or in 25- to 30-cm pots. Beds are prepared well and organic manure (farmyard manure [FYM], compost, or bone meal) is added ($155 \, g/m^2$). The rooted cuttings, tubers, or seedlings are then planted at $75–90 \times 45 \, cm$ depending upon variety and soil. When shoots are about 15–20 cm tall, the terminal buds are pinched off to promote branching (1).

Irrigation, Fertilization, and Interculture. Dahlias need frequent watering, although water logging should be avoided at all times. Heavy doses of nitrogen should also be avoided. Other operations like debudding, staking, and weeding are needed to obtain the best growth. The seed plot may need to be rogued for off-types and diseased plants, if any. Mildew and mosaic white are the most common diseases of dahlias, and aphids, thrips, and caterpillars are the important insect pests, which can be controlled chemically.

4. Harvesting and Postharvest Operations

Seeds are harvested when the pods turn yellow, which are dried on canvas or plastic sheets. Threshing or rolling extracts seeds. The threshed seeds must be cleaned immediately and stored in airtight containers.

F. Geranium

1. Introduction

Botany and Origin. The geranium (*Pelargonium hortorum*, L.H. Bailey), a member of the family Geraniaceae, has many forms. The leaves have zonal markings of band (blotch) variegation. Some cultivars have alternating colors of green and creamish yellow or white and sometimes red on the same leaf. The stem is fleshy but turns woody with age. The umbel-like inflorescence is borne on a long pedicel. The flowers are irregular with five petals, of which the upper two are relatively larger and more prominently colored. The most popular flower colors are pink, red, white, and purple. Geraniums are used as potted flowers and bedding plants.

Varieties. The present-day florist-type geranium is a hybrid derived from *P. zonale* × *P. inguinans* and *P. domesticum*. Based on the method of propagation, geraniums are grouped into seedling types and conventional types, which are propagated through cuttings. The seedling types take about 14–16 weeks from seed sowing to finish in 10-cm pots. Cultivars like Ireno, Sincerity, and Yours Truly are more popular than others. Some improved F_1 hybrids include the Ringo series, the Sprinter series, Mastang, Jack Pot, Tiffany Red, Show Girl, and Bright Eyes.

2. Breeding

Both seedling and vegetatively propagated geraniums have been available for commercial production since the 1970s. With the introduction of newer seed-propagated cultivars, several growers have begun growing seed geraniums that have shown better performance than the vegetatively propagated cultivars. The former are more tolerant to pests and diseases and have lower production costs (no need to raise stock plants). They can be made available in large quantities through breeding, since the exact time of flowering can be predicted more easily.

Marsolais et al. (5) produced somatic embryos from the petioles and hypocotyls of Zonal (*P.* × *hortorum*) geranium and from the petioles of Regal (*P.* × *domesticum*) geranium. Somatic embryos from both species were desiccated and subsequently germinated. Important factors that influence the rate of somatic embryo production, such as culture medium (auxin and auxin dosage, carbohydrates, amino acids, pH, and basal medium composition) as well as donar plant genotype, were found to have an effect on somatic embryogenesis and survival after desiccation. This study clearly showed that all 30 cultivars of Zonal geranium were capable of producing somatic embryos in the genotype survey (Table 1); 18 of these produced desiccated artificial seeds that turned green and began germination after rehydration. Most germinating artificial seeds, however, did not survive to develop into mature flowering plants, but were lost through the stages of radicle and shoot development and during transplanting to potting soil. About 41% of the seedlings derived from artificial (desiccated) Scarlet Orbit Improved seeds developed into mature plants (5).

3. Cultivation

Geraniums for seed are usually grown in pots under controlled conditions. Since plant height must be in proportion to pot size, growth regulators are generally used to produce plants of the desired size and shape. Chlormequat or chlorocholine chloride can be used as a foliar spray or soil drench to reduce plant height, to hasten flowering, and to increase the number of flowers per plant (1).

Geraniums are benefited by CO_2 enrichment in the greenhouse, especially during winter months, when ventilation is a problem. At a CO_2 of 500–700 ppm concentration plants grow faster and flower earlier. Wider plant spacings have been used to regulate plant height and produce better-shaped plants.

Seeds are harvested when the pods are fully ripe. They are dried, and then seeds are extracted, cleaned, and stored in a cool, dry place.

G. Marigold

1. Introduction

Popular as both cut flowers and bedding plants, marigold are easy to grow and are widely adapted to varied agroclimatic conditions. They produce flowers for a longer

Table 1 Somatic Embryogenesis and Germination After Desiccation of 30 Cultivars of Zonal Geranium

Cultivars	No. of somatic embryos per 100 hypocotyl sections	% Somatic embryos germinating after desiccation
Scarlet Orbit Improved	689 ± 79[a]	38
Merlin	649 ± 215	25
Picasso	589 ± 120	26
White Orbit	582 ± 277	40
Scarlet Eye Orbit	556 ± 184	19
Cherry Orbit	533 ± 164	0
Mustang	476 ± 125	48
Jackpot	382 ± 150	16
Cardinal Orbit	336 ± 45	28
Breakaway Salmon	327 ± 128	38
Cherie Light Salmon	256 ± 104	0
Ringo	211 ± 67	40
Pink Elite	176 ± 95	46
Ringo White	171 ± 65	0
Sprinter Scarlet	160 ± 27	0
Salmon Elite	158 ± 36	33
Ringo Salmon	140 ± 72	33
Bright Eyes	133 ± 54	30
Violet Orbit	118 ± 16	14
Ice Queen	104 ± 19	0
Orchid Orbit	87 ± 41	0
Deep Salmon Orbit	82 ± 47	0
Coral Orbit	80 ± 27	0
Pinto Rose	71 ± 27	0
Red Elite	71 ± 36	25
Ringo Rouge	67 ± 30	9
Sundance Orange Scarlet	31 ± 13	0
New Dawn	31 ± 11	75
Tiffany	31 ± 11	0
Ringo Rose	20 ± 10	0

[a] Values are the means of five replications ± SEM (standard error of means).
Source: Ref. 5.

time and the flowers last much longer than those of other bedding plant species. They are useful for herbaceous borders, hanging baskets, window boxes, garlands, rockeries, and edging owing to their wide variation in height, growth habit, and flower color. In the United States, marigolds are second only to petunias in popularity (1).

Botany and Origin.   In addition to *Tagetes erecta* (African marigold) and *T. petula* (French marigold), other species include *T. tenufolia* (*T. signata*), *T. lucida* (odorless), and *T. filifolia*.

The plants of Tagetes are quick-growing, bushy perennials but are grown as annuals. Marigolds vary in height from 10 to 100 cm or more. The leaves are dark green, deep cut, with a typical color. Flower heads vary in size, form, and color from small (1 cm) to large (15 cm). The African types are usually yellow, orange, yellow-orange, or greenish yellow, whereas French marigolds have a still wider color variation. The forms range from carnationlike peony and chrysanthemum-flowered, quilled, to fluffy heads.

Varieties. Important African marigold types are Cracker Jack (carnation-flowered), Guinea Gold, Gold Coin, Climax, Sun Giants, Smiles, Man-in-the Moon, and Yellow Supreme. The medium-sized varieties include the Jubilee series, with Happy Face, Gold Galore, and Lady. The dwarf African marigolds include the Space-Age series (carnation-flowered) with Apollo, Moonshot, Mariner, Aztec, Golden Age, Guys and Dolls, Spun Gold, Spun Yellow, Cupid, Happiness, Dolly, and Pot-O-Gold.

French marigolds are easier to grow and bloom earlier than African types. They are mostly dwarf and compact, producing single or double blooms, with flower colors including yellow-orange, primrose, tangerine, rusty red, bronze, mahogany, bicolors, deep scarlet, and combinations of colors. A new group of large-flowered French marigolds known as Super French have been developed. Some important French marigolds are Flame, Flaming Fire Double, and Rusty Red (all dwarfs). The Petile series are small dwarfs (15 cm) and include Gold, Gypsy, Orange Flame, Sunkist, and Tom Thumb. Some dwarf cultivars with single flowers are Naughty Marietta, Dainty Marietta, Ganuabar of India, and a tetraploid variety Tetra Ruffed Red.

The Super French group includes Sparky, Spanish Brocade, the Ole or Bonita series (Bolero), Golden Boy, Gypsy Dancer, Kind Tut, Aquaris, the Royal Crested series (Honey Comb), and others.

Some triploid marigolds (*T. erecta* × *T. petula*) are Seven Star and Showboat, with very attractive color combinations. Signet marigolds (*T. tenufolia pumila* or *T. signata pumila*) are very popular in Europe. These are compact, dome-shaped plants (15–25 cm) with fine lacy foliage; they include varieties like Golden Gem, Lemon Gem, Paprika, and Irish Lace, having compact growth with dense green foliage (30 cm high) often used for edging.

2. Culture, Interculture, and Harvesting

Marigold seeds (black, 300–350 seeds/g) remain viable for a couple of years. French seeds germinate to an extent of about 90–95%. Seedlings are grown on raised beds; seeds are sown in May–June, August–October, or February–March (in the plains) or March–April (in the hills). Seeds germinate best at 20°C within 8–10 days and are ready to transplant in 3 weeks.

Land is prepared well by plowing and two or three harrowings. The seed plot should be separated from other varieties by a minimum distance of 400 m to prevent contamination. Shivakumar et al. (5a) recommended that a NPK (nitrogen, phosphorus, potassium) level of 270:72:72 kg/ha with the spacing of 60 × 30 cm was optimum for quality seed production of marigold. The seed crop should be harvested at 28 days after opening.

Plant spacing varies with cultivar, growth habit, and soil type from 15 to 60 cm. Excessive nitrogen fertilization promotes vegetative growth at the cost of flowering.

Weeds must be controlled by frequent hoeings. French marigolds are pinched when the first flower appears to encourage branching.

Seeds are harvested when the crop is completely dried, threshed, cleaned, and stored in a cool, dry place.

H. Pansy

1. Introduction

The name *pansy* is derived from a French word, pensee, meaning thought. Pansy plants are most valued for garden decoration. Pansies are a perennial crop, but are mostly treated as an annual in the plains and a biennial in the hilly temperate regions.

Botany and Origin.   Pansies (*Viola tricolor*) belong to the Violaceae family and are native to Europe (southern France). The common florist-type pansy belongs to the species *V. tricolor* var. *Hortensis*, growing 15–25 cm tall. It has a compact, somewhat trailing habit. The leaves are long and cut at the margin. The flowers are borne singly on a long stalk and are attractively colored, which makes the plant an all-time favorite of gardeners. Flower colors include white, yellow, cream, orange, apricot, red, pink, rose-salmon, purple, primrose, blue, and mauve. Several varieties have flowers with wavy or frilled edges beautifully marked, striped, or blotched, veined, margined, or variegated in contrasting colors. Flowers also vary in size from miniature (2–3 cm) to large or giant (12.5 cm or more) types. Flower shapes vary widely too. They are sweetly scented, which can be noticed best in the early morning.

Varieties.   Important pansy varieties include the F_1 hybrids of the Universal and Vinking series, Magestic Giants, Golden Champion, Orange Prince, Mammoth Giant (all F_1), and Goldsmith Giant (F_2 mixtures), the Springtime series (F_1), the Imperial series (F_1), Ballering mix (formula mix F_2), Early Market mix (F_2), King Size mix (F_2), and some open-pollinated types.

2. Breeding

Breeding work in pansies has resulted in the development of several newer and more attractive varieties for the flower trade. The inheritance of flower color has been shown to be somewhat complex, indicating gene interactions. The prostrate growth habit is dominant to erect plant types. Variegation has been shown to be a non-Mendelian characteristic depending upon plastids in the cytoplasm. The development of velvety block color is controlled by the expression of three genes and five inhibitor genes suppressing the characteristic.

3. Cultivation

Soil and Climate.   Fertile, well-drained sandy loam soils are best suited for the growth of pansies. A high–organic matter (humus) content in the soil favors excellent growth and flowering. The soil pH should be around 7. Pansy is a cool-climate crop grown usually in winter; pansy seed production is thus concentrated in the hilly temperate regions. Warm climates adversely affect growth, flowering, and seed setting. The crop, however, must to be protected from frost and cold waves.

Land and Isolation Requirements.   The site selected for seed production should be open and protected from scorching sun. Although pansy is predominantly a self-pollinated crop, some cross pollination does occur. An isolation distance of 100–200 m from other varieties has been recommended for seed production.

Planting, Interculture, Irrigation, and Fertilization.   Pansies are propagated through seeds that are sown in September–October in the plains and a little earlier in the hills to enable the plants to become established before the cold weather sets in. The best time for planting in the hills is either March or August. Seeds are sown in raised beds or seed boxes well protected from direct sun. They are sown in shallow rows and covered lightly, and the beds are irrigated. Mild weather favors seed germination (20°C), and seedlings with three to four true leaves are used for transplantation. Seedlings are planted in rows, 15–25 cm apart, ensuringing adequate fertilization and irrigation. Weeds should be controlled by two or three hoeings. Apart from normal cultural practices, hand-pollination and use of growth regulator (naphthalene acetic acid) may be necessary to obtain good seed yields.

The common diseases of pansy are root rot, leaf spot, powdery mildew, rust, mosaic, and cottony mold. Aphids, wireworms, and sawflies are serious insect pests, which should be controlled using suitable measures.

4. Harvesting and Postharvest Operations

Seeds should be harvested before shattering begins. Since plants flower continuously, hand-picking of seed pods may be necessary. The capsules are collected in boxes and dried in a well-ventilated place. Seeds are usually extracted manually by rubbing or crushing. They are then cleaned, dried in the shade, and stored in a dry, cool place.

I. Petunia

1. Introduction

Botany and Origin. Petunias (*Petunia hybrida* L.), belonging to the Solanaceae family, are the most popular bedding plants worldwide and are also suitable for edging and pot culture. Today's petunias show a wide range of variation in color, size, and plant height (15–45 cm). The stems are herbaceous in nature with small, rounded to elongated, succulent to semisucculent leaves. In most cultivars, stems trail or creep on the ground. The trumpet-shaped flowers are borne in leaf axes. The flowers may be single, semidouble, or double, smooth, or frilled, having petals that are single-colored, bicolored, or tricolored. Petunia flower colors include all shades of white, pink, rose, red, scarlet, violet, blue, yellow, cream, salmon-rose, magenta, and lavender or combinations of these.

Varieties. Several F_1 and F_2 strains of petunia are available to the trade that have quality superior than straight puure-line varieties. Petunia cultivars can be grouped into the following six types:

1. *Dwarf and Compact Types*: suitable for edging and borders; most are early and flower readily (e.g., Blue Bird, Snowball, Rosy Mom, and Lady Bird).
2. *Grandiflora*: have large single or double blooms, with smooth-edged or ruffled petals, having attractive colors. The pure lines of this group include Bingo (smooth edged, wine red, and white), Dazzler (carmine), Purple Prince (smooth), Ramona, Snowstorn, and Defiance (all frilled). There are several F1 grandifloras on the market, which are classified on the basis of their petal colors; for example, grandiflora single blue and purple, grandiflora single pink and salmon, grandiflora single red, grandiflora single starred, and grandiflora ruffled (Can Can).
3. *Multiflora*: small-flowered types with either single or double flowers, with attractive colors; for example, multiflora single, blue and purple, multiflora single pink, multiflora single red, and multiflora white and yellow.

4. *F₂ Strains*: include both grandiflora (Carniwal) and multiflora (Colorama, Confetti) types.
5. *Hanging Basket or Trailing Types*: have long trailing stems with small flowers; for example, Ex Rose Wonder, Rose, and Blue Wonder.
6. *Doubles*: produce fully double, smooth or frilled flowers. Both grandiflora and multiflora are available.

2. Breeding

Both double-flowered forms and large-flowered grandifloras are products of hybridization, with the double form characteristic being dominant over the single form and linked to female sterility, producing only occasionally a functional pistil, which permits maintenance of the line. Using, seed of single-flowered parents and a homozygous double clone for pollen breeds double-flowered types. Double-flowered types are normally maintained vegetatively by cuttings.

Grandifloras with large flowers are also produced by breeding wherein the grandiflora characteristic is dominant to multiflora (small flowers) types. Crossing multiflora seed parents with grandiflora lines produces commercial F_1 grandiflora seeds.

Major efforts in petunia breeding are being made to develop varieties with better color, increased flower size, freedom from bloom, compact plant growth, and earliness. Resistance to *Botrytis* and petal spotting is also being sought through breeding. The use of cytoplasmic male-serile lines in petunia has been limited owing to factors such as production of inferior F_1 hybrids, increased flower bud blasting, and inferio flower quality (size and color). Genetic male-sterile lines and incompatibility systems are also being used to a limited extent because of the higher cost of maintaining these lines by vegetative means as compared to the cost incurred by pollination via hand emasculation.

3. Cultivation

Soil and Climate.   Light-textured, well-drained fertile soil with adequate organic matter favors petunia growth. Heavy soils with impervious substrata hinder growth and reduce the flower number and duration of flowering. Double varieties require soil with organic matter.

Cooler climates with moderate rainfall and 13–16°C soil temperatures are optimum for petunia. Petunia seed crops should be isolated, with a distance of at least 100 m from other varieties.

Planting Fertilization, and Interculture.   Petunia seeds are very small (about 9500–10,000 seeds/g). They are usually sown in September–October in the plains and in February–March in the hills. Seeds germinate well at 20–25°C and require light for better germination. Seedlings (5 cm) are ready to transplant about a month after sowing. All double-flowered varieties are propagated through terminal cuttings.

Uniform seedlings are transplanted in well-prepared fields enriched with FYM at 10 kg/m² at 30 × 30 cm spacings. The seedlings are pinched when they attain a height of about 15 cm.

Petunia seed crops may be sprayed with a solution of urea or top-dressed with nitrogenous fertilizers at 10 g/m². Adequate supplies of phosphorus and potassium should be ensured at the time of transplanting.

The field should be kept free of weeds and other pests with appropriate control measures. Seed crops are raised both under open-field conditions and under controlled greenhouse conditions. The latter is more effective in producing high-quality seeds. Pollination may be brought about either by hand emasculation or using certain equipment to facilitate the operation.

4. Harvesting and Postharvest Operations

The seed crop of petunia is harvested when it turns yellow or yellowish brown, before the seeds begin to shatter. The seeds are extracted from the pods, dried in the shade to reduce seed moisture content, and stored in airtight containers in a cool, dry place.

J. Snapdragon

1. Introduction

Snapdragons are a relatively recent introduction to the commercial cut flower group. Traditionally, they have been grown as perennial or annual flowers. Their production for cut flower purposes began with the introduction of a winter-flowering cultivar, Chevoit Maid, followed by several F_1 hybrids with desirable cut flower qualities.

Snapdragons are used in floral arrangements and for bouquets; the yellow and pink ones are highly preferred to others. The F_1 hybrids are grown year round.

Botany and Origin. Snapdragon (*Antirrhinum majus* L.), which belongs to the Scropulariaceae family, is a native of the Mediterranean region (6). The plant is a low-growing tender herb with side shoots arising very near to the ground. The leaves are usually opposing and entire. The corolla is gibbous or saccate at the base and personate or closed near the throat. The flowers are borne on spikes and florets in open succession.

Cultivars. Present-day greenhouse snapdragons are the F_1 hybrids arising from the crosses between the early-flowering winter varieties and the earlier F_1 hybrids. Two main types of snapdragon varieties are known: (1) cut flower, or florist, and (2) garden decoration types. Several varities are available, which can be used for both purposes.

In India, snapdragons are usually classified as tall, grandiflorum, or giant (1.0–1.2 m), medium (0.60–1.0 m), tomb or dwarf (0.3 m), and carpet types (0.10–0.15 m). Tetraploids with large flowers and fully double flowers are also available. More vigorous and floriferous F_1 hybrids and F_2 strains are more popular among commercial growers.

On the basis of their response to light and temperature conditions, cultivars of snapdragons can be grouped into four groups:

1. Early or winter groups flower readily in short photoperiods and at a relatively lower temperature (10°C), and take about 18–22 weeks to flower after sowing.
2. Late winter or spring group cultivars produce quality flowers at 10°C night temperature, and take 22–24 weeks for flowering.
3. Late spring or early summer cultivars take longer at 10°C, but flowering can be induced quickly (in about 16 weeks) by growing them at 15.5°C under long-day conditions.
4. Summer types perform much better at high temperatures (15.5°C or higher) and long days. Cultivars are selected depending largely upon their flowering period and adaptability to location.

2. Breeding

Present-day F_1 hybrids combine the winter-flowering and stout vegetative characteristics of their parents. Cross pollination in snapdragons occurs mainly through larger
insects like bumblebees. Plants resistant to rust (a devastating disease of snapdragons)
were found in early 1940s, which were selfed and coupled with selection over several
generations. The resistance to rust was found to be due to one dominant gene. The
segregation in the F_2 generation followed a typical Mendelian ratio of 3 (resistant) to 1
(susceptible).

Studies of genetics and inheritance in snapdragons indicated that magenta is a
dominant color and yellow a recessive one. The yellow varieties are, therefore, easy to
purify. Also, crimson has been found to be dominant to bronze, bronze to yellowish
bronze, magenta to rose, and rose to ivory. The delilah form of flower is recessive to the
nondelilah type; that is, colored tips and colorless tubes are recessive.

The postharvest quality attributes in snapdragons include stem length, total
number of florets, number of open florets, spike weight, freedom from abnormal
growth, crocked stems, susceptibility to diseases and pests and physiological disorders,
vase life, and retention of petal color (1).

3. Cultivation

Culture.

SOIL AND CLIMATE. Fertile but light-textured soils that are well aerated and well
drained favor the growth of snapdragons. The soil, however, should be at least 20 cm
deep. Severe soil compaction causes stunted growth, weaker root system, and delayed
flowering. Hanan and Langhans (7) reported that an increase in the bulk density of soil
from 1.089 to $1.479/cm^3$ reduced the growth of roots and shoots of snapdragons by
about 30%. Since the crop requires good soil aeration, an unamended field soil results
in poor performance with regard to quality. Soils with ample quantities of organic
matter and humus favor flower quality. Hanan and Langhans (7) recommended a ratio
of soil to sand or perlite mixture of 3:1 to produce a high-quality crop. The soil should
have adequate nutrient-supplying capacity for all necessary major and minor
elements.

Soil water is an important parameter determining productivity and flower
quality in snapdragons. Even short-term water logging during the initial stages of
crop growth can severely damage seedlings, decreasing the growth and quality of
spikes produced. Soil moisture stress is as damaging as excess water (8). An excellent-
quality crop can be produced by maintaining soil moisture and air at 24–34% by
volume and 45–55% of total pore space in the root zone, respectively (9). Water stress
also reduces the flower grades and the dry matter content of the flower tissue.

Snapdragons are long-day plants requiring ample sunlight for growth and
flowering. Both temperature and light should be considered together. The optimum
temperatures required for growth decrease with crop age. Whereas seedlings require
20°C (night) temperature, plants nearing flowering grow best at 13°C (night). Higher
temperatures adversely affect growth and flower quality.

SOWING AND TRANSPLANTING. Seeds almost exclusively propagate snapdragons.
Small seeds (6000 seeds/g) germinate about 80–85%. Seeds are sown in raised nursery
beds, which are prepared to a fine tilth with assured aeration and drainage. Seeds
germinate well at 18–21°C when supplemented with light. The recommended seedling

density is up to $3000/m^2$. Seeds are shallowly sown and lightly covered. The seedbeds may be covered with plastic sheets to maintain humidity. Seeds germinate within 5–7 days, and seedlings are ready to transplant in about 3–4 weeks. Seedlings can be stored for up to 6 weeks at 2–7°C with supplemental fluorescent light.

The seed crop needs to be isolated from other varieties by at least 100 m. Young seedlings with a pair of first true leaves may be transplanted in protected greenhouse conditions, whereas seedlings with four true leaves (1–3 cm tall) must be transplanted in open fields. In India, spacing of 20, 30, and 40 cm is recommended for dwarf, medium, and tall cultivars, respectively (about 115–125 cm^2/seedling). Young seedlings need to be protected from high winds, heavy rains, direct sun, and excessive drought conditions.

Irrigation, Fertilization, and Weed Control. Adequate soil moisture must be ensured throughout crop growth. Higher soil moisture levels decrease soil aeration and water use efficiency. It is advisable to grow snapdragons "on the dry side" to prevent the adverse effects of water logging, which may cause root rot.

Snapdragons have relatively low nutritional requirements, although high nitrogen levels produce stronger growth (6). Larson (10) has given a detailed account of the symptoms of nutritional deficiencies in major and minor elements. Boron deficiency inhibits the development of meristematic tissue, followed by death of growing points and development of axillary shoots. Calcium and boron activities are antagonistic to each other in soil.

Seed crops must be kept free of weeds, insect pests, and diseases. The common diseases of snapdragons include *Botrytis* blight, powdery mildew, downy mildew, and diseases caused by *Alternaria* and *Helminthosporium*. Plants or flowers infected by diseases and pests produce large amounts of ethylene.

Pinching and Roguing. Plants are pinched when they are 15 cm high to induce development of side shoots. Pinching delays flowering, however. Following pinching, the axillary shoots develop from leaf axils. Only the desired number of shoots should be maintained to produce a high-quality seed crop.

On the basis of morphological and flower characteristics, off-types as well as diseased plants should be rogued out before and after flowering to obtain pure seeds.

4. Harvesting and Postharvest Operations

Snapdragon cut flowers are very sensitive to ethylene exposure. In addition, they produce ethylene in sufficient quantities to cause a drastic reduction in their vase life and quality. Stem tip breakage due to rough handling during harvesting, grading, and packaging reduces the quality of cut flowers.

The spikes of seedpods are harvested when they are just become dry to minimize shattering losses of seeds. Spikes from lower branches mature earlier than upper ones. The seeds are usually extracted and cleaned manually, dried in the shade, and stored in airtight containers in a dry and cool place.

K. Sweet Pea

1. Introduction

Botany and Origin. Sweet pea is one of the most attractive annual climbers grown around the world. It originated in Sicily, and was introduced to the Netherlands and

England in the seventeenth century. It is valued for its attractively colored, fragrant flowers borne in a long panicle. Most present-day varieties have trailing stems requiring support. Plants attain a height of 0.5–2.5 m. The stem is rough and hairy, and the leaves resemble those of the garden pea. Two to four florets are borne on a long stalk. The flowers consist of a pair of "wings," a keel, and a hood; flower colors include hues of white, rose, scarlet, yellow, blue, crimson, red, purple, violet, cream-salmon, lavender, orange, and mauve. The form and size of flowers also vary greatly.

Cultivars.   The important sweet peas varieties are Eleanor (blue), Gloria (pink), Grace (lavender), Lily (white), Marylin (scarlet), Miranda (deep crimson), Patti (salmon pink), Ramona (orange), the Jet Set mixture, and the Royal Family mixture. Bijou is a bushy strain growing up to 30–40 cm tall.

2.   Breeding

About 500 varieties of sweet pea are available for both garden display and cut flowers. They have evolved through selection since the seventeenth century, with improved flower size, number of florets per spike, flower form, and growth habit. Hybridization of sweet peas began in England in 1880. Since then, many innovative hybrids have been introduced to the trade, adding to the present-day water-flowering group. A hybrid variety, Waved King Edward, was produced by crossing King Edward (plain red) with Countless Spencer (waved pink). The hybrid Primrose Spencer was bred by a similar procedure. Another hybrid, Black Knight Cupid, was synthesized by crossing Black Knight (tall, purple, dark wing) with Pink Cupid (cupid red, light wing). In hybridization, it is important to know the mode of inheritance of different characteristics. A single gene controls flower color. White color is dominant to cream. Similarly, colored flowers are dominant to white, purple to red, bright to dull colors, full to dilute colors, light to dark wings, and purple to copper and maroon. Two identical white parents give purple F_1 hybrids and nine purple to seven whites in F_2. There are also modifier genes for both purple and red. Tall, bush, cupid, and bush-cupid growth habits are found. A cupid × bush cross produces a normal plant in F_1.

3.   Cultivation

Soil and Climate.   Any well-aerated and well-drained garden soil is suitable for sweet pea. The optimum soil pH is 6.0–6.5. Soil should be medium textured. Sweet pea is a winter-flowering annual requiring a cool climate throughout its growth. It also requires ample sunlight.

Isolation.   Although sweet pea is a predominantly self-pollinated crop, some cross pollination may occur. It is therefore necessary to isolate the seed crop by a distance of 50–100 m from other varieties.

Planting and Interculture and Staking.   Bold seeds (12–15 seeds/g) are dibbled (at 3 cm) directly into the prepared field in September–October (on the hills) or in February–March (in the plains) in rows about 15–30 cm apart depending on the variety used. The plot is watered regularly and kept free of weeds by hoeings or weedings. A basal dose of 10 kg FYM/m² is recommended. Applying 300–400 g of fertilizer mixture per square meter must also ensure adequate supplies of phosphorus and potassium.

The plants need physical support, which is generally provided by bamboo sticks when plants are about 20–25 cm tall. In certain varieties, pinching may be necessary to encourage branching when plants have six or seven pairs of leaves. Flowering commences above 4–6 weeks after sowing and continues for more than one month.

4. Harvesting and Postharvest Operations

The seed crop is ready to harvest when the lower pods are completely dried. At this stage, whole plants are cut and dried in the shade before threshing to separate seeds. The seeds are then cleaned, graded, and stored in airtight containers.

III. BIENNIAL AND PERENNIAL FLOWER CROPS

Seed production for biennial and perennial flower crops is meager compared to that for annual flower crops. Most perennial flower crops are propagated vegetatively, and hence gardeners and florists do not depend on seeds for raising these crops. The initial planting material for biennial and perennial flower crops is costly. Also, it takes a long time for these crops to flower. Growers can, however, obtain seeds from a single crop for several years, with the length of time depending upon the species, management, climate, and soil fertility. Hardy crops can survive and produce enough seeds for several years.

Fields for the production of seeds of perennial flower crops should be located such that soils are fertile, well aerated, and well drained. Heavy clayey soils should be amended suitably by adding sand or coarse particles, and too light sandy soils may be improved by adding organic matter. The climatic conditions of the selected site should be able to meet minimum requirements of the flower crop. Perennial crops often form clumps, which should be divided and planted separately to maintain seed yields in later stages. The rows are dug out and clumps are separated every year, and can be used to establish new planting.

The time of planting depends on the crop species, location, and agroclimatic conditions of the area. Crops are generally planted in the fall (rainy season) or spring. In the plains, they are planted between August and November, whereas in the hills, spring (March) is the best time for starting a new seed crop.

A. Carnation and Pink

1. Introduction

Botany and Origin. Carnations (*Dianthus caryophyllus*) belong to the Caryophyllaceae family. Carnations and pinks are valued as a garden and florist crop for their attractive flowers. Several varieties of carnation are also grown under controlled conditions for cut flowers. In the garden, carnations and pinks are used for flowerbeds, borders, and cut flowers. Most garden types are hardier than florist types. Plants attain a height of 20–60 cm, having a tuft of grasslike leaves. Flowers are borne terminally either singly or in clusters. The petal margins are smooth, fringed, or frilled. Flower size varies from less than 2 to 10 cm or more. The flowers may be white, creamy yellow, pink, red-orange, mauve, or a combination of two or more colors. The florist-type carnation can be tinted to give any color.

Species/cultivars. The important species of carnation are *D. carophyllus* (carnation), *D. barbatus*, *D. chinensis* (pink), *D. deffoides*, (maiden pink), and *D. plumeris* (grass pink).

2. Cultivation

Soil and Climate.   Pinks and carnations prefer well-drained, well-aerated fertile soils. Sandy loam and alluvial soils with a pH of 6.3–6.8 are ideal. Plants are very sensitive to deficiencies of boron and other minor elements.

Although pinks and carnations are hardy, they require protection from frost and chilling temperatures during winter. The best growth occurs between 8.8 and 21°C. Plants need ample indirect sunlight.

Planting and Interculture.   Perennial carnations and pinks are started from seeds by direct sowing or transplanting seedlings. The seedlings are raised on flat or raised seedbeds, which are transplanted after 1 month. The best time for planting is September–October in the plains and February–March in the hills. The spacing can vary from 22.5 to 45.0 cm depending upon variety and species. The intercultural operations are similar to those for flowering annuals. The diseases of carnation, namely, rust, leaf spot, and wilt, and insect pests like aphids and thrips should be controlled using appropriate measures.

3. Harvesting and Postharvest Operations

Seeds are harvested before shattering begins. The pods are usually picked manually and dried in the shade. The seeds are extracted by hand or by threshing. After cleaning, seeds are stored in a dry, cool place in an airtight container.

B. Larkspur

1. Botany and Origin

Larkspur (*Delphinium grandiflorum*), belonging to the Ranunculaceae family, is popular in the trade for its attractive flowers that are borne on a long spike. It is used for flowerbeds and borders. The perennial plants grow to a height of 45–150 cm, with an average spread of 30–45 cm. The flowers have five sepals, which are petal-like, and the posterior petal is prolonged into a spur. The flowers at the bottom open first and proceed acropetally. The fruit has two to four carpels, with each follicle having many triangular and winged seeds.

2. Culture

Delphiniums require a fertile, light-textured, well-drained, and well-aerated soil medium for the best growth and flowering. A soil pH of 6.3–6.8 is optimum.

Larkspur is propagated by sowing seeds directly or by transplanting seedlings. Seeds are usually sown in August–September, and the seedlings are transplanted in October–November in the plains. In the hills, the seed crop is started in the spring. Seeds germinate best at around 15°C, and can be aided by placing seeds in moist blotting of filter paper for 7 days at freezing temperatures or for longer at 10°C as an after-ripening treatment. Light-textured, well-drained fertile soil is optimum for growth. Seeds germinate within 2–3 weeks, and seedlings with two to four true leaves are transplanted in prepared flat beds using wider spacing. The seedlings may be hardened before transplanting by exposing them to the sun or by reducing water so that they may more easily adapt to adverse field conditions.

Delphiniums can also be propagated by dividing the crown or by cuttings. The crowns are preferably divided in the spring after flowering is completed and the plants have put up new vegetative growth. Only the healthy part with one or two young vegetative shoots from the periphery of the crown is used for multiplication. Alternatively, cuttings about 5–10 cm long with one or two roots may be removed and planted in a suitable medium. These cuttings develop strong root systems within 2–3 weeks; at which time they can be transplanted singly in polyethylene bags or pots until they are finally set out in the field. Seedlings are usually spaced about 30–45/25–30 cm apart (1).

The crop must be ensured adequate fertilization. Dosages of 150 kg N, 100 kg P, and 50 kg K per hectare, depending upon soil fertility, may be used. A well-decomposed FYM at 5–10 kg/m^2 should also be applied at the time of land preparation. Diseases such as powdery mildew, black spot, blight, and aster-yellow and insect pests like cyclamen mites, thrips, and aphids can be damaging to the seed crop if not controlled.

3. Harvesting and Postharvest Operations

Seeds mature in June–July in the plains and August–September in the hills. The pods should be harvested before the seeds begin to shatter. The spikes are cut manually when the flower pods dry out and begin to open. The seepods are dried on canvas. The extracted seeds are cleaned and stored in a suitable container in a dry, cool place.

C. Hollyhock

1. Botany and Origin

Hollyhock (*Althea rosea*), belonging to the Malvaceae family, is an excellent garden plant used for borders, flowerbeds, and shrubbery. Although a perennial, it is mostly treated as an annual in India. Hollyhock plants grow to a height of 45–150 cm or more. The leaves resemble those of cotton or okra. The stems are glabrous and yellowish green in color. The flowers are perfect, sessile, and large and are borne singly in the leaf axils. Recent introductions to the germplasm are more floriferous double hollyhocks with frilled petals. Some have flowers in clusters of two or more per axil, which open in succession. The new types also have a longer flowering period and branching habit. The flower color varies from white, pink, orange, and red to bicolor types. The seeds are round, bold, and medium sized (15–20 seeds/g).

2. Culture

Hollyhock plants are raised from seeds sown directly in the field, which should be well prepared and leveled. Seeds are dibbled in rows, 60–90 cm apart, using about 2.5–3.0 kg seed/ha. Seeds germinate within about a week. Thinning may be required to orient seedlings about 60 cm apart from each other. The best time for sowing is May–June in the plains and February–March in the hills. Application of nitrogenous fertilizer at 10 g/m^2 with 10 kg FYM/m^2 accelerates crop growth and seed development.

In addition to the usual cultural practices, tall plants may need staking to keep them upright. Because hollyhocks are cross-pollinated crops, population of honeybees and other insect pollinators should be ensured.

The crop may be infested by pathogens causing leaf spot, root rot, and rust and insect pests like spider mites, which should be controlled from time to time.

3. Harvesting and Postharvest Operations

Seeds mature within 1 month after flowering. Mature seedpods may be harvested individually or by cutting whole plants. The pods are dried and seeds are extracted manually by hand or by threshing. After cleaning, seeds are stored in a dry cool place.

D. Phlox

1. Introduction

Botany and Origin.   Phlox (*Phlox* spp.), belonging to the Polemoniaceae family, can be grown as both an annual and a perennial crop in the garden. Because of its wide variation in flower size, shape, and color, phlox is an excellent plant for flowerbeds, edging, rock gardens, and flowering hedges.

Perennial phlox plants grow to a height of 60–150 cm and spread 30–45 cm. Flowers are produced terminally in clusters. Each flower contains three to five stamens and a three-celled ovary. The petals are attractively colored with shades of white, yellow, red, pink, orange, purple, and mauve or a combination of two or more colors.

Species/Varieties.   The annual varieties of phlox grown in the garden belong to *P. durmmondii*, whereas the perennial varieties are from *P. paniculata* and *P. maculata*. Other species like *P. subulata* (moss pinks), *P. multiflora* (alpine types), *P. pilosa*, and *P. divaricut* (early types) are also grown to a limited extent.

2. Culture

Seeds are sown directly in the field or seedlings may be raised for transplanting. Dividing the crown can also raise perennial types. The latter method ensures true-to-type and uniform progeny. The best time of planting is October–November in the plains and February–March in the hills. Seedlings are ready for transplanting about 1 month after sowing.

Annual varieties may be spaced at 25–30 × 20–25 cm, whereas perennials and tall types need wider spacings of 45–60 × 20–25 cm. The crop needs to be protected from freezing temperatures.

Adequate soil moisture and fertility should be ensured during crop growth. Application of organic matter at 10 kg FYM/m^2 at the time of land preparation and top-dressing with nitrogenous fertilizer at 10 g N/m^2 improves growth and flowering. Flowering begins about 45–60 days after transplanting, and seeds take another 30–40 days to mature. Diseases such as leaf spot and powdery mildew and other pests like nematodes and plant bugs should be controlled using appropriate measures.

3. Harvesting and Postharvest Operations

Mature pods should be harvested before seed shattering begins. Seed stems should, therefore, be harvested before the pods are completely dried. The harvested pods are then dried on canvas and threshed to extract seeds. Seeds are cleaned and stored in a dry, cool place using appropriate containers.

E. Minor Crops

In addition to the above perennial crops, other minor crops are also grown for seeds. The information regarding the botany and seed-production technology for these crops is summarized in Table 2.

Table 2 Botany and Seed Production Technology of Some Minor Perennial Flower Crops

Common name	Scientific name	Family	Plant height (cm)	Method of propagation	Growing conditions	Spacing (cm)	Harvesting	Remarks
Alyssum	Alyssum saxatile	Cruciferae	30	Direct seeding	Cooler temperatures	25 × 20	By cutting stem	Requires well-drained soil
Anchus	Anchus spp.		30–40	do	do	30 × 25	do	do
Columbine	Agilegia vulgaris	Renunculaceae	45–90	Transplanting	Milder climate	45 × 30	Picking pods manually	do
Coreopsis	Coreopsis lanceolata	Compositae	30–45	do	do	33 × 30	do	do
Campanula	Campanula spp.	Campanulaceae	60	do	do	45 × 30	do	do
Digitalis	Digitalis purpurea		90–120	do	do	45 × 45	do	do
Forget-me-not	Myosotis alpestois		30	Direct seeding		15 × 15	do	do
Gaillardia	Gaillardia aristata	Compositae	30–45	Transplanting	Warm climate	30 × 30	do	do
Linum	Linum perenne		30–45	Transplanting	Milder climate	20 × 20	Picking pods manually	Requires well-drained soils
Lupine	Lupinus polyphyllus		45–90	do	do	30 × 30	do	do
Iceland	Papaver nudicaule		30–90	do	do	30 × 45	do	do
Poppy, Oriental	Papaver orientale		33–90	do	do	45 × 60	do	do
Poppy, Shasta daisy	Chrysanthemum leucanthemum	Compositae	45–90	do	do	30 × 30	do	do

Source: Ref. 1.

The technique and benefits of seed priming, in which seeds imbibe sufficient water to begin the process of germination to a predetermined point and are then dried back prior to storage or packing, are employed widely. The advantages to this method include more uniform and rapid seedling emergence, greater tolerance of less than ideal environmental conditions at sowing, and an increase in the vigor of the developing seedlings. Primed seeds are presently being used commercially for the production of many ornamentals, including pot cyclamens and bedding plants of begonias, pansies (*Viola* sp.), polyanthuses (*Primula* × *polyantha*), and primroses (*Primula vulgaris*) (11).

IV. VEGETATIVELY PROPAGATED FLOWER CROPS

Several commercial flower crops are propagated vegetatively involving asexual reproduction through plant parts such as underground-modified structures, stem cuttings, budding, or grafts. The heterozygous types can be easily and quickly multiplied by this method to produce true-to-type plants. In recent years, application of the tissue culture method has opened a new horizon in plant propagation, especially in plants that do not set viable seeds.

The choice of the method of vegetative propagation depends largely on the plant species and objectives. Saluunkhe et al. (1) classified these methods as follows:

Use of polyembryonic seeds
Use of specialized plant parts: rhizomes, corms, bulbs, tubers, other structures
Cuttings and layering
Grafts and buddings
Tissue culture

A. Use of Polyembryonic and Apomictic Seeds

In some plant species, seed is formed from diploid cells, which may be either nonreduced megaspore mother cells or cells from ovular tissue. In these types, the usual meiotic cycle and fertilization are eliminated and as such they are asexual in nature. The seeds of these crops breed true-to-type. In a majority of cases, however, they are partially apomictic because of their association with sexual seeds. The seedlings from apomictic seeds are thus difficult to identify.

The development of polyembryonic or asexual seeds is probably controlled by recessive multiple genes regulating the synthesis of embryogenesis inhibitors in nucellar cells of monoembryonic species. The development of polyembryonic seeds from nucellar cells requires a stimulus provided by pollination, fertilization, and early development of zygotic embryo. It is possible to induce polyembryonic development of artificial seeds through tissue culture (e.g., citrus).

B. Use of Specialized Plant Structures

Certain plant species have specialized structures like modified stems or roots, which are primarily food-storage tissues and are often used to multiply the plants asexually. These specialized plant organs help them to withstand adverse climatic conditions such as very low or high temperatures by entering a dormant period. The shoots and roots in these structures are much reduced or modified to conserve maximum food material. The various specialized structures are known as rhizomes, corms, bulbs, tubers, runners, and offsets.

1. Rhizomes

Rhizomes are underground-modified cylindrical stems that grow horizontally. They are thick and fleshy and contain papery scales. They vary in size, branching, number of nodes, and length of internodes. The growth begins from terminal buds. The adventitious roots arise from nodes. It is normal practice to divide rhizomes into several pieces, each with one or more internodes, which are planted separately.

2. Corms

Corms are underground, highly compressed, solid, modified stems consisting of nodes and internodes. The topmost, or apical, bud usually gives rise to the main shoot, whereas the buds on the outer periphery produce multiple shoots. Whereas scaly leaves often cover the buds, thick leathery leaves cover the whole corm. The adventitious roots arise from the base of the corm. The planted corm is expended in flower production, and the new corm is formed at the base of the shoot above the old one. In a large corm, two or more flowering shoots may grow, each producing a new corm. The cormel or cormlets, which are fleshy in nature, have only one terminal bud growing in between old and new corms. These cormels produce single shoots, which produce a large corm at the end of the growing season. The cormels usually take two to three growing seasons to produce commercial-sized (or flowering-size) corms. The corms are graded according to their diameter or weight. Gladdiolus, crocus, freesia, and water chestnut are example of plants producing corms.

3. Bulbs

Bulbs are shortened, flat stems surrounded by fleshy leaves inside and leathery leaves outside. The nodes and internodes are much fewer in number. The buds present at the base of the scaly leaves produce bulblets or bulbils, which when grown to their full size are known as offsets. The bulbs are usually produced within one growing season, but increase in size if left undisturbed. In some plant species, the scales can be separated and used for propagation. Examples of plants propagated by bulbs include hyacinth, Easter lily, and tulip.

4. Tubers

Tubers are elongated, enlarged fleshy portions of underground stem. They may contain only one crown bud, as in dahlia, or several buds all over the surface, as in potato. The latter may be used as a whole or divided into pieces containing at least one "eye" or bud. In a whole tuber, the apical bud inhibits the sprouting of other buds. The seed pieces are often cured to heal the wounds before planting. Dahlia and tuberous begonia are commercially propagated through tubers.

5. Runners

Runners are aerial stems growing horizontally on the surface. There may be several runners arising from various nodes. Unlike corms, tubers, rhizomes, and bulbs, the runners do not serve as food-storage tissue. The new growth arises primarily from the apical bud, whereas roots grow from nodes. Plants can be multiplied 20- to 30-fold

using runners. Plants that are commercially propagated through runners are strawberry, *Saxifraga*, *Chlorophytum*, and *Ajuga* spp.

6. Offsets

The offsets are lateral shoots arising from the main stem, which are divided, rooted,
and used to grow whole plants. These shoots are also known as suckers, slips, or
divisions. Plants propagated through offsets are *Sansevieria*, pineapple, and certain
cacti. The offsets are usually first rooted in seedbeds prior to their planting in the field.

C. Cuttings and Layers

The plant's ability to produce shoots and roots under favorable conditions to develop
into a new plant has been utilized to propagate several plant species. The regenerated
plants that are induced to root while they are still attached to the stock plant are known
as "layers," whereas those that are detached from the parent plant and then induced to
root in the medium are called "cuttings." These plant parts have the ability to
regenerate the missing part(s). Some plants produce aerial roots while they are still
attached to the plants, but in the majority, etiolating, wounding, girdling, bending, and
reorientation of the stem encourage the root development. These practices probably
check the flow of hormones and carbohydrates, which accumulate at the cut end or
wounded portion to stimulate rooting.

　　The process of root development involves root initiation and root growth.
Whereas the first phase is characterized by the increased meristematic activity and
differentiation into root cells, the second phase involves cell division and cell
elongation. Certain plant species root more easily than others; the ability to root
depends on the interaction of internal factors present in the shoot cells. These
substances constitute carbohydrates, auxins, vitamins, nitrogenous compounds, and
other unidentified compounds, which are usually synthesized and transported besi-
petally and accumulate near the cut ends where they promote rooting. Several
compounds that either promote or inhibit rooting in different species have been
isolated. The fact that leaves and buds promote rooting suggests that they provide
auxins and other cofactors needed for rooting. In several species, higher carbohydrate
levels and relatively low-nitrogen levels promote rooting, increasing the number of
roots. However, nitrogen deficiency can also affect normal root development.

　　In most plant species, the ability to root is influenced by the age of the stock plant
(juvenile, reproductive, or mature), the type and location of cutting (soft or hard
wood), the stage of development at the time of cutting, and the time of the year, with
spring to late fall being optimal. In addition to internal factors, environmental
parameters like temperature, light, relative humidity, and type of rooting medium
also exert a significant influence on the rooting response. The plants differ in their
requirements for light, temperature, and rooting medium, and the conditions favoring
maximum metabolic activity in the cell have generally been found to promote rooting.

D. Grafting and Budding

Grafting and budding involve joining of plant parts or tissues for their physical union
and subsequent independent plant growth. The part that provides the root system is
known as the "rootstock" and the part that is united to the stock is called the "scion."
The scion may contain one bud (budding) or more (grafting). The graft combinatons
in some cases are made up of more than two parts, in which case the middle section is

called the "interstock." The strong and well-developed root system of one plant (rootstock) can be combined with the desirable qualities of other plant parts (scion) into a single plant through grafting and budding to enable the plant to withstand adverse soil and climatic conditions, achieve resistance against stresses, and yield better.

The scion may be either detached or attached to its parent (approach grafting) while the graft union is taking place. The approach graft is used in cases where it is difficult to obtain graft union. This method is most commonly used in roses, cacti, and other woody ornamental plants.

E. Tissue Culture

Plant tissue culture, also known as in vitro culture or microculture of plants, is the modern biotechnological method of growing, developing, and preserving plant cells, tissue, organs, or plants in aseptic and controlled conditions of a well-defined physicochemical environment. The material is cultivated within enclosed vessels and maintained in a sterile nutrient medium and gaseous environment. The small scale and limited internal space within the tissue culture vessels results in miniaturization of the material (microculture). Plant tissue culture involves a range of technologies to produce plants on a commercial scale. It is also an ideal medium for research investigation into crop improvement. It enables mass propagation of economically valuable clones and new crop introductions and production of high-health (virus-free) plants. It facilitates engineering and selection of new and elite genotypesof crops in addition to the synthesis of valuable secondary metabolites of agroindustrial importance. Plant tissue culture has also led to the production of artificial or synthetic seeds of several crop plants through somatic embryogenesis (see Chapter 23). The importance of this newer biotechnology is increasingly being recognized in the areas of agriculture, horticulture, and forestry (12).

F. Production of Vegetatively Propagated Seed Materials

The commercial production of planting materials in bulbous plants is a specialized operation involving several steps:

1. Harvesting and grading of bulbs into stocks and marketable bulbs
2. Preplanting storage of bulbs
3. Planting, rooting, and low-temperature mobilization for flowering or bulb growth
4. Flowering
5. Increasing the size and number of bulbs

In recent years, the tissue culture technique has found an important place in the commercial multiplication of bulbous plants, including those of flowers and ornamentals. The basic steps in the production of planting materials of these crops are given in the following sections. Detailed information can be found elsewhere (13–15).

1. Gladiolus

Botany and Selection of Cormels. Gladiolus (*Glodiolus* [*Toum*] L.), belonging to family Ixidaceae family (subfamily Ixioideae), is propagated through corms, which are graded according to their size. The corms are multiplied from cormels, which grow in clusters in between the old and new corms. The cormels can also be graded into different sizes for better results. The cormels selected for planting must be free from pests and diseases. The cormels are carefully selected from a mother block of corms and may be treated with hot water, with or without fungicide, to eradicate nematodes and other pathogens, if any. A 30-min dip of cormels in benomyl and captan (or thiram) solution held at 53–56°C has been recommended (16,17). The nineteenth century miniature cultivars (hybrids) of gladiolus are *G. colvillei* and *G. nanus*.

Field Selection and Fertilization.   The field selected for the production of gladiolus bulbs should be well aerated, well drained, and light in texture. Sandy loam or alluvial soils are best for this purpose. In heavy soils, the yielda and quality of bulbs produced are poor. The ideal soil pH is between 5.8 and 6.5. It may be necessary to fumigate the soil before planting to eradicate pathogens. Gladiolus needs ample sunlight and clear weather for its growth. An ample quantity of organic matter (5–10 kg FYM/m^2) and 30 g of N, 10 g of P, and 20 g of K per suqare meter should be applied at the time of planting (18). In addition to the basal dose, application of liquid fertilizers is known to encourage faster vegetative growth.

Planting, Harvesting, and Storage.   The bulbs are planted in rows about 25–60 × 10–15 cm apart. Adequate soil moisture should be ensured to facilitate early sprouting.

After flowering, when the leaves turn yellow and become dry, the corms and cormels are lifted from the field manually, and are then cleaned and dried in the shade. Precautions must be taken to avoid physical injury to corms during harvesting and postharvest operations. After adequate drying, the corms are stored in a cool and dry place. The use of fungicide like benomyl, captan, or thiram along with 5–10% DDT and suitable airtight containers such as polyethylene bags makes safe storage of corms possible for up to 6 months. Cold storage may be employed for longer storage periods.

2. Freesia

Botany and Cultivars.   Freesia (*Freesia refracta*), belonging to the Iridaceae family, is a bulbous ornamental plant of great value in the florist trade. Freesias are grown either in pots or in fields for cut flowers and are an excellent bedding material. The plants are low growing (30–40 cm) with linear dark green leaves. The flowers are trumpet shaped and brightly colored with shades of white, yellow, cream, pink, red, mauve, orange, bronze, and blue. The flowers are borne in recemes on a long stalk and are often highly scented.

Several varieties of Freesia known in the trade have mixed colors. The leading varieties include Pink Giant, White Swan, Sapphire, Gold Coast, and Blue Banner.

Freesia plants thrive best in open areas and dry conditions. A light-textured, well-aerated, and well-drained soil is ideal for growth. Poor drainage leading to water-logging conditions favor development of diseases, adversely affecting both the yield and quality of the seed crop.

Planting, Irrigation, and Harvesting.   Freesias are commercially propagated through corms, which resemble those of gladiolus but are smaller in size. Both corms and cormels can be used for multiplication.

The corms are planted at 15.0 × 2.5–5.0 cm; closer spacing may be used for cormels. Seed crops may be planted in October in the plains and from August–November in the hills, protecting the plants from cold waves and frosts.

Irrigation is withheld after flowering until the leaves turn yellow and dry. The corms and cormels are then lifted from the ground manually using forks, taking care not to injure them during harvesting. They are cleaned carefully to remove soil and other extraneous matter. The corms are dried in the shade for 2–3 days and stored in a cool, dry place after treating them with a suitable fungicide and insecticide.

3. Dahlia

Botany. Dahlias (*Dahlia variabilis*), members of the Compositae family, vary greatly in height (30–180 cm) and flower size (10–30 cm). The botanical details and varietal classification of dahlias are discussed earlier in this chapter (Sec. II.E).

The ideal site for dahlia seed crops should be sunny and well protected from cold and freezing temperatures. A light-textured, well-aerated, and well-drained soil favors growth (e.g., sandy loam or alluvial soil).

Planting, Harvesting, and Storage. Plants are propagated through seeds, cuttings, or tubers. Sexual propagation does not breed true-to-type progeny. The double, cactus, decorative, and pompom types are generally raised from tubers are generally planted in June in the plains and in March–April in the hill. Tubers in clumps are separated carefully with the help of a sharp knife, taking care not to injure the crown bud. The tubers are planted in rows 45 × 90 cm apart. The plants are pinched off when they are 15–20 cm high to remove the terminal portion of the main shoot to encourage branching. Interculture operations like irrigation, fertilization, and staking, are similar to those for the commercial flower crop.

After flowering, the plants are headed back to about 15 cm from the ground. The tubers are lifted from the ground carefully with the help of a fork, taking care not to injure the crown of the roots. The tubers are cleaned, dried in the shade, and treated with a suitable fungicide and insecticide before storing in a cool, dry place. Temperatures varying from 4.0 to 7.2°C with adequate air circulation and low humidity are best for storing tubers in good condition.

4. Narcissus and Daffodil

Botany and Cultivars. Narcissus (*Narcissus* spp) and daffodil, belonging to the Amaryllidaceae family, are similar to each other in all respects except that daffodils produce large trumpet-shaped flowers. The plants are low to medium growing (up to 60 cm) and have dark green, long, and narrow leaves. The flowers are borne on a long stalk and consist of a perianth tube having six segments at the top with an attractively colored corona projecting forward. The species and varieties differ in the size, shape, and color of the corona. The perianth tube is ether white or yellowish in color, whereas the corona tube may be yellow, white, red, pink, or a combination of different colors. The flowers are produced singly or in clusters of two or more flowers. Narcissus and daffodil are ideal for cut flowers, beds, and borders. The dwarf types are also used for rock gardens.

Planting and Harvesting. Narcissus can be multiplied by offsets. The best planting time is October in the northern plains and September–October in the hills. The bulbs are planted about 7.5–12.5 cm deep, spaced at 15 × 10 cm, and covered with soil. A sandy loam or alluvial soil with ample organic matter (humus) content is ideal for the seed crops. The site selected should also have ample indirect sunlight.

Irrigation is stopped after flowering is completed. The bulbs are carefully lifted with the help of forks when the leaves become yellow and dry. The bulbs are dried, cleaned, and stored in a cool, dry place at 17–20°C. The bulbs may be treated with hot water at 44°C and formalin or other pesticides to prevent infection with Fusarium and nematodes.

5. Hyacinth

Botany and Cultivars. Hyacinth (*Hyacinthus orientalis*), belonging to the Liliaceae family, is a native to Syria, Palestine, Turkey, and the Asia Minor region. It is ideal for pots, beds, window boxes, bowls, vases, indoor gardens, and hyacinth glasses. The flowers are borne in dense racemes, each consisting of about 50–60 bell-shaped florets, bluish in color. The leaves are dark green and strap-shaped. Several varieties available in the trade are used for forcing. Most of these thrive best in cool climates and require well-drained fertile soils.

Planting, Harvesting, and Storage. The bulbs are planted in October in the northern plains and in February in the hills 12.5 cm deep and spaced at 15 × 15 cm.

Hyacinth bulbs are generally multiplied in the hills only, which is done by offset (bulblet) reproduction. Because the crop is perennial, the production of offsets (bulblets) is slow. The bulbs are scooped to remove the basal place entirely, leaving only the scales, or they are scored to cut the basal three times in a pie-shaped pattern to facilitate rapid multiplication (19). Using this technique, adventitious buds are formed on the cut scale surfaces. The scooping method produces more bulbs, but it takes about 3 years to produce commercial-sized bulbs, whereas the scoring method produces fewer but large-sized bulbs in 1 year. The spikes produced are removed in the early stages to encourage bulblet production.

The bulbs are harvested when they are ready, cleaned, and dried in the shade. Hyacinth bulbs are best stored at 25°C. They may be treated with hot water (44°C) to control yellow disease (*Xanthomonas hyacinthi*). Under humid storage conditions, *Penicillium* is another prevalent fungal disease.

6. Iris

Botany and Cultivars. Iris (*Iris* spp.), belonging to the Iridaceae family, has several species. Most of the cultivated ones belong to one of two groups: (a) rhizomatous rooted (tall and dwarf all), bearded flag, or German types or (b) bulbous rooted Dutch, Spanish, and English types. The Dutch irises are more attractive and last longer than German types. The Dutch iris has resulted from a cross between *I. xiphium* Preacox, *I. tingitana*, and *I. lusitanica*.

The important varieties of Dutch iris are Wedgewood, Ideal, and Prof Blaauw (Blue Ribbon), primarily grown in the United States, The Netherlands, and France. Flower colors include white, red, orange, yellow, and blue. *I. reticulata* is also grown widely in gardens. *I. danfordiae* (yellow) and *I. reticulata* (blue) are grown in pots. Cantab, an outstanding variety of *I. reticulata* with bright blue flowers and ideal for growing in an alpine house, is shown in Figure 2.

The rhizomatous German irises are tall with beards on the top of three petals (fall petals). The plants grow to a height of 45–60 cm and have swordlike leaves. The flowers are borne on a long spike and are blue, purple, white, yellow, creamish, rose, pink, or bronze in color. Some Himalayan species (*I. napalensis* and *I. kashmeriana*) belong to this group. Irises are excellent for cut flowers, borders, and beds.

Figure 2 *Iris reticulata* Cantab, an outstanding variety with bright blue flowers. (From Ref. 26.)

Planting and Harvesting. The plants are propagated from rhizomes, with both types being reproduced by annual replacement or division. The corms are harvested and graded in July–August and stored at 16–18°C until the next planting. In Dutch irises, the stock plants that flower are usually considered to be nonmarketable, because they produce flat bulbs, whereas the large, round bulbs from vegetative plants produce the best planting materials.

7. Lilies

Lilies are excellent for pots, beds, and borders. They are generally multiplied by offset bulblet production similar to that of other bulbous crops. The bulbs are lifted after flowering, dried in the shade, and stored in a cool, dry place until the next planting.

Table 3 Prices for Rooted Plants Commercially Produced by Micropropagation

Plant name	Cost of acclimated plants ($)
Dieffenbachia amoena Schott ("Compacta")	0.51[a]
Nephrolepis exalta bostoniensis Davenport ("Boston fern")	0.40
Ficus benjamina L.	0.47
Hemerocallis thunbergii Baker ("Aztec Gold")	0.47
Spathyphillum patinii Patin ("Petite")	0.49
Syngonium podophyllum Schott. ("White Butterfly")	0.40

[a] Prices are an average from 14 tissue culture companies for 1987.
Source: Ref. 21.

V. PRODUCTION OF SYNTHETIC SEEDS OF ORNAMENTAL CROPS

Ornamental crops are presently valued at $2.4 billion or more (20). Many of these are laboriously micropropagated via tissue or organ culture where per plant production costs exceed $0.50 (Table 3). Most of this cost is a result of the manpower needed for multiple cultures and rooting steps. According to Gray and Purohit (21), micropropagation for such crops could be replaced by embryogenic culture to reduce labor costs, since somatic embryos can be mass produced on calluses, which can then be manually selected and placed directly into planting flats resulting in rooted plants. This could eliminate several labor-intensive steps. The implementation of artificial seeds for

Table 4 Somatic Embryogenesis in Ornamental Plants

Genus and species	Explant source
Foliage plants	
Cheiranthus cheiri L.	Seedling callus
Corton bonplandianum L.	Endosperm
Hedera helix L.	Stem section
Peperomia longifolia C. DeCandole	Stem section
Piper nigrum L.	Stem section
Yucca spp.	Lateral bud, shoot tip
Zamia integrifolia Ait.	Megagametophyte
Flowers and foliage plants	
Antirrhinum majus L.	Leaf protoplast
Calostama purpureum R. Br.	Nucellus
Euphorbia, pulcherrima Wild	Seed
Gentiana spp.	Embryo
Helleborus foetidus L.	Anther
Iris spp.	Shoot apex callus
Kalanchoe pinnata Pers.	Leaf sections
Lilium spp.	Embryo
Mesembryanthemum floxibundum Haw.	Hypocotyl, leaf, root, shoot tip
Narcissus biflorus Curt.	Anther
Paeonia hybrid Pall.	Anther
Paeonia lutea Pall.	Anther
Petunia hybrida Vilm	Leaf section, stem, intermode
Petunia inflata R.E. Fries.	Leaf section, stem, intermode
Spathyphillum patinii Patin	Nucellus
Tulipa gesneriana L.	Nucellus
Orchids	
Aranda spp.	Embryo
Cattleya spp.	Embryo
Cymbidium spp.	Embryo
Cypripedium spp.	Embryo
Dendrobium spp.	Embryo
Oncidium spp.	Embryo
Phalaenopsis spp.	Embryo
Vanda spp.	Embryo
Vanilla spp.	Embryo

Source: Ref. 21.

flowers and ornamental crops is compelling, because a relatively modest level of technological development would result in lowering existing costs of somatic embryogenesis (22). Somatic embryogenesis has been reported for a number of ornamental species (Table 4). Additional, more refined ornamental systems have also been envisaged with a decrease in the cost of production.

Ruffoni et al. (23) described the production technology of artificial seeds (encapsulated somatic embryos) of *Lisianthus* (*Eustoma grandiflorum*) and *Genista monosperma*, which included investigation to improve the germination and establishment of embryos. The addition of zeatin (0.5 mg/L) to the alginate encapsulation coating improved the shoot production, and sucrose (40 g/L) added to the MS agar germination medium improved the percentage root emergence of *Lisianthus*. Holding the sown embryos in the dark for the first 30 days also improved germination. Encapsulated *G. monosperma* embryos did not respond to the application of abscissic acid (2.5 mg/L) as a 1-hr soak prior to drying, and suffered severe damage from as little as 6 hr of desiccation in a laminar flow cabinet (23). Gill et al. (24) reported that somatic embryos of geranium (*Pelargonium* × *hortorum*) cv. Scarlet Orbit Improved germinated normally and produced flowering plants within 12–14 weeks when they were hydrogel encapsulated using 3% sodium alginate and 50 mM calcium chloride. The synthetic growth regulator thidiazuron at 1 µM was superior to 1 µM IAA (indole-3-acetic acid) + 8 µM benzl adenine for producing somatic embryos that confer viability to synthetic seeds following storage at 4°C for 45 days.

REFERENCES

1. D. K. Salunkhe, N. R. Bhat, and B. B. Desai, *Vegetable and Flower Seed Production*, Agricole Publishing Co., New Delhi, 1987.

2. H. Bodger, The commercial production of seeds of flowers, *Seeds: The Yearbook of Agriculture*, U.S. Department of Agriculture, Washington, DC, 1961, p. 216.

3. K. Redenbaugh, D. Slade, P. Viss, and J. A. Fujii, Encapsulation of somatic embryos in synthetic seed coats, *HortScience*, 22(5):803 (1987).

4. K. Redenbaugh (eds), *Synseeds: Application of Synthetic Seeds to Crop Improvement*, CRC Press, Boca Raton, FL, 1993.

4a. J.H. Zhang and M.B. McDonald, The saturated aging test for small-seeded crops, *Seed Sci. Technol.* 25:123 (1997).

4b. S.P. Poddagaudar, B.S. Vyakaranahal and M. Shekhargouda, Effects of growth regulators and boron on seed quality parameters of China aster (*Callistephus chinensis* L), XI National Seminar on Quality Seed to Enhance Agricultural Productivity, UAS, Dharwar, January 18–20, 2002 (A. Gaur, A.K. Vari, and J.L. Varshney, and K. Kant, eds) Indian Society of Seed Technology, Division of Seed Science and Technology, IARI, New Delhi, *Seed Technol News* 32:3, 2002.

5. A.A. Marsolais, D.P.M. Wilson, and M.J. Tsujita, Somatic embryogenesis and artificial seed production in Zonal (*Pelargonium* × *hortorum*) and Regal (*Pelargonium* × *domesticum*) geranium. *Can. J. Bot.* 69:1188 (1991).

5a. C.M. Shivkumar, B. Gouda, M. Shekhargouda, M.B. Kurdikeri, P.S. Dharmath, V.S. Patil and S.B. Negalur, Quality seed production in marigold (*Tagetes erecta* L) cv. Double orange, XI National Seminar on Quality Seed to Enhance Agricultural Productivity, UAS, Dharwar, Jan, 18–20, 2002 (A. Gaur, A.K. Vari, J.L. Varshney, and K. Kant, eds.), Indian Society of Seed Technology, Division of Seed Science and Technology, IARI, New Delhi, *Seed Technol News* 32:3, 2002.

6. M.N. Rogers, Snapdragons, *Introduction to Floriculture* (R.A. Larson, ed.), Academic Press, New York, 1980, p.107.
7. J. J. Hanan and R. W. Langhans, Soil aeration and moisture control snapdragon quality, *NY State Flower Growing Bull.* No. 210, 196 pp. 3–6.
8. J.J. Hanan and R.W. Langhans, Soil water content and the growth and flowering of snapdragons, *Proc. Am. Soc. Hort. Sci.* 84:613 (1964).
9. J.J. Hanan and R.W. Langhans, Efficiency and effects of irrigation regimes on growth and flowering of snapdragons, *Proc. Am. Soc. Hort. Sci.* 86:681 (1965).
10. R.A. Larson (ed.), *Introduction to Floriculture*, Academic Press, New York, 1980.
11. P. Jordan, Why use primed flower seed, *Plantsman* 14(4):247 1993.
12. M.A.L. Smith, Plant tissue culture, *Encyclopedia of Agricultural Science*, Vol. 3, Academic Press, New York, 1994, p. 369.
13. *Bulb Corm Production*, Her Majesty's Stationery Office, London, 1964.
14. C.J. Gould, *Handbook of Bulb Growing and Forcing*, Northwest Bulb Growers' Association, Washington, DC, 1957.
15. A.R. Rees, *The Growth of Bulbs*, Academic Press, New York, 1972.
16. R.O. Maggie, Effectiveness of treatments with hot water plus benzamidazole and ethephon on controlling *Fusarium* disease of gladiolus, *Plant Dis. Rep.* 55:82 (1971).
17. R.O. Maggie, The hot water treatment for gladiolus propagation, *Gladiograms* 17:4 (1975).
18. G.J. Wilfret, Gladiolus, *Introduction to Floriculture* (R.A. Larson, ed.), Academic Press, New York, 1980.
19. H.T. Hartman and D.E. Kester, *Plant Propagation: Principles and Practices*, Prentice-Hall, Eaglewood Cliffs, NJ, 1975.
20. Floriculture crops, 1989, Summary, U.S. Department of Agriculture, Washington, DC, 1990.
21. O.J. Gray and A. Purohit, Somatic embryogenesis and development of synthetic seed technology, *Crit. Rev. Plant Dev.* 10(1):33 (1991).
22. W.J. Florkowski, O. Lindstrom, C. Robacker, and W. Simonton, Biological, technical and economic aspects of micropropagation, *Georgia Agric. Exp. Sta. Res. Rep.* 556 Athens, GA, 1988.
23. B. Ruffoni, F. Massabo, and A. Giovannini, Artificial seed technology in the ornamental species, *Lisianthus* and *Genista*, International Symposium on Agrotechnics and Storage of Vegetable and Ornamental Seeds, Bari, Italy, June 14–16, 1993, *Acta Hort.* No. 362, 1994, p. 297.
24. R. Gill, T. Senaratna, and P.K. Saxena, Thidiazuron-induced somatic embryogenesis enhances viability of hydrogel-encapsulated somatic embryos of geranium, *J. Plant Physiol.* 43(6):726 (1994).
25. R. Reimann-Philipp, Heterosis in ornamentals, *Heterosis*, Vol. 6 (R. Frankel, ed.), Springer-Verlag, Berlin, 1983, p. 234.
26. I.G. Walls, *The Complete Book of the Greenhouse*, 3rd ed., World Lock, London, 1983, p. 354.

12

Sugar and Fiber Crops

I. SUGAR CROPS

The total world production of sugar (sucrose) is over 80 million tons per year. About 55–60% of this comes from sugar cane and about 40% from beets. Sucrose from all other sources, such as sugar palm, sweet sorghum, and maple trees, amounts to about 1% of the total. The major sugar-producing countries are Russia, Cuba, Brazil, the United States, and India.

A. Sugar Beet

1. Introduction

Botany and Origin. The sugar beet (*Beta vulgaris* L.) is an herbaceous dicotyledon belonging to the Chenopodiaceae family, which also includes spinach and goosefoot. Chenopods are characterized by small, greenish, bracteolate flowers, which are perfect, regular, and without petals. Sugar beet is a biennial plant, which stores a reserve food supply in a large fleshy root forming in the first growing season. The plant then overwinters, producing flower stems and seed the following year. The sugar beet seed of multigerm varieties is actually a fruit consisting of a cluster of seeds enclosed in the woody outer casing or husk. On germination, the cluster produces a number of seedlings that develop closely interwined and are difficult to thin. Monogerm varieties with good bolting resistance are available, but the seeds of monogerm varieties tend to have a flat disclike shape and are not suitable to precision drills (Fig. 1).

Unless the crop is grown for seed purposes, harvesting the fleshy roots during the first autumn when the amount of sugar in the root is at its peak interrupts the life cycle of the sugar beet. Sugar beets thus require two seasons to produce seeds. In the first season, the plant develops a rosette of leaves and a fleshy taproot, and in the second season it flowers and produces seeds (1). Cooler temperatures (>7.5°C) for 50–60 consecutive days are essential for inducing the reproductive phase of the crop growth (2). According to Rekhi (3), this period may be extended beyond 60 days depending upon the varietal character. In India, sugar beet can be produced only at higher altitudes with a cooler climate.

Wood et al. (4) reported that environmental factors such as temperature during seed ripening greatly influence yield and seed quality of sugar beet. Low temperatures

Figure 1 Sugar beet seeds of monogerm varieties.

gave greater yields of larger fruits but produced smaller true seeds with poor germination. Partial seed vernalization led to many bolters in the root crop. Sites with the most consistently favorable conditions, therefore, should be selected for seed production (4).

Species and Cultivars.   *B. vulgaris* includes four botanical groups or types: sugar beet, mangelwurzels or mangels, garden beets, and leaf beets such as chard and ornamental beets. A wild type, *B. maritima*, is thought to be the progenitor of the sugar beet.

Sugar beet varieties in Europe are usually classified into three groups (5):

1. Those producing large roots with a relatively low sugar content, often described as the "E" type
2. Varieties producing relatively smaller roots with a high sugar content, known as "Z" or "ZZ" types
3. Intermediate varieties with medium root size and sugar content, called "N" types

Since the range in the sugar content among these different types of sugar beet rarely exceeds 1%, this classification is not very useful (6).

2. Breeding

Sugar beet is normally a cross-pollinated crop, and the traditional methods of plant breeding are usually designed to secure maximum cross pollination between a number of carefully chosen parent lines. Sugar beet varieties are thus less stable than those of common cereals and unless maintained by constant breeding, selection, and trial, they will deteriorate. The discovery of male sterility in sugar beets has led to the development of several hybrid varieties (5).

Curtis (7) discussed the problems of breeding sugar beets under the following two categories: (a) the generation of progenies in sufficient quantity for evaluation and (b) the use of evaluation techniques to quantify selection criteria. The processes of evaluation open to breeders depend on the quantity and quality of seed produced. A balance must be struck, taking into account the number of progenies, seed quantities, and reliability (7).

Zakhariev (8) obtained seeds of hybrid sugar beet by crossing six diploid monogerm cytoplasmically male-sterile (CMS) lines with five tetraploid multigerm pollinators. Some CMS lines, irrespective of the pollinator, produced hybrid seeds with more than 5% hard seeds, whereas in others, its proportion was always below 5%. Some pollinators produced hybrids with a lower percentage of hard seeds (4%) when crossed with any of the maternal lines (8).

Ostrovaskii et al. (9) outlined a scheme for the production of heterotic sugar beet hybrids using CMS. Hybridization plots of at least 25 ha are needed for this purpose. The scheme requires the CMS forms and maintainers and the multigerm pollinators to be raised at the same site on different plots. The scheme also indicates the ratios of CMS to pollinator plants for different row spacing and the optimum regimens (9).

Two types of sugar beet hybrids are produced: diploid hybrids and polyploid hybrids. Utilizing cytoplasmic and genetic factors that condition sterility, as shown in Figure 2, produces the seed of diploid sugar beet hybrids. This is known as a four-way hybrid or a double-cross hybrid, which is comparable to that used in hybrid maize production. The four inbred lines are required for the production of the four-way hybrids. Lines A and B are monogem "O" types; line A is the male-sterile equivalent (Fig. 2). The hybrid of the two is male sterile, which serves as a seed parent in the commercial hybrid. Planting the two lines in alternate rows produces the hybrid between lines C and D. This hybrid is multigerm and is mixed with the monogerm A × B parent to constitute the stock seed for commercial production. Ten percent of the seed of the pollinator is usually mixed to raise a hybrid seed crop (10).

Hybrid seed production fields should be isolated from other fields by a distance of about 3000 m. All A × B (monogerm) seed, when harvested in a commercial field,

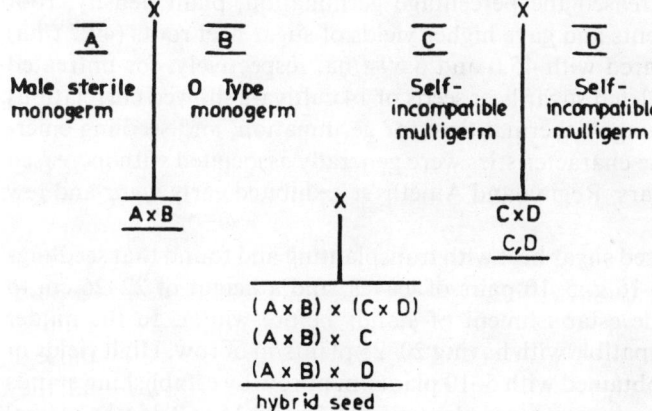

Figure 2 Double-cross hybrid seed production of sugar beet. (From Ref. 10.)

will be hybrid and can be separated from multigerm pollinator seed by using thickness grading with cylindrical slotted screens. Once the inbred lines are selected, the hybrid seed can be produced easily. Male-sterile diploid plants are interspersed with tetraploid pollinators to give rise to tripoid (polyploid hybrid) commercial seed (10).

According to Laby (53), the aims of sugar beet breeding remain the improvement of germinative capacity and bolting resistance through gene complementation between the two hybrid parents, especially for autuma- or winter-sown varieties in the Mediterranean. Owing to improved breeding methods on plantlets, very vigorous lines can be selected, which will have a higher storage for sugar in roots rather than in leaves.

3. Cultivation

Culture.

   ISOLATION REQUIREMENTS AND LAND PREPARATION.   The land selected for sugar beet seed production should not have been used for the growth of any *Beta* species within the last 4–6 years to avoid contamination from volunteer plants arising from fallen seeds of preceding crops. This is especially important in hybrid seed production.

Sugar beet is cross pollinated by wind or insects. Pollen is normally shed over a period of several weeks, prolonging the chances of contamination due to outcrossing. The seed crop of sugar beet should, therefore, be isolated from other fields of sugar beet and related species (mangels, garden beet, spinach beet, Swiss chard, and red garden beet) with a distance of 3000 m for the foundation seed class and 800–1000 m for a certified seed class (10). Land is prepared by one plowing, two to three harrowings, and leveling to bring it to a desired tilth.

PLANTING AND CULTIVATION.   The seed of sugar beet may be produced either by an in situ method (over wintering) or by transplanting (steckling). In the first method, the crop is overwintered in the field to allow it to flower in situ after the necessary thermal induction, whereas in the second method, the selected stecklings (roots) are taken out at the end of the first season and stored in shallow trenches for overwintering. These stecklings are replanted to produce seed the following season. One-third of the root and part of the top are chopped to retain about 10 cm of top leaves before planting.

Sugar beet seeds sown after pelleting with a polymer film containing fungicides and a growth regulator increased the percentage germination, plant density, root weight, and root sugar contents and gave higher yields of sugar beet roots (49.7 t/ha) and sugar (7.48 t/ha) compared with 45.0 and 6.59 t/ha, respectively, for untreated seeds (11). An analysis of pelleted sugar beet seeds of 14 cultivars showed correlations between germination percentages, thermal time for germination, and seedling emergence. Improvements in these characteristics were generally associated with increased bolting, although two cultivars, Regina and Amethyst, exhibited early vigor and few bolters (12).

Balan et al. (13) produced sugar beet with transplanting and found that seedlings required a root weight of 7–16 g, 5–10 pairs of leaves, and a height of 22–26 cm to become winter hardy for the establishment of stands before winter. In the milder southern areas, this was compatible with having 20–25 plants/m of row. High yields of good-quality seed could be obtained with 5–10 plants/m of row by establishing stands in the spring. Seed yields from overwintered plants varied from 2.15 to 2.36 t/ha at final plant densities of 107,000–199,000 plants/ha (13).

Adequate moisture in the field at the time of sowing is necessary to ensure good germination. It is necessary to plan sowing so that stecklings are fully developed before the onset of winter. In areas of severe winter with snowfall, the best time for sowing is mid-June, whereas in areas of low snowfall, sugar beet is ideally sown in July. Nucleus/ Breeder's seed (for foundation seed crops) and foundation seed (for certified seed crops) obtained from an approved source is sown in ridges and furrows, which are normally recommended for sugar beet. In areas where dry climates prevail, flat sowing may also be adopted. The recommended spacing for in situ crops is 60×15–20 cm, using about 5–6 kg seed/ha for monogerm varieties and 10 kg/ha for multigerm varieties. Spacing for the transplanted sugar beet crop is 45×10 cm for raising stecklings and 60×45 cm for transplanting stecklings, which requires about 10 kg seed/ha for monogerm varieties and 18–20 kg/ha for multigerm varieties. The stecklings from a 1-ha area are usually sufficient to plant about 7–8 ha of seed field (10).

Fertilization, Irrigation, and Weed Control. The commercial sugar beet seed crop requires 100–150 kg N, 50 kg P, and 150 kg K per hectare depending upon soil fertility. About 50–60% of these requirements are supplied before sowing the crop in the nursery. Beet seed crops need 125–150% of the commercial beet crop requirements of the fertilizer nutrients. A full dose of P and K and half of the N are applied at the time of planting the stecklings, and the remaining half-dose of N is given after 45 days at the time of leaf development. In addition, an application of 20–30 kg of borax per hectare has been found to be useful in preventing canker. Boron also helps in the development of good roots (10).

Treatment of sugar beet stecklings with 0.2% Ethrel (ethephon) 2 weeks before harvesting produced uniform maturation of seeds on all the shoots, increased total seed weight, and increased the production of large seeds (more than 5 mm), providing 82.7% germination as compared to 76.9% germination in seeds from untreated stecklings. The effect of application of benzyladenine (BA), gibberellins, and 2,4-D at the bud-formation or flowering stage of stecklings on the proportion of seed sizes and germination has been investigated (14).

Adequate soil moisture at the time of flowering of sugar beet is very essential. At other times, the seed plot may be irrigated as deemed necessary. Interculture operations such as weedings might be needed, especially during the first 60 days. Two herbicides, beetamin and pyramin, applied at 2 kg/ha 30 days after sowing and before emergence, respectively, can control weeds effectively (10). Ling et al. (55) developed a mathematical model for sugar beet seed yield in China. The effects of sowing date, plant density, and levels of N, P, and K were analyzed by multiple regressions. A computer-optimized combination of the five factors showed that for sugar beet yields of more than 225 kg/mu (1 mu = 0.067 ha), seeds should be sown on August 7, grown in a population of 2860 plants/mu, and fertilized with 54.1 kg ammonium sulfate, 51.4 kg calcium superphosphate, and 7.0 kg potassium chloride/mu. Of the five factors, N had the greatest effect on seed yields of sugar beet (55).

Control of Pests and Diseases. Sugar beet seed crops are often infested with a combined attack of curly top and yellow virus, which can be very damaging. Control of these two diseases depends upon the effective control of insect pests acting as vectors. Diseased plants should be removed and destroyed from time to time. A regular spraying of the seed crop with copper fungicide controls *Cercospora* disease.

The cutworms of sugar beet crops can be controlled by dusting with 5% benzene hefachloride (BHC) at 10 kg/ha during the early stages of crop growth.

Roguing.   Diseased and off-type plants as well as early bolters should be rogued out from the field from time to time. In the steckling method, undesirable stecklings such as diseased or plants malformed and those not conforming to the varietal characteristics should be rejected before transplanting.

4.   Harvesting and Threshing

The seed crop of sugar beet is ready to harvest when two-thirds of the seeds are ripe. At this stage, the seeds at the base of the branches mature and turn noticeably brown. Since the ripe seeds are prone to shattering, the crop should be harvested by pulling the plants early in the morning when moist to minimize seed losses. Harvesting by hand is preferred, since mowing is not satisfactory. Owing to uneven ripening of the seed, two or three hand harvestings are often necessary. In humid regions, plants are usually tied into bundles and cured in stacks and left to harden and mature. The seed may be discolored by rain.

The crop is threshed with an ordinary grain thresher. Combines may be used for a large-scale crop.

5.   Postharvest Operations

Cleaning, Drying, and Seed Treatment.   The threshold crop is cleaned with the help of a fanning mill to separate stalks, weeds, and shrunken seeds. The seed must be dried to 9–10% moisture content before its storage. If heated air is used, its temperature should not exceed 38–40°C. After the cleaning, seeds may be rubbed and graded with the help of suitable machines.

Busol et al. (15) described various methods of cleaning for different groups of sugar beet seeds, such as treatment of heavily soiled dry seeds, damp seeds, and heavily soiled damp seeds. The methods combine preliminary cleaning using airflow and sieves, drying, secondary cleaning, sorting, and weighing. Temperature regimens and timing of each method and equipment used for belt seed drying have been described (15).

Guidelines dealing with the evaluation of fungicidal seed treatment for the control of seedling diseases like damping-off due to soilborne fungi for beet root and certain large-seeded field and vegetable crops have been published (16) based on trials conducted under controlled conditions due to the difficulties of field testing. Special details are given for sugar beet, maize, and legumes, but the basic principle can be extended to other crops. In general, the method aims at controlling soilborne diseases, but it may also be applied to some seedborne diseases (e.g., *Pleospora betae* of beets). Guidelines for artificial inoculation, application of treatments, mode of assessment, recording, and measurements are also provided (16).

Seed Yields.   The average yields of sugar beet seeds vary from 8 to 10 Q/ha in India.

B.   Sugar Cane

1.   Introduction

Botany and Origin.   Sugar cane (*Saccharum officinarum* L.) belongs to the family Gramineae of the order Glumaceae and class Monocotyledon. Husz (17) and Barnes (18) have described sugar cane morphologically and anatomically. Sugar cane is closely related to the grasses and other members of the Graminae, of which it is a giant member. Stems or stalks develop from the bud of another stem following the usual vegetative propagation by cuttings, also known as "setts," "points," or "seeds." The cuttings consist of stem sections containing one or more buds. The secondary stalk develop underground from one bud of the primary stalk and produces tertiary stalks, introducing the process of tillering. The stalks (or stems) are differentiated by nodes and internodes.

A node is the base of the leaf, and depending upon the varietal characteristics and growth conditions, the internodes may be long or short, tumescent, bobbin-shaped, conoidal, obconoidal, curved, or cylindrical. The lateral buds appear at the nodes, one on each, normally on alternate sides of the stem. They usually lie flat in a longitudinal groove. The buds are shell shaped and clearly characteristic of each variety. Under favorable conditions, the buds germinate, thus ensuring the usual vegetative propagation and genetic constancy of the sugar cane varieties (17,18).

Unlike most grasses, sugar cane does not have a hollow stem, but one that is filled as in maize and sorghum. The stalk terminates at the upper end in a whorl of developing leaves and a growing point, which under certain conditions, develops into a slender arrow bearing a tassel of tiny flowers. Because sugar cane is a short-day plant, its photoperiodic flowering conditions can be attained only in the tropics. Although for crossing purposes it is necessary to cultivate perfect (fertile) blossoms, this is undesirable in agricultural cane production, because flowering (tasseling) reduces cane yields. Optimum growth conditions such as abundant supplies of nitrogen and moisture restrict inflorescence, whereas under more strenuous conditions, flowering is stimulated. The inflorescence is a open-branched panicle with thousands of blossoms which appear in pairs—one stalkless (sessile) and the other attached to a stalk (pedicillate). The flowers open in succession over a number of days. Flowers of all varieties have both male and female organs, but do not all produce fertile pollen. Sugar cane varieties may thus be classified as those having fertile pollen and eggs and those having sterile pollen and fertile eggs. Salunkhe and Desai (6) described the postharvest physiology and ripening process of sugar cane.

Species and Varieties. Jeswiet (19) divided the genus *Saccharum* into the following five species: *S. spontaneum*, *S. sinense*, *S. barberi*, *S. robustum*, and *S. officinarum*. Each of these can be subdivided into varieties with different genotypic (number of chromosomes) and phenotypic characteristics. *S. spontaneum* (wild cane), with its varieties having chromosome number 54, 112, or 118, is valued for hybridizing. *S. sinense* (Chinese cane) has chromosome numbers 116 or 118, whereas *S. barberi* has 82, 90, 116, or 124; it is also called the Indian species and is sturdier and more disease-resistant than *S. officinarum*. These qualities, together with higher sugar and ample fiber content, make it important for breeding purposes. *S. robustum*, with chromosome numbers 70, 80, and 84, is found in New Guinea and is similar to *S. officinarum*.

S. officinarum also know as noble cane, is rich in sucrose and relatively low in fiber content. Its sonatic cells have 80 chromosome numbers. *S. officinarum*, because of its vigorous and long stems, is used for commercial sugar production all over the world. Seven original (nineteenth century) groups of *S. officinarum* have been recognized (17):

1. Otaheiye or Bourbon or Lahina (from Hawaii)
2. Cherilon (from Indonesia, Java)

3. Preanger (from Java)
4. Tanna or Caledonia (from Australia, Mauritius, Fiji, and Hawaii)
5. Badilla (from New Guinea and from thence to Australia)
6. Black Borneo, Borneo
7. Creole or Criolle

None of these old groups is cultivated industrially at present. Old varieties deteriorate in productive capacity and are constantly replaced by more productive, newer varieties of sugar cane.

2. Breeding

Screening, selection, and hybridization of sugar cane aims at desirable characteristics such as vigor, cane tonnage, sucrose levels, sugar recovery, hardiness, disease resistance, early maturity, condition at harvest, tasseling, cold tolerance, suitability to mechanical harvesting, and rationing ability. Reasons for the deterioration of sugar cane varieties include declining fertility of sugar cane–producing soils, development of unfavorable soil physical conditions, the cumulative effect of pests and diseases, and the existence of symptomless or unidentified diseases (20).

Several high-yielding hybrid varieties of sugar cane have been developed at the Sugar Cane Breeding Center, Coimbatore, India, and at other centers around the world. Salunkhe and Desai (6) have listed the most promising varieties of sugar cane grown all over the world.

3. Cultivation

Irrigation, Fertilization, and Selection for Disease Resistance. Increasing doses of nitrogen application has been reported to advance the flag emergence of sugar cane seed crop by a few days and increase the weight of seeds produced (21). The frequency of irrigation, however, did not influence the date of flag appearance in the first year, but in the second year this was delayed 2 weeks by reducing the irrigation frequency from 5 to 15 days. The date of appearance of inflorescence was also delayed by infrequent irrigation in the second year. The number of germinated seeds per gram increased by reducing irrigation frequency in the first year of seed crop. Sugar cane flowering and seed production as influenced by photoperiod have been investigated (22). Malik (22) outlined the variety and clonal selection schemes along with approaches to screen varieties for pest and disease resistance.

Pollen Viability and Seed Production. Abou-Salama (23) investigated the effects of day length and relative humidity (RH) levels and duration on pollen viability and seed setting of sugar cane. Effects of declining day length rates of 30 sec and 1 min per day were studied on 354 crosses made during 3 years using 54 males and 92 females. Males experiencing 30 sec/day decreases had higher pollen viability. Seed production, however, was not influenced by the treatments; most of the variation was attributed to the varietal differences. The two rates investigated in this study could be used in a breeding program without any negative effect on seed production from the crosses. Altering the humidity level or duration did not influence pollen viability and seed production. Setting RH at 65–81% did not adversely affect pollen viability or seed production in sugar cane crosses.

Slow germination and seedling growth often reduce the effective growing season of sugar cane in parts of Australia. The effects on three sugar cane cultivars of presoaking sets with glycine betaine prior to germination and application of glycine betaine to the potting mix during early growth were investigated. Set germination was not significantly affect germination. Increases of 24–43% in plant growth were observed for cv. Q138 after treatment with 2–4 g glycine betaine/L, with no response to higher concentrations. Results for the other two cultivars (Q117 and Q124) were less promising, varying from no effect to apparent negative effects on growth after treatment with high (8–16 g/L) concentrations (23a).

Cleaning, Drying, and Storage of Seeds. In sugar cane breeding, fuzz found in the inflorescence causes problems in separating seeds and also causes absorption of moisture leading to loss of viability and reduces the storage life of seeds. Goonasekera et al. (24) designed a degluming machine using the principle of friction between sugar cane spikelets and sand, which are mixed vigorosly. After separating the fuzz from the seeds, the sand, seed, and fuzz mixture is separated using a shaker that sieves out the sand and fuzz.

Abayomi and Fadayomi (25) dried the sugar cane fuzz collected from five open-pollinated cultivars to 10.5% moisture, sealed in high-density polyethylene bags, and stored at the ambient conditions (25–30°C, 75–90% RH) as well as in a refrigerator (10°C, 65% RH) and in a freezer (−15°C, 50% RH). The germination tests carried out at 2-month intervals for 1 year revealed that seed viability was completely lost in fuzz stored at ambient conditions after 6–8 months, but it was best preserved under very low temperatures and relative humidity (10–66% after 12 months' storage in the freezer depending upon cultivar differences). The cultivars Co 443 and BR 6223 had high-germination capacity prior to storage.

C. Minor Sugar Crops

1. Sugar Maple

Sugar maples (*Acer* spp.) are relatively long-lived trees (70–100 years), sensitive to atmospheric pollution, and easily injured by city smoke and gas fumes. The following four important species of the genus Acer have been identified (26):

1. Sugar maple, *A. saccharum* Marsh—hard maple (lumber), rock maple, *A. saccharophorum* K. Koch
2. Black maple, *A. nigrum* Michx. f.—hard maple (lumber) black sugar maple, sugar maple, *A. saccharum*, var. *nigrum* (Michx. f. Britten)
3. Silver maple, *A. Saccharum* L.—soft maple (lumber), white maple, river maple, water maple, or swamp maple
4. Red maple *A. rubrum* L.—soft maple (lumber), water maple, scarlet maple, white maple, or swamp maple.

Salunkhe and Desai (6) have described the morphological characteristics of these four species.

Maples prefer well-drained fertile soils. They are drought resistant and respond well to irrigation. In the United States, they are mainly grown in the extreme eastern area of the western plains, where rainfall is heaviest. Maples are widely planted as shade trees in the Northeast. They are short-lived, subject to windbreak, and require moist rich soil.

Schwedler maple is a variety of Norway maple grown extensively in parts of the United States and Canada. It is less subject to leaf scorch than sugar maple. Maple varieties producing higher quantities of sap with an increased sugar concentration have been developed through hybridization and selection. The rate of annual growth, degree of apical dominance, and sugar content of the sap are the important parameters of genotypic evaluation. The characteristic of drought resistance can be used in the first year of plant growth (27).

2. Sweet Sorghum (Sorgo)

Sweet sorghum or Sorgo (*Sorghum bicolor* L.), or sorgo, has been used as a source of table syrup as well as for forage and silage. It is grown in India, Africa, parts of Central and South America, China, Pakistan, and the United States. It is adapted to a wide range of soil and climatic conditions, and tolerates moisture and other stress conditions fairly well. It is an excellent crop of varied utility grown under dry farming conditions, and has good potential as a supplementary source of sugar for the sugar cane and sugar beet industries, where, with slight modifications, the processing equipment used to produce sugar from these established crops can also be used to manufacture sugar from sweet sorghum. The amount of sugar produced from sweet sorghum could be great considering the fact that this crop can be raised throughout the year (three crop seasons) under tropical and subtropical conditions, and that its water requirements are considerably low compared those for either sugar cane or sugar beet (28).

The sweet sorghum cultivars used for sugar production should have a high juice purity (77.5%) and a low rate of sucrose inversion. Other desirable characteristics are higher stalk yields; erectness; resistance to pests and diseases; tolerance to drought, excessive water, and pesticides (weedicides); high degree of practically attainable sucrose extraction; and high juice brix. The higher starch and aconitic acid content of sweet sorghum can be troublesome in sugar production. A good variety for sugar production must produce a high yield of stalks per unit area, with a high percentage of extractable juice that is high in sucrose but low in invert sugars, dextrose and levulose. Starch should not be present in more than trace amounts, and the aconitic acid content should be low enough not to interfere with the crystallization of sucrose (29). The crop (stalk) of a typical sweet sorghum variety (SSV) grown at the Mahatma Phule Agricultural University, Rahuri, India, is shown in Figure 3.

A highly productive and disease-resistant sweet sorghum variety, Rio, was released in the 1960s for potential use as a sugar crop. Rio is a selection from the cross between MN 1048 and Rex and is highly resistant to leaf anthracnose, red rot, and rust. Brawley is another high-sugar sweet sorghum cultivar released for sugar production in areas where diseases are not a problem; it is a selection from the cross between two susceptible parents, Rex and White Seeded Collier. Coleman (30) has reviewed the breeding research work carried out on sweet sorghum in the United States.

Breeders have developed two types of sweet sorghum varieties: syrup types, which contain invert sugars in the juice to prevent crystallization, and sugar types, which contain mostly sucrose and very few invert sugars in the juice for crystallization. Salunkhe and Desai (35) have reviewed the breeding research work to evolve new improved cultivars of sweet sorghum that are being grown successfully in different parts of the world.

Figure 3 Sweet sorghum crop (stalk) of a typical SSV variety grown in India. (From Ref. 6.)

The land preparation, isolation requirements, and seed-production technology for sweet sorghum is similar to that described for sorghum seed production under cereal crops.

3. Sugar Palms

Palmyra Palm or Borassus Palm. *Borassus flabellifer*, known as Palmyra palm in India, is found growing in its wild state in West Africa, New Guinea, and Asia. It has many uses: food, drink, firewood, timber, clothing, writing material, toys, medicine, and intoxicants. The palm juice comes from tapping the inflorescences, the male plant yielding better than the female. The young inflorescence is beaten with a mallet for 3–4 days to encourage sap flow. The tip is then shaved off daily for 4 days. On the eighth day, clear, sweet liquor begins to flow from the wounded spadix, which is collected (about 2 L/day/inflorescence). The juice can be evaporated to make jaggery or palm sugar or fermented into an toxicant (32). Salunkhe and Desai (31) have discussed the cultivars, processing, and utilization of sugar palms.

Caryota Palm. Caryota palm is a fishtail palm like its close ally *Arenga*. Its leaf axis ends in a terminal leaflet. The tall trunks of *Caryota urens* (13–20 m high) begin to flower after 15–20 years. The starch in the trunk is converted into sugar, entering each inflorescence as it develops from the leaf axis in order from topmost to lowest. The carbohydrate is mobilized in stages node by node, down to within 1 or a few feet of the roots. Young inflorescences are tapped to produce sugar, wine, and a drink called arrack. They are pummeled with a wooden mallet for several days until bruised internally, and then sliced gradually down to the stalk using thin cuts once or twice a day. The large *C. urens* yields about 7–14 L of sap per inflorescence in 24 hrs, or if several inflorescences are tapped at the same time on one trunk, 20–27 L/day (31).

Nipa Palm. Nipa palm (*Nipa* [Nypa] *fruticans* Warmb.), also called sugar palm gulga, gubna, golphal, golpatta, or Poothada belongs to the Palmae family. It is an advanced genus of palms and is thornless, stemless, with unisexual flowers and capitate female inflorescence. Nipa is a prostrate palm with a stout, branched, creeping rhizome. Sweet sap is tapped from the spadix after second flowering when the plant is 5 years old, and tapping may be continued for 50 or more years. The average yield of sap per plant is 43 L during the season. Nipa sap contains about 17% sucrose and traces of glucose and fructose (33).

Phoenix Palm. The genus phoenix contains the date palm, *P. dactylifera*, the most historic of all palms. It has a rough trunk. The fleshy fruit is stoneless, for the "stone" is the seed. The wild date palm, *P. sylvestris*, provides mainly a source of sugar. The sugar is obtained not from the inflorescence but from the sap flowing out of wounds inflicted on the active part of the trunk within the crown. Tapping begins by cutting enough leaves on one side of the stem to expose a clean surface—about 30 cm each day. A V-shaped incision is made at the foot of the surface, about 7.5 cm wide and 1.5 cm deep. A bamboo spout is put at the base of the V to collect the sap into a vessel. Every evening for about a week, the cut is deepened slightly to renew the flow. A new cut is opened after 1 week, above the first, and so on, until by the end of tapping season (March–December), the whole of the exposed surface has been cut to a depth of about 7.5–10.0 cm. A new tapping panel is opened the next season above the first on the opposite side of the trunk. Thus, tapping may be continued for 40 years or more. A single tree may yield about 4–19 L per day or 250–1100 L per tree per season.

Raphia Palm. The genus *Raphia* originated in Africa and consists of about 20 massive species. *R. humilis* is a stemless species, and *R. sassandrensis* has a creeping trunk. Most appear to be solitary palms. Massive inflorescences emerge from among the leaves looking like enormous worms covered with spathes and branch into enormous centipedes with bracts set in two regular rows. Palm wine is made from the sap obtained from *R. vinifera* and *R. taedigera*.

Amazonian Fan Palm. The Amozonian fan palm (*Mayritia flexuosa*) needs to be felled and defoliated and scorched with fire to stimulate the flow of sap from its inflorescence stalks. The sap is evaporated into brown molasseslike palm sugar (jaggery), and then drunk as such or fermented into palm wine (34).

4. Jerusalem Artichoke

Jerusalem artichoke (*Helianthus tuberosus* L.) is an herbaceous perennial crop bearing oblong tubers. It is native to the Mississippi Valley and belongs to the Compositae family. It is used as feed for hogs and for human consumption. It contains high levels of a storage polysaccharide, inulin, which is a rich natural source of fructose. On hydrolysis, inulin yields fructose, a sugar having dietary value to diabetic patients. The large woody plants are cut and removed before the tubers are harvested with a plow or middle buster or by hand digging with forks. Good, sound, disease-free tubers can be stored for several months at 0°C and high humidity (35). The tubers are processed to produce several products such as fructose syrups, dried fructose powders, or artichoke flour (36,37).

II. FIBER CROPS

The important fiber crops cultivated in different parts of the world include cotton, sisal, jute, mesta or kenaf, flax, hemp, ramie, and manila fiber. Next to cotton, sisal is the most extensively used fiber in the United States for binder twine and for mixing with manila fiber in the manufacture of cordage of various sizes. Indian jute is used extensively to make gunnysacks and other containers.

A. Cotton Seed

1. Introduction

Botany and Origin.   Cotton (*Gossypium* spp.) is a member of the family Malvaceae. It is primarily grown for its fiber and yields about 180 kg of oil-containing seed for each 100 kg of fiber. Long fibers comprising staple or lint cotton are removed from the seed during the ginning operation. Seed is pointed and ovoid in shape (8–12 mm long), varying in color from brown to black. The seeds of chemically and mechanically delinted cotton are shown in Figure 4. The hull (seed coat) is relatively thin but tough. The mature kernel consists of a major portion of embryo, developed at the expense of reserve endosperm. Commercial cottonseed consists of about 10–15% lint, 35–40% hulls, and 50–55% kernels (38).

Species and Cultivars.   All cultivated cottons are members of four species of the genus Gossypium: *G. arboreum*, *G. herbaceum*, *G. hirsutum*, and *G. barbadense*. The first two are Old World diploids, whereas the last two are New World tetraploids, with $n = 26$, comprising a large number of types, often classified into groups, varieties, or races. In India, *G. barbadense* types are not grown on a commercial scale. Salunkhe and Desai (31) have listed the improved types of cotton grown in India.

A new cotton variety with low gossypol content, Zhemian 9, has been reported (39). It has early-intermediate maturity and is resistant to wilt. Its free gossypol content in seeds is 0.0041–0.0144%, which is lower than the accepted allowance of the International Health Standard. The seeds have an average protein content of 42.75% and fat content of 33.75%. Wu (40) reported a new low-phenol cotton variety, Xinluzao 3, derived from a hybridization program. It is high yielding and tolerant of *Verticillium albo-atrum* and resistant to *Xanthomonas campestris*. The seed phenol content is 0.006%; protein, 35.8%; and oil, 34.6%.

2. Breeding

Hybrid cottonseed is produced by hand pollination. The individual bud of the female parent is emasculated in the evening and pollinated the next morning with the pollen of the male parent. The stigmatic surface is covered with an envelope after emasculation and pollination.

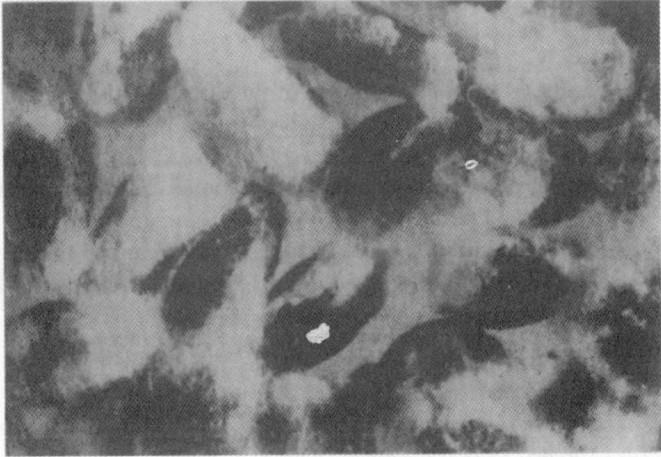

Figure 4 Chemically and mechanically delinted cotton seeds. (From Ref. 56.)

The planting ratio of female to male parents should be 4:1 or 5:1; that is, if there is a total of 50 rows, the first 40 rows should be planted with the female parent and next 10 rows with the male parent. Male parents must be sown at two or three times at 8- to 10-day intervals, since the flowering period in cotton is spread over a long time. A row-to-row distance of 150 cm and plant-to-plant spacing of 100 cm (in female) and 50 cm (in male) are recommended. Seed rates used are 3.75 kg/ha for female parents and 2.5 kg/ha for male parents. The cultural operations required for the production of hybrid cottonseed are similar to those described for normal cottonseed production.

Petalcorin et al. (41) evaluated two methods of emasculation and three pollination schedules on cotton in terms of percentage boll set, seed number and weight/boll, percentage seed germination, and germination rate. Emasculation using the thumbnail–soda straw technique, which completely removes the entire staminal column and corolla in one piece, has been found to be satisfactory whether done the day before flower opening or a few hours before flower opening. Pollination on the day of flower opening, preferably after sunrise, is recommended to produce good boll set, larger and heavier seeds, and a high percentage germination and germination rate.

Zhang and Pan (42) investigated the inheritance of genetic male sterility with the virescent marker characteristic and its cytology in *G. hirsutum* line 81A, having chromosome number 2n = 52. The pollen abortion occurs mainly in the late uninucleate stage. A single recessive gene controls male sterility and virescence in 81A. The close association between virescence and male sterility suggested that either close linkage or pleiotropy was involved. Allelic tests indicated that virescence in 81A is nonallelic to all the virescence genes identified previously in upland cotton and its male sterility is nonallelic. The development of the 81A genetic male-sterile line associated with virescence could markedly improve the efficiency of hybrid seed production in cotton.

According to Redenbaugh and associates (43,44), there is a strong technological and commercial basis for the improvement and multiplication of cotton by the production of synthetic seeds through somatic embryogenesis technology.

3. Cultivation

Culture.

LAND PREPARATION AND ISOLATION REQUIREMENTS. The selected land should be essentially free from volunteer cotton plants. The soil should be deep, well drained, moisture retentive, and fertile. It should also be free from hard layers of carbonates and hardpans, which may interfere with the proper development of the crop (10).

Land is prepared by deep plowing, two to three harrowings, and leveling to make the soil well pulverized and leveled.

Cotton is predominantly a self-pollinated crop, but natural cross pollination occurs to the extent of about 10–50% in *G. hirsutum*, 1–2% in *G. arboreum*, and 5–10% in *G. barbadense*. The isolation requirements vary according to the magnitude of cross pollination. Only one variety of seed crop should ideally be grown per farm. Isolation distances of 50 m (foundation seed) and 30 m (certified seed) from fields of other cotton varieties of the same or other species, as well as from fields of the same varieties not conforming to the varietal purity requirements for certification have been recommended (10). In the United States, where the discard row system is used, isolation distances varying from 45 to 90 m are employed for different seed classes.

PLANTING AND CULTIVATION. Cottonseed crops are sown in India 1 or 2 weeks before the onset of monsoon for the best results, using seed obtained from an approved source (nucleus or breeder's seed for raising foundation seed crops or foundation seed for raising certified seed crops). The seed is usually treated with organomercurials to prevent seedborne diseases. A basal dose of fertilizers along with 25 kg of 5% disulfoton or 20 kg of 10% phorate per hectare is applied at the base of each ridge in the ridges and furrows opened at distance of 120–150 cm. Seed is sown by dibbling, two or three seeds per hill, at a distance of 60–75 cm. The plot is irrigated immediately after sowing followed by a second irrigation 4–6 days after seeding. Gap filling may be necessary if germination is not satisfactory, using seedlings raised simultaneously in polyethylene bags. Thinning may be performed 20–23 days after planting to maintain one healthy seedling per hill. Seed rates vary from 8 to 10 kg/ha.

Fertilization, Irrigation, and Weed Control.   The nutrient requirements of cottonseed crops are relatively high depending upon the soil capacity and yield potential of varieties grown in different agroclimatic regions. In addition to about 5 tons of farmyard manure (FYM), the seed crop may require 50 kg N, 50 kg P, and 50 kg K per hectare as a basal dose given at the time of sowing followed by two top dressings of 25 kg N/ha after 60 days. A foliar spray of urea or diammonium phosphate (15–20 g/L) during the boll development stage has been found to be advantageous. Depending upon soil type and climatic conditions, the cottonseed crop may need irrigation once in 15–20 days, adopting the furrow irrigation method. Heavy irrigation may follow each picking.

Sawan and Sakr (45) reported that a foliar spray of mepiquat (1,1-dimethyl piperidinium chloride) increases the number of open bolls per plant, boll weight, seed index, and seed cotton yield. The crop matured faster with the increasing rate and number of applications of mepiqua.

Interculture operations (four to five times) such as hoeing or hand weeding are necessary to keep the field free of weeds. Weeds may be controlled chemically using herbicides like diuran, Toke-25, and Cotoran.

Control of Pests and Diseases.   A number of insect pests and a diseases affect cotton crops, the most important being aphids, jassids, thrips, mites, spotted bollworm, pink bollworm, American bollworm, and blackarm, and *Alternaria* leaf spot and *Helminthosporium* leaf spot diseases. Agrwal (10) has described a schedule of pesticides used to control the insects and diseases of the cottonseed crop.

Roguing.   Off-types and diseased cotton plants (seedling blight) should be rogued during the early stages of crop growth. Subsequent roguings for off-types and plants affected with several diseases should also be performed from time to time, especially at the square initiation and flowering stages.

4. Harvesting

Picking of cotton should commence soon after the bolls begin to open. Bolls ripen over 2–3 months, needing several pickings. The appropriate picking time is necessary to obtain high-quality viable seed. Cotton picked from late-formed bolls (last pickings) should not be used for seeds.

The bolls should not be picked when they are wet from dew or rain. Also, bolls spoiled due to rains or damaged (e.g., by insects) should be picked separately and discarded. The cotton picked should be clean, with a minimum amount of other material such as leaves and bark. The completely dried cotton should be stored in a dry place if not ginned immediately. Maiya et al. (45a) noticed that first-picking seeds had slightly higher germination, field emergence, root and shoot length, and seedling vigor compared to second- and third-pickings seeds of cotton. Seed grading with a 4.75-mm (R) sieve gave higher seed recovery (86.75%), germination (82.04%), field emergence (65.0%), root length (12.63%), shoot length (8.75%), and seedling vigor index (1752).

5. Postharvest Operations

Ginning.   Ginning operations significantly influence the quality of cottonseed. Ginning should be carried out on one variety of gin or on gins approved by a seed-certification agency. The machinery used should be clean to avoid any possible contamination. Care should be taken to ensure that seed is not damaged during the mechanical operation of ginning.

Cotton should be ginned at a 6–8% moisture content. The rate of ginning should not exceed 4.5–5.5 kg lint/hr. Agrawal (10) listed various precautionary measures to be taken to avoid mechanical damage during ginning.

Nicholas (54) reported on the dry gas delinting process for cotton in which dry HCl gas is injected in a revolving drum containing seed heated to 49°C. The hydrolyzed lint is then broken in a scalper and acidic traces are neutralized using ammonia gas. In a foam acid delinting process, expensive centrifugation involved in the dilute acid process is eliminated. In foam-acid delinting, using the principle of hydrolysis, H_2SO_4 is produced in the form of foam by applying a foaming agent followed by heat treatment. The lint gets brittle, and is then removed by abrasion. According to Nicholas (54), in the dry-gas delinting process, product standardization, scientific management, and cost reduction is achieved with 99.9% purity and 80% germination, and a better-quality seed can be made available to the growers as a primary input.

Storage. High-moisture cottonseed should not be stored for long. Seed should be dried with heated air (38–40°C) or combustion gases blown through it. The seed must be cooled before storage. Cottonseed containing up to 12–16% moisture can be stored satisfactorily at 21°C. Above this temperature, it is unsafe to store cottonseed in quantity when the seed moisture is above 11–12% (46). Heat is generated in the stored cottonseed because of the action of respiratory enzymes, and is not rapidly dissipated owing to the insulating effect of cotton fuzz unless adequate precautions are taken to prevent a rise in temperature. The glycerides present in the seed may be metabolized to yield free fatty acids, further deteriorating the seed quality. Storage houses, therefore, must be well ventilated or provided with forced air circulation to prevent a rise in storage temperature.

Seed Treatment. A spray of 4–5 kg of a mixture of propylene glycol dipropionate and bischloromethylene per ton of cottonseed prior to its storage has been recommended to prevent heat development and seed damage. Treatment with Nacconol (sodium alkylaryl sulfonate) also gives satisfactory results (47).

Seed Yield. Average seed yields of cottonseed vary from 300 to 600 kg/ha.

B. Jute

1. Introduction

Botany and Origin. Jute (*Corchorus* spp.) belongs to the family Tiliaceae. It is a natural inhabitant of the tropical and subtropical regions of the world. The fiber of commerce is obtained from the phloem tissue in the stem of the cultivated varieties of two species: *C. capsularis* L. and *C. olitorius* L. In both species, new leaves grow continuously as the plant gains height. Forking of the apex in two or more branches marks the transition to the reproductive phase. Inflorescence is extra-axillary, cymose with two to three flowers. Fruit is globose to pear shaped. Dehiscence is loculicidal with 30–48 seeds in each fruit of normal size (1.8 × 1.3 cm).

Unlike cotton or ramie, jute fiber has good lignin deposition (11–13%, as lignocellulose, with 80–87% cellulose). Lignin accounts for the rigidity and coarseness peculiar to jute and delays the action of retting (see Sec. II.D.1) microbes on the fiber.

Species and Cultivars.   Two species of *Corchorus*, *C. capsularis* L. (white jute) and *C. olitorius* L. (tossa jute), are distinguished on the basis of their fruit or pod shape, seed size and color, and tolerance to water logging. *C. capsularis* is more tolerant to water logging than *C. olitoris* in the later stages of crop growth. In India, 75% of jute is grown as *C. capsularis*. There are several varieties or strains of both cultivated species of jute.

2. Breeding

The normal methods of crossing of *C. capsularis* and *C. olitorius* have not been successful. When *C. capsularis* is used as the pistillate parent, it the flowers drop within 3 days after pollination, which does not happen when *C. olitorius* is used. Fertilization takes place without formation of seeds owing to premature abortion of the embryos. This is attributed to an impaired capacity for growth of the endosperm (48,49).

3. Cultitivation

Culture.

ISOLATION REQUIREMENTS AND LAND PREPARATION.   The field selected for the production of certified seed of jute should be one on which the same kind of crop has not been grown in the previous season or, if grown, the variety of the crop was the same and was raised using a certified seed. The land should be fertile, well drained, and of neutral character (pH – 7.0). It should also be free from noxious weeds and soilborne diseases. The land should be prepared well by deep plowing, and three to four harrowings followed by leveling to a fine tilth, which is necessary for good germination.

Jute is normally a self-pollinated crop; cross pollination does occur to the extent of about 2–17% in *C. capsularis*, but very little in *C. olitorius*. The seed fields should be isolated with a distance of 50 m for foundation seed and 30 m for certified seed from the fields of other varieties of the same species as well as from the same variety not conforming to the genetic purity of the seed variety being grown as required for certification.

PLANTING AND CULTIVATION.   Jute is best sown from April to May using nucleus/breeder's seed (for foundation crops) or foundation seed (for certified-seed crops) either by broadcasting or using a seed driller. Sowing in lines is preferred to facilitate interculture, rouging, and inspection of seed plot. Seed is generally sown about 3–5 cm deep, with spacing of 30×10 cm. The seed rates for *C. capsularis* and *C. olitorius* range from 8 to 10 and 4 to 6 kg/ha, respectively.

Fertilization, Irrigation, and Weed Control.   Depending upon the variety and soil fertility, jute crops require about 60–100 kg N, 20–30 kg P, and 30–40 kg K per hectare. The application of lime and potassium sulfate at 700–1500 kg/ha has been found to be beneficial when applied to acidic soils or in soils experiencing root rot or stem rot diseases. All of the phosphorus and potassium and half of the nitrogen should be applied before sowing, and the remaining half-dose of N should be split into two doses, one applied 3–4 weeks and another 6 weeks after planting. Adequate soil moisture content should be ensured at the time of application of the nitrogenous fertilizer. Two to three irrigations may be required during the crop growth depending upon rainfall.

Interculture operations such as hand weedings, mulching, and thinning are required to keep fields free of weeds and to maintain appropriate plant populations.

Control of Pests and Diseases. Agrawal (10) recommended the following methods of producing disease-free jute seeds:

1. Use of seed treated with organomercurials or captan
2. Application of lime and potassium sulfate to soil
3. Timely weeding and thinning operations
4. Crop rotation
5. Uprooting and destroying of infested plants

The insect pests and diseases of the jute seed crop can be controlled as follows:

Pests	Control measures
Insects	
Jute semilooper	Spray with folidol E 605 at 1 mL/5 L water or 20% endrin at 1.5 mL water
Indigo caterpillar	Dust crop with 10% BHC at 12–15 kg/ha
Hairy caterpillar	Spray with folidol E 605 at 102 mL/5L of water or endrin 20 EC at 7.5–10.0 mL/5L water
Yellow and red mites	Dust with lime-sulfur at 15–18 kg/ha or spray with lime-sulfur solution (1 in 49 parts of water) or folidol E 605 at 1 mL/5L water, covering underside of the leaves
Diseases	
Anthracnose	Spray the crop with copper oxychloride (phytolan)
Soft rot	Spray with 1% Bordeaux or copper oxide
Die back or black band	Irrigate crop when required and give a prophylactic spray of copper oxychloride
Leaf mosaic	Uproot the infected plants and destroy
Mildew	Dust lime-sulfur (3:1) in the morning hours when dew is on the leaves
Sooty mold of pods	Give a prophylactic spray with Bordeaux mixture

BHC = benzene hexachloride.

Rouging. Off-types, diseased, and pest-infested jute plants should be rogued out at three stages of crop growth. The first rouging should be carried out when plants are 30–40 days old followed by a second at the budding or flowering stage and a third at the capsule-formation stage.

4. Harvesting and Threshing

The jute seed crop is ready to harvest in October–November when the seedpods become brown. The proper stage of harvesting is important to minimize seed losses due to shattering. A good-quality seed of *C. capsularis* can be obtained from the crop when both stems and pods are not overripe. This may need several hand cuttings. The crop of *C. olitorius*, however, should be harvested when the stem and pods are dead ripe. The crop is usually cut with a sharp sickle in the morning hours without jerking or much bending of the plants. The crop should be placed in bundles on tarpaulin to collect the seed that shatters. The stalks are then dried under the sun before threshing. Threshing is usually done by beating the capsules with sticks.

5. Postharvest Operations

Cleaning and Drying.   Winnowing with wind or using an efficient air/screen cleaner cleans the seed. The cleaned seed is further dried to reach a moisture content below 9% before storage.

Seed Yields.   The seed yields of *C. capsularis* and *C. olitorius* vary from 3 to 4 and 1.0 to 1.5 Q/ha, respectively.

C. Mesta (Kenaf)

1. Introduction

Botany and Origin.   Mesta, (*Hibiscus cannabinus* L.), or kenaf, belongs to the Malvaceae family. It is indigenous to India; however, some authorities believe that it probably originated in tropical Africa.

Mesta is an herbaceous annual with a strait, slender, glabrous, or pricky stem that reaches a height of about 2.5–4.0 m or more. The lower leaves are cordate and the upper ones are palmated with five to seven lobes. The flowers are axillary, large (7.5–10.5 cm in diameter), and yellow with a crimson center. The capsules are globose, pointed, and bristly. The seeds are large, brown, and nearly glabrous. The fiber obtained from the stem of *H. cannabinus* is known by many names, such as ambaree, Deccan hemp, kenaf, mesta, dab, guinea hemp, and others. The plant has a taproot system penetrating deep into the soil.

Species and Varieties.   Crossing between different plants of *H. cannabinus* and other varieties or types occurs in nature. The mesta varieties are distinguished from each other on the basis of color and thickness of stem, leaf form, blossoms, seed type, and adaptability to different growth conditions. Of the cultivated species, those with green stems (i.e., *H. viridis* and *H. vulgaris*) are suitable for fiber production, whereas those with red or purple stems (e.g., *H. simplex*, *H. ruber*, and *H. purpureus*) are not suitable for fiber production.

2. Cultivation

Culture.

ISOLATION REQUIREMENTS AND LAND PREPARATION.   For the production of certified seed of mesta, the field selected should be such that the same kind of crop was not grown in the previous season or, if grown, the variety of the crop was the same and was raised by using a certified seed. The land should also be well drained and loamy with a high organic matter (humus) content. On poorly drained soils, plants usually die before flowers or seeds are produced, since the crop is highly susceptible to water logging. The land is prepared by plowing and three to four harrowings followed by leveling to provide a fine tilth.

PLANTING AND CULTIVATION.   Mesta crops are usually sown in India from mid-May to June before the onset of monsoon. The nucleus/breeder's/foundation seed obtained from a source approved by a seed-certification agency is sown in rows with a spacing of 30.0 × 7.5–10.0 cm, which requires about 10 kg seed/ha.

Fertilization, Irrigation, and Weed Control.   The seed crop of mesta requires about 70–80 kg N, 120–140 kg P, and 90 kg K per hectare depending on soil fertility status. Excessive nitrogen produces too much vegetative growth. One or two irrigations are required before the onset of monsoon and subsequent irrigations as and when required depending upon rainfall situation.

One or two hand weedings may be needed to keep the seed plot free of weeds. The first weeding and thinning may be carried out after 2–3 weeks followed by subsequent weedings as required. The field should be especially free from *Convolvulus* plants before harvest to avoid contamination of mesta seeds.

Roguing.   Rouging for off-types and plants affected with stem and root rot caused by *Macrophomena phaseol* may be carried out in the early stage of crop growth followed by subsequent rougings as and when required during the entire period of crop growth.

3.   Harvesting and Threshing

The mesta crop is ready to harvest when the seeds on the lower and middle branches are ripe. The crop is cut manually with sharp sickles and left in the field to dry for 3–4 days, allowing the seed on the top branches to mature. Delayed harvests result in considerable losses due to seed shattering.

The dried crop is threshed by hand by placing stocks on a tarpaulin and flailing with sticks, long poles, or planks wrapped in canvas. Alternatively, the crop may be passed through a stationary thresher.

4.   Postharvest Operations

Cleaning and Drying and Storage.   Winnowing in natural wind or by using a coarse sieve and fanning mill cleans the threshed crop. The cleaned seed should be dried to about 8% moisture before storage.

Seed Yields.   The average seed yields of mesta vary from 800 to 900 kg/ha.

D.   Minor Fiber Crops

1.   Flax

Common flax (*Linum usitatissimum* L.) is an annual plant with a single, upright branching stem about 45–90 cm or greater in height. It has a threadlike taproot sparingly supplied with tender branches. The leaves are simple, narrow, entire, and nearly sessile. It has perfect, symmetrical rather conspicuous blue flowers, all parts occurring in fives. The capsule is usually 10 seeded and the seed boll is about 1.5 cm long. The seeds are lenticular and compressed with a smooth, polished surface, varying in color from yellow to dark brown (Fig. 5). Flax is grown in the United States almost exclusively for the oil in its seed, which is highly prized for paint and varnish on account of its quick-drying quality.

In a flax stem, three zones may be recognized, namely, the pith, the wood, and the bark, which is further divided into four layers, the skin, or epidermis; the parenchyma; the bast, or flax fiber cells and the cambium layer. Since the tough bast cells lie between

Figure 5   Common flax seeds (*Linium usitatissimum* L.). (From Ref. 56.)

the tender thin-walled cells of the parenchyma and cambium, it is possible to clear the flax fiber from the adjacent cells by "retting" the stems. Retting is performed by exposing the stems to dew or rain or by placing them in pools of water, which causes the parenchyma and cambium cells to decay or become tender, permitting the tougher fiber cells to be separated. The fibrous cells of the bast layer are long but are so cemented together that continuous fiber along the length of the flax stem can be removed. The dry stems of flax contain about 20–27% bas, 58% of which is pure fiber. When dew retted, the fiber is silvery gray, whereas water-retted fiber is yellowish white. Two forms of commercial fiber are obtained: long, straight link, about 30–100 cm in lenght, and the short, tangled fiber (tow), which separates in dressings.

The seed-production technology of flax has been described under the oilseed crops.

2. Hemp

Hemp (*Cannabi's sativa* L.), closely related to the hop and ramie, belong to the mulberry family, Moraceae. It is a native of western and central Asia and has been cultivated in China from ancient times. Hemp is a rough, erect annual, usually about 8–10 feet but, in some cases, it may be 12–15 feet tall. It has staminated and pistillate flowers on separate plants. Pistillate plants are more branched than staminate ones and are less valuable for fiber. The seeds on the market consist of naked fruits or

Figure 6   Hemp seeds (*Cannabis sativa* L.). (From Ref. 56.)

achenes. The seeds are oval and about 3–4 mm long and 2 mm wide (Fig. 6). The seeds contain 30–35% oil, which, like olive oil, may be used for culinary or industrial purposes.

Hemp thrives best in temperate climates and can be grown on any soil adapted to maize. It is usually sown by broadcasting as a spring or winter crop. It is harvested when the first ripe seed is found in the head, which takes about 100 days. Hemp may be cut with a mower or self-rake reaper when not too large or manually by hand, as in the case of maize. It is allowed to lie on the ground until retted or rotted by dew and rains, after which it is tied in bundles and stacked. It may either be broken in the field (so as to add waste products to the soil) or transported to a central place where it can be processed mechanically. The fiber yields of hemp vary from 500 to 1500 kg/ha and seed yields from 10 to 30 bu/acre. When grown for seed, hemp is planted like maize at 25 kg/acre. When retted by dew and rain, the fiber is gray and somewhat harsh, but when retted in water, the fiber is creamy white, lustrous, soft, and pliable (50).

3.   Ramie

Ramie (*Bochmeria nivea* Gaud.) is a perennial shrub with herbaceous shoots belonging to the nettle family. It resembles hemp in its growth and appearance. It is an intertropical plant and grows readily in the Gulf. Seeds, cuttings, or root division propagates the plant. Plants started using seeds are grown in hotbeds. They are spaced about as densely as the hills of maize. Ramie is grown in eastern Asia, where the fiber is extracted by hand by a slow and tedious process and is used for cordage and for making textiles.

4.   Manila Fiber

Manila fiber (*Musa textilis* Nee.), or abaca, also known as Manila hemp, is a hard or structural fiber coming from one or more perennial species belonging to the same family as the common banana (Musaceae), occurring only in restricted areas of the

Philippines. The plant requires abundant rainfall, a moist atmosphere, and well-drained soil.

Manila plants are propagated from suckers (or seed) set in hills 5–8 feet apart and require little cultivation, since the rapid growth soon shades the ground. The plant is harvested as soon as the flower bud appears—about 3 years after planting when propagated by cuttings and about 5 years when grown by seeding. The plant attains a height of 3–6 m, and the leaf sheaves of which the stems are chiefly composed are about 1.5–3.0 m long. After cutting, the leaves are divided into thin strips and drawn by hand with a knife, scraping away the pulp. The yields of fiber vary from 350 to 500 kg/ha annually.

5. Sisal

Several species of the genus Agave of the Amaryllidaceae family are cultivated in Central America as a highly prized source of commercial fiber. Sisal (*A. rigida* Miller) is known in Spain as henequin. It is indigenous to tropical and subtropical America. The hard, structural fiber is obtained from the large thick leaves mechanically. About 1000 leaves produce 25 kg of fiber. Yields of 600–1200 kg/ha are obtained under favorable conditions. The fiber is yellowish white, 1.0–1.5 m long, harsh, lacking flexibility, and easily decomposed by saltwater. Sisal is a tropical plant growing on barren, rocky land that is useless for other agricultural purposes. It grows best in limestone soil and a comparatively dry climate. Its cultivation is confined to Yucatan, the West Indies, and Hawaii.

Breeding of sisal aims to raise a new and superior plant from a true seed through crossing selected parents combining as many desirable characteristics as possible in one individual. Once this stage is reached with a fiber agave, its multiplication by means of bulbils and suckers is easier (51). Improved types of agave must give high yields of good-quality fiber. They should make rapid growth without being too exacting to cultivate, and should be adaptable to different soils and a fairly wide range of climatic conditions besides being resistant to pests and diseases.

Important species of agave are *A. sisalana*, *A. fourcroydes*, *A. cantala*, *A. amaniensis*, and *A. angustifolia*. *A. sisalana* is not pure genetically, since it is a naturally occurring polyploid hybrid. *A. fourcroydes* sets fruits sparingly at the coast. Attempts to cross *A. cantala* with other species have failed. *A. angustifolia* is a small plant with a squat growth habit commonly planted for ornamental purposes. It is notable for its rapid and prolific leaf production. Its life cycle corresponds with that of sisal. It is a sexually fertile species, and self-fertilized seedlings of *A. angustifolia* hold promise as parent material in agave breeding. *A. amaniensis* is a blue sisal having characteristics that make it extremely valuable as a parent for crossing with other species, such as its long heavy leaves with a good yield of fine fiber. It is also resistant to sisal weevils.

Agave hybrid No. 11648 is a derivative of backcrossing a F_1 (*A. angustifolia* × *A. amaniensis*) hybrid with *A. amaniensis*. This has bluish green, waxy, smooth-edged rigid leaves with small tip spines. It is sensitive to overcutting.

The sisal flower is epigynous, since its floral parts are above the gynaecium or ovary. They are pale green with a regular perianth. The fruit or seed capsule is of walnut size (3 × 2 cm) and green and fleshy during the formative stage, but turns black on ripening and drying. The capsule has three loculi, each carrying seeds in two rows. The fruit ripens about 6 months after pollination of the flowers. Sisal rarely sets seeds in East Africa, although flowers are perfectly normal and the pollen is sound. It does, however, set seed in the Kenyan highlands at altitudes from 5000 to 6000 ft, where night temperatures are low (Amani and Pare districts of Tanzania). Even under these exceptional conditions, seed capsules will not be borne unless the pole is cut back when it is in the "giant asparagus" stage, that is, before any flowering branches have developed and only if flowers, instead of large bulbils, are produced from one of the buds on the mutilated pole.

The production of bulbils by sisal is a form of vegetative production. The bulbils arise from tiny buds, being protected by inconspicuous papery bracts or bracteoles found on each flower stalk. Each bulbil is a plantlet having six to eight reduced layers plus a rudimentary root system developed from the bases of the first leaves. A massive sisal pole with many flowering branches will bear about 2000–3000 bulbils of various sizes. Bulbils may appear on the bottom branches of a pole before flowering is over and take about 2–3 months to grow to a size of 6–10 cm depending upon the weather. When ready, they either fall off or may be easily shaken off the pole. During a drought, bulbils shrivel up, becoming a purplish gray color; although dehydrated, they will soon form roots upon being watered after planting (51,52).

REFERENCES

1. E. Artschwager, Anatomy of the vegetable organs of the sugar beet, *J. Agric. Res.* 33:143 (1926).
2. A. Madsen, Sugar beet in India, *Seed Technol. News* 4(3): 1 (1974).
3. S. S. Rekhi, Sugar beet seed production in India, *Seeds Farming* 2(1): 25 (1976).
4. D. W. Wood, R. K. Scott, and P. C. Longdon, The effects of mother-plant temperature on seed quality in *Beta vulgaris* L. (sugar beet), *Seed Production* (P. O. Hebbelethwaite, ed.), Butterworths, Boston, 1980, p. 257.
5. *Sugar Beet Cultivation Bulletin No. 153*, Ministry of Agriculture, Fisheries and Food, Her Majesty's Stationery Office, London, 1970, p. 33.
6. D. K. Salumkhe and B. B. Desai, *Postharvest Biotechnology of Sugar Crops*, CRC Press, Boca Raton, FL, 1988.
7. G. J. Curtis, Problems for the plant breeder in small-scale seed production in *Beta vulgaris* L. (sugar beet), *Seed Production* (P. D. Hebbelethwaite, ed.), Butterworths, Boston, 1980, p. 243.
8. A. Zakhariev, Effect of the maternal and pollen parent on the formation of normally developed but ungerminating (hard) seeds in some triploid hybrids of sugar beet, *Rasteniev'dnt Nauki* 27(5): 72 (1990) [Bulgarian].
9. L. L. Ostrovskii, V. A. Doronin, V. L Polishchunk, and L. I. Lishchinovich, Features of the seed production of heterotic hybrids using male sterility, *Sakharnaya Svekla Proizvodstvo i Peterabotka* (3): 28 (1990) [Russian].
10. R. L. Agrawal, *Seed Technology*, Oxford & IBH Publishing, New Delhi, 1980.
11. I. A. Abugaliev, S. S. Utebaev, and N. N. Bigaziev, Chemical methods of presowing seed treatment, *Sakharnaya Svekla* (2): 36 (1990) [Russian].
12. M. J. Durrant, S. I. Mashi, and S. McCullagh, Interactions between seed quality, emergence and bolting in sugar beet, *Seed Sci. Technol.* 18(3): 833 (1990).
13. V. N. Balan, N. M. Kirichenko, and L. Ya, Zhovtonozhuk, Stand density for seed production in sugar beet grown without transplanting, *Vestnik Sel'skohozyaistvennoi Nauki* (7): 97 (1991) [Russian].
14. V. B. Varshavs'ka and L. K. Lenchevs'ka, Effect of treatment of sugar beet stecklings with growth i regulators on the quality of ripening seeds, *Ukrainskii Bot. Z* 48(I): 67 (1991) [Ukrainian].

15. N. V. Busol, V. L. Polyvyannyi, and N. G. Digtyar, Postharvest seed treatment, *Sakharnaya Svekla No. 627* (1989).

16. Guideline for the biological evaluation of fungicides No. 125, Seed treatments against seedling diseases, *Bull. OEPP* 18(4): 743 (1988).

17. G. S. Husz, *Sugar Cane: Cultivation and Fertilization*, Ruhr-Stickstoff, A.G., Bochum, Germany, 1972.

18. A. C. Barnes, *The Sugar Cane*, 2nd ed., Leonard Hill, London, 1974.

19. J. Jeswiet, World material of Saccharum, Proc. Second Int. Congress of The Society of Sugarcane Technologists, Havana, 1927.

20. J. H. Buzacott, Cane varieties and breeding, *Manual of Cane Growing* (N. J. King, R. W. Mungomery and C. G. Hughes, eds.), Elsevier, New York, 1965, p. 220.

21. N. A. Noureldin, M. A. El-Moursi, A. A. Tabbakh, and I. H. El-Geddawi, Effect of some agricultural practices of flowering and seed setting of sugarcane in Egypt, *Pakistan Sugar J.* 3(3): 9 (1989).

22. K. B. Malik, Sugarcane flowering and variety development in Lousiana, USA, *Pakistan Sugar J.* 4(1): 10 (1990).

23. A. M. Abou-Salama, Sugarcane pollen viability and seed setting as affected by day length decline rates and relative humidity, *Dissert. Abstr. Int. B. Sci. Eng.* 52(3): 11348 (1991).

23a. J.A.Campbell, B.P.Naidu and J.R.Wilson, The effect of glycinebetaine application on germination and early growth of sugarcane, *Seed Sci. Technol.* 27:747(1999).

24. K. G. A. Goonasekera, A. Kumarisinghe, and T. B. J. Rajapakse, A low cost design of a sugarcane seed deglumer for sugarcane breeding, Proceedings of International Agricultural and Engineering, Conference and Exhibition, Bangkok, Thailand, December 3–6, 1990 (V. M. Salokhe, and S. G. Iangantileke, eds.), Asian Institute of Technology, Bangkok, Thailand, p. 139.

25. A. Abayomi and O. Fadayomi, Effect of short-term storage conditions on viability and seedling vigor of sugarcane fuzz (seeds), *Niger. J. Botany* 2:109 (1989).

26. E. L. Little, Important forest trees of the United States, *Trees: The Yearbook of Agriculture*, U.S. Department of Agriculture, Washington, DC, 1949, p. 763.

27. H. B. Krisbel, Some techniques for early diagnosis of genotypes in *Acer saccharum* L., *Zuchter* 6: 68 (1963).

28. R. Ferraris and G. A. Stewart, New options for sweet sorghum. *J. Australian Inst. Agric. Science* 45(3): 156 (1979).

29. J. H. Martin, History and classification of sorghum, *Sorghum: Production and Utilization* (J. S. Wall and W. M. Ross, eds.), AVI Publishing, Westport, CT, 1970.

30. O. H. Coleman, Syrup and sugar from sweet sorghum, *Sorghum: Production and Utilization* (J. S. Wall and W. M. Ross eds.), AVI Publishing, Westport, CT, 1970.

31. D. K. Salunkhe and B. B. Desai, *Postharvest Biotechnology of Oil seeds*, CRC Press, Boca Raton, FL, 1986.

32. E. B. H. Comer, *The Natural History of Palms*, Weidenfeld and Nicolson, London, 1966, p. 284.

33. H. F. Loomis, The nipa palm of the orient, *Principles* 1:41 (1956).

34. J. M. Dalziel, *Useful Plants of West Tropical Africa*, Crown Agents for the Colonies, London, 1937.

35. D. K. Salunkhe and B. B. Desai, *Postharvest Biotechnology of Vegetables*, Vol. 2, CRC Press, Boca Raton, FL, 1984.

36. B. B. Chubey and D. G. Dorrell, Jerusalem artichoke, a potential fructose crop for the Braines, *Can. Inst. Food Sci. Technol.* 7(2): 98 (1974).

37. H. J. Schaefer and J. W. Tintera, U.S. Patent No. 3,497; 360 (1970).

38. V. J. Godin and P. C. Spensley, *Oils and Oilseeds, Crop and Products Digest No. 1*, The Tropical Products Institute, London, 1971.

39. B. X. Yu, R. B. Xia. X. D. Wang, G. G. Zhu, and X. M. Qiu, A new cotton variety of low gossypol content-'Zhemian 9,' *China Cottons* No. 5:23 (1990).

40. Q. X. Wu, A new low-phenol cotton variety, 'Xinluzac 3' *Crop Genet. Res.* No. 2:46 (9990).

41. M. R. Petalcorin, R. P. Cabangbang, and R. F. Bader, Efficiency of some controlled pollination techniques in cotton (*Gossypium hirsutum* L.), *Philip. Agric.* 72(3): 291 (1989).

42. T. Z. Zhang and J. J. Pan, A male-sterile line with 'Vireslence' marker character in upland cotton, *Euphytica* 48(3): 233 (1990).

43. K. Redenbaugh, D. Slade, P. Viss, and J. A. Fujii, Encapsulation of somatic embryos in synthetic seed coats, *HortScience* 22(5): 803 (1987).

44. K. Redenbaugh, *Synseeds: Synthetic seeds to Crop Improvement*, CRC Press, Boca Raton, FL, 1993.

45. Z. M. Sawan and R. A. Sakr, Response of Egyptian cotton (*Gossypium barbadense*) yield to 1, 1-dimethyl pipridinium chloride (Pix), *J. Agric. Sci.* 114(3): 335 (1990).

45a. R.M.Maiya, B.Gouda, M.Shekhargouda, B.S.Vyakaranahal and B.M.Khadi, Seed technological investigations in naturally colored cotton, XI National Seminar on Quality Seed to Enhance Agricultural Productivity, UAS, Dharwar, January 18–20, 2002, *Seed Technol News* 32: (A.Gaur, A.K.Vari, J.L.Varshney, and K.Kant, eds.), Indian Society Seed Technology, Division of Seed Science and Technology, IARI, New Delhi, March 2002, p. 46.

46. Cotton seed drying and storage at cotton gins, Technical Bulletin No. 1262, Agriculture Research Service, U.S. Department of Agriculture, Washington, DC, 1962.

47. A. P. Nachaev, N. N. Busareava, V. D. Nadykta, and A. A. Yukasheva, Study of the lipid complex of cottonseeds during storage under controlled atmosphere conditions, *Maslozhir Promst.* 4: 13 (1983).

48. A. T. Ganesan, S. S. Shah, and M. S. Swaminathan, Cause *for* the failure of seed setting in the cross *C. olitorius* × *C. capsularis*, *Curl: Sci.* 26:292 (1957).

49. T. Ghosh, *Handbook on Jute*, The Food and Agriculture Organization of the United Nations, Rome, 1983, p. 38.

50. T. F. Hunt, *The Forage and Fibre Crops* in *America*, Orange Judd, New York, 1912.

51. G. N. Lock, Sisal: *Tropical Agriculture Series*, 2nd ed. Longmans Green, London, 1969, p. 246.

52. J. F. Osborne, *A Handbook of Sisal Planters*, Tanganyka Sisal Growers Association, Tanganyka, 1965.

53. H. Laby, The future varieties of sugar beet, *Seed Research*, Special Vol. No. 1 (S. P. Sharma, ed.), Indian Society of Seed Technology, New Delhi, 1993, p. 109.

54. I. Nicholas, Development of cotton seed processing in India, *Seed Research*, Special Vol. No. 2 (S. P. Sharma, ed.), Indian Society of Seed Technology, New Delhi, 1993, p. 818.

55. T. X. Ling, G. X. Tang, R. H. Cai, and H. F. Gu, Mathematical model on high seed yield of sugar beet and optimized combination of cultured measures, *China Sugar beet* No. 4:8 (1991).

56. *Seeds: The Yearbook of Agriculture*, U.S. Department of Agriculture, Washington, DC, 1961.

13
Grasses and Forage Legumes

I. INTRODUCTION

Unlike cereals and pulses, grasses and legume forages are not generally grown for their seed but to produce leafy forage for livestock. Consequently, seed crops of forage plants usually require different management from that required for pasture or hay crops. According to Kelly (1), this is a developing field of study, particularly with regard to the use of herbicides for weed control. Specialized harvesting equipment is also being devised, especially for the grasses and forage legumes grown in the tropics and subtropics. Agricultural chemicals in forage crops are being used to shorten the flowering stem to facilitate harvesting, to desiccate the matured crop to reduce moisture content before harvest (chemical desiccation), and to prevent shedding of ripe seed by the use of resins or polymers. Because research work on these aspects for most forages is still at an experimental stage, recommendations are not available for general or commercial use of these practices. Work in progress may provide additional tools of great benefit to forage seed growers.

Most grass species are cross fertilized, and many of the species of forage legumes are also cross pollinated, but, unlike grasses, the latter require insects to affect a good seed set. The flowers in several species have to be tripped, i.e., the weight of the insect alighting on the keel causes the stamens and, in some cases, the stigma to protrude out of the flower, facilitating pollination and seed set. Honeybees and bumblebees are the most effective pollinators for many species. The seed grower thus should ensure that the proposed area for seed production of forage crops supports an adequate population of suitable pollinators.

Owing to significant cross fertilization of grass species, plant-to-plant variation within a cultivar is relatively high. Environmental influences and management practices may, therefore, favor seed production of one kind of plant more than another, especially when the seed is produced in an area far removed from the place where the parents of the cultivar were selected and the foundation seed or breeder's seed was produced. In the United States, for example, herbage seed is produced in the West, where the climate is suitable for seed production of cultivars for use in the eastern states as well as in other countries where the climate is more suitable for hay or pasture. Kelly and Boyd (2) concluded that growing seed crops of forages for one generation in a different environment does not usually have a significant effect. The seed growers should have adequate knowledge of the variation that exists in the stability of cultivars.

Irrigation practiced in the production of forage seeds can be beneficial in producing seed in areas where the climate is most suitable for seed maturation but where seed yields may be affected by drought conditions. The seed crop can be irrigated during its establishment phase to encourage early tillering for improved seed production. Water stress induced at later stages encourages the production of flowering stems in a more uniform way, whereas subsequent further irrigation encourages the production of plump seed. Irrigation timings must be integrated with application of fertilizers, especially nitrogenous fertilizers, to prevent excessive leaching losses of valuable nutrients. Supplementary nitrogen dressings given in splits in addition to a basal application have proved to be very advantageous to seed crops. The species of forage crops that are adapted to drier conditions generally respond more poorly to irrigation than those that are less drought resistant (1).

The flowering phases of most forage legumes extend over a long period of time, prolonging seed ripening and making the harvesting of the seed crop difficult. Most species shed the seed freely, contributing to significant harvesting losses. It is, therefore, important to judge an appropriate time of harvesting of the seed crop to minimize these losses.

Various species of forage crops are grouped as perennials, biennials, and summer and winter annuals. The longer-lived species generally have the widest range. Alfalfa is the most widely grown of all cultivated forage species because of the success of plant breeders in developing numerous varieties that differ markedly in growth response, cold tolerance, and persistence (3).

Several hundred varieties of grasses and forage legumes comprising more than 125 species are grown all over the world. They differ in performance and in their requirements for soil, climate, irrigation, fertilization, and plant protection. At least one variety/species is adapted to each use and to each particular site or location. Grasses and forage legumes have thus become an important part of the cropping system on nearly every farm. A highly specialized seed-production industry is, therefore, essential to ensure domestic supplies of seed having the specific superior genetic characteristics that differentiate the recommended varieties one from another.

II. BREEDING

Great progress has been made in breeding improved varieties of grasses and forage legumes since 1940. The development of newer varieties with disease resistance and tolerance to a range of temperatures and improved management technology have permitted extension of some species of forage crops into new regions and soils previously unsuited to grassland agriculture.

Newer management practices such as row planting with proper field isolation; timely irrigation, fertilization, and the use of growth regulators, forcing plants to flower when temperatures are favorable; adequate control of insects, diseases, and weeds; satisfactory pollination; and timely harvest have advanced the production of forage seed from a minor to major farm enterprise. Seed production in the United States has also developed to include new forages like bluegrass, alfalfa, bent grass, clover, and lupines, and a range of other grasses.

Modification in alfalfa at first was achieved almost entirely by natural selection. Northern common types had fall dormancy and the ability to persist through cold winters. The important varieties developed through introduction and selection during the early period were Grimm, Baltic, Cossack, and Ladak. Grimm, a winter-hardy variety, became one of the most widely distributed varieties in the northern United States until bacterial wilt attacked it. Alfalfas introduced from Russian Turkestan, northern India, western China, and northwestern Iran were found to have some resistance to bacterial wilt. A few resistant plants also were found among some adapted varieties.

State and federal alfalfa-breeding programs were enlarged and coordinated to concentrate on wilt resistance and other breeding problems. From this work emerged improved varieties like Ranger, Buffalo, Atlantic, and Vernal, representing marked improvements in disease resistance and adaptation to areas for which they were recommended. These new synthetic varieties of alfalfa were produced by recombination of individually selected and tested clones or inbred lines. Efforts are being continued to produce newer hybrid alfalfa varieties.

Preliminary work in the improvement of grasses also began with selection, but due precautions were not taken to prevent cross pollination. A number of improved varieties were distributed, but most of them were adapted to a limited area (1).

According to Sprague (4), rust has been one of the major diseases of grasses. It was relatively easy to isolate and select resistant types of timothy, which is usually grown for hay with alfalfa or clover. Shelby and Marietta are the varieties of grasses ready for cutting when the legume crop is also ready for harvest. Types of grasses especially adapted for pasture have been developed. Progress was made in isolating prostrate types. The work with timothy has been carried out almost entirely by evaluating and propagating clones, but research work with other grass species has attempted to utilize hybrid vigor more completely. In types such as brome grass, orchard grass, and red canary grass, clonal selection has been followed by the recombination of a relatively small number of selected clones into synthetic cultivars (1). The level of hybrid vigor in such material is much greater than the vigor of the parent clones.

In Pensacola Bahia grass, about 20% of plants exhibit self-incompatibility. Such plants, when self-pollinated, set less than 2% seed. However, when grouped in pairs, seed setting may be as high as 90%. This method of hybridization has been used to produce Tifhi-I, having significantly higher yield of a better-quality grass per unit area than the conventional types. The same breeding method is used with several other perennial pasture plants.

Buffalo grass is dioecious, having staminate and pistillate flowers borne on separate plants. This phenomenon has been used to produce the F1 hybrid Mesa. Seed fields are prepared by vegetatively interplanting selected clones of the two dioecious parents in the seed fields.

Gahi-I pearl millet is a hybrid composite made from the random interpollination of four selected inbred lines. The seed produced represents a combination of F_1 hybrid seed and selfed or sibbed seed, resulting from interpollination among plants of the same parental line. The nonhybrid seed is selectively eliminated through competition in heavy seedlings; thus an essentially pure stand of hybrid plants remains. Yields of forage crops obtained by this breeding procedure have exceeded those of the parental check by more than 50% (1).

The cytoplasmic male sterility discovered in pearl millet and other crops has provided an alternate method of producing hybrid forage seed. As the research work on various forage species has progressed in the past few decades, growing emphasis has been placed on procedures that permit wider utilization of hybrid vigor.

Hybrids between *Lolium multiflorum* and *L. perenne* exhibit more or fewer of the characteristics of either species. The cultivars that are tetraploid have seeds almost double the size of those that are diploid, and moisture content of the plant is 1–2% higher, resulting in slower drying. Some tetraploids are less persistent than their diploid equivalents, but the more recent tetraploids have been greatly improved in this respect (1).

Seed production of grasses can be more appropriately discussed when they are classified as cool-season grasses and warm-season grasses. The latter can again be divided into great plain grasses and southern grasses. Copeland and McDonald (5) classified the grasses grown in the United States as shown in Table 1.

Cool-season grasses (most lawn and turf grasses) have been adapted to the northern regions of North America. These grasses grow well at cooler temperatures during the spring and fall, with their winter response varying according to severity of the climate. Whereas in the milder Pacific Northwest climate, they maintain their green color, under severe winter conditions, they go dormant, losing their green color. Dormancy is followed by very rapid growth in spring followed by flowering in early June, with seed becoming mature in July.

Warm-season grasses exhibit their best growth during the summer aided by adequate summer rainfall in the area of their adaptation. These grasses start their spring growth about 3 weeks later than the cool-season grasses and cease growing with the first hard frost in the fall. In the winter, they remain completely dormant.

Great plain grasses are mostly native grasses of the central and Great Plains (Texas, Oklahoma, Kansas, and Nebraska), and were originally harvested from native pastures for use in converting croplands to grass and for improving rangelands. Even today, seed of most of these grasses is harvested from native stands. However, a specialized seed-production industry is developing, owing to sporadic and undepen-

Table 1 Classification of Grasses of the United States

| | Warm season grasses | |
Cool season grasses	Great plain grasses	Southern grasses
Tall fescue	Grama	Burmuda grass
Red fescue	Bluestem	Dallis grass
Rye grass	Buffalo grass	Rescue grass
Tall oat grass	Indian rice grass	Bahia grass
Kentucky bluegrass	Lovegrass	Zoysia grass
Smooth brome grass	Blue panicum	Napier grass
Orchard grass	Switch grass	Pangola grass
Red canary grass	Dropseed	Centipede grass
Timothy	Buffel grass	Rhodes grass
Bent grass	Texas wintergrass	Carpet grass
Russian wild grass	Indian grass	Vasey grass
Wheat grass	Vine mesquite	St. Augustine grass
	Curly mesquite grass	
	Thatch grass	

Source: Ref. 5.

dable native stands occurring only when favorable seasonal conditions exist, and improved varieties of native grasses are available.

Southern grasses (like Bermuda grass) are adapted to southern regions of the United States (Arizona and southern California). Much of the southern forage grass seed production is harvested from pastures that are grazed during the summer, fall, and early spring. This dual use and early removal of grass material has stimulated seed production as compared to ungrazed stands. The availability of modern improved varieties of grasses has further stimulated seed production in rows rather than as a by-product of pasture production (5).

Seeds of the succulent annual halophyte *Zygophyllum simplex* germinate after rainfall during July or August. *Z. simplex* is one of the few annual halophytes found in subtropical salt marshes and deserts. Its leaves and seeds have medicinal properties, and shoots are grazed by animals. It has a life span of about 80 days. Khan and Ungar (5a) conducted experiments to determine the effects of NaCl and thermoperiods on the germination of seeds and their recovery responses after being transferred to distilled water. Cooler temperatures significantly inhibited germination at all treatments, and the highest temperature also caused some inhibition. Seeds were moderately tolerant to NaCl concentrations during germination. Highest germination percentages in all salinity treatments were obtained at a moderate (15–25°C) thermoperiod. Few seeds germinated at concentrations higher than 100 mM NaCl. When seeds were transferred to distilled water, after 20 days of salinity treatment and at various thermoperiods, there was some recovery. Recovery ranged from 0 to about 20% germination at extreme thermoperiods for seeds that were germinated at high salinity concentrations.

A study of tall fescue (*Festuca arundinacea*), Kentucky bluegrass (*Poa pratensis*), and creeping bent grass (*Agrostis stolonifera* var. *Palustris*) was undertaken to determine if reversed-phase high-performance liquid chromatography (RP-HPLC) of seed proteins could be adapted to the cultivar identification of turf grass species. Chromatograms of different cultivars of these species were all different and were used to characterize those cultivars. RP-HPLC was shown to be a quick, repeatable, and reliable method of turf grass cultivar identification for general screening of seed lots. Owing to the detection sensitivity of the apparatus used and the technical problems of working with small volumes, the analysis of single seeds for varietal identity was not possible. Mixtures and blends of seeds were analyzed and compared. The sensitivity of the detector was such that the changes of the area under the peaks of the chromatogram did not become obvious until proportions increased or decreased 20% in total composition (5b).

Freeman and Yoder (5c) determined seed esterase isoenzyme banding patterns of *P. pratensis* using polyaceylamide gel electrophoresis (PAGE) for 20 cultivars and for several mixtures of cultivars. Extracts of the individual cultivars were run in lanes adjacent to extracts from the mixtures in order to determine if the individual components could be identified in a mixture by the presence of bands unique to particular cultivars. The unique bands of individual components were only detected consistently in two-component mixtures. In mixtures made in the laboratory to match commercial mixtures, if a substitution of a component was made at a 20% or greater proportion, differences in the two banding profiles were detected. This technique could be useful in confirming that cultivars stated on labels of commercial and certified-seed mixtures were used.

III. SEED-PRODUCTION TECHNOLOGY

A. Tropical and Subtropical Grasses

The grasses of the tropics and subtropics are adapted to produce seed under short-day conditions, whereas some species are inhibited to produce seed under long-day conditions. Certain species prefer humid conditions, whereas others are adapted to drier areas. Some tropical grasses are propagated vegetatively, whereas others are normally grown from seed. Unlike the perennial temperate grasses, the seed of tropical and subtropical grasses is usually produced in the year of sowing, with lower harvested yields. It is often difficult to produce the seed of tropical grasses with regular viability. Boonman (6) employed the criterion of yield of "pure germinating seed," which may be less than the yield of clean seed, some of which may be infertile. Also, seed is usually produced two or three times a year depending upon the availability of irrigation.

Prolonged heading, both within and between plants, as well as prolonged flowering within each head resulting in an extended ripening period of tropical grasses is the primary reason for their low yields (7). According to Boonman (8) the commercial yields of about 100 kg/ha with a quality of 25% pure germinating seed have been reported. The following discussion on the production of seeds of tropical and subtropical grasses is mainly based on Hebblethwaite (9) and Kelly (1).

1. Rhodes Grass

Rhodes grasses (*Chloris gayana* Kunth.) comprise seven cultivars, some of which are diploid and others tetraploid (10).

Rhodes grass, being a cross-fertilized forage crop, needs to be isolated by a distance of 100–200 m from other grass fields depending upon whether or not the seed is to be used for further multiplication (in the case of smaller seed plots of 2 ha or less). For larger fields, an isolation distance of 100 or even 50 m may be sufficient. Diploid and tetraploid cultivars do not intercross, but some infertility may occur if anthesis coincides in adjacent crops. Kelly (1) advocated an interval of 2 years between crops of Rhodes grass (pasture, hay, or seed) unless the crop raised is of the same variety and the seed used is from an approved source. The land selected should be free from other grass species with similar characteristics to avoid any possible contamination.

Culture.   The seed of Rhodes grass is sown in May–June with an interrow spacing varying from 25 to 100 cm. Closer spacing helps to produce more uniform flowering and facilitates harvesting (11).

Rhodes grass responds well to the application of nitrogenous fertilizers and irrigation (11). Depending upon soil tests, the seed crop may require 100 kg N, 50 kg P, and 50 kg K per hectare. Tropical soils are usually deficient in micronutrients like boron, copper, manganese, zinc, or molybdenum as well as calcium and sulfur. Deficiencies of these micro- and macronutrients, if any, should be corrected to obtain good seed yields. The use of farmyard manure or compost using dung has proved to be advantageous provided these manures do not contain any viable seeds that might produce plants injurious in a seed crop.

Intercultivation.   The seed crop must be kept free of weeds and plant species with seeds with characteristics similar to those of *C. gayana*. Starting with a clean field and presowing cultivations (plowing, harrowings, and leveling) aided by application of paraquat is usually sufficient to control weeds. Intercultural operations like hoeing during crop growth in the seeding year will reduce weed populations. Weeds may also be hand rogued if possible. Thrips may be problematic in some areas, and should be controlled from time to time.

Harvesting. It is difficult to judge the correct time for harvest, since the crop bears inflorescences at various stages of maturity over an extended period of time. Early-developing inflorescences generally provide the highest yield of viable seed. Some seed shedding does occur from the earliest-maturing inflorescences before the crop is harvested; others will have seed in the hard dough stage or milky stage depending upon maturity of tillers (11). The crop may be harvested manually by beating or shaking the matured inflorescences against a container and then repicking the field after 7–10 days. The field should be cleaned up soon thereafter to prepare for the next seed crop.

Seed crops are not defoliated before the first seed harvest after sowing. Grazing, cutting, or burning to clean the field may remove crop residues after the harvest. There is no evidence to show that further defoliation is beneficial. Any subsequent defoliation should be well timed not to remove the developing inflorescence apices, which may delay anthesis to a less favorable period. Earlier defoliation may, however, help to synchronize inflorescence development and provide a more uniform crop at harvest. Seed crops sown in rows normally fill in the interrow spaces within a year and will not permit interrow cultivation after the first (sowing year) seed harvest.

Postharvest Operations. The hand-harvested inflorescences cut from the plants may be bound into sheaves and stocked and transported to a drying floor. The seed is then threshed either manually or with a stationary thresher. A combine can also be used. Swathing and picking up by combine, direct combining, or using a stripper to beat the seed heads to shake out ripe seeds are also employed in developed countries.

The handpicked seed is usually dry (14% or less) and can be stored as such. The combined seed will, however, need to be dried by spreading thinly on a well-ventilated floor and turning frequently. The dried seed is then cleaned by air/screen cleaner. The cleaned seed is usually about 30% fertile seed, or which about 70–80% is viable. Storing the seed under favorable conditions for a year may improve its germination by a decline in dormancy (12).

2. Sudan Grass and Proso Millet

Botany and Isolation Requirements. Sudan grass (*Sorghum sudanese* [Piper] Staph) or (*S. bicolor* × *S. sudanese*) and proso millet or common millet (*Panicum milaceum* L.) belong to a group of tropical annual grasses containing about 25 cultivars (10). The sorghums are cross fertilized and the millets are both self- and cross fertilized. These grasses, therefore, require sufficient isolation from other fields of forages when grown for seed. Seeds of sudan grass are shown in Figure 1.

Cultivation and Harvesting. Sudan grass seed crops are sown in rows up to 100 cm apart, using about 4 kg of seed per hectare. Millet is sown at a distance of 40–60 cm and requires about 14 kg of seed/ha. In areas where frosts occur, sowing may be delayed until the soil is warm enough to facilitate germination (23–27°C). *S. halepense* (Johnson grass), a perennial forage grass, is the most difficult weed to control. The seeds of this weed are slightly smaller than those of sudan grass.

Figure 1   Seeds of sudan grass (*Sorghum sudanese*). (From Ref. 3.)

Since the crop ripens unevenly, harvesting should be timed properly to obtain maximum yield from the main tillers. Swathing is preferred in areas where weather is favorable; otherwise the seed crop can be directly combined. The sorghum seed crop may be harvested two times depending upon seed maturity, although the best seed is generally obtained from the first harvest.

3. Sataria Grass

Botany.   Sataria grass (*Sataria sphacelata* [Schumach.]) also known as South African pigeon grass or golden timothy grass, belongs to a group of grasses containing both diploid and tetraploid cultivars. Some authors call the species *S. anceps* Staph. ex. Massey (1,10).

Cultivation and Harvesting.   Sataria grass is mainly a cross-fertilized crop, although some selfing does occur. The seed crop is sown with an interrow spacing of 30–50 cm (13). Anthesis extends over a long period of time, making the harvesting more critical. The spikelets of sataria grass are subtended by one or more bristles (sterile branches of the rachis), which remain attached to the rachis when seed sheds (1). The seed-cleaning operation is comparatively easier than in *C. gayana*, since there are no awns.

Other cross-fertilized species such as colored guinea grass or small buffalo grass (*Panicum coloratum* L.) and Columbus grass (*Sorghum* × *almun* Parodi) are grown in a manner similar to that of *C. gayana*.

4. Bracharia spp (Apomictic Grasses)

Signal grass, (*Bracharia decumbens* Staph.), or Surinam grass, Koronivia grass (*B. humidicola* [Randle] Schweichart), Ruzi grass (*B. ruziziensis* Germaine et C. Everardt), or signal grass, and paragrass (*B. mutica* [Forsk] Staph.) are all apomictic types and hence require an isolation distance of only 3 m to avoid physical mixing of the seed from crops with similar seed size. The seed is sown in lines (30–50 cm). The crop responds well to nitrogen up to 150 kg/ha. The seed sheds easily and is harvested when it becomes hard and difficult to mark with the thumbnail. Crops are usually harvested by direct combining with the table set low to pick up lodged panicles. Suction harvesters are used in Australia after combining to retrieve the shed seed. *B. mutica*, being a shy seeder, is often propagated vegetatively.

Other apomictic species of forage grasses include buffel grass (*Cenchrus cilaris* L.), jaragua or thatching grass (*Hyperrahania rufa* [Nees] Staph.), molasses grass (*Melina minutijlora* Beauv.), plicatulum (*Paspalum plicatulum* Michx), and *Urochloa mozambicensis* (Hack) Dandy. Being self-fertilized, none of these grasses requires an isolation distance of more than 3 m from other grasses. They respond to the application of nitrogenous fertilizers, especially when planted in rows up to 100 cm apart. Two of these, *P. plicatulum* and *U. mozambicensis*, are the easiest to harvest mechanically and may be combined directly. The other species are difficult to harvest because of a prolonged maturity period both within and between inflorescences. The seed crop may be harvested manually by hand or mechanically using a machine with a beating or stripping action to collect seed as it ripens. The seed of H. rufa has long twisted awns, making it awkward to handle. The *U. mozambicensis* seed crop can be harvested more than once (three to five times) a year.

5. Partially Apomictic Grasses

Botany.   Guinea grass (*Panicum maximum* Jacq.), Nallis grass or paspalum (*Paspalum dilatatus* Poiret), Bahia grass (*Paspalum notatum* Fluegge), and Kikuya grass (*Pennisetum clandestinum* Hochst) are the species belonging to a partially apomictic class of grasses. *P. maximum* is apomictic but cross-fertilizes about 5%. The Paspalum grasses are also apomictic, but have some cultivars that cross fertilize. Kikuya grass is a facultative apomict (1).

Cultivation and Harvesting.   These grasses are usually grown in rows (40–60 cm) and respond to grazing, which induces uniform emergence of inflorescence. The seed crops must be provided with gate isolation distances depending upon the mode of reproduction of a particular cultivar to be grown for seed (100–200 m).

High levels of nitrogen have been reported to encourage seed production. *Panicum maximum* and *Paspalum dilatatum* have wider maturity periods and are therefore difficult to harvest. Hand harvesting to remove ripe panicles is often followed. Bahia grass is comparatively easier to handle and, when grazed earlier, can be combined directly. Kikuya grass also responds well to hard grazing, but the seed heads are then very close to the ground. Seed is usually collected by cutting with a rotary mower set close (1 cm) to the ground. The material contains much leaf and stem and must be dried artificially or by spreading thinly on the floor. The dried material is treated in a hammer-mill to separate the seed and cleaned with an air/screen cleaner. Occasionally, the long-standing crop in the field can also be swathed and picked up with a combine (14).

B. Temperate Grasses

The grasses adapted to temperate, cooler climates include some annual and biennial forms, but most are perennial. Both biennials and perennials respond to lengthening days and temperature to induce a reproductive phase (flowering). They usually produce tillers in the autumn, which contribute to the seed yield in the following year.

1. Italian Rye Grass and Westerwold

Botany. *Lolium multiflorum* Lam., annual (westerwold) and biennial (Italian rye grass), can be distinguished from the perennial rye grass, *L. perenne*, by the presence of awns on the outer paleas. The rachilla is usually flattened in *L. multiflorum* but oval in *L. perenne*, the shoots of *L. multiflorum* are rounded but those of *L. perenne* are flattened, and the auricles are more predominant in *L. multiflorum*. Within *L. multiflorum*, cultivars that are short-lived and produce a seed crop in the sowing year (sown in early spring) are called westerwolds in Europe. The cultivars of *L. multiflorum* are distinguished on the basis of time of inflorescence emergence (early, late), plant growth habit at inflorescence emergence (erect, prostrate), flag leaf length at anthesis (short, long), flag leaf width (narrow, wide), and stem length (short, long).

Isolation Requirements. Italian rye grass is harvested only once for seed purposes because of the reduced seed yield of the second crop and contamination due to volunteer plants from the fallen seed. It is a crop cross fertilized by wind. According to the specifications of OECD (10), both *L. multiflorum* and *L. perenne* should be isolated by a distance of at least 200 m for small fields (2 ha) and 100 m for larger fields when the seed produced is intended for further multiplication, and 100 and 50 m for the production of certified seed or other use. Diploid and tetraploid cultivars do not cross fertilize, but the pollen may cause infertility, reducing seed yields. An isolation distance of 50 m is considered to be safe.

Owing to the persistence of most grasses in the field, it is advisable to allow fields an interval of about 2 years free of grasses of similar seed size before sowing Italian rye grass seed crops. Defoliants like paraquat may be used to eradicate established grass plants and shorten the interval needed between seed crops. Autumn-sown crops tend to encourage the build-up of volunteer grasses and grass weeds more than spring crops, like cereals or root crops.

Cultivation, Fertilization, and Weed Control. Italian rye grass is preferably sown in rows (10–20 cm) in early spring without a cover crop. Seed rates vary from 12 to 16 kg for diploid cultivars and 15 to 22 kg/ha for tetraploids depending upon the conditions of seedbeds.

The nutritional requirements for major and trace elements of rye grass are similar to those of wheat. The grass weeds are the most difficult to control and their seed the most difficult to remove after harvest. Kelly (1) described different species of weeds of the rye seed crop and their control measures. Most of these weeds can be effectively controlled by the application of MCPA, 2,4-D, mecoprop, ioxynil, and bromoxynil at the three- to four-leaf stage of the autumn crop.

Roguing. The seed needs to be inspected critically for off-types and diseased plants at the beginning of inflorescence emergence when the cultivar characteristics are most obvious. The standards for cultivar purity are often expressed as the maximum number of off-types permitted per unit area because of the difficulty in distinction of individual plants and assessment on a percentage basis.

The most serious seedborne disease of Italian rye grass is blind seed disease (*Gloeotina temulenta*), which is even more prevalent in perennial rye grass. There is no effective seed treatment to combat this disease, although a small quantity of seed can be immersed in hot water (50°C) for 30 min followed by cooling and drying. The best remedy is to use a clean, disease-free seed or a resistant cultivar. Also, a seed that has been stored for more than 2 years is usually free of the disease. Field sanitation is very important. Straw burning in situ also controls the disease but has been questioned on environmental grounds (15).

Harvesting. Italian rye grass seed crops are either swathed and picked up later or combined directly. The former method is preferred wherever the climate permits. However, where drying of the harvested seed is necessary, direct combining is advocated. As in the case of most grasses, shedding soon follows seed maturation. Inflorescences at different stages of ripening make it difficult to decide on a time of harvest. It has been observed that the greatest weight of seed is generally yielded by the longer inflorescences borne on the older tillers. It is, therefore, preferable to harvest the seed crop when the majority of these are ripe than to wait for the later, shorter inflorescences to mature (1). The ripe seed is doughy, changing its green color to yellow. Swathing may be begun when the seed moisture content decreases to about 43 and 45% and to 38 and 40% before direct combining of diploid and tetraploid species, respectively.

Drying, Cleaning, and Storage. The crop may be dried in swath and later picked up with a combine when seed moisture is 14% or below. When field drying is less certain, the crop is directly combined and the harvested seed dried immediately to safe storage moisture. Floor drying with ventilated air may be used to reduce seed moisture below 20% within 5 days. The seed depth should not exceed 55 cm when seed moisture content is above 35%. No heat should be used in the early stages and only very little in the final stages with 65% RH. The maximum temperatures for drying air are 38, 49, and 54°C for seed at moisture contents of 45, 40, and 35%, respectively. It is equally important to cool the seed that is dried in warm air before storage. The seed can be cleaned with an efficient air/screen cleaner or a gravity separator.

2. Perennial Rye Grass

Botany and Cultivars. Perennial rye grass (*Lolium perenne* L.) is grown for grazing or forage as well as for its amenity uses (e.g., to cover sports fields). The cultivars of *L. perenne* can be distinguished on the basis of their maturity group or inflorescence emergence (very early, early, intermediate, late, very late), length of flag leaf (short, medium, long), width of flag leaf (narrow, medium, wide), length of longest stem including inflorescence (short, medium long), as well as length of inflorescence (short, medium, long).

The cultivars of early perennial rye grass are usually harvested for seed in the year of their establishment. The later cultivars are managed so as to harvest them in the second or third year, although higher yields are realized from the first harvest year.

The range of flowering times in *L. perenne* cultivars is much greater to enable the seed grower to isolate the seed crop in time rather than space, between very early and very late cultivars, with a difference in their flowering time of about 1 month.

Cultivation and Fertilization. The *L. perenne* seed crop is managed in a manner similar to Italian rye grass during establishment and the first harvest year. During spring grazing, seed crops of early cultivars should be closed up early—by the middle

Table 2 Effect of Time of Application of Nitrogen Fertilizer on the
Number of Fertile Tillers at Final Harvest of S-24 Perennial Rye Grass

Time of application (120 kg N/ha)	Relative number of fertile tillers per m^2
Spikelet initiation	100[a]
30% ear emergence	92
70-80% ear emergence	74
Split between spikelet initiation and 30% emergence	96

[a] Values expressed are relative to the number of fertile tillers obtained when all N
 was applied at spikelet initiation, average of six experiments, 1971–1976.
Source: Ref. 17.

or end of March (in the United Kingdom). Late cultivars can be defoliated later, but
crops should be closed up before the end of April. If the crop is to be used for a second
or third seed harvest, the field should be cleaned up immediately after harvest by
burning, cutting, or trimming all material closely.

Nitrogen dressings are done at the same time as in the first year, with an increase
in rate of about 30%. However, later application of nitrogenous fertilizer in the spring
will reduce the number of fertile tillers (Table 2). The application of nitrogen in the
spring should, therefore, be aimed at satisfying the requirements of fertile tiller
production (16,17).

Phosphorus and potassium may also be needed to supplement soils low in these
nutrients. These deficiencies are best corrected before sowing the crop in the first year.
Autumn and spring grazing or defoliation follows the same pattern as in the first year.

Roguing. Seed crop inspection, harvesting, drying, and seed cleaning are the same as
for Italian rye grass. Ripeness in the early perennials resembles Italian rye grass, but in
late cultivars, the stems are still green when the seed is ready to harvest. Seed moisture
at harvest should be about 20% lower in *L. perenne* than in Italian rye grass.

Harvesting. Hebbelthwaite et al. (17) reported that delaying the harvest of S-24
perennial rye grass from 40 to 30% seed moisture content increased the 1000–seed
weight, although the seed yield and seeds per spikelet decreased significantly (Table 3).
This was attributed to either an indirect effect of shedding of the lighter seeds from the
distal florets within the spikelets, resulting in a higher mean weight per seed, or to an
increase in seed weight per se. An increased 1000–seed weight was insufficient to

Table 3 Effect of Time of Harvest on Seed Yield and Yield Components of S-24
Perennial Rye Grass

Harvest	Seed yield (kg/ha)	1000-seed wt. (g)	Seeds per spikelet
40% moisture content	1180	2.07	1.14
30% moisture content	880	2.18	0.78
SE[a]	3.6	0.017	0.043

[a] Standard error of a treatment mean.
Source: Ref. 17.

compensate for the reduction in the number of seeds harvested per spikelet. Hence, optimum harvest time for maximum seed yield depends upon the balance between loss of seed by shedding from overripe inflorescences and gain in seed yield from the later-maturing inflorescences.

Effects of 1, 2, or 3 kg tripenthenol per hectare on seed yields of *L. perenne* cvs. Frances, Talbot, and Melle (very early, intermediate, and late flowering, respectively) harvested on five dates from July 8 to 22 (Frances), July 15 to 29 (Talbot), and July 24 to August 7 (Melle) were investigated (18). The maximum seed yields did not differ between treated and untreated plots. Tripenthenol, however, increased seed yield as harvest was delayed, because it delayed crop maturity and prevented seed shedding (which occurred in untreated plots). Seed yield increase with tripenthenol was greatest in cv. Frances followed by Talbot and Melle.

Seed yield in perennial rye grass is low, and a large difference exists between potential and actual seed production. According to Elgersma (19), the seed yield of *L. perenne* depends largely on the degree of floret site utilization (FSU). A distinction is made between biological and economic FSU; the former is based on direct measurement and the latter on harvested seeds. The losses of potential seeds during seed development and seed losses during processing have been quantified and discussed (19).

L. perenne is more susceptible to blind seed disease. *L. rigidum* Gaud, known as wimmers rye grass, is an annual species grown for seed in southeast Australia. Its production technology is similar to that of westerwold rye grass.

Hybrid rye grass (*Lolium* × *Boucheanum* Kunth.) or hybrids between *L. multiflorum* and *L. perenne* resemble one or the other parent or may be intermediate in their characteristics. For seed growing, they are treated as either Italian or perennial rye grass based on which one they resemble.

3. Meadow Fescue

Meadow fescue (*Festuca pratensia* Huds.) cultivars are distinguished on the basis of length of flag leaf, width of flag leaf, and length of longest stem including inflorescence, as in the case of *I. perenne*. The requirements of seed crops are also similar to those for *L. perenne*. The seed rates, however, are similar to those for *L. multiflorum* (diploid), as the seed is somewhat larger. Defoliation is beneficial in autumn, but grazing or cutting in the spring must be completed before the end of March to prevent a decrease in seed yield. Spring defoliation may be skipped if the growth is not excessive. The crop is swathed when the seed turns brown but retains some green color. Stems at this stage begin to lose their green color. Early cultivars are more suitably direct combined than those maturing later.

Hybrids between *Lolium* and *Festuca* have been produced and are also known to occur in nature. These, however, have proved to be poor seed producers.

A few cultivars of *Alopecurus pratensis* L., commonly known as meadow foxtail, are grown in limited areas for seed, mostly in central Europe and the northwest United States. They are uneven seed ripeners having a shedding tendency. The crop prefers a short growing season to ripen more uniformly. Seed crops can be grown in a manner similar to that of perennial rye grass.

4. Orchard Grass (Cocksfoot)

Botany and Isolation Requirements. The cultivars of orchard grass (*Dactylis glomrata* L.), or cocksfoot, are distinguished on the basis of plant characteristics similar to those of perennial rye grass. Seed crops are usually harvested for seed for 3 years, thus occupying a field for 4 years (Fig. 2). The crop is cross fertilized by windborne pollen. The requirements for isolation, previous cropping, and weed management are the same as described for Italian rye grass. Seeds of orchard grass are shown in Figure 3.

Cultivation and Weed Control. The seed is sown in rows spaced widely apart depending upon growing conditions (45–60 cm). Narrower spacing may be used when soil moisture is not limiting, and spacing wider than 60 cm can be adopted under very dry conditions. Seed rate is about 8 kg/ha in wide hills (1). Spring sowing is preferred with or without a cover crop, which should not be too heavy to suppress the seed crop. First-harvest seed yields are generally higher from crops sown without a cover crop. Interrow cultivations suppress weeds and provide a beneficial environment for the seedlings to grow. Cultivation is especially beneficial after the removal of a cover crop to loosen the soil between rows. Weeds can also be controlled with herbicides more economically. The defoliation before the first seed harvest is not advantageous unless autumn growth is excessive. Excess leaf may cause some loss of seed yield if the crop suffers from winter frost. It should, however, be removed before the end of January when the crop should be closed up for seed. After the first seed harvest, the crop should be cleaned of excess herbage, and in dry areas, it may be burned to advantage. Interrow cultivation is beneficial if soil has become compacted, but must be performed without damaging the roots.

Fertilization. The crop responds well to nitrogen provided other nutrients like phosphorus, potassium, and calcium are in adequate supply. Some nitrogen is needed in the sowing year to stimulate quick establishment of the seedlings;

Figure 2   A field of orchard grass grown for foundation seed. (From Ref. 3.)

Figure 3 Seeds of orchard grass (*Dactylis glomerata*). (From Ref. 3.)

subsequently, the crop will need about 100–150 kg N/ha in February–March each harvest year. Early-flowering types respond better to split applications of nitrogen: half or more in autumn and the remainder in the spring (1).

Roguing. The seed crop requires roguing of off-types and diseased plants starting from inflorescence emergence. Ergot is not a serious disease in cocksfoot, but choke (*Epichloe typhing*) damages the crown of the plant, especially in older crops (from the third year onward). Two or three sprays of benomyl or tridemorph can control powdery mildew in the autumn followed by one or two sprays in the spring. Cocksfoot moth (*Glyphipterix cramerella*) can cause damage in the older stands. Burning of stubble after harvest and clearing field boundaries of roughage destroys the sites where larvae hibernate. Midge eggs are laid in the developing inflorescences and larvae feed on the seed, but can be easily controlled by insecticides.

Harvesting (Optimum Swathing Date), Drying, and Cleaning. Swathing and subsequent picking of the seed crop is preferred to the direct combine method in areas where good weather conditions prevail. Seed moisture content relates well to maximum harvested yield of pure live seeds. Klein and Harmond (20) established an optimum cutting time, which is the balance between the gain in seed weight and increased germination associated with maturity and the losses due to shattering. Given the seed moisture, it is possible to predict the optimum swathing date. The rate of decrease in seed moisture and the seed moisture at mowing time for optimum seed yield of orchard grass as compared to other grasses are given in Table 4. The seed can be combined at about 14% moisture content. This method of harvesting requires that the crop should not be grown in rows spaced too widely; otherwise it would be difficult to keep the swath off the ground. The seed crop can be swathed when the seed has about 44% moisture (Table 4) and is light brown in color. The stems below the inflorescence turn from yellow to brown. Some seed shedding does occur. Direct combining requires that the seed moisture has been reduced to 30% or below, which takes about 10 days after the swathing stage. Drying and cleaning are performed as described for Italian rye grass.

5. Timothy

Botany and Cultivation.   Timothy (*Phleum pratense* L.) cultivars are hexaploid and are distinguished on the basis of the length and width of flag leaf and length of the longest stem, including inflorescence. The seed crop is mostly managed like orchard grass. However, timothy requires narrower interrow spacing than orchard grass (10–20 cm), although weeds can be controlled more effectively with wider spacings of 45–60 cm. Seed rates vary from 6 to 8 kg/ha depending upon spacing. Since timothy is late maturing, spring defoliation may be performed about 2 weeks later than for orchard grass Timothy is susceptible to leaf and stem diseases that are rarely serious. The seed crop may have to be protected against aphids and midges in some seasons.

Harvesting and Threshing.   Timothy seeds turn gray with a brownish tinge when the crop is ready for swathing (Fig. 4). At this stage, the inflorescence may begin to lose seed from the tip. Seed threshes more easily from the swath after a period of alternate wetting and drying caused by light, rain, or dew followed by sun and wind. The crop can also be combined directly first with a wide clearance at the concave to cut only the ripest seed and then return to pick up and rethresh the crop at a normal combine setting.

Another species of timothy (*P. bertolonii* DC) is a diploid equivalent of *P. pratense* and is treated similarly in all respects. The plants are smaller, having a

Table 4   Seed Moisture as a Harvesting Index

	Seed moisture (%)	
Crop	At optimum mowing time	Daily decrease near maturity
Tall fescue (cv. Alta)	43	2.5
Orchard grass	44	1.0
Perennial ryegrass	35	3.0
Kentucky bluegrass cv. Newport	28	4.0
Chewing fescue	30	5.0
Red fescue	25	4.0

Source: Ref. 20.

Figure 4 Timothy in swath. (From Ref. 1.)

creeping habit. The seed is also used for amenity purposes and to provide a bottom grass in orchards.

6. Tall Fescue

Tall fescue (*Festuca arundinacea* Schreb.) cultivars are distinguished from each other, and the seed crop requirements are similar to those outlined for orchard grass. The crop normally differentiates inflorescences early and responds well to the application of nitrogen in autumn. It should be defoliated with care not too late in the spring. The seeds shed easily, and delayed harvests can result in significant seed losses. Burning of stubble after the crop harvest has shown improvement in seed yields in some areas. Tall fescue is rather difficult to establish in the first year, but once established, it can produce many seed crops—as many as 15 seed harvests can be realized from one crop stand. Seeds of tall fescue are shown in Figure 5.

7. Soft Brome (Brome Grass, Smooth Grass)

Soft brome, smooth brome, or brome grass (*Bromus inermis* Leyss) is recognized as cultivars of two regional groups. Southern U.S. cultivars with more vigorous underground rootstocks produce earlier spring growth than those from the northern United States. Some cultivars have been developed in Europe. The requirements of the seed crops are similar to those of tall fescue.

The chaffy seed is difficult to thresh, clean, and handle and may present problems at sowing time. The seed stores well for about 1 year.

Other *Bromus* species and their cultivars grown in Europe and parts of the United States and New Zealand include field brome (*B. avensis* L.), California brome

Figure 5 Seeds of tall fescue (*Festuca arundinacea*). (From Ref. 3.)

(*B. carinatus* Hook et Arun and *B. sitchensis* Bong), and rescue grass or prairie grass (*B. willdenowii* Kunth.).

8. Tall Oat Grass

Tall oat grass (*Arrhenatherum elatius* [L.] Beauv. ex. 1.Ś. et K.B. Presl.) cultivars are mainly grown in Europe. Seed crops sown in lines with wider spacing produce higher yields. Crops should be harvested by combining when the stems turn yellow but before the top half of the panicle has shed seeds to avoid shedding losses. This can be ascertained by striking a number of panicles on the palm of the hand; if the seed in the top half can be threshed in this manner, the crop is ready for combining. Like bromes, the crop is difficult to thresh and clean owing to its long, twisted awns on the dorsal paleas. If the seed is threshed too hard to remove pales, the naked caryopsis may easily lose its germinability.

C. Prairie Grasses

Prairie grasses are mainly grown in areas having a typical continental climate with relatively hot summers and cold winters.

1. Wheat Grass

Botany and Species. The species of wheat grass (*Agropyron* spp.) are as follows (10):

A. *desertorum* (standard crested wheat grass)
A. *cristatum L.* (fairway crested wheat grass)
A. *dasystachum* (northern wheat grass)
A. *elongatum* (tall wheat grass)
A. *inerme* (beardless wheat grass)
A. *intermedium* (intermediate wheat grass)
A. *riparium* (streambank wheat grass)
A. *smithii* (western wheat grass)
A. *trichophorum* (pubescent wheat grass)
A. *trachycaulum* (slender wheat grass)

All of these species, except *A. trachycaulum*, are cross-fertilized. OECD (10) lists six cultivars for *A. intermedium* and one or two cultivars for other species.

Isolation Requirements. The seed crops are raised by giving a sufficient isolation distance of 100–200 m for the cross-fertilized species when the seed to be produced is to be multiplied again and 50–100 m for certified seed. The self-fertilized *A. trachycaulum* may need to be physically separated with a 3-m–wide strip. An interval of 2 years between seed crops of wheat grasses is generally advocated, with longer intervals being needed for the earlier generation of seed. Chemical weed control can help to reduce the interval.

Cultivation and Weed Control. The seed crop is sown in lines about 60 cm apart; wider spacing may be adopted in drier areas when irrigation is not available. Depending upon crop species and planting width, seed rates can vary from 2 to 3 kg/ha.

The seed plots must be kept free of weeds with intercultivation or weedings and chemical control. *A. repens* (couch grass) or *Avena fatua* (wild oat) are the most difficult weeds of wheat grasses. The fields must be thoroughly cleaned of these weeds before sowing the seed crop, and plots should be kept free of these weeds during seed production by hand roguing, if necessary. Canode (21) found that light working with a rotary hoe to break the soil crust in the spring and to provide mulch was more advantageous to extensive interrow cultivation. Weeds can also be controlled economically by using suitable herbicides.

Fertilization and Roguing. A single dose of nitrogen at 100–120 kg/ha, depending upon soil fertility, applied in early spring yields a better seed crop, especially in the fourth harvest year (1). The old growth and crop residues like straw should be removed immediately after harvest by grazing, mowing, or burning, if necessary. The crop should be inspected for off-types and diseased plants from time to time and especially at the time of inflorescence emergence.

Harvesting, Threshing, Drying, Cleaning, and Storage. Swathing or combining may harvest the seed crop when the seed reaches the soft- to hard-dough stage. The combining stage is slightly later when the seed becomes harder. At this stage, seed should knock out easily when the inflorescence is struck against the palm of the hand. Widely spaced rows may cause difficulty in preventing swath from lying in the interrow space, resulting in nonuniform drying and uneven pickup by the combine. Wheat grass species such as *A. dasystachum*, *A. intermedium*, and *A. trichophorum* are comparatively more resistant to shedding and less troublesome to harvest.

Care must be taken while threshing to ensure that good seed is not blown away with the blast of air from the fan to the cleaning shoe. When picked up from swath, seed moisture content is usually below 14%, which is safe for storage. Direct combining may give seed with a higher moisture content and mixed with green vegetative debris. The seed must immediately be spread out to dry on a floor with good ventilation over it and turned frequently until the seed moisture has dropped below 14%. Seed can be cleaned with an efficient air/screen cleaner. Some species with awns may require deawning. The harvested and threshed seed usually contains unbroken spikelets or seed clusters, which should be broken up and cleaned to enable the seed to flow easily in a seed drill while sowing.

2. Bluestem

Botany and Species. Three species of bluestem (*Andropogon* spp.) are known (10)—*A. geradii* Vitm (big bluestem), *A. hallii* Hach (sand bluestem), and *A. scoparius* Michx. (little bluestem). All three species are cross fertilized.

Cultivation, Harvesting, and Cleaning. Most of the requirements of *Andropogan* are similar to those of *Agropyran* (wheat grass) species. However, adequate soil moisture must be ensured during the culm-extension and seed-filling phases of crop growth. At these stages, bluestem is susceptible to high temperatures. It is a vigorously growing grass, filling the interrow spaces within a short time. Some interrow cultivation is desirable to control weeds and loosen the soil; however, defoliation in the spring of the harvest year should be avoided. The seed crop responds well to nitrogen, which may be applied in suitable splits. The seed normally matures late in the season. The harvested seed consists of the fertile spikelets together with the stalks of the infertile spikelets and awns. The crop is thus difficult to handle and requires further breaking up in a deawner before cleaning. The debris, including the infertile spikelets, may be removed using an efficient air/screen cleaner.

3. Canary Grass

Two species of canary grass, *Phalaris arundinacea* L. (red canary grass) and *P. aquatica* L. (harding grass or phalaris) each with six cultivars, have been recognized (10). Both species are cross fertilized.

The requirements of *Phalaris* seed crops are similar to those of *Agropyron*. The seed tends to mature first at the tip of the spike and then progressively toward the base. Some loss of seed from the tip generally occurs before the optimum harvest maturity is reached when about half of the seeds turn brown. Cultivars resistant to seed shedding have been selected.

4. Grama

Grama (*Boutelua oligostachya* [Nutt.] Torr. ex. A. Gray), or side oats grama, with its six cultivars, is a cross-fertilized species. The requirements of a seed crop are similar to those of *Agropyron*. The seed matures on the spike rather irregularly, necessitating an appropriate harvest time to minimize seed losses due to shedding. The crop is generally harvested by the direct combine method. The harvested seed usually contains immature seed and green vegetative debris. It should, therefore, be immediately spread on a well-ventilated floor and turned frequently for effective drying. The dried crop is then cleaned with air/screen cleaner.

5. Wild Rye

Russian wild rye, *Elymus junceus* Fisch., with its five cultivars, and Canadian wild rye, *E. canadensis* L., are grown for seed in limited quantities in Europe, Canada, and Russia. Both species are cross fertilized.

The requirements of *Elymus* seed crops are similar to those of *Agropyron* (wheat grass). The seed is sown in rows with wide spacing (up to 2 m in very dry conditions). The crop responds well to nitrogen application. Russian wild rye sheds seed easily and is required to be harvested early when the seed is still in the dough stage for swathing or somewhat later when it is beginning to harden for direct combining. Canadian wild rye sheds seed less easily, allowing some latitude in harvesting. The seed is awned and requires deawning during its cleaning.

6. Other Minor Grasses

In addition to the grasses listed above, the OECD (10) includes the following species, whose seed-production requirements are very similar to those of *Agropyron* (wheat grass):

> *Buchloe dactyloides* (buffalo grass)
> *Eragrostis curvula* (weeping lovegrass)
> *Sorghastrum nutans L.* (Indian grass)
> *Panicum virgatum L.* (switch grass)
> *Stipa viridula L.* (green needle grass)
> *Trisetum florescens L.* (golden oat grass)

D. Amenity Grasses

These grasses are grown in temperate regions mainly for the amenity purposes, such as lawns, golf courses, sports fields, and play areas. Some are also used in agriculture in special situations; for example, creeping red fescue may be grown for hill grazing. Some grasses from the preceding section, like rye grass, can also be used for amenity areas such as football fields or on roadside verges, selecting special cultivars for this purpose. A high-quality pure seed must be produced for amenity purposes. Many cultivars are highly specialized and have been selected to provide color or other features for use in specific situations. The seed of red fescue should be absolutely free from the seed of perennial rye grasses, because even one or two rye grass plants will spoil red fescue turf. The fields selected for seed production must be totally free from any volunteer grasses or other weeds to avoid possible contamination.

1. Meadow Grass

Botany and Cultivars. Smooth-stalked meadow grass (*Poa pratensis* L.), or Kentucky bluegrass, constitutes an important group of meadow grasses grown for amenity purposes. They are used extensively to provide turf giving a hard-wearing, fine surface. They are also used in agriculture as pasture grasses (1).

The cultivars of *P. pratensis* are classified on the basis of their ligule pubescence, leaf sheath color in the young stage (green, red), and upper leaf surface pubescence, whereas they can be distinguished from each other on the basis of leaf width, hair on leaf surface, culm length at maturity, panicle color, rachis shape, seed size, date of panicle emergence, and other plant characteristics.

P. pratensis is generally apomictic with some degree of cross fertilization. If the species is not truly apomictic, some isolation may be required for the seed crop as per the advice of the plant breeder. The apomictic species, however, need to be physically isolated by a gap of 3 m or wider. The field selected for the seed crop should be free of weeds and volunteer grass plants, controlling them with different herbicides, if necessary. Seed crops of *Poa* spp. should be separated by an interval of at least 2 years or more for raising a foundation/nucleus or breeder's seed.

Cultivation.   The crop is sown in rows, about 30–60 cm wide, which requires about 1 kg seed/ha. Seed crops are normally sown in the spring without a cover crop. *Poa* seed becomes well established when sown under a protective covering of charcoal about 2.5 cm wide, which allows the use of overall sprays of herbicides, protecting the emerging *Poa* plants. Lee (22) developed a "carbon-banding" technique for establishing weed-free stands from autumn planting. This method controls grass weeds in the crop when there is no basis for herbicidal selectivity. Youngberg (15) described specially designed equipment (drill) to apply a band of activated charcoal directly over the seeded row in the planting operation. One kilogram of activated carbon is mixed with 16.7 L of water, and the carbon slurry is applied at the rate of 340 kg/ha carbon in a band 2.5 cm wide.

Fertilization.   *P. pratensis* seed crop requires adequate fertilization for establishment, with 40–60 kg N applied at the time of sowing. In the United States, seed crop responses up to 135 kg N/ha have reported. Meadow grass, or Kentucky bluegrass, responds well to nitrogen fertilization up to 90 or 112 kg N/ha (21) as compared to other cool-season grasses such as red fescue and smooth brome grass, which do not respond to N applications beyond 67 kg/ha (Table 5). After the sowing year, nitrogen is applied in the spring as early as practicable. There is no advantage to splitting the nitrogen application. Adequate supplies of phosphorus, potassium, and calcium must also be ensured.

Harvesting.   *P. pratensis* is very sensitive to shading from excess herbage in the seed crop. Cover-crop debris and excess herbage from *Poa* plants should be removed; the latter by a light grazing. The main seed yield comes from the tillers formed early in autumn. About three to four seed harvests can be obtained from a seed crop. After each harvest, excessive herbage must be removed immediately by burning the straw and stubble to reduce shed seed and weed seed.

Since burning causes pollution from smoke, mechanical removal of excess herbage by chopping the straw and stubble close to the ground and blowing the chopped material into trailers with a flail harvester have been proved to be very effective (1).

Plant Protection, Weed Control and Roguing.   There are no serious diseases or insect pests of meadow grass warranting chemical treatment. Most cool-season grasses, including Kentucky bluegrass, can be grown for seed only with minimum cultivation if weeds are controlled by herbicides. Canode (21) showed that additional cultivation with sweep shovels (conventional cultivation) actually reduced seed yields of Kentucky bluegrass (Table 6).

Table 5 Seed Production of Five Species of Cool-Season Grasses as Influenced by Rate of Nitrogen Application

Species and N rate (kg/ha)	Seed yield (kg/ha)[a]				
	1st crop	2nd crop	3rd crop	4th crop	5th crop
Kentucky bluegrass					
67	495b	692b	578c	499b	441b
90	529b	724b	623b	551a	493a
112	587a	792a	668a	575a	504a
Red fescue					
67	841a	625a	540a	533a	578a
90	895a	671a	564a	545a	597a
112	870a	627a	531a	528a	586a
Brome grass					
67	1250a	926a	792a	741a	
90	1241a	920a	750a	676a	
112	1217a	899a	746a	683a	
Crested wheat grass					
67	899a	710a	640a	651b	
90	853a	687a	625a	634bc	
112	867a	724a	655a	684a	
Orchard grass					
67	275b	214b	286b	262b	
90	304b	257a	349a	315a	
112	340a	268a	350a	328a	

[a] Means with in column of each species followed by the same letter are not significantly different at 5% level of probability.
Source: Ref. 21.

Table 6 Influence of Time and Method of Cultivation of Seed Production of Three Continuous Crops of Newport Kentucky Bluegrass

Cultivation[a]		Seed (kg/ha)[b]			
Time	Spacing	Second	Third	Fourth	Mean
April		916a	291ab	982a	730a
September	Wide	856a	337a	853bc	682b
April	Wide	863a	233b	808c	635c
September	Close	766b	353a	753c	624c
April	Close	642c	320a	768a	577d

[a] All plots received minimum cultivation with a rotary hoe in April. Wide spacing left a 46-cm and close spacing a 30-cm undisturbed row.
[b] Means within same column followed by the same letter are not significantly different at 5% level of probability.
Source: Ref. 21.

Broad-leaved weeds hinder seedling growth, and grass weeds compete with the crop for nutrients and water and cause problems later if allowed to produce seed. Bromoxynil sprays effectively control annual broad-leaved weeds in established stands (21). Perennial broad-leaved weeds can be controlled with dicamba, and young weeds at the tillering stage can be treated with 2,4-D or MCPA. Grass weeds should ideally be eliminated before the crop is sown using the "stale seedbed" technique. Propham or paraquat applications before sowing kill the grass weeds as soon as they appear. After sowing, interrow spraying with paraquat can be adopted. Stray grass weeds can be removed by hand roguing. The best time for seed plot inspection is after the emergence of inflorescence, but preferably when the early-emerging plants can still be distinguished.

Harvesting, Threshing, Drying, and Cleaning.   The seed crop is usually swathed and picked up later with a combine. For swathing, the inflorescences should be yellow to brown in color and seed should be firm, with a moisture content of about 28%. Swath can be picked when seed has dried to 14% moisture or less. The seed is difficult to thresh and may require double threshing. However, care needs to be taken not to injure the seed. In Oregon, combines are adapted for this purpose.

The seed is enclosed in the paleas, which are hairy at the base and tend to entwine the seeds into clusters. These need to be broken to get the free-flowing seed. If the harvested seed has a moisture content above 14%, it should be dried immediately by spreading out thinly in a well-ventilated storage area and turning it frequently. If heated air is used, its temperature should not exceed 49°C for a seed moisture content of 30% and may be increased by 5°C as the seed dries. The dried seed can be cleaned on an air/screen cleaner.

Some other species of *Poa* are as follows, in the order of their importance (1):

> *P. trivialis L.* (rough-stalked meadow grass or rough bluegrass)
> *P. nemoralis L.* (wood meadow grass or wood bluegrass)
> *P. palustris L.* (swamp meadow grass or fowl bluegrass)
> *P. compressa L.* (Canada bluegrass)
> *P. ampla L.* (big bluegrass)

2.   Red Fescue

Botany and Cultivars.   Red fescue (*Festuca rubra* L.), also known as Chewing's fescue or creeping red fescue, have diploid, tetraploid, hexaploid, and octaploid cultivars, which can be distinguished on the basis of time of inflorescence emergence, length and width of flag leaf, stem length including inflorescence, and length of inflorescence. Chewing's fescue is distinguished from creeping red fescue by the absence of rhizomes, whereas creeping red fescue may have slender or strong rhizomes.

Isolation Requirements.   Red fescue, being cross fertilized, must be isolated by distances of at least 100–200 m for fields of 2 ha or less and 50–100 m for larger fields. In each case, greater isolation would be needed if seed is being produced for further multiplication.

Cultivation and Harvesting.   Seed crops should be inspected somewhat earlier than for *Poa*, so that isolation can be checked before anthesis. The seed-production technology including postharvest operation is similar to that described for *Poa*. Red fescue seed yields do not seem to be influenced by row spacing, unlike other temperate grasses like Kentucky bluegrass, which produces more seed in rows of 30 and 60 cm, and crested wheat grass and smooth brome grass, which grow best with 60 cm between rows (Table 7), whereas orchard grass grows best with even wider spacing (21).

Threshing. The seed is relatively easy to thresh and handle and does not shed before being ready for harvest. The ideal seed moisture content for swathing is 25%.

Some other *Festuca* spp. include *F. ovina* (L.) *sensu lato* (sheep's fescue, fine-leaved sheep's fescue, and hard fescue) and *F. heterophylla* Lam. (shade fescue). The hard fescue does not respond to stable burning, therefore its postharvest residue should be removed by chopping and carting away.

3. Browntop

Botany and Cultivars. Browntop (*Agrostis tenuis* Tibth), or colonial bent grass, cultivars are distinguished on the basis of their ploidy (diploid, tetraploid, and hexaploid), growth habit (erect, medium, prostrate), leaf width (narrow, medium, wide), rhizomes and stolons (present, absent), leaf color (pale, medium, dark), time of inflorescence emergence (early, medium, late), length and width of flag leaf (short,

Table 7 Seed Production of Five Species of Cool-Season Grasses as Influenced by Row Spacing

Species and spacing	Seed yield (kg/ha)[a]				
	1st crop	2nd crop	3rd crop	4th crop	5th crop
Kentucky bluegrass					
30	603a	814a	669a	569a	499a
60	579a	784a	659a	568a	503a
90	429b	611b	541b	486b	436b
Red fescue					
30	836a	631a	557ab	549a	569a
60	924a	696a	580a	547a	578a
90	846a	597a	497a	509a	614a
Brome grass					
30	1200b	870a	705a	645b	
60	1305a	968a	792a	741a	
90	1203b	908b	792a	714ab	
Crested wheat grass					
30	889a	651a	602a	628b	
60	926a	771a	886a	704a	
90	805a	698a	643a	637b	
Orchard grass					
30	173c	154c	248c	226c	
60	342b	250b	322b	306b	
90	406a	336a	415a	377a	

[a] Means within column of each species followed by the same letter are not significantly different at 5% level of probability.
Source: Ref. 21.

medium, long, and narrow, medium, wide), and length of the longest stem, including inflorescence (short, medium, and long).

Isolation Requirements and Seed Production.   Because browntop is a cross-fertilized crop, its isolation requirements are similar to those of *Festuca rubra* (red fescue) The method of seed production is similar to that used for *P. pratensis*, Browntop, however, requires less nitrogen and responds well to stubble burning. Some other *Agrostis* spp. grown for amenity purposes include (10):

> *A. gigantea Roth* (redtop)
> *A. stolonifera L.* (creeping bent)
> *A. canina* L. subsp. *canina* (velvet bent)

Since *A. stolonifera* and *A. canina* do not tolerate stubble burning, their postharvest residues should be chopped and removed. Other requirements of seed crop are similar to those for *A. tenuis*.

E.   Tropical and Subtropical Forage Legumes

Seed production of tropical and subtropical forage legumes is comparatively recent, and their production technology is being developed in Australia, India, and other tropical countries. Seed of forage legumes produced in the United States has two distinct types of management systems and climatic regions: (a) the specialized seed-production industry, developed in the western United States and Canada and (b) a by-product of forage production scattered throughout North America.

1.   Centro

Botany and Isolation Requirements.   Centro (*Centrosema pubescens* Benth.) is a vigorously growing perennial climber, reaching a height of about 40–45 cm. It is a self-fertilized crop and only needs to be physically isolated by a gap of 3 m to safeguard the seed crop from mechanical mixture. Since viability of hard centro seed is comparatively long in the soil, it is advisable to allow an interval of 5–6 years free from centro before sowing a seed crop. This interval can be reduced with special meausures taken to control volunteer plants or if the same cultivar is to be sown again.

Cultivation and Roguing.   Centro seed crop requires fertile soil well supplied with Ca, P, K, and N. Seed is sown in prepared land in rows spaced 80–100 cm apart to allow intercultivation in the early stages for weed control. Wider spacing also allows support fences to be erected when the crop is intended to be harvested manually. Seed rates vary from 4 to 8 kg/ha. A light presowing irrigation will help to reduce hard seed proportion and facilitate germination. Seed should be inoculated with rhizobium before planting.

The best seed yields are usually obtained from the first growth, but grazing and cutting can be adopted to reduce excess foliage or to adjust the time of flowering in such a way that harvest time coincides with favorable weather conditions. The crop should be inspected critically at the full flower phase to remove off-types and diseased plants.

Harvesting, Drying, and Cleaning.   Centro seed ripens comparatively uniformly. Two manual pickings in a supported crop generally give the best yield. Crops should be harvested when the pods begin to open. In unsupported crops, the pods remain concealed in the foliage and a lot of vegetative material may be harvested with the seed. Windrowing for 5 or 6 days before picking up with a combine is, therefore, preferred to direct combining.

The harvested crop should be dried immediately to avoid spontaneous heating due to the presence of green matter (broken leaf, stem, and immature pods). Seed should be dried to an 8–10% moisture content for its safe storage. Spreading the produce thinly on a tarpaulin or on a concrete floor may carry out drying. In humid regions, forced air-drying may be needed, and the temperature of the drying air should not exceed 35°C. The dried produce is cleaned with an air cleaner to remove the rubbish, which may pick up moisture rapidly in humid climates.

Seed Treatment. Mechanical scarification to improve seed germination and rhizobium inoculation should be carried out as close to sowing as possible. Ants and bean flies generally attack the seed and young seedlings after sowing. Seed may be treated with an insecticide dust to control ants and bean flies. The seed, however, must first be calcium coated or pelleted to protect the rhizobia, especially in high-manganese soils [I].

2. Stylo

Botany, Cultivars, and Isolation Requirements. Stylo (*Stylosanthes guianensis* [Aubl.] Swtz) is a self-fertilized, small, generally insignificant perennial of the tropical and subtropical regions. The OECD (10) listed four cultivars of stylo. It usually has yellow flowers, but one cultivar (Cook) has been reported to have an orange standard with a central purple stripe (14). Although predominantly self-pollinated, a certain amount of cross fertilization can occur. Isolation by a physical barrier or a 3-m gap is generally sufficient.

Cultivation/Defoliation. A certain amount of physical scarification prior to sowing stylo seed is recommended to assist seed crop establishment. Inoculation with *Rhizobium* may be advantageous, although not always necessary. Seed crops are sown in rows 45–60 cm a part using 2–3 kg seed/ha. Crops should be defoliated 1 month before flower initiation to give a level crop canopy at harvest (1). Defoliation performed too late may decrease seed yield.

Harvesting. The seed ripens relatively uniformly and is prone to quick shedding. The crop should, therefore, be harvested as soon as seeds begin to fall from the pod when the plants are struck with the hand. The leaves produce a sticky exudate, which makes harvesting and handling of the seed difficult. Crops are normally combined directly, because manual harvesting is not economical. The combines work efficiently at the driest times of the day and should not be used when humidity is high. Vacuum harvesting can be employed with some cultivars effectively.

Other perennial species of *Stylosanthes* include *S. hamata* (L.) Tumb. (Caribbean stylo) and *S. scabra* Veg. (shrubby stylo), which have one and two cultivars, respectively, grown similarly to *S. guianensis* (10).

3. Townsville Stylo

Botany and Cultivars. Townsville stylo (*S. humilis* HBK) is an annual in contrast to other *Stylosanthes* species, which are perennial. Three Australian cultivars of *S. humilis* are known. Crops are adapted to a wide range of conditions in the humid tropics and warmer areas of subtropics. They will not produce seed if minimum night temperatures fall to 9°C or below (14). Townsville stylo is also critically photoperiodically, requiring 12–13 hrs of light per day, depending upon cultivars. It is a prolific seeder, but the disease caused by *Colletotrichum gloesporoides* can drastically reduce yields.

Cultivation, Harvesting, and Cleaning.   The seed crop is sown in rows 30–45 cm apart using 3–5 kg seed/ha. Seed normally does not need inoculation. Flowering extends over a long period, and seed sheds rapidly. Direct combining or flail harvesting for subsequent threshing is not always effective. Most seed is collected from the ground using a vacuum harvester or mechanical rotary brush or by hand brushing (1). Material swept or sucked from the ground should be cleaned first with a wide mesh screen to separate sticks and broken stalks and then on a fine mesh screener to remove soil. Hammer-mills can be used to break up the soil and dehull the seed.

4.   Siratro

Botany and Cultivation.   Siratro (*Macroptilium atropurpureum* [DC] Urb.) is a perennial climber with intermediate habits. The crop is self-fertilized, requiring isolation only to prevent physical admixture of seed with other types. A high proportion of hard seed occurs. Seed is produced best from crops sown in lines about 40–60 cm apart using 2–3 kg seed/ha. Seed should be inoculated with the cowpea strain of *Rhizobium*. Crops grown in dry areas respond well to irrigation with the best seed yields. A certain amount of moisture stress applied at the time of flowering has been found to benefit the crop (14).

Yadav and Sinha (23) investigated the effects of GA_3 and boron on the growth, flowering, and seed yield of siratro plants (Table 8). The height of siratro plants increased significantly with the foliar application of 50 ppm GA_3 and 1 ppm boron. Growth regulators also reduced days to flower and increased total number of flowers and pods (up to 25 and 50 ppm GA_3), which were decreased by higher concentrations of GA_3 (100 ppm). Similarly, lower concentrations of boron (0.5 and 1.0 ppm) promoted flower production, but the highest level (2 ppm) decreased it significantly. Lower concentrations of both GA_3 and boron increased the seed yield and 1000—seed weight of siratro plants (Table 8).

Harvesting, Drying, and Cleaning.   The ripe seed sheds very easily. Early ripened pods at the top of plants may be harvested with the combine set high and the rest combined later. The pods can be handpicked for maximum seed yield when the crop is grown on support fences; two to three pickings are needed with this method. Alternatively, a densely grown crop may be allowed to shed the first maturing seed into the foliage to combine it later. However, the herbage to be dealt with in postharvest operations like drying and cleaning can be very high. Vacuum harvesting at the end of the season may be employed to retrieve a worthwhile amount of fallen seed after cleaning.

5.   Leucaena (Jumbie Bean)

Leucaena (*Leucaena leucocephala* [Lam.] de Wit.) or, jumbie or white popinae, is an erect, perennial shrub. It is a self-fertilized plant, requiring minimum isolation. It is normally sown in lines about 60–120 cm apart using 15–40 kg seed/ha. The long, flat pods are usually hand harvested and are threshed and cleaned later. Leucaena must be

Table 8 Effects of GA₃ and Boron on Growth, Flowering, and Seed Yield of Siratro (*Microptelium atropurpureum*) Plants

Treatment (ppm)	Plant height (cm)	Days to initial flowering	No. of flowers per main shoot	Days to pod maturity	Seed yield (g/plant)	1000—seed weight (g)
0 (Control)	93.37	61	113.3	24	23.0	11.6
GA₃						
25	103.6	63	116.0	26	22.0	12.1
50	109.4	60	122.0	23	25.5	14.4
100	99.2	56	194.3	21	19.0	11.0
Boron						
0.5	98.9	61	116.0	25	26.0	13.0
1.0	105.7	59	120.3	20	29.7	16.2
2.0	98.5	57	96.2	17	10.5	11.3
CD at 5%	10.18	—	12.44	—	1.38	1.40

CD = critical difference.
Source: Ref. 23.

inoculated with a particular strain of *Rhizobium*. Physical scarification before sowing is generally advocated.

Other species of tropical legumes cultivated for forage are *Desmodium intortum* (Mill.) Orb. and *Glycine wightii* (R. Grah. ex. Wight et Am) Verdcourt. Both are partly cross and partly self-fertilized, requiring isolation. *G. wightii* has both diploid and tetraploid types (1).

F. Temperate Forage Legumes

1. Alfalfa (Lucerne)

Botany and Cultivars. Alfalfa (*Medicago sativa* L. and *Medicago × Varia* Martyn), or lucerne, is a generally erect annual forage crop growing well in cooler temperate climates. The cultivars are classified as early, medium, or late. The early types grow and flower early in spring and are generally erect in growth habit. Late cultivars grow and flower about 4 weeks later and are not completely erect. Some cultivars having a creeping habit with rhizomes are also available.

Alfalfa cultivars can be distinguished on the basis of length of the central leaflet on the third or fourth leaf below inflorescence (short, medium, and long), stem length (very short, short, medium, long, and very long), dominant flower color (white, yellow, light blue–violet, dark blue–violet, red-violet), and time of flowering (very early, early, medium, late, and very late). Alfalfa seeds are shown in Figure 6.

Alfalfa cultivars may be regarded as "synthetic," since they represent the controlled multiplication from selected parent lines (inbred lines). Cytoplasmic male sterility has been found in alfalfa and is used in the production of hybrid alfalfas, which have not been used largely because of their lower yields (1).

Isolation Requirements. Alfalfa is naturally cross pollinated by insects like honeybees, alkali bees, bumblebees, and leaf cutters, which "trip" the flowers; the extent of cross pollination depends on insect activity. The seed crop needs to be isolated by a

Figure 6 Seeds of alfalfa (*Medicago sativa*). (From Ref. 3.)

minimum distance of 400 m for the production of foundation seed and 100 m for certified seed from fields of the same variety not conforming to varietal purity requirements for certification (24). Alfalfa seed can live in the soil for several years. Kelly (1) recommended an interval of 4 years free from alfalfa before sowing a seed crop. In earlier stages of seed multiplication, a longer interval of up to 6 years is advised. These intervals can, however, be shortened if authentic seed of the same cultivar was used for preceding crops.

Land Preparation and Sowing. The land for seed crops should be prepared by one deep plowing and two to three harrowings followed by leveling to bring the soil to a desired tilth. In India, alfalfa is usually sown in October–November in lines 15–20 cm apart using about 12–15 kg seed/ha. In Europe, it is sown in the spring under a light cover crop (cereal), and in parts of the United States it may even be sown in autumn to harvest for seed the following year. It can also be sown early in the season to take the first growth for seed in the same season. Adequate soil moisture and fertility conditions, which encourage vigorous crop growth, as well as very dry areas, will require wider spacing (100–150 cm) for better seed yields. In the United Kingdom, 35–60 cm spacing between rows is considered optimum. For uniform stand of crops, seed rates in western countries vary from 1 kg/ha with rows 90 cm apart to 3 kg/ha for 35-

cm–spaced rows. Seed rates may be increased when conditions less than optimum prevail.

Alfalfa seed should be inoculated with the appropriate strain (*Rhizobium melilotii*) before sowing, especially when it is being raised for the first time in the field. In India, 1 kg of *Rhizobium* culture is mixed in 1 L of 10% sugar solution, and this mixture is uniformly applied to the seed. The seed should be dried immediately in the shade and used within 24 hr of its inoculation (24).

Fertilization and Irrigation. In India, depending upon the soil fertility level, alfalfa requires about 25–30 kg N, 50 kg P, and 25 kg K per hectare, applied as a basal dose at the time of planting. In Europe and the United States, much higher levels of fertilizers are used to ensure an adequate supply of macro- and micronutrients. Irrigation may be required to maintain enough soil moisture during crop growth. In temperate regions, the crop is trimmed before winter sets in without cutting it too close to the ground. This is done to remove excess herbage and get a uniform crop stand. In spring, the first growth is allowed to flower and set seed. The seed from the first growth can be harvested in time to allow late autumn forage cut and seed from the second growth. In subsequent years, the crop is cut in the late autumn to provide a uniform stand into the winter and the seed is taken from the first growth the following year. Manonmani et al. (24a) reported that an alfalfa crop sprayed with 0.3% of each zinc and boron (borax) produced the highest pod yield and seed yield per plot, 100–pod weight, 11–seed weight, with good germination potential and vigor index.

Control of Diseases and Pests. The crop needs adequate protection from insect pests and weeds during the seedling stage. Some insects attack flowers and seeds and require special control measures not usually required by forage crops. *Lygus* bugs, chalcids, and crickets are particularly serious pests of seed crop. Chemical control measures adopted should take into account the need to protect insect pollinators like honeybees by avoiding insecticide sprays at the time of flowering. Flowering periods can also be adjusted by taking appropriate forage cuts to avoid the most damaging insect attacks.

Table 9 Effects of B-9 on Growth, Flowering, and Seed Production of *Medicago sativa* L.

B-9 conc. (ppm)	Plant height (cm)	Days to flower	Seed weight (g/plant)	1000–seed weight (g)	Seed yield (q/ha)	% Increase or decrease in seed yield over control
0 (Control)	48.41	106	5.33	2.89	4.32	—
10	45.41	106	5.83	2.97	4.50	+4.16
100	40.62	108	6.56	2.96	4.83	+11.80
250	41.12	111	7.26	3.10	5.39	+24.76
500	37.62	118	6.90	3.13	5.02	+16.20
1000	36.49	122	6.43	2.93	5.01	+15.97
5000	35.41	125	4.66	2.82	4.60	+6.40
CD at 5%	2.56	—	0.17	0.08	0.17	—
CD at 1%	3.53	—	0.24	0.12	0.24	—

CD = critical difference.
Source: Ref. 23.

Pollination Aids.　Honeybees are the most important pollinators. However, flowers are not "tripped" by bees collecting nectar; "tripping" normally takes place by pollen-collecting bees. The pollen collectors are about 45% more efficient as pollinators than are nectar collectors. An insect density of one bee per square yard (0.84 m^2) and up to 10 bees per square yard when collecting pollen and nectar, respectively, have been recommended (25). Alternatively, other pollinators such as bumblebees (*Bombus terrestris*), alkali bees (ground-nesting species), and leaf-cutter bees should also be encouraged to develop (1,25).

　　Yadav and Sinha (23) reported the effects of growth regulators such as B-9 (N-dimethylaminosuccinamic acid), ethrel, and Planofix (naphthalene acetic acid) on the growth, flowering, fruiting, and seed production of three forage legumes: *Medicago sativa* L., *Meliotus parviflora* Desv. and *Vigna unguiculata* L. (Tables 9–11). The average plant height in *Medicago sativa* was found to decrease concomitantly with B-9 treatment. Flowering was delayed by 2–19 days. The seed yield per plant and total seed yield increased significantly at 250 ppm, followed by 500 ppm, but the highest concentration (5000 ppm) reduced seed yield remarkably (Table 9). A similar response was also noted in other crops (walp and cowpea) where lower concentrations of B-9, ethrel, and Planofix increased the seed production and higher concentrations were inhibitory and modified plant growth characteristics (Tables 10 and 11).

Roguing.　Seed crops need critical inspection for off-types and diseased plants, particularly when in full bloom. Off-types can be recognized on the basis of differences in flower colors.

Harvesting.　The period of flowering extends for some time, and hence all of the seed does not ripen at the same time. The crop can be harvested best when three-quarters of the pods have changed color to brown or dark brown but before many of them begin to shed seeds.

　　The crop may be windrowed or desiccated in the field. The crop is cut earlier to allow some immature pods to ripen after cutting. Desiccants like diquat allow the crop to mature to the optimum harvest stage before desiccation, after which it can be

Table 10　Effect of B-9 on Flowering and Fruiting of *Meliotus parviflora* Desv. Plants

Treatment concentration (ppm)	Days to flower		Seed yield (g/plant)		1000–seed weight (g)	
	Without cut	With cut	Without cut	With cut	Without cut	With cut
0 (Control)	46	68	6.23	5.06	1.28	1.21
10	48	70	5.70	5.33	1.34	1.31
100	51	72	7.10	6.26	1.47	1.41
250	53	75	7.96	7.10	1.53	1.52
500	55	77	8.63	7.43	1.62	1.57
1000	56	79	6.90	6.70	1.41	1.28
5000	59	84	5.43	5.13	1.35	1.16
CD at 5%	—	—	0.56	0.66	0.03	0.03
CD at 1%	—	—	0.78	0.93	0.04	0.05

CD = critical difference.
Source: Ref. 23.

Table 11 Effects of Ethrel and Planofix on Seed Yield of Forage Cowpea (*Vigna unguiculata* L.) Plants

Treatment concentration (ppm)	Seed yield (mean of 3 y)	% Increase over control
0 (Control)	4.41	—
Ethrel		
10	5.53	25.39
100	6.07	37.64
1000	4.86	10.70
Planofix		
10	6.38	44.67
50	5.52	25.17
100	4.85	9.97
CD at 5% level	0.941	—

CD = critical difference.
Source: Ref. 23.

combined directly. If the combining operation is delayed to more than 2–3 days after desiccation, seed shedding takes place, resulting in secondary growth. Combining should be adopted only when the leaves are dried to less than 20% moisture, although stems may still be green with a higher moisture content, especially in desiccated crops.

The chaff and straw after harvest may be removed and destroyed or even burned in situ to prevent the spread of some harmful insect pests. In situ burning will not harm alfalfa plants.

Drying and Cleaning.   Seed is dried to 12% moisture or less for medium-term storage and to 8% or less for long-term storage (over 1 year). The temperature of the drying air should not exceed 38°C when the initial moisture content of seed is below 20%; if it is higher, the air temperature should be lower than 30°C. The dried seed can be cleaned with an air/screen cleaner or a gravity separator. Roll separators and magnetic separators are also often used.

Seed Treatment.   Apart from treatment with *Rhizobium*, alfalfa does not require any other chemical treatment. Fumigation with an appropriate insecticide is sometimes needed to control seed from stem eelworm (*Ditylenchus* sp.).

2.   White Clover

Botany and Isolation Requirements.   White clover (*Trifolium repens* L.) cultivars are classified on the basis of their leaflet size. Generally, the larger the leaflet size, the shorter lived are the plants in the field. Cultivars with large leaflets are sometimes grouped as Ladino white clover, and those with the smallest leaflets are called wild white clover. The predominant distinguishing characteristics of white clover cultivars are frequency of white leaf marks (absent, or very low, low, medium, high, and very high) and time of flowering (early, medium, and late). Seeds of white clover are shown in Figure 7.

White clover is cross pollinated by bees and other insects, and small fields up to 2 ha should be isolated by a distance of 200 m from other clover fields when the seed is required for further multiplication and 100 m when the seed is needed for forage use.

Figure 7   Seeds of white clover (*Trifolium repens*). (From Ref. 3.)

For fields larger than 2 ha, the distances can be 100 and 50 m, respectively. Since clover seed can remain viable in the soil for several years, a period of about 4 years free from white clover is recommended before sowing a seed crop unless the same variety of clover was grown using an authentic seed (23).

Sowing.   The seed yields of white clover obtained from a pasture or meadow are relatively small. For higher yields, white clover must be raised as a seed crop. Good seed yields result from a pure white clover stand. Crops are sometimes sown with other, not very aggressive forage crops (grass), which provides a better pasture or forage when the field is not tied up in seed production. In the latter case, however, much lower seed yields are realized.

White clover is usually sown in the spring on bare ground, but it is more economical to sow under a cover crop, generally a stiff-strawed cereal. When cover crops are used, the nitrogen dose is reduced from that normally given to a cereal, so that the latter will not smother the young clover plants. Seed crops in some areas are also sown in late summer without a cover crop. Drilling in rows, about 15 m apart, using 1.5–3.0 kg seed/ha, sows seed. When a cover grass is sown, its seed rate is also reduced (2–3 kg/ha of meadow fescue is sufficient).

Using certain persistent wild cultivars, good yields of white clover seed can be produced for several years (10 years or more). However, with most bred cultivars and for routine seed production of local cultivars, three to four seed harvests are normally planned, after which seed becomes increasingly contaminated with weeds and other impurities, making the management of the seed crop difficult.

Fertilization and Defoliation.   The crop requires adequate supplies of calcium, phosphorus, and potassium during plant growth, and any deficiency of macro- or micronutrients should be corrected before planting. Sheep grazing helps the young plants to become well established, filling the gaps. Prolonged grazing and poaching in wet weather, however, should be avoided, The crop can also be alternatively trimmed with a mower to remove excessive herbage. A rolling may be needed under dry conditions (1).

White clover flowers freely if its stolons where the floral primordia develop are not very shaded. This necessitates removal of excessive herbage in the spring, which must be programmed carefully. Too late trimmings may destroy some of the earlier developing flowers, which give the best seed. Too early trimmings may also, however, cause excessive vegetative growth, having an undesirable shading effect on the stolons. In the United Kingdom, white clover usually begins to flower in May or early June, so that fields should be shut up for seed by mid-May. The exact best time for defoliation varies with cultivars and areas of crop production. Strong growth of early flowers should be encouraged for better seed yields. Strong growth of herbage in humid conditions will generally delay shutting up. Sheep grazing or a light cut for silage is preferred. The crop can also be trimmed frequently with a mower and the cut material dispersed in the field. The field will again be available for grazing or forage use after the first seed harvest and is treated like the first-year crop. A greater amount of herbage will be available from the second year onward, which should be removed before winter sets in. Excessive grass growth can be checked by a light application of contact herbicide like paraquat without damaging the clover crop. The crop in subsequent years is managed in the same way as in the first year. If the seed crop is not desired in any one year, the field should be grazed or cut for forage. When it is cut, excessive vegetative growth should not be allowed to remain too long, because it will drastically reduce the clover population (1).

Weed Control.   Dodder (*Cuscuta* sp.) is the most difficult weed of the white clover seed crop. It is, therefore, essential to use white clover seed free from dodder and choose clean land for seed crops. Other weeds, such as thistles, mouse-eared chickweed, fathen, cranebill, campion, broad-leaved plantain, self-heal, docks and sorrels, field madder, and chickweed, can be controlled with suitable herbicides. Two weeds, trefoil and suckling clover, should be weeded out physically, as they are difficult to control chemically (1).

Pollination Aids and Roguing.   One honeybee hive per hectare of seed crop is recommended for the effective pollination of white clover crop. The seed crop should be inspected critically at flowering time to check the cultivar purity and isolation.

Harvesting and Threshing.   White clover crop flowers and ripens over an extended period; the highest yield is usually obtained from earlier-formed flowers. The ripened seed heads turn brown, and seeds become hard and light yellow in color. Seed crops should be harvested when about 80% of the heads have changed color depending upon weather conditions. Because the crop is short, it is difficult to secure all seed heads during harvesting. It is, therefore, preferred to cut and windrow the crop to pick up later rather than direct combining. Lawnmower or forage harvesters are sometimes used to cut a very short crop. The cuttings are then collected for threshing in a stationary combine or huller. The product thus harvested contains a lot of green matter and plant debris, which needs to be separated before the seed is stored to avoid heating and consequent loss of germination. If the crop is desiccated, it should be threshed within 24 hr of treatment. Desiccation of the windrowed crop can also hasten the drying process, particularly when it contains a large proportion of green vegetative growth.

White clover seeds are tightly enclosed in the pods, making threshing and hulling operations difficult. Cylinder clearance and speed are required to be adjusted carefully to avoid physical damage during threshing. Stationary threshers are not generally satisfactory, and a huller specifically designed for use on clover crops with a double-threshing mechanism (with a second or hulling cylinder) may be required.

Cleaning, Drying, and Storage. The threshed material may be precleaned by aspiration to remove the green vegetative debris to reduce drying time. Spreading on a floor in a well-ventilated shed may dry it. The spread should be 10–12 cm deep and should be turned frequently. Forced air or ventilated bins can also be used to hasten drying. If heated air is employed, its temperature should not exceed 10°C above the ambient or higher than 35°C (38°C in case of continuous-flow driers). If the initial seed moisture content is high, the air temperature should be below 35°C (1).

The dried produce needs to be cleaned thoroughly by first using an air/screen cleaner followed by a velvet belt, roll cleaner, or magnetic cleaner. A gravity separator also can be used with satisfactory results. Cleaning of white clover seed is a specialized operation and needs skilled labor.

The short-term storage of white clover for some months can permit a seed moisture content of around 12%, but for long-term storage of 1 year or more, seed must be dried slowly to a 5% moisture content, sealed in suitable moisture-proof containers, and stored at 0–5°C (1).

3. Red Clover

Botany and Cultivar. Red clover (*Trifolium pratense* L.) cultivars have been traditionally classified as broad red, single-cut, and late-flowering types based on their earliness to start spring growth and flowering time. A more recent classification includes only two groups: early and late types. The former group can be cut two to three times in a season in northern Europe, whereas the late group cultivars produce most of their yield in the first cut. Within each group of red clover there are both diploid and tetraploid cultivars. Seeds of red clover are shown in Figure 8. The cultivars of red clover are distinguished on the basis of their stem length at flowering time (very short, short, medium, long, very long), length of central leaflet (short, medium, long), width of central leaflet (narrow, medium, broad), and time of flowering (very early, early, medium, late, and very late).

Isolation Requirements. Red clover is cross fertilized, mainly by bees, and needs to be isolated from other fields as described for white clover. Seed setting does not take place when diploid cultivars are pollinated by tetraploid cultivars or vice versa. According to

Figure 8 Seeds of red clover (*Trifolium pratense*). (From Ref. 3.)

Kelly (1), an isolation distance of at least 50 m is required to avoid reductions in seed yields.

Since red clover remains viable in the soil for long periods and is prone to attack by soilborne diseases like clover rot (*Sclerotina* sp.) and eelworm (*Ditylenchus* sp.), it is advisable to observe an interval free from red clover of at least 6 years, which may be shortened to 2 years when the field is free from soilborne infection and the seed produced is not required for further multiplication, or when the variety grown was the same using an authentic seed.

Early types of red clover are mostly used for the production of seed, but late-flowering cultivars give better yields when raised without a cover crop. However, in some areas, grass is grown as a companion, which is a not very aggressive and vigorously growing crop. Timothy is used most advantageously in England for this purpose, because it produces seed at the same time as does the late-flowering clover.

Sowing.   The seed of red clover is usually grown in spring under a cover crop (cereal) that would not suppress development of young clover plants. Seed may be broadcast or, preferably, sown in rows about 15 cm apart. Late-sowing cultivars are placed wider (60 cm) than early types provided lodging is not a serious problem. Seed rates vary from ≤1 kg/ha for wider-spaced crop to 5 kg/ha for closely spaced early cultivars.

Weed Control and Defoliation.   The field should be kept free from weeds. Dodder (*Cuscuta* spp.) is a problematic weed and very difficult to control; fields free from dodder should be selected for seed crops. Other weeds that may cause seed contamination are thistles (*Cardenus* sp. and *Cirsium* sp.), carnesbill (*Geranium* sp.), trefoil (*Medicaqo lupilina*), campion (*Melandrium* sp.), ribgrass (*Plantago lanceolata*), and docks and sorrels (*Rumex* spp.). All these weeds, except trefoil, can be controlled using suitable herbicides.

The first crop growth of early-flowering types of red clover sown early in the season is allowed to grow unchecked, and seed is harvested at the first opportunity. Fields should be trimmed of excessive herbage at the end of active growth to keep the crop uniform. Cover crops, if grown, should be removed in time to allow the development of clover seedlings. High seed yields can be obtained from the first cut of the early-flowering types raised for seed purpose. However, foliage growth is generally excessive with severe lodging, making harvesting difficult. They are, therefore, preferably harvested for seed from the second cut in the year after sowing. The first cut for hay or silage should be timed properly to allow the second growth to flower and set seed at a time when favorable weather conditions prevail. In the United Kingdom, the first cut is normally taken in late Mayor to early June. Late-flowering types are best raised for seed with their first growth of the season when they are not likely to lodge. The intermediate types benefit from a light early grazing or cutting to remove excess herbage.

Pollination Aids.   Bees mainly pollinate red clover. The length of the corolla tube and the height of the nectar in the tube determine the efficiency of pollinators. Honeybees effectively pollinate early-flowering types, but their tongue is too short to reach the nectar in the longer corolla tubes of late-flowering red clovers. The bumblebees (*Bombus terretris*) take nectar by piercing holes at the base of corolla, which are also used by honeybees to collect nectar. This avoids both tripping of flowers and pollination. Other species of bumblebee (*B. hortorum*, *B. ruderatus*, and *B. subterraneus*) having longer tongues are effective pollinators. One hive of honeybees/3 ha provides sufficient effective pollinators (1).

Roguing.   Seed crops should be inspected for off-types and diseased plants, if any, at the beginning of flowering.

Harvesting and Threshing.   Early-flowering red clovers are usually harvested for seed only once from each crop, whereas late-flowering cultivars may be kept for a second year, although the seed yields are usually less than in the first year. Occassionally, a third seed harvest is also taken in some areas. Since flowering and seed ripening extends over several weeks; the appropriate time of harvest is difficult to judge. The earlier flowers generally produce good viable seed, and harvesting should concentrate on these. The crop is ready to harvest when the majority of heads turn brown and seed becomes dark colored and hard. The proportion of fully ripe heads varies from 60 to 90% depending upon cultivars and areas.

Harvesting by direct combining is preferred, although the amount of green vegetative matter is usually too great. It may then be necessary to windrow the crop before picking up or to desiccate it with a suitable desiccant like diuran before combining. In the latter case, a few days of dry weather should be foreseen to prevent regrowth of the crop during rains. A fully ripe crop can be desiccated, whereas windrowing can be carried out a few days before the full-ripe stage, because seed matures further m the windrow. Sealing them properly can minimize loss of free-flowing red clover seed through combines and other equipment. Threshing needs careful setting of combines, with cylinder clearance being somewhat greater than that required for white clover. If threshed using a stationary thresher, a clover huller or a hulling attachment to an ordinary thresher will be required. Physical injury to the seed by too hard threshing must be avoided.

Drying and Cleaning. Red clover seed is dried and cleaned in a manner described for white clover. A bulk quantity of seed may be handled during these postharvest operations. The seedborne pest stem, eelworm (*Ditylenchus* sp.) can be controlled by fumigation. Other temperate clovers include *T. hybridum* L., *T. fragiferum* L., *T. alexandrium* L., *T. incarnatum* L., *T. resupinatum* L., *T. semipilosum* Fres., and *T. vesiculosum* Savi, whose seed crops are treated like other annual or perennial clovers (1).

4. Vetches

Botany and Cultivars. Vetches (*Vicia* spp.) are annual leguminous forage crops commonly grown for seed in temperate regions. Three species of *Vicia* with a number of cultivars in each are listed by the OECD (10). These are *V. sativa* L. (41 cultivars), *V. villosa* Roth (16 cultivars), and *V. pannonica* Crantz (2 cultivars), which are also known as common vetch, hair vetch, and Hungarian vetch, respectively. Some cultivars, especially *V. villosa*, that can be sown in autumn are winter hardy. Cultivars are distinguished on the basis of a number of plant characteristics (1).

Isolation Requirements. Whereas *V. sativa* is almost entirely self-fertilized, *V. villosa* is largely cross pollinated. Isolation requirements, therefore, differ from 3 m for *V. villosa* to 200 and 100 m for *V. sativa* crops of 2 ha or less depending upon whether the seed produced is required for further multiplication (nucleus/breeder's seed/ foundation) or to sow crops for fodder production (certified seed). Vetches have a high proportion of hard seed, which can remain viable in the soil for several years. It is, therefore, necessary to observe an interval of 6 years free from any vetch crop while selecting land for seed production. Adopting special measures to eliminate volunteer plants can shorten intervals or when the same cultivar is to be grown again.

Cultivation, Harvesting, and Threshing. The seed is sown without a cover crop in lines about 20 cm apart using 20–40 kg seed/ha. The first growth of a seed crop is allowed to mature to seed before harvest. *V. villosa* crops will need one or two honeybee hives for pollination. Other cultural requirements are similar to those of other forage crops. Docks, sorrels (*Rumex* spp.), and some clovers can be difficult weeds of the vetch crop. Severe drought conditions during flowering can inhibit seed setting. *V. sativa* and *V. villosa* cultivars shed seeds easily, and the time of harvest should be carefully adjusted to reduce seed losses. The lower pods, which ripen first, have the high-quantity viable seeds contributing toward seed yield. When these seeds are firm and have changed color, the crop is directly combined in 3–4 days. Alternatively, crops may be windrowed and picked up later, after 5–6 days, at a somewhat earlier stage than for the direct combine method to reduce the seed loss. *V. pannonica* is comparatively resistant to seed shedding and can be harvested when about 90% of the pods are mature with ripe seed. The crop is difficult to thresh and will need an appropriate combine setting to avoid loss of seed in pods.

Drying, Cleaning, and Storage.   Seed from combining normally contains green vegetative debris and requires immediate drying, as described for white clover. Other postharvest requirements of cleaning and storage are similar to that of white clover.

5.   Sainfoin

Botany, Cultivars, and Seed Production.   Sainfoin (*Onobrychis viciifolia* Scop.) has 13 cultivars (10), which can be grouped in two main types: short-lived biennials surviving not more than 3 years and longer-lived perennials. Seed crops require calcareous soils containing a high proportion of calcium. The crop establishes rather slowly, but once established, it resists drought well. Sainfoin is a cross-fertilized crop requiring good isolation. Honeybees are the main pollinators, and 2–10 hives per hectare have been recommended to achieve good seed set.

Sowing and Harvesting.   Seed is usually undersown in a cereal nurse crop in lines 15–20 cm apart using about 50 kg hulled seed/ha. The crop should not be grazed or cut after removal of the nurse crop unless the growth is excessive. The biennial type is first cut for hay in the second year, and seed is harvested from the second growth. The seed of the perennial type, on the contrary, must be harvested from the first growth. The second and third harvests of perennials generally give the best seed yields. The harvesting and postharvest operation are similar to that for red clover.

6.   Serradella

Serradella (*Ornithopus sativus* Brot.) is mainly grown in coastal areas of Spain, Portugal, and Morocco. It is predominantly self-fertilized, requiring minimum isolation. The seed crop is grown as an annual with a cover crop like spring rye to provide support. The *Rhizobium*-treated seed is sown in rows 15–20 cm apart using about 20–40 kg of unhulled seed per hectare. Seed sheds easily, and the crop is usually harvested by windrowing and subsequent picking by combine. It is susceptible to a fungal disease caused by *Colletotrichum trifolii* and requires seed treatment.

G.   Forage Legumes Adapted to Hot Summers

These are grown in North America and the warmer Mediterranean areas. Some are used for soil conservation.

1.   Korean Lespedeza

Botany and Cultivars.   Korean lespedeza (*Lespedeza stipulacea* Maxim.) is an annual forage legume with two cultivars grown extensively in the southeastern United States. The crop has two types of flowers: one is completely self-fertilized and the other can be self-fertilized, but can be cross pollinated with bees and requires adequate isolation. An interval of about 4 years free from lespedeza is required before sowing a seed crop unless the same variety is to be planted again. Seed stored longer than 2 years is not good for sowing. Seeds of Korean lespedeza are shown in Figure 9.

Cultivation.   Seed is sown by drilling in lines 15–20 cm apart using about 25 kg of unhulled seed/ha. Excessive herbage is removed and a second growth is taken for seed

Figure 9   Seeds of Korean lespedeza (*Lespedeza stipulacea*). (From Ref. 3.)

production. The seed crop is inspected at flowering time for roguing off-types. Dodder (*Cuscuta* spp.) can be a serious weed of the lespedeza, necessitating selection of clean land for the seed crop. Other common weeds are Johnson grass and ragweed, which can be controlled with herbicides.

Harvesting, Drying, and Cleaning.   Lespedeza seed sheds easily when ripe. The harvesting must not be delayed unduly in order to avoid seed losses. Seed is usually combined directly. If the seed is not harvested at a low enough moisture content for storage, it should be dried immediately. Precleaning helps to reduce drying time. Seed is further cleaned on an air/screen cleaner. A velvet roll mill is required to remove dodder. The seed is usually left unhulled because of its short dormancy period.

      Other species of *Lespedeza* include *L. striata* (Thunb.) Hook et Com (striata lespedeza) and *L. cuneata* (Dum.) G. Don (sericea lespedeza).

2.   Bird's-Foot Trefoil

Botany and Isolation Requirements.   Bird's-foot trefoil (*Lotus corniculatus* L.) is a self-incompatible perennial forage leguminous crop cross fertilized by bees. It requires about 200 m isolation for small fields up to 2 ha and 100 m for larger fields of seed crop meant for further multiplication. An interval of 5 years free from trefoil is recommended for sowing, since trefoil seed remains viable in the soil for several years. Seeds of bird's-foot trefoil are shown in Figure 10.

Sowing and Roguing.   Seed is sown by drilling in rows 15–60 cm apart using 3–5 kg seed/ha. Closer spacing tends to produce a crop with a shorter range of flowering time and a more uniformly ripened seed. Seed inoculation with *Rhizobium* can be advantageous. Cover crops and companion grasses enhance the forage value of the crop, but generally decrease seed yields; hence, they are not usually recommended. Seed crops should be inspected critically for off-types at the flowering time when it is also possible to check isolation.

Weed Control.   Because of its open growth habit, bird's-foot trefoil has to compete with weeds. Bindweed (*Convolvulus arvensis*), docks, and sorrels (*Rumex* spp.) are the difficult weeds associated with trefoil. Other forage legumes like clover and lucerne also occasionally cause contamination. Most of these weeds can be controlled by herbicides.

Figure 10   Seeds of bird's-foot trefoil (*Lotus corniculatus*). (From Ref. 3.)

Harvesting. Bird's-foot trefoil is a perennial crop, and several seed harvests can be taken from one crop. Although crops can continue for much longer, about five seed harvests are considered to be optimum owing to build-up of volunteer plants causing increasing contamination. The crop establishes rather slowly, and therefore it is preferably sown in spring in colder areas. It may be sown in certain favored areas if irrigation is available.

The first seed harvest is taken in the year of its establishment. The crop is trimmed of excessive herbage to leave a uniform stand into the winter, and in the following spring, the first growth is considered to be best for seed. However, if later seed harvests are desired, the crop should be trimmed by grazing or cutting, which should be timed to obtain the following harvests at favorable times. It takes about 2 months for the crop to flower after defoliation. Flowering is also controlled by day length. After the first harvest, the seed crop of trefoil is managed in much the same way in subsequent years, alternating judicious grazing or cutting to obtain seed harvests at the best time of year.

Very high soil fertility may produce excessive vegetative growth, leading to lower seed yields of trefoil. However, it is also necessary to ensure that soils are not deficient in major and minor essential nutrients, especially phosphorus and potassium. Bees being the best pollinators, two hives of bees per hectare are recommended for obtaining uniform seed setting and good seed yield.

Trefoil is difficult to harvest on account of seed shedding and uneven ripening over an extended period, although some cultivars with more uniform ripening are now available. The seed crop is generally harvested by windrowing rather than by direct combining. The most appropriate time for this would be when plants remain green. As the seed ripens, the pods turn dark to black and seed becomes firm and yellow to brown in color when ripe. The use of defoliants like diquat reduces the amount of green material to be harvested; however, crops must be harvested a few weeks after its application, as regrowth appears quickly. Windrowing followed by picking of the crop when it is damp helps to minimize seed-shedding losses.

Drying and Cleaning. The green matter present in the threshed produce makes it necessary to dry it immediately to prevent heating and seed deterioration. The drying and cleaning operations are similar to that for white clover seed crops. Because of the high proportion of hard seed, physical scarification of seed is recommended before sowing. Inoculation of seed with rhizobium can also be advantageous.

Some other species of *Lotus* include *L. tenuis* Waldst et kit ex. Wild (slender bird's-foot trefoil) and *L. uliginosus* Schk. (greater bird's-foot trefoil), with one cultivar each. *L. tenuis* is similar to *L. corniculatus*, whereas *L. uliginosus* is somewhat different, requiring a different strain of *Rhizobium* for seed treatment.

3. Sweet Clover

Botany and Cultivars. Sweet clovers (*Meliotus* spp.) are both annual and biennial in growth habit, with a wide range of maturity periods, cultivars differing in ripening times by about 2 months. Two species of *Meliotus* are known: *M. alba* Med. (white sweet clover with two cultivars) and *M. officinalis* (L.) Pall (yellow sweet clover with three cultivars) (10). Seeds of sweet clover are shown in Figure 11.

Isolation Requirements. Owing to a high degree of cross fertilization by bees (greater in white types than in yellow), sweet clover seed crops need to be isolated like other

Figure 11   Seeds of sweet clover (*Melilotus* spp.) (From Ref. 3.)

cross-fertilized crops; for example, *L. corniculatus*. Similarly, an interval of 5 years free from sweet clover and 2-3 years from other seed crops would be needed.

Sowing.   Seed is preferably sown in spring by drilling rather than by broadcasting with a spacing of 15--100 cm between rows—the wider spacing is favored in drier areas. Seed rates can vary from 3 to 4 kg/ha with very wide rows to 15 kg/ha when closely spaced. The best seed yields are obtained from stands that are not very thick. Very tall growth prevents full pollination and makes it difficult to harvest the crop. Companion grasses or cover crops can be used to improve the fodder value of the crop.

Harvesting and Roguing.   Seed crops sown in spring are harvested the following year. Crops grazed in the establishment year produce the best seed yields. Crops should be shut up for winter with about 15 cm growth and trimmed uniformly. The seed yield from the first growth is generally the highest. Sweet clover crops grow to a height of 100–120 cm, causing difficulty in harvesting. Hence, grazing early or takin an early light cut helps to reduce the amount of herbage at harvest. Two late cuttings or grazing may reduce seed yields significantly.

As with most forage legumes, seed crops can be inspected most effectively when they are in full bloom. Sweet clover grows vigorously after its establishment and competes well with weeds. Weeds, however, should be controlled before sowing and during the crop establishment using suitable herbicides.

Sweet clover is prone to quick seed shedding. It is, therefore, better to harvest crops by windrowing than by direct combining, which also allows herbage to dry before threshing. The crop is ready to harvest when about 60% of the pods have turned brown or black, with the stems of plants still sappy, and when only a few of the leaves have fallen. The crop may be direct combined within 3 days of desiccation with diquat for easy harvesting. Harvesting the crop in the early morning hours when it is damp may minimize shedding losses. It may be necessary to harvest the crop before its 60% color change stage in very dry areas (1).

Drying and Seed Treatment. The harvested produce usually contains green vegetative material and needs to be dried and cleaned immediately. The combine will generally leave a high proportion of unhulled seed. Scarification may be necessary to remove hulls effectively as well as to reduce the hard seed content.

4. Chickpea Milk Vetch

Chickpea milk vetch (*Aestragalus cicer* L.) is very slowly established and needs careful seedbed preparation. Seed is sown in lines 60–90 cm apart using 3–6 kg seed/ha. The cross-fertilized crop requires adequate isolation. It is rather resistant to shedding but is difficult to thresh. Seed crops may be harvested by windrowing and picking up by combining after thorough drying.

5. Crown Vetch

Crown vetch (*Coronilla varia* L.) with its three cultivars is mainly grown for soil conservation. It grows and sustains well even under adverse conditions. It is a perennial, growing to a height of 90 cm, with the herbage below forming a dense thick mat of 30 cm. It is cross fertilized by bumblebees and honeybees and needs adequate isolation. Seed sheds easily, making its harvesting difficult. Direct-combining has been found to be the best method. Scarification is necessary for dehulling the harvested seed and to reduce the content of hard seed, which is done taking care not to damage the seed physically.

6. Black Medic Trefoil

Black medic trefoil (*Medicago lupilina* L) is a biennial that is sown in the first year for seed production in the following year. Although grazing increases herbage production, it should not be allowed to continue late in the spring. The plant is generally low growing (8–10 cm), but may become as tall as 40 cm. Flowers appear on the lower parts of the stem first, and these seeds may shed before the majority of the pods are ready for harvest when the ripe pods turn black. Being small, the seed is difficult to handle. The seed crop is normally windrowed and picked up within 1 or 2 days of drying. The optimum harvest period is short–not lasting more than a week. Scarification is necessary for dehulling the seed before cleaning.

H. Forage Legumes Adapted to Very Dry Conditions

The forage legume species that are adapted to very dry conditions were developed mainly in south and western Australia and are also grown in South America and other drier regions. These are characteristically difficult to harvest, because the seed sheds easily to the ground as it ripens. Once sown, the crop reseeds itself each year. The seed varies in dormancy period, and it is not possible to identify the generations. All cultivars are self-pollinating and can be identified from the marker genes (10). The plant breeder produces the basic seed, followed by the production of certified seed, which may be from fields originally sown with the basic seed but subsequently reseeded naturally. A field may thus continue to produce certified seed indefinitely provided the crop has not more than 5% off-types not conforming to the genetic character of the species (1).

Subterranean clover (*Trifolium subterranean* L.) is an annual whose ripe seed heads characteristically turn downward and are pressured into the soil by the toughened peduncle. The crop requires cooler climates for the induction of flowers and longer days for flowering. It is not generally frost hardy. Once established, the crop is reseeded, because the seed is buried in the soil. It is therefore difficult to eliminate and change seed production from one cultivar to another on the same farm. Also, the seed harvested from the same field in successive years will be a mixture of generations, and it is not possible to operate a system of generation control as an aid to preserve the genetic purity of the cultivar.

Subterranean clover is a self-fertilized crop and needs minimum isolation (3 m) to prevent physical mixing of seed at harvest. New cultivars are preferably sown on fields not having grown subterranean clover; otherwise an interval of about 6 years free from the plants of the species is advisable to maintain cultivar purity of the seed.

Subterranean clover requires a good seedbed having a flat and firm surface on the field to discourage the plants from burying their seed heads in the soil. Grazing and cultural operations can also reduce this tendency. The soil should be supplied with calcium, phosphorus, and potassium if found to be deficient. Good seed yields are obtained from an autumn-sown pure crop stand obtained by drilling seed closely spaced at a rate of 6-10 kg/ha. Companion grasses may improve grazing but are likely to suppress the subterranean clover and reduce seed yield unless carefully controlled. The weeds described under red clover can also be problematic in subterranean clover.

The crop should be inspected for off-types and diseased plants at the time of flowering. Seed is normally harvested from the first growth in the following year. If a cover crop is grown, it should be controlled by grazing. The field is cleaned after harvest to prepare it for further seed harvest. Occasional grazing may be possible, taking care not to move sheep too quickly from one cultivar to another to prevent seed transfer.

Dense and thick stands are easier to harvest than sparse ones in which the soil surface is exposed and seed gets lost in the cracks. The seed is ready to harvest when the plants are dry and dead and seeds are hard. The crop is mowed close to the ground and picked up by combine. After combining, it may be useful to recover the seed from the soil surface with a suction harvester. A tinted windrower may be used to swath subterranean clover (Fig. 12). The harvested seed is normally dry but may contain plant debris, soil, and stones. Precleaning with aspiration followed by air/screen cleaning is required as described for other clovers.

Some other species of similar type are (10) *Medicago littoralis* Rhode ex. Loisel (strand medic), *M. tornata* (L.) Mill. (disc medic), and *M. truncatula* Gaertn (barrel medic). All these species shed seeds freely but do not bury seed in the soil. The seed is harvested in a manner described for subterranean clover by recovering a high proportion of it from the soil surface with suction. Kelly (1) mentioned two other

Figure 12 A tinted windrower for swathing subterranean clover. (From Ref. 3.)

species of *Medicago* grown for their seed in Australia: *M. rugosa* Desr. (gama medic) and *M. scufeilata* (L.) Mill. (snail medic).

IV. ARTIFICIAL SEEDS

According to Redenbaugh and associates (26,27), forage crops such as alfalfa, orchard grass, *Panicum*, and *Pennisetum* have a strong technological basis for improvement and multiplication by producing synthetic/artificial seeds through somatic embryogenesis technology. High-quality somatic embryos are currently produced in these crops.

Table 12 Germination of Dehydrated Orchard Grass Somatic Embryos After Storage at 23°C for 0, 7, and 21 Days[a]

Response[b]	Days of dehydrated storage		
	0 days	7 days	21 days
No germination	126 (28)[c]	333 (74)	396 (88)
Germination—no further growth	180 (40)	81 (18)	36 (18)
Germination—viable plants	144 (32)	36 (8)	18 (4)
Total	450 (100)	450 (100)	450 (100)

[a] Embryos were imbibed on solidified medium after test storage periods.
[b] Embryos that produced root hairs, roots, coleoptiles, and/or shoots but failed to develop further were scored as germinated—no further growth. Those that produced green leaves and continued to grow were considered to be viable.
[c] Figures represent total number, and those in parenthesis are % of total.
Source: Ref. 29.

Synthetic cultivars of seed-propagated self-incompatible crops like alfalfa and orchard grass are developed laboriously by selecting phenotypically uniform but genetically distinct lines (31).

These lines are then cross pollinated to produce seed. Such seed is, however, not genetically uniform. This type of breeding does not allow incorporation of specific new genes into existing lines. The use of synthetic seed enables single outstanding hybrids to be utilized as cultivars, because self-fertilization is not required to increase seed. Excellent somatic embryonic culture systems exist for both alfalfa and orchard grass, which are limited by low per plant value and low cost of existing seed (see Table 4 in Chapter 23). According to Gray and Purohit (31), an intermediate use of synthetic seed for these forage crops may be for an increase in parental lines prior to establishment in open-crossing blocks.

Plant recoveries from dehydrated somatic embryos were first obtained in orchard grass (28,29). The embryos were exposed to 70% RH air and stored up to 21 days at 23°C. Under these conditions, embryos dehydrated rapidly, became discolored, and decreased in size, and their outer cell walls collapsed and the water content dropped from 83 to 13% within 24 hr. When rehydrated, the embryos regained their normal white color and germinated. However, only well-developed, white, opaque somatic embryos were responsive, which were mature and contained starch and lipid storage compounds (30). When these embryos were stored in a dehydrated state for 21 days, 4% germinated and produced plants after imbibition (Table 12). This study demonstrated the occurrence of quiescence in somatic embryos (31).

REFERENCES

1. A. F. Kelly, *Seed Production of Agricultural Crops*, Longman Scientific and Technical, New York, 1988.
2. A. F. Kelly and M. M. Boyd, The stability of cultivars of grasses and clovers when grown for seed in different environments, *Proc. 10th Int. Grassland Congress*, Helsinki, 1966.
3. H. O. Graumann, Our sources of seeds and grasses and legumes, *Seeds: The Yearbook of Agriculture*, U.S. Department of Agriculture, Washington, DC, 1961, p. 159.
4. G. F. Sprague, Breeding for food, feed and industrial uses, *Seeds: Yearbook of Agriculture*, U.S. Department of Agriculture, Washington, DC, 1961, p. 119.
5. L. O. Copeland and M. B. McDonald, *Principles of Seed Science and Technology*, 3rd ed. Chapman & Hall, New York, 1995.
5a. M.A.Khan and I.A.Ungar, Germination response of the subtropical annual halophyte *Zygophyllum simplex*, *Seed Sci. Technol. 25*: 83 (1997).
5b. G.W.Freeman, C.A.Wagg and M.M.Mileva, Identification of turfgrass cultivars using reversed-phase high-performance liquid chromatography, *Seed Sci. Technol. 24*: 495 (1995).
5c. G.W.Freeman and F.A.Yoder Jr., Examination of Kentucky bluegrass blends using esterase isoenzyme electrophoresis, *Seed Sci. Technol. 19*:553(1991).
6. J. G. Boonman, General introduction and analysis problems (experimental studies on seed production of tropical grasses in Kenya), *Neth. J. Agric. Sci. 19*:23 (1971).
7. J. G. Boonman, Tillering and heading in seed crop of eight grasses (experimental studies on seed production of tropical grasses in Kenya). *Neth. J. Agric. Sci. 19*:237 (1971).
8. J. G. Boonman, Seed production of tropical grasses in Kenya, *Seed Production* (P. D. Hebblethwaite, ed.), Butterworths, Boston, 1980, p. 215.
9. P. D. Hebblethwaite (ed.), *Seed Production*, Butterworths, Boston, 1980.

10. *List of Cultivars Eligible for Certification*, Organization for Economic Cooperation and Development, Paris, 1984.
11. J. W. Purseglove, *Tropical Crops: Monocotyledons*, 5th ed., Longman, London, 1985.
12. J. G. Boonman, Effect of nitrogen and planting density on *Chloris gayana* cv. *Mbarra* (experimental studies on seed production of tropical grasses in Kenya), *Neth. J. Agric. Sci. 20*:218 (1972).
13. J. G. Boonman, The effect of row width on seed crops of *Sataria sphacelata* cv. Nande (experimental studies on seed production of tropical grasses in Kenya), *Neth. J. Agric. Sci. 20*:22 (1972).
14. L. R. Humphreys, Plant Production and Protection, *Tropical Pasture Seed Production*, Food and Agriculture Organization, Rome, 1979.
15. H. Youngberg, Techniques of seed production in Oregon, *Seed Production* (P. D. Hebbelethwaite, ed.), Butterworths, Boston, 1980, p. 203.
16. P. D. Hebbelthwaite and J. O. Ivins, Nitrogen studies in *Lolium perenne* grown for seed. II. Timing of nitrogen application, *J. Bl: Grassland Soc.* 33: 159 (1978).
17. P. D. Hebbelthwaite, D. Wright, and A. Noble, Some physiological aspects of seed yield in *Lolium perenne* L. (perennial rye grass), *Seed Production* (P. D. Hebbelthwaite, ed.), Butterworths, Boston, 1980.
18. J. J. J. Wiltshire and P. D. Hebblethwaite, Harvest timing in *Lolium perenne* L. seed crops treated with the growth regulator tripenthenol, *J. Appl. Seed Prod. 8*: 12 (1990).
19. A. Elgersma, Floret size utilization in perennial rye grass (*Lolium perenne* L.), *J. Appl. Seed Prod.* 9(suppl.): 38 (1991).
20. L M. Klein and J. E. Hammond, Seed moisture, a harvest timing index for maximum yields. *Transact. ASAE* 14:124 (1971).
21. C. L. Canode, Grass-seed production in the inter-mountain Pacific Northwest, USA, *Seed Production* (P. D. Hebbelthwaite, ed.), Butterworths, Boston, 1980, p. 189.
22. W. O. Lee, Clean grass seed crops established with activated carbon bands and herbicides, *Weed Sci.* 21:537 (1973).
23. R. B. R. Yadav and N. C. Sinha, Effect of plant growth substances on seed production in forage legumes, *Seed Research*, Spec. Vol. No. 1 (S. P. Sharma, ed.), Indian Society of Seed Technology, New Delhi, 1993, p. 182.
24. R. L. Agrawal, *Seed Technology*, Oxford and IBH Publishing, New Delhi, 1980.
25. V.Manonmani, D.Raja and A.Gopalan, Effect of foliar spraying with nutrients on seed production and seed quality of Lucerne (*Medicago sativa* L.), XI National Seminar on Quality Seed to Enhance Agricultural Productivity, UAS, Dharwar, January 18-20, 2002, *Seed Tech News* 32: (A.Gaur, A.K.Vari, J.L. Varshney and K.Kant, eds.), Indian Society Seed Technology, Division of Seed Science and Technology, IARI, New Delhi, March 2002, p. 34.
26. C. H. Hanson, (ed.). *Alfalfa Science and Technology*, American Society of Technology, Madison, WI, 1975.
27. K. Redenbaugh, D. Slade, P. Viss, and J.A. Fujii, Encapsulation of somatic embryos in synthetic seed coats, *HortScience* 22 (5): 803 (1987).
28. K. Redenbaugh (ed.), *Synseeds: Application of Synthetic Seeds to Crop Improvement*, CRC Press, Boca Raton, FL, 1993.
29. D. J. Gray and B. V. Conger, Quiescence in somatic embryos of orchard grass (*Dactylis glomerata*) induced by desiccation, *Am. J. Bot.* 72:816 (1985).
30. D. J. Gray, B. V. Conger, and D. D. Songstad, Desiccated quiescent somatic embryos of orchard grass for use as synthetic seeds, *In Vitro Cell Dev. Biol.* 23(29): 1987.
31. D. J. Gray and B. V. Conger, Somatic embryo ontogeny in tissue culture of orchard grass, *Tissue Culture in Forestry and Agriculture* (R. R. Henke, K. W. Hughes, M. J. Constantin, and A. Hollaender, eds.), Plenum Press, New York, 1985, p. 49.
32. D. J. Gray and A. Purohit, Somatic embryogensis and development of synthetic seed technology, *Crit. Rev. Plant Sci. 10*(1): 33 (1991).

14
Drying, Cleaning, and Upgrading

I. INTRODUCTION

After harvesting and threshing, partially field-dried seeds must be further processed by drying to an optimum moisture content to prevent seed germination, retain maximum quality (genetic purity, viability and germination, and analytical, physical, and storage quality), prevent bacterial and fungal growth, and retard infestation by mites and insects. Seeds of different crop plants like cereals, oilseeds, legumes, and grasses differ considerably in their physical and biological make-up. The moisture content in seeds varies according to their grain type, chemical composition, moisture at harvest, harvesting methods, relative humidity of the atmosphere, and seasonal fluctuations (1,2).

The normal sequence of operations included in the processing of seed after its harvest include receiving–(drying)–precleaning–cleaning–separation–blending–treatment–packaging–storage/dispatching (Fig. 1). In the progression of a seed lot through a processing plant, drying and storage operations can be inserted at any specific point depending upon the kind of seed, the condition of the seed lot, and the workload (3). A seed lot remains within the processing plant for weeks or months, but only a fraction of this time is spent actually moving through the machines. The seed is stored for most of the time provided it is dry enough. It can be stored in bulk or in bags as received, in bins or large boxes between operations, or in bags after packaging. It may have to be stored temporarily because the next stage in the processing sequence is full to capacity, or it may be moved out of storage because space is required for another seed lot. Similarly, drying may also need to be carried out at different stages in the sequence and may not be completed in one operation. Seed may be partially dried on the farm, but it may need immediate drying if received with a moisture content more than 3% above a safe storage level. Seed that is close to the safe moisture content may be cleaned wholly or partially before a second drying, but it will require precleaning to remove coarse trash. In this way, seed is less susceptible to mechanical damage before it is completely dried. Seed destined for sealed polyethylene bags may need a final drying before packaging (3).

A. Seed-Processing Plant

Seed processing mainly consists of seed cleaning, pesticide treatment, and packaging together with drying and storage under closely integrated control. For carrying out these operations efficiently, a seed-processing plant must be constructed carefully according to a planned layout. All operations like cleaning, upgrading, and treatment

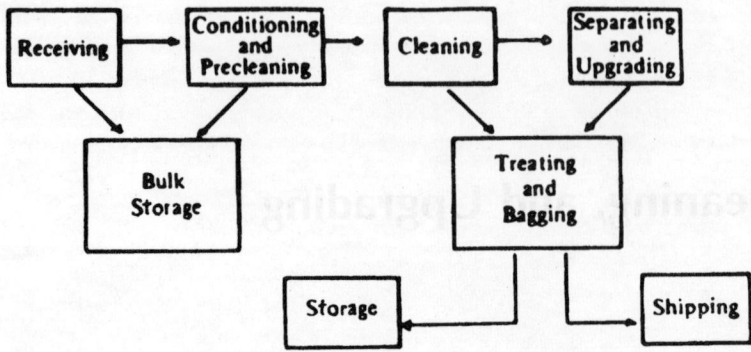

Figure 1   Basic flow diagram showing essential steps in seed conditioning. (From Ref. 12.)

are carried out without mixing or damaging the seed. Agrawal (4) listed the following factors to be considered in planning and designing a seed-processing plant:

1. Kinds of crop seeds to be handled and type of contamination present in seed lots: for example, weed or plant debris
2. Size of operation
3. Essentiality of drying facility
4. Selection of suitable equipment
5. Location of the plant
6. Source of power (electricity) for the processing equipment
7. System of seed delivery to the plant
8. Availability of manual labor

The efficiency of the plant depends on a thorough knowledge of the general sequence of the seed processes involved as well as an intimate knowledge of the seed to be processed. The sequence of operations depends upon the kind of seed, its initial quality, types of the contaminants present, and seed moisture content. The processing machinery also needs to be arranged in a sequence so that seed flows continuously through the separators, elevators, conveyers, and storage bins, which should be flexible enough to change the sequence of operations if necessary. The size or capacity of the equipment must also be adequate to handle the overall operating capacity of the plant. The conveying system employed should also be able to handle the needed capacity and fully adapted to the seed to be handled.

Three types of layouts, multistory, single story, and combined designs, are used to construct processing plants. In multistory layouts, the seed is carried by elevators to the top floor and emptied into large bins, where from it goes to the cleaners arranged in a vertical series on the lower floors by gravity. In a single-story plant, the seed moves from one machine to another through elevators placed between the machines. One person can supervise all the operations of such a processing line. A compromise between single and multistory systems also can be adopted.

In planning the layout and setting up machines, it is important to remember that all seeds are to some degree fragile—some (e.g., pulses) more than others. The consequences of breakage and physical damage to seed are quite obvious. The cracks and abrasions in the seed coat, although they may not be apparent to the eye, may indirectly provide access for microorganisms and toxic chemicals, leading to loss of seed viability. The germination capacity of snap beans, for example, declined from 95 to 55% when the seed was dropped onto a hard floor from a height of about 2 m, and after nine such drops, the seed failed to germinate (3). The physical damage to the seed is commonly caused during processing by conveyors and drops; for example, during threshing and shelling operations owing, to the faulty adjustment of machines or unsuitable moisture content.

B. Seed Moisture Content

A high moisture content of harvested seeds is one of the main reasons they lose their ability to germinate during storage. The moisture affects the respiration rate, and microorganisms and moisture levels above 20% may produce heat rapidly enough to kill seed or start fires in a seed mass. Seeds may suffer mechanical damage in handling and processing if their moisture content is too high. Fungi or molds tend to grow in moist seed lots, especially through the cracked or damaged seed coats. Most weevils and insects breed rapidly at a seed moisture content above 8%. Fumigation used to control insects may also cause injury to the seed if it has a high moisture content. The damp seeds easily stick together, interfering with the processing machinery (5).

The moisture in the seed has a definite vapor pressure, as does the water present in the atmosphere. Drying of seeds, which is basically the evaporation of moisture, takes place only when the vapor pressure of the seed moisture is greater than the resisting vapor pressure of the surrounding air. The rate of drying is high when the difference in these two pressures is high, and it declines as the difference in pressure lessens; drying will stop when equilibrium is reached between the two vapor pressures.

Under proper conditions, air readily absorbs moisture present on the outer surface of the seed. The internal moisture distributed throughout the cellular structure of the seed, however, either involves capillary action or diffusion of moisture to the surface where it can evaporate.

Drying can be either natural or artificial. In natural drying, the atmospheric air moves naturally around damp seed spread on trays, canvas, or floors or in the fields, whereas artificial drying, also known as dehydration, uses heated or unheated air that is forced mechanically through a drier. All drying operations thus involve some movement of air through the seed.

Evaporation (escape of water molecules from seed into the air) and absorption (entry of water from the atmosphere into the seed) go on continuously. When evaporation exceeds absorption, the seed dries out, and when the reverse happens, seed imbibes water from its surrounding. When both processes are equal, the seed is in equilibrium with the humidity of the air. When water evaporates, it changes from the liquid state to a vapor. The number of freely vibrating water molecules that can be contained in a given volume of air is limited; the saturated air carries the maximum amount of water vapor. However, the hotter the air, the more water it can carry (Table 1).

The water content of air is expressed as relative humidity (RH), which is the amount of water expressed as a percentage of the amount that the same air can carry when saturated. At the saturation point, RH is 100. Both the water-carrying capacity

Table 1 Maximum Amount of Water
Held by Air at Different Temperatures

Temperature (°C)	Mass of water (g/kg air)
0	3.9
10	7.6
20	14.8
30	26.4
40	41.4

Source: Ref. 6.

and the RH of air vary with temperature. If, for example, air at 20°C has a RH of 90, it contains 13.3 g (90% of 14.8) of water per kilogram of air. If this air is cooled to 10°C, the amount of water remains the same, but it is now in excess of the air's holding capacity of 7.6 g/kg and the excess water appears in the liquid state; that is, some of the water condenses. Similarly, if air is heated, its water content remains the same, but its moisture-carrying capacity increases and its RH falls.

For drying to take place, the air must have a RH below the equilibrium point (seed moisture content in equilibrium with the RH of the air). The drying process can be hastened by blowing air through the seed mass. However, if the RH of air is close to or above the equilibrium point, it must be heated to increase its water-carrying capacity and lower its RH (3).

II. SEED DRYING AS RELATED TO HARVESTING AND THRESHING

During seed development and accumulation of storage material, seeds contain more than 50% moisture on a wet weight basis. The farmer aims to have this seed in storage with a moisture content of less than 15% just a few weeks later. In the beginning, the sun and wind naturally promote drying while the seed is attached to the mother plant. Harvesting operations interrupt this process, and subsequent drying may be either natural or artificial depending upon the harvesting system used. Nevertheless, the loss of moisture from seed, whether slow or fast, is regarded as a single process of evaporation (that may even go in the reverse direction) determining the ultimate quality of the seed.

Preharvest drying depends on the prevalence on natural dry weather conditions in the field. Seed production is normally carried out in the part of a country that provides the best weather at the time of seed ripening and harvest, and it may be possible to adjust the date of sowing so that ripening and harvesting operations coincide with dry weather. This may necessitate the use of presowing irrigation, may involve a serious reduction in yield, or may even be impossible for the rain-adapted or day-length–adapted cultivars.

Harvesting of most crops constitutes two major operations, cutting of plant parts and separation of seeds from the cut material (threshing), each being performed either manually or mechanically by machine. A machine can perform both operations in one pass over the crop (combine-harvester). Mechanical harvesting, although not necessarily more effective than manual methods, can deal with a large crop area in a brief spell of dry weather. Manual methods may be preferred for economic or social reasons. The final drying of the seed by artificial means depends on the moisture content of seed after harvesting.

Thomson (3) described three types of harvesting systems used for seed crops: handpicking, cutting, and threshing.

A. Handpicking

In handpicking, ripe seeds or fruits are picked by workers walking through the crop and collected in baskets. This is practiced in small vegetable crop areas and for species in which all seed does not ripen simultaneously (cotton, pigeon pea, carrot, some cultivars of *Pennisetum*), and therefore requiring two or more pickings as more seed ripens. At the final picking, plants are cut and harvested.

Apart from handpicking, there are two common harvesting systems: cutting and threshing as two separate operations and as one combined operation. The effectiveness of one or the other system of harvesting depends on utilization of natural drying, shedding losses of seed, and the extent of mechanical damage to the seed. The latter may be caused by faulty setting of the threshing mechanism, but the nature and extent of damage is greatly influenced by the moisture content of the seed. If they are too wet, seeds are crushed and the tissues bruised, and if too dry, they may be liable to fracture.

B. Cutting and Threshing

Plants are usually cut and left in the field to dry before threshing. The cutting may be performed with a machine or manually. At this time, seed is generally too wet to separate readily from the plant and is very susceptible to mechanical damage. Threshing must therefore be performed only when seed has dried sufficiently. Plants are cut before the seeds begin to shed and are tied into bundles or sheaves. After a period of natural drying, they are transported to a nearby threshing floor or static machine for immediate threshing or may be stacked for a time before threshing. Shedding losses can be minimized by cutting the crop in the early morning hours while the humidity is still high or by fitting trays under the cutting machine to catch the shed seed, as is traditionally done for small-grain cereals, some pulses, and erect-growing grasses.

In some crop species, while the seeds ripen, the stems and leaves remain green. These are usually cut and left lying in a window or in small heaps or may be hung on tripods or racks in wet situation. When dry enough for threshing, they are either carried to a static threshing machine or picked up from the window and threshed by a moving combine-harvester, as is normally done for herbage crops and most pulses. This system can be used as a fallback method for any crop in favorable weather. The crop is cut as high as possible to minimize the amount of stem and leaf that must be dried and fed into the thresher as well as to hasten the drying process.

Alternatively, the crop may be cut and carried to a drier, followed by threshing. This technique is normally used for peas and the tropical grass *Setaria anceps*. In addition to cutting and threshing, maize requires picking (removal of ears from the plants) and shelling (separation of seeds from the ear). The crop is normally picked up early while the seed moisture content is quite high to prevent infestation by weevils and molds under warm and humid conditions. Shelling is carried out later when the ears have dried. For grain production, picking and shelling can be performed at the same time, an operation corresponding to combine harvesting of small-grain cereals, but this may cause physical damage to the kernels. Maize cobs are picked up by hand or machine, leaving some husk to provide enough protection. Mechanical picking may cause severe damage to the seed if it has a moisture content above 30%; 20–25% is considered appropriate. The ears are dried in open-sided cribs (Fig. 2) or in heaps of 12–14% moisture content before shelling. Alternatively, they can also be dried artificially after removing the husk.

Shelling of maize also may be carried out by hand or mechanically after the removal of husks. The latter facilitates examination of ears for mold infestation or damage by insects. The off-type and diseased ears can be removed to aid the roguing operation carried out earlier in the field.

Groundnut is either dug up manually or by machine or left on the ground to dry in windrows or small heaps. The pods are then separated from the plants in a stationary thresher or by a combine-harvester picking up from the windrows. Peanut flowering and seed ripening are very uneven (extending over a couple of months in some cultivars) and make it difficult to judge the appropriate harvest time. Early harvesting produces many empty pods with immature and shriveled seeds, whereas late harvests will produce some swollen and germinating seeds. The swelling may cause splitting of pods and free access of soil fungi to the seeds through ruptured seed coats. The inside of a pod provides a suitable environment for the growth of fungi both before and after harvesting (3).

1. Direct Combine Harvesting

In this method, the machine cuts off the inflorescences with a variable amount of stem and leaf and threshes the crop directly. Most cereal crops grown on a medium to large scale are harvested by directly combining. It is suitable for any crop whose seed is

Figure 2   Open-sided cribs used to dry maize ears in East Africa. (From Ref. 3.)

retained on the standing plant until it is dry enough to thresh. Some crops like pulses, grasses, and forage legumes (lucerne, clover) are sprayed with chemical agents to accelerate drying of the standing plant about 3–10 days prior to combining. Three types of chemical agents have been used, each functioning in a different way:

1. Desiccants (e.g., diquat, paraquat), which kill the green tissue by damaging the cell membranes
2. Defoliants (e.g., ethephon), which cause abcission of leaves by release of ethylene gas
3. Soil sterilants, which stop water intake by damaging the root tissue

Desiccants, although most effective, are costly and may cause discoloration and injury to seed. Spraying of the inflorescences shortly before maturity with a water-soluble plastic resin has been reported to prevent physical shedding of seeds as they ripen; thus harvesting can be delayed until all seeds are fully ripe (3). However, this technique, tried on an experimental scale, needs further investigation.

The seed of many species of tropical pastures also flower and ripen over an extended period of time and fall to the ground before harvesting. Conventional harvesting methods are able to secure only a small proportion of the total seed produced (ripe or ripening seed on the crop). In one of the improved harvesting methods, a machine moves through the crop beating or rubbing the inflorescences to detach the ripe seeds, leaving the unripe ones to ripen on the plant and be collected later (e.g., *Cenchrus ciliaris* and *Paspalum dilatatum* grasses). The other method allows the seed to fall to the ground to be picked up later by a suction harvester (e.g., *Stylosanthes humilis*).

The moisture content of the seed, however, can be used as a precise indicator of the seed crop's harvest maturity for cutting or threshing. Thomson (3) has summarized the available information for several species of crops as follows:

Mechanical damage to seed can be minimized when threshed or combined within the following limits:

Small-grain cereals	
Wheat	16–19%
Barley	17–23%
Oats	19–21%
Rice	17–23%
Maize	Ideally picked at 20–25%, but some cultivars may be picked up to 35%; ears from the pollen plants are picked earlier at 30–35%
Soybean	Harvested at not more than 14%, very sensitive to damage, beyond the 13–15% range
Groundnut	Shelled at 10%, very sensitive to physical damage
Phaseolus peas	Harvested at 14–20%
Flax and linseed	Cut at not more than 35%
Rye grass	Cut at 40–50%, thresh or combine at not more than 35%, depending on cultivars
Dactylis	Cut at less than 40%, threshed at 30% or less, liable to shed seed if combined direct

The electrical moisture meters are not reliable at high moisture contents, and they are therefore not suitable to judge the harvest maturity of the crop, especially of the grasses, the seeds of which may contain more than 40% moisture shortly before cutting. In such situations, infrared moisture meters are more reliable. The crop's readiness to harvest is also indicated by its color and hardness (3).

2. Threshing

Threshing consists of:

1. Beating or rubbing the plant material to detach the seeds
2. Windrowing to remove chaff, straw, and other light material from the seed
3. Sieving to remove heavier material like stones, soil, and seeds of different size

All of these operations are traditionally carried out on a smooth floor. Seed may be separated manually by beating the plants with a stick (flail), trampling by animals, or with a rubber-tired tractor. A stoneroller or a sledge on a set of metal discs like a disc harrow drawn by animals may also be used. Windrowing by wind can be supplemented by use of flexible trays of plaited bamboo sheets. Finally, hand sieves are used to separate heavy particles that are smaller and larger than the seed.

Power-driven mechanical threshers are used to handle a large-scale crop. Beaters, corresponding to flails, revolve at the specified speed within a drum into which plant material is fed. The threshed material is then subjected to a strong air blast and shaken over sieves of different sizes. The emerging produce is fairly clean.

The machines are adjustable to handle different crop species and different conditions of the seed; the main consideration is to obtain clean produce that is subject to minimum physical damage and loss of seed. The type of seed and its moisture content mainly govern the seed's susceptibility to mechanical damage. Oilseeds (groundnut, sesame, linseed), for example, are more susceptible than cereals. Also, wet seed tends to be crushed or skinned, and too-dry seed tends to be broken. In a combine-harvester, the plant material is lifted from the knife blades into the threshing mechanism. In addition to the adjustments necessary in a static threshing machine (beater speed, width of the gap between beaters and the concave, airflow, and sieve sizes), it is also necessary to adjust cutting height and forward speed and the pick-up reel to prevent speed loss and mechanical damage. The moisture content of the seeds is liable to be greater than for a stationary machine, necessitating a lower beater speed to prevent crushing. Too slow a speed will not detach seed completely. Combine-harvesters can handle much more variable material than a static machine in respect to crop uniformity and seed moisture. Combine contractors may be paid by the hour to discourage too rapid harvesting (3).

Misra (7) investigated the effects of seed moisture, seed temperature, cultivar, and conditioning equipment on soybean seed quality during processing. Seed lots below 10% moisture declined in germination during processing. Temperature influenced the amounts of splits produced as a result of processing operations; low-moisture seed lots processed at -8 to $-3°C$ declined in germination. Also, variety Beeson was more susceptible to damage during processing than were Corsoy or Williams. The air/screen cleaner and gravity separator improved the quality of soybean seeds. Seed lots with low moisture and conditioned at cold temperatures had the lowest average germination (Table 2).

Table 2 Effects of Moisture, Temperature, and Conditioning Sequences on Soybean Seed
Germination

	Germination (%)			
Sequence	M: 12–15% T: 18°C	M: 12–15% T: −8° to −3°C	M: 3% T: 18°C	M: 3% T: −8° to −3°C
Initial	96.2	95.8	94.8	95.3
After auger	95.2	93.8	93.7	93.0
After cleaner	95.3	94.8	94.5	94.2
After spiral	95.7	95.0	93.3	93.0
After gravity	96.0	96.5	94.0	92.8
After bagging	95.7	95.8	94.3	91.2

M = Moisture content; T = temperature.
Source: Ref. 7.

Nascimento and West (7a) investigated effects of drying conditions during seed
priming on muskmelon seed quality either immediately after seed priming or after 12
months of storage. Seeds of muskmelon cv. Top Net SR were primed for 6 days in
darkness at 25°C in KNO$_3$ (0.35 M) aerated solution. Seeds were rinsed in running tap
water and redried at 18, 28, and 38°C during 24, 48, and 72 hrs. Seeds were stored at
10°C and 45% RH for 12 months. The beneficial effects of seed priming were
maintained after dehydration the drying temperature did not affect seed germination
after priming. An interaction effect on germination between temperature and drying
duration was observed at low temperature. Seed germination and vigor of primed
seeds decreased after 12 months of storage. Both temperature and duration of
drying affected seed vigor after storage. A better management of drying conditions
during muskmelon seed priming should be focused on in order to achieve the benefits
of seed priming and to maintain seed quality, especially if primed seeds are to be
stored.

III. SEED-DRYING TECHNOLOGY

Seed drying should reduce the seed moisture content to safe limits to maintain
its viability and vigor during storage, which may otherwise deteriorate quickly owing
to mold growth, heating, and enhanced microbial activity. Seed drying also permits
early harvesting, long-term storage of seeds, more efficient use of land and manpower,
the use of plant stalks as green fodder, and production of high-quality seed (8).

Depending upon the climate and the method of harvesting adopted, the threshed
seed may or may not be dry enough for safe storage. In a dry climate, seed from the
combine-harvester may be stored directly either temporarily or for a longer term based
on seed moisture content.

Under les favorable conditions, threshed seed almost always needs further
drying. If the seed is not dry enough after threshing, it must be dried to within 3%
of the moisture content required for storage within 24 hrs, followed by a second
drying. Seed at this stage may contain impurities, for example, plant debris or weed

seeds, which should be removed to facilitate further drying. This may be achieved by passing the seed through a simple precleaner before artificial drying. This is a partial cleaning, and the main cleaning operation occurs later when the seed has been dried and stored for some time (9).

The moisture content limit at this stage (i.e., 3% above storage requirement) is absolute and must be taken into account in planning any seed-multiplication program. In drier regions, seed may require minimum drying (blowing air heated to about 5°C above the ambient temperature) with simple driers installed on medium- to large-sized farms. In the humid tropics, however, large-scale drying plants may need to be located in each seed-producing district to avoid delays in drying. Mechanical transport to bring the seed rapidly from the producing centers to the processing plants must also be ensured.

The seeds of most agricultural crops can retain their viability even if they are dried to as little as 5% moisture content, whereas seeds of some tropical rain forest species will lose their viability if the moisture content falls below 15% (3). Certain ill effects of artificial drying on the germinability of seeds of agricultural crops have been noticed and attributed either to rapid drying or exceeding the limits of temperature and time to which the seed is exposed for drying. Soybean seed dried by air with a relative humidity (RH) of less than 40% loses viability even at low temperatures. Sometimes seeds are dried in two stages to minimize the adverse effects (e.g., rice).

Drying air temperatures of up to 45°C are generally safe (3), although higher temperatures may be used in continuous-flow driers than in batch driers, because the time of exposure in the former is shorter. It is not possible to give maximum drying temperatures guaranteed to be safe. The seed temperatures given in Table 3 provide a general guide. A range of temperatures for some seeds indicates that the higher temperatures are applicable only to relatively dry seed in a continuous-flow drier, whereas the lower ones are recommended for drying. Kelly (10) also suggested maximum limits of drying temperatures for a range of seeds based on their initial moisture content. It is very important that after drying the seed is cooled by forced ventilation before it is stored.

Table 3 Recommended Maximum Drying Temperatures of Seeds Generally Considered Safe

Crop	Temperature (°C)
Wheat, barley, oats	43–65
Rice	50–60
Maize	40–45
Pea	30–50
Groundnut	36
Beet	45
Soybean	30–55
Rye grass	40–70
Clovers and brassicas	27–40
Onion	21–33

Source: Ref. 3.

A. Methods of Seed Drying

The methods of seed drying can be grouped into traditional sun (and wind) drying and force-air drying (batch drying or continuous-flow drying).

1. Traditional Sun Drying

Farmers traditionally dry small quantities of seed, utilizing the sun and natural wind, by spreading the threshed seed in a thin layer on a smooth earthen floor or on straw matting. Ventilation can be improved by stretching the matting on a horizontal framework supported on stakes above the ground to allow the wind to blow through the seeds. Unthreshed inflorescences can be hung on frames or placed in cribs with open sides.

Sun drying requires no additional expenditure or special requirements. The major limitations of sun drying, however, are delayed harvest, risks of weather damage, and the possibility of mechanical admixtures. Direct sunlight also can adversely affect seed germinability owing to high temperature and ultraviolet radiation, especially if the moisture content of the seed is high. Prolonged exposure of seed to direct sun must, therefore, be avoided.

2. Forced-Air Drying

In forced-air drying, natural air or air supplemented with heat is blown through a layer of seed until drying is completed. Two types of driers are used: batch driers and continuous-flow driers (11).

Batch Driers.   In a batch drier, relatively dry air is blown through a layer of seed until the seed is dried completely, after which it is removed and replaced by another batch of seed. The method is simple and well suited to small quantities of seed, allows easy cleaning, and is recommended for farm drying. In a horizontal drier, the seed is contained in a box or chamber with a perforated or slatted floor through which the air is blown. Air ducts can be installed in a barn floor and the seed to be dried piled over them. In a modified sack drier, seed contained in a woven sack is placed on top of a grid through which air is blown (Fig. 3). A cylindrical storage bin with a raised perforated floor arranged to blow air underneath the floor can also be used (Fig. 4). A vertical

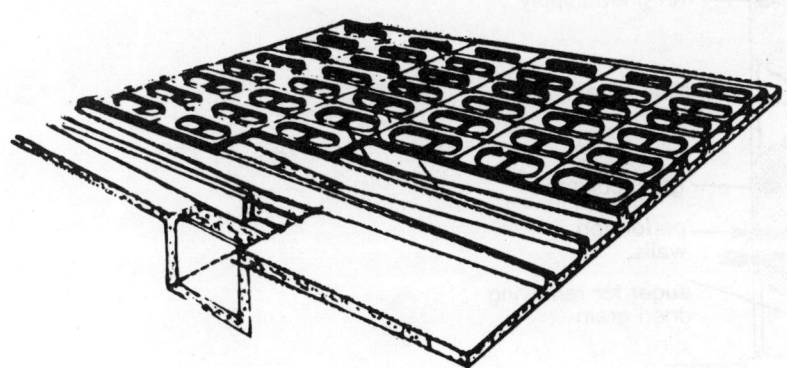

Figure 3   Sack drier. (From Ref. 3.)

Drying, Cleaning, and Upgrading

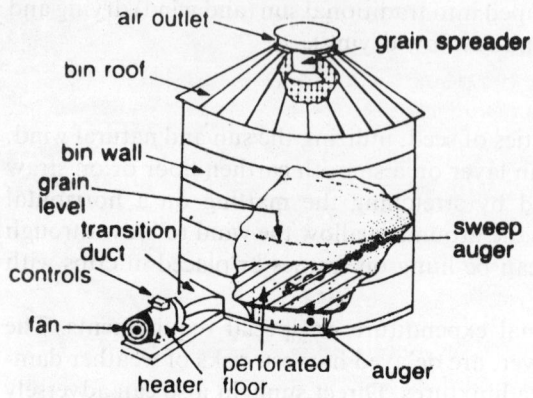

Figure 4   Bin drier. (From Ref. 3.)

batch drier consists of two concentric perforated cylinders. The space between the two cylinders is filled with seed, and air is blown into the inner cylinder from where it passes outward through the seed (Fig. 5). The size of the batch determines the drying rate (3,11).

In a horizontal batch drier, the seed at the bottom dries first, with the dry zone extending gradually upward. The drying of the uppermost seed may be delayed unduly if the seed layer is too thick or the airflow is inadequate. The seed layer should not exceed a depth of 3 m, and for high-moisture forage seed, it should be reduced to 1 m or less. If the seed is dried in a storage bin, a layer of undried seed can be added on top of the dried batch and drying continued, but only if the seed is already fairly dry and the air is not too hot. Seed loss also can be avoided by drying in two stages. After the first batch has partially dried, the emerging air is passed through a second batch held in another chamber, repeating the process with the second and a third batch, and so on.

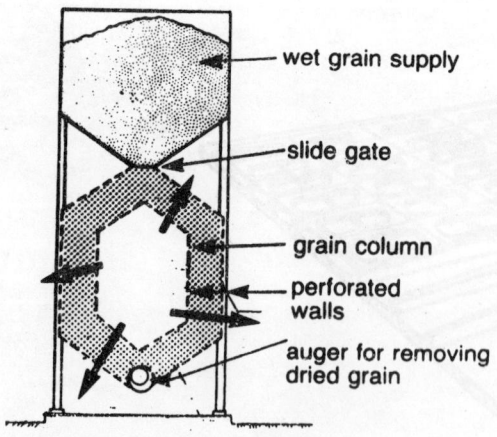

Figure 5   Vertical drier. (From Ref. 3.)

The air blown through a batch must not be too hot, because the seed at the bottom may be overheated by direct exposure to the entering hot air. It is often not necessary to heat the air at all, and heating to less than 10°C above the ambient temperature can be very effective, but on a hot humid day in the tropics even a few degrees above ambient temperature can harm the seeds. Dehumidifiers may need to be used under these circumstances (3).

An appropriate drying rate is very important. Too raid drying can harm the seed because of the high drying temperature or a quick loss of water from the seed. Slow drying may mean maintaining a high-moisture seed at a higher temperature for a prolonged time, resulting in deterioration of seed (11).

Continuous-Flow Driers. In this type of drier, the seed moves horizontally or vertically through a stream of hot air and then into a cooling chamber (Fig. 6). It is a continuous drying process on a factory scale suitable to handle large quantities of seed. These driers are, however, difficult to clean when there is a change of cultivar. Continuous-flow driers can use air temperatures higher than those of batch driers, because the seed is heated for a much shorter time (3).

B. Storage Structures for Seed Drying

Building requirements for a seed-drying system depend upon the size of operation, the number of different seeds to be dried, the level of mechanization desired, and future expansion. Different types and forms of storage structures can be built for handling

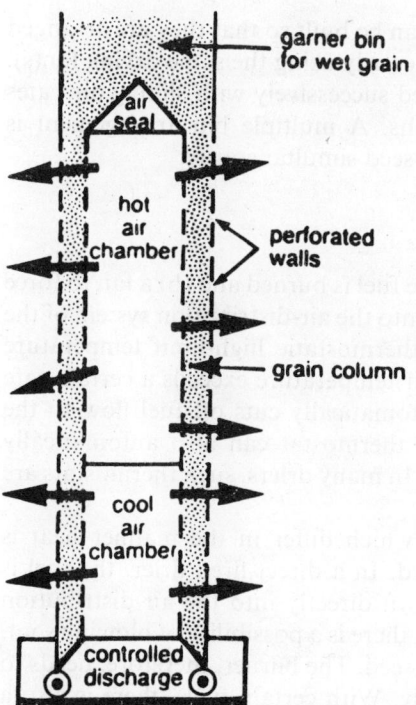

Figure 6 Continuous-flow drier. (From Ref. 3.)

seeds to be dried with forced air. These may be made of steel, wood, plywood, or concrete, and they may be cylindrical or rectangular in shape (12).

Regardless of the type of structure, all storage bins used for forced-air drying in storage of seeds must have the following features (4,12):

1. Small-grain seeds in bulk exert large pressures against the sidewalls. The side pressures are converted to a vertical load on the foundation, which should be strong enough to hold the seed lots.
2. The roof and walls of bins must be airtight for drying (and fumigation) to proceed satisfactorily.
3. The openings for filling and removal of seed should be large and convenient to use. A full-size entrance door is desirable.
4. A hand space of about 1 m should be provided for easy inspection of seed. Cleaning and spraying operations should be convenient. For fumigation, the structures should be airtight, with a provision for temporary sealing of all openings.
5. The structure should be able to dry and store more than one kind of seed.
6. The drying air should be uniformly distributed through all portions of the seed lot for efficient drying.
7. The flow of air leaving the seed should proceed rapidly so that back pressures do not hinder the flow of drying air into the seed.

Agrawal (4) described three types of air-distribution systems used for seed drying: (a) the main and lateral duct system, (b) a single central perforated duct, and (c) the perforated false-floor system.

Multiple bin storage structures for drying can be built so that they are arranged to enable the drying of several seed lots simultaneously using the same drying fan(s). Alternatively, different seed lots can also be dried successively with sliding air gates controlling the flow of air to the respective bins. A multiple bin arrangement is particularly useful to dry more than one kind of seed simultaneously.

C. Heated Air-Drying System

Heated air driers consist of (a) a heater unit where fuel is burned and (b) a fan to force the heated air through a canvas connecting duct into the air-distribution system of the drying bin. Safety features such as automatic thermostatic high-limit temperature control, which cuts off the burner flame if the air temperature exceeds a certain safe maximum, and flame-failure control, which automatically cuts off fuel flow to the burner if the flame goes out, are provided. A thermostat can also automatically maintain the air temperature at a desired setting. In many driers, such thermostats are provided as a standard feature.

Two main types of driers are available, which differ in the manner heat is supplied to the air: direct fired and indirect fired. In a direct-fired drier, the fuel is burned and the hot combustion gases are thrown directly into the air-distribution system. Although the heat is used very efficiently, there is a possibility of blowing soot, unburned fuel, and objectionable fumes into the seed. The burner, therefore, needs to be adjusted properly to burn the fuel completely. With certain fuels, there is also a danger of blowing small sparks into the seed.

In indirect-fired driers, the hot combustion gases pass into a chamber. The drying air circulates around this chamber (the heat exchanger) and picks up heat as in a hot-air furnace. The drying air thus does not include combustion gases, sparks, soot, or fumes. These driers are less efficient in the use of heat, but are safer than direct-fired types.

The driers are designed to burn various types of fuels (e.g., liquid propane) or butane, natural gas, fuel oil, and coal). Both liquid propane and natural gas burn readily with minimum soot and are the best fuels for direct driers, and kerosene (rock) oil is better for indirect-fired driers.

Drier fans may be driven by either an electric motor or a gasoline engine and power take-off. Electrical motors are most efficient and economic, and they are commonly used.

Two important aspects that must be considered while calculating the requirements of a suitable to crop drier are the required airflow volume and the heat capacity (BTU/hr) for drying seeds at the specified desired rate. The fan requirements can be computed by knowing the total airflow at the static pressure of the seed at a given drying depth, and heater requirement are estimated by calculating the amount of water to be removed from the seed per hour. Based on these calculations, a suitable crop drier can be selected to provide a minimum required airflow volume (bin capacity × air flow rate) and heat capacity in BTU/hr. Agrawal (4) categorized types of heated driers as layer-in-bin, batch-in-bin, batch, and continuous driers and described their functions.

Stirring devices keep the seed in a loose fill condition, allowing easy airflow through the bottom layers. Such mechanisms alleviate the problems of uneven drying (or over drying) by breaking up pockets of fires and trash and blending the seed by constant mixing.

Large differences in the degree of drying between the top and bottom (air inlet) layers of seed have been noticed during drying by heated air. It is, therefore, advisable to dry seed at shallow depths to minimize these differences and avoid overheating of the bottom layer. Agrawal (4) recommended maximum seed depths and temperatures for batch drying of seeds of different crop species in bins (Table 4).

Table 4 Recommended Maximum Depths and Temperatures for Heated Air-Drying of Some Crop Seeds in Bins

Crop seed	Maximum depth (cm)	Recommended maximum temperature (°C)
Shelled corn	50.8 (20 in.)	43.3 (110°F)
Wheat	50.8 (20 in.)	43.3 (110°F)
Barley	50.8 (20 in.)	40.4 (105°F)
Oats	91.4 (36 in.)	43.3 (110°F)
Rice	45.7 (18 in.)	43.3 (110°F)
Soybean	50.8 (20 in.)	43.3 (110°F)
Peanuts	152.4 (60 in.)	32.2 (90°F)
Grain sorghum	50.8 (20 in.)	43.3 (110°F)

Source: Ref. 4.

Heated air drying requires higher rates of airflow, because water is evaporated faster and more air is needed to carry it away. The higher airflow rate also ensures more uniform drying of the top and bottom layers of the seed, completing the drying much faster at the recommended temperatures.

The general procedure for bin drying of seeds with heated air consists of charging seed into the bin to the recommended depth, maintaining a uniform distribution of trash and broken kernels. The drier is operated at the recommended temperature of the seed using either manual or thermostatic controls to set the desired temperature. After the drying is completed, blowing of air through the seed is continued for some time without heat to bring the seed to an ambient temperature, which may take from 30 to 120 min depending upon the quantity of seed and the air temperature. The maximum moisture contents for safe storage of seeds of different crop species are given in Table 5.

Some variations of batch drying with heated air include wagon drying, bag drying, and box drying. In wagon drying, the seed is loaded directly from the combine onto a wagon especially constructed for drying. The wagon is then drawn to the drier unit and connected with a canvas distribution duct. Forcing heated air up through the perforations in the wagon floor dries the seed. After drying, the wagon is disconnected from the canvas duct and the seed is cooled with a fan and towed to the storage bins. Wagon drying provides continuous drying, versatility, easy cleaning, and low initial cost.

Bag drying is another suitable variation to handle several varieties of smaller quantities of seed simultaneously. Seed received in jute bags is exposed to airflow with minimum static pressure, because the drying bed is only one sack deep. Typical design criteria provide 25–40 m^3 of air/min/m^3 seed at a static pressure of 3 cm or less. The construction is simple and inexpensive (4).

A box drier is a modified bag drier well adapted to dry smaller quantities of basic or foundation seeds. With box driers, it is possible to maintain the identity of small seed lots despite bulk handling. The boxes can be constructed of locally available materials, which are fitted with perforated metal or woven wire bottoms.

The design of seed-drying facilities must take into consideration both size and type or components as dictated by the drying system. Radial axial fans with

Table 5 Maximum Moisture Content for the Safe Storage of Seeds of Some Crops

Crop	Maximum moisture content (%)
Wheat	12
Oats	13
Barley	13
Grain sorghum	12
Shelled corn	13
Soybean	11
Rice	12

Source: Ref. 4.

backwardly curved blades are normally recommended for seed-drying systems. The fan should be large enough to deliver 8 m³ of air/min/m³ of seed against a static pressure of 8.75 cm of water for drying deep-bed seed in bins.

Heaters that maintain the drying temperature at 43°C should be used, because higher temperatures are detrimental to seed viability (13,14). The overall management of the seed-drying operation requires that records be kept of the temperature, airflow, drying time, seed moisture content, depth of drying, and different types of seeds handled.

IV. CLEANING AND UPGRADING

Seed as it comes from a field is often mixed with impurities (e.g., seeds of weeds and other plants, plant debris, soil, and stones), which must be separated to obtain pure live crop seeds for replanting. Seed from each crop is basically different in physical make-up from others and can be identified easily. The physical differences in size, shape, weight, surface area, specific gravity, color, texture, stickiness, pubescence, and electrical properties can be measured or sensed by mechanical devices called separators, which cull unwanted impurities from the seeds on the basis of one or more of these physical differences. Seed separators can remove dirt, leaves, stems, and chaff. Cleaning thus reduces the bulk to be handled and stored and removes moist and green plant material that may cause heating in storage. All seed crops such as grasses, legumes, vegetables, flowers, fibers, trees, and shrubs require some cleaning.

During processing, seeds first go through a precleaning operation, which may include (a) scalping, the removal of material coarse enough to be easily separated by screens; (b) hulling, a completion of the field threshing operation; and (c) scarifying, the scratching of hard-coated seed. The processing of seeds of native grasses and other seeds that have awns and appendages may require additional precleaning to remove awns and beards.

The seed is then processed on an air/screen cleaner in which the bulk of the foreign material is removed by screens and air. The final separation is achieved on one or more finishing machines, which normally separate only one type of contaminating seed from the desired clean product. Specific gravity separators divide seeds according to their weight and size. Indent disc and cylinder separators remove long seeds from short ones, whereas pneumatic and aspirator separators work on the basis of differential resistance to airflow. Velvet roll separators remove smooth seeds from rough seeds, and spiral, inclined draper, timothy bumper mill, vibrators, and horizontal disc separators divide seed according to shape. Electronic separators sense a difference in the electrical properties of seeds, whereas magnetic separators and the buckhorn machine separate rough or sticky-surfaced seeds from smooth seeds. Color separators can distinguish light-colored seed from dark ones.

Klein et al. (15) cited the following factors to be taken into consideration when designing a seed-cleaning plant:

1. Handling and cleaning of the seeds should be possible without mixing or damaging seed with a minimum of equipment, personnel, and time.
2. Seed separators, elevators, conveyors, and storage bins should be arranged so that seeds can flow continuously from beginning to end, yet be flexible enough to bypass a machine or return part for recleaning.

3. Other factors to be considered are kinds of crop seeds to be cleaned, nature of contaminants and weed seeds, volume of seed to be handled, method of handling (bulk or sacks), type of conveying system (pneumatic or mechanical), and location of shipping and receiving facilities.

Cleaning usually requires a succession of operations, which can be regarded as proceeding in three stages (3,16): conditioning or precleaning, basic cleaning, and separation and grading.

Brassica juncea cv. Pusa Bold seeds were graded into three size categories: > 2.4 mm, 1.8–2.4 mm and <1.8 mm. Initial germination was 90.3, 89.3, and 81.6% in the three sizes as listed, whereas field emergence was 45.4, 40.6 and 30.4%. Seedling length and storability of seeds also decreased with decreasing seed size (16a).

A. Precleaning/Scalping

Conditioning, or "scalping," is a rapid precleaning process consisting essentially of an air blast and large-meshed screens or cylinders to remove the most bulky material and the rubbish most liable to choke up conveyors and sieves. Thus, the purpose of scalping is to facilitate movement of the seed through the machines in subsequent cleaning operations. The material falls through an upward air stream that removes dust and other light material. Vibrating screens or revolving cylinders allow the seed-size particles to pass through, retaining chaff, pods, awns, stem, leaves, and any other large-sized particles, which are shaken off to the side. This process is colloquially known as "scalping." Some machines also remove some smaller material.

Although scalping and blowing operations are common to all seed crops, "hulling" and "shelling" are required by particular crops only. In crops like clover, for example, the fibrous fruit wall (hull) adheres closely to the seed and is not always removed by the threshing machines. Such seeds are fed into a cylinder containing revolving arms that rub the seed against the internal concave surface of the cylinder. The machine requires careful adjustment to avoid mechanical damage to the seed. In addition to the hulling operation, this machine partially scarifies hard seeds, rendering their seed coats permeable to water and facilitating germination. Hullers and scarifiers usually abrade the seeds between two rubber-faced surfaces or fling seeds against roughened surfaces, such as sandpaper. The severity of abrasion or impact must be controlled accurately to prevent damage. Seeds with a high moisture content are harder to hull or scarify than seeds with less moisture. The seed moisture should be determined before hulling or scarifying, because a huller or scarifier adjusted for moist seeds may damage dry seeds. Based on seed moisture, necessary adjustments can be made. Hulling and scarification may be performed separately or jointly depending on the presence of unhulled or hard seed or both. Some seeds that may require hulling are Bermuda grass, Bahia grass, buffalo grass and koreagen, koloe, and common and bicolor lespedeza, whereas seeds that require scarification include wild winter peas, hairy indigo, alfalfa, crotalaria sub cloven, and suckling clover. Some seeds such as sweet clover, sericea, lespedeza, crown vetch, black medic and sour clover require both hulling and scarification.

Removal of groundnut seeds from their fibrous pods and separating maize kernels from their cobs, although different operations, are both known as shelling. In shelling maize, the cobs are pressed against revolving cylinders with projecting lugs, which detach the caryopses from the tough fibrous axis with a thumblike action. Delaying this operation until after drying can minimize mechanical damage to maize seeds. The testa of the groundnut kernel is too thin to provide any protection against mechanical damage, and the radicle is so prone to injury that groundnut shelling is mostly performed by hand, and in some countries, seed is even sold to farmers unshelled.

Scalpers are of many types. One type consists of a reel of perforated metal screen, which is inclined slightly and turns on a central shaft. Seeds fed into the higher end tumble inside the reel until they drop through the perforations, whereas larger particles continue through the reel and are discharged separately. Another type of scalper operates similarly with a single, flat, perforated screen that is shaken mechanically. These simple devices remove only large trash. An air/screen cleaner separates light chaff and dust with a controlled air current, large trash with a large-hole screen, and small foreign matter through a small-hole screen. Most scalpers are arranged to accomplish air separation before the seeds reach the screens.

Many native grasses and small grains produce seeds that have awns, beards, hairs, glumes, and other appendages, which make the difficult to handle in processing and planting operations, because they tend to interlock and cause undesirable clustering. They can be removed by precleaning treatments, which improve their flow properties and quality. Species of *Stipa* (needle grasses) and *Elymus* (wild rye) and grasses like bluestems and gramas are especially troublesome awned seeds.

High-speed threshers, hammer-mills, and debearding machines and tumbling pebble mills are used to remove awns and appendages. All employ a vigorous abrading action and must be operated carefully to ensure removal of awns without causing damage to the seed. Barley seed awns and fibrous tips of oats are removed by a machine with a similar abrading action (16).

Precleaning facilitates mechanical handling with more even subsequent flow of eed and reduces the time required for artificial or natural drying.

B.   Basic or Secondary Cleaning

Basic cleaning is a second stage of cleaning carried out with air blasts and vibrating screens and is applicable to all kinds of seeds. It is essentially the same as scalping but more refined, carrying the cleaning process a stage further. Basic cleaning is performed mostly by one machine known as an air/screen cleaner (Fig. 7). The air blast removes lighter material, and a series of screens separate particles larger and smaller than the crop seed, which may be seeds of weeds or other crops or broken seeds. An element of grading can be introduced at this stage by including screens that can separate seeds on the basis of their size. Screens to separate seeds according to their shapes also can be fitted. The success of this operation depends on selection of appropriate screens out of about 200 available screens and adjustment of the air blast to suit the species of seed being cleaned and the material that is to be removed.

Almost every kind of seed must be cleaned over an air/screen cleaner before any other separation can be attempted. Many kinds of seeds can be cleaned completely on this machine to a finished product. The air/screen cleaner, therefore, is known as the basic equipment in cleaning plants. In this unit, screens take advantage of a difference in size and shape and moving air senses a difference in surface area and density in seeds to effect separation.

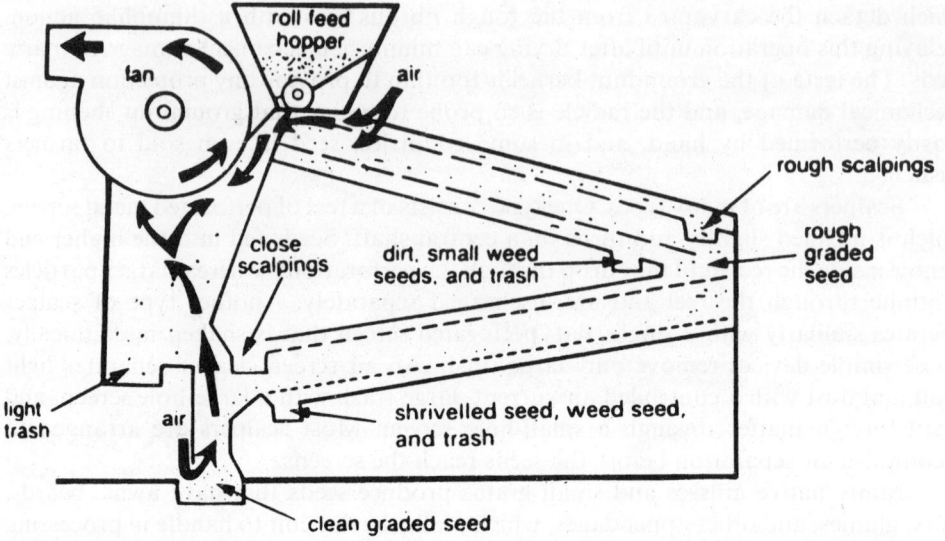

Figure 7   Air/screen cleaner. (From Ref. 3.)

A seed mixture obtained directly from the combine or from any of the precleaning units flows by gravity from a hopper to the feeder, which meters it into an air stream. Light, chaffy matter is blown out by air, and the remaining seeds are distributed uniformly over the top screen. In a typical operation of a four-screen machine, the top screen scalps large material, the second screen sizes or drops particles smaller than the seeds, the third screen scalps the seeds more closely, and the fourth screen performs final grading. The graded seeds then pass through a second air stream, which lifts light seeds and chaff into the trash bin and drops the plump, heavy crop seeds into a clean chute. The size of air/screen cleaners varies from the small, two-screen farm models to the modern precision unit that is arranged to use several top and bottom screens in one cleaning operation (15). Large cleaners, used in commercial seed-cleaning plants, subject seeds to as many as seven screens and three air separations in one pass through the machine, with capacities up to about 3000 kg of seed/hr.

Air/screen cleaners have been provided with perforated metal or wire mesh screens. The former are available with round, slotted, or triangular openings. Wire mesh screens are woven with square or rectangular openings. The top two screens generally have one round hole and one slotted screen. The first bottom screen is slotted and the second has round, square, or triangular holes. The screen perforation sizes most commonly used are given in Table 6.

Each screen is slightly slanted to enable the seed to roll or slide downward over the openings. The pitch of each screen is adjustable to facilitate accurate separation. Screens can be adjusted to shake seeds slowly or rapidly so that seeds slide smoothly over the screen or are agitated by the screen motion as desired. The feed rate of seed can also be adjusted to keep the screens operating at nearly full capacity. The airflow in each air separation is normally regulated with the help of dampers provided in the air ducts. Commercial machines have cleaning brushes that travel under the screens to

Table 6 Screen Perforation Sizes (mm) Commonly Used for Some Seed Crops

Crop	Removal of oversize material	Removal of undersize material
Cereals	20 × 4.5 or 20 × 4.0 (slot)	20 × 1.5 to 20 × 2.6 (slot)
	6.0 or 5.5 (round hole)	
Peas	9.5 (round hole)	25 × 5.5 (slot)
Rapeseed	3.5 (round hole)	20 × 1.2 (slot)
		1.2 (round hole)

Source: Ref. 10.

prevent seed from lodging in its openings, and some have mechanical screen bumpers to assist in dislodging seed.

The air/screen cleaners generally have a much higher capacity than the length-separation machines (indented cylinder and disc separator). The latter are usually mounted in groups of two or three in sequence so that different indent sizes can be used in the same cleaning line.

Kausal et al. (17) attempted to standardize the sieve sizes for grading sorghum, green gram, black gram, red gram, and soybean. About 10–11% good-quality seed can be saved by replacing the present grading sieve by 8/64-inch, (3.17 rom) and 7/64-inch (2.78 rom) sieves in sorghum and green gram. Germination, physical purity, and 100–seed weight were found to improve significantly with an increase in sieve size, whereas processing recovery in all crop varieties decreased. Kausal et al. (17) stressed the need for revision of sieve sizes for grading new hybrids and high-yielding varieties to minimize the processing cost as well as to meet the demand for good-quality seeds.

Pandya (18) reported on optimizing the capacity for desired grading efficiency of air/screen cleaner–grader by microprocessor technology. Since the proportion of undersize materials varies with seed lot and even within a seed lot, Pandya (18) recommended continuous adjustment of grading time available to seed according to the proportion of the undersize materials in the seed lot being processed to obtain accurate grading. Reprograming to suit any specific requirements of the breeder, foundation, and commercial seed processors can modify the system developed.

In a new ultrasound and computer vision technology, seed is subjected to sound waves, and a portion of the waves is transmitted, absorbed, or reflected. The frequency components of the sound wave are analyzed and correlated to various seed-quality parameters such as weight, size, mechanical damage, and seed health (19).

C. Separation and Upgrading Techniques

After the basic cleaning operation removes most of the impurities that can be removed by a simple combination of air blast and screens, some seed lots require further cleaning treatment to remove adulterants that have remained too close to the pure seed in size and shape to be separated by the air/screen cleaner.

Mechanized planting requires seed of a uniform size, which can be obtained by more precise sizing and grading techniques. Sizing is the operation of removing seeds larger or smaller than the required size using special machines. Grading consists of removal of cracked, damaged, or otherwise defective seeds, which may have reduced germinability and vigor.

Seeds of any one crop species are not uniform in physical characteristics and vary in size and weight. When two crop species are very similar, their dimensions may overlap, making separation difficult. Some of the pure seed often is separated along with the impurity, and seed loss of this type may be as high as 20% (3). The separation process, therefore, needs continuous scrutiny to regulate the fine adjustments provided in separating machines to ensure more efficient cleaning.

A seed has three dimensions—length (the greatest dimension), width (the greatest dimension of all possible cross sections taken at right angles to the length), and thickness (the greatest dimension of all possible cross sections taken at right angles to both length and width).

1. Separation Techniques Based on Length, Width, Thickness, and Shape

Ordinary sieves cannot separate seeds having a similar girth but different lengths (e.g., wheat and oats), because the longer seeds can be tilted from the horizontal position and fall. They can, however, be separated by an indented cylinder, which has numerous pockets in its inner surface. It revolves around a sloping axle and has a fixed trough along the upper half of its length. The seed mass fed into the higher end of the cylinder slowly descends to the lower end. As the cylinder revolves, seeds are carried upward in the pockets, and the short seeds carried into the upper half of the cylinder's rotation fall out of the indents into the trough, and the long seeds fall out in the lower part of the rotation, remaining in the cylinder. The short seeds are thus removed by way of the trough, whereas the long ones move along the length of the cylinder and emerge from the lower end. The indented cylinder can also be used to separate seeds of different girths, as shown in Figure 8.

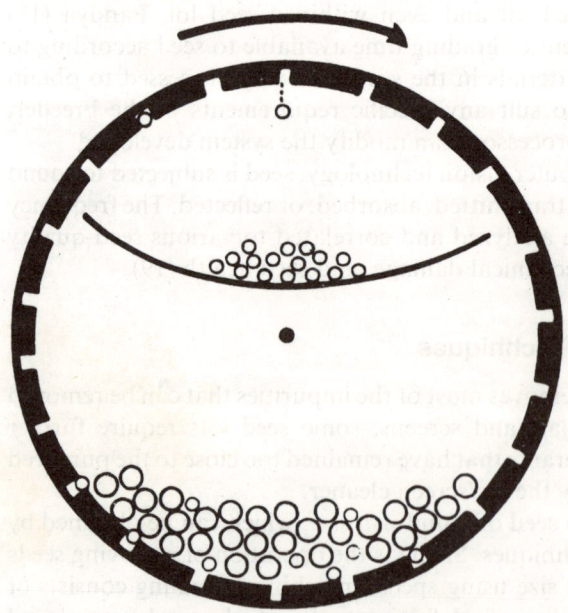

Figure 8 Indented cylinder. (From Ref. 3.)

The seeds of pulses damaged by weevils can also be separated by cylinders adapted to projecting pins (instead of indents) that can enter the holes left by the weevils and lift the damaged seeds out of the cylinder. As a seed impaled in this way enters into the upper half of the cylinder's revolution, it falls into the trough and is separated. The indented cylinder separator can separate slough grass from meadow foxtail, long barley from oats, and dock or sorrel from orchard grass or fescue. It can also be used for sizing oats and rice and grading hybrid corn by length (16).

A variant application of the same principle has pockets cut in the faces of discs revolving at 30–40 rpm inside a static cylinder. At the top of the revolution the short seeds are thrown out into shallow troughs placed parallel to the discs (Fig. 9).

Spiral separators can separate round seeds from flat seeds when the seeds when the seeds are fed on the top of a spiral chute; flat seeds slide down the chute relatively slowly and tend to follow the inner edge of the spiral, whereas the round ones roll down more quickly and can then be separated by the centrifugal force carrying them over to the outer edge. The purple moonflower seed, shaped like an orange segment, can be separated from the almost spherical seeds of soybean in this way.

Size grading uses techniques of separation by width and thickness in which holes with a certain cross section are set in screens or cylinders (e.g., rectangular holes for

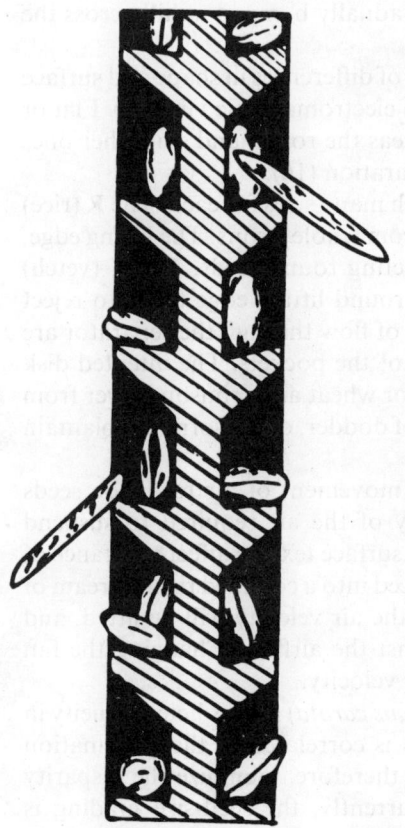

Figure 9 Pockets cut in the face of a Carter disc. (From Ref. 3.)

maize and round holes for barley). To achieve this, holes can be set in grooves for flat maize or in round pits for barley.

Indented disc and indented cylinder separators take advantage of a difference in seed length for separation. Both types use pockets or indentations to lift short seeds from a mixtue, rejecting the long ones. The indented disc separator consists of a series of discs revolving together on a horizontal shaft inside a close-fitting housing. Each disc has many pockets (cups) on each face. As the discs revolve through a seed mixture, the recessed pockets lift out short seeds and drop them in a trough at the side of the machine. The rejected long seeds are conveyed through the disc spokes to the end of the machine and discharged through the openings.

When seeds are to be graded by length, the mixture first encounters discs with small cups and then discs with progressively larger cups from inlet to discharge when only one separation is required. Many discs with the same indent size may be used in a machine to increase capacity.

The timothy bumper mill separates seeds on the basis of a difference in shape. It is a machine developed to remove weed seeds. The timothy seed mixture is fed onto the lower corner of the upper end of a small rectangular metal plate, which is inclined slightly in both directions, but more in the lengthwise direction. A back-and-forth movement of the plate, parallel with its short side, is smooth in one direction but is stopped with a bump in the other. Round timothy seeds roll downhill between bumps and fall off the end. Irregularly shaped seeds are gradually bumped uphill across the plate and fall off the side.

A vibrator separator divides seeds on the basis of differences in shape and surface texture. The deck in this machine is activated by an electromagnetic vibrator. Flat or rough seeds clim to the high side of the deck, whereas the round and smoother ones roll, tumble, and slide to the low side, effecting separation (16).

Disc pockets are made in two basic shapes, with many sizes for each. The R (rice) pocket was designed to remove broken rice grains from whole grains. The lifting edge, being flat, lifts out cross broken or flat seeds, rejecting round seeds. The V (vetch) pocket was designed to remove round seeds. The round lifting edge tends to reject tubular or elongated seeds. The disc speed and rate of flow through the separator are important adjustments to obtain proper emptying of the pockets. The intented disk separators can be used to separate vetch from oats or wheat and crimson clover from rye grass and fescue seed. Small particles like seeds of dodder, dock, sorrel, or plantain can also be separated from fescues and rye grass.

Pneumatic and aspirator separators use the movement of air to divide seeds according to their terminal velocities (the velocity of the air required to suspend particles in a rising air current). Density, shape, and surface texture affect resistance of a particle to airflow. When a seed mixture is introduced into a confined rising stream of air, all particles with a terminal velocity less than the air velocity will be lifted, and seeds with higher terminal velocities will full against the airflow. Changing the fan speed or the size of the air inlet can regulate the air velocity.

The variability of the size of carrot seeds (*Daucus carota*) causes heterogeneity in the crop. Some studies show that the size of seeds is correlated to the germination process. Having lots of seeds of the same size will, therefore, minimize the disparity among seeds during sowing and germination. Currently, the seed-size grading is evaluated by sieving, which takes a long time and can induce errors due to the precision and the accuracy of the sieves. Anouar et al. (19a) used a vision system to measure the size and shape of seeds. Compared to the sieving method, they evaluated the individual seed size, so the grading evaluation was more precise. Seed grading could be easily represented as a histogram. This technique offers an automated and reliable tool for seed-size grading. The results of this study are also likely to be useful and generalized to other species of seeds.

2. Separation by Surface Texture

Seeds with rough seed coats adhere to velvet cloth and can be separated from smooth seeds (e.g., rough weed seeds like dodder and cranesbill can be separated from smooth seeds of clovers and lucerne). The velvet roll separator classifies seeds according to a difference in the texture of the seed coat. The separator consists of two parallel, inclined velvet-covered rolls in contact with each other that revolve outwardly. When a seed mixture is fed onto the upper end of the rolls, the smooth seeds travel downhill betwen them and are discharged at the lower end. Rough-coated seeds caught in the velvet take a bouncing path between the shield and rolls and are thrown over the sides and separated.

Spiral separators divide seeds according to shape or the degree of the ability to roll. A mixture fed onto the spiral at the top slides or rolls down the inclined surface. The fast-rolling seeds gain speed and are thrown by centrifugal force into an outer housing, which directs them to a chute below. The sliding or slow-rolling seeds remain on the inner inclined surface and enter a second chute at the bottom, effecting separation.

The inclined draper separator senses a difference in shape and surface texture to separate seed on an inclined plane. A seed mixture is placed onto the center of an inclined draper belt traveling in an uphill direction. Round or smooth seeds, which roll or slide down the draper faster and drop off, are caught in one hopper. Flat and rough-coated seeds are carried to the top of the incline and dropped into a second hopper (Fig. 10).

The horizontal disc separator takes advantage of the difference in shape and surface texture to determine whether seeds slide or roll when subjected to centrifugal force.

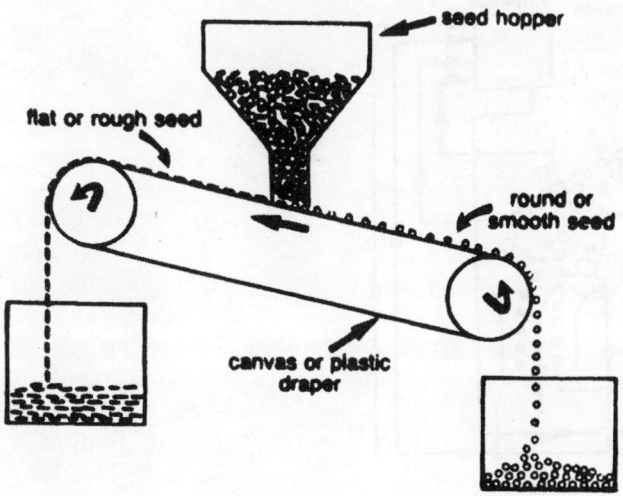

Figure 10   Inclined draper separator. (From Ref. 3.)

Magnetic separators also take advantage of the surface texture and stickiness of seed to effect a separation. A seed mixture and a proportioned amount of water and finely ground iron powder are mixed in a screw conveyor or other mixing device. In the presence of moisture, the powder will adhere to rough, cracked, and sticky seeds. When the mixture is fed onto the top of a horizontal revolving magnetic drum, smooth or slick seeds relatively free from powder fall from the drum in a normal manner and rough-textured or sticky seeds coated with iron powder are attracted by the magnetic drum and stick to it until removed by a rotary brush or a break in the magnetic field. Cleaning success depends largely on the extent of the difference in seed coats and proper mixing of water, iron powder, and seed. This is an excellent device for removing seeds of weeds like dodder, buckhorn, plantain, and mustard from alfalfa and clovers. Dirt and cracked seeds can also be removed.

3. Separation by Color

Color separators divide seeds on the basis of a difference in color or brightness. They ·are often used to remove discolored seeds of pulses, which might be diseased. In one type, the machine picks up the seeds on a series of suction fingers and carries them past a phototube where they are judged for color or brightness and ejected into a separate container one at a time (Fig. 11). In another type, the variant seed is given an electric charge; the seed then falls through an electric field, which diverts the charged seeds.

Color separators are practical only for larger seeds like those of beans and peas. They have not been used for smaller seeds because of the low capacity of the separator involved in scanning the particles individually.

4. Separation by Specific Gravity

Seeds of similar size and shape but slightly different in weight can be separated by specific gravity separators. These are often used to separate empty seeds of flax and various pulses from full seeds. The seed is spread in a layer over a perforated plate (or

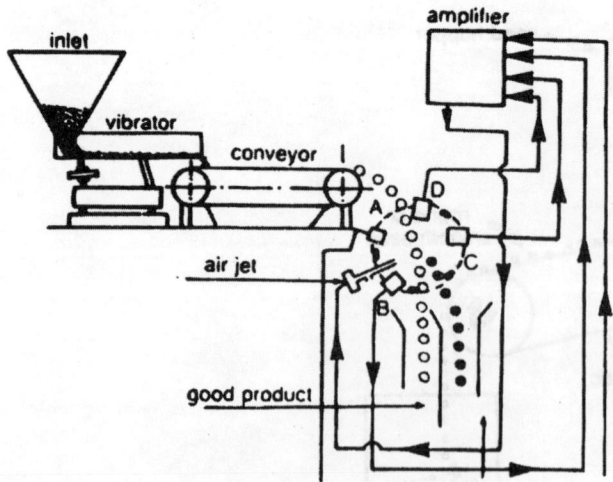

Figure 11   Electronic color separator (A, B, C, and D represent photo cells). (From Ref. 3.)

deck) set at a slope. A current of air, which is strong enough to lift the light seed slightly off the surface, is blown vertically through the deck, leaving the heavy seed on the deck. The deck shakes back and forth with sudden jerky movements, each jerk moving the heavy seed slightly up the slope. The lighter seed, floating on air and unaffected by the movement, tends to move down the slope under the influence of gravity, effecting separation.

The oscillating table separator consists of a table with zigzag partitions along the length. The table is inclined longitudinally and is shaken with a jerky movement from side to side. The seed flows slowly onto the table and tends to slide downward, but is rhythmically struck against the partitions by the lateral movements of the table. In addition to lateral forces, the seeds are subjected to two forces acting up and down the slope—the impact of diagonal faces of the partitions acting upward and the gravity acting downward. The impact of the partition is sufficient to throw a light seed across to the opposite diagonal, which gives it another thrust upward, causing the seed to move upward. The heavy seed, under the force of gravity, moves down the slope, missing the opposite diagonal and striking the next lower one. The heavy seed thus tends to move down the slope. The slope of the table and speed of oscillation must be adjusted according to the size of the seeds and differences between them to effect complete separation.

Harikumar et al. (20) studied *Gymnema sylvistre* seeds to standardize air pressure for seed density grading. The seeds were graded in an air blower with 0.5 inches of water pressure for 1 (T_1), 2 (T_2), and 3 (T_3) min. Among the grading levels, T_1 gave the highest values for recovery (88%), 100–seed weight (5751g), seed germination (47%), root and shoot length (4.43 and 4.875 cm, respectively), and vigor index (436). Jamun (*Syzygium cumini*) seeds lose their viability below 17% within a month under ambient conditions of Coimbatore, Tamil Nadu, India. Hence, attempts were made to upgrade the quality of seeds through physical grading during the period of deterioration. Seeds were collected at different storage periods after extraction (1, 2, and 3 weeks) and were graded through water flotation. The results revealed that seed quality decreases with an increase in seed age. The rate of reduction of seed germinability was 27, 32, and 69% at 1, 2, and 3 weeks after extraction, respectively, compared with fresh seeds that recorded 100% seed germination. At each age group, the performance of sinkers was higher in the separation of seeds through water flotation, suggesting the efficacy of the technique in increasing jamun seed quality (20a).

5. Separation by Electrical Conductivity

Electronic separators divide seeds on the basis of differences in their electrical properties. The degree of separation depends on the relative ability of seeds in a mixture to conduct electricity or to hold a surface charge. The seeds are passed through an electrical field immediately after being charged. The charged seeds are diverted, whereas seeds that lose the charge fall straight through. Alternatively, the seeds may fall onto a conveyor belt to which only the charged seeds adhere. Docks can be separated from clovers in this way.

Yadav and Gupta (21) have successfully designed and developed a viable seed-sorting machine, working on the basis of seed viability using an electrostatic charge method. The machine effectively sorts and counts the viable and nonviable seeds of bold, medium-, and small-sized seeds of soybean, maize, and okra. More extensive tests on various other crops are being conducted to make it a multicrop seed-sorting machine on the basis of seed viability; an important seed quality that can be used to prevent pests and diseases being spread to the fields through seeds.

D. Blending

Some variations have been noticed in seed sample taken from different bags within the same seed lot and even from different positions within a bag in respect to seed purity, weed content, germination capacity, and so forth. Since perfection is never achieved, some variation within a lot is acceptable, but investigations indicate that variation may be excessive in about 10% of the seed lots sold to farmers (3).

The variability within the crop from which the seed is harvested contributes significantly to the observed differences in the samples. Seed harvested from different parts of a field varies in maturity, germination capacity, weed content, and so forth. The seed lots from different fields or from different locations are put together without thorough mixing. Seed lots of any particular cultivar arriving at a processing plant differ greatly in size and viability.

The purpose of blending is to pool all the variable seed lots into a larger bulk and mix the constituents so thoroughly that when the composite lot is packaged, each bag has a uniform quality seed up to the required standard. The extent to which this can be achieved in practice depends on the diversity of the constituents, seed quality, and the method of blending. The greater the difference between the constituents, the smaller the chance of achieving uniformity. The blended seed lots should have maximum uniformity. A smaller seed lot is more likely to be uniformly mixed than a larger one. The international rules for seed testing prescribe a maximum seed lot size: 10 tons for seeds smaller than wheat and 20 tons for larger seeds.

Despite various machines and methods used for blending, uniformity is not always attained. Mixing involves movement of particles induced mechanically or by gravity. Introducing the components simultaneously into a bin and circulating the bulk two to three times between bins usually blend the large seed lots of one species. For smaller lots, the seed is run into a mixing chamber and stirred around by a revolving screw or by rotating the whole chamber. Although the latter is a more effective way of blending, revolving screws may damage seeds.

Experimentally designed blending bins have been found to be more effective. The bin is filled with the seed to be blended, and seed is drawn from different levels at the same time, so that seed from each level flows into another bin along with the seed from every other level. The four-side bin has an external duct running from top to bottom along with the center of each side (Fig. 12). In each duct, at equal intervals along its length, are circular ports through which seed flows from the bin into the duct. The bin is filled with the ports closed, which are then opened simultaneously to allow the seed to run down into a hopper at the base of the bin.

From the hopper, the seed is carried to a second similar bin to repeat the operation. After the seed lot has passed three blending bins, the composite lot is sufficiently uniform in quality.

According to Thomson (3), sometimes a mixture of seeds of different crop species or of different cultivars is required by the farmer for sowing. Because the physical properties of the seeds of different species differ, seeds are liable to segregate after blending. The contents of each bag are, therefore, mixed independently in a small mixing chamber. The constituents held in large storage bins are measured out

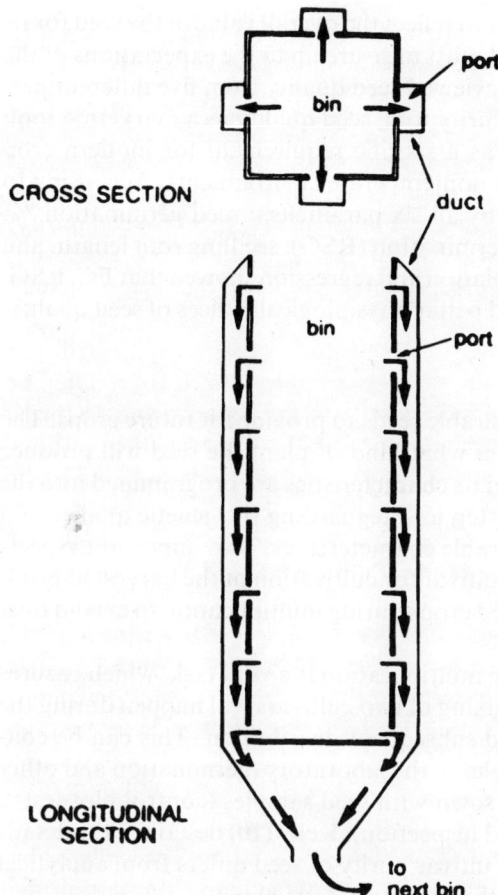

CROSS SECTION

LONGITUDINAL
SECTION

Figure 12 The blending bin. (From Ref. 3.)

electronically in required proportions into the mixing chamber. The mixed seed is then transferred automatically to a bag.

V. WHAT IS SEED QUALITY?

Thomson (3) described seed quality as consisting of several aspects, such as seed health, germination capacity, analytical purity, cultivar purity, freedom from weeds, seed vigor, and uniformity. According to Kelly (10), the genetic purity of the seed is the most important aspect of its quality, because it ensures that the plants making up the crop possess the desired characteristics.

In addition to having high viability, germination rate, and vigor, high-quality seeds must also have good storage quality to ensure that it maintains their characteristics until they is sown.

If quality is defined as the degree or standard of excellence, then seed quality can be viewed as a standard of excellence in certain characters or attributes that will determine the performance of the seed when sown or stored (22). In practice, the expression of "seed quality" is used loosely to reflect the overall value of the seed for its intended purpose, the performance of seed must measure up to the expectations of the end user of that seed. Hampton (22) has reviewed seed quality from five different perspectives: seed quality as a misused biosecurity tool, seed quality as a marketing tool; seed quality as a nonsense, seed quality as a specific requirement for modern crop production; and seed quality as needed in nontemperate environments. According to Jain et al. (22a), seed quality is reflected by all six parameters: seed germination %, electrical conductivity (EC), rate of seed germination (RSG), seedling root length, and seedling shoot/coleoptile length. The correlation and regression showed that EC, RSG, and seedling root length were sensitive and better physiological indices of seed quality.

A. Genetic Purity

The seed producers aim at multiplying desirable seeds to provide for future crops. The genetic background of the seed determines what kind of plant the seed will produce within the next crop. The kind of plant and its characteristics are programmed into the seed at the time of fertilization. The first step in safeguarding the genetic quality of a seed is to identify cultivars that show desirable characteristics. Two important aspects of cultivar purity are (a) the value of the cultivar for cultivation of the harvested product and (b) identification of seed lots or seed crops during multiplication to extend over several generations (10).

Assessment of cultivar purity during multiplication is a vital task, which ensures the seed producers that nothing such as mixing of two cultivars will happen during the growing, harvesting, seed processing, and subsequent distribution. This can be controlled in three ways: (a) testing seed samples in the laboratory (germination and other tests), (b) testing growing plants in plots sown with seed samples (control plot tests), and (c) inspecting growing seed crops (field inspection). Kelly (10) described the details of assessing the cultivar purity of seed. (Cultivar purity of seed differs from analytical purity in that it cannot be determined with laboratory tests alone.)

B. Analytical Purity

The analytical purity of seed is an indication of how much of the material in the bag is actually intact seed of the species named on the label. This can be determined by analyzing a small quantity of the seed in the laboratory. During analysis, the impurities are separated from the pure seed of the named species. The impurities include seeds of other crop species, weed seeds, and inert matter such as, for example, broken seeds, chaff, plant debris, and soil particles. After separation, the pure seed is weighed, and analytical purity is expressed as the percentage by weight of the whole sample. It is not sufficient, however, to state only the percentage of analytical purity, because the nature of impurities is also vital. A high purity percentage depends on the success of the cleaning operations after harvest.

Isoenzymes are valuable markers for cultivar identification and varietal purity tests in seed lots. Choer et al. (22b) investigated the possibility of using these markers in *Phaseolus vulgaris*. Horizontal polyacrylamide gel electrophoresis (PAGE) was used with different buffer and staining systems, extraction techniques, gel concentrations, and types to analyze esterase, leucine aminopeptidase, aspartate aminotransferase, and 6-phosphogluconate dehydrogenase isoenzymes from seeds. Cv. BR-IPAGRO 1-Macanudo did not show a definite isoenzymatic pattern. It was concluded that only esterase patterns (pH 8.3 and 6.5) allowed cultivar identification and detection of intravarietal polymorphism. Only the cultivars Irai, Carioca, and Rio Tibagi showed 100% purity.

C. Germination Capacity and Vigor

High analytical purity means nothing if the seeds do not germinate and produce vigorous seedlings in the field. The germination capacity of the seed lot is the percentage by number of pure seeds that produce normal seedlings in a laboratory test. Weak and abnormal seedlings are rejected. This figure indicates the potential of a lot for establishing seedlings under field conditions. According to Thomson (3), analytical purity and germination capacity can be combined and expressed as pure live seed, which is calculated by the formula:

$$\frac{\% \text{ Analytical purity} \times \% \text{ germination capacity}}{100}$$

This indicates the proportion of the seed lot consisting of seeds of the named species capable of germinating to produce robust seedling.

Whereas the germination capacity of a seed lot indicates its ability to establish seedlings in good field conditions, seed vigor indicates its ability to do so in poor growth conditions. The germination figure may, therefore, include seeds of insufficient vigor for good farm establishment. High germination capacity may be associated with high vigor, but it does not follow that seed lots with the same high germination capacity will have equal vigor.

Seed vigor has no precise physiological definition, although some useful vigor tests have been developed for certain crop species using physiological properties; for example, the cold test for maize and the conductivity test for peas.

Vigor can be affected by damage to the embryo incurred during harvesting or seed processing. Other factors include the environment and nutrition of the mother plant, stage of maturity at harvest, seed size, senescence, and pathogens. Low seed vigor may be associated with changes in the permeability of cell membranes arising from either immaturity at harvest or poor storage conditions. Seedling vigor is a genetic character often expressed by the hybrid cultivars (3).

Steiner and Stahl (23) determined seed vigor of hybrid varieties, population varieties, cytoplasmic male sterility (CMS) single cross lines, and CMS inbred lines of rye by a controlled deterioration technique. Seeds were deteriorated at 45°C for 24 hr at 20% m.c. Vigor decreased in the following order: hybrid varieties > population varieties > CMS single cross lines > CMS inbred lines. This vigor rating corresponded with the previously determined storage potential rating of the same seed lots. Therefore, in rye, vigor determination by controlled deterioration testing can be used to indicate the storage potential of seeds. Veselova (24) subjected pea seeds to accelerated aging for 10 days at 40°C and 80% RH. An aging-dependent increase in seed heterogeneity followed from greater dispersion of the average values of the germination capacity, room temperature phosphorescence of air-dried seeds (an indicator of seed moisture content), and electrolyte leakage (an indicator of seed intactness). Veselova examined the seed distribution by the level of phosphorescence and electrolyte leakage, as well as the distribution of seedlings by the length of their axial organs. The increase in seed heterogeneity upon aging was caused by the appearance of a new fraction in an initially uniform seed lot. The air-dried seeds of this fraction had a lower moisture content and differed from other seeds by twofold phosphorescence intensity and a twofold electrolyte leakage. These seeds germinated earlier but grew slower and gave rise to many abnormal seedlings. This new fraction contained seeds of low vigor. The vigor of an individual seed, that is, the ability of seed to produce a normal seedling, supposedly declines in a stepwise manner. The gradual vigor reduction in the total seed population upon aging seems to result from stepwise transitions of individual seeds to a subpopulation with low vigor but from a continuous vigor decline in all seeds. Thus, the phosphorescence of air-dried seeds can be used as a tool to descriminate low-vigor seeds (24).

Parmeswari et al. (24a) graded tamarind seeds into yellowish brown, brown, and blackish brown categories and tested for germination and seedling growth. Brown seeds exhibited the highest germination percentage and vigor and greatest seedling growth followed by the blackish brown seeds.

D. Seed Size and Uniformity

Seed size includes actual size and uniformity of size. The harvested seed lots have seeds of different sizes, the differences arising from seeds harvested from different plants or due to differences in the maturity of seeds borne on the same plant.

Large seed size is an indication of vigor. Seeds may be embryonic alone or may also contain endosperm ready to be absorbed into the developing seedling. Hence, the bigger-sized seeds normally produce larger seedlings with a larger area of green leaf capable of photosynthesis. The seedling's emergence from the soil is also facilitated by the larger seed size. Compared with small seeds, large seeds produce seedlings that grow more readily in the field, and a greater proportion of them emerge through the soil surface. Very small, shriveled seeds have no practical planting value. (Seed size is normally expressed as the weight of 1000 seeds.)

Studies were conducted on yam bean (*Pachyrhizus erosus*) for the selection of the best seed quality for further handling of the seed with reduced seed polymorphism. Bulk seeds of yam bean were size graded using 8.20-, 7.8-, 7.41-, 7.02-, and 6.63-mm round, perforated metal sieves and were designated as 21R, 20R, 19R, 18R, and 17R, respectively. The seeds that were passed through 6.63-mm round, perforated metal sieves were designated as 17P and bulk seeds were designated as UG (ungraded). In a related experiment, another set of seeds was acceleratedly aged at 40 + 1°C and 100% RH for 7 days. Seed recovery was maximum in 20R (31.46%) and was followed by 18R (25.7%) and 19R (22.96%), whereas the undersized seed recovery (17P) was 12%. The 100–seed weight was positively correlated with seed weight, which showed a decreasing order from 21R to 17R. A linear relationship was also evident with size and germination wherein the 21R seeds recorded the highest germination (97%) and reduced to 95, 94, 90, 85, and 80%, respectively, with 20R, 19R, 18R, 17R, and 17P seeds. In the accelerated aging test, large seeds performed better in maintaining higher variability and vigor in storage than the smaller seeds. The seed-quality characters of different size grades after accelerated aging revealed higher rates of reduction in seed-quality performance with reduction in seed size (24b).

Uniformity of seed size influences the effectiveness of seed-cleaning operations, which were dealt with in previous sections of this chapter. In addition to size uniformity, uniformity of the seed lot is also important. Every seed lot is to a certain extent a mixture of pure seed, inert matter, crop seeds, and weed seeds, as well as some live and dead seeds. Seed lots, therefore, need thorough blending before the seed is packaged. Bonow et al. (25) used isoenzymes to compare the efficiency and cost of PAGE and isoelectric focusing to identify the genotypes and to detect possible seed mixtures in irrigated red and black hulled ecotypes and to observe intercultivar and intracultivar variation. Both techniques were efficient to detect varietal mixtures. Owing to the high cost of isoelectric focusing, this technique should only be used if PAGE was not efficient to differentiate genotypes. Pamuk et al. (26) developed a method to evaluate the thickness and uniformity of polymer coating (Cepril) on pine seeds, using scanning electron microscopy (SEM). This method is especially thought to be useful for evaluating seed coating uniformity and thickness when the intention of the coating is to control sed hydration.

E. Seed Health and Moisture Content

Seed health is vital in controlling crop diseases and ensuring good field establishment. Parasites and disease-producing organisms may survive the dead season between crops, carrying the infection from one year to the next throughthe seeds (seedborne diseases). The smut and bunt diseases of cereals do not produce spores or any symptoms of diseases until the flowering stage. Many flat-stripe and root-rot diseases of crops are also borne through seeds. Certain viruses are carried within the embryos of infected seeds, causing mosaics on mature plants. The seedborne diseases can be controlled by chemical treatment of the seed. However, sowing seeds harvested from healthy crops better prevents these diseases. Many seed-certification schemes impose standards of seed health. The laboratory tests identify the pathogens present and estimate the percentage of seeds infected.

The moisture content of the seed is more important to the processor and the store manager than to the farmer. It is the key parameter in determining whether or not seed will retain its germination from harvest to the next sowing. It can be easily measured in a laboratory or, more quickly but less accurately, on the spot by moisture meters.

Thomson (3) concluded that a high-quality seed should have the following characteristics:

1. High analytical species and cultivar purity
2. Freedom from weeds
3. High germination capacity and vigor
4. Uniformly large size
5. Freedom from seedborne diseases
6. Fairly low moisture content

REFERENCES

1. U. S. Deshpande and S. S. Deshpande, Legumes, *Foods of Plant Origin: Production, Technology and Human Nutrition* (D. K. Salunkhe and S. S. Deshpande, eds.), Van Nostrand Reinhold, New York, 1991, p. 137.
2. L. O. Copeland and M. B. McDonald, *Principles of Seed Science and Technology*, 3rd ed., Chapman & Hall, New York, 1995.
3. J. R. Thomson, *An Introduction to Seed Technology*, Leonard Hill, London, 1979.

4. R. L. Agrawal, *Seed Technology*, Oxford & IBH Publishing, New Delhi, 1980.
5. N. R. Brandenberg, Why and how are seeds dried? *Seeds: The Yearbook of Agriculture*, US Department of Agriculture, Washington, DC, 1961, p. 295.
6. J. F. Harrington, *Seed Ecology* (W. Heydecker, ed.). Cited by R.L. Agrawal (Ref. 4).
7. M. Misra, Maintaining soybean seed quality during conditioning, *Seed Research*, Special Vol No. 2 (S. P. Sharma, ed.), Indian Society of Seed Technology, New Delhi, 1993, p. 841.
7a. W. M. Nascimento and S. M. West, Drying during muskmelon (*Cucumis melo* L.) seed priming and its effects on seed germination and deterioration, *Seed Sci Technol*. 28: 211 (2000).
8. J. E. Douglas, *Successful* Seed *Programs: A Planning and Management Guide*, West view Press, Boulder, CO, 1980.
9. B. R. Gregg, A. G. Law, S. S. Virdi, and J. S. Balis, *Seed Processing*, Mississippi State University, National Seed Corporation and United States Agency for International Development, New Delhi, 1970.
10. A. F. Kelly, *Seed Production of Agricultural Crops*, Longman Scientific & Technical, New York, 1988.
11. ISTA, Drying and storage, *Proc. Int. Seed Testing Assoc*. 28 (4) (1963).
12. C. E. Vaughan, B. R. Gregg, and J. C. Delouche, *Seed Processing and Handling Handbook*, No. 1, Seed Technology Laboratory, Mississippi State University, State College, MS, 1967.
13. J. F. Harrington, Drying, storing and packaging seeds to maintain germination and vigor, *Seedsmen's Digest* 11(1): 16 (1960).
14. R. K. Matthews, G. B. Welch, J. C. Delouche, and G. M. Dougherty, Drying, processing and storage of com seed in tropical and subtropical regions, Proc. Am. Soc. Agric. Eng. St. Joseph, Mich. Paper; Nos. 69–77, 1969.
15. L. M. Klein, J. Henderson, and A. D. Stoesz, Equipment for cleaning seeds, *Seeds: Yearbook of Agriculture*, U.S. Department of Agriculture Washington, DC, 1961, p. 307.
16. J. E. Harmond, N. R. Brandenberg, and L. M. Klein, *Mechanical Seed Cleaning and Handling: Agriculture Handbook* No. 354, ARS, US Department of Agriculture in Cooperation with Oregon Agricultural Experiment Station, Washington, DC, 1968.
16a. K. Kant and S. R. S. Tomar, Effect of seed size on germination, vigor and field emergence in mustard (*Brassica juncea* L.) cv. Pusa Bold, *Seed Res*. 23: 40 (1995).
17. R. T. Kausal, S. P. Changade, and U. N. Patil, Standardization of sieve sizes for grading crop seeds, Seed *Research*, Special Vol. 2 (S. P. Sharma, ed.), Indian Society of Seed Technology, New Delhi, 1993, p. 826.
18. L. S. Pandya, Optimizing the throughput capacity for desired, efficiency of air screen cleaner-cum- grader by microprocessor technology, Seed *Research* Special Vol. 2 (S. P. Sharma, ed.), Indian Society of Seed Technology, New Delhi, 1993, p. 833.
19. M. K. Misra, B. Koerner, and Y. Shyy, Ultrasound and computer vision technology for determining seed quality. Seed *Research*, Special Vol: No. 2 (S. P. Sharma, ed.), Indian Society of Seed Technology, New Delhi, 1993, p. 809.
19a. F. Anouar, M. R. Mannino, M. L. Casals, J. A. Fougereux, and D. Demilly, Carrot seed grading using a vision system, *Seed Sci. Technol*. 29: 215 (2001).
20. C. Harikumar, K. Malarkodi and P. Srimathi, Density grading on seed quality and seed recovery *Gymnema sylvistre*, *Madras Agric. J*. 87: 166 (2001).
20a. P. srimathi, K. Malarkodi, K. Parameswari, and G. Sasthri, Water floatation technique to upgrade the quality of jamun (*Syzyqium cumini*) seeds, *Progr. Hort*. 31: 20 (1999).
21. B. G. Yadav and C. P. Gupta, Viable seed separator, *Seed Research*, Special Vol. 2 (S. P. Sharma, ed.), Indian Society of Seed Technology, New Delhi, 1993, p. 899.
22. J.G.Hampton, What is seed quality? *Seed Sci. Technol*. 30: 19 (2002).

22a. S. K. Jain, A. Sharma, and A. K. Singh, Electrical conductivity, rate of seed germination and seedling root length are better physiological indices of seed quality, XI National Seminar on Quality Seed to Enhance Agricultural Productivity, UAS, Dharwar, Jan, 18–20, 2002, *Seed Tech News* 32: (A. Gaur, A.K. Vari, J.L. Varshney, and K. Kant, eds.), Indian Society Seed Technology, Division of Seed Science and Technology, IARI, New Delhi, March 2002, p. 82.

22b. E. Choer, E. Augustin, I. F. Antunes, J. B. da Silva, and J. A. Peters, Cultivar identification and genetic purity analysis of beans (*Phaseolus vulgaris* L.) through isoenzymatic patterns, *Seed Sci. Technol.* 27: 517 (1999).

23.   A. M. Steiner and M. Stahl, Vigor rating of rye varietal categories (*Secale cereale* L.) using controlled deterioration testing, *Seed Sci. Technol.* 30: 219 (2202).

24.   T. V. Veselova, Assessment of individual seed vigor and seed lot heterogeneity by room temperature phosphorescence, *Seed Sci. Technol.* 30: 187 (2002).

24a. K. Parmeswari, P. Srimathi, K. Malarkodi, Chintala-Suresh and C.Suresh, A note on the influence of seed coat color on seed quality of tamarind (*Tamarindus indicus*), *Orissa J.Hortic.* 27: 97 (1999).

24b. P. Srimathi, K. Malarkodi, R. Geetha and K. N. Navamaniraj, Influence of size grading and standardization of sieve size for seed selection in yam bean (*Pachyrihizus erosus*), *Orissa J.Hort.* 28: 45 (2000).

25.   S. Bonow, E. Angustin, D. F. Franco and A. L. S. Terres, Detection of varietal mixture in rice through isoenzyme, *Seed Sci. Technol.* 30: 197 (2002).

26.   G. S. Pamuk, T. Olsson, U. Bergsten and H. Lindberg, Evaluation of polymer coating on Scots pine (*Pinus sylvestris*) seeds using scanning electron microscopy (SEM), *Seed Sci. Technol.* 30: 167 (2002).

15
Seed Treatment

I. INTRODUCTION

After cleaning, seed must be treated for several different purposes:

1. Seed disinfection/disinfestation to combat seedborne diseases and insect pests
2. Protection of seeds against diseases and pests that may be present in soil or be airborne when seedlings emerge
3. Specialized seed treatments such as, for example, coating, pelleting, scarification, decortication, irradiation, blending, and delinting (cotton), to protect seeds against pests or aid in germination

II. DISINFECTION AND DISINFESTATION

Seed treatment commonly refers to the application of pesticides (fungicides, insecticides, or a combination of both) to seeds to disinfect and disinfest them from various seedborne and soilborne pathogenic organisms and storage insect pests. The term *disinfection* refers to the eradication of fungal spores established within the seed coat or in the inner tissues, whereas the term *disinfestation* is generally used to mean destruction of surface organisms (fungi, bacteria, insects) that have contaminated but not infected the seed surface. Simple chemical dips, soaks, and fungicides applied as dust, slurry, or liquid have been found to be quite satisfactory for this purpose. The major diseases and insect pests of seeds that can be controlled by seed treatment are (1):

1. Systemic diseases infecting the seed during harvesting or storage; for example, bunt or stinking smut of wheat, loose kernel or covered kernel smuts, Helminthosporium stripe of barley, loose and covered smut of oats, head and kernel smut of rye, loose kernel and covered kernel smuts of sorghum, and kernel and head smuts of millets. Appropriate seed treatment can prevent the spread of these diseases.
2. Systemic diseases that infest seeds during the flowering stage to become established within the seed and the resulting plant; for example, loose smuts of barley and wheat. These diseases are generally treated with fungicides, like Vitavax, to check their spreads.
3. Nonsystemic diseases that infest seed during harvesting and storage; for example, *Helminthosporium* spots; blotches or blights of barley, oats, rice, rye, sorghum, and wheat; bacterial blights of barley, oats, and sorghum;

anthracnose; bacterial and *Aschochyta* blights of cotton, rust; and anthracnose and pasmo disease of flax.

4. Seed rots and seedling blights—*Rhizoctonia* and *Pythium* spp.—present in soil may rot the seed before germination or may kill young seedlings before their emergence. Appropriate seed treatment forms a protective coating around the seed and acts as a barrier to ward off attack by seedborne and soilborne organisms.

5. Storage and soil insects—Most storage insects (weevils, beetles, grain borers, bruchids, and moths) and certain soil insects like wireworms and seed corn maggots can be effectively controlled by seed treatment; for example, a combination of insecticide and fungicide as a means of giving limited protection to the seed and seedling until it becomes resistant to attack or can survive limited attacks.

A. Essential Situations Needing Seed Treatment

Agrawal (1) listed the following situations essentially requiring seed treatments:

1. Injured seed coats provide an excellent opportunity for fungi to enter the seed, which can eventually kill the seed or weaken the seedling. Seeds suffering physical or mechanical injuries during harvesting and postharvest handling are also prone to fungal attack and are required to be protected by seed treatment.

2. Seed may be infested by disease at the time of harvest or may become infected or infested during its processing.

3. Seeds must occasionally be planted under unfavorable conditions, like cold, damp soils, or extremely dry soils, which favor the growth and development of certain fungal spores that may attack and damage seed sown under these conditions.

4. Seed treatment also provides good ensurance against diseases and soilborne organisms, affording protection to seeds during germination and early establishment. Seed treatment often improves germination and crop establishment by controlling surface molds. The latter are not generally considered to be pathogenic, but may infest seed following moist harvesting and storage conditions.

According to Copeland and McDonald (2), seed treatment is a sound agronomic practice and a routine part of seed conditioning to reduce, control, or repel seedborne, soilborne, or airborne pathogens causing various diseases in field and garden crops. The history of seed treatment dates back to the seventeenth century, when salt water was accidentally discovered to control bunt and stinking smut of wheat seed. In 1755, Matthieu du Tillet, a French botanist, recommended the use of lye and lime as a chemical treatment for wheat seed. Some 50 years later, Prevost, a Swiss botanist, discovered the use of copper fungicides (copper sulfate) as seed treatment. Newer concepts of treating seeds with organomercurials like caresan and semesan were introduced in 1920. These and several other toxic pesticides have been recognized to pollute the environment, necessitating other effective and safer measures for treating seeds (2).

B. Methods of Seed Treatments

Seed treatments are used to prevent or reduce losses from diseases caused by organisms associated with seed or present in the soil. These organisms may be associated with seed in the form of sclerotia, smut balls, nematode galls, and infested plant parts. The pathogens may be present in or on seeds.

Hanson et al. (3) divided seed-treatment methods into three groups; mechanical methods, physical methods, and chemical methods.

1. Mechanical Methods

Mechanical methods of seed treatment are designed to remove infectious materials mixed with seeds. Seeds can be mechanically cleaned thoroughly before seedling to remove most pathogenic organisms from the seed surfaces. Mechanically treated seed is not completely free from pathogens and often requires further treatment.

2. Physical Methods

Physical methods of seed treatment are used primarily to kill pathogens rooted deep into the seeds. Some, such as loose smuts of barley and wheat, cannot be inactivated effectively in any other way. Physical methods include hot-water and water-soak treatments and ultraviolet, infrared, x-ray, and other types of irradiation. Dry heat also has been used occasionally. However, only the hot-water and water-soak treatments have been shown to be more practical. Physical methods, however, do not protect seeds against soilborne organisms; they are effective only against pathogens present on or in the seeds.

Although hot-water treatment was the most commonly recommended physical method before the 1950s, it has not been used extensively because of difficulties encountered in controlling the temperature and duration of treatment and with little margin of safety. The method also requires adequate supplies of steam or hot water, accurate thermometers, water tanks or vats, and drying facilities. The use of this method, therefore, has been restricted mostly to disinfecting small seed lots and batches of small-seed crops that require low seeding rates per acre (e.g., grasses). Modified water-soak methods are safer and less critical in their requirements than the traditional hot-water treatments. These have been used mostly to control loose smuts of wheat and barley, but are also effective against other pathogens.

In all water-soak methods, seeds are soaked in water for about 2 hrs and subsequently kept under anaerobic or near-anaerobic conditions for 1 or more days. In some cases, seeds are soaked for 64 hrs in water at about 72°F (22.2°C) and then dried; sometimes seeds are soaked for only 2 hrs and then placed in airtight containers at 80°F (26.3°C) for 48 hr before being dried. Other effective modifications differ in the temperatures used and the duration of the treatment. The higher the temperature, the shorter is the time required. Crop varieties differ considerably in their sensitivity to injury from soaking for longer periods, which can be reduced by adding either 1% common salt or 0.2% Vaccine 51 to the soaking water.

Tawfik and Noga (3a) showed that priming of cumin seeds with water was a satisfactory alternative to osmotic priming using polyethylene glycol (PEG). Presowing priming for 3 days with water and planting the primed seeds within 30 days after initial drying for 2 days is beneficial. Rowse (3b) described a new seed-priming method in which seeds were hydrated to a predetermined water content over a 1-day period by placing them inside a horizontal rotating drum into which water vapor was released. The drum was mounted on an electronic balance, linked to a computer which monitored the seed water content, and controlled the production of water vapor so that at no time were the seeds visibly wet. Hydrated seeds were kept in a rotating drum for a further 14 days before being either sown immediately (nondried seed) or dried to facilitate storage. Drum-primed seeds of 10 leek cultivars germinated faster and more uniformly than seeds primed for the same period by placing them on paper saturated with a solution of PEG, MW 20,000 with an osmotic potential of -1.5 MPa. A similar difference was observed when seeds primed by both methods were germinated at all combinations of four temperatures (5, 10, 15, and 20°C) and five osmotic potentials (0, 0.1, −0.2, −0.3, −0.4 MPa). Drum priming enables large quantities of seed to be primed without the use of large quantities of PEG solutions.

3. Chemical Methods

Chemical methods of seed treatment are the most commonly used means of treating seeds because of their effectiveness and ease of handling. Several excellent chemicals are available for seed treatment. Organic fungicides are used more than inorganic ones, although the latter are preferred for specific purposes. Fungicides can be applied as dusts, liquids, or suspensions with or without equipment. The recommended dosages vary with the chemical, the crop, the length of storage period after treatment, and the method of application. Wet treatments involve soaking seeds in a water solution of a fungicide for a prescribed time, after which the seeds are removed and dried before they are used or stored. This method is time consuming, needing extra work and space for drying. Today, wet treatments are applied mostly by the slurry method or by quick-wet procedures in which no drying is necessary, because the treatments add less than 1% water to the seeds. In the slurry method, seeds are completely coated with a thick suspension of the chemical in water. A special machine, a slurry treater, applies the suspension. This method eliminates flying dust during treatment and is safer. Also, more accurate and uniform dosages of chemicals can be applied to most kinds of seeds. In the quick-wet method, a concentrated solution of a volatile fungicide is added to the seeds and thoroughly mixed with them (e.g., Panogen 75 and Ceresan 75). The use of volatile liquid fungicides has increased since the 1950s, especially for treating small grains like sorghum, cotton, flax, and rice.

Pelleting is another method of applying chemicals and is mostly used as a protectant against soilborne organisms and as a repellent against birds and rodents. The seeds of pine and other conifers are particularly treated this way. It is also used to treat seeds of other crops like onion to control smuts. Veasey and Teixeira De Freitas (3c) recommended the concentrated sulfuric acid treatment for 40 min. or, as a cheaper and safer alternative, sandpaper scarification to treat seed dormancy of *Sesbania sesban*, *S. rostrata*, and *S. virgata*.

Cowpea, black gram, green gram, and pigeon pea seeds dried to safe moisture levels were treated with a mixture of $CaOCl_2$ and $CaCO_3$ (3 g kg^{-1}) and kept for 1 week before artificial aging. Seed treatment increased germination, seedling growth, dry matter production, and vigor index in all the pulse seeds studied. There was also a significant reduction in bruchid infestation in treated seeds compared with controls (3d).

Takaki and Gama (3e) described the effect of scarification with sodium hypochlorite on light sensitivity in seeds of lettuce cv. Grand Rapids. Normally light-

requiring lettuce seeds germinated in both dark and continuous light after scarification. After preincubation at 36°C, scarified seeds behaved similarly to intact seeds,
with far-red light inhibiting germination and germination being poor in the dark.
Dose-response curves indicated that chemical scarification induced a change in the
control of seed germination from the low fluence response to the very low fluence
response. Preincubation at 36°C returned the control to the low fluence response of
phytochrome action.

4. Other Methods

Apart from mechanical, physical, and chemical methods, still other methods are used
to control certain diseases. The best method of destroying the pathogen causing
bacterial canker of tomato is to ferment seeds and pulp at about 70°F (21.1°C) for 72 hr
before extraction. Storage of seed for one or more seasons destroys some pathogens;
for example, the fungus causing late blight of celery and the virus causing mosaic of
tomato. Certain insecticides applied to seeds may increase the need for fungicidal
treatment, because they tend to predispose seeds and young seedlings to attack by soil
fungi. However, some compatible fungicidal and insecticidal treatments such as
captan, dieldrin, and thiram-dieldrin are also available.

Siddiqui and Agrawal (4) brought to light a new material for seed treatment,
ABCD (attapulgite-based clay dust), developed at the Indian Institute of Chemical
Technology (CSIR, Hyderabad, India) for seed treatment and safe storage of seeds and
food grains. This nonconventional dust was eveloped especially by modifying attapulgite clay for antifungal and antipest treatment of seeds. Its efficiency was tested on a
variety of seeds and seedborne diseases in comparison with organic fungicides and
pesticides for the control of bioinfection and maintenance of high seed germination.
ABCD specifically controlled *Fusarium, Alternaria, Drechslera, Curvularia, Macro-
phomina phesolina, Aspergillus flavus,* and *Phoma.* It was also very effective in
controlling the major stored grain pests, causing almost 100% mortality of the
bruchids and sustaining its effect through 1 year of the prestorage treatment of seeds.
Of all the treatments tested, ABCD was unique in providing 0% oviposition. It
protected sorghum, wheat, paddy, groundnut, and pulses for over 10 months with the
least damage to seeds. Data presented in Table 1 show that the efficacy of ABCD was at
a par with the best fungicidal treatment, that is, bavistin + thiram (1:1), in maintaining
high germination (76–78%) and lowering the infection level of fungi (1–5%). ABCD
was also found to be effective in reducing seedborne infection of soybean to about 5%,
maintaining a high level of seed germination (>85%) (Table 2).

C. Chemicals Used for Seed Treatment

Various types of chemicals used for seed treatment may be classified as organic or
inorganic mercurial or nonmercurial and metallic or nonmetallic depending upon their
chemical nature. They can also be grouped as fungicides, bactericides, nematicides,
insecticides, and so forth, based on the pathogen they control.

The ideal chemical would be the one that is highly effective in controlling the
pathogen, harmless to the seed (even at higher than recommended dosages), economic
to use, easy to apply, noncorrosive to machinery, adapted to use in planting equipment
so as not to interfere materially with uniform seed flow, stable for long periods, and

Table 1 Effects of Seed Dressing Fungicides and ABCD on Germination and Percentage Infection of Fungi in Sorghum

| | Mean performance as recorded at two locations[a] | | | | | |
| | *Alternaria* spp. | | *Fusarium* spp. | | *Curvularia* spp. | |
Treatment particulars	Germination (%)	Incidence (%)	Germination (%)	Incidence (%)	Germination (%)	Incidence (%)
Mean of fungicides (bavistin, bavistin + thiram, topsin, captafol, thiram, MBC, delsan, bavistin 50 WP, and JK Stem), excluding ABCD	76	5.2	76	2.8	76	5.7
Bavistin + thiram (best fungicide treatment)	76	4.5	76	2.0	77	1.0
ABCD	78	3.5	78	2.0	78	4.5
Control	66	14.0	66	10.0	66	12.5

[a] APAU, Hyderabad, and MPAU, Rahuri.
Source: Ref. 4.

relatively harmless to animals and nontoxic to people. It is difficult to find a chemical that meets all these requirements.

The choice of chemical depends on the type of seed, nature of the pathogen, the condition of the seeds, the relative cost, availability of acceptable chemical and treating equipment, and the weather conditions expected after seeding.

Fungicides are relatively ineffective in preventing the mold growth that occurs when seed with high moisture content is stored. They can, however, control to some extent fungal growth in seeds prone to mechanical injury (e.g., groundnut).

Organomercurial fungicides are most effective agaist infection carried on the seed surface. These are applied in powder form, but they are volatile and diffuse

Table 2 Effects of Fungicides Against Seedborne Infection of Anthracnose of Soybean by *Colletotrichum truncatum*

Treatment	Germination (%)	Infection (%)
Delsan	>85	7.5
ABCD	>85	5.0
Captan	>85	2.5
Emisan	>85	1.5
Carbendasim	>85	1.0
Diathane M-45	>85	0.0
Control	68%	22

Source: Ref. 4.

throughout the seed mass—seeds are kept in closed containers for 24–48 hr after the treatment. These fungicides are commonly used on small-grain cereals, but are phytotoxic and require careful dosaging to avoid harmful effects. They have a bactericidal effect and, therefore, should not be applied to seeds treated with a bacterial inoculant.

Organomercurial are generally used to treat small grains, flax, cottonseed, and safflower, whereas inorganic mercurials like mercuric chloride, mercurous chloride, and mercuric oxide are used to treat seeds of tuber and root crops as well as garden and vegetable crops. Inorganic nonmercurials like copper fungicides and zinc oxide and zinc hydroxide are used to treat bunts of wheat and damping-off in vegetables. However, the nonmercurial organic fungicides, principally captan and thiram, are used most extensively for seed treatment of many types of agricultural crops (Table 3). They are applied as either dusts or slurries. Captan is superior at low dosages and on old seed, especially when conditions after planting are unfavorable. Both chemicals provide adequate protection for good seed when applied at the recommended rates. The insecticide dieldrin can be combined with either captan or thiram when protection against certain soil insects is anticipated. Heavy insect infestation should, however, be controlled by elaborate soil treatment. Thiram and captan are nonvolatile and hence are not effective against soil fungi. They are nonphytotoxic and commonly applied to beans, groundnut, vegetable, and maize seeds (5).

Seedborne pests and diseases may be carried within some seeds (e.g., loose smut of wheat) or on the seeds (e.g., spores of bunt of wheat), or they may accompany the seed as free-living organisms or in debris. Systemic fungicides, which can be translocated to the infected site, are able to control difficult diseases, like loose smut of wheat, which earlier required hot-water soaks for their control. External infections can be attacked directly, and can be effectively controlled by organomercurial dressings. Because of the toxicity of these compounds, they are, however, being replaced by other effective formulations such as dithiocarbamates (6).

Protecting seed and seedlings against possible attack in the soil or against aerial plants after emergence has been successful in some cases using systemic insecticides or fungicides. Carbofuran applied to sorghum seed gives some protection against shoot fly. Similarly, benomyl can control powdery mildew of barley after seedling emergence.

Seed coating or pelleting has been widely adopted to treat seeds of sugar beet and forages, although it is little used in other agricultural crops in general. An irregularly shaped seed can be converted to a regular shape by pelleting, which facilitates sowing operation.

Insecticides have also found increasing use in seed treatment and are often marketed in combination with the fungicides. They offer protection against storage insects of cereals, pulses, and other crops and soil insects that attack seed and seedlings during germination (e.g., sorghum, corn, beans, and seeds of vine crops). The insecticides most commonly used for seed treatment are DDT, aldrin, lindane, dieldrin, heptachlor, chlordane, and methoxychlor. Proper dosage of seed treatment with insecticides is very important, because a narrow range exists between the amount of insecticide that can be used without causing injury to seeds and the amount that will effectively protect the seeds from insect attack in soil (1).

Thiamethoxam 70 has been extensively tested under field conditions in Brazil on crops such as wheat, barley, maize, soybean, dry bean, rice, and cotton, and has shown an outstanding levels of insect control (6a). Thiamethoxam at rates varying from 17.5

Table 3 Schedule of Chemical Seed Treatment Followed by N.S.C. and T.D.C. in India[a]

Crop	Name of chemical	Quantity of chemical (g/Q of seed)	Quantity of water slurry for treating 1 Q of seed
1. Cereals			
Paddy	2.15% wet Ceresan	60	Amount sufficient to immerse seed completely
Wheat	75% Thiram, WDP or	100	500 mL
	1% Agrosan GN dust	250	Dry dressing
Oats	1% Agrosan GN dust	250	Dry dressing
Sorghum	75% Thiram, WDP	85	500 mL
Pearl millet	75% Thiram, WDP or	75	500 mL
	1% Agrosan GN dust and	250	Dry dressing
	5% brine solution[b]		To immerse seed completely
Maize	75% Thiram, WDP or	70	500 mL
	75% Captan, WDP	70	500 mL
2. Legumes			
Pigeon pea	75% Thiram, WDP	75	500 mL
Green gram	75% Thiram, WDP	75	500 mL
Black gram	75% Thiram, WDP dust	250	Dry dressing
Khesari, lentil	1% Agrosan GN dust	250	Dry dressing
Soybean	75% Captan dust or	300	Dry dressing
	75% Thiram dust	(1:1)	
3. Oilseeds			
Sesame	1% Agrosan GN dust	250	Dry dressing
Sunflower	75% Thiram dust	250	Dry dressing
Rape, mustard	1% Agrosan dust	250	Dry dressing
Groundnut	75% Captan dust or	225	Dry dressing
	75% Thiram, WDP	125	500 mL
Castor	1% Agrosan GN	250	Dry dressing
4. Fiber crops			
Cotton (ginned, delinted)	75% Captan dust or	250	Dry dressing
	75% Thiram, WDP or	225	500 mL
	25% Ceresan, wet	2 g in 1000 mL	To immerse seeds completely
Jute	75% Captan, WDP	80	500 mL
Mesta	75% Captan, dust	250	Dry dressing
Sunhemp	75% Thiram, WDP	75	500 mL
5. Forage Crops			
Guar	75% Thiram, WDP	75	500 mL
Lucerne	75% Thiram, WDP	75	500 mL
Teosinte	75% Thiram, WDP	75	500 mL
6. Sugar Crops			
Sugar beet	75% Thiram, dust	250	Dry dressing
7. Vegetable Crops			
Peas, beans, cowpea	75% Captan, WDP or	100	500 mL
	75% Thiram, dust	125	Dry dressing
Bhindi	75% Thiram, WDP or	100	500 mL
	75% Captan, dust	250	Dry dressing

Table 3   Continued

Crop	Name of chemical	Quantity of chemical (g/Q of seed)	Quantity of water slurry for treating 1 Q of seed
Brinjal (eggplant)	75% Thiram, dust or	250	Dry dressing
Chilis	75% Captan, dust	250	Dry dressing
Tomato	75% Thiram, dust	335	Dry dressing
Cole crops (cabbage, cauliflower)	75% Thiram, dust	85	Dry dressing
Cucurbits	75% Captan, dust or	250	Dry dressing
	75% Thiram, dust	250	Dry dressing
Leafy vegetables	75% Thiram, dust	335	Dry dressing
Roots and bulb vegetables	75% Thiram, dust	250	Dry dressing

Agrosan GN and Ceresan wet are mercurial fungicides; seeds treated with these should not be stored for more than 1 month, and they should be applied to seed no earlier than 1 month before sowing.
[a] DDT 50% (WP) should be added to the fungicide at 100 g/kg of fungicide and mixed well.
[b] 5% brine solution (5 g salt/100 mL) should be used only when ergot infection is to be controlled.
Source: Ref. 1.

to 210 g a.i./100 kg seed provided a high level of control of *Metopolophium dirhodum* and *Diloboderus abderus* on wheat; *Dichelops furcatus* and *Elasmopalpus lignosellus* on maize; *Oryzophagus oryzae* and *E. lignosellus* on rice; *Aphis gossypii* and *Frankliniella shultzei* on cotton, and *Sternechus subsignatus* on soybean. Thiamethoxam 70 was safe to all crops at the tested rates, and it can be used in integrated pest management (IPM) and integrated resistance management programs (6a). According to Hofer et al. (6b), thiamethoxam (CGA-293 343) is a novel insecticide discovered and currently under worldwide development by Syngenta Crop Protection. It belongs to a new class of chemicals—the neonicotinoids—and is the first representative of the thianicotynyl subclass—chlorothiazole ring responsible for the broader activity and higher control potential. A good crop start is ensured by efficient control of all important soil-dwelling and early leaf-feeding and sucking insects like wireworms, false wireworms, flea beetles, pea weevils, Colorado potato beetles, aphids, white flies, thrips, and different other bugs. By controlling sucking insect pests, it also prevents the transmission of insect-vectored viruses. The long-lasting and reliable activity of the compound is based on low use rates, systemicity, and robustness of performance under different environmental conditions (6b).

Brandl (6c) provided a wide-ranging survey of new developments and trends in seed-treatment technologies during the last decade and identified future directions. The major crops that benefit from seed treatment are cereals, maize, cotton, potato, oilseed rape, and sugar beet. Seed treatments are being transformed from commodity to high-value status. Active ingredients such as tebuconazole, triticonazole, fludioxonil, silthiofam, imidacloprid, thiamethoxam, and fipronil are providing a broader spectrum of activity and longer lasting control of diseases and pests in early crop growth stages and better toxicological and ecotoxicological profiles. Modern seed-treatment products demand accurate application techniques and quality assurance systems to optimize efficiency, crop safety, and the cost/benefit ratio for the grower. Further research is needed on germination-enhancement techniques and the role of the seed as the delivery vehicle for additional crop inputs. These developments, along with changes in crop production systems and genetic technologies for environmentally friendly crop production methods, including nonsynthetic crop-protction agents, are the needs of the future.

Maude (6d) developed an image analysis system for quantitative analysis of seed-treatment coverage on treated seed. Using this technique, the effects of application parameters and formulation component change on seed coverage could be investigated on maize cv. Silverio. This technique has provided a useful tool for the optimization of formulation and application parameters for seed-treatment products. Rose and Oades (6e) noted that imidacloprid as a seed treatment gave good protection against *Agriotes* spp. (wireworm) in the United Kingdom, and was equivalent to gamma HCl both in improving crop emergence and reducing *Agriotes*-damaged plants. Slug feeding on germinating wheat was reduced by an average of 68% when seed were treated with imidacloprid. In field trials, these effects translated into an improved crop stand, but slug grazing on emerged foliage was not reduced much.

Two seed samples of China aster (*Callistephus chinensis*) were primed in PEG 8000 at an osmotic potential −1.25 MPa for 7 days in darkness at 15°C. In the primed seeds, a considerable increase in seed infestation with most fungi (such as Cladosporium sp., *Fusarium* sp., and *Rhizopus nigricans* [*R. stolonifer*]) was observed. In order to control the growth of fungi, PEG priming was combined with a fungicidal treatment. Rovral [iprodione] 50 WP at 0.2% concentration was used in the experiment. In combined treatments, the fungicide was applied before, during, and after PEG priming. Mycological and germination tests were performed in all combinations. Combining a fungicidal treatment with priming, in general, caused improvement of seed health compared with priming in a PEG solution alone. However, none of the combined treatments was fully effective against the fungi. Priming seeds in PEG alone and combined with a fungicide significantly increased the speed of germination and germination capacity (6f).

Zhang and Hampton (6g) investigated the effects of four systemic fungicide-based products, Aliette Super (fosetyl-aluminium 528 g/kg + thiram 172 g/kg + thiabendazole 129 g/kg), Apron 35SD (metalaxyl 350 g/kg) and Apron TZ (metalaxyl 450 g/kg + thiabendazole 240 g/kg) plus captan (800 g/kg), on the conductivity of garden pea (*Pisum sativum*), soybean (*Glycine max*), French bean (*Phaseolus vulgaris*), and broad bean (*Vicia faba*). The four seed treatments were applied by hand at the recommended product application rate (Aliette Super, 290 g/100 kg seed; Apron 35SD, 200 g/100 kg seed; Apron TZ = 150 g/100 kg seed; captan, 250 g/100 kg seed) and at double this rate. At the recommended application rates, the conductivity of the fungicide-treated garden pea, soybean, French bean, and broad bean seeds did not differ from that of the untreated control. However, at double these fungicide application rates, Apron SD significantly increased the conductivity of garden pea and soybean, Apron TZ significantly increased the conductivity of soybean, and Aliette Super significantly increased the conductivity of garden pea, soybean, and broad bean. A comparison of eight commercially treated (Apron 35SD or TZ, Aliette Super) garden pea seed lots with untreated seed from the same lots also provided no significant changes in conductivity for seven of the eight seed lots. It has been concluded that for the products tested, and provided the manufacturer's recommended application rates are strictly adhered to, fungicidal seed treatment does not affect conductivity results in seed vigor testing for large seeded legumes.

D. Dosages of Seed Treatments

Dosages of fungicides vary with the crop species to be treated and the chemical used. Lower dosages are required for large-seeded crops than for small-seeded species and for seeds sown in the springtime than for fall-sown seeds. Approximately twice as much captan as thiram is required for seed crops like red pine (*Pinus resinosa*).

A schedule of chemical seed treatment being followed by the National Seeds Corporation and Tarai Development Corporation in India is given in Table 3. It specifies the quantity of the chemical applied and the mode of application for different seeds. According to Agrawal (1), seed should be treated with fungicide-insecticide mixtures only when it is required to be planted in a soil infested with insects that are likely to attack the seeds or young seedlings.

E. Major Seed Crop Diseases

1. Cereal Crops

Maize is an extensively treated crop, and all hybrid corn seed is treated chemically to prevent seed rots and seedling diseases. Sorghum is almost always treated to control kernel smuts, seed rots, and seedling blights, using both mercurial and nonmercurial fungicides (Ceresan, Panogen, and Chipcote). Wheat is treated to control bunt, loose smut, seed rots, and seedling blight, especially with organic mercurials (Ceresan, Panogen, Chipcote). Loose smut of wheat can be controlled by hot-water or water-soak treatments.

The application of seed treatment on the basis of the percentage of seed infection was demonstrated only in the case where control of loose smut of wheat was carried out through seed treatment with Vitavax (7–10). The Uttar Pradesh Seeds and Tarai Development Corporation in India conducted testing of wheat seeds for loose smut in 1976. It not only restricted the increase in the incidence of loose smut, but also saved unnecessary treating of seed lots with less than 0.5% infection. According to Agrawal (11), this resulted in a net saving of approximately 45000 U.S. $ during 1975–1978. The National Seeds Corporation (NSC) also undertook the testing of wheat seeds for loose smut in India in 1977. This testing scheme has saved approximately 700,000 U.S. $ on seed treatment alone (11).

Effects of various physicochemical seed treatments such as scarification over sandpaper, sulfuric acid treatment for 30, 60, and 120 secs, and hot-water (80°C) soaking for 2 mins on the breaking of hard-seededness in 10 faba bean genotypes were investigated (12). The scarification over sandpaper resulted in the maximum germination percentage. The treatment with sulfuric acid for 30, 60, and 120 secs was not effective in breaking the seed hardness of faba bean cultivars (Table 4).

Barley diseases that respond well to seed treatment are covered smut, loose smut, stripe, seed rot, and seedling blight. Loose smut of barley is more effectively controlled by the water-soak method than that of wheat (3). All organic compounds used on wheat can control the diseases of barley. Oats are treated with the same compounds against loose smut, covered smut, *Helminthosporium* blight, seed rots, and seedling

Table 4 Effects of Different Treatments on Breaking Hard Seededness in Faba Bean, 1987–1988

	Seed treatment									
	Control		Scarification		H_2SO_4 treatment					
					30 sec		60 sec		120 sec	
Genotype	G	H	G	H	G	H	G	H	G	H
VH 130	39.0	58.3	79.7	0.0	37.3	35.6	40.3	48.6	32.6	41.6
VH 131	56.7	39.7	85.0	0.0	37.6	58.0	39.6	50.3	37.6	2.86
VH 132	43.9	53.6	78.0	0.0	30.3	67.3	36.3	54.6	25.6	41.3
VH 133	46.0	50.7	81.5	0.0	31.0	53.3	39.3	56.0	34.6	40.3
VH 134	31.6	67.0	85.0	0.0	35.3	52.0	32.3	51.6	46.5	46.5
LMI	44.6	52.0	87.0	0.0	47.6	30.6	43.6	46.0	37.3	34.3
Local	44.0	54.0	84.5	0.0	45.3	40.6	35.0	50.6	51.3	0.3
JV 1	29.0	67.0	82.7	0.0	34.3	48.3	30.6	49.6	26.3	52.6
JV 2	42.6	54.0	45.7	0.0	44.3	50.3	36.0	72.0	32.0	52.6
DB 6	39.2	58.7	84.2	0.0	35.3	59.6	34.0	53.6	29.0	45.0
Total	416	5545.5	793.8	0.0	378.3	495.6	367.0	532.7	352.5	390.8
Mean	41.6	55.5	79.4	0.0	37.8	49.5	36.7	53.3	35.3	39.1

G = % germination; H = % hard seeds left over.
Source: Ref. 2.

blights. Rye is treated like wheat with organomercurials for bunt, stalk smut, and other seedborne pathogens causing seed rots and seedling blights. Rice is treated with captan or thiram dusts or slurries against soilborne pathogens and with organomercurials against *Helminthosporium*, *Piricularia*, and other seedborne fungi. Seeds containing nematodes are fumigated with methyl bromide. This chemical has been banned in some countries.

2. Fiber and Sugar Crops

Cottonseed is treated to achieve protection against angular leaf spot, anthracnose, sore shin, seed rots, and seedling blights. Seed is normally delinted before treatment, either mechanically or chemically using acids. The method of delinting may influence the choice of fungicide used. Organomercurials are most effective against seedborne pathogens and nonmercurials against soil organisms. Flax seed is commonly cracked in threshing, letting fungi enter, which cause seed rots and seedling blights. They also kill seedborne pathogens and are generally applied as liquids, dusts, or slurries. Heavier doses are required for flax than most other field crops. Sugar beets are best treated with caplan, thiram, dexon, or dichlone dusts to control seed rots and damping-off.

3. Oilseeds

Groundnut may be treated with thiram, chloranil, and 2% Ceresan dusts to control seed rotting. Soybeans are rarely benefited by seed treatment, but may be treated with captan, thiram, and chloranil if required.

4. Forage Crops

Small-seeded forages (alfalfa, clovers, vetches, lespedezas, and trefoils) usually do not respond to seed treatment. Forage grasses differ in disease problems and in their responses to seed treatments. Using captan or thiram can reduce losses from seed decays and damping-off. Damaged seeds usually benefit more from the seed treatment than do sound ones (e.g., sudan grass seeds). Smuts that are common on millets, slender wheat grass, Canada wild rye, and sudan grass can be controlled by any of the organomercurials, as in sorghum and other small grains.

5. Vegetable Crops

Most vegetable crops benefit from seed treatment, especially by controlling the most common problems of seed rots and damping-off. Seed treatment also prevents seedborne pathogens. Asparagus seeds are treated for damping-off with mercurials like Calogreen or Ceresan M dusts. Beans are usually treated with a combination of fungicide and insecticide applied as a dust or slurry (captan or thiram with dieldrin or lindane).

Beets, spinach, and Swiss chard can be protected from seed rotting and preemergence damping-off with captan, thiram, or dichlone. Carrots are treated with a combination fungicide-insecticide dust to control damping-off and rust fly in first-generation (thiram or dichlone with lindane). Carrot seeds are often treated with hot water before other seed treatments to free them from bacterial blight.

Celery seeds are also soaked in hot water at 110°F (43.3°C) for 30 min to destroy pathogens of early and late blights. Seeds stored for 2 years or more do not require this treatment. Crucifers (cabbage, cauliflower, broccoli, and Brussels sprouts) are usually infested by internal seedborne diseases and require hot-water treatment at 122°F (50°C) for 20–30 min. Other crucifers like radishes, rutabags turnips, kale, kohlrabi, and mustard also benefit from this seed treatment.

Cucurbits (cucumbers), muskmelons, watermelons, squashes, and pumpkins may be disinfected with mercuric chloride to protect them from anthracnose and angular leaf spot diseases. Seeds are soaked for 5 min in a solution of mercuric chloride (1 oz in 7.5 gal of water), rinsed thoroughly in running water, and dried before sowing. Protectants like captan or thiram alone or in combination with dieldrin or lindane are used afterward.

Onion seeds are protected against decay with captan or thiram, which also control seedborne smuts. Pelleting seeds with thiram or captan has replaced the old formaldehyde-drip method. An insecticide like aldrin may be included in the pelletant to control maggots.

Peas are treated against seed rots and seedling blights with captan and thiram alone or in combination with aldrin. Tomato, eggplant, and pepper seeds are disinfected by soaking in water at 45°C for 20 min and then treated with the protectants like mercuric chloride solution in hot water (pepper) or Ceresan M (tomato). Tomato seeds treated with mercurials should not be treated with any other chemical; again to prevent injury to seeds. Protectants are beneficial to the seeds treated with hot water as well as those that have not been previously treated.

6. Fruit Crops

Seeds of apple, pear, cherry, peach, plum, almond, and apricot are commonly soaked in water for several hours before planting and normally do not require treatment with fungicides. Treating with a suitable protectant can reduce occasional losses caused by damping-off.

7. Flowers and Ornaments

Seeds of most ornamental plants benefit from seed treatment, especially by controlling losses caused by seed rots and damping-off diseases. Thiram, captan, chloranil, and Semesan are the most commonly used protectants applied as dusts. The hot-water and chemical-soak treatments are also used to eliminate seedborne pathogens. Seeds of China aster are sometimes treated with mercuric chloride or Semesan soaks to prevent the introduction of *Fusarium* or *Verticillium* wilts into clean soil. This treatment will not control wilt if the soil is already contaminated with the fungus. To control wilt, seeds are soaked for 30 min in 1:1000 mercuric chloride solution (7.5 grains in 1 pint of water). The treated seeds are rinsed immediately for 5 min in running water and dried at room temperature. Instead of mercuric chloride, a 0.25% solution of Semesan can also be used.

Sweet peas are frequently treated with captan, thiram, or chloranil to prevent seed rots and seedling blights. Seeds of forest trees are also treated before sowing in nurseries to control seed rots and preemergence and postemergence damping-off using thiram and captan. Other chemicals may be added to protect seeds from insects, birds, and rodents. Seeds of hardwoods and some large-seeded southern conifers (e.g., loblolly and slash pine) can be protected by dusting with thiram or captan, and conifers with smaller seeds (> 40 seeds/g) are usually pelleted (3).

III. SEED-TREATING EQUIPMENT

The equipment used to treat seeds with chemicals consists of mainly two types: slurry treaters and direct treaters.

A. Slurry Treaters

Slurry treatment involves preparation of a suspension of wettable powders in water, which is applied to seeds. The material to be treated is accurately metered through a simple mechanism composed of a slurry cup and seed dump pan. The cup introduces a given amount of slurry with each dump of seed into a mixing chamber, where they are blended together. The slurry treaters are adapted to all types of seeds and rates of treating. The small amount of moisture added to seeds (0.5–1.0% of the seed weight) does not affect seed in storage, since the moisture is added to the seed surface and is soon evaporated.

The equipment designed for treating seed is available for placing it in the seed-cleaning line. Mixing is usually achieved either by an auger or by a rotary action. It is important that the seed receive the correct dose, so that both seed and chemical have to be metered into the mixing chamber. Some machines work on the batch system, where a weighed quantity of seed is released into the mixing chamber together with an appropriate amount of chemical. Other machines are of a continuous-flow type, with both seed and chemical being metered continuously.

Powders are the most difficult to handle and to mix, although they are easy to transport. Slurries and liquids are easier to mix, but may cause damage to the seed if not handled carefully. Most seed-treatment machines are now designed to handle all forms of chemicals.

B. Direct Treaters

Direct seed treaters are comparatively newly designed seed-treating equipment. Two models, Panogen and Mist-o-matic, are widely used (1). The Mist-o-matic treater applies chemical as a mist directly to the seed. The metering operation of the treatment cup and seed dump is similar to that of other treaters (Panogen); cup sizes are designed according to the quantity of chemical they actually deliver (i.e., 2.5, 5, 10, and 15 mL). The machine is equipped with a large treatment tank, a pump, and a return that maintains the level in the small reservoir from which seed is fed. After metering, the treatment material flows to a rapidly revolving fluted disc mounted under a seed-spreading cone. The disc breaks drops of the chemical into a fine mist and sprays this outward to coat the seed falling over the cone through the treating chamber. Just below the seed dump are two adjustable retarders designed to give a continuous flow of seed over the cones between seed dumps to ensure continuous misting of the material from the revolving disc. Selecting the appropriate cup size and adjusting the seed dump weight can obtain the desired treating rate.

C. Other Miscellaneous Equipment

Running a pipe through a drum at an angle can make a simple homemade drum mixer. The drum is mounted on two sawhorses. The seeds and chemical are placed in the drum, which is rotated slowly until the seeds are covered with the chemical.

Liquid chemicals can be made to drip to the seed as they enter a grain auger or screw conveyor. By the time the seeds leave the auger, the liquid is spread well over most of the seeds. Dusts and slurries can be applied, but only with some difficulty.

Seeds are spread about 10–15 cm deep on a clean, dry surface. An appropriate amount of chemical is diluted with water and sprinkled evenly over the seed. Mixing is then accomplished with the help of a shovel or scoop, turning the seed at least 15–20 times.

IV. SPECIAL SEED-TREATMENT PROCESSES

In addition to the treatment of seeds with hot water and chemicals to control numerous seedborne and soilborne diseases, seeds are also required to be treated for specific purposes; for example, delinting of cotton, scarification of legumes, decortication of sugar beet seeds, and irradiation of hard seeds to facilitate germination and blending, pelleting, and coloring of seeds (13).

Taylor and Harman (14) have recently reviewed and discussed the use of seed-treatment technologies such as seed coating, seed pelleting, physiological seed treatments like fluid drilling, and seed priming, as well as seed treatment with beneficial microorganisms (amendments that enhance biological seed treatment) to protect sown seeds from soil fungi.

A. Seed Coating and Pelleting

Seed coating is a process of direct addition or application of materials to the seed. Roos and Moore (15) defined "coated seed" as seed that has been pelleted, tableted, or taped. The term *seed coating* is generally used to denote application of useful materials to the seed without changing its general size or shape, whereas the term *pelleted seed* refers to the addition of inert fillers to increase the apparent seed size and weight (14).

Dry powders, which are widely used as planter box treatments, do not adhere well to the seed surface and result in nonuniform application and dust problems. Active agents may be dispersed or suspended in water to form slurry for uniform application. Adhesives (stickers, glues, or binders) such as methylcellulose, dextran, gum arabic, and vegetable or paraffin oils improve retention of the applied materials (16). Jeffs and Tuppen (17) reviewed the seed-coating equipment developed for laboratory and commercial applications. In the film-coating technique, active material is dispensed or dissolved in a liquid adhesive and applied to seeds either with a fluidized-bed treater or pharmaceutical coating drum. This method permits application of multiple coatings with a 1–10% increase in seed weight (16). Seed coating has been utilized to ameliorate environmental stresses such as drought and flooding. Hydrophilic polymers like hydrolyzed starch-graft polyarylonitrile I (H-SPAN) maintain a high water potential around germinating seeds (18). Polymer coatings have also been reported to retard imbibition rates when seeds are sown in a wet soil. Seed coatings with peroxide compounds provide oxygen to seeds under anoxic soil conditions (19). Similarly macro- and micronutrients applied in seed coatings have been found to improve early plant growth (20,21).

Many crop seeds are small and irregular in shape, posing difficulty in precision seeding and uniform plant spacing of high-value horticultural crops. Pelleting increases seed size and weight and alters seed shape for precision planters. The binders for pelleting include gum arabic, gelatin, starch, methyl cellulose, polyvinyl alcohol, polyoxyethylene glycol–based waxes, and carboxymethyl cellulose, and the commonly used fillers or particulate matter for pelleting, are calcium carbonate, limestone, gypsum, talc, vermiculite, diatomaceous earth, kaolin clay, bentonite, zeolite, and peat (14). The pellet may act as a physical barrier to water and oxygen diffusion to the seed and can provide a mechanical barrier to radicle protrusion.

B. Encapsulation of Seeds

Encapsulation technology was developed to form capsules of somatic embryos by gelation to enhance their stability. Natural seeds also can be encapsulated using hydrogels. Redenbaugh et al. (22) mixed propagules with sodium alginate solutions and then transferred the coated propagules to a calcium salt to form a soft capsule.

C. Fluid Drilling and Seed Priming

Fluid drilling, or gel seeding, consists of germinating seeds in aerated water until radicle emergence. The seeds are then mixed in a viscous gel and sown with an appropriate drill (23). The gel prevents seeding injury to the emerging radicle and maintains seed moisture. The nonviable or slow-to-germinate seeds can be separated from the germinating seeds on the basis of density. Gels also have been used to deliver pesticides to control soilborne diseases. Seed priming, or osmoconditioning, is a presowing hydration treatment developed to facilitate seedling establishment. Unlike the fluid drilling of germinating seeds, the primed seeds can be handled like conventional dry seeds. The osmotic priming of seeds with liquids has been discussed and reviewed extensively (24–26).

Seeds may be primed in a controlled environment, regulating the water potential of the priming media, aeration, temperature, and duration of treatment. Osmoticums such as PEG (MW = 6000–8000) and inorganic salts have been used as priming agents. Alternatively, seeds can also be primed in a solid medium; for example, onion seeds can be primed in slurry of PEG −6000 and vermiculite (27). The solid matrix priming (SMP) process thus involves mixing of seeds with a solid material and water in known proportions. Bioprotectants and/or chemical pesticides can be used in conjunction with SMP (14,28).

D. Biological Seed Treatment

Seed treatment with beneficial microorganisms like *Rhizobium* spp. to treat leguminous seeds for nitrogen fixation and *Azospirillum* and other nitrogen-fixing bacteria has long been known. The ability of certain fungi and bacteria to control plant diseases and other microorganisms or to colonize plant parts and deliver specific beneficial genes within the plants is currently being investigated. Fathey (29) noted that an endophytic bacterium, *Clavibacter xylii*, when introduced into corn seeds, is transformed to produce an insecticidal protein (δ-endotoxin) from *Bacillus thuringensis*. The toxin produced by the transformed bacterium is active against the European corn borer.

According to Taylor and Hannan (14), the seed treatments that alleviate stress associated with the soil environment as well as those that directly increase or improve plant growth will continue to play a significant role in seed-production technology. Advances to refine existing seed-treatment technologies and strategies are needed.

E. Delinting of Cotton

Whereas 1 ton of undelinted cottonseed will be needed for about 60 acres of land, the same quantity when delinted gives 1700 lb of seed, which can plant 210 acres, allowing more uniform planting. The delinted seed is free from certain lintborne pathogens and permits better coverage with chemical treatments (1).

1. Mechanical Delinting

After ginning, cottonseed still has a small amount of lint or fuzz on the surface, making it difficult to plant, because seeds stick together and do not flow easily. Mechanical delinting is more like ginning, with closer saws and finer teeth. Lint is cut as close as possible to the seed without breaking the hull, although a small amount of fuzz remains on the seed. Flash or flame processing removes more of the fuzz on mechanically delinted seed. Flash furnaces have a vertical duct at the bottom. Seeds are metered onto a vibrating conveyor, feeding them into the furnace operated at 2400°F (1315.5°C) to remove lint. The seeds then pass over an air/screen cleaner to remove immature seeds and foreign material.

2. Chemical Delinting

Chemical delinting with inorganic acids removes all of the lint. In the "dry process," hydrochloric and sulfuric acids are mixed to form a gas, which is piped into a revolving tank. A reaction between the acid gas and fuzz, hastened by heat, crystallizes the fuzz. The seeds are dumped into a revolving perforated cylinder to remove the crystallized lint and an air/screen cleaner removes immature seeds and other debris. Anhydrous ammonia neutralizes acid on the seed in a revolving tank. In the "wet process," an acid solution is applied to seed until fuzz is crystallized, and seeds are rinsed to remove the crystallized fuzz. Treating seeds with an alkali neutralizes the acid. The seeds are dried and cleaned as in the dry process.

F. Decortication of Sugar Beet Seeds

Planting sugar beet seeds involves the problems of multiple germs and irregular size, shape, and density of seeds. Decortication makes it possible to obtain seeds of more uniform size and fewer germs in a seed ball. Earlier attempts to obtain single-germ seeds by a segmenting process injured the seeds, resulting in poor germination and damaged seedlings. A less drastic method of decortication involves reduction in the size of seed units by removing the outer corky parts of the seed balls, which decreases the number of germs in the larger seed balls. A cleaner and grader separate the mixture of seeds and dust. The seeds are passed on to a gravity table or aspirator to remove light or incomplete seed units. The decorticated seeds have a greater density than whole seeds (about one-third the volume of whole seeds) and are smoother and more uniform in size and easy to plant.

G. Scarification of Hard Seeds

Some hard seeds such as legumes, asparagus, and okra contain a high proportion of hard seeds, and need to be scarified or scratched to break the impermeable layer of surface cells forming a barrier to water. Hard seeds do not readily germinate to give a uniform crop stand. Several methods are used to scarify hard seeds, such as buffing after treating seeds with a special oil, treating seeds with heat or acid, irradiating them electrically, and abrading them mechanically. Care must be taken, however, that seeds are not damaged during the process. The seed industry mostly uses mechanical scarification, although the chemical method (acid) is used extensively for cottonseed. Irradiation has been experimentally used on seeds like corn and on leguminous seeds. Mechanical scarifiers pass the seeds over some abrasive surface like sandpaper or carborundum stone. Air streams separate hulls and chaff.

Seeds of tamarind (*Tamarindus indica*), collected in November 1998 in western Sudan, were germinated following seven pretreatments: control, mechanical scarification by scratching the region near the hilum on coarse sandpaper or nicking it with a knife and acid scarification by immersion in concentrated sulfuric acid (H_2SO_4) for 15, 30, 45, and 60 min. Final emergence was 100% for all pretreatments, but days to first emergence (Eist), days to 50% emergence (E50), and mean emergence time (MET) varied significantly among them. The best germination response occurred with acid pretreatment followed by mechanical scarification. Eist, E50, and MET decreased as acid pretreatment time increased. Scratching on a coarse sandpaper advanced the E50 by 4.5 days and shortened the MET by 4.1 days compared to nicking (29a).

H. Irradiation of Seeds

Compact x-ray equipment adapted to grain inspection has been used for routine seed inspection. It requires little technical skill and minimal preparation of the sample. Radiographs of grain samples are easy to interpret and give rapid indications of the degree of internal infestation of insects. They are also used to determine the amount of checkling in rice, the effectiveness of fumigation, and the selection of grain for processing. Seeds are also treated electrically to increase water absorption and reduce the proportion of hard seeds.

Irradiation can be accomplished experimentally in two ways. Placing seeds in an evacuated glass tube and applying about 1000 V electricity to electrodes on the tube ends (10-50 mA) causes gases in the tube to glow and affect seeds (glow-discharge treatment). In another method, seeds are subjected to high frequencies using radio frequency equipment.

Irradiated cottonseeds sink in water immediately, whereas nonirradiated ones will float 24 hr or longer before they absorb enough water to sink. Irradiated seeds of alfalfa, corn, and clovers absorb water more rapidly and show improved germination. Irradiation may sterilize some seeds. In one irradiation test, turnip seeds were found to lose their germinability after irradiation, whereas red clover and mustard seeds were not affected.

I. Coloring of Seeds

Coloring agents of dyes are often added to the treating mix to indicate specific treatments; for example, green for seeds treated with wireworm repellant fungicide, red for seeds treated with wireworm repellant; and yellow for fungicide-treated seeds. The colors also indicate the uniformity of the chemical coverage. The high price and machine capacity may, however, limit any general use to processors that handle large amounts of seeds. Coloring of seeds prevents the inadvertent use of treated seeds for food or feed purposes.

V. PRECAUTIONS FOR SEED TREATMENT

Most of the chemicals used to treat seeds are highly toxic to humans and animals. In addition, they can be harmful to seeds. Extreme care is, therefore, necessaryto ensure that human beings or animals do not consume treated seeds. The treated seeds must be properly labeled, giving appropriate warnings of their dangers when consumed. Care should also be taken to treat only the quantity of seed for which sales are assured or the quantity of seeds immediately required for planting to avoid the temptation to use the unsold or unplanted seed for human or animal consumption.

The use of appropriate dosages of chemicals cannot be overemphasized. Applying too much or too little chemical can be as damaging as not treating the seeds at all. Seeds with a high moisture content are very susceptible to injury, especially when treated with certain concentrated liquid compounds. Colors or dyes can serve to warm against inadvertent contamination of the treated seeds with food grains meant for human and animal consumption. Any seed treatment must be effective for its intended purpose over a wide range of field conditions and not be deleterious to the seed. Also, the technique must be economical and practical for the specific crop, and above all, the material(s) used should be environmentally safe.

According to Cockrell et al. (30), several significant changes have taken place in the UK cereals industry in the last few decades. Lower cereal prices have forced growers to seek input cost savings. Increased consumer concerns for the environment and food safety and the introduction of more rigorous quality standards imposed by the end-use markets all require that pesticide use is according to need. The diversity of new seed-treatment products also encourages growers to target seed treatment to the range of pathogens present such as *Tilletia tragic, Microdochium nivale*, yellow rust, bunt, loose smut, and barley yellow dwarf virus. Winter wheat being the largest cereal crop in the United Kingdom, requiring over 300,000 tons of seed each year, in low disease seasons it may be possible to sow over half of the winter wheat. The current status of wheat seed production in the United Kingdom and the research work being carried out to support and promote better targeting of seed treatments through improved seed testing technology while maintaining a high level of seed health have been reviewed.

The efficacy of presowing seed management on the seed-yield potential of tomato cv. PKM-1 upon the addition of macro- and micronutrients to the seed pellet were studied in a field experiment conducted during the kharif season of 1997 in Tamil Nadu, India. The treated seeds were also evaluated for storability to determine the period of usage of seeds after treatment. The seeds were divided into five treatments and pelleted using 10% maida (200–300 mL/kg) and arappu (*Albizia amara*) leaf powder (200–300 mg/kg). The treatments were 2 g diammonium phosphate (DAP)/kg, 0.250 g zinc sulfate/kg, 0.100 g borax/kg, simple pelleted seeds without nutrients, and unpelleted seeds (control). The initial germination were insignificant owing to the pelleting treatments, but the treatments exerted significant differences in field emergence wherein zinc sulfate–pelleted and simple-pelleted seeds had 14 and 7%, respectively, higher germination than the control (57.41%). The nutrient-pelleted seeds had approximately 50% higher 100–seed weight than the control and highest plant height (59 cm). Zinc sulfate–pelleted seeds had the highest seed yield (174.4 g) followed by DAP (173.4 g) and borax (140.3 g). The storability of seeds tested after 3 months showed a reduction of germination in pelleted seeds compared to the control and was more reduced (31%) in seeds pelleted with borax. However, seeds pelleted with zinc sulfate retained the viability potential higher than simple pelleting, borax and DAP pelleting which were only 8% lower than the control, but higher than the certification level (70%)(31).

Bitter gourd (*Momordica charantia*) requires high temperature (between 25 and 28°C) for successful seedling emergence, and poor emergence is common at suboptimum temperature. Lin and Sung (31a) evaluate the effect of suboptimum temperature on seedling emergence and several physiological characteristics related to seedling growth in bitter gourd cv. Special Six. Priming was achieved by mixing the seeds with moist No. 3 vermiculite, incubating at 25°C for 36 hr, and then air drying to the original moisture level. Soaking the seed in water at 50°C for 60 min and then air drying to the original moisture level achieved warm-water soaking. Seedling emergence from vermiculite was determined at 25, 20, and 15°C. The emergence of nontreated seeds at 25°C was 50%. No seedling emerged at 20 or 15°C. However, both priming and warm-water soaking improved the emergence response of bitter gourd seeds at 25 and 20°C. The observed decrease and delay in emergence at suboptimum temperature were linked to the reduced activity of enzymes (i.e., isocitrate lyase, malate synthase, and malate dehydrogenase) involved in lipid and sucrose conversion. Both priming and warm-water soaking improved the low-temperature (20°C) seedling emergence. These improvements were attributed to the increased enzyme activities. Nevertheless, the morphological changes and softening in seed coat and seed-treatment-stimulated embryo growth might also play crucial role in speeding up the seedling emergence.

Molecular marker technologies are currently being used as an alternative strategy to identify population dynamics and environmental activity profiles of bacterial antagonists in the soil environment. Leifert (32) has recently reviewed the opportunities and problems associated with the use of molecular marker–based assays, (e.g., use of lux marker technology) for population and activity profiles of bacterial antagonists in soil. Quinn (33) has outlined main data requirements for pesticide seed treatments within the European Community and Great Britain (UK) to ensure safety to humans and the environment. Leuenberger (34) employed an optional tool for the uniform color distribution to measure the treatment uniformity at the point of application. The colorimeter Minolta CR-331C provided the data processed by a computer. Uniformity information was obtained rapidly for dressed seeds. This technology has proved to be very promising to assess the treatment quality seed-treatment dressings.

Mani et al. (35) pelleted *Acacia leucophloea* seeds with diammonium phosphate (30 g kg^{-1} of seed), commercial micronutrients mixture (19.7 g kg^{-1} of seed), *Rhizobium* (50 g kg^{-1} of seed), sevin (carbaryl) (2 g kg^{-1} of seed), and *Trichoderma viride* (4 g kg^{-1} of seed). The pelleted seeds along with unpelleted control were evaluated in calcareous, sandy loam, acidic, and sodic soils. Pelleted seeds registered significantly higher germination and seedling vigor compared to unpelleted control under all soil types. However, higher germination and seedling vigor were recorded in calcareous soil. In the acidic soil, pelleted seeds also recorded significantly higher germination and seedling vigor than the unpelleted control. Pelleting of seeds thus could be recommended for augmenting germination and seedling vigor under adverse soil conditions.

REFERENCES

1. R.L. Agrawal, Seed *Technology*, Oxford and IBH Publishing, New Delhi, 1980.
2. L.O. Copeland and M. B. McDonald, *Principles of Seed Science and Technology*, 3rd ed., Chapman & Hall, New York, 1995.
3. E.W. Hanson, Hansing, and W.T. Schroeder, Seed treatments for control of diseases, *Seeds: The Yearbook of Agriculture*, US Department of Agriculture, Washington, DC, 1961, p. 272.
3a. A.A.Tawfik and G. Noga, Priming of cumin (*Cuminum cyminum* L.) seeds and its effects on germination, emergence and storability, J. Appl. Bot. *75*:216 (2001).
3b. H.R. Rowse, Drum priming—a non-osmotic method of priming seeds, *Seed Sci. Technol. 24*:281(1960).
3c. E.A. Veasey and J.C. Teixeira De Freitas, Breaking seed dormancy in *Sesbania sesban, S. rostrata* and *S.virgata*, Seed Sci. Technol. *30*:211 (2002).
3d. C. Dharmalingam, R. Visantha, K. Malarkodi and S. Lakshmi, Halogenation treatment to safeguard pulse seeds in storage under ambient conditions, *Seed Res. 28*:42 (2000).
3e. M. Takaki and L.H.P. Gama, The role of the seed coat in phytochrome-controlled seed germination in *Lactuca sativa* L. cv. Grand Rapids, *Seed Sci. Technol. 26*:355 (1998).
4. M.K.H. Siddiqui and P.K. Agarwal, Attapulgite based clay dust (ABCD) for seed treatment and storage of food grains, Seed *Research*, Vol. 1 (S.P. Sharma, ed.), Indian Society of Seed Technology, New Delhi, 1993, p. 568.
5. J.R. Thomson, *An Introduction to Seed Technology*, Leonard Hill, London, 1979.
6. A.F. Kelly, *Seed Production of Agricultural Crops*, Longman Scientific and Technical, New York, 1988.
6a. O. De C. Lette, F. Brandl, D. Hopper, P. Aranmaki, K. Gehmann and J. Weissenberg, Seed treatment- an emerging technology in agriculture in Latin America, demonstrated by the technology developed for thiamethoxam, Proc. Int. Sym. Wishaw, UK, February 26–27, 2001, *Seed Abstr. 25*(6):288 (2002).
6b. D. Hofer, F. Brandl, B. Druebbisch, F. Doppmann and L. Zang, Thiamethoxam (CGA 293343)—a novel insecticide for seed delivered insect control, Proc. Int. Sym. Wishaw, UK, February 26–27, 2001, *Seed Abstr. 25*(6):287 (2002).

6c. F. Brandl, Seed treatment technologies: evolving to achieve crop genetic potential, Seed Treatment: Challenges and Opportunities, Proc. Int. Sym. Wishaw, UK, February 26–27, 2001, *Seed Abstr.* *25*(6):253 (2002).

6d. S.J. Maude, The development of an image analysis technique for the quantitative analysis of seed treatment coverage on seed, Proc. Int. Sym. Wishaw, UK, February 26–27, 2001, *Seed Abstr.* *25*(6):254 (2002).

6e. P.W. Rose and L. Oades, Effects of imidacloprid cereal seed treatment against wireworms and slugs, Proc. Int. Sym. Wishaw, UK, February 26–29, 2001, *Seed Abstr.* *25*(6):288 (2002).

6f. H. Dorna, K. Tylkowska and X. Zhao, Effects of osmo priming and Rovral on health and germination of China aster seeds, *Phytopatholgia-Polonica 21*:34 (2001).

6g. T. Zhang and J.G. Hampton, Does fungicide seed treatment affect bulk conductivity test results? *Seed Sci. Technol. 27*:1014 (1999).

7. V.K. Agarwal, H.S. Verma, and S.B. Singh, Technique for the detection of loose smut infection of wheat seeds, *Seed Tech. News. 8*(3):1 (1978).

8. V.K. Agarwal, Testing wheat for need of seed treatment in India, *Seed Pathol. News. 12*:1 (1979).

9. V.K. Agarwal, Assessment of seed-borne infection and treatment of wheat seed for the control of loose smut, *Seed Sci. Technol. 9*:725 (1981).

10. V.K. Agarwal, Quality seed production at Pantnagar, India, *Seed Sci. Technol. 11*:1071 (1983).

11. V.K. Agarwal, Present status of seed pathology in India, *Seed Research*, Vol. 1 (S.P. Sharma, ed.), Indian Society of Seed Technology, New Delhi, 1993, p. 549.

12. P. Kumari, R.P.S. Tomer, C. Ram, and T.P. Yadava, Effect of location and physico-chemical treatments on hard seeds in faba bean (*Vicia faba* L.), *Seed Research*, Special Vol. No. 2 (S.P. Sharma, ed.), Indian Society of Seed Technology, New Delhi, 1993, p. 914.

13. L.H. Purdy, J.E. Harmond, and G.B. Welch, Special processing and treatment of seeds, *Seeds: The Yearbook of Agriculture*, US Department of Agriculture, Washington, DC, 1961, p. 322.

14. A.G. Taylor and G.E. Hannan, Concepts and technology of selected seed treatments, *Annu. Rev. Phytopathol. 28*:321 (1990).

15. E.E. Roos and F.D. Moore, Effect of seed coating on performance of lettuce seeds in green house soil tests, *J. Am. Soc. Hortic. Sci. 100*:573 (1975).

16. P. Halmer. Technical and commercial aspects of seed pelleting and film-coating, *Application to Seeds and Soil* (T.J. Martin, ed.), British Crop Protection Council, Thornton, Heath/Surrey, England, 1988, p. 191.

17. K.A. Jeffs and R.J. Tuppen, Application of pesticides seeds, requirements for efficient treatment of seeds, *Seed Treatment* (K.A. Joffs, ed.), British Crop Protection Council, Thornton, Heath/Surrey, England 1986, p. 17.

18. J. Baxler and L. Waters, Effect of hydrophilic polymer seed coating on the imbibition, respiration and germination of sweet corn at four matric potentials, *J. Am. Soc. Hortic. Sci. 111*:517 (1986).

19. J.P. Liver and E.H. Roberts, Peroxides in seed coatings, *Outlook Agric. 13*:147 (1984).

20. D. Scott and U.J. Archie, Sulphur, phosphate and molybdenum coatings of legume seed, *NZ J. Agric. Res. 21*:643 (1978).

21. R.G. Silcock and F.T. Pruith, Seed coating and localized application of phosphate for improving seedling growth of grasses in acid, sandy red earths, *Aust. J. Agric. Res. 33*:785 (1981).

22. K. Redenbaugh, D. Slade, P. Viss, and J.A. Fujii, Encapsulation of somatic embryos in synthetic seed coats, *HortScience 22*:803 (1987).

23. D. Gray, Fluid drilling of vegetable seeds, *Hortic. Rev. 3*:1 (1981).

24. W. Heydecker and P. Coolbear, Seed treatments for improved performance-survey and attempted prognosis, *Seed Sci. Technol. 5*:353 (1977).

25. A.A. Khan, N.H. Peck, and C. Samimy, Seed osmoconditioning, physiological and biochemical changes, *Isr. J. Bot. 29*:133 (1980–1981).

26. K.J. Bradford, Manipulation of seed water relations via osmotic priming to improve germination under stress conditions, *HortScience 21*:1105 (1986).

27. J.R. Peterson, Osmotic priming of onion seeds-the possibility of a commercial scale treatment, *Sci. Hortic 5*:207 (1976)

28. G.E. Harman and A.G. Taylor, Improved-seeding performance by integration of biological control agents at favorable pH levels with solid matrix priming, *Phytopathology 78*:520 (1988).

29. J.W. Fathey, Endophytic bacteria for the delivery of agrochemicals to plants, *Biologically Active Natural Products: Potential Use in Agriculture* (H.G. Cutler, ed.), American Chemical Society (Symp. Ser.), Washington, DC, 1988, p. 120.

29a. K. El-Siddig, G. Ebert and P. Ludders, A comparison of pretreatment methods for scarification and germination of *Tamarindus indica* L. seeds, *Seed Sci. Technol. 29*:271 (2001).

30. V. Cockrell, V. Mulholland, M. McEwan, N.D. Paveley, W.S. Clark, S. Athony, J.E. Thomas, J. Bates, D.M. Kenyon and E.J.A. Taylor, Seed treatment according to need in winter wheat, Seed Treatment: Challenges and Opportunities, Proc. Int. Sym. Wishaw, UK, February 26–27, 2001, *Seed Abstr. 25*(6):254 (2002).

31. P. Srimathi, K. Malarkodi, R. Geetha, and V. Krishnasamy, Influence of presowing pelleting treatment on seed yield and storability of tomato, cv. PKM-1, *Orissa J.Hort. 28*:33 (2000).

31a. J.M. Lin and J.M. Sung, Pre-sowing treatments for improving emergence of bitter gourd seedlings under optimal and sub-optimal temperatures, *Seed Sci. Technol. 29*; 39 (2001).

32. C. Leifert, Improving bacterial seed treatments-advantages and problems with use of molecular marker technology, Seed Treatment: Challenges and Opportunities, Proc. Int. Sym. Wishaw, UK, February 26–27, 2001, *Seed Abstr. 25*(6):254 (2002).

33. J.O, Quinn, The regulation of seed treatments in the European Community and Great Britain, Proc. Int. Sym. Wishaw, UK, February 26–27, 2001, *Seed Abstr. 25*(6):254 (2002).

34. A. Leuenberger, New technologies for seed loading and seed-to-seed distribution analysis-the critical parameters for treatment quality, Proc. Int. Sym. Wishaw, UK, February 26–27, 2001, *Seed Abstr. 25*(6):254 (2002).

35. G. Mani, A.S, Ponnuswamy and K.Vanangamudi, Performance of seed pelletization in Acacia leucophloea (Roxb.) under different soil types, *Tropical-Agricultural Res. Extention 2*:30 (1999).

16
Seed Packaging and Handling

I. INTRODUCTION

After seeds are dried, cleaned, and treated with suitable chemicals, they must be packaged in containers specifying their net weight. Packaging is the last operation of the seed-processing line, in which seeds are packed into bags of uniform size. Packaging consists of the following operations:

1. Filling of seed bags to the specified weight
2. Placing leaflets in the seed bags regarding improved cultural practices pertaining to the cultivar
3. Attaching labels and seed-certification tags to the seed bags and sewing of the bags
4. Transportation or storage of the bags

Transfering the cleaned seed from the processing plant to the field where it is to be sown is neither a simple nor a speedy operation (1). The seed may have to travel long distances by a variety of means of transport during which there must be no leakage of the seed or the pesticide with which it has been treated. Thomson (2) regarded packaging as:

1. A convenient unit for handling, transporting, and storing
2. A protection against contamination, mechanical damage, and loss
3. A suitable microenvironment for storage
4. A barrier against loss of seed and escape of pesticides
5. A sales promoter

Seed handling consists of receiving, elevating, and conveying operations in the processing plant. Seed-processing plants should have adequate equipment for receiving and conveying seed throughout the plant. A well-equipped plant has a pit area where incoming trucks can be unloaded quickly. From here seed is conveyed further to the processing units in the plant vertically, horizontally, or on an inclined plane as required.

II. SEED-PACKAGING MATERIALS

Modern packaging materials and methods maintain seeds at their original quality from the time of their packaging to the time they are used for planting. The best way to maintain the viability and vigor of many kinds of seeds to store them in a dry, cold place. Many kinds of seeds maintain their germination properties for several years even at quite high temperatures if they are kept very dry. But when dry seeds in porous containers (burlap, cotton, paper) are removed from refrigerated and dehumidified storage, they absorb moisture rapidly from the atmosphere, which can impair their viability in a few days or weeks. Very dry seeds (3–8% moisture) kept in moisture-proof containers retain good viability and vigor in different conditions of temperature and humidity (3).

Packages designed to protect most physical qualities of seeds, such as weight, size, color, moisture content, and purity (freedom from weeds, inert matter, and disease organisms and damage), as well as their physiological aspects, like viability, vigor, and dormancy, are made up of materials that have sufficient tensile strength, bursting strength, and tearing resistance to withstand normal pressures and handling procedures. Such materials do not normally protect seeds against either insects, rodents, or changes in moisture unless special protective qualities are built into them.

Packaging materials used for storing processed seeds can be made of burlap, cotton cloth, paper, films, metal, glass, fiberboard, or various combinations of ma-.terials. Some offer protection against moisture, whereas others do not. Each material has characteristics that make it suitable for a particular type of package (4).

A.　Burlap or Gunny Sacks

Burlap is a low-cost fabric woven of good-quality jute yarn. Both cotton and burlap as well as cloth bags are made in a variety of fabric construction to conform to the buyers' specifications. Burlap is strong in tensile strength and tear resistance, and burlap bags can be stacked high in storage and will withstand rough handling in distribution. The bags can be reused many times and still adequately protect most of the physical properties of seeds. Another minor advantage of cloth bags (cotton, jute, or synthetic fibers) is that samples for tests can be drawn with a spear without damaging the seed material. However, they are not suitable for seed treated with a highly poisonous pesticide.

Burlap cloth is used in laminations with various other flexible materials like asphalt, films, and paper. Laminated burlap bags are resistant to moisture transmission, insects, and rodents, and some types can retain the gases used for fumigation and seed treatment. They are used for both their strength and their barrier properties.

B.　Cotton Bags

Cotton bags used for packaging seeds are made from sheeting, print cloth, drill, osnaburg, and special seamless material. Cotton fabrics are produced in different sizes, each designed with a special purpose. Osnaburg, which is stout and coarse, and seamless fabrics having the greatest tensile strength and tear resistance of the cotton materials are most widely used. Seamless cotton bags can be reused many more times than ordinary cotton bags.

Cotton fabrics can be coated and laminated. Selected laminating materials are bonded together by asphalt or vegetable or compounded latex adhesives; the cotton cloth provides the required strength and protective properties.

Although cotton bags are good protectants of the physical properties of seeds, they are not resistant to moisture, insects, or rodents unless special barrier properties are built into them. According to Bass et al. (4), the two most widely used laminated moisture-proof bags consist of one or two sheets of paper attached to a fabric with asphalt or compounded latex adhesive. Other moisture-proof packages utilize such barrier materials as vegetable parchment, pliofilm, polyethylene, and rubber coatings. Moisture-proof liners are sometimes used inside cotton bags.

C. Paper Products

Products made of paper are widely used for packaging seeds. Small packets are made of bleached sulfite or bleached kraft paper, which is coated with white clay to facilitate printing. The packets are designed to contain a measured amount of seed without loss, but they do not normally protect seed viability under unfavorable conditions.

Many paper seed bags have multiple plies or layers. Such bags can be constructed of various thicknesses of smooth or crinkled paper. Regular multiwall bags contain two or more plies of kraft paper; the outside ply is heavier to take wear, and special plies are hidden among the layers of the multiwall. A special barrier material like asphalt, polyethylene, or aluminum foil is included when moisture protection is desired.

Ordinary multiwall bags have poor bursting strength, and bottom bags may burst when they are piled high. The top bags in high piles often slip. They also tend to dry out in dry climates and become brittle along folds and on the corners at wear points. If a sample is drawn through a spear from paper bags, the hole can be repaired with an adhesive patch.

D. Elastic Multiwall Paper Bags

Elastic multiwalled paper bags have several walls of crinkled paper; the number of plies depending on the weight of the product to be packaged. Two outer plies are often laminated together with asphalt to provide a moisture-proof barrier, which also protects the inner layers of paper from damage by rain.

Duplex-ply paper bags are considered to be airtight, but there may be some interchange of gases through the top closure sewing holes. The thickness, toughness, and stretch of the outer ply provide the elastic multiwall paper bag with excellent puncture resistance. The resilience of the packaging material absorbs the shock of impact, keeping the bags from splitting.

Some bags are made of laminates of paper, polyethylene, and aluminum foil, which afford better protection against moisture than foil or polyethylene layers used alone with paper. The paper can also be treated to repel insects and rodents.

Multiwall paper and laminated cotton and burlap bags are designed to meet the specific needs of a variety of conditions of weather, shipping, handling, and storage.

E. Cellophane-Polyethylene Laminate Bags

Films of cellophane, pliofilm, polyester, polyvinyl, aluminum foil, and polyethylene are used alone or in various combinations to make suitable bags for seeds. Cellophane made of regenerated cellulose is produced in more than 100 varieties, each designed for a specific purpose. Moisture-proof types, having low moisture vapor–transmission rates, are used for small packages. Cellophane alone may become brittle with age and break easily, but combinations of cellophane and polyethylene do not normally become brittle and offer good moisture protection to seeds. Polyethylene-cellophane laminates heat seal easily and perform well on automatic packaging machines.

Pliofilm is a thermoplastic rubber hydrochloride plastic film that resists ripping, tearing, and splitting. It also seals well at low temperatures, has good moisture-barrier properties, and can be laminated to itself, paper, foils, and other films. Pliofilm is used on most packaging machines designed for flexible film packaging. It deteriorates in strong light.

Polyester films are heat-sealable, transparent, flexible plastics with low moisture vapor–transmission, carbon dioxide–transmission, and oxygen-transmission rates. They have great tensile strength, and do not dry out or become brittle with age, because they do not contain plasticizers. Polyester film can be laminated to itself and practically any other material. Its flexible laminates are commonly used in a variety of flexible packages. One novel construction utilizes a base of light cotton fabric and metalized polyester film, which offers easier fabrication, stronger seals, and better resistance to flex damage, rough handling, and pinholes (4).

Polyvinyl films are heat sealable, deteriorate slowly in sunlight, and have outstanding tensile strength and tear resistance. They provide only moderate moisture protection, however, unless laminated with a good moisture-barrier material. They laminate well to paper, foil, or other films.

F. Annealed Aluminum Foil

Annealed aluminum foil has a high tensile strength and low moisture vapor–transmission rate even for films of the lowest thicknesses, which have tiny perforations called pinholes. Aluminum foil alone does not make good seed packaging, but it can be bonded to other materials to produce combinations with almost any desired characteristics. Despite the pinholes in thin foils, combinations with paper or plastic films offer an effective barrier to moisture and gases. Laminations such as (a) aluminum foil–glassine paper–aluminum foil–heat-sealing lacquer, (b) aluminum foil–tissue paper–polyethylene, and (c) paper–polyethylene–aluminum foil–polyethylene have been used most satisfactorily.

G. Polyethylene Bags

Polyethylene is the most extensively used thermoplastic film. It is made from aliphatic hydrocarbon resins (polymers of ethylene). Commercially made polyethylene resins are of three types—low-, medium-, and high-density films differing in their molecular and physical structures. Conventional low-density films are generally used in seed packages, but a special new medium-density film shows promise. Medium- and high-density films are progressively less permeable to moisture vapor and gases. A special medium-density polyethylene film with a specific gravity of 0.938 has better tensile properties and greater elongation than conventional films. Because of its high percentage of stretch, it has good resistance to puncture by rodents. With a tight closure and a heat seal, these medium-density polyethylene bags can be made completely insect proof.

Polyethylene films can be laminated to themselves, other films, foil, paper, textile fabrics, and fiberboard. Laminations further improve such film properties as moisture barriers and other physical properties, some being practically completely impervious to moisture vapor and gases. The disadvantage of plastic film bags is their smooth surface, which allows bags to slide over each other so that stacks collapse, and the difficulty of drawing seed samples for testing. This packaging material is ideal for providing sealed-storage conditions in a humid climate, giving protection against high humidity in a seed store, during transport, in the merchant's premises, on the farm, and even against rain when left outdoors. Packaging in sealed bags, however, requires additional drying, which involves increased cost (5).

Packaging seeds in moisture-resistant (hermetically sealed) containers and the effectiveness of several types of packaging materials in preventing moisture uptake and maintaining seed viability of creeping red fescue seeds was investigated (6). The results of this study (Figs. 1 and 2) indicated that ordinary paper and cloth containers were least effective, whereas various laminate and polyethylene materials were moderately effective, and metal cans were completely effective in maintaining seed moisture at the initial 5% level. Such completely moisture-proof containers hermetically seal the seed and effectively preserve seed quality for 5–10 years or longer. The effectiveness of packaging materials is directly associated with their ability to resist moisture (7). Doijode (8) reported that papaya seeds need greater air circulation during storage and could be kept viable for 2 years in paper bags at 10°C without a significant decline in vigor. Low temperatures reduced the viability of papaya seeds remarkably (Figs. 3–6).

Pandey (9) evaluated the prospects of conserving French bean seeds by embedding them in an organic liquid (diethylene glycol) medium and aging at 58°C. The diethylene glycol (DEG) dynamically quenched seed moisture to about 5% and enhanced seed longevity about 18-fold over the seeds hermetically sealed and similarly aged. The reduction in the rate of aging of the seeds embedded in DEG was evident from the decrease in membrane disruption, dehydrogenase activity decline in embryonic axes, and loss of vigor and viability over the seeds hermetically sealed and aged. Quenching of moisture from the seeds and elimination of oxygen from the seed environment were the decisive factors in retarding the rate of aging by embedding

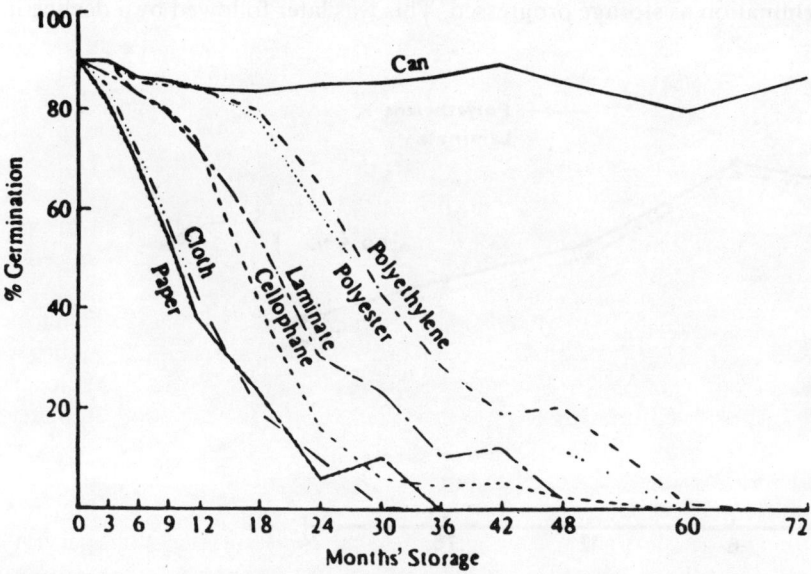

Figure 1 The effects of different packaging materials on the germination of creeping red fescue seed. (From Ref. 6.)

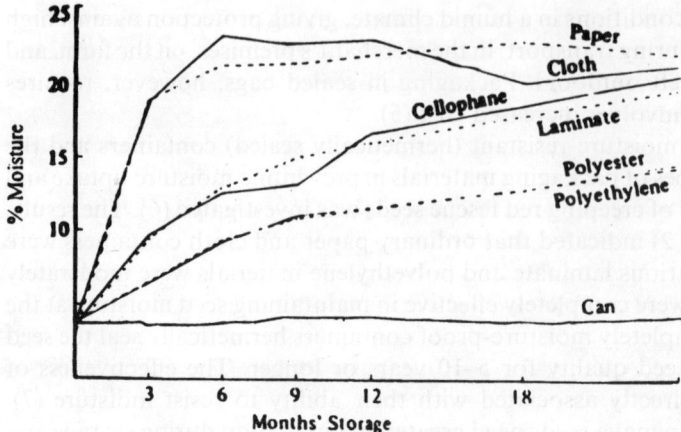

Figure 2 The effects of different packaging materials on the moisture content of creeping red fescue seed. (From Ref. 6.)

seeds in DEG. About 14% of seeds absorbed the chemical, which invariably resulted in loss of viability. This study indicated that orthodox seeds could be preserved for long periods by embedding them in a suitable organic liquid medium at ambient conditions, avoiding costly refrigeration (9).

Freshly harvested pepper (*Capsicum annuum*) cultivars Tatashe and Rodo seeds were packaged in laminated aluminum foil, polythene, and paper envelopes at moisture contents of 9.2% (Tatashe) and 9.3% (Rodo), respectively, and stored for 24 weeks at 30°C and at about 90% RH. Seeds were drawn at intervals for germinability and seedling development tests. There was an initial improvement in total and rate of germination as storage progressed. This was later followed by a decline in

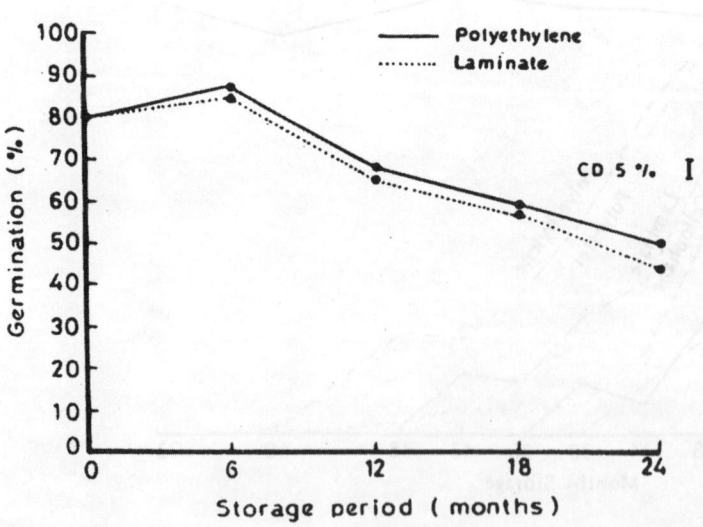

Figure 3 Viability of papaya seeds stored at ambient temperature. (From Ref. 8.)

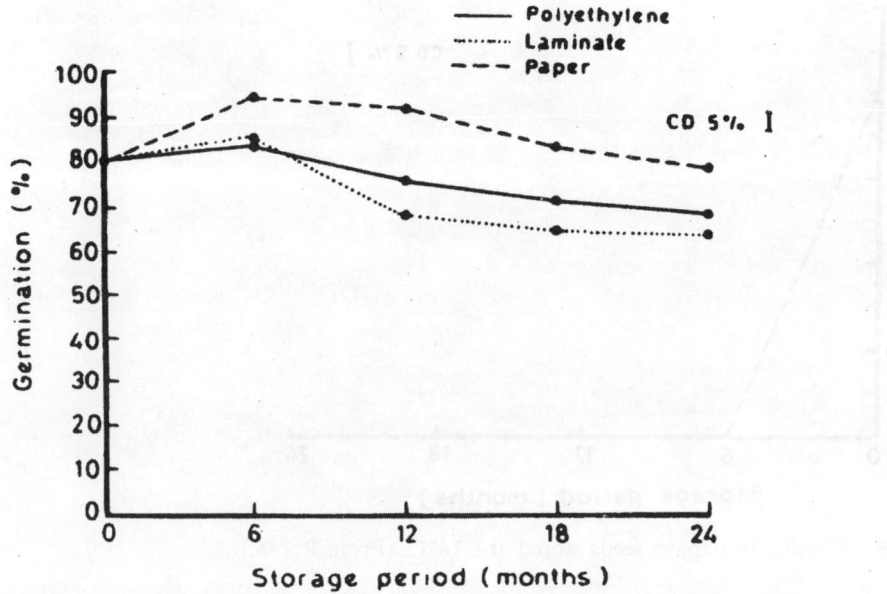

Figure 4 Viability of papaya seeds stored at 10°C. (From Ref. 8.)

Tumaized lids and other volatile substances originating from aluminum containers or glue on seed packets may cause a rapid decrease of seed germinability.

the three containers. Measurements were best in aluminum foil packets and worst in polythene toward the end of storage. Serious fungal growth was observed on seeds packaged in polythene. Marked improvements were observed in seedling growth with storage in all containers up to certain ages beyond which a decline set in. The improvement in seed germination and seedling development from some aged seeds was taken to be an indication of the need for after-ripening of pepper seed (9a).

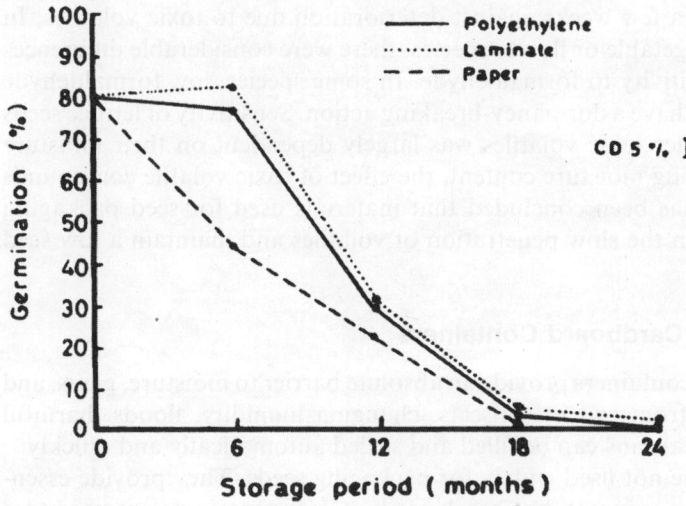

Figure 5 Viability of papaya seeds stored at 5°C. (From Ref. 8.)

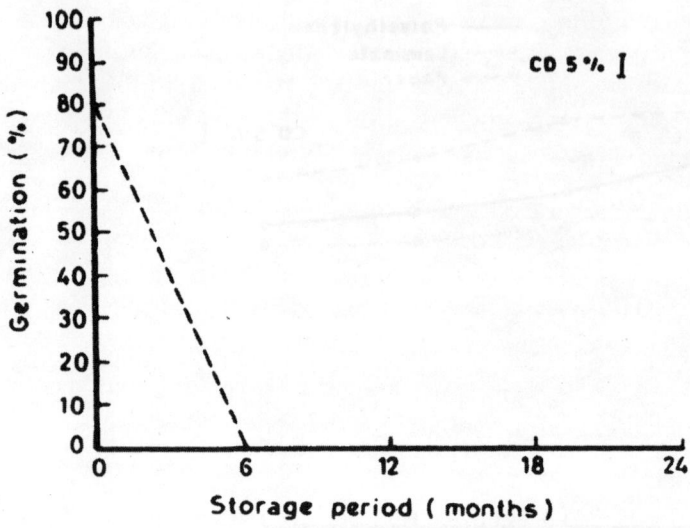

Figure 6 Viability of papaya seeds stored at −18°C. (From Ref. 8.)

Formaldehyde and other volatile substances originating from printing ink, paper, or glue on seed packets may cause a rapid decrease of seed germinability. Lettuce seeds appear to be particularly sensitive. At 20°C and 75% RH, lettuce seeds (cv. Prado) lost germinability within 1 month when stored in a desiccator together with five recently printed paper bags as a source of toxic volatiles. In the absence of toxic volatiles, germination capacity had not decreased after a storage period of 12 months under the same conditions of temperature and RH. Various packaging materials (based on polypropylene or polyethylene) tested maintained seed germinability for a period of 4 months in the presence of toxic volatiles, but germination capacity started to decrease after 8 months. At 30°C and 75% RH, the packaging materials tested protected seeds for only a few weeks against deterioration due to toxic volatiles. In tests with a total of 48 vegetable or flower species, there were considerable differences with respect to seed sensitivity to formaldehyde. In some species, low formaldehyde concentrations appear to have a dormancy-breaking action. Sensitivity of lettuce seeds to formaldehyde and other toxic volatiles was largely dependent on their moisture content. With an increasing moisture content, the effect of toxic volatile compounds increased markedly. It has been concluded that materials used for seed packaging should protect seeds from the slow penetration of volatiles and maintain a low seed moisture content (9b).

H. Metal, Glass, and Cardboard Containers

If properly sealed, metal containers provide an absolute barrier to moisture, gases, and light and protect seeds from rodents, insects, changing humidity, floods, harmful fumes, and so forth. Metal cans can be filled and sealed automatically and quickly.

Glass containers are not used widely for packaging seeds. They provide essentially the same protection as metal, but break easily. Glass containers are used to a limited extent in research laboratories and occasionally as display receptacles in stores where bulk sales are made. Cardboard boxes and cans are used extensively to store seeds. Cardboard cans have metal lids and bottoms. Conventional cardboard (boxboard) is laminated with polyethylene, aluminum foil, or other material to achieve a moisture-barrier quality. Sometimes cartons are overwrapped with wax paper, aluminum foil, or polyethylene to get the same effect. Cardboard containers protect the physical qualities of seeds and are well adapted to automatic filling and sealing (4).

III. SEED-PACKAGING TECHNIQUES AND EQUIPMENT

Seed packages are filled in many ways, and the equipment used may range from a simple spoon (or seed scoop) to a high-speed, completely automatic small packet, metal can, or plastic bag filler. Gravity flow from a manually controlled bin is also used. Most filling equipment has a measuring device, which may be controlled manually or automatically by a signal from a weighing machine.

Practically all seeds except those in small packets are sold on a weight or volume basis. Seeds sold on a volume basis are associated with weight; for example, a bushel of corn or wrinkled peas weighs 56 lb, and a bushel of smooth peas weighs 60 lb. Weighing devices range from an ordinary beam scale to an elaborate scale that activates a pneumatic or electrical device to shut off the flow of seed when a certain weight or volume is reached. Rigid containers always have the same volume, but the seeds needed to fill them may vary in weight if the seeds are not vibrated properly while filling. The vibrator may be attached to the platform that supports the container while it is being filled.

Some seeds are sold on the basis of number; either actual or adjusted for percentage of pure live seed (% of pure seed × % germination). In the United States, some seed companies package hybrid corn seed in acre units—each package contains the correct amount of seed for planting a specified acreage.

Except in small operations, seeds to be packaged are delivered to hopper bins above the filling machines. Seeds may come to the hopper from bulk storage bins by gravity flow through pipes, airlift, belt conveyors, or storage boxes or in bags handled by elevators, forklifts, or manually.

IV. SEED HANDLING

All handling of seeds during packaging operations must be done with care to avoid any possible effect on the physical quality of the seed through impact or undue pressure. Heavy seeds like those of peas, beans, corn, and soybean can be fractured if they strike or are struck by a hard object or film surface. The breaks may not be visible in the dry seeds. The severity of injury is related to the moisture content and the force of impact. Seeds that contain too much or too little moisture are normally damaged when they are forced through a restricted opening (4,5).

Several types of conveyors are used for moving seed into, through, or away from the processing plant in a vertical, horizontal, or inclined direction. Selection of conveyors of adequate capacity avoids damage to seeds and increases the overall efficiency and effectiveness of seed handling during packaging.

Agrawal (10) described the following types of conveyors used at seed-processing plants:

1. Bucket elevators
2. Belt conveyors
3. Vibrating conveyors
4. Pneumatic conveyors
5. Screw conveyors
6. Chain conveyors
7. Lift trucks

Selection of the proper conveyors for receiving seed in the plant, moving it from driers, shellers, and processing machine to another and, finally, moving the filled seed bags to storage has an important bearing on the efficiency of seed handling during processing. The kinds of seeds handled, the direction and length of conveying, and equipment capacity are other important considerations in eliminating as much manual handling as possible. In sizable operations, dependence on manual handling limits the capacity and efficiency of different processing steps. Conveying equipment should always have a greater capacity than the equipment to or from which it is intended to carry seeds.

Some miscellaneous accessory equipment needed during packaging operations include blowers, vacuum cleaners, and different types of bins for storage of seeds. Rough handling of packages and planting seed with equipment that squeezes the seeds as they are fed into the planting spout can also cause injury to seeds. Such damage may kill the embryo or cause weak or abnormal seedlings if germination takes place. Mechanical damage shortens the storage life of seeds even when held under most favorable conditions of storage (3).

The types and sizes of seed packages used for wholesale distribution are often quite different from the ones used for retail sales. Processors usually package seeds in burlap, osnaburg, or seamless and multiwall paper bags holding 25–50 kg of seed. A number of companies use moisture-proof packages or burlap or cotton bags with polyethylene liners for cereal grains, soybean, hybrid sorghum, corn, peas, beans, and other seeds. A valve-type polyethylene bag prevents loss of material while filling and can be sealed more easily than the conventional bag. Both paper and polythene bags can be closed by a valve, which is shut automatically by the weight of the seed when the bag is turned over after filling. For certified seed, an official sealing device is incorporated in whatever method used (2). Fumigants and inert gases can be easily introduced into the filled bags. Some alfalfa seed is packaged in large cardboard cartons, whereas flower seeds are sometimes packaged in cans. Large nonrigid containers (burlap, cotton, multiwall bags) are held in place by hooks or clamps or manually by hand during filling.

Hand trying of the open ends of cotton and fiber bags has largely been replaced by sewing machines. Multiwall paper and laminated bags are closed by sewing or sewing and taping, and polyethylene bags are closed by heat sealing. Heat sealers include small hand irons or rollers, hand- or foot-operated jaws or clamps, and elaborate automatic machines for forming, filling, and sealing bags and pouches. Some sealers use thermostatically controlled bars, bands, or rollers.

A wide variety of materials and package sizes are used to prepare seed for retail sales. Most field seeds are sold at retail in the original wholesale packages, but seeds of vegetables, flowers, and lawn grasses are packaged for various types of retail customers using multiwall paper, cloth, and plastic bags, cardboard boxes and drums, and metal cans of 1–5 kg capacity. Small paper, foil, and plastic packets and cardboard boxes containing few seeds (10–50 g) are used for mail order and store sales of vegetable and flower seeds (11).

V. LABELING AND STENCILING

Filled seed packages must be labeled appropriately to show the species, variety, grade and lot number, percentage of live seed, purity, content of noxious weeds, and seed treatment if any. All this information may be printed on a tag attached to the bag or on a label that is glued to the container. It may also be printed or stamped directly on the container. Seedsmen usually print their own tags and labels. A packaging bag of any suitable material should also be regarded as a means of promoting sales, and the minimum amount of information, as required by law, should be supplemented by a bold distinctive design and, for high-technology cultivars, some advice on cultural methods. If the seed has been chemically treated, some dramatic emblem should indicate this.

Rotary printers may carry out stenciling on seed bags either manually or automatically as the container passes a point on an assembly line. The closing machine normally does embossing. Special labeling machines can apply glue to the can or label and wrap the label around the can as it rolls through the machine. A special printer may imprint information on the can before the label is applied, so that it can be identified even if the label is removed.

VI. ASSEMBLING AND STACKING

The final packaging operation is the assembling of packages and stacking. Larger seed containers are usually brought together by belt or roller conveyors or placed manually by hand on pallets handled by forklift or using hand trucks. Smaller seed units are frequently placed in larger cartons by hand or automatically with a machine built to assemble a selected number of units and place them in cartons. The seeds are then ready for transportation and distribution to farms or warehouses for temporary storage. An industrial forklift truck is used with pallets to handle bagged or packaged seed in warehouses that have suitable floors and column arrangements. This method is adapted to picking up and stacking unit loads—groups of bags or packages—rather than single bags.

REFERENCES

1. L. E. Holman and J. R. Snitzler, Transporting, handling and storing seeds, *Seeds: Yearbook of Agriculture*, US Department of Agriculture Washington, DC, 1961, p. 338.
2. J. R. Thomson, *An Introduction to Seed Technology*, Leonard Hill, London, 1979.
3. C. E. Vaughan, B. R. Gregg, and J. C. Delouche (eds.), *Seed Processing and Handling*, Handbook No. 1, Seed Technology Laboratory, Mississippi State University, State College, MS, 1967.
4. L. N. Bass, T. M. Ching, and F. L. Winter, Packages that protect seeds, *Seeds: Yearbook of Agriculture*, US Department of Agriculture, Washington, DC, 1961, p. 330.

5. J. E. Douglas, *Successful Seed Programs, A Planning and Management Guide*, Westview Press, Boulder, CO, 1980.
6. D. F. Grabe and D. Isely, Seed storage in moisture-resistant packages, *Seed World 104(2): 4* (1969).
7. L. O. Copeland and M. B. McDonald, *Principles of Seed Science and Technology*, 3rd ed., Chapman & Hall, New York, 1995.
8. S. D. Doijode, Influence of storage temperatures and packaging on the longevity of papaya (*Carica papaya*) seed, *Seed Research*, Special Vol. No. 1 (S. P. Sharma, ed.), Indian Society of Seed Technology, New Delhi, 1993, p. 288.
9. D. A. Pandey, Prospects of seed conservation by embedding in organic liquid medium-possibilities and implications, *Seed Research*, Special Vol. No. 1 (S. P. Sharma, ed.), Indian Society of Seed Technology, New Delhi, 1993, p. 262.
9a. J. A. Oladiran and S. A. Agunbiade, Germination and seedling development from pepper (*Capsicum annum* L.) seeds following storage in different packaging materials, *Seed Sci. Technol.* 28: 413 (2000).
9b. H. L. Kraak and J. G. van Pijlen, Packaging materials and seed viability: effect of formaldehyde and other toxic volatiles, *Seed Sci. Technol. 21:* 463(1993).
10. R. L. Agrawal, *Seed Technology*, Oxford and IBH Publishing, New Delhi, 1980.
11. J. E. Harmond, N. R. Brandenberg, and L. M. Klein, *Mechanical Seed Cleaning and Handling*, Agricultural Handbook No. 354, ARS, US Department of Agriculture in cooperation with Oregon Agricultural Experiment Station, Washington, DC, 1968.

17
Seed Transportation and Storage

I. INTRODUCTION

Seed is required to be transported from the place of its production to the place where it is processed, from processing plants to warehouses where it is stored, and again from these places to farms for planting. The transfer of the raw seed to the place of processing as well as the transfer of a clean, processed seed to the place of planting is not a simple and speedy operation; its transit involves a variety of modes of transportation, such as carts, trucks, railrods, and ships. The seed may have to travel long distances by a variety of means, and it can be subjected to jolting and rough handling during transportation. Its journey may be interrupted by periods of storage prior to sale. Even after its arrival at the farm, seed may need to be kept for a time in a barn or outbuilding before planting.

Maintenance of high germination and vigor from a seed's harvest until it is planted in soil is the primary aim of good seed-production technology. Seeds are practically worthless if upon planting they fail to germinate and give a healthy and vigorous crop stand. The safe handling, transportation, and storage of seed throughout its journey is a basic requirement of seed production.

II. SEED TRANSPORTATION

A. Modes of Transportation

Trucks, railroads, and ships constitute the major modes of seed transportation all over the world. The primary users of these facilities are middlemen and seed merchants, who assemble, process (condition), store, and ship the seed to large-scale seed growers or to other middlemen, who sell it to farmers for planting. A substantial quantity of seed is transported on a global basis; however, seed still accounts for only a fraction (1–3%) of the total agricultural produce transported by the three major types of carriers (1).

All modes of transport are used for hauling seeds. In India and the United States, transportation of seeds by railroads predominates (>6% of the total), representing primarily the long-haul movement of seeds; that is, from the country assembly points through the various wholesale trade channels to the retail stores. Trucks mostly accomplish the short-haul movement from the farm to the country assembly point and from the retail outlet to the farm. In the developed countries, the long-haul movement of seeds also involves, to a limited extent, the use of boats and airplanes. The seed shipped by air is primarily experimental and high-value seed for which the buyer is willing to pay the high cost of this premium service. Postal and package services are used extensively for small shipments of vegetable and flower seeds, utilizing mainly railroads and trucks.

Privately owned and leased trucks are generally used to transport short-haul seed shipments. These trucks are either owned or leased (a) by seed producers and wholesale buyers who haul the seeds from the farm to the country assembly point for reshipment, (b) by farmers who purchase the seed from the local retail seed dealers and haul it to their farms, and (c) by the retailers who provide delivery service to the farm.

According to Holman and Snitzler (1), two types of leased carriers engage in seed transportation in the United States: exempt carriers and regulated carriers. The distinction is based on the Motor Carrier Act of 1935, which provides exemptions for vehicles hauling nonmanufactured agricultural commodities such as fresh fruit and vegetables, poultry, eggs, livestock, grain, and seeds. Exempt carriers haul only exempt commodities and are subject to the rules of the Interstate Commerce Commission (ICC) as to safety and hours of service of drivers. The regulated carriers are authorized by the ICC to transport other than exempt commodities; they may haul exempt commodities, but then are not subject to regulation by the ICC. Seeds serve to balance out return trips for many regulated motor carriers. Both types of motor carriers haul seeds from the production centers or warehouses of seed wholesalers to their retailers.

Seedsmen can save money using rail transportation, although sometimes it is cheaper to use trucks for either inbound or outbound trips. Convenience of loading and unloading is another advantage of rail transportation. The shipper or receiver of seeds has 48 hr free of charge for loading or unloading the rail car after it has been placed at his disposal; any time beyond the 48 hr is chargeable as per the published demurrage rates. The rail carriers can handle large shipments on long hauls at low rates.

Trucks usually deliver seeds from the warehouse of a wholesaler to customers in less time than it would take to move them by rail. Speed of transportation is important, particularly late in the planting season when retailers run out of certain varieties of seeds and need refill orders immediately or during emergency conditions such as floods and droughts that may have ruined a farmer's first planting, making replanting necessary. Speed is also important in the servicing of the supermarket trade by wholesale seed dealers. Large chains operate on the basis of fast turnover, needing replenishment of seed stocks several times during the season. The date and hour of delivery is often specified and mandatory, making the fastest mode of transportation important.

A large portion of the seed-marketing business, however, does not require fast delivery service. The seeds of grasses, legumes, vegetables, and flowers are harvested in the late summer and fall, and several months elapse before seed is needed for planting. During this interval, the seed is generally stored in the wholesalers' warehouses. In order to reduce storage risks and to ensure more orderly distribution, shipments of seeds to independent seed retailers are begun as early as December, using both rail and truck transportation.

Trucks also provide pickup and delivery services, which are particularly important to customers who are not near railroads. Both rail cars and trucks can be partially loaded at one place, completing the loading somewhere else. They may also be stopped for unloading at more than one place. Stopping in transit to load or unload some of the seed allows buyers to obtain the benefit of lower carload rates usually available on heavier shipments.

B. Losses During Shipments

Both the shippers and carriers are responsible for taking protective measures to ensure that seed lots arrive at their destination in satisfactory condition. The shippers are responsible for seed being properly loaded and protected from ordinary transportation hazards, and carriers must deliver it to the destination in the same condition in which it was received.

A major cause of loss and damage to seed shipments is torn sacks or bags (normally associated with railway shipment) causing loss of both seed and container. Protruding nails and bolts and loose or splintered boards are common causes of torn bags during transport. Water may cause the containers to split apart, damaging the seed's germinability. Leaky roofs, loose-fitting doors, and worn tarpaulins may lead to such damage, which occurs more frequently in truck shipments.

Industrial chemicals and oil residues not removed from the rail car or truck before loading may cause the seed bags to disintegrate. The container may soak up some of the residues or take up an odor from them.

Freezing injures some seeds such as potato seeds. Portable heaters may need to be placed in the truck or car for shipments in winter, taking care that they do not overheat.

Grain that is moved in bulk into the wholesaler's plant for further processing may suffer damage and loss during transit because of loose doors, loose or broken floor boards, or cracks in the floor.

C. Precautions and Protective Measures

The shipper must critically inspect the carrier's equipment to ensure that it is fit to haul the seeds in cool and dry conditions. Such inspections should reveal the presence of loose or broken wallboards or floorboards, protruding nails or bolts, broken pieces of wire strapping, chemical or oil residues, or other material that might damage the seed or its containers. The shipper should ask the carrier to replace the equipment if it is unsatisfactory.

Before loading, cars or trucks should be swept carefully and the floor, sidewalls, and ends lined with heavy paper. Special precautions need to be taken for bags or cartons stacked near the doorway of a rail car, using strips made of heavy paper, reinforced at regular intervals with steel strapping, and nailed to the doorposts through prepunched holes in the strappings. For bulk shipments of seeds, one-piece wooden doors or heavy-duty water-repellent paperboard, reinforced with steel strapping, are placed inside the regular car doors for additional protection.

The heaviest bags should be placed on the bottom to prevent other bags from splitting open from the overhead weight. The bags should also be stacked tightly together in an interlocking pattern to reduce chances of shifting the load during transit. While unloading the shipment, bags must not be dragged over the car floor and the bottom and outside layer of bags must be removed with care to avoid any possible damage. Care should be taken to remove all nails and bolts before loading, but some may work out in transit.

Seeds should be handled with great care during loading and unloading. Modern automated equipment is available for safe handling of most kinds of seeds: for example, bucket elevators to move seeds vertically in bulk and self-cleaning vertical elevators for handling bulk seed; pneumatic conveyors to carry materials through a pipe in a high-velocity stream of air; high-pressure fluidized conveyors to avoid physical damage to seeds during handling; belt conveyors to move seeds in bags or in bulk in a horizontal or inclined direction; flat-belt conveyors for bagged or packaged seeds; and troughed-belt conveyors to move seed in bulk. Portable belt conveyors with platform elevators are used in warehouses for piling bagged seeds, removing bags from the piles, and moving bagged seed into and out of the warehouse.

The industrial forklift is used with pallets to handle bagged or packaged seed in warehouses having suitable floors and column arrangements. Unit loads (group of bags or packages) can be picked up and stacked by this method, enabling the transport of loads of 150 m or more. Forklift trucks with a capacity of 1500–2000 kg are suited to handle bagged seeds in many warehouses. Smooth, level floors and runways speed the movements of such trucks. Pallets form a natural base for transporting unit loads. The wooden pallet (1.2 × 1.2 m) is used widely because of its low cost, light weight, and fair durability. More costly metal skids of various types and sizes are also available. A pusher bar installed on the front end of a forklift is used to push the load off of a pallet onto the floor of a rail car or truck. If pallets accompany rail and truck shipments, the charges for their return shipment can be sizable. Expendable, one-trip paper pallets are often used to save the cost of the return freight.

III. THE IDEAL SEED-STORAGE ENVIRONMENT

The principal agents of deterioration of seed during storage are fungi and insects, whose development is influenced by the seed moisture content and temperature of the storage facility. Important fungi and insect pests associated with some stored food legumes and grains are given in Tables 1 and 2, respectively.

Beetles such as the granary weevil (Fig. 1) bore into the seed and lay their eggs internally; the larvae then feed on the endosperm and the embryo. In some crop species, the eggs are laid on the surface and the larvae bore into the seed (Fig. 2).

The limiting relative humidity (RH) for the growth of fungi is about 65–70%; corresponding to the equilibrium moisture content of about 14–15% in starchy seeds and 8–9% in oilseeds (Fig. 3). Both seed moisture and humidity within the seed airspaces determine the growth of insects (3). Insects generally do not develop at seed moisture contents equivalent to 40% RH: that is, about 10–11% in starchy seeds and 5–7% in oilseeds, but there is considerable variation between different crop species (Table 3).

Table 1 Important Fungi Associated with Some Stored Food Legumes

Legume crop	Fungal species
Cowpea	*A. niger, Rhizoctonia baticola, Cladosporium herbarum, A. flavus, Absidia* spp., *Rhizopus* spp.
Pea	*Aspergillus* spp.
Peanut	*A. flavus, A. parasiticus*
Phaseolus spp.	*A. glaucus, A. restricus, Penicillium* spp.
Soybean	*A. flavus, A. glaucus, A. restrictus, A. ochraeus, A. niger, A. funigatus, A. repens*

Source: Ref. 2.

Table 2 Important Insect Pests of Some Stored Food
Grains

Order and pest	Common name
Order Coleoptera	
Sitophilus granarius	Granary weevil
S. oryzae	Rice weevil
S. zeamais	Corn weevil
Tribolium castaneum	Red flour beetle
T. confusum	Confused flour beetle
Ginathocerus cornutus	Broad-horned flour beetle
Trogoderma granarium	Khapra beetle
Stegobium paniceum	Drugstore beetle
Oryzaephilus surinamensis	Saw-toothed grain beetle
Achanthoscelides obtectus	Common bean weevil
Callosobruchus spp.	Cowpea weevil
Bruchus pisorum	Pea weevil
Bruchus rufimanus	Bean weevil
Order Lepidoptera	
Ephestia kaehniella	Mediterranean flour moth
Cadra cautella	Almond moth
Plodia interpunctella	Indian meal moth
Nemupogon granellus	Grain moth

Source: Ref. 2.

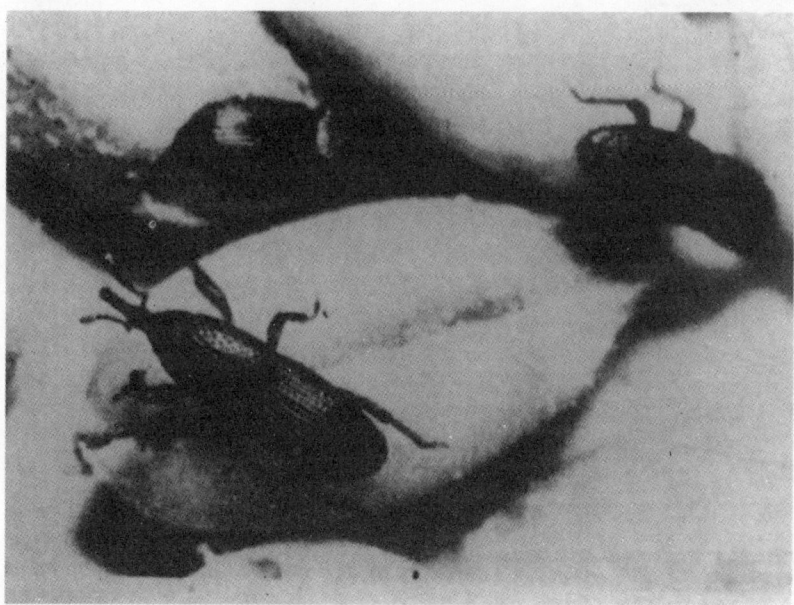

Figure 1 Granary weevil on wheat (Ministry of Agriculture, Fisheries and Food). (From Ref. 3.)

Figure 2 Lentils damaged by beetles. Egg capsules are on the surface and adults are ready to emerge (Ministry of Agriculture, Fisheries and Food). (From Ref. 3.)

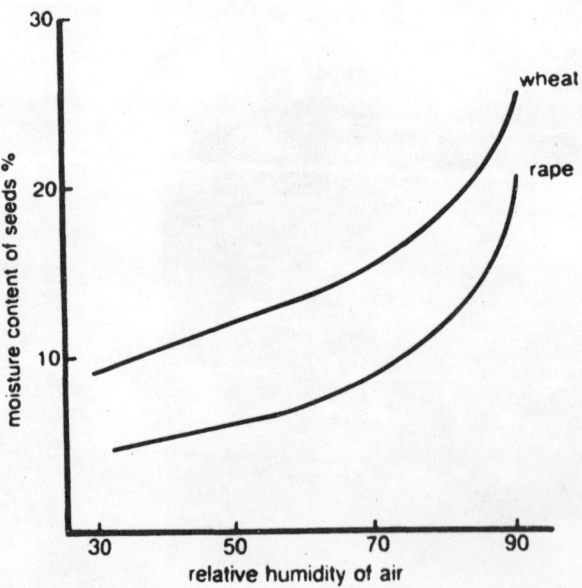

Figure 3 Curves showing how the equilibrium moisture content of wheat and rape seed changes with the relative humidity of the ambient air (J. Kreyger, IBVL, Wageningen, 1972). (From Ref. 3.)

Table 3 Typical Moisture Content (%) of Seeds in Equilibrium with a Range of Relative Humidities (RH)

Crop	Percent moisture at:				
	30% RH	40% RH	50% RH	60% RH	70% RH
Wheat, rye	9	10	12	13	15
Barley, oats	8	10	11	12	14
Maize	9	11	12	13	14
Sorghum	9	10	11	12	14
Rice	8	9	11	12	13
Rye grass	9	10	11	12	14
Lucerne	8	9	10	12	14
Phaseolus	7	10	12	14	16
Pea	8	10	12	13	15
Beet, onion	7	8	10	11	13
Soybean	7	7	8	10	12
Cotton	6	7	8	9	10
Brassica	5	6	7	8	9
Groundnut	4	5	6	7	9

Source: Ref. 3.

Even if fungi and insects do not develop because of an unfavorable storage environment (too dry and too cold), seeds can still deteriorate with age and lose viability. Deterioration due to aging alone becomes significant only in long-term storage; it does not affect short-term (season-to-season) or reserve storage of most kinds of farm seeds.

Taking into consideration all the causes of seed deterioration, two of the most important factors determining the storage life of seeds are the seed moisture content and the temperature at which seeds are held. Availability of oxygen plays some part because of the role played by seed respiration. However, oxygen becomes limiting only in hermetically sealed storage systems, and can be ignored in the storage of large seed lots.

Barring extreme conditions, the effects of seed moisture and temperature can be summarized in terms of Harrington's rules (4); namely, "The storage life of seeds is halved for each 1% increase in moisture content and for every 5°C rise in storage temperature." These rules operate with seed moisture contents ranging from 5 to 14% and temperatures from 0 to 50°C and are independent of one another. Above 14% moisture, molds can grow on seeds, and heating can occur above 18% moisture. Insect activity is greatest between 21 and 27°C, and therefore seeds should be stored below 20°C whenever possible (5).

The ideal storage environment to maintain seed viability can be expressed simply as "dry and cool," but the material and mechanical resources required to provide this ideal storage environment vary enormously according to the prevailing climate. Duration of storage is another important variant that must be taken into account. Most seeds required to be stored from harvest to planting (one season or less than 1 year) do not require elaborate storage conditions, but seed stocks to be stored for more than a year will need more stringent storage conditions. A rule of thumb would be that the temperature (in °F) added to the relative humidity should not exceed 100 (3).

Because seed is a poor conductor of heat, seed lots tend to maintain their original moisture content and temperature for a considerable time depending upon the weather. In drier and temperate regions, a well-dried and cooled seed can store well without elaborate efforts. In temperate climates, seed deterioration is mainly caused by fungi, and hence it must be dried to a moisture content in equilibrium with 65% RH or less (see Table 3). Stored seeds can be kept cool simply by blowing ambient air through the seed lot. Seed moisture content can be adjusted slightly depending on local factors, such as the kind of seed container, the structure and site of the warehouse, and the duration of storage desired. Deterioration occurs mainly during the warm summer months in seeds stored for longer periods. This can be checked to some extent by aerating the seed lot with cool air of low humidity (3).

Longevity and storage capacity differ greatly between seed lots and even within the same seed lot. The typical differences in the longevity of three seed lots each of garden beans and sorghum under open storage conditions are shown in Figure 4 (6). The curves shift from high to low seed quality, and the range becomes wider with an increase in storage time. All seed lots are composed of individual seeds, each possessing its own unique capacity to perform in the field. A hypothetical distribution of loss of seed quality over a period of 3 years is shown in Figure 5. The proportion of high-quality seeds within a population decreases significantly with increasing storage time (7).

In the hot dry tropics and subtropics, insects are the principal cause of seed deterioration. Because temperature control of large warehouses is costly and difficult, the most effective preventive measure is to dry seeds to a moisture content in equilibrium with 50% RH or even 35% RH in some cases (see Table 3), which is not very difficult. Fumigants or contact insecticides may be used for carryover seed stocks.

Of all regions, the humid tropics present the most difficult situation. It is possible to store most seeds between harvest and their resowing without air conditioning if the temperature and RH remain below 30°C and 70%, respectively. Natural drying to a moisture content low enough for storage is possible only if the seed is exposed to the sun. Sun drying is limited to small quantities of seeds; it is slow and erratic, and the germination capacity of the seed is liable to be affected. To guard against both fungi and insects under the prevailing temperatures, the seeds must be dried to a moisture content in equilibrium with 40–50% RH or less than 40% for carryover stocks (see Table 3). The high ambient humidity will tend to increase the moisture content of the stored seed; when the humidity is not too high, cool night air can be circulated to lower the temperature. The refrigeration and dehumidification necessary for long-term storage in the humid tropics are prohibitively expensive for large quantities of seeds. On a medium scale, insulated rooms fitted with a capacity of about 100 tons of seed in bags have been successfully operated at 22°C and 50% RH. Since dehumidification creates heat, the refrigeration unit used should be powerful enough to counteract it.

Alternatively, seeds can be stored in sealed airtight, plastic containers in the wet tropics using a high-quality polyethylene film. For this kind of storage, the seed must be drier than for open storage, with a moisture content in equilibrium with 30–40% RH or even lower for long-term storage. A slight rise in the humidity during storage (due to seed respiration) is not enough to permit mold development. According to Thomson (3), the extra cost of drying can be weighed against the decreased risk of seed loss through the deterioration.

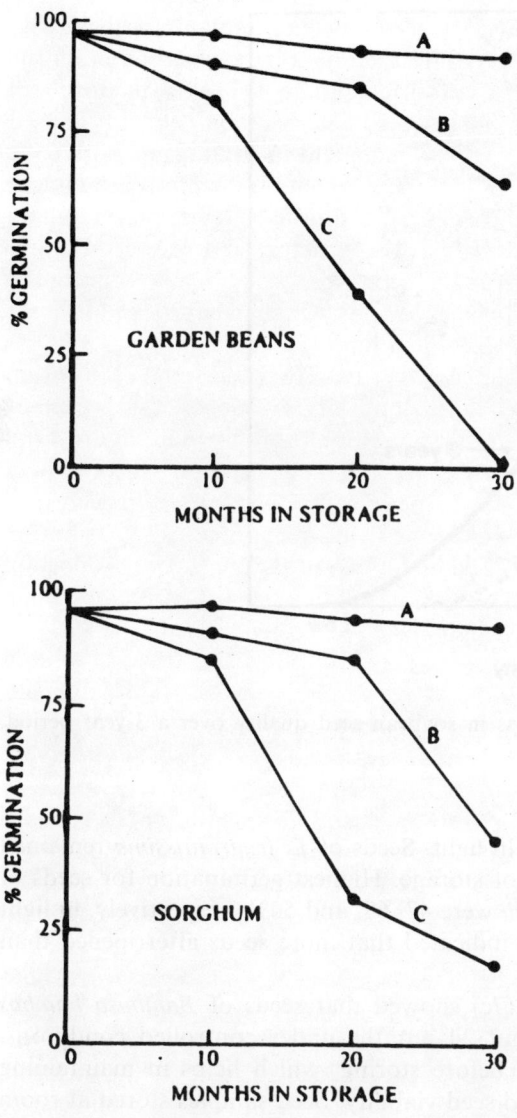

Figure 4 Differences in longevity of three seed lots each of garden beans and sorghum under open storage conditions. (From Ref. 6.)

Varghese et al. (7a) performed desiccation and storage experiments on mahua (*Mathuca indica*) seeds to determine their storage behavior and suggested that the storage behavior of *M. indica* seeds is true tropical recalcitrant; seeds becoming non-viable when desiccated below a 9.4% moisture content. The strong positive correlation was obtained between decline of viability and rate of dehydration. Hidayatiet al. (7b) tested the effects of dry storage under ambient laboratory conditions on germination and survivorship seeds of four *Lonicera* species. Fresh seeds of *L. fragrantissima* and *L. japonica* were dormant, whereas those of *L. maackii* and *L. morrowii*

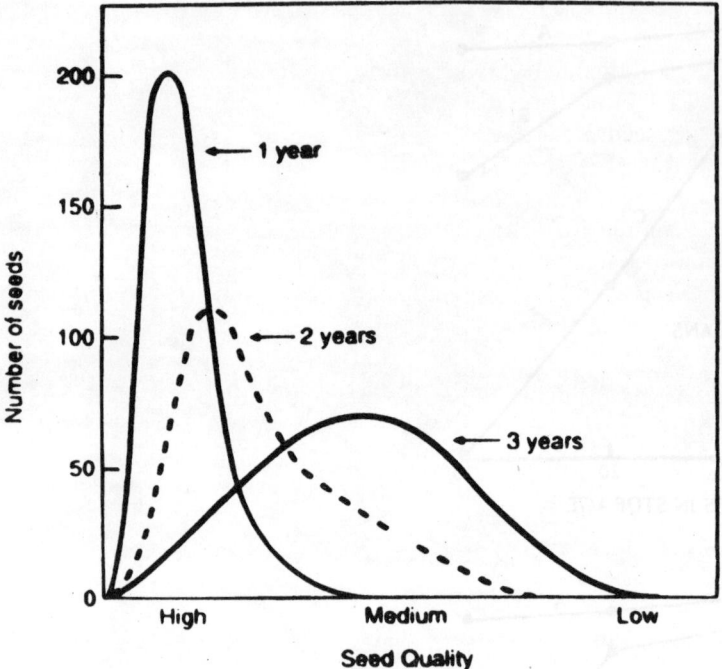

Figure 5 Hypothetical distribution of loss in soybean seed quality over a 3-year period. (From Ref. 6.)

germinated to about 30% at 25–15°C in light. Seeds of *L. fragrantissima* remained dormant during the entire 24 months of storage. Highest germination for seeds of *L. japonica*, *L. maackii*, and *L. morrowii* were 47, 68, and 50%, respectively, in light after 6 months of storage. The results indicated that more seeds afterripened than lost viability.

Shanmughvel and Peddappaiah (7c) showed that seeds of *Bambusa bambos* (giant grass) could be stored for about 24 months under controlled conditions. Reducing the initial moisture content before storing, which helps in maintaining viability for up to 9 months, can extend seed viability. Seed samples stored at room temperature lost their viability within 21 months in comparison to seeds stored for 24 months under low temperature, such as in a cold room (2–4°C) and deep freezer (−5°C) can maintain a high percentage of viability (89.2–92.5%), seeds germinating within 3–4 days.

Gamene et al. (7d) studied the storage behavior of a neem (*Azadirachta indica*) seed lot from Burkina Faso, West Africa, by varying both seed the moisture content and temperature during storage. The initial germination capacity, expressed as percentage normal seedlings, was 87%. To determine the effect of moisture content, seeds with intact endocarp were stored in drums with air of a constant temperature (20°C) and different relative humidities. During 9 weeks of storage, the germination capacity remained highest (70% or more) at RHs of both 55% (moisture content about 9% on a fresh weight basis) and 75% (moisture content about 13%) and was markedly lower at either lower (32 and 20–25%) or higher (95%) RHs. The effect of storage temperature was examined in two experiments. In the first one, seeds with 8.9% moisture were stored at 3, 20, 30, or 50°C for 2 weeks. The germination capacity of these seeds was lowered at the highest temperature only. In the second experiment, seeds with 4.5, 9.2, or 12.9% moisture were stored at both 3 and 30°C for up to 5 weeks. Irrespective of the seed moisture content, the germination capacity was lost most rapidly at the lowest temperature. The results for this neem seed lot are consistent with the intermediate category of seed-storage behavior. Ellis et al. (7e) concluded that storage of carrot, groundnut, lettuce, oilseed rape, and onion at −20°C rather than 20°C was beneficial to seed survival, and that hermetic storage at 20°C of seeds first dried at 20°C to moisture contents in equilibrium with about 10% RH provided greater longevity than a 5.5–6.8% moisture content in these five species.

In order to identify the volatile compounds that were evolved from five types of dry seeds (carrots, lettuce, soyabeans, peas, and rice) under different storage conditions and also released from them after heating, experiments were carried out using a gas chromatograph–mass spectrometer (GC-MS) fitted with a cryocondensation system. Fifty-nine types of volatile components were identified, of which the major components were methanol, ethanol, acetone, isopropanol, 2-butanone, and the aldehydes acetaldehyde, 2-methylpropanal, 2- and 3-methylbutanal, pentanal, and hexanal. Regardless of the storage temperature, the seeds stored at high RH contained more iso-type aldehydes, such as 2-methylpropanal and 2- and 3-methylbutanal, whereas n-type aldehydes, such as pentanal, hexanal, and heptanal, were evolved abundantly from seeds stored at low RH. Acetaldehyde, the most universal and abundant aldehyde in the seeds, was emitted at different levels depending on the seed species and the storage RH. The endogenous aldehydes of dry seeds appeared to be causally related to seed aging during storage (7f).

Sun-dried *Phaseolus vulgaris* cv. Selection 9 seeds at a 10% moisture content were kept in closed polyethylene bags at room temperature (28±4°C) for up to 3 years and tested for germination, solute leakage, and dehydrogenase activity in embryonic axes. After 1 year of storage, vigor had declined as indicated by a 7 and 16% reduction in seedling fresh weight (FW) and germination speed (GS), respectively. After 2 years, both vigor and viability declined. Reductions in percentage germination, seedling FW, germination rate (GR), GS, and standard germination (StG) were 9, 22, 38, 52, and 39%, respectively. Aging for 3 years further accentuated the loss of vigor and viability with reductions in germinability, seedling FW, GR, GS, and StG of 38, 50, 80, 90, and 74%, respectively, compared with fresh seeds. Vigor evaluation parameters such as electroconductivity, concentration of ultraviolet (UV)–absorbing sustances, sugars, amino acids, P, and K in aqueous seed extracts and dehydrogenase activity in the embryonic axes gave comparable results. The correlation coefficient between these parameters and germination, seedling FW, GR, GS, and StG were all highly significant except for percentage emergence and concentration of phenolic substances in aqueous seed extract, which was significant at the 5% level (7g).

IV. METHODS AND STRUCTURES OF SEED STORAGE

Seeds are stored at the farm, trade, and government levels in various types of storage structures until the following season's seed is available for planting. Methods of seed storage can be divided into two types: traditional methods and modern methods using improved storage structures.

A. Traditional Methods and Storage Structures

Seeds have been traditionally stored by farmers using different methods adapted to local conditions (climate). Storage structures suitable for farm, urban, and commercial storage of seed grains used in tropical and subtropical countries have been described (2,8–11).

The following traditional methods and storage structures are used to store grains and seeds in India and other tropical countries (9):

1. *Pusa Bin*—This is a conventional mud bin using a polyethylene sheet to make the structure airtight and moisturetight. It is simple in design, easy to construct, and inexpensive to suit the needs of average farmers. When filled to capacity and properly closed, the bin becomes an airtight, damp-proof structure preventing conductivity.
2. *RCC Bin*—The reinforced cement concrete (RCC) bin is circular, sturdy, weather-proof, and suitable for both indoor and outdoor storage. It is similar in construction to the circular steel bin.
3. *Circular Steel Bin*—This bin has an opening at the top for filling and a spout at the bottom to remove seeds. The bottom is constructed of plain sheets of mild steel. This structure can be easily assembled on the farm site and can be taken apart when not in use. It has a built-in arrangement for preventing the development of uneven temperatures, which lead to moisture migration and deterioration of seed quality. Seed grains stored in the bin can be fumigated whenever required.
4. *Plastic Bin*—This low-cost structure, suitable for indoor storage of seeds, has a tube-shaped metal base with a provision for placing bamboo sticks vertically around the side of the metal drum. A cylindrical rubberized fabric is hung inside, into which the seeds are loaded. They can be taken out from the top or through a sliding door at the bottom of the metal base. This is a compact and stable storage structure that can be dismantled when not in use.
5. *Aluminum Bin*—This is an outdoor storage structure consisting of a cylindrical body built of several corrugated aluminum curved sheets and a conical roof of flat aluminum sheets. The bin is constructed on a 60-cm-high platform and has a spout embedded in the platform and a manhole in the roof, both provided with a locking arrangement.
6. *Prefabricated Steel Bin with Hopper Bottom*—This is a strong and durable outdoor structure with a sloping roof, a manhole for filling, and a hopper bottom with a sliding door for discharging seeds. The bin is provided with natural aeration to prevent the heating and subsequent problems of translocation of moisture and resultant deterioration of seed quality. The bin stands on a firm support, creating a clearance of 60 cm at the bottom below the hopper. Both the manhole and the hopper gates have locking arrangements, and a metallic ladder and pulley are provided to facilitate filling of the bin.

B. Improved Storage Structures

Pingale (8) recommended for urban areas a rectangular steel structure that is rodent and moistureproof with an arrangement for controlling both aeration and fumigation. This structure is suitable for the compact storage of cereals and pulses in one place. It is made of galvanized iron sheets and is often partitioned into four or five compartments of different sizes.

Copeland and McDonald (12) recommended four principal types of seed storage: conditioned storage, cryogenic storage, hermetic storage, and containerized storage.

1. Conditioned Storage

Conditioned storage aims at storing seeds by carefully controlling the temperature and relative humidity. This method may be too costly to store most seed lots of agricultural crops, but it is extremely valuable to preserve germplasm and other high-value seed stocks. According to Harrington (13), in the tropics, conditioned storage is necessary to maintain seed viability. Welch and Delouche (14) considered the following four factors while evaluating the economics of seed storage: (a) the type of seed to be stored, (b) the length of storage, (c) the quality of seed stored, and (d) the loss of seed weight during storage.

2. Cryogenic Storage

In cryogenic storage, seeds are placed into liquid nitrogen at $-196°C$ (seeds are actually placed into the gaseous phase of the liquid nitrogen at about $-150°C$ for easy handling and safety). At these temperatures, there is little detrimental physiological activity, prolonging the storage life of seeds. This method is safe and cheaper than conditioned storage, but it is limited in capacity to the amount of storage space available in the cryogenic tanks. Hence, it is not practicable for most commercial seeds, but is useful to maintain a valuable germplasm over a prolonged period of time (15).

3. Hermetic Storage

Hermetic storage refers to packaging seeds in moisture-resistant or hermetically sealed containers for storage and marketing. Grabe and Isley (16) studied the effectiveness of several types of packaging materials in preventing moisture uptake and maintaining viability of creeping red fescue seed (see Chapter 16). Ordinary paper and cloth containers were least effective, whereas various laminate and polyethylene materials were moderately effective. Metal cans were completely effective in maintaining seed moisture at the initial 5% level. Using this method, seeds can be stored up to 10 years or longer. The ambient air from seed can be removed and replaced with specific gases like CO_2 and N_2, decreasing O_2 concentration. Pea seeds thus could be stored at 18.4% moisture at 25°C for 11 weeks when O_2 was decreased from 21 to 1.4% and CO_2 was increased from 0.03 to 12% (17). The extreme case of hermetic storage is to store seeds in vacuo at low moisture contents. Justice and Bass (18) and Bass and Stanwood (19), however, reported no major benefits from vacuum storage. Ambient air could be replaced with pure gas in sealed containers such as carbon dioxide (20) and inert gases like nitrogen, argon, and helium (21,22). Evidence available indicates that seeds stored in oxygen at low moisture contents deteriorate more rapidly (17,21,23).

4. Containerized Storage

In containerized seed storage, humidity is regulated in a closed container by the use of chemical desiccants like sulfuric acid, saturated salt solutions, or silica gel treated with cobalt chloride, which serves as an indicator dye, turning from blue to pink when relative humidity exceeds 45%. The desiccant storage system is cheaper and insect-, rodent-, and moistureproof. The seeds remain undamaged by storage fungi, since they are maintained at 45% RH (12).

C. Commercial Large-Scale Storage

Commercial large-scale storage structures are designed to store large quantities of seeds by government seed agencies that have adequate economic and technical support. More sophisticated improved structures can be constructed for bag and bulk storage of seeds.

1. Bag Storage

Bag storage warehouses should be constructed with waterproof walls, roofs, and floors and sealable openings for controlled ventilation and fumigation. The openings can be used for both natural and fan-controlled aeration. If a metal roof is used, it should be insulated to minimize temperature build-up. The seed bags should be placed on wood planking erected above the floor to provide insulation against floor temperature and moisture. The building also must be rodent-, bird-, and insectproof.

2. Bulk Storage

A bulk storage structure should meet basic requirements similar to those of a bag storage structure. However, sidewall insulation is more important in a bulk storage structure, since the seeds are in direct contact with the bin walls. Depending upon the temperature, the metal walls of these bins readily transfer heat inward or outward. The metal also absorbs radiation in varying degrees depending upon the reflectivity of the exterior surface and its capacity for absorption of infrared radiation (heat energy). Aluminum, bright steel, and white-painted roofs are good reflectors of sunlight and radiators of low-temperature heat energy. The surface temperature of the white-painted sheets is generally lower than most unpainted steel surfaces under bright sunlight conditions.

Studies carried out by Holman and Snitzler (1) in Illinois showed that soybeans stored in bulk at 12.0–12.5% moisture maintained their viability for 175 days and those at 8–9% for more than 650 days. Soybean seeds stored at 15% moisture retained their viability for less than 50 days.

Bulk-stored seed often needs to be turned or moved from one bin to another to break up any undesirable hot spots and to equalize the temperature of the seed to prevent translocation of moisture. Better results are obtained by aerating the stored seed (moving the air through the seed rather than the seed through the air). A motor-driven fan with a suitable duct system supplies the small amount of air needed for aeration. Automatic controls permit the fan to operate only when the air humidity and temperatures are within a selected range.

Sizes of bulk storage structures can vary depending upon the requirements. Large elevators may be flat or upright, whereas upright storage bins are mainly large, cylindrical silo types. Painted metal or concrete silos can be built singly or in rows (2).

V. FACTORS INFLUENCING SEED-STORAGE QUALITY

In addition to the seed moisture content and storage temperature, other factors influencing the quality of seeds during storage include initial seed quality, damage caused by rough handling, and the extent of cleaning. Seeds damaged during harvesting and processing and containing a high proportion of rubbish—broken pieces of straw or leaf—will not store well. Seed storability is also influenced by the kind of seed and its variety—some kinds being naturally short lived (onion, soybean, peanut). Sarwad et al. (27), employing the technique of accelerated aging, showed that ground variety, JL-24 was better storer compared to R-8808 and ICGS-11, and could be held for longer time in godowns. These authors further reported that in view of the scarcity of space and for immediate use, ground kernels could be stored with halogenation treatment with bleaching powder (5g/kg seed) or coated with ash + captan or captan + malathion (28).

The amount of moisture content in the seeds is the single most important factor influencing the seed viability during storage. Over the most commonly encountered seed moisture ranges, cereal seeds stored at temperatures not exceeding 90°F (32.2°C) will have storage life as shown in Table 4.

If seeds are stored at a higher moisture content, losses due to mold growth (12–14% moisture) and heating (18–20% moisture) are very rapid. Within the normal moisture range, the biological activity of seeds, insects, and molds increases with rises in temperatures. A very low moisture content (below 4%) may also damage some seeds severely owing to extreme desiccation and seed hardness.

Since the life span of seeds largely revolves around the moisture content, seeds should be dried to their safe moisture content, which depends upon kind and variety of seed, the length of storage period desired, and the type of storage structure and packaging material employed. The relative humidity and temperature during storage most significantly influence the viability and germinability of the stored seeds, as described in Section III. The effective storage life of seeds thus depends on three important factors; namely, the kind of seed, its preharvest history, including harvesting and drying, and the actual storage conditions employed (24–26).

The fungicides and insecticides used to treat seeds to protect them from pathogenic organisms during storage involve an element of phytotoxicity, which may depress the germination capacity of the seeds. The phytotoxic effect is exacerbated by cracks and fractures in the seeds, overdosage, inadequate drying, and prolonged storage. The treated seeds, therefore, should not be stored for long periods of time.

Soyabean (*Glycine max* UFV-10) seeds with a water activity (aw) varying from 0.66 to 0.86 (moisture content 11.3 to 17% wet basis) and infested with conidia of

Table 4 Moisture Content and Storage Life of Cereal Seeds[a] at Temperatures Not Exceeding 90°F (32.2°C)

Seed moisture (%)	Storage life
11–13	6 months
10–12	1 year
9–11	2 years
8–10	4 years

[a] Seeds of high germination and high vigor.
Source: Ref. 4.

Aspergillus ruber were stored at 25°C for 140 days. Practically all seeds were colonized by the fungus in less than 20 days independent of the aw. The free fatty acid (FFA) content increased linearly with an increase of the aw and storage time. The rate of FFA accumulation was slow at the lower aw and increased linearly with an increase of the aw. Seedling emergence started to decrease more slowly at lower than at the higher aw, and its decline followed the pattern of the FFA increase. All seeds lost viability in 140 days of storage independent of the aw. Regression equations showed strong negative linear relationships between the FFA accumulation and emergence percentage and positive relationships with an increase of abnormal seedlings characterized by negative geotropism. Seed colonization by *A. ruber* accompanied by an increase of the FFA content and a decrease of seed viability at an aw of 0.66 warrants revisions of the present recommendations regarding soybean seed moisture for inter season storage (29).

Storage behavior and longevity of seeds of lemon (*Citrus limon*) and sweet orange (*C. sinensis*) were tested following desiccation to between a 14 and 4% moisture content (fresh weight basis, wb) and subsequent hermetic air-dried storage for up to 914 days at temperatures between 15 and −20°C, and up to 6 days in liquid nitrogen (−192°C) (30). The results confirmed that both species showed intermediate seed-storage behavior. Air-dried storage environments, therefore, require careful optimization to obtain maximum seed longevity. Seeds of sweet orange were less tolerant of desiccation, and less than 20% of seeds survived more than 210 days at 5°C and 8.7% moisture content (63% equilibrium RH at 20°C). The optimum air-dried environment for the medium-term storage of the longer lived lemon seeds was 5°C and 7.5% moisture content (48.3% equilibrium RH at 20°C).

The seeds of five sunflower restorer lines (RHA-857, RLC-4, RLC-2, VI-46, and V-94) were treated with two fungicides (thorax and bavistin [carbendazim]), two halogens (potassium iodide and calcium oxychloride), and two plant products (*Pongamia pinnata* leaf powder and sweet flag [*Acorus calamus*] powder). The treated seeds were subjected to accelerated aging at 45 ± 1°C and 95 ± 1% RH for 4 days and were then tested for germination using the germination paper method. Percentage germination, root and shoot lengths, and seedling vigor were determined before and after the aging treatment. A decrease in percentage germination resulted when the seeds were subjected to accelerated aging. The highest percentage reduction in germination was observed in seeds of RHA-857 and those seeds treated with sweet flag powder. The seeds treated with potassium iodide and calcium oxychloride maintained significantly higher seed germination (98.5 and 97.4%, respectively), root and shoot length, and vigor index compared with the other treatments (31).

Daniel et al. (32) determined the longevity of botanical seeds of a West African yam species, *Dioscorea rotundata* cv. Obiaoturugo, under a number of dry-cold storage conditions to preserve yam genetic resources. Seeds were dried to 14.5%, 10.1%, and 5.0% seed moisture content and stored at −20, 5, 15, and 25°C, respectively, for 3 years. The half-life (P50) values derived by probit analysis of viability decline curves during storage of seeds for 12 dry-cold storage treatment combinations suggested that seed longevity could be prolonged by desiccation and cold storage. The highest seed storage life of 21 years was derived for seeds dried to a 5% moisture content and stored at −20°C. This was followed by 19 years with the seeds dried to a 5% moisture content and stored at 5°C. The shortest seed-storage life of about 7 years was estimated for seeds of 14.5% moisture content at 25°C. The storage of seed using the conventional dry-cold gene bank conditions offers good potential for long-term yam germplasm maintenance.

The influence of liquid nitrogen (LN2) on seed germination at different seed moisture contents and after 1,7, or 30 days of storage was studied in rice, oats, lettuce cv. Regina and Valdor, celery cv. Tall Utah 52–70R, *Betula celtiberica, Colutea atlantica, Datura ferox, D. stramonium, Halimium atriplicifolium, Lavandula stoechas* subsp. *pedunculata, Onobrychis eriophora (O. peduncularis), Onopordum nervosum,* and *Spartium junceum.* In most species, there were no effects of moisture content or storage period in LN2 on percentage germination. In *D. ferox,* percentage germination was lower after 1 day of storage than in control seeds or in those stored for 7 and 30 days. In *D. stramonium,* percentage germination decreased during storage in LN2. In *S. junceum* and *L. stoechas* subsp. *pedunculata,* percentage germination was higher in seeds with higher moisture contents but unaffected by storage (33).

Phaseolus vulgaris cv Calima seeds were stored at seven different seed moisture levels, ranging from 10.3 to 14.2% in hermetic containers at 30°C during a 32-week period. Seed moisture, percentage germination, and percentage emergence were evaluated every 2 weeks by sampling independent lots of each seed moisture level. A multiple regression analysis of the percentage germination and emergence data produced a determination coefficient of 0.85 for percentage germination and 0.93 for percentage emergence. Results indicated that bean seeds having initial high germination and emergence rates and a maximum moisture content of 11.5% could be hermetically stored for up to 8 months at 30°C without suffering significant losses in germination and emergence percentages. This moisture content could be achieved by small farmers using natural sun drying and by hermetic storage used to protect seeds from insects (34).

Carrot (cv. Pusa Kesar), onion (cv. Pusa Red), and tomato (cv. S. 120) seeds were either stored in cloth bags under ambient conditions or dried to a moisture content of 5%, spread in a single layer in a wire basket, and kept over silica gel in desiccators at room temperature. The percentage germination and moisture content of the seeds were determined at intervals of 6 months for up to 48 months of storage. In a second experiment, seeds of vegetable pea (Arkel and Bonneville) and grain (KPMR-10, Pant-4, and T-163) cultivars were stored under ambient conditions. Percentage germination was evaluated for up to 40 months of storage for the grain and up to 22 months of storage for the vegetable pea cultivars. Under ambient storage conditions, the loss of seed viability was greatest in onion with percentage germination decreasing from 97 to 3% after 28 months. Carrot seed germination rate decreased from 59 to 22% after 34 months. The loss of viability was least in tomato, with percentage germination decreasing by only 19% after 33 months of storage. When stored under reduced moisture conditions, the seeds of all three species in the first experiment showed little loss of viability even after 36 months of storage. After 22 months of storage, the two vegetable peas, Arkel and Bonneville, showed 50 and 36.14% losses of viability, respectively, whereas the loss in viability for all three grain types was negligible (35).

Thomson (3) stated the following general rules to be observed in the management of stored seed:

1. The seed stocks should be kept under continuous observation, especially looking for hot spots.
2. Seeds should be fumigated between outgoing and incoming stocks.

3. Floors should be swept thoroughly and rubbish burned.
4. Ventilation should be encouraged within and between stocks.
5. Seed should not be piled against a wall.
6. The storage building should be repaired and kept in good condition.
7. Seeds showing the best storage potential (as shown by accelerated aging test) should be selected for carrying over from one season to the next.
8. Seeds stored in bulk require frequent turnings to prevent deterioration due to heating.

VI. STORAGE REQUIREMENTS OF WAREHOUSES

The main requirements of the single-story or multistory warehouses used to store bagged seeds are weathertight roofs and walls; strong, smooth floors and properly spaced columns that permit the efficient use of forklifts and other equipment; and ceiling heights that permit bags to be stacked 5 m or higher.

Bulk seed is stored in bins or tanks separate from or within warehouses. The strength required for these containers varies with the size and kind of the seed stored. Bin walls and floors must be strong enough to support both the lateral and vertical pressures exerted by the stored seed.

Rodents are a problem in seed-storage facilities and warehouses. Multistory warehouses should ensure that they are completely rodentproof. In one multistory bag storage structure in Maryland, each floor is isolated so that no mice or rats can enter a floor except through incoming seed bags. An open space about 0.5 m wide is left between the stacked seed bags and the wall; this strip of floor is painted white to show rodent tracks and excreta. Regular fumigation should prevent any possible entry of rodents and other organisms.

REFERENCES

1. L. E. Holman and J. R. Snitzler, Transporting, handling and storing seeds, *Seeds: The Yearbook of Agriculture*, US Department of Agriculture, Washington, DC, 1961, p. 338.
2. U. S. Deshpande and S. S. Deshpande, Legumes, *Foods of Plant Origin: Production, Technology and Human Nutrition* (D. K. Salunkhe and S. S. Deshpande, eds.), Van Nostrand Reinhold, New York, 1991, p. 137.
3. J. R. Thomson, *An Introduction to Seed Technology*, Leonard Hill, London, 1979.
4. J. F. Harrington and J. E. Douglas, *Seed Storage and Packaging: Applications for India*, National Seeds Corporation, New Delhi, 1970.
5. A. F. Kelly, *Seed Production of Agricultural Crops*, Longman Scientific and Technical, New York, 1988.
6. J. C. Delouche, Precepts of seed storage, Proc. Mississippi State Seed Processors, Short course, 1973, p. 93.
7. M. B. McDonald and D. O. Wilson, ASA-610 ability to detect changes in soybean seed quality, *J. Seed Technol.* 5:56 (1980).
7a. B.Varghese, R.Naithani, M.E.Dulloo and S.C.Naithani, Seed storage behavior in Mathuca indica J.F.Gamel, *Seed Sci. Technol.* 30: 107 (2002).
7b. S.N.Hidayati, J.M.Baskini and C.C.Baskin, Effects of dry storage on germination and survivorship of seeds of four *Lonicera* species (Caprifoliaceae), *Seed Sci. Technol.* 30:137 (2002).
7c. P.Shanmughvel and R.S.Peddappaiah, Techniques for seed storage of giant grass (*Bam-*

mbusa bambos), *Seed Technology and Seed Pathology* (T.Singh and K.Agrawal, eds.), Pointer Publishers, Jaipur, India, 2002, pp.200–204.

7d.   C.S.Gamene, H.L.Kraak, J.G.van Pijlen, and C.H.R. de Vos, Storage behavior of neem (*Azadirachta indica*) seeds from Burkina Faso, *Seed Sci. Technol.* 24: 441(1996).

7e.   R.H.Ellis, T.D.Hong, D.Astley, A.E.Pinnegar and H.L.Kraak, Survival of dry and ultra-dry seeds of carrot, groundnut, lettuce, oilseed rape, and onion during five years' hermetic storage at two temperatures, *Seed Sci. Technol.* 24: 347(1996).

7f.   M.Zang, H.Yajima, Y.Umezawa, Y.Nakagawa and Y.Esashi, GC-MS identification of volatile compounds evolved by dry seeds in relation to storage conditions, *Seed Sci. Technol.* 23: 59(1995).

7g.   D.K.Pandey, Aging of French bean seeds at ambient temperature in relation to vigor and viability, *Seed Sci. Technol.* 17: 41(1989).

8.   S. V. Ping ale, *Storage and Handling of Food Grains*, Indian Council of Agricultural Research, New Delhi, 1976.

9.   B. R. Birewar, B. K. Verma, C. P. Raman, and S. C. Kanjilal, *Traditional Storage Structures in India and Their Improvements*, Indian Grain Storage Institute, Hapur, India, 1980.

10.   D. K. Salunkhe, J. K. Chavan, and S. S. Kadam, *Postharvest Biotechnology of Cereals*, CRC Press, Boca Raton, FL, 1985.

11.   D. K. Salunkhe, S. S. Kadam, and J. K. Chavan, *Postharvest Biotechnology of Food Legumes*, CRC Press, Boca Raton, FL, 1985.

12.   L. O. Copeland and M. B, McDonald, *Principles of Seed Science and Technology*, 3rd ed., Chapman & Hall, New York, 1995.

13.   J.F. Harrington, Biochemical basis of seed longevity, *Seed Sci. Technol.* 1:453 (1973).

14.   C. B. Welch and J. C. Delouche, Conditioned storage of seed, Proc. and Rep. of the Southern Seedsmen's Association, 1974, pp. 30–40.

15.   P. C. Stanwood, Cryopreservation of seed germplasm for genetic conservation, *Plant Cryopreservation* (K. Kartha, ed.), CRC Press, Boca Raton, FL, 1985, p. 199.

16.   D. F. Grabe and D. Isley, Seed storage in moisture resistant packages, *Seed World* 104 (2):4 (1969).

17.   E. H. Roberts and F. H. Abdalla, The influence of temperature, moisture and oxygen on period of seed viability in barley, broad beans and peas, *Annal. Bot.* 32:97 (1968).

18.   O. L. Justice and L. N. Bass, Principles an practices of seed storage, *USDA Agricultural Handbook No. 506*, US Department of Agriculture, Washington, DC, 1978.

19.   L. N. Bass and P. C. Stanwood, Long-term preservation of sorghum seed as affected by seed moisture, temperature, and atmospheric environment, *Crop Sci.* 18:575 (1978).

20.   A. Bennici, M. B. Bitonti, C. Floris, D. Gennai, and A. M. Innoceti, Ageing in *Triticum durum* wheat seeds: Early storage in carbon dioxide prolongs longevity, *Environ. Exp. Bot.* 24:159 (1984).

21.   B. J. Harrison, Seed deterioration in relation to storage conditions and its influence upon germination, chromosomal damage and plant performance, *J. Natl. Inst. Agric. Bot.* 10: 644 (1966).

22.   G. Quaglia, R. Cavaioli. P. Catani, J. Shejbal, and M. Lombardi. Preservation of chemical parameters in cereal grains stored in nitrogen, *Controlled Atmosphere Storage of Grains* (J. Shejbal, ed.), Elsevier, New York, 1980, p. 319.

23.   J. B. Ohlrogge and T. P. Kernan, Oxygen-dependent aging of seeds, *Plant Physiol.* 70:791 (1982).

24.   R. L. Agrawal, *Seed Technology*, Oxford and IBH Publishing, New Delhi, 1980.

25.   C. M. Christensen (ed.) *Storage of Cereal Grains and Their Products*, 2nd ed., American Society of Cereal Chemists, St. Paul, MN, 1974.

26.   1ST A, Seed storage and drying, *Seed Sci. Technol.* 1(3): (1973).

27.   R.K.Sarwad, V.K.Deshpande and M.Shekhargouda, Prediction of storability of groundnut through accelerated aging, XI National Seminar on Quality Seed to Enhance Agri-

cultural Productivity, UAS, Dharwar, January 18–20, 2002, *Seed Tech News* 32: (A.Gaur, A.K.Vari, J.L.Varshney and K.Kant, eds), Indian Society Seed Technology, Division of Seed Science and Technology, IARI, New Delhi, March 2002, p. 115.

28. R. K.Sarwad, V.K.Deshpande and M.Shekhargouda, Storability of groundnut kernels as influenced by seed treatment, XI National Seminar on Quality Seed to Enhance Agricultural Productivity, UAS, Dharwar, January 18–20, 2002, *Seed Tech News* 32: (A. Gaur, A.K.Vari, J.L.Varshney and K.Kant, eds), Indian Society Seed Technology, Division of Seed Science and Technology, IARI, New Delhi, March 2002, p.116.

29. O.D.Dhingra, E.S.G.Mizubuti, I.T.Napoleao and G.Jham, Free fatty acid accumulation and quality loss of stored soybean seeds invaded by *Aspergillus ruber*, *Seed Sci. Technol.* 29: 92001).

30. T.D.Hong, N.B.Ahmad and A.J.Murdoch, Optimum air-dry storage conditions for sweet orange (*Citrus sinensis* L. Osbeck) and lemon (*Citrus limon* L. Burm.) seeds, *Seed Sci Technol.* 29: 183 (2001).

31. B.S.Vyakarnahal, M.Shekhargouda, A.S.Prabhakar and S.A.Patil, Efficacy of halogens, plant products and fungicides on storage potentiality of sunflower restorer lines, *Karnataka J. Agric. Sci.* 13: 36 (2000).

32. I.O.Daniel, N.Q.Ng, T.O.Tayo and A.O.Togun, West African yam seeds stored under desiccated and cold storage conditions are orthodox, *Seed Sci. Technol.* 27: 969(1999).

33. J.M.Iriondo, C.Perez and F.Perez-Garcia, Effect of seed storage in liquid nitrogen on germination of several crop and wild species, *Seed Sci. Technol.* 20:165(1992).

34. R.Aguirre and S.T.Peske, Seed moisture content required for short-term hermetic storage of beans, *Seed Sci. Technol.* 19:117(1991).

35. A.Varier and P.K.Agrawal, Long term storage of certain vegetable seeds under ambient and reduced moisture conditions, *Seed Res.* 17:153(1989).

18
Seed Testing

I. INTRODUCTION

The success of a seed-production program depends on the availability of breeder's seed of high genetic purity for multiplication. It is necessary to maintain the genetic purity of the breeder's seed in self-pollinated crops, to prevent any possible multiplication of contaminants, and to avoid the chance of rejection of foundation and certified seed of the improved cultivars of agricultural crops. It also saves on the cost and effort involved in the country's breeding programs.

A newly bred crop variety is normally expected to be pure, homogeneous, and free from weed seeds and seedborne diseases. However, as a result of bulking of seed from early segregating generations, the new variety is often found to be contaminated with off-types upon further multiplication. Although off-types could be removed by negative selection during mass increase, it is preferable to select genetically pure plants, check their progenies for purity, and then bulk the seeds that are pure and true-to-type. The seed thus obtained is called nucleus or breeder's seed. Since the residual heterozygosity in the variety is more visible through segregation within the progeny, this method of nucleus/breeder's seed production has a distinct advantage over mass increase by negative selection (1).

II. SEED IDENTIFICATION AND SAMPLING

A. Concept of Cultivar/Variety

The concept of cultivar or variety of a crop is not as precise as that of a species. Barring a few exceptions, one species of crop does not cross pollinate with another, and intermediate forms are generally not produced. The crop species, therefore, can be identified beyond any doubts. Within a crop species, however, many cultivars exist, which are often interfertile, and crossing between the varieties can produce several intermediate types. The capacity of a variety to maintain its distinctive genetic feature depends greatly on its mode of propagation and the method of breeding.

Some crops, like potatoes, yams, and bananas, are propagated vegetatively through a piece of stem taken from the mother plant. From one such original plant, thousands of plants true to the cultivar can be produced, as though they were branches of a large tree growing separately with their own root system instead of being connected to the main trunk. Such a cultivar is called a clone, and all plants belonging to the same clone are genetically identical. Mutations do occur occasionally, which may change the cultivar's genotypic and phenotypic characters.

A seed develops in the ovary of a flower after the fusion of male and female gametes either coming from the same plant (self-pollination) or from different plants (cross pollination).

In self-pollinated species (e.g., wheat, barley), all plants are descended from one common ancestor by a repeated process of selfing (self-fertilization) and therefore are homozygous; that is, each plant is genetically identical to other plants within the cultivar, which will produce identical progeny in the next generation. Such cultivars are known as pure lines, and can maintain their distinctive genetic characters almost indefinely provided crossing does not occur. Nevertheless, irregularities do occur in the pairing and separation of chromosomes, resulting in off-types over several generations.

In the cross-pollinated cultivars, on the contrary, individual plants are genetically heterozygous and not identical. The cultivar is a mixture of plants differing in genetic characteristics like growth habit, flowering, and maturity, a range varying from very narrow in some cultivars to large in others. Because plants interpollinate, each plant is not the same as either of its parents, but the progeny from a group of plants show a range of plant characters roughly similar to those of all the parents, and the cultivar as a whole can remain broadly unchanged provided the pollen from another cultivar of the species is not introduced (2).

The concept of cultivar purity, therefore, cannot be the same for both the self-fertilized and cross-fertilized cultivars. Whereas for a pure line, a precise identity of cultivar can be defined and any plant that does not fit the description can be regarded as an impurity, for a cross-pollinated cultivar, the description of a cultivar is more general, allowing for variation between individual plants. A well-defined boundary cannot be set to exclude a plant as being impure. An individual plant can, however, be regarded as being impure if it is clearly and significantly outside the range of the accepted variation.

According to the standards of the European Economic Community, a cereal crop grown to produce basic seed must conform to a cultivar purity of 99.9% which is calculated by the number of plants. Forage crops do not have a precise purity standard, but a crop should have satisfactory trueness to its cultivar, apart from the gross off-types.

In a seed-multiplication program, cultivar purity must be confirmed for each generated by examining seed samples in the laboratory using a variety of tests or by growing plants in the field or in a check plot. The essential features that distinguish cultivars are the properties of the crop in which the farmer or ultimate user is interested; for example, yield, cooking quality, palatability, and straw strength. These characters are generally not tested in the laboratory or by field inspections. The cultivar should have characteristics by which it can be clearly and simply distinguished to control its genetic purity. Agronomic characteristics, such as growth habit and flowering time, as well as morphological characteristics like rachilla heirs (barley) and glume shape (wheat), although not of any significance to the user, are employed by breeders as markers of cultivar identity. The most useful are "two-state" markers, having only two alternate conditions; for example, colored or colorless, awned or awnless.

Most crop species are both self-fertilized and cross fertilized, in which some particular characteristic may predominate. The amount of variation within a cultivar of such species depends on the degree of self-fertilization. In a highly self-fertile species, a uniform cultivar can be produced by the method of selection as a nearly pure line.

Although maize is a self-fertilized crop, cross pollination predominates, because male and female flowers, produced in different inflorescences, do not open at the same time. It is possible to produce seed by artificial self-pollination (inbred lines), which are genetically pure but lack vigor and yield poorly. The seed of an F_1 hybrid cultivar is produced by crossing two of these inbred lines, the resultant hybrid being vigorous, high yielding, and completely uniform. However, if the grain harvested from these hybrid plants is sown as seed, uniformity is lost, and the next generation of plants will show various characteristics of the original inbred lines in different combinations, with a loss in vigor and yield.

A cultivar that is a mixture is sometimes well adapted to adverse soil and climatic conditions, with at least some plants growing well to produce a reasonable yield, although the yield may not be as good in more uniform conditions. To meet such needs, "synthetic" cultivars (composites) have been developed and have been successful in crops like maize and, to a lesser extent, in cotton. Breeders mixing together the seeds of 10–15 pure or inbred lines produce a synthetic cultivar. The mixed stock is then multiplied over a few generations; during this time the plants intercross, eliminating the distinctions between constituent lines. At the end of a controlled multiplication program, the seed is sold to farmers, who grow it for several generations, using the same seed again and again.

Thus, there is no simple comprehensive definition of a cultivar, but the International Code of Nomenclature of Cultivated Plants defines it as, "an assemblage of cultivated plants that is clearly distinguishable by certain characteristics and which, when reproduced retains these distinguishing characteristics." This definition emphasizes the reproducibility of a cultivar. However, under various breeding systems and natural conditions, the distinguishing characteristics of cultivars may be easily lost. The task of the seed producer is to control and modify the breeding system in order to retain cultivar purity (1,3).

Recently used techniques like electrophoresis of seed proteins and enzymes have proved to be versatile and useful in the identification of seeds of crop varieties (4) and are described by McDonald and Payne in the Cultivar Purity Testing Handbook (5). According to Copeland and McDonald (6), this technique fails to identify a number of varieties in some crops. A newer and powerful technique known as randomly amplified polymorphic DNA (RAPD) is used to examine the base sequence of the DNA molecule using techniques similar to the electrophoresis of seed proteins (7).

Standardized testing of seeds required the development of a compendium of test procedures. This objective was met by the rules for testing seeds published by the Association of Official Seed Analysts (8).

B. Seed Sampling

When seeds of the same lot are tested in different laboratories, the results vary significantly. This variability is normally attributed to variation in sampling, experimental error, interpretational variation, and lapse of time. The variation occurs largely because the sample used in a test is very small and does not truly represent the seed lot from which it is drawn; the latter being a mixture of pure seed and impurities like dead seeds. In a purity test, the analyst examines every seed and particle of the sample (e.g., germination capacity). The results can be applied to the seed lot as a whole only if a representative sample has been taken from a uniform seed lot. In practice, however, a seed lot is never completely uniform for the following reasons (2):

1. Segregation by gravity of heavy and light seeds within the bulk or within a seed bag
2. Differences within the crop from which the seed was harvested owing to field variation in crops maturity, lodging, disease, or occurrence of weeds
3. Harvesting variations caused by interruptions by bad weather
4. Lack of uniformity in postharvest operations like threshing, drying, cleaning, and storage of seed from the same crop, using different machines, variation in their adjustments, and different storage conditions
5. Inadequate blending of seeds of two or more crops to form one lot before packaging

Some variation is inevitable, because blending does not ensure perfect uniformity, which would imply an orderly distribution of impurities like weed seeds and dead seeds throughout the seed bulk. A visibly uniform-appearing seed lot may have unevenly distributed seeds of different crops as well as dead and diseased seeds that are not apparent to the eye.

C. Sampling Procedure

One method of obtaining a representative sample from a seed lot is shown in Figure 1. Numerous small quantities of seed (primary samples) are taken randomly from

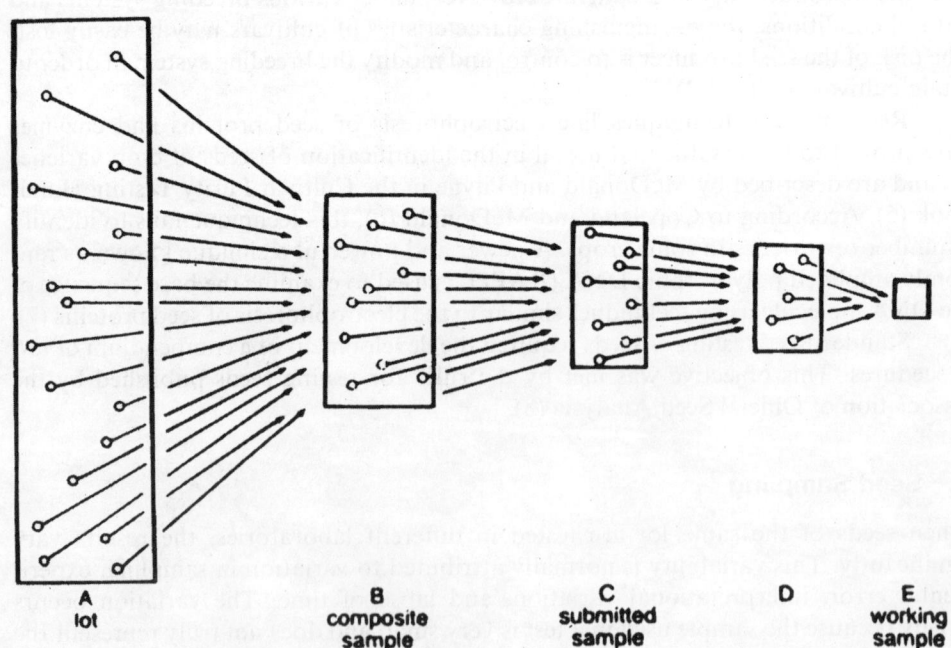

A	B	C	D	E
lot	composite sample	submitted sample		working sample

Figure 1 Diagrammatic representation of successive stages in taking a sample from a seed lot. (From Ref. 2.)

different points in the lot (A) and mixed thoroughly to form a composite sample (B). Small quantities are then taken from the composite sample and mixed to form a submitted sample (C). This process is repeated until a sample is obtained of the correct size for testing (E); known as a working sample. The sample size may be reduced by half at each stage after the second stage instead of taking a number of small quantities (successive halving).

It is advisable to carry out sampling in two stages because of the difficulty involved in drawing a small representative sample from a seed lot. In the first stage, a large seed sample (the submitted sample), usually about 10 times more than is required for testing, is taken in the warehouse and sent to the seed-testing laboratory. In the second stage, the working sample is drawn in the testing station using more refined methods (9).

1.　Warehouse Sampling

The size of the sample representing the seed lot depends upon the uniformity of the lot. Because the latter is an unknown variable, the International Seed Testing Association (ISTA) has set limits based on seed size, which are generally followed by seed-certification agencies and authorities enforcing seed laws (Table 1).

The number of primary samples to be taken depends on the size of the seed lot; the following are the minimum numbers:

Size of seed lot (kg)	Minimum number of samples
Up to 50	3
51–500	5
501–3000	One for each 300 kg, but not less than 5
3001–20,000	One for each 500 kg, but not less than 10

The primary samples should be taken at different depths from different locations in the bulk lot. The bags to be sampled are picked randomly from the whole lot, and a sample is taken from each bag, the number of primary samples being determined by ISTA rules. Bags are usually sampled with an instrument that leaves a hole in the seed bag. If the bag is made of cloth, the threads are not broken an can be returned to their original place, but if they are made of paper or plastic, the hole can be repaired with an adhesive patch.

The primary samples are thoroughly mixed in a suitable container to form a composite sample, which should be at least four times larger than the submitted sample and can be reduced mechanically using a suitable apparatus. If too large, it may be sent as is for reduction in the laboratory. Sample weights of seeds depend on lot size, varying from 25 to 1000 g (see Table 1).

The submitted samples are sent to the laboratory carefully closed and labeled in small bags bearing the name of the species, cultivar, reference number, size of lot, date, place of sampling, and name of sampler. If the seed sample is meant to be used for the determination of moisture content, it should be carried out quickly with minimum exposure to the atmoshere. Primary samples for this purpose should be packaged in polyethylene bags and submitted in sealed moistureproof containers.

Table 1 Size of Seed Samples Representing Seed Lots According to ISTA Standards for Seed Testing

		Minimum sample weight	
Crop	Maximum weight of seed lot (kg)	Submitted sample (g)	Working sample for seed testing (g)
Agrostis spp.	10,000	25	0.5
Medicago sativa	10,000	50	5
Pennisetum typhoides	10,000	150	15
Cannabis sativa	10,000	600	60
Sorghum bicolor	10,000	900	90
Triticum aestivum	20,000	1,000	120
Dolichos lablab	20,000	1,000	500
Zea mays	20,000	1,000	900
Vicia faba	20,000	1,000	1,000

·*Source*: Ref. 2

To use the various devices available to take samples from bags, from bulk, or from a packaging machine, it is necessary for seed to be clean and free flowing. A handful of seed can be withdrawn manually by inserting a hand into the bag or bulk.

Samples can be drawn from seed lots using a suitable instrument, such as a stick, thief, spear, or bulk sampler (Fig. 2). The stick is a double metal tube with a pointed end, both tubes having coinciding slots, which can be opened or closed by turning the inner tube within the outer one. The stick is inserted into the seed mass with the slots closed; the seed flows into the inner tube when the slots are opened, facilitated by one or two opening and shutting movements. The seed is withdrawn with the slots closed and the contents of the primary sample is transferred to a suitable container. Sticks are made in various diameters to suit seeds of different species depending upon size and flowing property. They vary in length from 75 cm for samplng bags to 3–4 m for sampling bulks. The stick is inserted diagonally from top to bottom while sampling a seed bag. The depth of sampling seed lots in bulk, bins, and boxes is limited by the length of the stick, which is inserted diagonally downward at a number of points depending upon the number of primary samples required.

A spear is a metal tube with a sharp point and an oval hole just behind the point that is suitable only for sampling seed in bags. It is thrust upward into the sack at an angle of about 30 degrees, with the hole facing downward, until the point reaches the opposite side; it is then turned until the hole faces upward, allowing the seed to flow down the spear into a small bag or container held by the sampler. A bag is usually sampled at one point, drawing primary samples at different levels (top, middle, and bottom) throughout the lot. Like sticks, spears are also made in various diameters to suit different seed sizes and lengths sufficient to penetrate to the farther side of the bag. The short spear, called a thief, can draw seed only from the outermost parts of the bag and is not very useful to draw a representative sample.

Bulk sampling devices can draw samples from very deep bulks. One type is a large version of the stick sampler. Another consists of a heavy metal cup with a close-fitting lid attached to the end of a metal rod to which additional lengths can be

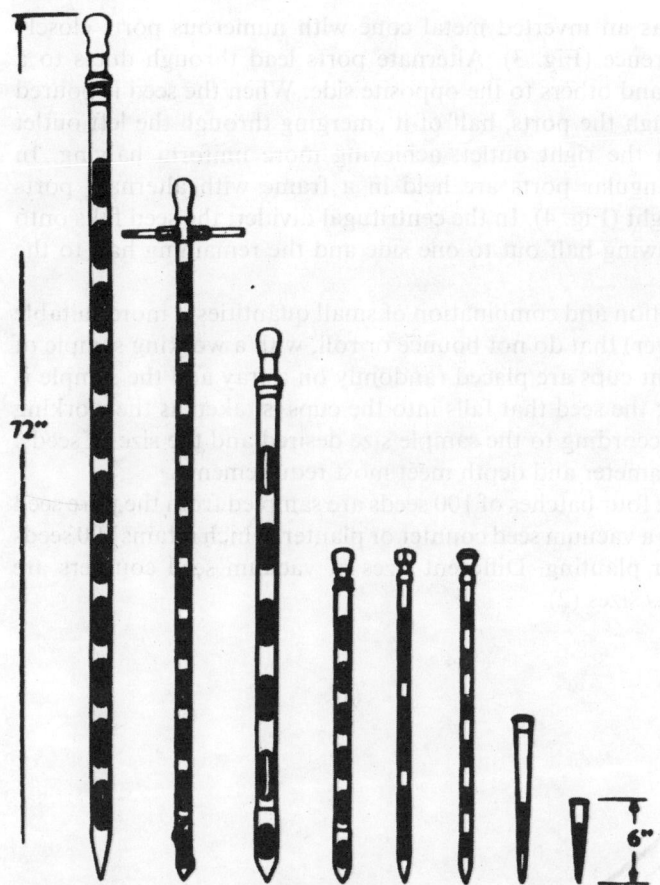

Figure 2 Probes used for seed sampling; on the far right is a "thief probe." (From Ref. 6.)

added according to the depth to be sampled. The cup is pushed vertically downward into the bulk; the pushing action keeps the lid pressed to the cup. When the rod is pulled to withdraw the cup, the lid is opened for the cup to fill with seed, constituting a primary sample.

Primary samples can also be drawn automatically from seed as it passes alog a conveyor or down a chute toward a packaging machine.

2. Laboratory Sampling

The sample arriving at the laboratory from the warehouse must be reduced to about one-tenth of its weight for the purity test, which is performed by successive halving or extraction and combination of small quantities with thorough mixing.

The principle of successive halving consists of dividing the sample into two approximately equal portions, with one being discarded and the other halved again. The process is repeated until the desired weight of working sample is obtained. Before each halving, the seed material should be mixed thoroughly. Various types of apparatus—conical, multiple-slot, and centrifugal dividers—are used for halving.

The conical divider has an inverted metal cone with numerous ports closely spaced around its circumference (Fig. 3). Alternate ports lead through ducts to a common outlet on one side and others to the opposite side. When the seed is poured over the cone, it flows through the ports, half of it emerging through the left outlet and the other half through the right outlet, achieving more uniform halving. In multiple-slot dividers, rectangular ports are held in a frame with alternate ports leading to the left and the right (Fig. 4). In the centrifugal divider, the seed falls onto a rapidly rotating disc, throwing half out to one side and the remaining half to the other side.

The principle of extraction and combination of small quantities is more suitable for small seeds (grasses, clover) that do not bounce or roll, with a working sample of about 109. About six to eight cups are placed randomly on a tray and the sample is poured evenly over the tray; the seed that falls into the cups is taken as the working sample. Cup size can vary according to the sample size desired and the size of seeds. Cups about 10–20 mm in diameter and depth meet most requirements.

For a germination test, four batches of 100 seeds are sampled from the pure seed fraction of a purity test using a vacuum seed counter or planter, which retains 100 seeds and spaces them readily for planting. Different sizes of vacuum seed counters are available for seeds of various sizes (2).

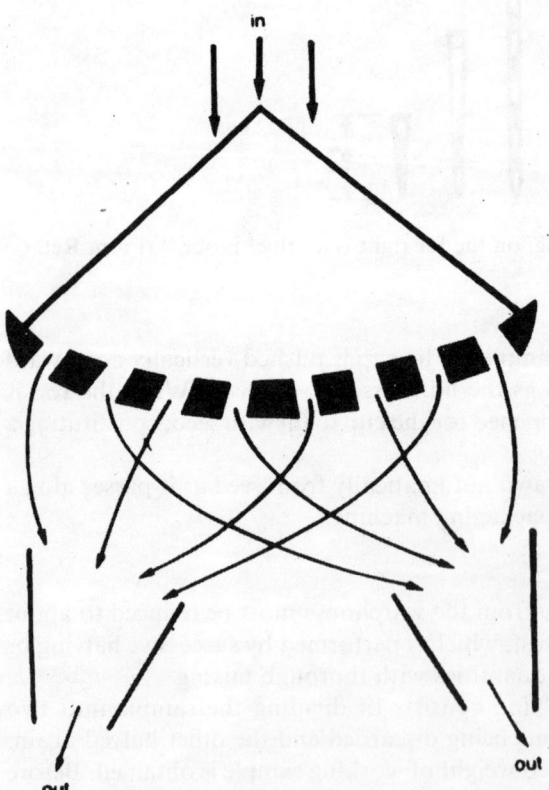

Figure 3 Conical divider with an inverted metal cone with numerous ports. (From Ref. 2.)

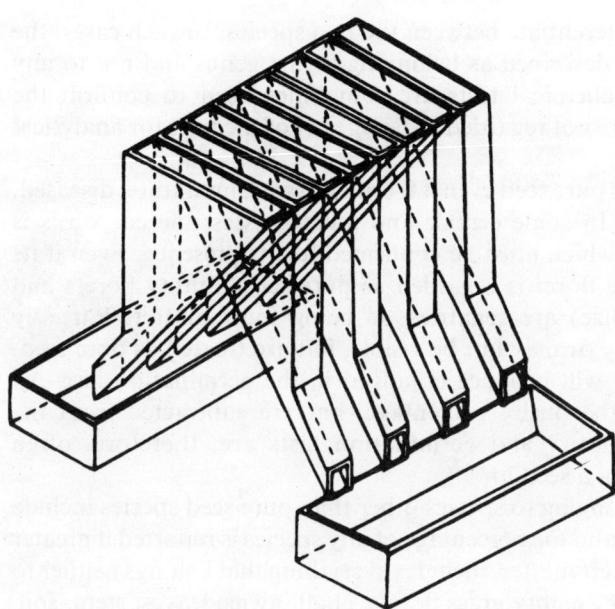

Figure 4 Multislot divider used for drawing laboratory samples (International Standards Organizations). (From Ref. 2.)

III. SEED TESTING FOR PURITY

Seeds can be tested for their analytical purity, species purity, freedom from weeds, and so forth (7,9).

A. Analytical Purity

To test the sample for analytical purity, it must be separated into three fractions: pure seed, seed of other species, and inert matter. The sample contains about 2500–3000 seeds. From the sample submitted to the laboratory, a sample is taken, either by successive halving or the random-cups method, and it is transferred to a dark smooth surface. The analyst examines every seed and particle, dividing the sample into the above-mentioned three fractions. After separation, each fraction is weighed and reported as a percentage of the total weight. Analytical purity is the percentage of pure seed present in the sample.

Pure seed is seed of the species that the sample purports to be without distinguishing between cultivars within the species. The analyst is faced with such difficulties as identification, the nature of the seed, and categorization of immature and broken seeds.

Most laboratories maintain authentic samples for identifying seeds of crop species and their cultivars by comparison. In some crop species, it is very difficult to distinguish with certainty between seeds of two species belonging to the same genus. Two species of rye grass, *Lolium perenne* and *L. multiforum*, for example, can be distinguished only by the presence of awns on the seeds of the latter. The awns are, however, liable to be broken off during postharvesting operations (threshing, cleaning), making it impossible to differentiate between the two species. In such cases, the pure seed can be identified and described as belonging to the genus and not to any particular species. Specialized chemical tests are sometimes used to confirm the authenticity of the species, but are not regarded as being part of the tests for analytical purity.

A whole seed is regarded as pure seed even if it is undersized, immature, diseased, or has germinated in storeage. In some cereals and most grasses, the caryopsis is hidden within the chaffy floret, which must be confirmed for its presence. Even if its development is only partial, the floret is regarded as pure seed. Empty florets and broken seeds (less than half size) are regarded as being inert matter. Partially developed and broken seeds may or may not be visible, but are treated as pure seed. The value of the seedlings they will produce is judged in the germination test. All nonviable seeds contribute to the purity percentage, but are subtracted from the germination percentage. The purity and germination tests are, therefore, often considered together in evaluating a seed lot.

The seeds identified as belonging to species other than pure seed species include both crop seeds and weed seeds, and the percentage of any species is reported if greater than 0.1%. The third fraction, inert matter, includes everything that belongs neither to pure seed nor to other seeds (e.g., empty grass florets, chaff, awns, leaves, stem, soil, sand, stones). The inert matter also includes fungal bodies like, for example, ergot and insect galls, although these are not really inert.

Seeds of pulses and cereals are identified by visual characteristics seen by the unaided eye or supplemented by a hand lens or occasionally by binocular microscope. Analysis of forage seeds may require mechanical aids. A machine has been developed that moves the seeds single file under a binocular microscope; the analyst can stop the flow, subjecting seeds to prolonged scrutiny. It is necessary to confirm that each floret contains a caryopsis, especially in grasses, which can be done either by touch or by transmitted light. By placing florets on a sheet of ground glass with a light underneath, full seeds can be distinguished by the dark shade of the caryopsis showing through the floret. Also, by feeling with a fingertip or spatula, an experienced analyst can tell whether or not a floret is empty. A blower also can be used to separate lighter, chaffy material from heavy seeds. The air current can be adjusted to blow the material uniformly to separate it into full, empty, and questionable seeds; the latter can be examined more closely to confirm. A sample may be blown uniformly for 3 min at a defined pressure that will separate empty and full seeds, leaving no questionable seeds. Many species cannot be separated by this blowing method; grasses like *Poa pratensis*, *Chloris gayana*, and *Dactylis glomerata* each require different air pressures. The method considerably reduces the analyst's workload and increases uniformity in test results. Different laboratories using the same standardized blower can obtain precise results in their tests of the same seed. Precise resultscan also be achieved using the same sieves to make a preliminary separation between pure seed and other seeds or inert matter, the separation being confirmed later by visual examination.

Fitzgerald et al. (9a) carried out image analysis of seeds, cotyledons, and leaves in two-dimensions and seeds in pseudo–three dimensions in selected populations of Brussels sprouts and cabbages to determine the amount of sib contamination in F_1 hybrid seed production. A novel cluster analysis was applied to the data and individual aberrants were selected. Comparative studies using acid phosphatase isoenzyme analysis, germination tests, and visual grading were carried out to characterize the selected aberrants. Image analysis offers the potential economic advantages of being less demanding of skilled labor, of reduced overheads, and of being carried out in real time. It also has the potential to select individual aberrants in large populations, allowing upgrading of seed quality. Criteria for seed quality control are discussed based on the results obtained.

Sousa-Santos et al. (9b) applied a polymerase chain reaction–based method for the detection and identification of *Clavibacter michiganensis* subsp. *michiganensis* in tomato seed. Two primers, CM3 and CM4, were used to amplify a specific 645-bp DNA fragment when the target bacterium was present in the amplification step. This fragment was produced with DNA from all *C. michiganensis* subsp. *michiganensis* strains and contaminated seed extracts tested from as few as 40 cells mL^{-1} but not from DNA of other plant pathogenic bacteria or tomato saprophytes. The results were confirmed by isolation, IF, pathogenicity tests and restriction fragment length polymorphism (RFLP) analysis of the amplified fragments by the endonuclease.

B. Freedom from Weeds

Sometimes a seed sample must be examined for the presence of seeds of other species, including weeds. The sample size in this case should be about 10 times greater than that used for analytical purity. This large sample is examined critically for seeds other than pure seed. Such tests may be required for certification or for law enforcement purposes. The result is expressed as the number of seeds found in the weight of seed material examined, e.g., one seed of wild oat found in x grams of barley seed. A certification standard for weed seeds is normally in the form of number of seeds of a species (or weed) in 1 kg of seed, using a standard weight of seed sample (10).

IV. TESTING FOR SEED GERMINATION AND VIGOR

A. Germination Testing

Seeds require oxygen, water, and a suitable temperature to maintain respiration (or biological oxidation) to provide energy for various metabolic processes. To test the germination capacity of seed, a small sample, usually 400 seeds, is provided with optimum conditions for germination, which are maintained for an appropriate duration of time, and the number of seeds producing normal seedlings is counted. When a viable, nondormant seed is given a wetted substratum, oxygen, and a suitable temperature, the seed imbibes water, respiration and metabolic events increase, and after some time the radicle emerges from the seed. In addition to the three essential factors for seed germination (moisture, temperature, and oxygen), some seeds like Lachica sativa may also require light for germination (11).

According to seed quality control programs, germination tests are performed at a stage where the presence, absence, and formation of essential structures of a seedling can be assessed properly to ascertain satisfactory development under favorable soil conditions. Since not all germinable seeds necessarily produce normal seedlings, the germination percentage is usually expressed on the basis of normal seedlings only.

Germination can be of two types: (a) epigeal, where cotyledons emerge above the soil surface and usually become green and photosynthetic, and (b) hypogeal, in which cotyledons do not emerge above the soil surface. Some endospermic and nonendospermic seeds differing in germination type are listed in Table 2 (13).

Table 2 Endospermic and Nonendospermic Seeds Showing Hypogeal or Epigeal Germination

Type of germination	Type of seed	
	Endospermic	Nonendospermic
Hypogeal	Hevea spp.	Zea mays
	Hordeum vulgare	Pisum sativum
	Tradescantia spp.	Vicia faba
Epigeal	Allium cepa	Arachis hypogaea
	Ricinus communis	Cucumis sativus
	Rumes spp.	Cucurbita pepo
		Lactuca sativa
		Phaseolus vulgaris

Source: Ref. 13.

The duration of the germination test, giving complete germination at the test temperature, varies with the type of seed species from 5 to 28 days. Dormat seeds require longer. The seed sample is usually a mixture of dormant and nondormant seeds, resulting in nonuniform germination requiring more than one count.

Among grasses, dormancy is common and light is necessary to break it. Germination tests for such seeds are carried out at a constant temperature under strong light. Some seed samples fail to respond to standard test conditions because of dormancy and require special measures. The test is repeated either after pretreatment or with special conditions to overcome dormancy. Some cereals may need chilling for a few days at 5°C (on a moist paper in a refrigerator).

Pretreatment with gibberellic acid can break dormancy in cereals. The grass seeds can be induced to germinate by soaking the blotters in a dilute solution of potassium nitrate. Washing seed in running water before the germination test can remove a soluble inhibitor present in best seeds. Dormancy in clover seeds is broken by enriching the environment with carbon dioxide, which may be achieved by enclosing the moistened sample in an impermeable polyethylene envelope; the carbon dioxide produced by the respiration of the nondormant seeds is sufficient to induce seed germination (2).

Temperatures needed for germination range from 15 to 35°C, with the optimum for most seeds being 20°C. The temperature may be steady or variable, alternating between relatively; high by day and low by night. A common regimen is 20°C for 16 hr and 30°C for 8 hrs/simulating natural fluctuations of night and day. Dormant seeds may require alternating temperatures, and in some species of grasses, dormancy is so common that the use of alternating temperatures is the common testing regimen.

Seeds do not germinate well when immersed in water, probably owing to insufficient oxygen for respiration or leaching of soluble substances from the seed. In germination tests, water is supplied by absorption through a medium or substratum such as sand or absorbent paper. Small pieces of paper (blotters) and large sheets in the form of "towels" are commonly used. Provided ventilation in the seed germinator is adequate, there is normally sufficient oxygen in the atmosphere and no special arrangements are necessary.

The seed analyst's ability to discriminate between normal and abnormal seedlings is one of the most subjective aspects of seed testing. The Association of Official Seed Analysts (AOSA)(12) therefore developed a Seedling Evaluation Handbook in 1992, which is now recognized as a formal component of the rules for testing seeds. This handbook has drawings depicting the differences between normal and abnormal seedlings to help analysts in discriminating among questionable seedlings (Fig. 5).

In a seed germination, temperature is controlled by a thermostat set to maintain the required range of temperature. Both heating and cooling arrangements may be required depending upon the climate and time of year. An arrangement for air circulation may also be provided to maintain uniform temperature. Fluorescent tubes generate less heat than other light sources and are preferred to maintain temperature control. High humidity is required for germination.

A small air-conditioned room with heat-insulated walls and a tight-fitting door maintained at a steady temperature at high humidity (by humidifier) and fan for air circulation meets most of the requirements of germination. A cabinet germinator is a cubical unit with internal lighting and shelves for the samples. Thomson (2) described the Jacobsen germinator consisting of a shallow tank holding water kept at a controlled temperature. It is covered by glass strips by a perforated metal plate, or by a grid through which water circulates. The blotters placed on it are kept moist by wicks dipping into the water in the tank. Each blotter is covered by a transparent bell jar, which reduces water loss but allows ventilation through a hole at the top. Water is kept at the required temperature and fluorescent tubes fitted above the tank supplement daylight.

The international rules for seed testing (10) specify requirements with respect to temperature, light, duration, and methods of breaking dormancy for the seeds of different crop species (Table 3).

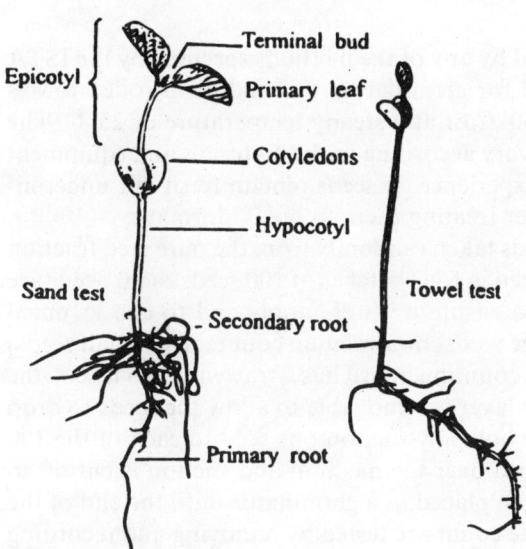

Figure 5   Seven-day-old seedlings of soybean under sand test and towel test. (From Ref. 12.)

Table 3 Germination Requirements for Seeds of Some Important Crop Species According to International Rules for Seed Testing

Crop species	Substrata	Temperature (°C)	Light	Duration	Methods of breaking seed dormancy
Avena sativa	S, BP	20	—	10	Prechill, GA, test at 10–15°C
Hordeum vulgare	S, BP	20	—	7	Prechill, predry, GA test at 15°C
Oryza sativa	BP, TP, S	20–30, 30, 25	—	14	Presoak
Pennisetum typhoides	SP	20–30	—	28	Light
S. vulgare	BP	20–30, 20–35	—	10	Prechill
T. aestivum	S, BP	20	—	8	Prechill, predry, GA test at 15°C
Zea mays	BP, S	20–30, 25	—	7	—
P. vulgare	BP, S	20–30, 25, 20	—	9	Light
Pisum sativum	S, BP	20	—	8	Light
Vicia faba	BP, S	20	—	14	Light, prechill
Lolium perenne	TP	15–20, 20–30	—	14	Light prechill, KNO₃, test at 10–30°C
Poa pratensis	TP	15–25, 15–30 10–30	L	28	Prechill, KNO₃
Medicago sativa	TP, BP	20	—	10	Prechill
Trifolium pratense	BP, TP	20	—	10	Prechill, test at 15°C
Brassica oleracea	TP, BP	15–25, 20–30 20	—	10	Light, prechill, KNO₃
Lactuca sativa	TP, BP	20	—	7	Prechill, predry
Helianthus annuus	BP, S	20–30, 25, 30	—	7	Prechill, predry
Corchorus capsularis	TP, S	30	L	5	—

ᵃ When more than one method is indicated, more than one may be used, but the first is to be preferred. TP = top of paper; S = sand; GA = gibberellic acid; BP = between paper, including rolled towels; L = light essential.
Source: Ref. 2.

Seeds of crop species may be tested by any of the methods specified by the ISTA (10). Maize, for example, can be tested for germination on sand or in rolled towels at alternating temperatures of 20 and 30°C or at a steady temperature of 25°C. The medium and temperatures selected can vary according to the materials and equipment available as well as the analyst's past experience. If seeds remain fresh but ungerminated, they are tested a second time after treating them to break dormancy.

The test sample consists of 400 seeds taken randomly from the pure seed fraction (from test for analytical purity) and placed in four batches of 100 seeds each. Seeds are spaced out evenly on the substratum to ensure a water supply and to check fungal development. Counting boards (for large seeds) or a vacuum counter (for small seeds) is often used to facilitate operation. The counting board has a tray with two layers: the upper layer has 100 holes and the lower layer is removable to allow the seeds to drop onto the medium. The vacuum counter holds by suction one seed to each of the 100 holes in a metal plate. The plate is placed over the medium and suction is cut off to place the seeds. The four batches are then placed in a germinator until the end of the test period. Seeds needing more than one count are tested by removing and recording the seedlings developed in the first count; the dead seeds are also removed and counted. The remaining seeds are again placed in the germinator to allow them to develop further for subsequent counting and recording. At the end of the test period, each seedling is evaluated and classified as normal or abnormal and the numbers of dead and dormant seeds are counted. The latter may be of two kinds—fresh ungerminated seeds and hard seeds. The hard seeds found commonly in pulse do not imbibe water and probably will not germinate when sown in the field. If the numbers of normal seedlings in four replicates agree within the statistical limits and if the sample does not contain more than 5% fresh ungerminated seeds, the germination test is considered complete; the percentage of normal seedlings being considered the germination capacity of the sample tested (2).

A normal seedling has the following essential structures, which are often lacking in abnormal ones:

1. A well-developed root system
2. A well-developed and intact hypocotyl or epicotyl and a normal plumule or a well-developed first leaf (in cereals and grasses)
3. One cotyledon (in monocot seedlings) and two cytoledons (in dicot seedlings)

Abnormal seedlings may be damaged, deformed, or decayed.

Scrupulous cleanliness must be ensured while carrying out germination tests. Seeds are often infested with molds and other microorganisms, which can become a permanent source of contamination in the laboratory. All germinated and dead seeds as well as towels and glassware should be destroyed after the test. Blotters must be cleaned thoroughly to avoid the spread of microorganisms. Germinators should also be cleaned periodically by washing with a mild disinfectant such as formalin or potassium permanganate. Sand used repeatedly must be sterilized by dry heat before each test.

B. Biochemical (Tetrazolium) Test

The germination capacity of a seed sample can also be tested biochemically using the reagent tetrazolium (2,3,5-triphenyltetrazolium chloride or bromide). It is a colorless soluble salt whose aqueous solution easily diffuses through living plant tissue, where it is enzymatically reduced to triphenyl-formazan by a dehydrogenase. The reduced compound is red in color and remains inside the cells where it was formed, being nondiffusable. Since enzyme dehydrogenase is active only in living tissue, the compound stains only living seeds, with the dead ones being unaffected. This reaction is also brought about by light, alkaline solution, and ascorbic acid. The pH of tetrazolium solution is therefore adjusted to between 6.5 and 7.0, and it is kept in darkness both before and during the test. The results of the tetrazolium tests are interpreted on the basis of the knowledge of seed and seedling structures, understanding of the test mechanism, and its limitation. The result of staining patterns should also be interpreted in the light of other visible aspects of seed quality and the experience of the analyst (13).

The seeds are immersed in a 1% tetrazolium solution to allow the solution to penetrate, after which the straining of the embryo is examined. The embryos that are completely stained and completely unstained indicate, respectively, the presence of living and dead seeds. Some seeds may be partially stained; indicating that the plumule is viable but the root is dead. In this test, the embryo is evaluated based on the staining of essential parts. The minimum staining required has been determined experimentally and is not the same for all seed species; for example, all the root tissue should be stained in sorghum, but only a small part may be stained in barley and an intermediate proportion in wheat. The analyst with experience learns the minimum staining pattern of each species.

The entry of tetrazolium solution into the embryo is facilitated by presoaking the seed in water, removing the outer layers, or cutting each seed into halves. At high germination capacity, tetrazolium tests generally agree with conventional tests, but a lower germination, the biochemical test is not as reliable; mainly because dormant seeds are not distinguished from nondormant seeds by the latter test. Abnormal growth is not detected, and embryos that develop into deformed seedlings stain normally. Living microorganisms present in decayed embryos may also cause staining.

The biochemical test is, however, very rapid and is of value in processing establishments where it is necessary to assess incoming seed quickly. The test takes about 24 hr to complete, but time can be reduced by vacuum infiltrating the solution into the seed, raising the temperature slightly during immersion, or using the iodide salt of tetrazolium. This test is also employed to determine whether the seeds left ungerminated at the end of the conventional germination test are dead or dormant. The tetrazolium test is a versatile tool to investigate the causes of poor seed germination and low vigor.

Kuo et al. (13a) studied procedures and evaluation criteria used in the tetrazolium test on the seeds of *Salvia splendens* and *S. farinacea*. Decoating was found to be necessary for satisfactory staining. Soaking the naked embryo in 0.5% triphenyltetrzolium chloride solution and incubating at 37.5°C for 30 min yielded good results. Six staining patterns were recognized. The viable categories of each pattern were determined by the root mean square method. Three patterns represented viable and three nonviable seeds.

C. Testing Seeds for Vigor and Viability

According to Isley (14), vigor is the sum total of all seed attributes that favor stand establishment under unfavorable field conditions. The AOSA (15) defined seed vigor as the sum total of all those properties in seed which, upon planting, result in rapid and uniform production of healthy seedlings under a wide range of environment, including both favorable and stress conditions. According to Perry (16), seed vigor is the sum total of all those properties of the seed that determine the level of activity and performance of a nondormant seed or seed lot during germination and seedling emergence.

Seed vigor or viability is thus the ability of a seed to germinate and grow into a plant in adverse field conditions. Adverse field conditions not being the same in different areas, seeds face a variety of stresses like cold, heat, drought, acidity, and salinity, and it is difficult to express vigor as a precise numerical value. A vigor test is a supplement to the standard germination test and can at most differentiate between seeds that will perform relatively well in poor soil conditions and those that will perform relatively poorly among seed lots of the same species with approximately the same germination capacity (17).

1. Direct Vigor Tests

In direct vigor tests, the adverse conditions expected in the field situation are simulated in the laboratory and applied to the seeds, allowing them to germinate and develop in the artificial environment, followed by a performance evaluation. It is difficult to standardize adverse conditions, and significant variations in results between laboratories and even within the same laboratory have been reported from time to time. It is therefore necessary to plant a sample of a known high-vigor crop as a control for comparison.

2. Indirect Vigor Tests

In indirect vigor tests, a physiological activity known to be associated and to vary with vigor, such as respiration, enzyme activity, or permeability of the cell membrane, is measured to assess vigor indirectly. A possible cause of low vigor, for example, bruising of seed tissue during processing or senescence, can also indirectly reflect vigor. Speed of germination observed in a germination test is not a reliable indication of seed vigor, because it is significantly influenced by dormancy.

The most common vigor test are the cold test (maize), the Hiltner test (wheat, rye), and the conductivity test (peas). The first two are direct, whereas the third is indirect. In the cold test, seeds are planted in moist soil containing *Pythium* and other fungi and kept at 10°C for 1 week followed by 3 days at 30°C to allow the seeds that are still viable to germinate. In this test, seeds are subjected to two adverse factors: parasites and cold. In the Hiltner test, seeds are deeply sown in tightly packed brick dust. The physical force required for the seedling to reach to the surface is tested. In the conductivity test, seeds are immersed in water and soluble substances like sugars and amino acids exuded into the water through faulty permeable membranes are measured by testing the solution for its electrical conductivity.

Shen and Oden (18) assayed germination and fumerase activity using untreated, primed, and artificially aged seeds of *Picea abies* L. Karst, *Pinus contorta* Engelm, and *Betula pendula* Roth. Priming at a restricted and controlled moisture content significantly decreased the mean germination time of all these species, whereas artificial aging significantly increased the mean germination time of *P. abies* and *P. contorta* seeds. The correlation between mean germination time and fumerase activity was significant for all these species. The result indicated that the relationship between the mean germination time and fumerase activity can be developed into a reliable method of predicting seed vigor of *Picea abies* L. Karst, *Pinus contorta*. Engelm, and *Betula pendula* Roth seeds using fumerase activity. Germination and fumerase activity were assayed also using untreated, stratified, and hormone-treated *Fagus sylvatica* var. *latifolia* seeds. The fumerase activity in nondormant seeds was about two times higher than in dormant seeds; indicating that it may be possible to use fumerase activity as a measure of dormancy degree. Ranbo (19) has summarized the current understanding of seed vigor and seed vigor testing covering purity, health, and viability.

According to Sako et al. (20), seed vigor testing provides valuable information for assessing seed lot quality. However, vigor testing has not experienced widespread use because of its labor intensiveness, high cost, and variability in test results from laboratory to laboratory. Sako and associates have presented an automated seed vigor assessment system that is objective, economical, and easy to perform. The system interfaces an imaging device that captures digital images of germinating seedlings to a computer. These images are processed by a computer to generate numerical values that collectively represent the quality of a seed lot (vigor index) based on various statistics acquired from morphological features of the imaged seedlings, indicating the sample mean of hypocotyls and radicle lengths and sample standard deviation of the hypocotyl length, radicle length, total length, and radicle to hypocotyl length ratio that indicates the speed and uniformity of seedling development. This system was tested on lettuce seedlings grown for 3 days in the dark. The results indicated that the imaging system accurately quantified these parameters to yield reproducible, objective vigor assessments. Employing inexpensive flatbed scanners as an approach to capture high-quality seed and seedling images further enhances the standardization of seed testing (21).

Twelve rapeseed lots (three cultivars each of four lots) and nine pea seed lots (three cultivars each of three lots) of various ages were used in laboratory vigor test and field trials in Denmark (22). Plants of older rapeseed lots performed poorly throughout the growing season, and final seed yield was significantly lower compared to that of young seed lots. The germination percentage of rapeseeds in a standard test correlated poorly with field performance, whereas germination after controlled deterioration for 24 hr correlated significantly with plant growth in both sowings and with final yield in one sowing. The mean germination time in the standard test and after controlled deterioration had highly significant correlations with field performance and seed yield for rapeseed in both sowings; the applicability of a vigor test for *Brassica* species based on the germination rate is discussed. Seedlings produced by the older pea seed lots emerged more slowly than plants from the young seed lots, but differences in growth characteristics diminished throughout the growing season. Differences in seed yield were only slightly significant in one sowing. Controlled deterioration improved the correlation of the germination percentage and the mean germination time for pea seeds with field performance, but overall the correlation with plant growth and yield was poor. Larsen et al. have discussed the significance of these findings.

Strydom and Venter (23) correlated the emergence of cabbage (*Brassica oleracea* var. *capitata*) seedlings in four trials with the results of the standard germination test (at both 20 and 20/30°C), germination at nonstandard temperatures (10, 30, and 35°C), the controlled deterioration test, and Ki values determined from seed survival curves. Ki, determined on the basis of the percentage normal germination, produced highly significant correlations with probit emergence in all four trials and was the best indicator or relative emergence. In the germination tests, the highest number of correlations with emergence was in the case of the 20/30°C temperature regimen. Counts of normal seedlings correlated significantly with emergence in three of the trials, whereas radicle emergence counts after 3 days were significantly correlated with emergence in all four trials. The controlled deterioration test did not show greater potential as a seed vigor test than the above-mentioned ones.

Differences in field emergence of commercial combining pea seed lots (>80% germination) were revealed in field trials in 1989 (5 sowing dates, 29 seed lots, 5 cultivars) and 1991 (2 sowing dates, 26 seed lots, 3 cultivars) indicating differences in seed lot vigor (24). Powell et al. suggested that although vigor tests can identify low-vigor seed lots of combining pea, it might be difficult routinely to apply a single seed-vigor test to seed lots from all cultivars.

Seeds of goldpoppy (*Eschscholtzia californica*), China aster (*Callistephus chinensis*), garden balsam (*Impatiens balsamina*), and stock gillyflower (*Matthiola bicornis*

[*Matthiola longipetala* var. *bicornis*]) were subjected to accelerated aging at 42°C and 100% RH, and then three methods of assessing seed vigor were used to compare vigor in these seeds and in untreated ones. Seeds were tested in the laboratory using the tetrazolium test or Germ's method and in a greenhouse by monitoring seedling emergence at 20/15°C. Accelerated aging reduced seed viability by 19–42%. The tetrazolium test gave good results, but great care in interpretation was needed during seed assessment. Germ's method was the least useful, and gave good results only for goldpoppy. Seedling emergence in the greenhouse gave the best results but takes the longest and requires the greatest care in controlling environmental factors (25).

Duczmal and Ratajczak (25a) determined the germination capacity, Pieper's index (mean germination time), and electroconductivity of aqueous extracts for 10 seed lots of wrinkled-seeded peas of different cultivars. Germination was 56–85% at the first count and 85–98% at the final count. Pieper's index ranged from 4.2 to 5.5 days and the mean conductivity for 50 seeds from 160 to 333 µS/cm. Emergence at 15, 8, or 4°C in sand at 50 or 80% of the total moisture capacity ranged from 28 to 92%. From correlation studies, it was concluded that the use of the conductometric test to estimate seed vigor in peas is potentially unreliable.

Minicka et al. (25b) conducted studies to explain the significant differences between the results of the laboratory estimation of pea seed viability and their field emergence reported by many authors. Results showed that conductivity test results were better indicators of field emergence than germination test results but were of limited value. Seed age should be taken into account when predicting pea seedling emergence and excessive or insufficient precipitation, and low temperatures were the main factors affecting field emergence.

Holubowicz et al. (25c) conducted studies on the ethylene production of imbibed and germinating seeds of the radish cultivar *Saxa* in order to use this character as a biochemical marker for seed vigor. The seeds selected for use in the experiments had both high vigor and germination capacity. Those of low vigor were obtained by subjecting the given seed lot to accelerated ageing (AA). Determination of ethylene production was carried out 24, 38, 48, 72, and 96 hr after placing the seeds on wet blotting paper in Petri dishes. The intensity of ethylene production was found to depend on seed vigor and the stage of germination. The highest ethylene production (15.94 nL/hr for 50 seeds) was found after 38 hr. Seeds subjected to AA attained a peak of ethylene production (12.92 nL/hr for 50 seeds) after 72 hr. Those seeds with a higher ability to produce ethylene emerged faster, when sown in a greenhouse, than seeds with a lower ability. Plants from seeds subjected to AA had shorter aerial parts, a smaller hypocotyl diameter, and lower air-dried weights. The results showed that estimation of ethylene production could be used to determine seed vigor.

The slant-board test is a potentially useful predictor of seed vigor in small-seeded field vegetables. However, in its standard form, the test is time consuming. Image analysis has already been used to characterize and measure seeds, but the resolution of the very fine roots produced in early seedling growth has proved to be difficult. Modifications to the slant-board test necessary to achieve automated measurements by image analysis have been described. In particular, a novel growth-support medium has been described consisting of a thick paper blotter overlaid with black fabric to provide an optically favorable background. Problems of phytotoxicity, poor color contrast, dye leaching, root penetration, and growth distortion have been reported for many of the support media tested. The automated slant-board test offers

Table 4 Initial Germination, Accelerated Aging, and Storage Potential of Maize
Seeds of Two Different Lots

Test results	Seed lot A	Seed lot B
Initial germination, %	97	96
Germination % after accelerated aging	92	47
Germination % after 18-month open storage	92	20

Source: Ref. 2.

the prospect of more accurate and less expensive predictions of seed vigor in field
vegetables (25d).

D. Storage Potential of Seeds

Seeds with a high germination capacity generally maintain this capacity better for
longer periods than seeds of poor germinability. Seed lots of equal germination
capacity, however, may not necessarily be store with equal success. The storage
potential of seed is influenced by several preharvest and postharvest factors, such as
weather conditions during seed ripening and harvesting, treatment during processing,
and storage conditions.

Accelerated aging tests are used to forecast the probable storage life of a seed. In
these tests, the stresses of prolonged open storage are concentrated into a few days by
exposing the seed sample to the worst possible conditions that will not actually kill the
embryo; for example, 40–45°C at 90% RH maintained for 2–8 days. At the end of this
period, the sample is tested for germination capacity. A higher germination capacity
indicates the ability of the seed to survive prolonged normal storage conditions. This is
illustrated by the results from an experimental test of two contrasting samples of maize
(Table 4).

V. SEED TESTING FOR MOISTURE CONTENT

The moisture content of seeds greatly influences their storage potential and how
long they will maintain their viability. Seeds harvested at a high moisture content
are prone to damage during harvesting and threshing and are susceptible to attack
by molds and insects. It is, therefore, essential to know the moisture content of seed
at various stages during seed production. In the seed trade, a certain percentage of
moisture is allowed by law depending upon the species. The object of moisture analysis is to determine the moisture content of a seed lot at the time it is sampled. The
sample must, therefore, be handled carefully to retain the initial moisture content
by packing it in a moistureproof (metal or plastic) container, submitting the sample to
the testing center without delay, and analyzing the sample promptly after its arrival.
Even during moisture determination, the seed should be minimally exposed to the
laboratory atmosphere. Seed species that do not require grinding should not be exposed more than 2 min between removal from their containers and placement it in
a drier (26).

The moisture content of the seed can be thought of as existing in three states that
overlap and are not sharply distinguished from each other. Water is present as an

essential part of the living protoplasm and organic material constituting endosperm. This is chemically bound water and is virtually independent of the seed's atmosphere. In its second state, electrostatic forces associated with the molecular structure of the seed also hold water. This water is also held firmly, but to a lesser extent than in the first state. The third state of water, called "free" water, is rather loosely bound. It is the free water that easily exchanges itself with the seed's atmosphere, depending on ambient conditions, and affects the seed's longevity in storage. Theoretically, *moisture content* refers to the amount of free water present in the seed and is expressed as the percentage of the total weight of the seed at the time of its determination.

Determining the moisture content of seed in the laboratory is accomplished by expelling free water from a weighed seed sample by placing seed in a hot-air oven maintained at a constant temperature. The consequent loss in weight is taken as the moisture content of the seed.

The ideal drying temperature and time, which may vary with seed species, should be such that only free water is driven off, nothing more than free water (volatiles) is driven off, and no chemical change takes place (e.g., oxidation) as a result of heating (2). Ideal methods take too long, and methods recommended for routine laboratory use are not quite perfect, although they give accurate and precise results. Thomson (2), therefore, prefers to define the seed moisture content as the loss in weight when a sample is dried under standardized conditions.

The International Rules for Seed Testing (10) prescribe two oven-drying methods for agricultural seeds:

1. Low–Constant Temperature Method—This method involves drying seeds at 103°C for 17 hr. It is suitable for oilseeds like soybean, cottonseed, and sesame.
2. High–Constant Temperature Method—Nonoily seeds like cereals, pulses, grasses, and clovers can be dried more quickly by the high-temperature method—130°C for 4 hr (maize), 2 hr (other cereals), or 1 hr (for other seeds). The higher temperature used in this method may kill the embryos in the process of drying, which does not matter.

To ensure complete drying, large seed samples (cereals, pulses) are usually ground to a powder. Very wet seeds are dried in two stages. In the first, a large sample is weighed, spread out in a thin layer, and dried overnight in a dry, warm place. After this predrying, the sample is weighed and the percentage of water lost is calculated. This sample is then ground and a subsample is subjected to the normal constant-temperature method. The moisture content of the seed is calculated by combining the water losses obtained in the two drying stages.

Precautions should be taken to prevent the moisture content changing on exposure to air. After drying and before weighing, the sample should be cooled in a desiccator kep dry with desiccant such as phosphorus pentoxide or calcium chloride crystals.

Sometimes electrical moisture meters are used to obtain instant results. The meters measure the electrical property of the sample (the conductance or capacitance), which varies with the moisture content. The meter is calibrated for the species being tested so that the dial reading directly gives the moisture content. The calibration must be checked periodically against laboratory tests. The meters are most reliable in the moisture content range of 10–25%, but are not completely dependable in any circumstances and may have error rates up to 1%; temperature may also affect the reading. Meters are essential tools for plant operators, for whom speed is more important than absolute accuracy.

Seed moisture can also be determined by using infrared moisture meters; in this equipment, an infrared lamp is used as a source of heat for the quick results. However, infrared meters are not suitable for seeds with a high oil content because of their high temperature.

As for all other seed attributes, the result of the moisture test applies to the sample and is only applicable to the seed lot in so far as the sample represents the lot. Within a seed lot there may be a considerable variation in the moisture content, and if the test is to be of any value, the sampling procedure must be followed scrupulously. Even when the moisture content of the sample truly represents the average moisture content of the seed lot, the latter may contain spots with a higher than average moisture content.

VI. SEED TESTING FOR HEALTH AND UNIFORMITY

A. Testing for Seedborne Diseases

Different seedborne diseases caused by microorganisms like fungi, bacteria, viruses, and eelworms require different laboratory methods for detection. A pathologist cannot identify all diseases by one test, and although a seed sample cannot be tested for freedom from all diseases, if certain pathogens are found in the sample, it may be possible to report the type and extent of their presence.

Some diseases are detected by examining dry seeds with low magnification or even the unaided eye. Fungi producing hard dark-colored sclerotia or fruiting bodies can be detected in this way; for example, ergot of cereals, Sclerotinia on clover seeds, and Septaria on flax and celery seeds. Discoloration caused by bacterial diseases of pulses and galls produced by eelworms can also be identified.

A small seed sample is shaken in a small quantity of water and the suspension is examined under the microscope for the presence of microorganisms that have been washed off the seed surface. This quick technique is particularly useful to identify certain smut and bunt diseases of cereals like wheat. The spores of smuts and bunts are generally found on the seed surface and are easily washed out in an aqueous extract.

The mycelium of loose smut is often present within the embryo tissue, which requires separate microscopic examination of seed tissue. If it is present in the superficial layers of the seed, it can be detected by less elaborate methods.

The blotter test provides favorable conditions of temperature and humidity for the growth of both the seed tissue and the pathogens present on it. The seeds are placed on wet blotters (or filter paper) at 20–30°C for a few days. Pathogens can be recognized by the symptoms appearing on the seedlings or by the emerging fungal growth; for example, oat seeds infected with stripe disease (*Drechslera* spp.) give rise to seedlings with brown-colored stripes on the first leaf and cereal seeds infected with root rot (*Fusarium* spp.) produce seedlings with brown roots. Alternatively, the emerging seedlings can be dipped in a herbicidal solution (2,4-D) and deep frozen for 24 hr to arrest the plant growth. The pathogens is not affected by this treatment and becomes visible on the seed surface. Examining them under ultraviolet radiation can identify the spores of the fungi.

In another method, seeds are placed on the surface of a sterile nutrient agar gel in a Petri dish. Any fungus present on the seed grows and forms a colony on the agar, and is then identified by its color and type of growth. The seeds are washed with a dilute hypochlorite solution before being placed on agar gel to kill the surface spores accidentally attached to the seed and to ensure that the fungus growing on the agar was an internal pathogen.

Some diseases require elaborate and time-consuming techniques to identify them by growing plants under favorable conditions in a greenhouse and examining the development of any pathogenic symptoms on the emerging plants (e.g., lettuce virus). Such tests may take several weeks to complete. Bacterial diseases require special phage and serological techniques for identification.

B. Seed Testing for Uniformity

Bigger seed lots containing more than 15–20 tons of seeds are required to be sold in packages of 50 kg or less. It is desirable that within a lot, the contents of each package are uniform in quality. In practice, significant variations between packages have been noticed, which can be tested for uniformity.

For testing uniformity, several seed samples are taken separately from a number of small containers (up to 30) depending upon the size of the lot. Each sample is then tested for one attribute like the percentage purity, germination capacity, or the number of weed seeds per kilogram of seed. The amount of variation observed in numerical values is expressed statistically as the variance (V). This is then compared with the variance that would be theoretically expected if the variations between sample values were distributed at random throughout the entire seed lot (W). The standad for comparison is the ratio:

$$\frac{V}{W} = \frac{\text{actual variance found in sample}}{\text{expected variance in random state}}$$

This ratio would be unity if pure seed, viable seed, or weed seeds were distributed at random. In practice, the ratio is often found to be greater than unity, indicating heterogeneity in distribution, H, which can be obtained as follows:

$$H = \frac{V}{W} - 1$$

The standard of uniformity is taken to be that in which variation is at random, and the H value indicates the excess of the actual variation over this standard. Since randomness is not uniformity, the actual variance may be less than the standard, in which case H will have a negative value. Whereas a low H value does not necessarily indicate uniformity of the seed lot, a high H value surely denotes that blending has been inadequate (27).

Seed quality control must be closely associated with seed production. National seed quality control must be suited to the seed-production systems nationwide to find both large-scale and small-scale solutions. Seed legislation should provide options both for seed certification and some simplified quality declaration. The seed-testing capacity must be reliable and suited to cover different farming needs over time and territory (28). Roberts et al. (29) investigated the effects of seed tests for bacterial pathogens with varying soak times followed by two redrying regimens and two different storage conditions on germination of naturally infected seeds of peas and inoculated seeds of brassicas, tomatoes, and maize. The pathogens were detected in all soak times, and although greater numbers were generally detected at the longer soak times, these differences were not always significant. The effect of soak time on germination varied for the different crops, but in general germination was lower with longer soak times and it was better after redrying in the airflow of a laboratory fume hood at ambient temperature (20–28°C) than after drying in controlled conditions at 15°C and 30% RH. Germination also tended to be better when seed was stored in sealed foil packets at –20°C than in an ambient temperature store. It is, therefore, feasible to use short soak times with rapid redrying of seed as a means of nondestructively testing a range of different seed types for a variety of bacterial pathogens with only minimal or even no reduction in germination. Such an approach will be most useful for assessing the health status of small quantities of highly valuable seed, such as in germplasm collections or breeding material.

Lakshmana et al. (30) employed phenol color reaction to assess 15 pearl millet genotypes, including 6 varieties, 5 cyloplasmic male sterility (CMS) lines, and 4 restorer lines. This study showed that the variation in phenol color reaction in pearl millet could be used as a quick and simple method precisely to characterize genotypes. It was noticed that male sterility was not associated with the phenol color reaction trait in pearl millet and was not cytoplasmically controlled. Hang et al. (31) have developed a simple, accurate, and rapid method to identify malting barley cultivar, usng the polymerase chain reaction–randomly amplified polymorphic DNA (PCR-RAPD) technique with a selected set of 10-mer primers and DNA that was extracted from mature in bibed embryos. This technique can be adapted by industry to maintain cultivar purity and to check integrity of purchased seed lot.

Deswal and Sheoran (32) developed a simple method for measuring the leakage from single seeds of any size, which can help in predicting seed vigor, viability, and germination. The optical density of seed leachate was measured at 260 nm with a spectrophotometer and data compared with an electrical conductivity test. The method was tested using aged and unaged samples of chickpea cv. Gaurav, pea cv. HFP-4, cotton cv. H777, *Brassica juncea* cv. RH-30, and *Trifolium alexandrinum* cv. Mescavi seeds. Optical density and electrical conductivity data were highly positively correlated. It was concluded that the optical density method is very sensitive, rapid, and simple to use with a large number of samples.

Results of a survey of seed analysis equipment used in 51 Society of Commercial Seed Technologists and 43 Association of Official Seed Analysts affiliated seed-testing laboratories in the United States have been reported. Particular reference was made to the use of personal computers and adoption of computer technology, availability of basic and more specialized seed analysis equipment, purity and germination problems which present equipment cannot handle, inventions to make tedious tasks easier, and equipment seed analysts were interested in seeing demonstrated and/or discussed (33).

Pandey (34) investigated a short-duration accelerated aging technique consisting of incubating *Phaseolus vulgaris* cv. Selection 9 seeds in water at 58 ± 1°C for different lengths of time to achieve a wide range of stocks differing in viability. Membrane integrity loss as manifested by electrolyte and solute leakage increased consistently with an increase in the duration of aging. Aging in the method used was a function of temperature, duration, and dynamically increasing the moisture content. The germination performance of seeds aged for 5, 10, 15, and 20 min by this method was comparable with the performance of seeds aged for 0.16, 1, 2, and 3 years at ambient temperature.

According to Thomson (2), it is not possible to set minimum standards of uniformity. The test can be used to ascertain the efficacy of blending methods, but is not practicable for routine purposes.

REFERENCES

1. S. P. Sharma, Breeder seed production in self-pollinated crops, *Techniques in Seed Science and Technology* (P. K. Agrawal and M. Dadlani, eds.), South Asian Publishers, New Delhi, 1987, p. 1.
2. J. R. Thomson, *An Introduction to Seed Technology*, Leonard Hill, London, 1979.
3. R. C. Payne, Variety testing by official AOSA seed laboratories, *J. Seed Sci. Technol. 10*:24 (1986).
4. ISTA, International rules. for seed testing, *Seed Sci. Technol. 21*:1 (1993).
5. M. B. McDonald and R. Payne (eds.), *Cultivar Purity Testing Handbook*, Contribution No. 33 to the *Handbook on Seed Testing*, Association of Official Seed Analysts, Lincoln, NE, 1991, pp. 1–76.
6. L. O. Copeland and M. B. McDonald, *Principles of Seed Science and Technology*, 3rd ed., Chapman & Hall, New York, 1995.
7. M. B. McDonald, L. J. Elliot, and P. W. Sweeney, DNA extraction from dry seeds for RAPD analyses in varietal identification studies, *Seed Sci. Technol. 22*:171 (1994).
8. M. B. McDonald, H. Danielson, and T. Gutormson, *Seed Analyst Training Manual*, Association of Official Seed Analysts, Lincoln, NE, 1992.
9. O. L. Justice, Essentials of seed testing, *Seed Biology*, Vol. 3, Academic Press, New York, 1972, p. 301.
9a. D.M.Fitzgerald, D.Barry, P.R.Dawson, and A.C.Cassells, The application of image analysis in determining sib proportion and aberrant characterization in F1 hybrid Brassica populations, *Seed Sci. Technol. 25*:503 (1997).
9b. M.Sousa-Santos, L.Cruz, P.Norskov, and O.F.Rasmussen, A rapid and sensitive detection of *Clavibacter michiganensis* subsp. Michiganensis in tomato seeds by polymerase chain reaction, *Seed Sci. Technol. 25*:581 (1997).
10. ISTA, International rules for seed testing, *Seed Sci. Technol. 13*(2): 307 (1985).
11. P. K. Agrawal, Germination test under controlled conditions and its evaluation, *Techniques in Seed Science and Technology* (P. K. Agrawal and M. Dadlani, eds.), South Asian Publishers, New Delhi, 1987, p. 54.
12. AOSA, *Seedling Evaluation Handbook*, Contrib. No. 35 to the Handbook on Seed Testing, Association of Official Seed Analysts, 1992, p. 101.
13. P. K. Agrawal and M. Dadlani, Tetrazolium test for seed viability and vigor, *Techniques in Seed Science and Technology* (P. K. Agrawal and M. Dadlani, eds.), South Asian Publishers, New Delhi, 1987, p. 79.
13a. W.H.J.Kuo, A.C.Yan, and N.Leist, Tetrazolium test for the seeds of *Salvia splendens* and *S. farinacea*, *Seed Sci. Technol. 24*:17 (1996).
14. D. Isley, Vigour tests, *Official Seed Analysts*, 1957, pp. 177–182.
15. Association of Official Seed Analysts, *Vigour News Letter*, 1975.
16. D. A. Perry, Report of the vigour test committee, 1974–1977, *Seed Sci. Technol. 6*:159 (1978).
17. P. K. Agrawal, Concept seed vigor and its measurement, *Techniques in Seed Science and Technology* (P. K. Agrawal and M. Dadlani, eds.) South Asian Publishers, New Delhi, 1987, p. 90.
18. T.Y.Shen and P.C.Oden, Relationship between seed vigor and fumerase activity in

Picea abies L. Karst, *Pinus contorta* Engelm, *Betula pendula* Roth and *Fagus sylvatica* var. latifolia, *Seed Sci. Technol. 30*:177 (2002).

19. P.Ranbo, What is seed vigor? *Rasen-Turf-Gazon, 31*(4): 62 (2000), Seed Abstr. 25 (4):158 (2002).

20. Y.Sako, M.B.McDonald, K.Fujimura, A.F.Evans, and M.A.Bennett, A system for automated seed vigor assessment, *Seed Sci. Technol. 29*:625 (2001).

21. M.B.McDonald, A.F.Evans, and M.A.Bennett, Using scanners to improve seed and seedling evaluations, *Seed Sci. Technol. 29*:683 (2001).

22. S.U.Larsen, F.V.Povlsen, E.N.Eriksen, and H.C.Pedersen, The influence of seed vigor on field performance and the evaluation of the applicability of the controlled deterioration vigor test in oil seed rape (*Brassica napus*) and pea (*Pisum sativum*), *Seed Sci. Technol. 26*:627 (1998).

23. A. Strydom and H.A. van de Venter, Comparison of seed vigor tests for cabbage (Brassica oleracea L.) var. capitata L., *Seed Sci. Technol. 26*:579 (1998).

24. A.A.Powell, A.J.Ferguson and S.Matthews, Identification of vigor differences among combining pea (*Pisum sativum*) seed lots, *Seed Sci. Technol. 25*:443 (1997).

25. R.Holubowicz, K.Tylkowska, and E.Jankowska, Methods of investing seed vigor of selected ornamental plant species, *Folia-Hortic. 9*:15 (1997).

25a. K.W.Duczmal and K.Ratajczak, Emergence and the conductometric test for single seed peas, *Folia-Hortic. 5*:39 (1993).

25b. L.Minicka, K.W.Duczmal, D.Come, and F.Corbineau, Correlation between pea seed laboratory testing methods and field emergence, Proc. Fourth Int. Workshop on Seeds; Basic and applied aspects of seed biology, Angers, France, July 20–24, 1992, Vol. 3, 1993, pp. 1003–1007.

25c. R.Holubowicz, K.Tylkowska, and L.Gruchala, Ethylene production as a biochemical marker for radish seed vigor, *Folia-Hortic. 5*:19 (1993).

25d. A.C.McCormac, P.D.Keefe and S.R.Draper, Automated vigor testing of field vegetables using image analysis, *Seed Sci. Technol. 18*:103 (1990).

26. C. White, Moisture determination, *Techniques in Seed Science and Technology* (P. K. Agrawal and M. Dadlani, eds.), South Asian Publishers, New Delhi, 1987, p. 47.

27. J. H. Khare Thomson, The heterogeneity test, *Proc. Int. Seed Testing Assoc. 37*:669 (1972).

28. L. Khare, Centralized-decentralized seed quality control, *Seed Research*, Special Vol. No.2 (S. P. Sharma, ed.), Indian Society of Seed Technology, New Delhi, 1993, p. 877.

29. S.J.Roberts, J.Brough, and S.Chakrabarty, Non-destructive seed testing for bacterial pathogens in germplasm material, *Seed Sci. Technol. 30*:69 (2002).

30. D.Lakshmana, R. Vasudeva, and T.R.Radhamani, Phenol colour reaction (PCR) in pearl millet (*Pennisetum glaucum* L.), R.Br. J. Res. ANGRAU 28: 84 (2000), *Seed Abstr.* 25 (4):159 (2002).

31. A.Hang, C.S.Burton, D.L.Hoffman, and B.L.Jones, Random amplified polymorphic primer-generated embryo-DNA polymorphisms among 16 North American malting barley cultivars, *J.Am.Soc.Brew.Chem. 58*:147 (2000).

32. D.P.Deswal and I.S.Sheoran, A simple method fo seed leakage measurement: applicable to single seeds of any size, *Seed Sci. Technol. 21*:179 (1993).

33. W.C.Ebener and M.Misra, Census of seed laboratory seed analysis equipment, *Newsletter of Assoc. Off. Seed Analysts, 64*:97 (1990).

34. D.K.Pandey, Short duration accelerated aging of French bean seeds in hot water, *Seed Sci. Technol. 17*:107 (1989).

19
Seed Marketing and Distribution

I. INTRODUCTION

Seeds of agricultural crops, whether produced by farmers or seed companies in special areas, follow many paths from the producer to the planter. The marketing of field crop seeds may be simple or complex with one or several contributive functions. The simplest marketing cycle is completed when one farmer sells seeds to a neighbor. Millions of pounds of seed move in this way. Most seed distribution, however, calls for a more sophisticated chain of events. Much of the seed produced outside its major area of use is not sold directly and must enter the marketing channel similar to that illustrated in Figure 1.

Seeds of field crops follow no normal route from producer to user, but most seed movement involves a grower, a wholesaler, and a retailer. The activities of each vary widely depending on the item being handled. The size and location of any of these three segments will modify the pattern of their operation (1).

II. MARKETING REQUIREMENTS

The principles of seed production apply when the seed is produced by farmers for their own use or when it is produced for marketing on a wider scale. There is little point in producing seed for marketing unless the producer knows where and when it will be used. Seed is a perishable commodity of high value, and it is expensive to produce, store, and transport. Seed production, therefore, must be geared to realistic marketing and distribution targets.

According to Kelly (2), in assessing marketing requirements, one must answer the following questions:

1. How much, where, and when is the seed needed?
2. What standards of quality are required?
3. At what price can it be sold?

After getting answers to these questions, one can plan both the scale and the timing of seed production to meet the expected market demand.

Agrawal (3) listed the following factors that influence seed marketing:

1. *Clear-Cut Policy*—This is necessary to develop the seed industry, defining the tasks and responsibilities of the official, semiofficial, and private sectors to foster seed marketing on a sound footing.

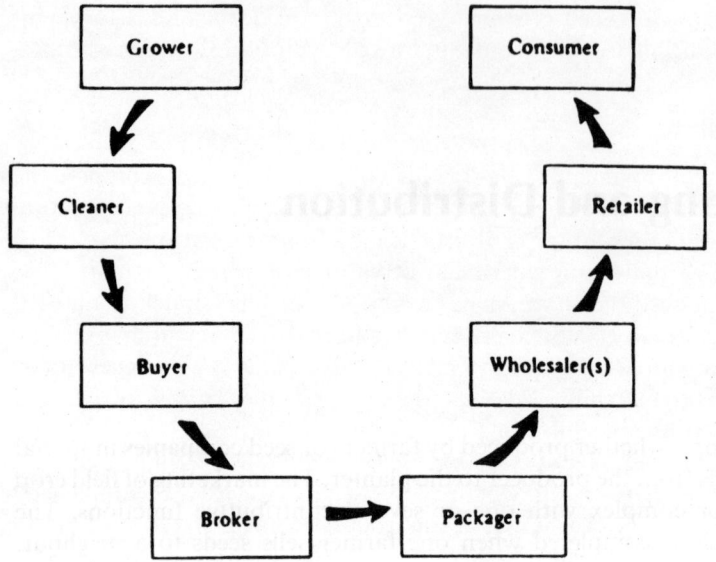

Figure 1 A generalized seed-marketing scheme. (From Ref. 1.)

2. *Availability of Well-Identified and Well-Adapted Varieties*—Newly developed superior varieties of crops must be introduced in seed programs to promote seed trade.
3. *Adequate Production, Storage, and Testing Facilities*—These are necessary to produce and maintain high-quality seeds in accordance with established standards, which are in turn vital to seed marketing.
4. *Official Program*—Initiation and promotion of new varieties and hybrids by government agencies is important for seed trade development. The government should also provide market information to set targets and to regulate and control seed agencies and enterprises.
5. *Demand Forecast*—Realistic assessment and targets of seed demand are necessary to put the seed trade on a sound footing.
6. *Market intelligence*—Comprehensive market intelligence to indicate the needs of farmers, location of production areas, size of market demands, and marketing costs influence seed marketing.
7. *Transport and Storage Facilities*—Adequate storage and transport arrangements must be ensured for timely supply of seeds to the end users.
8. *Nature of Product*—Seeds need to be handled with care at all stages of its marketing chain. Seed marketing is seasonal; unsold seed stocks, if carried over, will bring losses.
9. *Quality Control*—Effective, legally enforced control of seed quality as per internationally established standards are necessary to prevent unhealthy competition among seed companies and corporations.
10. *Publicity*—The value, availability, and returns from certified seed need adequate publicity.

11. *Financial Rewards*—Seed growers, seedsmen, and dealers need to be encouraged financially for their outstanding contributions to the seed trade.

A. Marketing Research

Based on crop statistics, cropping area, and local seed rates, it is easy to compute the total seed requirement for a particular crop in a specific country. However, the total seed requirement of a state or country is of little use by itself. It is also necessary to know what proportion of that requirement can be supplied through the market, as opposed to farmers using seed they have saved themselves, and how much of the seed can be supplied by the cultivator(s). Such estimates require market research involving questioning the farmers who are expected to acquire the seed. It is also necessary to know what motivates farmers to buy seed, what their purchasing power is likely to be, and how many potential customers live in a particular area.

Convincing the farmer to purchase seed can be effectively achieved by conducting field-demonstration trials in areas where they can be shown to potential seed buyers, convincing them of the advantages of a particular cultivar in terms of better yields, improved quality, or other cultural advantages. Practical on-farm demonstrations have been recognized to have the greatest value in addition to conventional promotion programs such as newspaper articles, radio talks, television shows, and catalogs. These activities announce a forthcoming product, which evokes feedback of information to permit a realistic assessement of the quantity of seed that may be disposed of and probable locations of the market outlets.

Seed must be multiplied through several generations before a sufficient quantity of seed is obtained for marketing. The timing of a marketing operation will, therefore, depend on the seed-multiplication rates. The quantity of seed that is available from the cultivar maintenance program and rate of seed multiplication must be known to plan a production program to meet a predetermined quantity of seed. It is then possible to estimate the probable number of growing seasons needed to reach the quantity of seed required for market, making due allowance for unforeseen failures or other losses during multiplication. The target quantity of seed required each year is translated into areas of seed crop required. The initial multiplication steps are taken while the market research is still proceeding. The initial quantities should be small enough to avoid overproduction. Advance publicity is inevitable in estimating market potential, which should be followed soon thereafter by the actual availability of seed. Once farmers are convinced of the potential value of a new cultivar, they want to try it as soon as possible and will lose interest if kept waiting too long (4).

Seed production should also be planned to make seed available continuously. The stimulated demand should be satisfied in each growing season as long as the cultivar remains popular. It is, therefore, necessary to estimate the demand for the next growing season and plan production to satisfy that demand. The production of seeds of new and improved cultivars of crops should be integrated into the regular program as and when they are available, dropping obsolete cultivars. Kelly (2) described a hypothetical calculation for the production of a new wheat cultivar. For a self-fertilized crop like wheat, it is possible to plan and continue seed multiplication for long periods of time by retaining part of the produce for further multiplication. This is not possible with cross-fertilized crops, where generations are limited.

Seed production of a hybrid cultivar must be planned in such a way that sufficient quantities of each seed of the parents are made available in the year in which the hybrid is to be produced to meet the market needs. Since less seed of the male parent is usually required than of the female parent, separate calculations for male and female lines are usually needed (5,6).

B. Quality Control During Marketing

Both the customers' needs and the difficulty and cost involved must be taken into consideration when deciding upon the standards of seed quality required. The quality standards should have a real practical value for the farmer. Standards that are too high increase the cost of seed production unnecessarily, whereas those that are too low serve no useful pupose to the user. The desire to make claims for a high-quality standard as an aid to sales usually conflicts with the realistic situation in relation to its actual needs, often necessitating a compromise between the two. Kelly (2) listed the following quality standards to establish:

1. Authenticity and purity of the cultivar
2. Analytical purity of the seed
3. Content of weed seeds (count per unit weight)
4. Germination capacity
5. Any other important quality parameter specific in the area where seed is to be marketed; for example, disease resistance, freedom from seedborne disease, or moisture content.

The germination capacity is the most difficult standard of quality to achieve, because, unlike other quality parameters such as cultivar purity, analytical purity, or weed seed content, the germination capacity can deteriorate during storage and transportation, especially when it is exposed to adverse climatic conditions. The test results indicated on the package or container can thus be valid only for a limited period (6–12 months).

The health standards for seed are also difficult to establish. The number of seeds that can be examined for a particular disease is very limited, and it is difficult to relate the test result to the actual situation existing in seed lot. A negative test for a disease examined may not necessarily prove that the complete seed lot is free from infection. Standards of seed quality, therefore, must represent a compromise between what is desirable or ideal and what is practical and achievable.

The control of seed quality by the government during marketing is rather different from seed certification. Whereas seed certification may be regarded as a service to seed growers, providing expert advice on seed quality and cultivation, seed quality control during marketing may be likened to consumer protection legislation (2). Minimum standards of quality should be specified for seed offered for sale, and seed sellers are required to ensure that the seed being sold is at least of this quality.

The truth-in-labeling concept requires the seller to make certain statements about the seed quality on offer, but does not set minimum standards. In such cases, the government is able to take seed samples in the marketplace and make independent quality tests. The system relies upon random checks being made by government inspectors at appropriate times of the year; a check of about 10% of the seed on offer is regarded as being satisfactory. If these tests show that the seed being sold is below the required standards, the person(s) responsible can be prosecuted in the courts and, if found guilty, would be liable for a penalty. Quality tests made in a laboratory with a reasonable time scale can be completed in time to prevent poor-quality seeds from being sold. However, for more important aspects of genetic quality requiring control-plot tests, such action cannot be taken to prevent the sale of poor-quality seeds (7,8).

In a seed market survey, 207 seed samples of summer/kharif vegetable crops (cucurbits fruit vegetables, and others) collected from 21 seed markets in Madhya Pradesh (Jabalpur, Raipur, Indore, and Gwalior) were studied at the Indian Agricultural Research Institute, New Delhi, India, for seed-quality parameters in relation to the minimum standards prescribed by the Indian government. Four samples (1.9%) were found to be substandard for physical purity and 68 (32.9%) for germination. Loose seed samples were the most likely to be substandard (42.6%) for germination, followed by packed seed without a test report (33.3%), and packed seed with a test report (30.2%), respectively. Thus, packing helped to improved seed quality. It was evident that seedsmen were not aware of the provisions of the Seeds Act, 1966, as 91 seed samples (44.0%) did not carry any label or were offered for sale after the expiration date, suggesting an urgent need for strengthening the seed law enforcement program and educating seedsmen (8a).

C. Selling Price of Seed

The cost of seed usually constitutes about 10% of the cost of growing a crop, and traditionally seed has been relatively cheap considering the benefits a high-quality seed of an improved cultivar offers to the grower. Seed price is subject to severe competition, especially for self-fertilized crops, for which farmers have the option of saving seed from the preceding crop for sowing.

The price of seed should include all direct and indirect costs of its production plus a reasonable profit, with due regard to what the customer will pay. The user must be able to perceive some benefit at the price being asked.

Calculations of the selling price of seeds of food crops like cereals, pulses, and oilseeds usually start by taking the current market price of the product as the basic production cost to which can be added the following costs (2):

1. Seed grower's premium to cover the extra expense such as cleaning equipment before sowing and harvest,
2. Cost involved with safeguarding isolation, roguing, and so forth
3. Seed drying, cleaning, treatment, and packaging charges
4. Storage and transport costs
5. Allowances for seed loss during postharvest operations and depreciation on plant and buildings and interest on capital invested
6. Costs of quality tests, advertisement, and other promotional costs
7. A reasonable profit, along with a plant breeder's royalty, if any.

This will normally give a seed price at least 50% above the selling price of the product or as high as 400% when the cultivar showing marked advantages has been introduced to the market. For other crops like grasses and forage legumes where the seed is not usually the product for which the crop is grown, there is no overall market price on which to base calculations. The seed grower's remuneration in these cases must be based on an estimate of the costs of seed production, but all other cost elements in calculating the price of seed will be the same as for food crops.

III. SEED-MARKETING SERVICES

Various services available for seed marketing include the Seed Varification service, seed testing and certification, and cooperatives for seed distribution.

The Seed Verification service is primarily concerned with the origin of seed; that is, the locality where it is produced. In the United States, it is a voluntary service conducted by the Seed Branch of the Grain Division, Agricultural Marketing Service, and covers shipments of seeds within and between states. This service enables the seed dealers and farmers to buy seeds of alfalfa and red clover with a positive assurance of origin.

The purpose of seed certification is to make available to the public seed of superior varieties of known genetic purity and identity. The Seed Verification service facilitates the marketing of seed under its true origin, although it does not supplant the state seed-certifying agencies. Misrepresentation of the origin of alfalfa seed in the United States before the Seed Verification Service existed resulted in unfair competition among dealers and in heavy losses to farmers who unknowingly purchased nonhardy seed that was killed in winter because it was not adapted to the localities where it was planted (2).

Seed dealers apply for enrollment in the Seed Verification Service each year, agreeing to comply with the instructions and procedures for verification of seed origin, to keep prescribed records, and to confine purchases of seed to lots eligible for verification. A comprehensive system of supervision provides for inspection in surplus-producing districts, at offices and warehouses of verified-origin dealers, and in consuming sections. Records alone will not accomplish little unless provisions are made to supervise them.

The tests are made in accordance with the method of seed testing prescribed under the Federal Seed Act or in accordance with the International Rules for Seed Testing when requested by the applicant. Samples are tested and certified for one or more factors of quality as requested by the applicant for inspection.

Two types of inspection are made: (a) sample inspection certificates apply only to the sample tested, and this service is available only on seeds in or destined for foreign commerce; and (b) lot inspection certificates apply to the lot of seed sampled and tested. This service is available on seeds tested for other government agencies and departments provided the Grain Division officially samples the seed lot.

Service testing is not available for seeds in interstate or intrastate commerce. State, commercial, and private laboratories perform such services (9).

Testing of seeds for purity and germination is carried out routinely in many countries. There has been growing realization among seedsmen, seed growers, and scientists that testing for purity and germination is not enough, and that seeds should be tested for health and freedom from various seedborne diseases using standardized methods. Laboratories for testing seeds may be state run, commercial, or private. Organizations that help in testing seeds are the Association of Official Seed Analysts, the International Seed Testing Association, whose secretariat is in Europe and whose members are national governments, and the Society of Commercial Seed Technologists. These organizations develop and adopt standard methods of testing seeds, promote research leading to the improvement of seed testing, and exchange information through meetings and publications.

IV. SEED TRADE ASSOCIATIONS

Like most other professions, the seed trade is highly organized. Seed trade organizations provide their members an opportunity to meet and associate on a professional and social level with persons having mutual interests. They also provide the seed trade with an identity and power in influencing public opinion as well as state and national seed legislation.

The trade associations keep the producers, conditioners or processors, and distributors of seeds informed of newer developments that bear on their business. Each of four types of seed trade association (state, regional, national, and specific interest) serves its own purpose and deals with its own problems.

A. United States

The seed trade within the United States is organized into state, national, regional, and specific interest associations. Although most states have their own seed dealer associations, the national seed industry is organized into the American Seed Trade Association (ASTA) organized in 1983. It is divided into five regional associations, representing the Pacific, Western, Northern, Atlantic, and Southern regions. It is divided by commodity interests into four divisions: (a) the farm seed division, (b) the garden seed division, (c) the hybrid corn division, and (d) the lawn and turf division. Each division has its own staff and committees.

ASTA has many varied activities, which are geared to the interests of the seed industry. It has been highly effective as a lobbying organization, influencing state and federal seed legislation (e.g., ASTA was influential in the passage of the Plant Variety Protection Act of 1970). It also sponsors educational meetings such as its annual Farm Seed Conference and the Hybrid Corn and Sorghum and Soybean Research Conferences. Through the American Seed Research Foundation, ASTA sponsors seed research projects in public institutions, the results of which are published and made available to the entire seed industry. The Association of Official Seed Analysts (AOSA) is an organization of seed analysts from official state, federal, and university seed laboratories throughout the United States and Canada, which was established in Washington, DC, in 1908. The Society of Commercial Seed Technologists (SCST) is an organization of seed analysts from private or commercial seed laboratories in the United States and Canada.

B. Canada

In Canada, the Canadian Seed Trade Association (CSTA), established in 1923, represents the seed industry. This group is active as a lobbying agency in the interest of the seed trade of Canada and to foster professional associations within the Canadian seed industry as well as with that of the United States and other parts of the world.

C. India

In India, the National Seeds Corporation (NSC) was established in 1963 to promote healthy development of the seed industry, initiate measures leading to the production of high-quality seeds, and to produce, process, and market hybrid maize. The establishment of the Tarai Development Corporation (TDC) in 1969 with the assistance of the World Bank was another important landmark in the development of the Indian seed trade. The Indian Society of Seed Technologists (ISST), established in 1971, provides an opportunity for exchange of information to persons engaged in the seed industry through its publications; for example, Seed Research and Seed Technology News.

D. International Organizations

The Federation Internationale du Commerce des Semaces (FIS) is an international seed trade organization that fosters cooperation among different nations and facilitates the international seed trade. Although it has no policymaking powers, the FIS is effective in enforcing national and international policies affecting movement of seed from one country to another.

The International Seed Testing Association (ISTA) is a worldwide organization dealing with seed testing. ISTA develops rules for seed testing, standardizes testing techniques promotes seed research, and fosters cooperation between international seed agencies for seed improvement (1).

V. SEED DISTRIBUTION

After deciding the quantity of seed to be marketed, the area of sale, the growing season, and the time of sale, as well as the quality of the seed to be aimed at and the seed price to be offered for sale, it is necessary to decide how seed should be distributed logically. Seed is normally stored in a more sophisticated central storage system until the last possible moment, because farm storage facilities are not usually adequate to protect seed quality for long and seed could deteriorate if not sown soon after delivery unless farms are large and well organized. In situations where a large central store is expected to serve a very large area, it would be convenient to set up a network of subcenters with short-term storage facilities for effective distribution.

The seed must be transported quickly and efficiently to the planting area with reasonable care, especially in the humid tropics. Supply also must be flexible enough to transport surplus seed quickly to where it is needed.

Delivering the required quantity of seed of the desired quality in time for sowing when conditions are at their best can ensure an efficient distribution of seed. It is therefore necessary to plan the delivery system carefully to function smoothly in the limited period available for seed movement. Because seed distribution is a seasonal activity, it is usually not economical to hire a transport system solely for seed for the entire year. The system may be used to haul other goods like fertilizer or pesticides during the off-season, or transport could be hired on a contract basis. Trading functions and seed distribution may be performed by a government agency, by farmers' cooperatives, or by private or commercial seed companies (10).

Duczmal and Tylkowska (11), discussing carrot seed production in Poland, estimated that the needs of the country's seed market were about 160–180 t/year. The yields of roots and seeds, their germination capacity, and infection by *Alternaria dauci* and *A. radicina* were considered. Regions most conducive to acquiring healthy seeds were investigated, with central regions providing to be the best. The seed-to-seed method of production was studied in the field during 1980–1992; of the five cultivars examined at Boguslawice, southeastern Poland, Jawa and Koral (with optimum sowing dates of August 20) and Perfekcja (optimum sowing date August 10) were most suited to the seed-to-seed method. Ways of improving seed quality by grading fungicidal treatment, and conditioning have been mentioned.

VI. PROMOTION OF SEED SALES

Seed does not sell itself. For most self-fertilized crops, farmers try to save seed for their own use. The seed growers and seedsmen must try to convince farmers of the advantages of buying seeds of newly introduced varieties in terms of higher yields, better quality of the produce, and better financial return from the crops the new seed will produce. Kelly (2) divided the seed trade into the following categories:

1. The plant breeder/geneticist, who evolves new cultivars and limited quantities of seed from a cultivar maintenance program
2. The wholesaler, who multiplies plant breeder's seed by contracting with farmer/seed grower to a marketable quantity
3. The retailer, who buys seed from the wholesaler and sells it to the farmers
4. The brokers acting as intermediaries between the wholesalers and retailers

It is possible to combine two or more of these functions in a single organization or company; some large commercial enterprises combine all four categories. Many specialized wholesalers deal with a limited number of crops, but retailers cover a wide range of crops and often supply to farmers other commodities in addition to seeds.

Seed sales can be promoted through (a) demonstrations, organizations of farmers' rallies, trade shows, and so forth, (b) advertising in the press, on radio and television, and through posters, and (c) individual contact of farmers by salespersons.

Government subsidies can encourage the use of good-quality seeds and may be offered in the form of cheap credit facilities so that farmers are not required to pay for seed until the crop grown from it is harvested. The farmer can also exchange grain for an equivalent amount of good seed. The seed must be sold at a price that will yield a reasonable return to the grower and those concerned with its processing and marketing. The government extension or advisory services should also be fully aware of the benefits of using good seed and of the proven attributes of the cultivars that have been newly introduced to the market.

Practical demonstration can be combined with cultivar trials in the area of expected seed sales. Alternatively, demonstrations may be laid out on the farmer's fields to convince others of the practicality of the new cultivars. In India, "mini-kit" demonstrations have been successful, where enough seed and other necessary inputs like fertilizer/pesticides for a seed plot are supplied in a package with instruction to farmers in each area where the cultivar is to be sold.

Advertising should be factual and should rely upon evidence of trials conducted to speak for cultivars. Seed may be delivered in an attractive package or bag with an appeal as a part of the advertisement. Salespeople approaching individual farmers should know all the details of the seed being sold and should be able to discuss the relative merits of the cultivars, if asked for.

The success of an organization selling seeds to farmers depends greatly on the reputation it has been able to build up over the years by supplying good-quality seeds of reliable cultivars. The consistency of performance of the cultivars from year to year is more important to farmers than an outstanding performance in any one year followed by a partial failure or disaster in another. Like any other trade, the seed trade should aim at constant improvement and perfection to raise the standard through purity of stocks, honesty of representation, carefulness of obligations, and promptness and efficiency in execution (12).

REFERENCES

1. L. O. Copeland and M. B. McDonald, *Principles of Seed Science and Technology*, 3rd ed., Chapman & Hall, New York, 1995.
2. A. F. Kelly, *Seed Production of Agricultural Crops*, Longman Scientific & Technical, New York, 1988.
3. R. L. Agrawal, *Seed Technology*, Oxford & IBH Publishing, New Delhi, 1980.
4. A. G. Law, B. K. Gregg, P. B. Young, and P. R. Chetty, *Seed Marketing*, National Seeds Corporation, New Delhi, India, 1971.
5. B. R. Gregg, Seed marketing in the tropics, *Seed Sci. Technol. 11*:129 (1983).
6. A. J. G. Van Gastel, Seed marketing, *Seed Production Technology* (J. P. Srivastava and L. T. Simarski, eds.), ACAR-DA, Aleppo, Syria, 1986, p. 232.
7. K. P. Wagner, H. F. Creupelandt, and W. H. Verburgt, Seed marketing, *Cereal Technology* (W. P. Feistntzer, ed.), Food and Agriculture Organization, Rome, 1975, p. 108.
8. J. E. Douglas (ed.) *Successful Seed Programs: A Planning and Management Guide*. Westview Press, Boulder, CO, 1980.
8a. K.Kant, M.M.Verma, Sukhvir Singh and S.Singh, Studies on the quality of vegetable seeds in the markets of Madhya Pradesh, *Seed Res.* 27:1(1999).
9. W. H. Crispin, Seed marketing services, *Seeds, The Yearbook of Agriculture*, US Department of Agriculture, Washington, DC, 1961, p. 470.
10. D. K. Cristensen, E. Sieveking, and J. W. Neely, Handling seeds of the field crops, *Seeds: The Yearbook of Agriculture*, US Department of Agriculture, Washington, DC, 1961, p. 409.
11. K.W.Duczmal and K.Tylkowska, Carrot seed market and prospects for carrot seed production in Poland, *J. Appl. Genet. 38A*: 5 (1997).
12. Memorandum of the Seed Trade Association of the United Kingdom, Central Office of Information, London, 1961.

20

Seed Certification and Legislation

I. INTRODUCTION

Seed certification is a legally sanctioned system designed to control and maintain high-purity seed and propagating material of genetically distinct crop varieties. It allows one to check on the origin of seed and trueness to its cultivar purity, to evaluate the growing crop, and to supervise the preharvest, harvest, and postharvest operations during seed production and processing, as well as conduct sample inspection (laboratory test), bulk inspection for homogeneity, and controlled plot testing.

Certified seed is produced by some outstanding farmers and seedsmen using careful quality-control measures, pedigreed plating stock, and critical field and seed inspections during the entire process of seed production. Seed certification is the only method of maintaining the varietal identity of seed on the open market, and it is particularly important for field crops whose varieties are released and sold on the open market. It is of less significance for other types of agricultural crops, varieties of which are often released privately and seed production of which is controlled by private seed companies.

A. History of Seed Certification

Seed certification in the United States and Canada began early in the nineteeth century when new varieties were developed by state land-grant colleges and government experimental stations. Prior to this, most field crops had originated through introductions from other countries. Newly produced crop varieties were distributed to farmers haphazardly and inefficiently, resulting in contamination and loss of varieties. During 1900–1920, organizations were set up in different states through which the newly evolved varieties were distributed to farmers more systematically. These organizations, which were an outgrowth of state experimental stations, soon became known as crop-improvement associations or seed-certification agencies. They were administered by the experimental stations or by the extension service staff of the land-grant institutions. Under the guidance and influence of the universities, seed certification turned into an established institution to increase the production of high-quality seeds.

Today, seed certification in the United States is the responsibility of each state. The basic authority for seed certification is derived from the seed law of the individual state. Several states administer certification programs through their departments of agriculture or cooperative extension services.

In Canada, seed certification is administered by the Canadian Seed Growers Association, which represents pedigreed (certified) seed growers of Canada. In India, the Central Seed Committee was established under the Seed Act of 1966, followed by the Central Seed Certification Board, which coordinates the activities of the state seed-certification agencies to ensure uniform application of seed-certification standards (1).

B. Objectives

The objective of seed certification is to ensure genuineness and quality of seed to the seed grower. A well-organized seed-certification program may be regarded as a guardian of the pure seed-quality supply.

Seed certification is built around three primary concepts (2,3):

1. Superior variety
2. Genetic purity
3. High seed-quality standards

Over the years, these fundamental concepts have become synonymous with seed certification. However, since the late 1950s, this philosophy has greatly changed with the concept and practices of seed certification. The advent of certification interjected a third, unbiased party between seedsmen and their customers to attest to seed quality, which was welcomed by buyers and sellers alike.

The Association of Official Seed Certifying Agencies (AOSCA) is an organization of certification agencies in the United States and Canada with the following purposes:

1. To establish minimum standards for genetic purity and to recommend minimum standards for the classes of certified seed
2. To standardize seed certification regulations and procedures
3. To encourage cooperation with all individuals, agencies, groups, and organizations to accomplish these purposes
4. To assist its member agencies in seed promotion, production, and distribution

In 1919, the International Crop Improvement Association (ICIA) was established to promote the agricultural interest of the various states of the United States and the provices of Canada, especially emphasizing the improvement of field crops in general and seed improvement in particular. The main objectives of the ICIA are:

1. To encourage the breeding and improvement of field crops and seeds
2. To husband, propagate, and disseminate elite, registered, certified, and improved seed
3. To create a more active interest in better seeds through circulars, reports, and other publicity measures, as well as by encouraging local, state, national, and international shows
4. To assist in the standardization of seed improvement and certification work being done by the member agencies

The ICIA has a major influence on certification and has been instrumental in enunciating the fundamental concepts of seed certification, establishing field and laboratory inspection standards, and encouraging uniformity in certification procedures among its member agencies. In 1968, the ICIA was renamed the Association of Official Seed Certifying Agencies (AOSCA).

According to Donglas (4), a well-organized seed-certification system should have the following three objectives:

1. A systematic increase in superior varieties
2. The identification of new varieties and their rapid increase under appropriate and generally accepted names
3. Provision of a continuous supply of comparable seed material by its careful maintenance

In addition to varietal purity and identity, seed certification endeavors to maintain reason-able standards of other quality parameters such as freedom from weeds and diseases, seed viability, mechanical purity, and grading (5).

II. SEED-CERTIFICATION SCHEME

The seed-certification system aims at keeping pedigree records for crop varieties and making available sources of genetically pure seed and propagating material for general distribution. This is achieved by means of the inspection of fields and seeds and regulations for checking the production, harvesting, and cleaning of each seed lot. Such a system ensures that both seedsmen and farmers get genetically pure seeds when they distribute or use certified seeds. Without this the seeds of varieties would tend to become contaminated and mixed and lose their identity (6).

The organization of seed certification necessitates enacting of certain statutory rules and regulations (Sec. V). The following three important steps are needed to organize seed certification:

1. Establishment and operation of a seed-certification agency
2. Establishment of minimum seed-certification standards
3. Establishment of procedures for field and seed inspections

The seed-certification agency must be established under the statutory regulations of the country. Agrawal (2) stated the following broad principles for the functioning of a seed-certification agency:

1. A seed-certification agency should not be involved in the production and marketing of seeds.
2. A seed-certification agency should have autonomy.
3. The seed-certification standards and methods adapted by seed-certification agency should be uniform throughout the country.
4. A seed-certification agency should be closely associated with the technical institutes on clearly defined terms.
5. The agency should operate on a no-profit, no-loss basis.
6. The agency should have adequate technical staff well trained in seed certification.
7. The agency should have adequate facilities to ensure timely and thorough inspection.
8. The seed-certification agency should serve the interest of both seed producers and buyers.

Careful planning is necessary to organize a seed-certification agency based on pertinent data regarding the anticipated acreage for certification of different crops, area of operation (i.e., distance to be covered), farm size, nature of crop and varieties, educational status of people in the area, seed growers' experiences, and special problems. A typical organization has a board of directors, secretariat, technical and other staff, and operating facilities. The agency may or may not require a seed-testing laboratory and may have seed samples tested through a state seed laboratory.

A. The Generation Scheme of Certification

The concept of seed certification has a generation system whereby the pedigree of superior crop varieties is maintained through subsequent seed production. A four-generation scheme has been devised for this purpose, the seed of each generation being identified by a special color-labeling system (1).

1. *Breeder's Seed*—Produced under the direct supervision of the plant breeder and represent the true pedigree of the variety. The containers of breeder's seed are labeled with white certification tags.
2. *Foundation Seed*—The first generation of seed from the breeder's seed, usually produced under contract by a foundation seed organization. It is also labeled with white tags.
3. *Registered Seed*—The seed from the foundation stock intended for multiplication before the production of a large quantity of certified seed. It is not a seed of commercial class and is designated by purple certification tags. In some U.S. states (e.g., Michigan and Wisconsin), all certified seed is the direct progeny of the foundation seed, eliminating the registered seed class, as is done in many cross-pollinated crops, especially with crop species whose seed is produced outside the area of adaptation.
4. *Certified Seed*—Produced from foundation or registered seed and represents the final product of the seed-certification program. It is labeled with a familiar blue tag.

The foundation seed is a vital link between the breeder's seed produced under the control of the plant breeder and the certified seed produced by the seed grower. It is the seed stock from which registered and certified seeds are produced by a foundation seed organization, which may be a private association of seed growers, a special project within an experimental station or university, or a private seed company. Only the best seed growers with the right combination of experience, appropriate land, facilities, and ability are accepted as foundation seed growers. The supply of foundation seed should not exceed the demand, which calls for advance planning. If produced in excess, it must be carried over at extra cost and sold as commercial grain or even destroyed at a considerable loss.

B. Release of Varieties

Although most agronomic varieties are released by state agricultural universities and state department experimental stations, they are also released by private seed companies and research foundations (e.g., vegetables, corn, sorghum, cotton, and other field crops). The plant breeder submits to the appropriate review boards a description of his or her new variety and its identifying characteristics, along with the performance data. A commodity committee, consisting of persons closely concerned and familiar with the crop (plant breeder, plant pathologist, entomologist, extension worker, agronomist, and other trained persons) may first assess the merits of the new variety. After favorable action, the proposal goes to a larger more formal committee composed of experimental station personnel from different disciplines responsible for the release of new crop varieties. Whereas the first group provides expert knowledge and involvement with the specific crop area, the second observes uniformity of release procedures and evaluates the variety to be released more objectively and advises on specific release conditions and seed increase matters.

To be eligible for certification, a variety must be properly named and described. The term *variety* has been defined as a specific close subdivision of a kind with definite distinguishing genetic characteristics that can be maintained or inherited when plants are produced over a period of years. It is rather difficult to ascertain whether many candidates for certification actually qualify as varieties because of the different kinds of crops and germplasm available for certification. In the United States, an ad hoc committee representing the U.S. Department of Agriculture (USDA), the Association of Official Seed Certifying Agencies, the American Society of Agronomy, and the American Seed Trade Association has developed a comprehensive consensus definition for different varieties, which have been published and made available to all the concerned organizations (7).

Individual certification agencies are aided in deciding the eligibility of varieties for certification by national variety review boards; established by the AOSCA. Four review boards, representing alfalfa, grasses, soybeans, and small grains, were set up in the United States by 1973. Each board is composed of six members representing the ASTA, AOSCA, National Council of Commercial Plant Breeders, USDA, and the Agricultural Research Service (ARS) (USDA). These boards review and evaluate information presented by plant breeders requesting certification of new varieties and advise the AOSCA on their acceptability as bona fide varieties (8).

C. Certification Procedure

Appropriate planting stock is essential for seed certification. It provides a pedigree and forms the basis for certification. Certified seed is usually produced from the registered seed, but it may also be produced using either foundation or breeder's seed. Similarly, other classes of seed may be produced from earlier seed generations.

A seed grower submits an application for certification to the appropriate state seed-certifying agency requesting certification as foundation, registered, or certified seed. The application is accompanied by at least one official tag substantiating the class of the seed planted (breeder, foundation, or registered). Each agency has its own application procedure to be followed. Certification procedures generally followed are shown in Figure 1.

The technical staff of the agency scrutinizes the applications, and field inspections are performed on all fields for which applications are received. Field inspections are timed such that the varietal off-types and other crop and weed contamination are detected easily. In crops like clover and alfalfa, seedlings may also need to be inspected for volunteer crop plants a few weeks after the seedling stage. Small-grain crops are usually inspected after the chaff color has changed and during the hard-dough stage when off-types with different chaff color are easily noticeable. Oats are often inspected while plants are still green and the seed is in the soft-dough stage. Grasses and legumes are usually inspected during pollination when the off-types and weed plants are most easily detected and when isolation from adjacent cross-fertile fields is apparent.

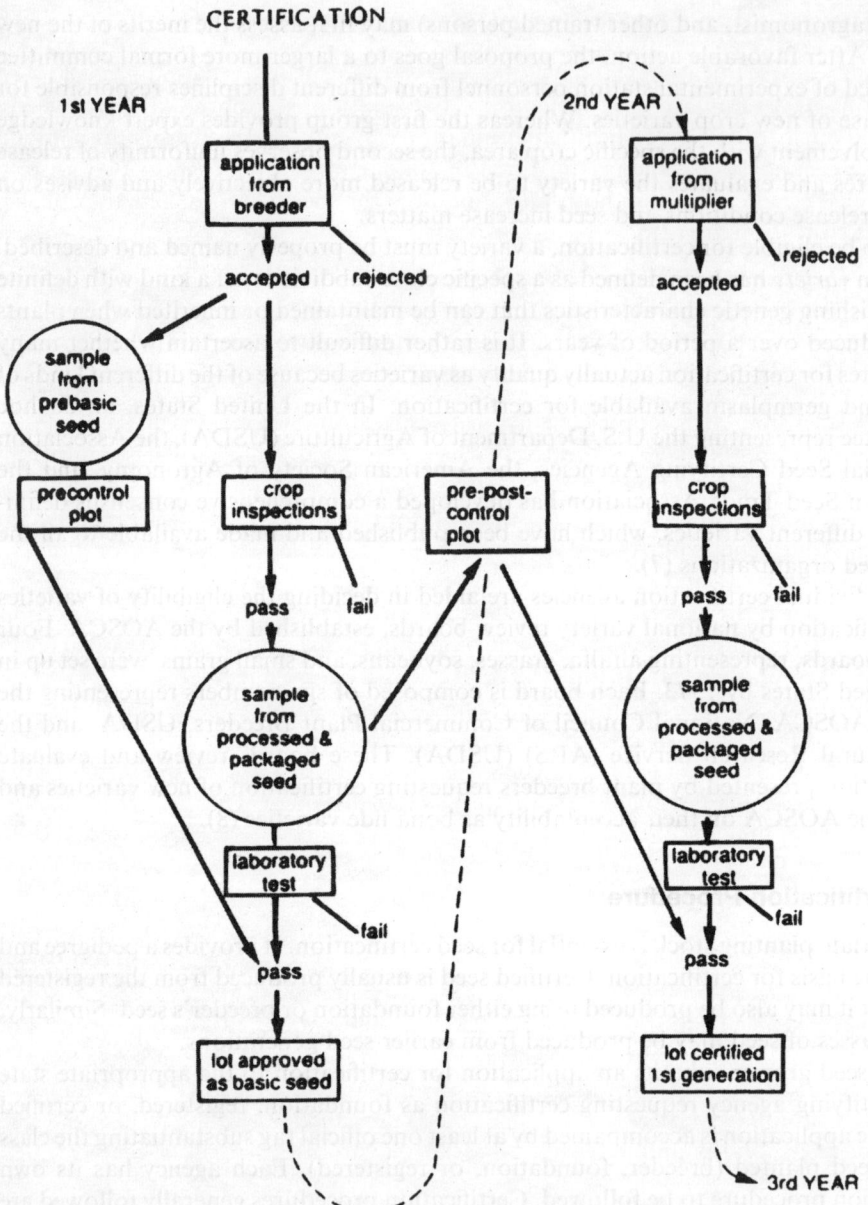

Figure 1   Seed-certification procedures. (From Ref. 13.)

III. MINIMUM SEED-CERTIFICATION STANDARDS

Minimum seed-certification standards are necessary to ensure seed quality. Field seed inspections have standards for varietal purity, isolation, seedborne diseases, weeds, analytical purity of seed, seed health, germination capacity, and seed moisture content.

The certification standards in force in India can be divided into two groups: general seed certification standards and standards for specific crops. The four phases of seed certification are (2):

1. Verification of seed source
2. Field inspection to verify conformity to the prescribed field standards
3. Seed analysis to verify conformity to the prescribed seed standards
4. Tagging and sealing of the container

The seed standards set for genetic purity of the foundation and certified seeds in India are 99.5 and 99.00%, respectively. Seeds not conforming to these minimum standards may be rejected by the certification agency.

In addition to the general seed-certification standards, specific standards for maintaining the genetic purity and quality of seeds have been established for different crops field standards and seed standards. Field standards have been prescribed for factors that affect the genetic and analytical purity and seed health of standing crops. Various field standards, such as for land requirements, isolation requirements, off-types, diseases, and weeds, for different seed crops have been described (2,3,9).

The AOSCA has recommended genetic standards of isolation (time and distance) and tolerances of off-types permitted for different classes of seeds of some important crops species (Table 1). The data presented in Table 1 show that cross-pollinated crops must be isolated! (both for time and distance), whereas self-pollinated crops require minimal separation.

Crops to be certified must be harvested and processed more carefully than con-ventional grain crops to prevent mechanical damage to the seeds. The seed crop must also be harvested at its appropriate moisture content to maintain maximum quality during postharvest operations, including storage. The seeds must be thoroughly dried and cleaned to meet the purity standards for certification. All the processing equip-ment used must also be thoroughly cleaned to prevent contamination with the seeds of other crops.

The processed and blended seed is then sampled, employing the standard methods of sampling to represent the seed to be marketed (see Chapter 19) and tested to determine its quality and acceptance for certification. If the seed is treated with a pesticide, the sample should be drawn from the treated lot.

The sample is normally tested for genetic purity, germination capacity, and free-dom from weed seeds. Occasionally, phytosanitary tests for diseases are also required by the certification agency. The analysis performed either by the certification agency or state seed laboratory is used to decide whether the seed lot qualifies for certification or not. Apart from genetic purity, seeds are examined for other parameters, such as the presence of inert matter, and weed seeds, although this is not mandatory for cer-tification. Some nongenetic seed standards suggested by the AOSCA for three classes of alfalfa seeds are given in Table 2.

Adequate seed tagging is compulsory in seed certification. Some agencies have a one-tag system, in which analysis information (purity, germination) is printed on the

Table 1 Minimum Genetic Standards for Three Classes of Seeds of Some Important Crop Species Recommended by AOSCA

Crop species	Foundation seed				Registered seed				Certified seed			
	Land[a]	Isolation[b]	Field[c]	Seed[d]	Land[a]	Isolation[b]	Field[c]	Seed[d]	Land[a]	Isolation[b]	Field[c]	Seed[d]
Alfalfa	4	600	1,000	0.1	3	300	400	0.25	1	105	100	1.0
Barley	1	000	3,000	0.05	1	000	2,000	0.1	1	000	1,000	0.2
Hybrid	1	660	3,000	0.05	1	600	2,000	0.1	1	300	1,000	0.2
Trefoil	5	600	1,000	0.1	3	300	400	0.25	2	165	100	1.0
Clover	5	600	1,000	0.1	3	300	400	0.25	2	165	100	1.0
Corn												
Inbred lines	0	660	1,000	0.1	–	–	–	–	–	–	–	–
Foundation (single cross)	0	660	1,000	0.1	–	–	–	–	–	–	–	–
Hybrid corn	–	–	–	–	–	–	–	–	0	660	–	0.5
Open pollinated	–	–	–	–	–	–	–	–	0	660	200	0.5
Sweet corn	–	–	–	–	–	–	–	–	0	660	–	0.5
Cotton	0	0	0	0	0	0	35,000	0.01	0	0	7,000	0.1
Cowpea	1	10	1,000	0.1	1	10	500	0.2	1	10	200	0.5
Crambe	1	660	2,000	0.05	1	660	1,000	0.1	1	660	500	0.25
Crown vetch	5	600	1,000	0.1	3	300	400	0.25	2	165	100	1.0
Field and garden beans	1	0	2,000	0.05	1	0	1,000	0.10	1	0	500	0.20

[a] Number of years that must elapse between destruction of a stand of variety and establishment of a stand of a specific class of a variety of the same crop kind.
[b] Distance in feet from any contaminating source.
[c] Minimum number of plants or heads permitted per plant or head of other variety or off-type.
[d] Maximum percentage seed of other varieties or off-types permitted.
Source: Ref. 1.

Table 2 Nongenetic Seed Standards for Three Classes of Alfalfa Seeds Suggested by AOSCA

Factor	Standards for each class		
	Foundation	Registerd	Certified
Pure seed (minimum %)	99.00	99.00	99.00
Inert matter (maximum %)	1.00	1.00	1.00
Weeds seeds (maximum %)	0.10	0.20	0.50
Objectionabl or noxious weed seeds[a] (maximum %)	None	None	None
Other crop seeds (maximum %)	0.20	0.35	1.00
Other variety seeds (maximum %)	0.10	0.25	1.00
Other kinds[b] (maximum %)	0.10	0.10	0.50
Germination and hard seed (minimum %)	80.00	80.00	80.00

[a] Include bindweed, Canada thistle, dodder, dogbane, Johnson grass, leafy spurge, perennial sow thistle, quack grass, Russian knap weed, and white top.
[b] Include sweet clover seed, which should not exceed 9 per lb for foundation, 90 per lb for registered, and 180 per lb for certified seed.
Source: Ref. 1.

certification tag. Most other agencies adopt a two-tag system, where the analysis tag and certification tag are different. With the latter, the seed-quality information can be changed or updated without removing the official blue certification tag. The tag is attached in such a way that it will reveal evidence of any opening, reclosing, or tampering with the contents of the container, because it is not possible to open the container without breaking or defacing the tag. This is accomplished by sewing the tag into the seam or by attaching the tag to the stitching with a metal seal.

Davidson (10) described the pattern of labeling adopted in the United States and the average quality standards needed for different types of seeds.

IV. FIELD AND SEED INSPECTION

A. Field Inspection

Inspection of the standing seed crop is an essential step in verifying the conformity of seed crops to the prescribed certification standards. Field inspections are carried out to verify the origin (source) of seed and identity of the variety, to collect the cropping history of the seed plot, to verify land requirements, and to check the crop and cultivation conditions, isolation distance, and freedom from impurities like weeds, off-types, and seedborne diseases. The field observations recorded are compared with the set prescribed standards specified for each crop type.

Field inspections are carried out by well-qualified and trained personnel who know the prerequisites and standards of seed production and who are familiar with the characteristics of the varieties to be inspected. The field inspections are generally made with prior notice. However, the examinations of cross-pollinated crops during flowering and those of self-pollinated crops infected by seedborne diseases (e.g., loose smut of wheat during flowering stage) should be performed by surprise inspections without prior notice. Seed growers should fully cooperate with field inspectors in cases of rejections and remain present at the time of inspection. The inspector checks all the pertaining information about the species, variety, origin of seed, cultivated area, seed class, cropping history of the field being inspected and the adjacent fields of the same species, and isolation requirements of the seed crop. The field inspector should cover the seed plot thoroughly, walking in a schematic pattern to cover the maximum possible area (11). Heavily lodged (one-third or more) seed crops, which are difficult to inspect, are liable to be rejected. If at a given inspection the seed crop does not conform to the prescribed standards, further inspections are not made unless the seed crop is eligible for reinspection after removal of contaminating factors. Reinspection to confirm the removal of contaminants and conformity to the standards is permitted only once over and above the minimum number of inspections prescribed for the crop. Reinspection is carried out only at the request of the seed grower and upon the payment of fees prescribed for reinspection.

Field inspections are made at the time of sowing, during the preflowering or vegetative stage, the flowering stage, the postflowering and preharvest stages, and at the time of harvesting. The various type of contaminants normally encountered during field inspection at various stages of crop growth include off-types (plants of the same crop differing in morphological characters), pollen shedders (in hybrid seed production), shedding tassels (maize seed production), inseparable other crop plants, objectionable weed plants, and diseased plants.

It is also necessary to inspect the seed producer's or contractor's premises, such as threshing place, seed-drying and storage premises, seed containers, and sampling arrangements. Inadequate facilities and poor postharvest handling of seeds may disqualify the seed lot from certification.

B. Seed Inspection

The seed-certification inspector as per ISTA rules should sample seed after ensuring that the seed moisture content is not notably higher than the maximum standards and that the seed lot is acceptably uniform. The sampling report contains information regarding the name and address of the seed lot owner, seed producer and location, species and cultivar, lot number or marking, weight of seed lot, number and kind of units, origin of seed, certification class, date of sampling, and signature of the inspector.

Seeds are sampled at least twice—once before the seed is processed to test it for purity, viability, germination capacity, and seed health followed by a second sampling after cleaning. The seed is analyzed completely for germination, purity, moisture, other crop seeds, weeds, and seed health to ensure that it meets the requisite seed-certification standards.

The seeds sampled are then tested in the laboratory for various parameters of quality using the methods described in the preceding chapter.

V. LEGISLATION AND SEED LAWS

The basic purpose of seed legislation and the enforcement of seed laws is to control the quality of seed that is produced and sold to the farmers, which has become especially important with technologically and scientifically advanced agricultural practices all over the world. Two types of seed legislation are available: sanctioning legislation and control legislation. Sanctioning legislation authorizes, establishes, and accords legal sanctions for the formation of advisory bodies, seed-certification agencies, seed-testing laboratories, foundation and certified seed programs, and so forth. The control legislation regulates the quality of seed produced and sold in the market, including the establishment of suitable agencies for regulating seed quality.

Seed laws are designed to aid in the orderly marketing of the seed. They establish regulations governing seed sale, thereby providing legal protection to both the buyers and sellers of seeds. In most countries, seed legislation exists at both the state and federal levels. Seed laws ensure truth-in-labeling.

In India, the Seeds Act was passed in December 1966, and came into force on October 2, 1969. It is applicable only to notified varieties of seeds and vegetatively propagating materials used for sowing crops. The Act also provides for the formation of an apex advisory body, namely, the Central Seed Committee, the Central Seed Certification Board, and establishment of seed-certification agencies and central and state seed testing laboratories, as well as for notification of the kinds/varieties under the purview of the Act.

On September 16, 1988, the government of India announced a new seed import policy. Officially called the New Policy on Seed Development, it was implemented on October 11, 1989. It has modernized Indian agriculture by making available to farmers the best seed and planting materials available in the world to enhance productivity, farm incomes, and export earnings. The policy takes a particular approach to importing various kinds of seeds and planting materials, keeping in view the indigenous capabilities. The quarantine policy regarding import of seeds also has been changed suitably. According to the policy, importers can make available small quantities of seeds of coarse cereals, pulses, oilseeds, flowers, and vegetables as well as tubers/bulbs and cuttings/saplings to the Indian Council of Agricultural Research (ICAR) for testing and accession to the National Gene Bank.

Incentives provided to the seed industry and importers, such as reducing import duty on seeds from 95 to 15% as well as a proposed reduction in the duty on equipment and machinery used in seed production, have been envisaged to play an important role in the development of the Indian seed industry (12).

A. Regulatory Legislations

1. Notification of Kinds or Varieties

In consultation with the Central Seed Committee, the Central Government can issue notification for kinds and varieties of crops for which it has become necessary to regulate the quality of seed sold for the purpose of agriculture, bringing the varieties under the purview of the Seeds Act. Different kinds or varieties may be notified for different states or areas within the state. The seed is sold in containers only truthfully labeled or certified to meet the minimum prescribed requirements or standards.

2. Requirements for Selling Seeds

The seeds of notified varieties must be truthfully labeled for sale and should meet the prescribed quality standards. The minimum standards for purity and germination for the purpose of labeling different field crops are given in Table 3. In addition, the label should provide information regarding kind, variety, lot number, date of testing, germination percent purity, inert matter, seeds of other crops and weeds, net weight, seller's name and address, and warning against food use of treated seed. The certified

Table 3 Minimum Seed Standards for Labeling

Crop	Purity (%)	Germination (%)
Field crops		
Barley	97	75
Castor	98	70
Cotton	95	55
Groundnut	97	70
Bengal gram	97	75
Maize	97	80
Paddy	97	70
Pearl millet	96	70
Finger millet (Ragi)	97	70
Sorghum	97	70
Soybean	96	70
Wheat	97	80
Vegetable crops		
Cowpea	97	70
Garden pea	96	70
Cabbage	97	65
Cauliflower	97	60
Knol-khol	97	65
Methi	96	65
Spinach beet	93	55
Carrot	94	55
Beet root	94	55
Onion	97	65
Radish	97	65
Turnip	97	65
Brinjal	97	65
Chilli	97	55
Okra	98	60
Tomato	97	65
Muskmelon	98	65
Watermelon	98	65

Source: Ref. 2.

seed offered for sale should also bear the certification tag and label of the seed-certification agency (9).

The seed must be sold only during the period of its validity recorded on the container. The seller should keep a complete record of the notified seed lot sold for a period of 3 years. The Seed Act also provides for import or export of seed of any notified kind or variety, provided (a) it conforms to the minimum limits of germination and purity specified for that seed and (b) container bears, in the prescribed manner, the mark or label with all the particulars as specified for that seed. These regulations do not apply to farmers who grow seeds of notified kinds and varieties for their own use and have sold to or bartered with other growers on their own premises.

The seed inspectors, possessing the prescribed qualifications appointed by state governments through notification in the *Official Gazette*, are responsible for the enforcement of seed laws in their areas. The state governments also notify state seed laboratories and state seed analysts where the samples taken by the seed inspectors can be analyzed or tested under the technical supervision of state seed analysts.

Persons contravening the provision of the Seeds Act or persons who prevent a seed inspector from taking a seed sample or exercising any other power conferred on him by the Seeds Act are liable for punishment; the penalty varying from a fine to up to 6 months imprisonment or both depending upon the nature of the offense. In certain cases, contravention of the Act may result in forfeiture of the property of the person committing the offense. If a seed company has committed the offense, all persons responsible for the conduct of the business and the company are deemed to be guilty and liable to be punished.

B. Statutory Regulations

The seed inspectors are responsible for enforcing various statutory regulations provided by the Act. They are required frequently to inspect all places and premises used for growing, processing, and selling seeds of any recognized kind or variety to ensure that the provisions of certification/labeling are observed scrupulously. The inspectors can procure seed samples and send for analysis in the case of suspected contamination or other contravention of the provisions of the Act. The inspector also investigates any complaint made in writing in respect to contravention of the Act. The records of all such inspections and seizure of stocks and the actions taken thereupon are maintained and copies sent to the Director of Agriculture or the certification agency. Agrawal (2) has described a detailed procedure to be followed for instituting any legal action; giving levels of tolerance for comparing the results of a purity test with that of the seller's label on the container in respect to seed purity, germination, and foreign seed number present in the sample. Seed legislation and enforcement of the seed laws in developed parts of the world like the United States, Canada, and Europe have been described (1).

A typical labeling tag (analysis tag) by the Perfection Seed Company (U.S.A.) listed the following details:

Variety	Vernal
Kind	Alfalfa
Lot No.	307-31
Pure seed	98.90%
Inert matter	01.05%
Other crop seed	00.00%
Weed seed	00.05%
Noxious weeds	00.00%
Germination	90.00%
Hard seed	05.00%
Date tested	Jan. 1974
Net weight	60 lb

Most countries have patent laws to protect inventions, with the protection covering the method and the product. Similarly, the concept of plant breeder's rights has been accepted, resulting in the formation of the International Convention for the

Protection of New Varieties of Plants in 1961. Under this convention, the signatory countries have introduced legislation to enable breeders to secure legal rights to their cultivars. The United States has not signed the convention, but it has introduced a system of rights that differs in some respects; for example, the breeder is required to submit a detailed description of the cultivar on a prescribed form and a sample of seed which may be used to confirm the description (1,13).

The International Union for the Protection of New Varieties of Plants, an organization based in Geneva, coordinates the administrative and technical procedures followed by the member countries for the granting of rights. The introduction of plant breeders' rights has given a tremendous impetus to plant-breeding research in Europe and other parts of the world (14).

VI. SEED FRAUD

Compliance with seed laws requires careful technical work. Violations may be due to faulty organization, careless procedure, inexperience, or incompetence. Errors may also occur through circumstances not under the control of seedsmen. All these may cause loss to the farmers or seed growers.

Sometimes seed fraud is a result of dishonesty on the part of persons engaged in the seed trade. Such fraud can be avoided by proper implementation of legislation. It is important to realize the basic differences in the types of violations of the law to prevent unjustful condemnation of legitimate seedsmen as well as condoning seed fraud. Clark (15) described the unusual conditions of supply and demand that tempt some unscrupulous dealers who may occasionally cause enormous losses to the farmers and gardeners through seed frauds.

REFERENCES

1. L. O. Copeland and M. B. McDonald, *Principles of Seed Science and Technology*, 3rd ed., Chapman & Hall, New York, 1995.
2. R. L. Agrawal, *Seed Technology*, Oxford & IBH Publishing, New Delhi, 1980.
3. J. R. Cowan, Seed certification, *Seed Biology*, Vol. 3 (T. T. Kozlowski, ed.), Academic Press, New York, 1972, p. 371.
4. J. E. Douglas, Opening opportunities for seed technologists, *Seed Technol. News* 2(4):2 (1972).
5. J. E. Douglas (ed.), *Successful Seed Programs: A Planning and Management Guide*, Westview Press, Boulder, CO, 1980.
6. F. G. Parsons, C. B. Garrison, and K. E. Beeson, Seed certification in the United States, *Seed: The Yearbook of Agriculture*, US Department of Agriculture, Washington, DC, 1961, p. 394.
7. A. L. Larsen, J. H. Wiersema, and T. Handwerker, Uniform classification of weed and crop seeds, *Contribution No. 25 to The Handbook on Seed Testing*, Association of Official Seed Analysts, Lincoln, NE, 1993, pp. 1–137.
8. S. F. Rollin and F. A. Johnston, Our laws that pertain to seeds, *Seeds: The Yearbook of Agriculture*, US Department of Agriculture, Washington, DC, 1961, p. 482.
9. E. R. Clark and C. R. Porter, The seeds in your drill box, *Seeds: The yearbook of Agriculture*, Department of Agriculture, Washington, DC, 1961, p. 474.
10. W. A. Davidson, What labels tell and do not tell, *Seeds: The Yearbook of Agriculture*, US Department of Agriculture, Washington, DC, 1961, p. 462.

11. O. Z. Svenson, H. Al-Jibouri, and E. J. Fuentes, *Seed Certification in Cereal Seed Technology*, FAO Agriculture Development Paper No. 98, Rome, 1975, pp. 163–202.
12. N. S. Randhawa, Seed programmes in India, *Seed Research*, Special Vol. No. 1 (S. P. Sharma, ed.), Indian Society of Seed Technology, New Delhi, 1993, p. 9.
13. J. R. Thomson, *An Introduction to Seed Technology*, Leonard Hill, London, 1979.
14. A. F. Kelly, *Seed Production of Agricultural Crops*, Longman Scientific & Technical, New York, 1988.
15. E. R. Clark, Sometimes there are frauds in seeds, *Seeds: The Yearbook of Agriculture*, US Department of Agriculture, Washington, DC, 1961, p. 478.

21
Seed-Production Industry

I. INTRODUCTION

Food, feed, and fiber are the basic and essential requirements of humans and animals and have made agriculture one of the most important primary industries in many countries. The seed industry is a part of this great primary industry of agriculture, comprising all of the complex interlocking operations necessary to ensure a regular supply of uniformly high-quality seeds to farmers and horticulturists. According to Thomson (1), the seed industry cannot be studied in isolation, because its different operations bear a complex relationship to the social, economic, and political structures of any country. The rapid increase in agricultural productivity to meet an ever-increasing demand for food, feed, and fiber necessitates adequate input, both in terms of materials (seeds, fertilizers, pesticides, growth regulators, irrigation water, farm machinery and equipment, and energy) and services (research, extension, training, marketing, transport, and credit). High-quality seed is the most critical ingredient upon which all the others depend for their full effectiveness. The seed industry is a specialized business and undertaking operated by experienced, knowledgeable personnel who have been specially trained. The farmer or grower is only one member of the seed industry, which consists of home gardeners, market gardeners, truck growers, florists, vegetable processors, food technologists, and merchandisers.

II. SPECIAL FEATURES OF GROWING SEEDS

Cultivation of agricultural crops for the production of seeds demands more special care and continuous attention that required for growing crops. The culture of all seed crops, especially vegetables and flowers, is a specialized business requiring trained skills, knowledge, and vigilance. Unlike the growing crops, the culture of most vegetable and flower crops for home or market use or market does not involve the production of inflorescence, fruit, and seed. Many vegetables for market are allowed to grow until the edible plant portion attained the desired stage of maturity and is ready for harvest. Growing of seed crops, on the contrary, requires completion of further stages of crop growth, necessitating familiarity with methods and techniques not generally known by vegetables gardeners or florists. Production of seed of most agricultural crops requires elaborate techniques with increased risk of crop loss, and great skill and technical knowledge are needed to carry the crops through the final essential stages of flowering and seed development. High-quality seeds can greatly increase the profits of farmers and growers. Although the cost of seed represents a mere fraction of the total crop production, the seed quality plays a decisive role in influencing profitability. High-quality seed, therefore, is the basic requirement of growers of agricultural crops. Seed producers thus have a great responsibility and need a thorough scientific knowledge of the biology, production, storage, and processing of seeds. The costs of seed production of agricultural crops are most always higher than for crops raised for home or market use, because additional precautions must be taken to prevent contamination and loss of germination capacity.

III. FUNCTIONS OF THE SEED INDUSTRY

Thomson (1) listed the following functions of the seed industry:

1. Plant breeding—including genetic research
2. Cultivar assessment
3. Multiplication—growing of seed crops on farms
4. Processing—including drying, storage, and packaging
5. Marketing and procurement
6. Legislative control—seed certification and testing
7. Quarantine
8. Extension activities

Plant breeding, cultivar assessment, quarantine, and extension activities are specialized subjects of study and are essential parts of the seed industry. Seed multiplication, processing, and marketing are the technological areas, requiring special skills, training, and experience. In order to ensure a supply of high-quality seed for farmers, it is necessary to exercise some control over the industry. Special powers enforced by the government in the form of seed acts and legislation are essential steps in the development of a seed industry.

In addition to a regular research program in plant genetics, a research service unit with scientists and laboratories must be provided to solve the day-to-day problems as they arise as well as to foresee and forestall difficulties before they become acute.

Development of a seed industry should be regarded as part of a general development plan of any country having agriculture as its economic base. The capital requirement of fixed assets (buildings, equipment, and machinery) is less than for other industries. The main requirements of the seed industry are working capital and a high level of technical and commercial management Seed production is not a routine process and is far removed from the assembly line concept of modern industry. The situation may change from year-to-year with unpredictable weather, changes in the patterns of pests and diseases, and introduction of new crops and their cultivars. The product of the seed industry is a "living package," always at risk, and surpluses are a dead loss. Skilled managers able to adapt and cope with critical situations and emergencies as and when they arise are typically needed by the seed industry. Also, demand for better seeds must be created by first producing the seeds of improved crop cultivars and then demonstrating their value to farmers and growers (1).

Agrawal (2) distinguished between grain and seed, the latter being a living organism (embryo) embedded in the supporting storage (food) tissue. The seed-technology business seeks to protect this biological entity and looks after its welfare, whereas the field of food technology focuses on the supporting food tissue. Seed technology thus consists of seed production and maintenance and preservation of seed quality and encompasses the methods for improving the genetic and physical characteristics of seeds, involving variety development, evaluation, and release and seed certification and quality control.

Seed technology performs the following important roles in seed improvement (3):

1. Carrier of new technologies
2. Basic tool for secured food supply
3. Principal means to secure crop yields in less favorable production areas
4. Medium for rapid rehabilitation of agriculture in cases of natural disaster

National seed stocks enable countries to produce food grains rapidly during emergency periods by providing improved seeds. Regional disasters can be circumvented in a similar way.

The major goals of the seed-technology business to increase agricultural production through the spread of high-quality seeds are (2):

1. Rapid multiplication
2. Timely supply
3. Assured high quality of seeds
4. Reasonable price

Describing the role f a well-organized seed industry, Baird (4) stated the following important components: research, production, quality control, and marketing. Involving both the private and the public sectors, each of the four components should be given appropriate attention; the components should also be linked and integrated. The public sector institutions are directly involved in the development and maintenance of quality standards in interstate marketing and in farmer education, and research is an important function of national, state, or regional agricultural institutions. Both public and private sector institutions concerned with seed may be involved in research, extension, and production and marketing of foundation, certified, and commercial seeds. Conditions and an atmosphere that stimulates the production and use of high-yielding and high-quality seeds must be encouraged (12).

Freeman (5) listed the various functions of the seed industry and its associated agencies (Table 1). A quality seed-production program requires step-by-step implementation of these functions. The European experience of developing a hybrid seed industry showed that a lack of competition created by cartels and government interference can retard development of the industry. Production of seed by the department of agriculture and its distribution through an extension agency as a community contribution do not affect the production and marketing of hybrid seeds. Public agencies should take primary responsibility for quality control and extension aspects, whereas private agencies may be primarily involved in the production and marketing aspects. The state and/or central government has a continuing obligation to maintain its leadership in adaptive research, development of foundation seed stock organization, preservation of the identity and high quality of seed through certification, and seed regulatory work. In addition, it has the vital functions of providing an agricultural extension service and credit to seed producers. Programming of a growth pattern based on rates of acceptance and on normal growth rates in seed and fertilizer

Table 1 Various Functions of the Seed Industry and Its Associated Agencies

Function	Agency
1. Research (central, regional, state, coordinated)	Government or private
2. Variety release	State and/or central variety release committee (for products of public breeding research operations)
3. Foundation seed production	Foundation seed stocks agency (private producers have their own hybridization programs)
4. Seed production and distribution	Private producers marketing seed to cultivars
5. Seed certification	Recognized agencies to undertake seed certification
6. Seed regulation	District, province, and central departments of agriculture
7. Extension	State government departments, agricultural universities, community development blocks, private seed companies
8. Credit facility	Banks and cooperatives

Source: Ref. 5.

utilization could act as a guide to seed producers in gauging seed consumption and could serve as a production pattern. Governments can provide incentives to private producers if total needs exceed a nation's production capacity (5).

Isolation problems, weather conditions, water supply, and other factors determine the suitability of an area for seed production. If free movements of seed are allowed, seed growers will produce seed in the areas that are most economic for production and distribution. State line boundaries that restrict seed movement will tend to retard the growth and development of a seed industry.

A seed industry requires men and women with technical knowledge. Training is therefore necessary to make technically skilled personnel available. This phase of a seed program can be greatly accelerated by establishing in-country training programs.

Seed producers and seedsmen have the important tasks of mobilizing seed growing, seed drying, seed processing, seed storage, and seed distribution. Seed producers need to be encouraged to discharge these responsibilities. They cannot be expected to assume these responsibilities if fear of changing government policy discourages substantial investment or if there is a threat of price control, forcing them to sell seed at below-cost prices.

A mature seed industry asks little from the government and assumes many risks, which will be covered by profits when production is good. It also takes responsibility for the carryover of seed down to the retail level. All seed is returned for proper storage, retreatment, and rebagging before again being offered for sale. Unsalable seed can be used as grain to salvage some of its costs. A properly phased seed program will enable the seed industry to assume all the responsibilities to ensure the production of the needed quantity of high-quality seed.

IV. GROWTH AND DEVELOPMENT OF THE SEED INDUSTRY

The seed industries in different countries have developed to varying levels in step with the overall agricultural development of each country. In highly developed countries with higher levels of productivity, less than 10% of the human population is involved with agriculture and food production, whereas in a developing country like India, about 70% of the people depend upon agriculture for their income.

A. American Seed Industry

The seed industry in the United States began in a small way in the later parts of the eighteenth century. It developed into a great industry based on modern scientific methods of plant breeding and cytogenetics, labor-saving devices, mechanical farming procedures, and improved methods of sowing, intercultivation, harvesting, handling, storage, seed processing, and distribution. Trained geneticists and plant breeders supervise the day-to-day activities of crossing, roguing, selection, and hybridization and are engaged in extensive long-range plans for crop improvement, producing crop varieties that are resistant to pests and diseases or tolerant of certain stress conditions such as salinity, extreme temperatures, and environmental pollution.

World War I had far-reaching effects on the seed industry in most countries, especially in North America and Europe (6), forcing them to seek other seed sources during the war. When the seed supply was critical, production was increased rapidly, forcing the seed trade to enter a period of tremendous activity and expansion. Seedsmen were encouraged to increase acreage in hitherto untried areas. Within the trade a great deal of experimentation and research began on cultural methods under new situations. Efforts were made by reputed seedsmen within the trade to maintain the high quality of seeds. The curtailment of European supplies stimulated American growers to expand the production of vegetables and flowers for seeds. The global effect of World War II also steadily increased the importance of seed production resulting in greater food supplies. During the World War II, the U.S. Department of Agriculture (USDA) under the Lend-Lease Act entered into numerous contracts with established seedsmen as well as a number of newcomers for the production of large quantities of vegetable seeds. The state agricultural experimental slations conducted extensive research in collaboration with the USDA further to expand the production of seeds of crop plants. The American seed trade has had a unique experience in the production maintenance of quality distribution of seeds. The American seed trade is not likely ever to depend upon imported seed. Today's seedsmen handle many times the number of varieties of crops known in the early nineteenth century. Modern plant-breeding methods thus have significantly influenced the growth and development of the world seed trade.

Copeland and McDonald (7) listed the following factors responsible for the development of the seed industry in North America:

1. Increased number of new available varieties
2. Development of seed-certification and seed law enforcement programs
3. Development of a seed cleaning and conditioning technology
4. Better knowledge of seed quality
5. Emergence of the seed grower as a specialist

Seed production is a complex business involving many different and integrated operations, which have been reviewed by Hebblethwaite (8), George (9), Kelly (10), and McDonald and Copeland (11).

Plant breeders today use huge collections of seeds of cultivated crops and related wild species to select germplasm for improving the existing crop varieties and to search for new ones. Modern scientific tools and technical knowledge permit plant scientists, geneticists, and plant breeders to probe deeper into the plants' composition. Changing needs and an expanding seed industry economy will give rise to increasing demands for new crop varieties with different characteristics in the world market.

B. International Seed Industry

The need for uniformity in seed testing was met by the formation of the International Seed Testing Association (ISTA) in 1924. The purpose of the ISTA was to discover causes of serious discrepancies between the results of tests performed by laboratories in different countries and to eliminate them by devising procedures and techniques that could be followed by every seed-testing laboratory in the world. The Food and Agriculture Organization (FAO) of the United Nations, established in 1948, realized that one of the most effective means of increasing the world agricultural production is through promotion of improvement in seed quality and distribution. In the 1950s, the proliferation of national certification schemes with different nomenclature and control systems seriously impended the free flow of good seed across international borders. Setting up the FAO evolved a plan for harmonization for Economic Cooperation and Development (OECD) in Paris in 1958 to promote the postwar economic recovery of European countries, but its seed scheme is open to all member countries of the United Nations.

The initial development of a seed industry in terms of the investment of capital finance requires government sponsorship with low interest rates and subsidies if required; a well-founded and efficiently managed seed industry will not require governmental help for operational purposes, since these will be subsidized by other sectors of the community. The seed industry requires, large sums of working capital compared to the capital required for nonrecurring assets such as buildings, equipment, and machinery, since the harvested seed is not sold until the following season, and even then sales may be credited until the next harvest (1).

In some countries where seeds are produced under planned projects, they may be sold below cost to encourage farmers to buy seed, artificially creating a high demand for high-quality seeds, which can be grown under the conditions of traditional crop husbandry. Good seed may be wasted if high-technology cultivars are sown using traditional crop husbandry. Such cultivars should not be released until the research organization or seed industry has developed new cultural methods allowing these cultivars to develop to their full potential. The introduction of new or improved cultivars of crops thus must be coordinated with the development of new husbandry methods and techniques. Only growers who are prepared to husband these cultivars will buy them, even if the seed is expensive.

Farmers should be charged a price that covers at least the operational cost of seed production. If growers are not willing to pay this price, either the cultivar is not good enough and the grower does not expect a fair return on the investment, or the methods of seed production are not efficient. Plant variety rights and the sale of basic seed can recover the cost of plant breeding for improved cultivars. Thomson (1) recommended that subsidies from taxation should generally be directed to long-term objectives such as basic research, technical training of the personnel involved, season-to-season storage, and building up and maintenance of reserve seed stocks. It may be in the public and national interest to subsidize the sale price of seed if, for example, vegetables and flowers are involved in the export trade. Trade in such products can be encouraged by selling the uniform type of vegetable and flower seeds of high quality, which will automatically "flush out" the seed of inferior cultivars traditionally propagated by growers and home gardeners.

The development of the seed industry may be coordinated with other sectors of agricultural and general economic developmental programs by state and national governments. Past experience in several countries has shown that it is often not possible to implement an entire seed-improvement program at once. The initial steps to increase the demand for good seeds of newer cultivars would include production of seeds of improved cultivars, demonstration of their value to the growers, and leading farmers through extension programs, with encouragement of use by national extension agencies. Scientific breeding, demonstration, and seed production on experimental farms run by state departments and agricultural universities are the important initial phase of the development of a nationwide, large-scale seed industry. The growth and development of seed industries in different countries have rendered a great service by enhancing the overall prosperity and progress of plant and animal life and human societies on this planet.

Pasichnick et al. (12) provided information on the amount of seed required by the sugar beet industry for 2002, possibilities for increasing seed production, state regulations on seed production, and seed-certification systems. Karim (13) presented an overview of the methods that can be used to establish and evaluate the profitability of seed programs in developing countries and described the cost and benefit items of individual seed programs, discussing the problems of reducing costs and improving the efficiency of seed production as well as the use of seed-pricing policy. Karim (14) further examined the current status of seed programs in developing and developed countries and showed that many seed technologies which are used in almost all developed countries are lacking in the developing countries. Elements of seed program management such as varietal development, organized multiplication, quality control, processing, packaging, and marketing are always present but the level of efficiency differs according to the stage of the develop of a country. Some policy suggestions for developing countries have been outlined.

The use of certified seed or improved seed in many African countries is largely restricted and limited to a few crops, such as maize, and to a limited extent, imported vegetables. For other crops, the majority of smallholder farmers in Africa use on-farm saved seed and, in some cases, seed from relief agencies. The reasons for the poor supply of improved seed include the high cost (both real and perceived) of seed and associated inputs and the unreliability of returns. More importantly, administrative or institutional inadequacies limit both the production and the supply of improved seed. These inadequacies (e.g., seed monopolies that tend to limit private sector participation, lack of policy attention to minor crops, which are essential for seed security and hence household food security) must be resolved through discussions involving both producers and users of seed. Owing to these lack of a coherent seed policy in many of the African countries, seed insecurity and widespread food insecurity have become an important phenomenon. Besides structural rigidities, a number of regulations need to be reviewed and simplified. Variety and certification procedures are often cumbersome and expensive. Seed regulations also need to be harmonized across different countries to ease variety movement and, in certain instances, seed imports. Seed information exchange is also needed among different seed experts on the African continent on a number of issues including the performance and potential suitability of newly developed and local landraces and rainfall patterns. Seed quality control systems must be modified and, in some instances, simplified to reduce delays and costs and ensure that certification standards are not excessively strict. These challenges could be faced squarely by establishing an international seed fund to finance regular seed policy reviews, seminars, and seed networks to provide support to emerging seed entrepreneurs, such as individuals, family seed companies, NGOs, cooperatives, and farmers' groups, and to improve Africa's capacity for seed policy formulation and implementation (15).

C. Indian Seed Industry

Agrawal (2) reviewed the early developments of the seed industry in India before and after independence through its five Five-Year Plans (FYP) from 1951 to 1974; coordinated crop-improvement schemes; National Seeds Corporation Ltd. (NSC), founded in 1963; High Yielding Varieties Programs (HYVP), launched in 1966; development of private seed industries; Annual Plans of 1966–1969, establishment of the Tarai Development Corporation (TDC) in 1969, with the assistance of the World Bank; and the formation of Indian Society of Seed Technology (ISST) in 1971, with its two publications, Seed Research and Seed Technology News, India has succeeded in establishing a well-organized seed industry within a short period of time. A seed law has been enacted and the Certified Seed Producers Association has been formed. The National Commission on Agriculture (1976) estimated the following magnitude of the Indian Seed Industry by the year 2000: 70 breeder institutions, 140 processing plants and storage facilities (5–10 tons) for breeder's seed, 75 foundation seed agencies, 150 processing plants and storage facilities (200 tons) for foundation seed, 360 certified seed agencies, 12 million small farmers participating in the annual production of certified seed, 3000 processing plants (1000 tons) for certified seed, 3000 storage facilities (1000 tons) for certified seed in towns and 30,000 primary market centers (10 tons) and 30 million seed boxes for villages, 12,000 village carpenters for seed boxes, and 10,000 graduates and postgraduates in the seed-technology field. The area needed under seed crops and nurseries to provide seed and planting material for a gross cropped area of about 200 million hectares would be 2.4 million hectares (16).

The Genetic Enhancement Division (GED) was created in 1994 by amalgamating the erstwhile cereal- and legume-breeding units of the International Crops Research Institute for the Semi-Arid Tropics (ICRISAT), Patancheru, India. The creation of GED has facilitated increased disciplinary interaction among breeders based at different ICRISAT locations. The GED serves as the hub of crop-improvement activities, and its scientists interact closely with scientists from other divisions at ICRISAT and National Agricultural Research Systems (NARS) in their region. The ICRISAT gene bank possesses a collection of 18,761 seed species, including 5014 sorghum, 5483 pearl millet, 4470 chickpea, 1694 pigeon pea, and 3000 groundnuts, with an average viability of all lines stored being 96% (17).

D. Vegetable and Flower Seed Industry

The success of private breeding programs with selected agronomic crops might suggest that the private seed industry should take responsibility for development of cultivars of all crops. The basic differences between vegetable and field crops, however, make it difficult for private industry to assume responsibility for improvement of all types of crops. The volume of the field crop seed sold provides profits that cannot be matched by the sale of vegetable seeds, thereby reducing motivation for extensive breeding programs. Also, seedsmen and growers of feed grains and oilseeds are concerned with the same end product—seed. For most vegetable crops, the seedsmen produce and sell seed, but the growers' end products may be leaves, petioles, bulbs, inflorescence, fruits, pods, or roots. These different end products may involve different objectives in plant-improvement programs. Seedsmen need quantities of seed to maintain sales, but the grower is mainly concerned with crop quality, not seed production. Much private vegetable breeding is carried out in seed-producing areas, often geographically remote from major vegetable-producing areas, creating problems of local adaptation of vegetable and flower cultivars.

Vegetables and flowers consisting of several species cannot be considered as being one crop; each species requires individual attention. With many vegetable species, the acreage, crop value, and urgent problems may be quite inadequate to support the needed research. According to Vest (18), the private seed industry alone cannot assume the responsibility for genetic improvement of these valuable crops. Profit motivations are feared to eliminate research in marginal crops, leaving only a few cultivars of vegetable and flower species from which to choose. Citing an example of the private seed industry's focus on major crops, such as maize (one species), Kalton and Richardson (19) stated that 155.10 Ph.D. scientific year (SY) equivalents were employed in the case of one maize crop for breeding, whereas only 96.36 Ph.D. SY equivalents were employed for all vegetable and fruit crops together, containing more than 30 plant species.

Small seed companies can rarely afford competitive research programs, and they may be forced out of business without a strong public plant-breeding effort (18). Many of these companies have played a vital role in the vegetable and flower seed industry by producing and merchandising publicly developed varieties and hybrids of crops. Plant breeders in the public (government) sector can be involved in long-term breeding programs without a need to demonstrate cost effectiveness in the short term. They can also be involved in the enhancement of germplasm—a costly and long-range effort. Constant basic and applied research efforts are needed to develop the biotechnology fundamental to continued progress of the seed industry. Many current developments in the seed industry, including breeding for crop improvement, are the result of the vision and needs of consumers, growers, packers, shippers, processors, and seedsmen as well as the significant contribution made by many federal and state breeding programs, initiated and sustained with public funds. It takes a long time to develop a successful breeding program, but the benefits of a public breeding program exceed their costs by a wide margin. Both public and private breeders can cooperate professionally. The material released and ideas generated by public breeders have been utilized successfully by the seed industry.

The estimated value of internal commercial seed markets for the world's top 16 countries ranges from $4500 million for the United States to $300 million for the

Netherlands. These sixteen countries represent more than two-thirds of the world's commercial seed market. The 10 largest seed companies in the world have a total turnover of $4.33 billion, but this represents only 15% of the total world commercial seed market, proving that the international seed market, is very fragmented. The value of international trade is seed is estimated at $2.9 billion: $1.9 billion of which is for agricultural crops. Globally, the most important products are maize, vegetable, flower, herbage, and beet seeds. International seed trading has been increasing, but within Europe, trading accounts for 45% of this global exchange. Trade in the future may be restricted by governmental regulations (which are replacing tariff barriers), national listing of seeds organizations, regulations for genetically modified organisms and failure to protect intellectual property (20).

V.   NEWER BIOTECHNOLOGY FOR CROP IMPROVEMENT

According to a recent report from ICRISAT (21), biotechnology is now on the cutting edge of plant science—offering new techniques, new applications, and new opportunities for crop improvement. Scientists are now using a variety of techniques to track genes that determine specific traits (e.g., drought tolerance or disease resistance). They make crosses between species previously believed to be incompatible and produce improved genotypes much faster than was possible using traditional breeding methods. One of the most far-reaching applications of recent years is restriction fragment length polymorphism (RFLP) mapping the use of DNA fragments as genetic markers to follow chromosomal segments through segregating generations. Using DNA markers (practically unlimited numbers of markers are available in virtually all plant species), scientists can now directly follow chromosomal segments during recombination and create genetic maps far more quickly and accurately than before. Rather than selecting for a particular trait (which can be tricky, because expression of the trait may depend not only on genetic factors but also on environmental conditions and genotype-environment interactions). they can now select for the presence or absence of molecular markers linked to genes controlling that trait.

Van Geyt (22) reviewed the impact of biotechnological techniques in the seed industry and agriculture. In the 1980s, the boom of newly created seed companies concentrated on very specific areas of biotechnological research. The main goal was to provide breeders with new tools and gradually to incorporate biotechnology in the process of the development of a new variety, whereas allowing classic agriculture, breeding efforts, and phytosanitary treatments to coexist with biotechnological solutions to agronomic problems. Van Geyt (22) reviewed the actual and expected influence of newer biotechnologies. Various applications such as in vitro clonal propagation and dihaploids are already being used routinely. In the short term, applications based on DNA markers (RFLPs) are expected to emerge. Transgenically introgressed resistance or modified genes are believed to be of medium- or long-term importance, mainly due to regulatory legislation and the field tests needed before commercial release is possible. Various other possibilities such as the se of plant cells or whole plants to produce high-value products, artificial seeds, growth-promoting organisms, and biological control agents are still at a developing stage (22). Glimelius (23) considered the following biotechnological tools for breeding: micropropagation embryo culture, haploid production, somatic embryos, somaclonal variation, protoplasts (somatic hybrids or cybrids), cybrids, and combination of cell culture and molecular biological techniques.

REFERENCES

1. J. R. Thomson, *An Introduction to Seed Technology*, Leonard Hill, London, 1977.
2. R. L. Agrawal, *Seed Technology*, Oxford and IBH Publishing, New Delhi, 1980.
3. W. P. Feistrizer, The role of seed technology for agricultural development, *Seed Sci. Technol. 3*:415 (1975).
4. G. B. Baird, The role of a well-organized seed industry, *A Guide to Sorghum Breeding*, 2nd ed. (L. R. House, ed.), International Crops Research Institute for Semi-Arid Tropics, Patancheru, India, 1985, p. 152.
5. W. H. Freeman, Developing a seed industry, *A Guide to Sorghum Breeding*, 2nd ed. (L. R. House, ed.), International Crops Research Institute for Semi-Arid Tropics, Patancheru, India, 1985, p. 162.
6. L. R. Hawthorn and L. H. Pollard, *Vegetable and Flower Seed Production*, Blakistan, New York, 1954.
7. I. O. Copeland and M. B. McDonald, *Principles of Seed Science and Technology*, 3rd ed., Chapman & Hall, New York, 1995.
8. P. D. Hebblethwaite (ed.), *Seed Production*, Butterworth, London, 1980.
9. R. A. T. George, *Vegetable Seed Production*, Longman Press, New York, 1985.
10. A. F. Kelly. *Seed Production of Agricultural Crops*, Longman Press, New York, 1988.
11. M. B. McDonald and L. O. Copeland. *Principles and Practices of Seed Production* Chapman & Hall, New York, 1995.
12. P.K. Pasichnick, V.V. Kuyanon and V.B. Khikhlovskli, Regulations on sugar beet seed supply for sugar beet sowing economics of the Ukraine in 2001, Tsukor Ukraine No. 1/2, 12-14, 2001, *Seed Abstr. 25*(6):251 (2002).
13. M.R. Karim, Financial and economic analysis of seed programs for developing countries, *Bangladesh J. Train Dev. 13*:93 (2000).
13a. H. G. Vest Jr., The vegetable seed industry and public plant breeding: some concerns, Hort Science *19*(2): 1984.
14. M.R. Karim, Problems and issues in seed program in developing and developed countries, *Bangladesh J. Train Dev. 13*:181 (2000).
15. S.W. Muliokela, The challenge of seed production and supply in Africa, *Seed Sci. Technol. 27*:811 (1999).
16. N. C. A., *Report of the National Commission of Agriculture*, Government of India, New Delhi, 1976.
17. *Asia Region Annual Report*, 1994, International Crops Research Institute for the Semi-Arid Tropics, Patancheru, India, 1995, pp. 71–91.
18. H.G. Vest Jr., The vegetable seed industry and public plant breeding programs: a major thrust in U.S. Agriculture, Diversity *5*:16 (1983).
19. R. R. Kalton and P. Richardson, Private sector plant breeding programmes: a major thrust in U.S. Agriculture. *Diversity 5*:16 (1983).
20. B. le Buanec, Globalization of the seed industry: current situation and evolution. *Seed Sci. Technol. 24*:409 (1996).
21. Report 1994, International Crops Research Institute for the Semi-Arid Tropics, Patancheru, India, 1995, p. 9.
22. J. P. C. Van Geyt, Impact of biotechnological techniques in the seed industry and agriculture, *Rijksaniv. Gent. 55*(4):1587 (1990).
23. K. Glimelius, Utilization of biotechnology for breeding in agriculture and forestry L. Agriculture, *Kungl. Skogs-och Lantbruksakad. Lidske. 130(1-2)*:37 (1992).

22
Micropropagation

I. INTRODUCTION

Micropropagation, also knows as in vitro or clonal plant propagation, constitutes a range of tissue and cell culture techniques to propagate true-to-type plants. Micropropagation is one of the most commercially efficient plant biotechnologies employed in agriculture as an alternative to plant reproduction by seeds (sexual or generative propagation). This practically oriented plant biotechnology has formed a basis for the development of other useful biotechnologies such as somatic embryogenesis leading to synthetic seed biotechnology and genetically modified (GM) or transgenic seeds, plants, and crops through genetic engineering and recombinant DNA technology. However, the latter have not yet been commercialized to the same extent as micropropagation.

In addition to the traditional breeding where micropropagation is required to produce the first parent, a fertile plant that will be crossed and selected for eventual seed production, it is the ultimate need for large-scale regeneration of GM plants produced by in vitro techniques. These are normally cultivated by vegetative or clonal propagation: for example, cutting grafting, division and separation (1). Altman and Loberant (1) have described the basic principles and science of commercial practices of micropropagation followed successfully in agricultural and horticultural crops. In addition to this comprehensive review, the science and technology of micropropagation has been dealt with extensively by other authors (2–7).

Plants are being cloned using in vitro methods; that is, whole plants have been regenerated from cells, tissues, and organ explants since the 1940s (8), although under small-scale laboratory conditions. It was not until the early 1970s that agriculturally important crops, such as flowers and ornamentals, were produced on a large scale by tissue culture techniques (9,10). Micropropagation techniques are now being used on a variety of plant species, and in vitro methods are being practiced on a commercial scale globally in more than 500 million plants annually, the majority of which constitute flowers and ornamental crops (2,11–13). Cassells (7) has reviewed the developments in the use of plant tissue culture techniques to produce new plants with improved agricultural traits; that is, in vitro production of pathogen- and contaminant-free plants.

II. SEXUAL (SEED) VERSUS MICROPROPAGATION

Seeds through sexual or generative propagation routinely reproduce most of the agriculturally important field and vegetable crops. In homozygous and annual crop plants, this is easily done by producing pure seeds through backcrossing in a short time. However, most perennial heterozygous crops, especially the horticultural crops such as flowers, ornamentals, and plantation and fruit trees, are normally propagated vegetatively by a range of techniques—cuttings, layering, separation, division, grafting, and budding—resulting in true-to-type clonal plant material. In this respect, micropropagation is synonymous with vegetative or clonal propagation. Altman and Loberant (1) differentiated micropropagation from vegetative propagation such as cuttings on the basis of following:

1. A very small plant part (explant), usually a few millimeters or less, is used as starting material in micropropagation.
2. The explant is maintained in vials in a defined and balanced culture medium.
3. Micropropagation is essentially carried out under aseptic conditions.
4. In vitro or micropropagation normally results in numerous clonal propagules per unit of initial (stock) plant material (mass propagation) in a much shorter time, thus enabling very efficient plant propagation.

Micropropagation offers several qualitative, quantitative, and economic advantages for many plant species over the conventional vegetative propagation (1).

1. *Mass plant propagation in short time-span*—A very large number of clonal propagules can be produced in a short time. Thus, the plant tissue culture enables us to produce millions of plants rapidly from a relatively few selected source plants depending upon the multiplication rate.
2. *Production of disease-free plant material*—It has been possible to eliminate fungal, bacterial, and viral contamination of crop plants and produce totally disease-free plants material, which is a serious limiting factor in the conventional plant propagation (7,14). In vitro and micropropagation techniques minimized infection and contamination by horticultural crops. In addition to the production and maintenance of pathogen-free stock plants, other implications of micropropagation include long-term in vitro conservation of germplasm and selection and generation of transgenic plants (1) (see Chapter 24).
3. *Large-scale production of true-to-type plant material from a minimum plant source*—A commercial micropagation guarantees mass production of a consistently high degree of likeness (true-to-type) between the template (plant source) and the product for each selected plant type.
4. *Economic transportation and shipping*—A large quantity of micropropagated plant material can be air shipped quickly, efficiently, and relatively inexpensively. As many as 30,000–50,000 in vitro (in closed vials or small containers) or 3000–10,000 in vitro (hardened) plants may be transported in 1 cm of shipping space (7).
5. *Improved marketability*—It is possible to produce large quantities of newly bred plant crops and selections quickly. Selection of commercial plant products through conventional breeding is a lengthy process. Micropropagation techniques enable the plant breeder to track and test new genotypes quickly and with more efficient qualitative resolution. Molecular biology techniques have offered the possibility of patenting cultivars for the mass market (13).

Some of the limitations and disadvantages of in vitro and micropropagation include endogenous and environmentally induced contamination of plant cultures leading to great economic losses, higher than acceptable levels of somatic variation, losses incurred during transfer of plant material from in vitro conditions to the acclimatization stage, and high production costs owing to expensive technology and the labor-intensive nature of tissue culture (1).

III. BASIC PRINCIPLES AND TECHNOLOGY

A. Tissue Culture and Cellular Totipotency

The term *plant tissue culture* broadly encompasses protoplast, cell, tissue, and organ (anther, embryo) culture carried out under controlled and aseptic conditions. This techniques has revealed a unique capacity of living cell "totipotency," or total genetic potential, which is the inherent capacity of a plant cell to give rise to a whole plant. This capacity is often retained even after a cell has undergone final differentiation in the plant body. For a differentiated cell to express its totipotency, it must first undergo dedifferentiation, followed by redifferentiation. The former often involves embryonization of cells leading to callus formation, but in some instances, redifferentiation may occur directly from the dedifferentiation cell without going through the callusing phase (15). All in vitro techniques of plant propagation, including micropagation, rely on the unique totipotency of plant cells; that is, the regeneration of whole plants from individual cells, a tissue, or an organ, expressing the full plant genome. This is normally achieved after the plant tissue or organ has been excised and placed on culture medium (barring young seedlings, which can be cultured intact without prior excision), and involves the following consecutive events (1):

1. Dedifferentiation of the source tissue or organ, resulting in the activation of the physiological mechanisms that lead, under appropriate culture conditions, to cell division
2. Active cell division in the entire cut surface, or localized meristematic activity in specific regions of the explant, or both, often leading to proliferation of callus tissue
3. Organization of defined meristems, which occurs within the zones of active cell division, and results in formation of shoot or root meristems or both
4. Regeneration and differentiation of new organs—organogenesis (formation of new shoot buds or new roots) or somatic embryogenesis (the bipolar differentiation of somatic embryos)

Organogenesis and somatic embryogenesis may occur directly from the explant without involving callus formation or indirectly from a callus tissue, which reflects in the genetic stability of the resulting plantlets. Meristem organization from cell calluses often leads to mutant genetic aberrations compared to direct regeneration. The entire micropropagation process needs to be carefully controlled and monitored so as to generate true-to-type plantlets. Some micropropagation stages do not necessarily involve a regeneration process, for example, culture of shoot tips may result into proliferation of already-existing axillary buds, but their further development into plantlets does involve a regeneration process (i.e., root formation) (1).

Plant tissue culture has become an invaluable tool in the field of experimental botany to study basic problems related to growth and differentiation phenomena under highly precise conditions. Micropropagation is an efficient, safe, and often economic method of propagating horticultural and agricultural crops to raise healthy, pathogen-free plants, bypassing time-consuming quarantine requirements. It is possible to store valuable plant germplasms in cultures under low temperature more economically and safely than using conventional methods. It may also be possible through micropropagation to grow single cells and the fuse isolated protoplasts (cells without cell walls) of completely unrelated species of crops plants to develop new crops with a higher yield potential and better qualities. Above all, in vitro plant cell and tissue culture techniques have paved the way for the production of synthetic/artificial seeds through somatic embryogenesis (see Chapter 23) as well as the GM or transgenic seeds/plants/crops (see Chapter 24).

B. Axillary Bud Formation and Organogenesis

According to Altman and Loberant (1), in vitro axillary bud proliferation is usually considered to be a convenient route for micropropagation. It is considered to be "safer" for the preservation of the clonal characteristics of plants, because it does not include a callus formation stage. Bud meristems existing in the axils of leaves do not develop in plants because of apical control until the stem elongates and grows. The short stem tip contains many axillary buds at different stages of development condensed in a small explant. Thus, a very large number of otherwise quiescent axillary shoot buds proliferate extensively when shoot tips, or even smaller apical meristems, are cultured in an appropriate medium, usually containing high concentrations of cytokinins. The excised explant from the shoot tip or apical meristems, when cultured in cytokinin-rich medium, "activate" the bud, growing into many side shoots, which are separated for further culture after the induction phase.

Shoots and roots are formed de novo from explants during organogenesis. These two events may take place simultaneously during culture, but frequently either shoots or roots (usually shoots) are formed first. Complementary organs regenerate later either in the same medium or often subculturing to another under a different environment. Organogenesis begins with the distinction of a root meristem within the explant or from the callus and is controlled by the type of explant, composition of the culture medium (especially the balance of growth hormones), and the environmental conditions during culture. After the organization, the shoot and root meristems begin to form small shoot and roots. Proper organization and differentiation, including the formation of functional vascular connections between the developing shoots and roots, finally produce plantlets in vitro. The new plants are then acclimatized, hardened, and grown under greenhouse or field conditions.

Depending upon the plant genotype, in some cases, organogenesis may lead to the formation of modified shoots and roots, usually storage organs like bulbs and tubers. The shoot or the root meristems formed soon develop into minibulbs, tubers, or corms instead of further growth and elongation. Culture conditions determine whether the resulting storage organs grow further or go into dormancy. Axillary buds also can give rise to minibulbs, tubers and corms, especially when primary explants are obtained from basal plates of bulbs, tubers, or corms (1).

IV. ESSENTIALS OF MICROPROPAGATION

A. Experimental Micropropagation

Altman and Loberant (1) have listed the following basic and support facilities and activities to carry out experimental micropropagation:

1. Basic Facilities for Direct Production

Plants are prepare initially in an area adjacent to an aseptic room (clean). Media are prepared in a space having easy access to all parts of micropropagation laboratory. Plants are produced in a clean, aseptic room containing laminar flow hoods where most plant manipulation and replanting activities on new growth media take place. Plants are grown in a clean growing room for longer periods of time, usually lasting from 1 to 2 years. Plantlets are required to be grown and developed for 1–2 months after each manipulation, employing appropriate environmental conditions. An average-sized micropropagation laboratory may hold as many as 50,000 plantlets (1).

2. Allied Support Facilities

In addition to basic facilities required for direct plant production, around 30–40% of time, space, and technical personnel in a commercial micropropagation laboratory may be devoted to provision of allied support facilities and activities, such as quality assurance, an information system, and packing and shipping.

High-tech commercial micropropagation requires mass production of uniform plantlets to be delivered in a short time. Quality assurance being the most important central element in micropropagation, bringing commercial firms in line with total quality management (TQM) and the International Standards Organization (ISO 9000) (16,17), requires professional technical personnel and laboratory facilities that are appropriate for pathological and horticultural testing. A trained staff is required to maintain and control systems in plant-growing rooms, examining all media and plant material on a regular basis, and serving as a link between production and quality assurance staffs. Despite this, agricultural problems, such as plant source diversity, occurrence of pests and diseases, and the weather conditions occasionally complicate placing a living plant into the production line. The uncertainty principle underlying plant micropropagation is sometimes exacerbated by certain unforeseen issues, such as the use of potentially dangerous chemicals in plant media and pathological contamination of the system. Thus, the success of the micropropagation industry should heavily depend on quality assurance and plant efficiency.

The biological nature of micropropagation requires a computer-aided information system to monitor long production processes with multiple manipulations, identification of inventory, and quality control. The labor-intensive nature of commercial micropropagation assures the presence of many technical and nontechnical workers and individuals tasks performed by them. Ancillary space for dressing rooms, eating and rest areas, and office space also need to be provided. The ready-to-ship growth containers are moved to a space designed for sorting, final quality checking, and packing. Aseptic and environmental control are essential for the entire process in micropropagation starting from the in vitro production of plants to their transportation by air freight. Survival of micropropagated plants on arrival to the customer followed by their successful acclimatization and planting are necessary (1).

3. Cleanliness and Asepsis

Clean, aseptic environment and axenic culture media are the essentials of in vitro micropropagation, which can be accomplished by surface sterilizing the explant source (plant material) and planting and subculturing under strict aseptic conditions. Presterilized culture media and vials are used to maintain continuous culture in a clean environment. Aseptic rooms are buffered and isolated from other spaces that are progressively cleaner and require more stringent operations as the central, clean transfer room is approached.

"Clean room technology" provides work spaces with a minimal level of particulate matter, including spores. Air under positive pressure is directed through high-efficiency particulate air (HEPA) filters and discharged into the laboratory at central points. Similarly, and out-flowing gradient of clean air prevents the passive entrance of contaminated "fresh" air and particulate matter. Plant material is sterilized under laminar flow in a clean transfer room and stored aseptically in plant-growing rooms. Filters and levels of asepsis are examined regularly, employing standard methods and equipment (1).

Sterilization and disinfection constitute an internal part of cleanliness in creating an aseptic environment and a contamination-free product. Autoclaves are used to sterilize most media, containers, and tools in commercial laboratories. Additional large-scale sterilization may be performed by gamma irradiation or by ethylene oxide gas treatment, whereas most tools may be disinfected by glass lead sterilizers, electric incinerators, and Bunsen burners. Cleaning work surfaces, floors, and equipment are usually disinfected chemically. Plant material is first cleaned by scrubbing with a detergent and rinsing with clean water, followed by surface sterilization, and additionally with fungistats, fungicides, or antibiotics, if required.

4. Controlled Atmosphere

Maintaining an optimal controlled atmosphere in terms of temperature, relative humidity (RH), and light throughout the handling and development of in vitro plant cultures is a key factor in micropropagation. A constant balance between complex and expensive electrical, air-conditioning, and filtration systems, especially in plant-growing rooms, is required. The ambient temperatures for most plants may vary from 22 to 27°C, although temperature regimens can be altered slightly depending upon the plant species, stages of growth, or storage conditions.

The RH in the plant growth container is normally 98–100%. However, Tanaka and associates (18) have shown that the net photosynthetic rates of potato plantlets in vitro were higher when maintained at RH of 88–94%. The ambient RH in the growth room is much lower (50–80%). However, RH below 40% can result into plant desiccation, enhanced salt concentration, and drying of plant material, and RH levels higher than 85% often increase the incidence of microbial contamination. Thus, stringent humidity control measures may be necessary in areas with extreme climates and those with significant variations in weather conditions (1).

Despite some photosynthetic activity, in vitro plantlets in micropropagation do not rely on photosynthetically fixed carbon for growth. Hence, conventional in vitro growth media provide cultures with a ready-made carbon source. Fluorescent radiation with a photon density of 20–200 $\mu M \ m^{-2} \ s^{-1}$ and a standard photoperiod of 16 hr is normally used in micropropagation (1). However, exceptional environmental conditions are often required depending upon different plants, species, in vitro techniques and stages of plant growth. This may necessitate establishment of laboratories equipped with multiple growing rooms and options for changing standard parameters of light, RH, and temperature. Recent research indicates that many of the standard procedures described in the literature are not optional (19).

5. Growth Media

Initiation and sustenance of in vitro plant growth takes place in a sterile nutrient medium in an appropriate growth container. Synthetic growth medium is the primary source of nutrition for plants and plant tissues in vitro. Since the time when. While (8) demonstrated the continuous growth of tomato roots in vitro there has been constant improvement in the understanding and effectiveness of the use of synthetic plant growth media (1,6).

The components of a basal media are water, sugar(s), as a primary carbon source, inorganic salts, providing macroelements and microelements, vitamins (some essential, others beneficial), and growth hormones. The discovery, isolation, and synthesis of plant growth hormones (auxins, gibberellins, cytokinins, inhibitors, and ethylene) have provided the basis for the control of plant growth development and regeneration. Certain additional, undefined factors, such as coconut milk, yeast extract, or protein hydrolysate, have been shown to be beneficial for some plants. Growth media are normally gelled with agaragar or its substitute such as gelrite (2). Murashige and Skoog (MS) medium is the most widely used standard growth medium for in vitro techniques (20). Micropropagationists and in vitro technologists today use a vast range of synthetic, analytical-grade growth media and media components available in the market. Smaller quantities of nonautoclavable materials can be sterilized using microwaves or filtration. The preparation of media in a micropropagation laboratory requires a range of equipment, including a pH meter, analytical balances, measuring containers, cooking stirring, and pouring equipment, detonizers and distilled water units, a washing area for apparatus, and storage facilities.

6. Bioreactors, Containers, and Other Apparatus

A stringent control of the microenvironment during micropropagation, including generation and absorption of gases by the plantlet and culture media, needs additional specifications. The latter are usually dictated by factors such as availability, low unit cost, uniformity, nonphytotoxicity, and ease in sterilization and handling. The glass test tubes, vials, flasks, and jars are standard containers, but these have largely been replaced by more convenient plastic (polypropylene or polycarbonate), often disposable, wares, including petri dishes, test tubes, tubs, boxes, and flexible wall, and disposable bioreactors. Limitations in production procedures often require that most containers be small, holding from 1 to 100 plantlets, because storage for containers at various stages of micropropagation is an important issue in commercial laboratories. New in vitro techniques in liquid culture media often use bioreactors, holding from 1 to 2000 L of growth medium and plant material, thus saving a significant amount of storage space (1).

B. Phases of Micropropagation

Altman and Loberant (1) have described various stages involved in plant micro-propagation starting from isolation of cells, tissues, or organs of selected plants, surface sterilization, and incubation in a growth-promoting, sterile environment to produce many clonal plantlets. Micropropagation involves the following five critical and ordered operational phases, with the first three of them being in vitro (1–3,21):

1.   Phase 0: Selection and Preparation of Explant Source

The quality of an explant source (mother plant) is a deciding factor in the success of micropropagation. The effective selection and maintenance of an explant source should ensure that the selected material represents a certified, horticulturally true-to-type plant of the desired species and cultivars. It should be free from infectious disease, or can be made pathogen free using specific in vitro methods, and it should be viable and vigorous, that is, potentially able to respond to division and plant regeneration.

For compliance with the above requirements, selected explant sources are usually "preconditioned" by a variety of specific growth regimens and horticultural procedures, including nutrition and irrigation, optimal environment conditions of photoperiod, light quality, and temperature, treatment with growth regulators, and pest control. The explants from mother sources range from large corms (bananas, dates) to seeds or fragile stems and leaves (1).

2.   Phase 1: Establishment of Viable Explants in Culture

The initial explant, ranging in size from 0.1 mm (meristems) to around 1 cm (bulbs, stems), is established in culture, leading to tissue activation and multiplication. Explants are usually established in agar-based media although liquid media can also be used. The choice of basal media and growth hormones at this phase of micro-propagation is very important and depends on the type of plant tissue and method of multiplication. Microbial contamination and horticultural traits are monitored closely by visual observation, which may necessitate specific indexing and treatment for pathological contamination.

3.   Phase 2: Rapid Regeneration of Propagules

Primary explants, after successful passing through phase, are aseptically transferred to phase 2 for their rapid multiplication into numerous clonal propagates. Tissue masses are repeatedly manipulated by subculture onto new culture media to encourage proliferation of propagates. The method of regeneration and proliferation is largely governed by the combination of growth hormones. Whereas a high proportion of cytokinins normally stimulates continued multiplication of axillary or adventitious shoots, a higher proportion of auxins, such as 2,4-D, is required for callus proliferation vis-à-vis somatic embryogenesis (1). Higher multiplication rates and proliferation of high-quality new plant propagules can be obtained by combining, optimizing and appropriately adjusting growth hormones, composition of basal media, and environ-mental conditions. The rapid multiplication phase may last from several months to 1 or 2 years or more. The stock cultures are required to be renewed from time to time so as to prevent possible mutations, loss of vigor, and regeneration potential. Although basic media formulations do not vary significantly, extensive experimentation and research may be necessary to reach commercially efficient multiplication rates with specific crops and their varieties.

4. Phase 3: Establishment of Complete Plantlets/Plants

Followed by repeated subcultures and screening, the established plantlets are transferred to the final in vitro phase. The latter is aimed at arresting rapid multiplication, including the development of the plantlet fully, lending to shoot elongation, root formation, and formation of storage organs (bulbs, corns, and tubers) as independent propagation units. This phase provides conditions to stimulate photosynthesis and other autotrophic physiological changes required for ex vitro plant growth during the acclimatization phase. This can be accomplished by modifying the culture media; for example, decreasing or totally eliminating cytokinin concentration, sometimes increasing auxin levels, decreasing levels of carbon sources, and by modifying environmental conditions (e.g., increasing intensity of light) (1). The efficiency plantlet establishment at this phase can be improved and production costs can be reduced by adopting micropropagation systems that shorten the process or allow much of it to take place ex vitro or in a less aseptic, in vitro environment (2,19,22).

5. Phase 4: Plant Acclimatization ex Vitro

The healthy emerging plantlet from the micropropagation laboratory cannot normally grow under natural field or greenhouse conditions. It requires 4–8 weeks of greenhouse hardening and acclimatization; a weaning process. For the first couple of weeks, plants remain under low-light, high-temperature, and high-humidity conditions, which may be provided by fogging or an incubator (21). The plants' photosynthetic activity is improved and they become autotrophic with the development of cuticular waxes, stomatal function, and new, functional roots. During the following few weeks, the light intensity is raised, and the ambient temperature and humidity are regulated to the natural growing conditions of the field. The use of rooting hormones may stimulate root development.

V. COMMERCIAL MICROPROPAGATION

A. Economics and Financial Aspects

The production of in vitro plantlets continues to expand worldwide. According to Pierik (3), the Netherlands, a founding center for commercial micropropagation, produced 20 million plants annually in the early 1980s, and they were producing 50 million by the close of the 1985. With its beginning in the 1970s, commercial micropropagation saw a particularly sharp rise in 1980s, with an increase in the number of laboratories in the United States, Europe, and Israel, producing in vitro plantlets beyond the market demand. With cheaper manpower and overheads, Eastern Europe, Southeast Asia, and South America had entered the market by the mid to late 1980s, providing significant competition in the production of orchids, cut flowers, and house plants (13).

According to the European Community survey (12) of professional and commercial micropropagation laboratories (172 commercial of 501 responding laboratories), the commercial production of plants in Europe, Great Britain, and Ireland was over 80–100 million plants annually. Another survey of 113 commercial laboratories in the United States in 1996 showed that the production of around 120 million plants annually will almost certainly continue to grow in the next decade (4). Israel with a very active micropropagation industry had 10 commercial laboratories in 1996, producing 20–25 million plantlets per year (1).

Research and development in biotechnology present financial risks. According to Loberant (21), most micropropagation companies, of all sizes, types, and locations, have failed to meet the demands of a market-driven economy, with only a few having become financially successful and independent. During the 1970s, there were over 15 million hectares planted with food and forest crops worldwide. Many of these crops and areas were suffering from disease, poor cultivation techniques, and abiotic stress conditions, resulting in poor yields and quality of crops. Thus, plant biotechnologists have postulated that these conditions could benefit both from crop improvement programs that use disease-free plants produced by in vitro technologies as well as from the mass production of quality propagation material. This could, and still can, become a reality if more efficient and relatively inexpensive production processes are developed (1).

According to Loberant (21), the establishment of a micropropagation facility and the allied agricultural, marketing, and management support system costs over a million U.S. dollars. Also, conservative, agriculturally based farming populations are not willing to abandon traditional methods and to invest in new ideas of farming. Micropropagation companies are required to develop expensive, long-term marketing strategies, such as pilot farms, education, and technical after-sale-service networks. The labor-intensive nature of plant micropropagation being highly cost intensive, a large-sized commercial micropropagation company may employ hundreds of workers for in vitro subculture, transplantation, and other activities. In highly developed industrial countries and in an increasing number of developing countries, labor, which may account for 40–60% of production costs, remains a major limiting factor in the success of a micropropagation company. Export-oriented commercial tissue culture companies have begun operation in many developing countries to exploit cheap labor (23). Countries like China, India, and Indonesia are looking for labor-intensive industries to employ ever-enlarging masses of workers. Most of the production work in micropropagation being relatively simple and repetitive, automation or robotic solutions have been sought in recent years. The issue of automation versus inexpensive foreign labor has created a potential dichotomy in the micropropagation industry.

Despite major scientific advances in agricultural biotechnology, micropropagated plants are sometimes subjected to unforeseen, inexplicable problems associated with agriculture. For example, significant production loss may occur because of microbial contamination of plants, which is reflected by economic and marketing ramifications. Export-oriented micropropagation companies must be equipped to deal with international marketing, monetary issues, and plant-protection policies. National policies often limit marketing efforts, reducing profitability. However, many of these complications can be dealt with by persistent and consistent production of high-quality plant material and improved services offered to dedicated customers.

B. Micropropagation Industry

Altman and Loberant have defined the following characteristics of the micropropagation industry (1):

1. Has developed a specialty in certain plants or product types, and markets over 2 million micropropagated plants annually
2. Has become relatively self sufficient in research and development, production professional manpower, facilities and sources of plant material
3. Engages in broad-scaled, vertically integrated agrotechnology projects

A successful industrialized micropropagation company must have achieved expertise in most areas of growth and development of a particular product, coupled with a business-like approach to marketing, management, and customer service.

1. Dynamics of Scaling-Up and Cost Effectiveness

With an enlargement of the micropropagation industry, the dynamics of scaling-up and cost effectiveness must be harnessed to stringently engineered inventory control and quality assurance programs. Problems associated with plant expansion include an initial loss of the ability to detect subtle changes or bottlenecks in production without timely solutions to overcome them. Expanded inventories (liability) along with the perishable nature of the product are rather difficult to manage in an exporting company. Also, owing to the seasonal nature of field crops, there are peaks in production schedules, but commercially viable companies must employ staff and facilities on a year-round basis; maintenance of start-up culture stocks help to minimize the time lapse between order and delivery.

2. Automated Mass-Propagation System

As indicated by some surveys, the demand for high-quality plant material for transplants has increased rapidly worldwide in respect to plant material required for reforestation, food, fiber and forage crop production, landscape and indoor horticulture, and global environmental protection (1,2). However, the relatively high production costs of micropropagated plants have restricted the widespread use of major crops in agriculture and forestry (24,25). Kurata (26) showed that major efforts are being devoted to develop automated, robotized, and more efficient transplant production methods. Developing an automated mass-propagation system to produce in vitro plantlets on a large scale at lower costs will become increasingly important in both the agricultural and horticultural industries. Such automated systems are in various stages of development and testing. An automated system working in Holland is based on image analysis, computer-controlled laser cutting, and robotized planting of explants (27). Liquid culture, employing 1000- to 2000-L containers with plant material immersed in liquid media, is proving to be a very cost-effective form of micropropagation, because such a system leads itself to automation more easily than cultures on semisolid media. Automated systems in micropropagation are being used by several commercial and research laboratories in Israel and Japan, which have developed production lines and prototypes based on liquid culture (28–31), with others being reported in Spain (32) and the United States (33,34) Kozai (22) has claimed "a novel approach to multiplication of in vitro plantlets" in reduce production costs significantly by means of a deliberate multifactorial system with robotics.

3. Problems Facing Automated Mass Micropropagation

Altman and Loberant (1) categorized issues facing automated micropropagation and mass-production systems into technical, horticultural, and economic problems. Despite its several advantages, liquid culture continues to be a subject of intensive research. There are numerous physiological limitations in propagating many plant types in immersed culture; for example, vitrification, deformation, and somaclonal variation due to uncontrolled multiplication (35,36). Although such problems do occur in agar-based micropropagation, the smaller number of plants in a given batch or container does not cause heavy losses. In contrast, economic losses of contaminated or damaged large-scale liquid culture systems are potentially enormous.

The problems facing the systems using semiautomatic or robotized cutting tools are twofold (1):

1. Tools are usually engineered for one standard cutting mode (stem sections), making them useless for plants manipulated differently.
2. The use of computer-aided cutting systems and visual analysis is technically complicated.

The combination of these two issues makes the systems undesirable for many commercial micropropagation laboratories. High costs incurred on establishment of a micropropagation laboratory, research and development, and automated systems limit their use for most commercial laboratories. Only companies established in a high-volume, single-type high-value (elite) plant product might consider such an option in addition to these, many pitfalls and unsolved problems remain to be solved in the scale-up and fully automated micropropagation plants. Thus, many commercial laboratories have adopted partially automated or semiautomated systems; for example, developing a prototype for the automatic manipulation of growth containers with in vitro plants. This robotic system performs the sterile exchange of liquid medium at an accelerated rate with minimal labor (37). Computer-aided processing systems are used in which media components are measured, mixed, processed, and poured into containers automatically with minimal human support (1). Other similar automations employed in micropropagation include systems for inventory and production task control and computer-aided environmental control systems to monitor, store, report regularly on environmental data, and set off alarms when standards are not met.

Commercial micropropagationists continue to sock any cost-effective, reliable system that can be used routinely to minimize human labor and to guarantee the production of high-quality material uniformly. Altman and Loberant (1) have envisaged that a high-quality, moderately priced, maximally engineered and automated system for selecting, manipulating, or planting micropropagated plantlets could provide solutions to most production problems that are currently faced by micropropagation laboratories.

VI. MICROPROPAGATION RESEARCH ON CROP APPLICATIONS

Plant micropropagation is yet to become practical and economically feasible. Also, the number of plant species that can be potentially regenerated by in vitro techniques far exceeds the number of plants that are being propagated in vitro on a commercial scale. The practical application of in vitro techniques to plant propagation depends on two major factors (1):

1. The current status of a particular agricultural or horticultural plant relative to its market value and critical issues associated with its production and marketing
2. The cost-benefit ratio of micropropagation relative to conventional propagation methods

The first consideration is whether there is a technology available for micropropagation, and if so, is micropropagation expected to provide solutions to problems that arise in conventional propagation? The application of micropropagation to economically important crops has been discussed as follows.

A. Vegetables and Field Crops

Most vegetable and traditional field crops are being grown and improved successfully for generations through conventional seed technology and breeding. However, certain crops and situations, including varieties of vegetables that have been specially bred or altered genetically, crops demanding exact and uniform harvesting times to meet market demands, and production of disease-free propagation material, have successfully used micropropagation and in vitro technologies.

1. Potato

Potato (*Solanum tuberosum* L.), an annual plant belonging to the Solanaceae family and originating at the high altitudes (2000 m) of the Andean region of Peru, Bolivia, Ecuador, and Columbia, is an important source of food worldwide. Potato tuber ranks fourth in terms of global food production (38) and accounts for 90% of micropropagation of vegetable crops in the United States, with over 11 million plantlets being produced annually (11). Under the traditional cultivation of potato by planting tubers gathered from the previous season, many tubers get infected by various viral diseases while in the field and/or when replanted and pass the infection to future generations; a problem especially severe in tropical climates (39). According to Bajaj (38), potato is one of the crop plants on which in vitro propagation techniques have been applied most successfully to overcome the problems of time and quantity associated with conventional breeding, to select and propagate virus-free planting material to produce seed tubers, and to establish in vitro cryopreservation of seed germplasm.

Cultivation of potato worldwide has changed significantly with an advent of virus-free stocks of potato germplasm (40) and development of mass-propagation programs for plantlets and seed tubers (41,42), followed by reports of the successful research on the production of disease-free in vitro potato plants from several countries (43–50). Potato tubers and plants are now being produced in vitro as nuclear material and grown subsequently for multiple seasons in disease-free conditions The resulting certified seed stock is then distributed to farmers for general cultivation.

With increasing demand for large quantities of virus-free planting material globally, issues of somaclonal variation (51,52) and the costs of maintaining and running micropropagation laboratories and greenhouse facilities have achieved significance. However, the use of new in vitro propagation technologies in potatoes has led to significant increases in the yields of disease-free potatoes due to selection of quality traits and establishment of multiple, international storage banks for valuable germplasm.

International Potato Center (CIP).   A major research effort has been launched at the CIP in Lima. Peru, to develop an effective means to produce potatoes from "true potato seeds" (TPS) that are infection free. It has been reported that the crop yield from 50 g of TPS is equivalent to that obtained from 3 tons of tubers (39). A new program to produce disease-free potatoes in Denmark in 1977 envisages that all potatoes grown in this country from the late 1980s should have originated from meristem culture (43). A field study in Taiwan has shown that the disease rate for a virus-free clone was 0.65% compared to 16.8% for standard, virus-infected farmers' seed stocks (some farmers' stock showing a 100% disease rate), and yield of the virus-free clone was 26% higher than that of the farmers' seed stock (46).

2. Sorghum

Sorghum, an important crop of tropical and subtropical areas of the world, has received considerable attention in tissue culture research. Masteller and Holden (53) first reported plant regeneration from the primary callus of seedling tissue. Callus induction and plant regeneration in sorghum have been reported from immature embryo (54–56), mature embryo (57), young inflorescence (58,59), and young leaf segments (60–62) as well as from mature seed (63–65). In a somaclonal variation study, Bhaskaran et al. (63) obtained sodium chloride–tolerant callus derived from mature seed, whereas Waskom et al. (57) reported increased tolerance to acid soils and drought stress in the field evaluation of mature-embryo culture-derived sorghum. In the case of sorghum seed, only a small number of genotypes have been investigated. Hagio (66) examined the response of mature seeds of numerous varieties of sorghum, consisting of six variety groups to in vitro conditions and identified varieties for further studies, such as medium improvement, somaclonal variation, and genetic trans-formations. Five varieties (CK602, CKW5809, PE932 203, PE954 068, and PF954 110) showed a relatively high ability for plant regeneration. Among the variety groups, the Kafir group varieties had a relatively high ability for callus formation and plant regeneration. These five varieties and Kafir group varieties may be useful for the in vitro study of sorghum requiring plant regeneration (66). In a more recent study, Hagio (67) examined 11 sorghum genotypes for their response to tissue culture and optimized the system. The cultures were initiated from immature embryos taken 2 weeks after flowering with the response varying with the genotype. High-frequency regeneration was obtained with cytokinin in the medium devoid of growth regulators. The addition of proline and polyvinylpyrrolidone (PVP) also enhanced shoot forma-tion. Plants could be regenerated on the revised culture medium up to 100% of sorghum immature embryos (67).

3. Cotton

Cotton (*Gossypium* spp.) is the most important source of natural fiber used in the Indian textile industry. Although conventional breeding methods have contributed extensively to the improvement of crop varieties, biotechnological approaches have widened applications and have significantly enhanced crop productivity (68). In cotton, the need to develop reproducible and efficient tissue culture protocols has prompted research in the area of ovule culture (69–71) and somatic embryogenesis (72). Genetic transformation has been achieved (73) and transgenic plants have been developed for resistance to insect pests (74). However, most of these studies have been limited to Coker cultivars of *G. hirsutum*, which are known to be highly responsive to plant tissue culture. Until recently, success with plant regeneration of Indian cotton cultivars either through somatic embryogenesis or multiple shoot induction was limited. Agrawal et al. (75) have reported multiple shoot induction in cotton (cv. LRK 516), and Gupta et al. (76) have done the same in 10 other cultivars. Kumar and Penial (77) achieved induction of somatic embryogenesis and plantlet formation in cv. MCU-5. Mhatre et al. (78) developed a protocol for the production of high-frequency multiple shoots and plantlets using seedling spices with cotyledonary nodes of six Indian cultivars of cotton. An average 7–10 shoots per explant were produced directly from the shoot spices excised from 10-day-old seedlings on MS medium. Rooting of isolated shoots was achieved on MS medium containing naphthalene acetic acid (NAA). Rooted plantlets exhibited 95% survival during hardening in the greenhouse, and regenerated plants grew to maturity with normal flowering and boll set (78).

4. Rice

The three-line system used to utilize heterosis in rice has been replaced by a more advantageous two-line system, with the availability of a photoperiod–temperature-sensitive genic male sterile (PTGMS) rice to produce hybrids. However, owing to the drift in the response of PTGMS rice to photoperiod-temperature conditions that is accumulated from one generation to the next, the use of a popular PTGMS line has become unreliable to produce high-purity F_1 hybrid seeds. To deal with this problem, a strategy called the fundamental seed technique, that is, selecting typical individual plants under critical conditions and using the seed-propagated progeny seeds only for a limited number of generations has been proposed (79). In vitro plant regeneration culture is a vegetative means of propagating a large number of plants that can be recovered without increasing the generation number to lessen the extent of drift in PTGMS rice. Zheng et al. (80) have evaluated a newly established culture method for mass propagation of PTGMS rice (cv. N195). The new culture method has been found to be very efficient to regenerate plants from the callus of this strain. Also, a high concentration of 6-benzyl adenine (BA) that is necessary for efficient plant regeneration did not affect some agronomic traits of the regenerated plants and their progenies. By the new culture method, a large number of plants could be regenerated in a relatively short time (around 8×10^{12} plants within 300 days) from one naked seed explant. Thus, the prominent drift of the PTGMS strain, accumulated gradually during the course of propagation of the seeds through generation, could be avoided when in vitro culture protocol is employed instead of seed propagation. Differences in the races of pollen fertility and seed setting and between their respective first and second progenies noticed between regenerated and seed-grown plants have an important application value in two-line rice production (80). An improved procedure has recently been reported for high-frequency androgenesis in indica × Basmati rice hybrids using liquid culture media (80a).

B. Fruit and Plantation Crops

Although fruit crops (apple, pear, citrus, olive, peach) have not been subjected to plant micropropagation significantly, their ability to produce micrografts, rooted cuttings, and rooted plants for rootstock or grafts from selected, disease-free mother plants under in vitro conditions has created a good marker (1). Some micropropagation laboratories have even specialized in fruit trees and woody plants. Most of the plantation crops, such as pineapple, palms, and bananas, which require large quantities of planting material on a regular basis, have been cultivated for generations by conventional methods. The incidence of disease, lower yields, and demand for both quantity and quality of the planting material as well as the final product are some of the major issues in fruit and plantation crops that can be effectively addressed by micropropagation.

1. Banana

According to Altman and Loberant (1), varieties of dessert bananas with a triploid genome (e.g., Cavendish, AAA) are produced extensively for commercial markers using micropropagation technologies. As an international cash crop, bananas provide dessert fruit for the tables of developed countries and are also a primary staple food (plantain) for populations of Africa, South America, the South Asian Pacific region, and the Indian subcontinent. During the second half of the twentieth century, commercial banana production was plagued by pests and diseases, depleted and contaminated soils, increasing market demands for quality and uniformity, rising prices, and world policies. Costly, long-term breeding programs, underway for decades, have had a minimal effect on commercial production, field selection is still the primary source for potential new varieties of banana (1).

Ma and Shii (81) first reported the in vitro culture of banana meristems in 1972. Since then, large-scale commercial in vitro production of banana plants has began in many countries, including Taiwan, France, South Africa, and Israel (21,82). Standard shoot-tip tissue culture is used to produce micropropagated banana plantlets. In vitro, hardened plantlets are shipped to their destination, acclimatized in shade nurseries, and planted in fields. According to Loberant (21), micropropagation companies in 1990s were providing millions of banana plantlets annually for commercial plantation in South America, Africa, and Southeast Asia with nearly 100 million micropropagated banana plants having been planted on South America alone. Israel, with a highly developed micropropagation industry, an academic infrastructure, and experienced nursery, is in the forefront of providing disease-free, selected in vitro clones of dessert banana plantlets to the international market.

The primary advantage of micropropagation in banana is its ability to provide large quantities (millions per year) of uniform, disease-free plantlets from selected clones timely to fit predetermined planting regimens as determined by the climate and market forces. However, in vitro banana propagation is associated with somaclonal variation (82–84), leading to genotypic or epigenetic changes that are inconsistent with commercial production. Significant research efforts have been diverted toward cultural protocols, multiplication schedules, selection and quality of source plants, recognition and control of endogenous (nonpathogenic) bacteria, and adherence to slander commercial practices (1).

In vitro mutagenesis of multicellular meristems of banana leads to a high degree of chimerism. Roux et al. (85) noted that colchicine treatment induced ploidy chimerism (mixoploidy), and assessed chimera dissociation using three different micropropagations systems: shoot-tip culture (ST), multiapexing culture (MA), and corn slice culture (CS). The average percentage of cytochimeras decreased from 100 to 36% after three subcultures in ST culture, from 100 to 24% in CS culture, and from 100 to 8% in MA culture technique. However, all propagation systems failed to eliminate chimerism completely (85).

2. Grape

Apices and axillary buds have been used for the in vitro propagation of various species and cultivars of grape (*Vitis* spp.) (86,87), and micropropagation protocols have been reported for muscadine grapes (88,89). Compton and Gray (87) evaluated the potential of micropropagation as a means of rapidly propagating the popular Southern Home, a complex interspecific hybrid grape plant. This study indicated that shoot-tip micropropagation would be an excellent way rapidly to increase the number of Southern Home plants for wholesale and retail growers. Salunkhe et al. (90), using tendril explants of grape cultivars, reported somatic embryogenesis with low-frequency conversion of embryos to plants. However, Mhatre et al. (91) recently developed an improved micropropagation protocol, using nodal explants bearing a single axillary bud from three grape cultivars: Thompson seedless, Sonaka and Tas-e-Ganesh. These authors have described a culture procedure for enhanced multiple plant production (Figs. 1 and 2) and recovery of complete plants, and discussed the use of this protocol for the commercial exploitation of cultivated grapes.

3. Apricot

Apricot (*Prunus armeniaca* L) is an excellent candidate for in vitro propagation, because apricots are currently budded onto apricot seedlings (92); therefore, growing apricots on their own roots appears to be logical. Successful embryo culture of immature apricot seed has been achieved (93). Adventitious shoot generation was obtained from immature embryos of Sundrop (94) and from embryos and embryo-derived callus of Royal (95) cultivars. Only a small percentage of the buds developed into shoots, which produced roots spontaneously (95) or after transfer to a rooting medium (94). Recent reports on in vitro plant propagation of apricot deal with the development of a culture medium to supply the excised plant part with all necessary nutrients, vitamins, and growth regulators (96–99). Perez-Tornero et al. (100) reported data on the preservation of apricot germplasm by in vitro slow-growth storage and determined the effect of temperature on regrowth of stored apricot plantlets, measured as shoot production and shoot length during repropagation. Shoots stored at 3°C had the highest regeneration rates and shoot lengths following transfer to stand proliferation conditions. These authors also explored the conditions required for induction of adventitious buds and regeneration of shoots from explants of mature apricot trees propagated in vitro (101).

4. Pineapple

Pineapple (*Ananas comosus* L. Merr.) cultivation in recent years has suffered significant losses (over 30%) due to a devastating disease, fusariosis or gomosis (*Fusarium*

Figure 1 Multiple shoots and plantlets in grape: (a) tuft of multiple shoots obtained on GM 2 medium, (b) rooting of in vitro shoots on filter paper bridges in liquid GR 1 medium, and (c) in vitro–produced rooted plantlets established in plastic cups with Soilrite. (From Ref. 91.)

subglutinans WR.), that is disseminated through the cutting type slips and suckets. This disease can be effectively controlled by the use of plantlets free of pathogens or resistant cultivars (102,103). Thus, micropropagation techniques are valuable tools for mass clonal propagation of superior clones or varieties. Micropropagation in pineapple was first described in 1960 (104) and followed by a report of a protocol for multiple proliferations established from lateral buds (105). Several other explant sources have been tested successfully in several regenerative protocols. Wasaka (106) reported an alternate pineapple micropropagation protocol with etiolated shoots, resulting in high multiplication rates. Besides the explant source, performance of micropropagation is influenced by variations in culture medium composition, environment (CO_2 enrichment), and light intensity (107).

Direct organogenesis in pineapple follows through a general protocol: (a) selection and preparation of explant donor plant, (b) establishment of aseptic culture and induction, (c) multiplication, (d) rooting, and (e) establishment of plantlets in soil. Dal Vesco et al. (108) reported an improved protocol through manipulating both explant size and medium composition, as well as acclimation substrates and plantlet size, aiming for the mass clonal micropropagation of pineapple plantlets free of fusariosis. Using this protocol, it is possible to obtain 1 million in vitro plantlets after

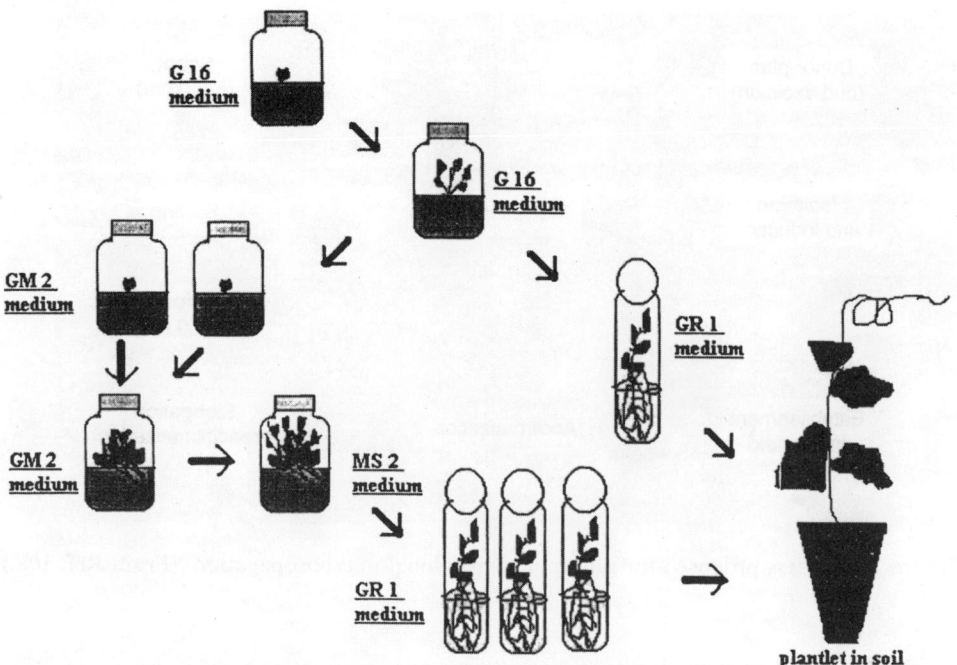

Figure 2 Schematic representation of the procedure employed in grape micropropagation using axillary buds. (From Ref. 91.)

9 months from a single bud, with a 45-day subculture interval and an average multiplication rate of 10 shoots per bud (108). An improved in vitro regenerative protocol for pineapple has the following stages (Fig. 3) (108):

1. Stage Zero: The selection and maintenance of mother plants in greenhouse conditions
2. Stage 1: Bud isolation on test tubes over bridge fitter paper in MS liquid culture medium supplemented with NAA (2 μM) and BAP (4 μM) for 6 weeks
3. Stage 2: Multiple bud induction in 500-mL flasks containing 30 mL of the same medium composition as in stage 1, with 5-week subculture intervals
4. Stage 3: Sorting of the shoots higher than 2 cm and subcultures in an elongation MS medium free of growth regulators. Shoots with a size less than 2 cm should return to the multiplication phase (stage 2)
5. Stage 4: Acclimatization of plantlets higher than 70 cm, after 5 weeks in stage 3
6. Stage 5: Transfer of the plants with a height between 15 and 20 cm to the field

Soneji et al. (109) recently reported high-efficiency regeneration of plantlets and somaclonal variants obtained in Micropropagated dormant axillary buds of pineapple. Phenotypic variants such as albinos, white-streaked shoots, and shoots with elongated internals were observed in in vitro cultures. Around 520 in vitro–propagated plantlets were established in the field, which exhibited somaclonal variation.

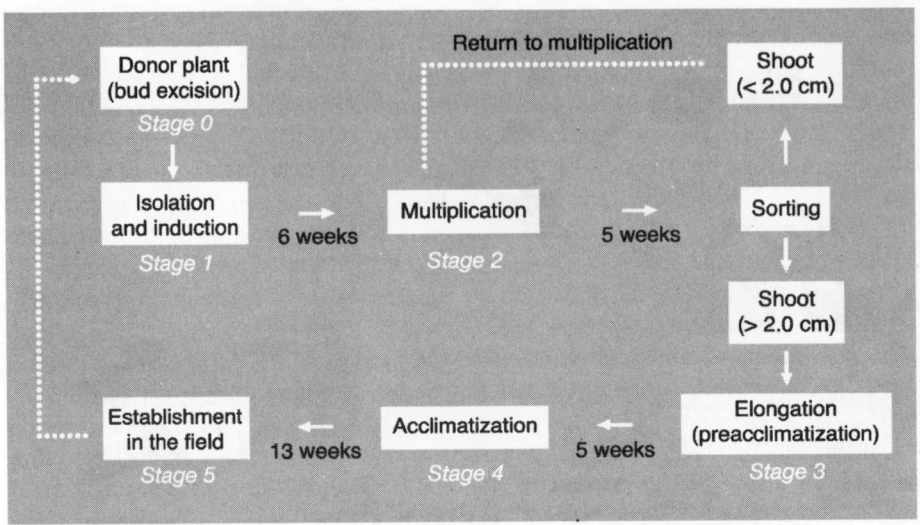

Figure 3 Process proposed for pineapple mass clonal micropropagation. (From Ref. 108.)

5. Strawberry

Strawberry (*Fragaria indica* Andr.) is one of the economically important edible fruits in India and other subtropical regions. Seeds and runners propagate the genus *Fragaria*. However, propagation by seeds is not adopted commercially, as seedlings take a longer time to fruit than plants grown by runners. Plant propagation through runners, adopted in several species of the genus, produces a limited number of propagules. The production of propagules through runners contributes 90% of the total Dutch strawberry production, but the product in the Elsanta cultivar was found to be susceptible to several fungal diseases (110). Bhatt and Dhar (111) developed an efficient method of micropropagation based on an increased percentage survival of explants and reduced phenol-induced browning in wild strawberry. Rooted shoots with fully explanted leaves (Fig. 4) acclimatized successfully, and around 70% of plantlets survived ex vitro. Jemmali et al. (112) reported carryover of certain morphological and biochemical characteristics associated with hyperflowering of micropropagated strawberries.

6. Fruit-Bearing Shrubs

The native North American fruit-bearing shrubs, chokecherry and pincherry (*Prunus* spp.), are well adapted to severe winter conditions of the Canadian prairies. With new commercial interest in these fruit crops, there is an opportunity for genetic improvement and utilization, requiring an efficient multiplication method to produce superior clones on large-scale basis. Micropropagation of these species could satisfy the current market demand. Such a method has been successfully utilized to produce other fruit-bearing, native prairie shrubs, for example, the Saskatoon cherry (*Amelanchier alnifolia* Nutt.) (113,114) and the sour morello cherry (*Prunus cerasus caproniana* L.), and well adopted to prairie conditions, the Nanking cherry (*Prunus tomentosa* L.) and Mongolian cherry (*P. × fruticosa*) (115). Micropropagation protocols have also

Figure 4 Stages of micropropagation of *Fragaria indica*. (A) multiple shoots in MS medium + 4.0 μM BA + 0.1 μM NAA; (B) rooted shoot; (C) plant raised in vitro 5 months after transplantation to potting mixture; and (D) micropropagated plant producing fruits.

been described for other Prunus species: *P. cerasus* L (116), *P. tenella* (117), *P. insititia* L. (118), *P. avium* (119), *P. armeniaca* (120), several cherry rootstocks, including *P. cerasifera* × *P. munsoniana* (121), *P. persica* × *P. amygdalus*, peach rootstock (122), and cherry and peach rootstocks (123). Pruski et al. (115) recently reported in vitro multiplication of *P. pennsylvanica* (pincherry) and *P. virginiana* (chokecherry). The ex vitro–rooted plantlets did not require any additional acclimatization prior to transplanting to the regular greenhouse conditions.

C. Flowers and Ornamental Crops

Initial research in commercial micropropagation was concentrated on flowers and ornamental crops. These crops lend themselves to newer biotechnological advances owing to their great commercial value as traditional consumer products. Laboratories associated with established nurseries began experimental production of cut flowers, ornamental potted plants, and foliage crops, serving as production models for in vitro problem-solving techniques, such as somaclonal variation (4,82–84), endogenous contamination, and detection and prevention of infectious diseases (10,14). These efforts led to the setting of international standards for establishing "mother block nurseries" with elite, tested, disease-free plants (124), the use of enzyme-linked immunosorbent assay (ELISA) to detect plant disease, and the establishment of germplasm banks for the maintenance, breeding, and protection of selected, disease-free cultivars (1). Cut flowers and ornamentals account for the majority of commercial micropropagation endeavors (11–14).

In vitro techniques of cut flowers and ornamental crops have certain advantages over traditional propagation; a relatively low frequency of off-types for many species, high uniformity, year-round availability of plant material to market, elimination or significant reduction of infectious diseases, and the ability to bring large quantities of plant material of newer varieties to market quickly. However, the high frequency of off-types in certain cultivars and control of endogenous bacterial contamination contributing to significant production losses still remain major limiting factors and challenges to overcome (1).

1. Carnation

Carnation (*Dianthus caryophyllus*) lends itself excellently to the use of in vitro techniques of micropropagation for commercial purposes. Carnation is a highly popular cut flower. During the 1960s and 1970s, the carnation industry suffered significant losses due to disease. Advances in micropropagation techniques during this period have led to improve protocols to combat viral and bacterial diseases and to provide higher yields of uniform, better-quality products. The following protocol developed in 1970s is used today with some modifications (1,21,125,126):

1. Selected cultivars are grown under meristem culture and the resulting material is further developed in vitro.
2. Clones of successful plants grown in vitro are then tested with ELISA to detect the presence of virus.
3. Virus-free clones are transferred as prenuclear stock plants to an approved, insectproof, quarantined greenhouse for further growth and testing.
4. Virus-free plants from this group are used as a source of cutting for similarly maintained and tested nuclear stock.
5. Cuttings from virus-free nuclear stocks are transferred to a nursery where plants are maintained and tested to provide cuttings for use as mother plants by commercial producers of carnation flowers.

This procedure, known as the mother block system, uses the micropropagation technique to eliminate pathogens and to clone initial plant material. It is a less expensive, protected plant propagation (using cuttings) and has served as a model for other crops, such as gypsophila, phlox, strawberry, and sugar cane (21).

Syngonium and *Spathyphillum* species, belonging to the Araceae family, are the world's most popular potted houseplants. Poor quality, low uniformity, and disease have remained problems to be overcome (127,128). According to Altman and Loberant (1), by using meristem culture of selected mother plants and micropropagation to produce millions of plants annually, today's market offers a wide variety of high-quality houseplants throughout the year at affordable prices.

D. Forest Trees, Medicinal Trees, and Minor Crops

1. Forest Trees

Although micropropagation is feasible in forest trees (129–131), barring few exceptions, in vitro propagation has not yet become practical or economically viable for most forest trees; thus, improvement in traditional propagation methods and their scaling-up is needed. It is commonly believed that, from the environmental perspective, the genetic diversity of forest plantation should be conserved, using traditional mixed populations of seedlings for forestation. Altman and Loberant (1), however, have pointed out the following three exceptions for which the application of micropropagation can be considered to be must relevant:

1. Breeding to accelerate the process of provenance selection and establish selected mother trees for further breeding.
2. Uniformity of the trees (clonal propagation) is desired where forestation is practiced for industrial applications; for example, for production of paper pulp or specialized timber or rubber.
3. Rapid, large-scale production of selected trees for forestation, whenever a need exists. Micropropagation of certain poplar and eucalyptus genotypes can compete with traditional vegetative propagation.

During the last two to three decades, a large number of coniferous species, such as Norway spruce, white and black spruces, radiata pine, and *Pinus pinaster*, have been propagated by in vitro-techniques (130–133). Much work has been done on southern pines, including loblolly, slash, longleaf, and Virginia pines, generating numerous transplantable materials. However, studies on other pines, such as Mediterranean species (Aleppo and brutia pines), are limited.

Four major types of micropropagation methods that have been employed in conifers are (1): (a) direct organogenesis of adventitious buds from organized explants, (b) fascicular and axillary bud breaking, (c) organogenesis by callus, and (d) somatic embryogenesis and artificial seeds.

2. Medicinal Trees

Pittosporum napaulensis (DC) Rehder & Wilson, belonging to the family Pittosporaceae, is a rare and endemic medicinal tree of the Himalayas, generally propagated by seeds. Although different presowing treatments have improved germination up to 83% (134), desirable characters may not be transmitted when propagated through seeds. Micropropagation therefore assumes importance, especially in view of the rarity of the species and the need to produce a large number of planting material for rural inhabitants of the region. Dhar and Bhatt (135) described the procedure adopted for micropropagation of five healthy genotypes of *P. napaulensis* using nodal segments.

Fritillaria thunbergii Miq, commonly known as fritillary, is a perennial plant of medicinal importance. Although this plant sets seeds, seed propagation has no practical value, because seedlings are week and the bulb development is too slow (5–6 years to develop a bulb to commercial size). In vitro propagation is an effective means for rapid multiplication of species. Earlier reports on tissue culture of this species (136,137) had a low multiplication rate. Paek and Murthy (138) have recently analyzed the role of growth regulators, method of culture, effect of dark and light incubation, and temperature on bululet regeneration of fritillary using bulb scale sections. They also established transplantation feature of in vitro–grown bulblets.

Alractylodes lancea (DC), belonging to the Asteraceae family, is a perennial medicinal plant originating in China. *A. lancea* has been propagated in vitro from flower buds and shoot tips. Hiraoka and Tomita (139) described the botanical and chemical evaluation of micropropagated plants grown under field conditions as compared to plants propagated by the conventional methods.

Soneji et al. (140) reported enhanced regeneration of multiple shoots and plantlets in young cotyledons of neem, *Azadtrachta indica* A. Juss. The in vitro method has made possible a cyclic production of a large number of shoots and plantlets in young embryo explants of neem.

Dhar et al. (141) reported in vitro regeneration protocols of Himalayan multipurpose trees (MPTs) to meet the bonafide needs of rural people for fodder, fuel wood, timber, and other products. Of the total 260 MPTs reported in India, 114 occur in the Himalayas, of which only 37 are Himalayan native species and the rest are either naturalized or have been introduced in the region. Propagation protocols of several MPTs have been envisaged to ensure mass scale availability of planting material of selected species, including several medicinal plants, which are in great demand, in addition to conservation of an important genetic resource.

Eucalyptus camaldulensis can be micropropagated through meristematic agglomerates (MAs) (4–6 mm diameter), dense shoot clusters initiated by the outgrowth of numerous successive buds (142). Viable plantlets can be raised from such MAs. This MA technique can be related to the nodule systems developed by Aitken-Chritie et al. (143) on radiata pine, by McIown et al. (144) on poplar, by Ito et al. (145) on eucalyptus, and cormlet propagation systems proposed by Ziv (146) on *Gladiolus*.

3. Minor Crops

Acacia mangium Willd, is a pioneer legumenous tree that has gained increasing interest for use in (re)afforestation in many tropical countries, especially for pulp wood production in Southeast Asia. Micropropagation is likely to play an important role in the improvement *A. mangium*. If the shoot tips used as scions are small enough, in vitro micrografting may overcome problems associated with micropropagation of *A. mangium* by regenerating pathogen-free and rejuvenated tissues (147). Other recent reports of micropropagation of *A. mangium* include plant regeneration through organogenesis in callus cultures (148), auxin metabolism in relation to adventitious root formation (149), and the effects of micronutrient solutions and growth regulators on micropropagation potential (150).

Panaia et al. (151) developed a micropropagation protocol for the critically endangered western Australian shrub *Symonanthus bancroftii*. Using seedling explants, an improved regeneration protocol has been developed for *Bauhinia vahlii*, a multipurpose leguminous plant of the tropics (152). *Corydalis ambigua*, a perennial herb of the family Papaveraceae, has been micropropagated through somatic embryogenesis starting from sliced tubers (153).

Medicago truncatula Gaertn is an annual pasture legume (family Fabaceae) grown globally. Simple and effective micropropagation methods of regeneration can successfully modify the agronomic characteristics of this plant. Neves et al. (154) have established a reliable method for the maintenance of a permanent stock of *M. truncatula* genotypes selected from a general seed stock by their in vitro culture amenability and embryogenic capacity. Populations of *Vriesea friburgensis*, commonly found in Brazil, are facing increasing dangers of deforestation. Micropropagation tools may be applied to mass propagate and conserve these important species Micropropagation for mass propagation and conservation of *V. friburgensis* has recently been reported (155).

Sapium sebiferum Roxb, is an oil-yielding tree crop of the tropics and subtropics valued for its waxy fruit, yielding vegetable fallow and stillingia oil as by-products. The former is used for human consumption and in soap manufacture and the latter in the formulation of paints and varnishes. Whereas seed propagation does not preserve desirable genetic characteristics, vegetative propagation via root suckers is slow and time consuming and provides a limited number of propagates. In vitro cloning of selected germplasms is highly desirable for optimum utilization of this nontraditional energy crop. Siril and Dhar (156) have described micropropagation of an "elite" 15-year-old *S. sebiferum* through in vitro shoot proliferation from nodal explants.

A micropropagation system using regeneration via somatic embryogenesis from immature inflorescences of reed grass (*Phragmites australis* [Cav.] Trim) has been optimized (157). This system has been proposed for the production of reed grass to construct wetlands used in waste water purification. Borthakur and associates reported protocols for the micropropagation of several traditional medicinal plants: *Eclipta alba* L. and *Eupatorium adenophorum* L. (158), *Alpinia galanga* Willd. (159), and *Diascorea floribunda*, medicinal yam (160).

Tetraploid plants of watermelon (*Citrullus lanatus*) have been produced by treating diploid seedlings with an aqueous solution of colchicine, which results in a mixed population of diploids, tetraploids, aneuploids, and sectoral and periclinal chimeras (161). However, the production of tetraploids in vitro through adventive tissue culture offers a more efficient alternative method to obtain tetraploid plants. Compton et al. (162) demonstrated that in vitro tetraploid induction could be used to produce high-quality tetraploids for use in triploid hybrid seed production. Another study indicated that watermelon cultivars that respond poorly to in vitro procedures might have fewer cells competent for shoot regeneration, requiring special care during explant preparation (163). The highly prized seedless watermelons are produced from triploid plants that arise from seed obtained by crossing tetraploid and diploid plants. In vitro techniques have been used successfully in this venture (164–170). Compton et al. (171) recently reported a micropropagation protocol that allows for the efficient cloning of *Cucumis hystrix*.

VII. MICROPROPAGATION OF DISEASE-FREE PLANTS

The use of disease-free seed or planting material is a prerequisite for the production of higher yield and a good-quality crop. Both in vitro and in vivo mass clonal propagation of plants pose great risks for the spread of pathogenic microorganisms in planting material. Cassells (172) has proposed a series of options that exist to minimize or eliminate pathogens from planting material (Fig. 5). This figure indicates several alternative in vivo and in vitro, or hybrid, options that are today available, both for the elimination of pathogens and for the multiplication of elite stock, the choice depending upon the technical and economic feasibility for the specific crop and circumstances (172). These options when combined with efficient pathogen detection and identification procedures can ensure the high quality of planting material. However, the disease-free mother stock must be multiplied on a large scale to provide an adequate supply of propagules for crop planting. In vitro cloning of disease-free material avoids the risk of reinfection during multiplication in the environment,

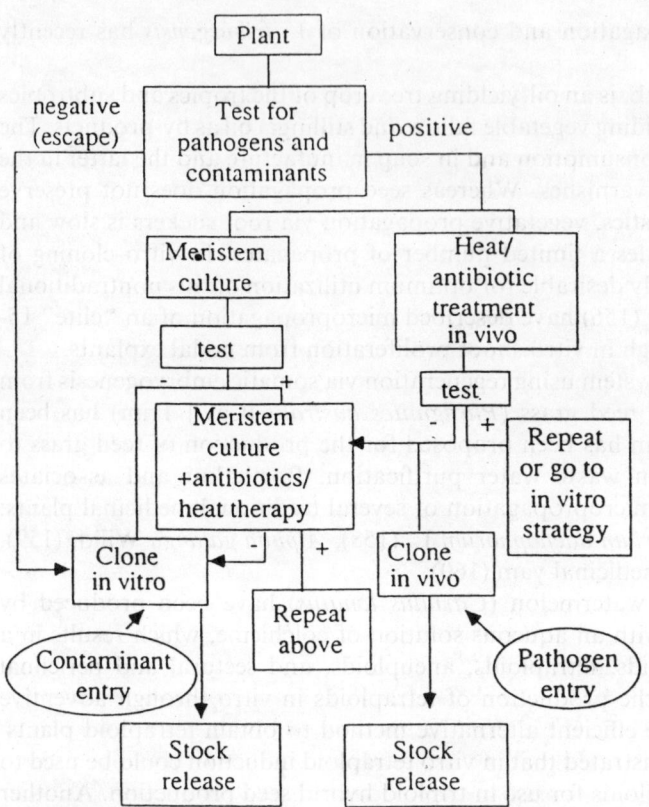

Figure 5 In vivo, in vitro, and hybrid strategies for the production of pathogen- and contaminant-free plants. (From Ref. 172.)

although in vitro cultures are also vulnerable to laboratory contamination, which in turn can be avoided by careful monitoring and maintaining clean cultures under slow-growth conditions (173). It may be advisable to use only officially certified or "classified" disease-free another stocks to initiate cultures, especially the major vegetatively propagated crops, such as potato and tuber crops, bananas, and other plantation crops, and in the absence of such cultures, the cultures may be sent to specialized laboratories for pathogen testing and elimination. However, owing to the diversity and limited scale of production of micropropagation crops, in-house pathogen and contamination detection, characterization, and elimination may be a carried out more economically. Cassells (172) recommended that such work might be subcontracted to specialist laboratories without taking the risks of the development of resistant microorganisms due to the use of bacteriostatic chemicals.

Significant developments in micropropagation techniques in recent times have helped to solve the duel problems of efficient pathogen elimination and of propagation of clean clonal stock; for example, meristem-tip culture could result in the elimination of pathogens from infected plants with pioneering developments in plant tissue culture (174,175). Romano and associates (176) have recently reported rapid clonal multiplication of *Lavandula* medicinal plants; the essential oils of this important species are used in perfumes, cosmetics, flavoring, and pharmaceutical industries and the plants are also used as ornamentals. They have also developed an in vitro propagation protocol for a carob tree (*Ceratonia siliqua* L.), belonging to the family Leguminosae, known for its industrial gum that is rich in galactomannas, an important food additive with a water-retention property (177).

VIII. CONCLUSIONS

A number of well-defined problems in physiological, epigenetic, and genetic quality are associated with the culture of plant cell tissue and organs in vitro; that is, the absence or loss of organ genetic potential (recalcitrance), hyperhydricity (vitrification), and somaclonal variation (176). The latter are known to be genotype and environment dependent. These phenomena affect the practical application of micropropagation and plant genetic manipulation. Cassells and Curry (178) have hypothesized that much of the variability expressed in microplants may be the consequence of, or related to, oxidative stress damage caused to the plant tissues during micropropagation due to media and environmental factors. Controlled stress in vitro may help to overcome weaning stress through cross tolerance, and monitoring of stress parameters may help to characterize the level of in vitro stress which is beneficial for high postvitrum performance and quality.

REFERENCES

1. A. Altman and B. Loberant, Micropropagation: Clonal plant propagation in vitro, *Agricultural Botechnology* (A. Altman, ed.), Marcel Dekker, New York, 1998, p. 19.
2. P. Debergh and R.H. Zimmerman, eds. *Micropropagation: Technology and Application*, Kluwer Academic, Dordrecht, the Netherlands, 1990.
3. R.I.M. Pierk. *In Vitro Culture of Higher Plants*, Martinus Nijhoff, Dordrecht, the Netherlands, 1987.
4. R.H. Zimmerman, R.J. Griesbach, P.A. Hammerschlag, and R.J. Lawson, eds., *Tissue Culture as a Plant Production System for Horticultural Crops*, Martinus Nijhoff, Dordrecht, the Netherlands, 1986.
5. T. Mrashighe and F. Skoog, Plant propagation trough tissue culture, *Annu. Rev. Plant Physiol. 25*:135 (1973).
6. H. Hartman, D. Kester, and F.T. Davies, Jr., eds., *Plant Propagation Principles and Practices*, 5th ed., Prentice-Hall, Englewood Cliffs, NJ, 1990.
7. A.C. Cassells, In vitro production of pathogen and contaminant-free plants, *Agricultural Biotechnology* (A. Altman, ed.) Marcel Dekker, New York, 1998, p. 43.
8. R.R. White, *The Cultivation of Animal and Plant Cells*, 2nd ed., Ronald Press, New York, 1963.
9. R.A. De Fossard, *Tissue Culture for Plant Propagators*, University of New England, Armidale, 1976.
10. J.F. Knauss, A tissue culture method for producing *Dieffenbachia picta*, cv. Perfection, free of fungi and bacteria, *Proc. Fla. State Hortic. Soc. 89*:293 (1976).
11. R.H. Zimmerman, Commercial application of tissue culture to horticultural crops in the United States, *J. Korean Soc. Hort. Sci. 37*:490 (1996).
12. F. O'Riordain, Compiler, Cost'87 Directory of European Plant Tissue Culture Laboratories, Commission of the European Communities, Division of Biotechnology, Brussels. 1993, pp. 1–172.

13. A. Sasson, ed., *Biotechnologies in Developing Countries: Present and Future*, Vol. 1, *Regional and National Survey*, UNESCO, Paris, pp. 1–764.

14. A.C. Cassells, ed, Bacterial and bacteria-like contaminants of plant tissue culture, *Acta Hortic. 225*:225 (1988).

15. S.S. Bhojwani and M.K. Razdan, *Plant Tissue Culture: Theory and Practice*, Elsevier, Amsterdam, 1983.

16. C. Shore, Total Quality Management (TQM), Quality Control and Design for Quality. Tel Aviv, 1992.

17. J.T. Rabbit and P.A. Bergh. *The ISO 9000 Book*, Tel Aviv, 1993.

18. K. Tanaka, K. Fusiwara, and T. Kozai. Effects of relative humidity in the culture vessel on the transpiration and net photosynthetic rates of potato plantlets in vitro in Japan, International Symposium on Transplant Production Systems: Biological, Engineering and Socioeconomic Aspects, Yokohama, 1992.

19. P. Kristensen and H. d'Allesandro, A method and a system for growing and cultivating plants. Patent Application No. 100159, Israel, 1991.

20. T. Murashige and R. Skoog, A revised medium for rapid growth and bioassay with tobacco tissue cultures, *Physiol. Plant. 15*:473 (1962).

21. B. Loberant, Industrial plant micropropagation, First International Conference on Biotechnology, Jerusalem, 1994.

22. T. Kozai, High technology in protected cultivation from environment control engineering point of view. Proceedings of an International Symposium on High Technology in Protected Cultivation. Collected Papers 1986, Tokyo, Japan, 1989.

23. V. Dhawan and A. Kapaki, Greenhouse in semi-arid region for handling tissue culture propagated plants. India, International Symposium on Transplant Production Systems: Biological, Engineering and Socioeconomic Aspects, Yokohama, 1992.

24. J. Aitken-Christie, Development of semi-automated micropropagation systems, *Acta Hortic. 230*:81 (1988).

25. J. Aitken-Christie, Large-scale production in plant tissue culture, an overview, Proceedings of the 8th International Congress, IAPTC, Florence, 1994.

26. K. Kurata, Transplant production robots in Japan, International Symposium on Transplant Production Systems Biological, Engineering and Socioeconomic Aspects, Yokohama, 1992.

27. D.P. Holdgate and E.A. Zandvoort, Automated micropropagation and the application of a laser beam for cutting. International Symposium on Transplant Production Systems: Biological, Engineering and Socioeconomic Aspects, Yokohama, 1992.

28. D. Redenbaugh, *Scale up and Automation in Plant Propagation*, Academic Press, Orlando, FL, 1991.

29. R.C. Harrell and C.V. Cantliffe, Automatic identification and separation of somatic embryos in vitro in USA, International Symposium on Transplant Production Systems: Biological, Engineering and Socioeconomic Aspects, Yokohama, 1992.

30. P.A. Cooper, J.T. Grant, L. Kerr, and Genghis, Development of a prototype automated cutting system for in vitro multiplication in New Zealand, International Symposium on Transplant Production Systems: Biological Engineering and Socioeconomic Aspects, Yokohama, 1992.

31. R. Levin, V. Gaba, B. Tai, S. Hirsch, D. Denola, and I.K. Vasil, Automated plant tissue culture for mass propagation, *Biotechnology 6*:1035 (1988).

32. J. Majada, R. Sanchez-Tameo, A. Fal, R. Ibarra, and F. Mateos, an automated system for vitro-ponic culture in Spain, International Symposium on Transplant Production Systems: Biological, Engineering and Socioeconomic Aspects, Yokohama, 1992.

33. R.E. Young, S.A. Hale, J.W. Adelberg, R.J. Keese, and N.D. Camper, Bioreactor development for continued-flow, liquid plant tissue culture in USA, International Symposium on Transplant Production Systems: Biological, Engineering and Socioeconomic Aspects, Yokohama, 1992.

34. R.E. Young, S.A. Hale, N.D. Camper, R.J. Keese, and J.W. Adelberg. An alternative, mechanized plant propagation approach, ASAE/CSAE Paper No. 896092, ASAE. St. Joseph, MO, 1989.

35. M. Ziv, Vitrification: morphological and physiological disorders of in vitro plants, *Micropropagation: Technology and Application* (P. Debergh and R. Zimmerman, eds.), Kluwer Academic, Dordrecht, the Netherlands, 1990, pp. 45–69.

36. M. Ziv, Morphologic control of plants micropropagated in bioreactor cultures and its possible impact on acclimatization, International Symposium on Transplant Production Systems: Biological, Engineering and Socioeconomic Aspects, Yokohama, 1992.

37. L.D. Gautz and C.K. Wong, Automation of the handling and manipulation of vessels during micropropagation in USA, International Symposium on Transplant Production Systems: Biological, Engineering and Socioeconomic Aspects, Yokohama, 1992.

38. Y.P.S. Bajaj and S.K. Sopory, *Biotechnology in Agriculture and Forestry*. Vol. 2, Crops I (Y.P.S. Bajaj, ed.), Springer-Verlag, Heidelberg, 1986.

39. E. Galun, *Weizmann Institute Cell Fusion Techniques May Lead to Virus-Free Potatoes*, The Weizmann Institute of Science, Rehovot, Israel, 1990.

40. A. Cassells, In vitro induction of virus-free potatoes by chemotherapy, *Biotechnology in Agriculture and Forestry*, Vol. 2, Crops I (Y.P.S. Bajaj, ed.), Springer-Verlag, Heidelberg, 1987, pp. 46–49.

41. L.G.H. Silva, In vitro storage of potato tuber explants and subsequent plant regeneration, *HortScience 20*:139 (1985).

42. T. Kzai, Y. Koyama, and I. Watanabe, Multiplication of potato plantlets in vitro with sugar free medium under high photosynthetic photon flux, *Acta Hortic. 230*:121 (1988).

43. H. Kristensen, Potato tissue culture, Proceedings of FAO/Norway Symposium on Plant Tissue Culture: Technology and Utilization, Norway, 1984.

44. K.C. Tao, W.T. Yin, H.Y. Cheng, and K.P. Kung, Meristem culture of potatoes and the production of virus-free seed potatoes, Proceedings of a Symposium on Plant Tissue Culture, Peking, 1979.

45. N. Van Uyen, From test tube to transplants in Vietnam, Proceedings of the 28th CIP Congress, Lima, 1984.

46. P-J Wang and C. Hu, In vitro mass tuberization and virus-free seed potato production in Taiwan, *Am. Potato J. 59*: (1982).

47. G. Mok, S.Y. Kim, and K.K. Kim, Use of rapid multiplication techniques by the Korean National Programs, Proceedings of the 29th CIP Congress, Lima, 1984.

48. D. Levy, Propagation of potato by direct transfer of in vitro proliferated shoot cuttings in the field, *Sci. Hortic. 26*:105 (1985).

49. P. Vander Zaag, V. Escobar, M.G. Torio, and I. Graza, Rapid multiplication of the potato (*Solanum* spp.): Research and use in the Philippines, Proceedings of the 28th CIP Congress Lima, 1984.

50. T. Kozai, K. Tanaka, K. Fujiwara, K. Watanabe, and S. Kira, Effect of relative humidity on the shoot elongation of potato plantlets in vitro in Japan, International Symposium on Transplant Production Systems: Biological, Engineering and Socioeconomic Aspects, Yokohama, 1992.

51. J.H. Dodds, Tissue culture propagation of potatoes: Advantages and disadvantages, Proceedings of the 28th CIP Congress, Lima, 1984.

52. A. Karp, M.G.K. Jones, D. Foulger, N. Fish, and S.W.J. Bright, Variability in potato tissue culture, Germplam Symposium Papers, *Am. Potato J. 66*:1 (1989).

53. V.M. Masteller and D.J. Holden, The growth of and organ formation from callus tissue of sorghum, *Plant Physiol. 45*:362 (1970).

54. E. Thomas, P.J. King and I. Potrykus, Shoot and embryo-like structure formation from cultured tissues of Sorghum bicolor, *Nuturwissen 64*:587 (1977).

55. O.L. Gamborg, J.P. Shyluk, D.S. Brar, and P. Constabel, Morphogenesis and plant

regeneration from callus of immature embryos of sorghum, *Plant Sci. Lett. 10*:64 (1977).

56. H. Ma, M. Gu, and G.H. Liang, Plant regeneration from cultured immature embryos of *Sorghum bicolor* L. Moench Theoret. *Appl. Genet. 73*:389 (1987).

57. R.M. Waskom, D.R. Miller, G.E. Hanning, R.R. Duncan, R.L. Voigt, and M.W. Nabors, Field evaluation of tissue culture derived sorghum for increased tolerance to acid soils and drought stress, *Van. J. Plant Sci. 70*:997 (1990).

58. R.I.S. Brettell, W. Wernicke, and E. Thomas, Embryogenesis from cultured immature inflorescences of *Sorghum bicolor, Protoplasma 104*:141 (1980).

59. M.C. Lusardi and F. Lupotto, Somatic embryogenesis and plant regeneration in *Sorghum* species, *Maydica 35*:59 (1990).

60. W. Wernicke and R.I.S. Brettell, Morphogenesis from cultured leaf tissue of *Sorghum bicolor*–culture initiation. *Protoplasma 111*:19 (1982).

61. T. Cai, B. Daily, and L. Butler, Callus induction and plant regeneration from shoot portions of mature embryos of high tannin sorghums, *Plant Cell Tiss. Org. Cult. 9*:245 (1987).

62. S. Bhaskaran, R.H. Smith, and K. Schertz, Control of morphogenesis in sorghum by 2,4-dichlorophenoxyacetic acid, *Ann. Bot. 64*:217 (1989).

63. S. Bhaskaran, R.H. Smith, and K.F. Schertz, Sodium chloride tolerant callus of *Sorghum bicolor* L. Moench Z, *Pflanzenphysiol. Bot. 112*:459 (1983).

64. S. Bhaskaran, S. Paliwal, and K.F. Schertz, Somaclonal variation from *Sorghum bicolor* L. Moench cell culture, *Plant Cell Tiss. Org. Cult 9*:189 (1987).

65. R.H. Smith, S. Bhaskaran, and K.F. Schertz, Sorghum plant regeneration from aluminum selection media, *Plant Cell Rep. 2*:129 (1983).

66. T. Hagio, Varietal difference of plant regeneration from callus of sorghum mature seed, *Breeding Sci. 44*:121 (1994).

67. T. Hagio, Adventitious shoot regeneration from immature embryos of sorghum, *Plant Cell Tiss. Org. Cult. 68*:65 (2002).

68. P.S. Rao, P. Suprasanna, and T.R. Ganapathi, Plant biotechnology and Agriculture: prospects for improving and increasing productivity, *Sci. Cult. 62*:185 (1996).

69. M.D.J. Stewart and C.L. Hsu, In-ovulo embryo culture and seeding development of cotton (*Gossypium hirsutum* L.), *Planta 137*:1 (1997).

70. Y.P.S. Bajaj and M.S. Gill. Micropropagation and germplasm preservation of cotton (*Gossypium* spp.) through shoot tip and meristem culture, *Indian J. Exp. Biol. 24*:561 (1986).

71. M.C. Petrs, K. Williams, and R. Swennen, Protoplast to plant regeneration in cotton (*Gossypium hirsutum* L.), cv. Coker 312, using feeder layers, *Plant Cell Rep. 13*:208 (1994).

72. J.J. Finer, Plant regeneration from somatic embryogenic suspension cultures of cotton (*Gossypium hirsutum* I.), *Plant Cell Rep 7*:399 (1988).

73. J.J. Finer and M.D. Mullen, Transformation of cotton (*Gossypium hirsutum* L.), via particle bombardment, *Plant Cell Rep. 8*:586 (1990).

74. F.J. Perlak, R.W. Deaton, T.A. Armstrong, R.L. Fuchs, S.R. Sims. J.T. Greenplate, and D.A. Fischhoff, Insect resistant cotton plants, *Bio/Technology 8*:39 (1990).

75. D.C. Agrawal, A.V. Kulkarni, S.M. Nalawade, S.M. Hazra, and K.V. Krishnamurthy. In vitro induction of multiple shoot and plant regeneration in cotton (*Gossypium hirsutum* L.), *Plant Cell Rep. 16*:647 (1997).

76. S.K. Gupta, A.K. Srivastava, P.K. Singh, and R. Tuli. In vitro proliferation of shoots and regeneration of cotton, *Plant Cell Tiss. Org. Cult. 51*:149 (1997).

77. S. Kumar and D. Pental, Regeneration of Indian cotton variety MCU-5 through somatic embryogenesis, *Curr. Sci. 74*: (1998).

78. M. Mhatre, P. Suprassana, R. Jaiswal, V.V. Kumar, and P.S. Rao, Production of high frequency multiple shoots and plantlets in six Indian cultivars of cotton, (*Gossypium hirsutum* L.), Asia Pacific, *J. Mol. Biol. Biotechnol. 7*:73 (1999).

79. L.P. Yuan, Purification and production of foundation seed of PGMS and TGMS lines, *Hybrid Rice* 6:2 (1994).

80. G-Z. Zheng, Y-S Yang, X-H. Chen, and B-H. Wan. Evaluation of a new culture method for mass-propagation of a photoperiod-temperature sensitive genic male sterile rice strain N19S, *Plant Cell Tiss. Org. Cult.* 68:195 (2002).

80a. U.S. Bshnoi, R.K. Jain, K.R. Gupta, V.K. Chowdhury, and J.B. Chowdhury, High frequency androgenesis in indica × Basmati rice hybrids using liquid culture media, *Plant Cell Tiss. Org. Cult.* 61:153 (2000).

81. S. Ma and C. Shii, In vitro formation of adventitious buds in banana shoot apex following decapitation. *J.Chin. Soc. Hortic. Sci.* 18:135 (1972).

82. B. Loberant and A. Duvdevani, Somaclonal variation of banana in Israel, First Workshop on Banana Biotechnology, Tenerife, 1993.

83. Y. Israeli, O. Reuveni and E. Lahav, Quantitative aspects of somaclonal variations in banana propagated by in vitro techniques, *Sci. Hortic.* 49:71 (1991).

84. O. Reuveni, Methods for detecting somaclonal variants in "Williams" bananas, *The identification of Genetic Diversity in the Genus Musa* (R.L. Jarret, ed.), INIBAP, Montpelier, 1990, pp. 108–113.

85. N. Roux, J. Dolezel, R. Swennen, and F.J. Zapata-Aras, Effectiveness of three micropropagation techniques to dissociate cytochimeras in *Musa* spp., *Plant Cell Tiss. Org. Cult.* 66:189 (2001).

86. D.J. Gray and L.C. Fischer, In vitro shoot propagation of grape species, hybrids and cultivars, *Proc. Fla. State Hortic. Soc.* 98:172 (1985).

87. M.E. Compton and D.J. Gray, Micropropagation of 'Southern Home' hybrid grape, *Proc. Fla. State Hortic. Soc.* 107:308 (1994).

88. K.L. Thies and C.H. Graves, Jr., Meristem micropropagation protocol for *Vitis rotundifolia* Michx. *HortScience* 27:447 (1992).

89. L. Torregrossa and A. Bouquet, In vitro propagation of *Vitis* × *Muscadinia* hybrids by microcuttings or axillary budding, *Vitis* 34:237 (1995).

90. C.K. Salunkhe, P.S. Rao, and M. Mhatre, Induction of somatic embryogenesis and plantlets in tendrils of Vitis vinifera L., *Plant Cell Rep.* 17:65 (1997).

91. M. Mhatre, C.K. Salunkhe, and P.S. Rao, Micropropagation of *Vitis vinifera* L.: towards an improved protocol, *Sci. Hortic.* 84:357 (2000).

92. H.T. Hartmann and D.E. Kester, *Plant Propagation: Principle and Practices*, Prentice-Hall, Englewood Cliffs, NJ, 1975.

93. L. Burgos and C.A. Ledbetter, Improved efficiency in apricot breeding: Effects of embryo development and nutrient media on in vitro germination and seedling establishment, *Plant Cell Tiss, Org. Cult.* 35:217 (1993).

94. W.D. Lane and F. Cossio, Adventitious shoots from cotyledons of immature cherry and apricot embryos, *Can. J.Plant Sci.* 66:953 (1986).

95. R.E. Pieterse, Regeneration of plants from callus and embryos of 'Royal' apricot, *Plant Cell Tiss. Org. Cult.* 19:175 (1989).

96. O. Perez-tornero, J.M. Lopez, J. Egea, and L. Burgos, Effect of basal media and growth regulators on the in vitro propagation of apricot (*Prunus armeniaca* L.), cv. Canino, *J.Hort. Sci Biotechnol.* 75:283 (2000).

97. O. Perez-Tornero and L. Burgos, Different media requirements for micropropagation of apricot cultivars, *Plant Cell Tiss. Org. Cult.* 63:133 (2000).

98. O. Perez-Tornero, J. Egea, E. Olmos, and L. Burgos, Control of hyperhydricity in micropropagated apricot cultivars, *In Vitro Cell Dev. Biol. Plant* 37:250 (2001).

99. O. Perez-tornero, L. Burgos, and J. Egea, Introduction and establishment of apricot in vitro through regeneration of shoots from meristem tips, *In Vitro Cell Dev. Biol. Plant* 35:249 (1999).

100. O. Perez-tornero, F. Ortin-Parraga, J. Egea, and L. Burgos, Medium-term storage of apricot shoot tips in vitro by minimal growth period, *HortScience* 34:1277 (1999).

101. O. Perez-Tornero and J. Egea, A. Vanoostende and L. Burgos, Assessment of factors affecting adventitious shoot regeneration from in vitro cultures leaves of apricot, *Plant Sci. 138*:61 (2000).
102. J.R.S. Cabral, A.P. Matos, and G.A.P. Cumba, Selection of pineapple cultivars resistant to fusariose, *Acta Hortic. 334*:53 (1993).
103. F. Leal and G. Coppens d'Eckenbrugge, Pineapple, *Fruit Breeding—Tree and Tropical Fruits* (J. Janick and J.N. Moore, eds.), Wiley, New York, 1996, pp. 515–557.
104. D. Aghion and G. Beauchesne, Utilisation de la technique de culture sterile d'organes pour obtenir des clones d'ananas, *Fruits 15*:464 (1960).
105. T.S. Rangan, Pineapple, *Handbook of Plant Cell Culture* (P.V. Ammirato, D.A. Evans, W.R. Sharp and Y. Yamada, eds.), Macmillan, New York, 1984, pp. 373–382.
106. K. Wasaka, Pineapple (*Ananas comosus* [L.] Meri.). *Biotechnology in Agriculture and Forestry-Trees II* (Y.P.S. Bajaj, ed.), Springer-Verlag, Berlin, 1980, pp. 13–29.
107. S. Mayak, T. Tirosh, A. Ilan, A. Duvdevani, and E. Khayat, Growth and development of pineapple (*Ananas comosus* L.) plantlets cultured in vitro at enriched and ambient CO_2 environments, *Acta Hortic. 461*:225 (1998).
108. L.L. Dal Vesco, A. de A. Pinto, G.R. Zaffari, R.O. Nodari, M.S. dosReis, and M.P. Guerra, Improving pineapple micropropagation protocol through explants size and medium composition manipulation, *Fruits 56*:143 (2001).
109. J.R. Soneji, P.S. Rao, and M. Mhatre, Somaclonal variation in micropropagated dormant axillary buds of pineapple (*Ananas comosus* L., Merr.), *J. Hortic. Sci. Biotechnol. 77*(1):28 (2002).
110. J. Dijkstra, Research on strawberries focuses on healthy plant material. Expensive cultural method requires excellent material, *Fruiteelt Den Haag 83*:14 (1993).
111. I.D. Bhatt and U. Dhar, Micropropagation of Indian wild strawberry, *Plat Cell, Tiss. Org. Cult. 60*:83 (2000).
112. A. Jemmali, P. Boxus, C. Kevers, and T. Gaspar, carry-over of morphological and biochemical characteristics associated with hyper flowering of micropropagated strawberries. *J. Plant Physical 147*:435 (1995).
113. K. Pruski, J. Nowak, and G. Grainger, Micropropagation of four cultivars of Saskatoon berry (*Amelanchier alnifolia* Nutt.), *Plant Cell Tiss Org Cult. 21*:103 (1990).
114. K. Pruski, M. Mohyuddin, and G. Grainger, Saskatoon (*Amelanchier alnifolia* Nutt.). *Biotechnology in Agriculture and Forestry*, Vol. 6, *Tress III.* (Y.P.S. Bajaj, ed.), Springer Verlag. Heidelberg, 1991, pp. 164–179.
115. K.W. Pruski, T. Lewis, T. Astatkie, and J. Nowak, Micropropagation of chokecherry and pincherry cultivars, *Plant Cell Tiss. Org. Cult. 63*:93 (2000).
116. B. Borkowska, Micropropagation of sour cherry cultivar-Schattenmorelle, *Fruit Sci. Rep. 10*:59 (1983).
117. P.G. Alderson, M.A. Harbour, and P.A. Patience, Micropropagation of *Prunus tenella*, cv. Firehill, *Acta Hortic. 212*:463 (1987).
118. F. Loreti, S. Morini, and P.L. Pasqualetio. Effect of alternating temperature during proliferation and rooting stages of GP 65512 and GF 677 shoots cultured in vitro, *Acta Hortic. 227*:467 (1988).
119. N. Hammatt and N.J. Grant, micropropagation of mature British wild cherry, *Plant Cell Tiss, Org. Cult. 47*:103 (1996).
120. Y. Murai and H. Harada, In vitro propagation of apricot (*Prunus armeniaca* L.), cv. Bakuoh-Junkyou, *J. Jpn. Soc. Hortic. Sci. 66*:475 (1997).
121. A. Dalzotto and D.M. Docampo, Micropropagation of rootstock from the Marianna-2624 plum (*Prunus cerasifera* × *Prunus munsoniana*), and the pixy plum (*P. insititia* L.) under controlled conditions, *Phyton Int. J. Exp. Bot. 60*:127 (1997).
122. G. Marino, The influence of ethylene on in vitro rooting of GF-677 (*Prunus persica* × *Prunus amygdalus*) hybrid peach rootstocks, *In Vitro Cell Dev. Biol. Plant 33*:26 (1997).

123. S. Radice, P.E. Perelman, and D.H. Caso, Clonal propagation of three rootstocks of the genus Prunus for the 'flooding Pampa', *Phyton Int. J. Exp. Bot. 64*:149 (1999).

124. B. Loberant and Y. Alon, The interaction between plant protection authorities, academia and industry, Proceedings of the 9th International Symposium on Virus Diseases of Ornamental plants. *Acta Hortic. 432*:218 (1996).

125. W.P. Hackett and J.M. Anderson, Aseptic multiplication and maintenance of different carnation shoot tissue-derived from soot spices, *Proc. Am. Soc. Hortic. Sci. 90*:365 (1972).

126. M. Hollings and O.M. Stone, Productivity of virus-tested carnation clones and the rate of reinfection with virus, *J. Hortic Sci. 47*:141 (1972).

127. W.R. Jarvis, *Managing Diseases in Greenhouse Crops*, APS Press, St. Paul, MN, 1992, pp. 112–120.

128. A. Fonnesbech, In vitro propagation of *Spathiphyllum, Sci. Hortic. 10*:21 (1979).

129. J.M. Bonga, Clonal propagation of matured trees: Problems and possible solutions, *Cell and Tissue Culture in Forestry*, Vol. I (J.M. Bonga and D.J. Durzan, eds.), Martinus Nijhoff, Dordrecht, the Netherlands, 1987, pp. 249–271.

130. B.E. Haissig, N.D. Nelson, and G.H. Kidd, Trends in the use of tissue culture in forest improvement, *Biotechnology 5*:52 (1987).

131. I.S. Harry and L.A. Thorp, In vitro culture of forest trees, *Plant Cell and Tissue Culture* (L.K. Vasil and T.A. Thorpe, eds.). Kluwer Academic, Dordrecht, the Netherlands, 1994, pp. 539–560.

132. D.I. Dunstan, Prospects and progress in conifer biotechnology, *Can. J. Forest Res. 18*: 1497 (1988).

133. M.S. Greenwood, G.S. Foster, and H.V. Amerson, Vegetative propagation in southern pines, *Forest Regeneration Manual* (M.L. Dryea and P.M. Dougherty, eds.), Kluwer Academic, Dordrecht, the Netherlands, 1991, pp. 75–86.

134. S. Airi, R.S. Rawal, S, S. Samant, and U. Dhar, Treatments to improve germination of four multipurpose trees of central sub-Himalaya, *Seed Sci Technol. 26*:347 (1998).

135. U. Dhar, J. Upreti, and I.D. Bhatt, Micropropagation of *Pittosporum napaulennsis* (DC) Rehder & Wilson—a rare, endemic Himalayan medicinal tree, *Plant Cell Tiss. Org. Cult. 63*:231 (2000).

136. C.S. Sun and D.Y. Wang. *Fritillaria* spp. (Fritillary): In vitro culture and the regeneration of plants, *Biotechnology in Agriculture and Forestry*, Vol. 15, Medicinal and Aromatic Plants III (Y.P.S. Bajaj, ed.), Springer Verlag, Heidelberg, 1991, pp. 258–269.

137. J.H. Seon, K.Y. Paek, W.Y. Gao, C.H. Park, and N.S. Sung, Factors affecting micropropagation of pathogen free stocks in *Fritillaria thunbergii, Acta Hortic. 502*:333 (1999).

138. K.Y. Paek and H.N. Murthy, High frequency of bulblet regeneration from bulb scale sections of *Fritillaria thunbergii, Plant Cell Tiss. Org. Cult. 68*:247 (2000).

139. N. Hiraoka and V. Tomita, Botanical and chemical evaluation of Atractylodes lancea plants propagated in vitro and by division of the rhizome, *Plant Cell Rep. 9*:332 (19990).

140. J.R. Soneji, P.S. Rao, and M. Mhatre. Enhanced regeneration of multiple shoots and plantlets in young cotyledons of neem, *Azadirachta indica* A Juss. *Physiol. Mol. Biol. Plants 7*:175 (2001).

141. U. Dhar, I.D. Bhatt, and S. Airi, *In Vitro Regeneration Protocols of Himalayan MPTs—Its Only the Beginning*, Gyanodaya Prakashan, Nainital, India, 2002, pp. 53–71.

142. O. Arezki, P. Boxus, C. Kevers, and T. Gaspar, Hormonal control of proliferation in meristematic agglomerates of *Eucalyptus camaldulensis* Dehn. *In Vitro Cell Dev. Biol. Plant 36*:398 (2000).

143. J. Aitken-Christie, A.P. Singh, and H. Davies, Multiplication of meristematic tissue: a new tissue culture system for Radiada pine, *Genetic Manipulation of Woody Plants* (J.W. Hanover and D.E. Keathley, eds.), Plenum Press, New York, 1998, pp. 413–432.

144. B.H. McCown, E.L. Zeldin, H.A. Pinkalla, and R.R. Dedolph, Nodule culture: a devel-

opmental pathway with high potential for regeneration, automated micropropagation, *Genetic Manipulation of Woody Plants*, (J.W. Hanover and D.B. Keathley, eds.), Plenum Press, New York, 1998, pp. 149–166.

145. K. Ito and K. Doi, Y. Tatemichi and M. Shibata, Plant regeneration of eucalyptus from rotating nodule cultures, *Plant Cell Rep. 16*:42 (1996).

146. M. Ziv, Enhanced shoot and carmel proliferation in liquid cultured Gladiolus buds by growth regulators, *Plant Cell Tiss. Org. Cult. 17*:110 (1989).

147. O. Monteuuis, In vitro shoot apex micrografting of mature Acacia mangium, *Agroforestry Systems 34*:213 (1996).

148. D. Xie and Y. Hong, In vitro regeneration of *Acacia mangium* via organogenesis, *Plant Cell Tiss. Org. Cult. 66*:167 (2001).

149. O. Monteuuis and M-C Bon, Influence of auxins and darkness on in vitro rooting of micropropagated shoots from mature and juvenile Acacia mangium, *Plant Cell Tiss. Org. Cult. 63*:173 (2000).

150. M-C Bon, D. Bonal, D.K. Goh, and O. Monteuuis, Influence of different macronutrient solutions and growth regulators on micropropagation of juvenile *Acacia mangium* and *Paraserianthes falcataria* explants, *Plant Cell Tiss. Org. Cult. 53*:171 (1998).

151. M. Pannaia, T. Senaratna, E. Bunn, K.W. Dixon, and K. Sivathamparam, Micropropagation of species *Symonanthus bancroftii* (F. Muell) L. Haegi (Solanaceae), *Plant Cell Tiss. Org. Cult. 63*:23 (2000).

152. I.D. Bhatt and U. Dhar, combined effect of cytokinins on multiple shoot production from cotyledonary node explants of *Bauhinia vahlii*, *Plant Cell Tiss. Org. Cult. 62*:79 (2000).

153. N. Hirooka, Y. Kato, Y. Kawaguchi, and J.I. Chang, Micropropagation of Corydalis ambigua through embryogenesis of tuber sections and chemical evaluation of the ramets, *Plant Cell Tiss. Org. Cult. 67*:243 (2001).

154. L.O. Neves, L. Tomaz, and M.P.S. Fevereiro, Micropropagation of *Medicago truncatula* Gaertn cv. Jamalong and *M. truncatula* spp. Narbonensis, *Plant Cell Tiss. Org. Cult. 67*:81 (2001).

155. G.M. Alves and M.P. Guerra, Micropropagation for mass propagation and conservation of *Vriesea friburgensis*, var. Paludosa from micro buds, *J.Bromeliad Soc. 51*:202 (2001).

156. E.A. Siril and U. Dhar, Micropropagation of mature Chinese tallow tree (*Sapium sebiferum* Roxb.). *Plant Cell Rep. 16*:637 (1997).

157. D. Lauzer, S. Dallaire, and G. Vincent, In vitro propagation of reed grass by somatic embryogenesis. *Plant Cell Tiss. Org. Cult. 60*:229 (2000).

158. M. Borthakur, K. Datta, S.C. Nath, and R.S. Singh, Micropropagation of *Eclipta alba* and *Eupatorium adenophorum* using a single step nodal cutting technique, *Plant Cell Tiss. Org. Cult. 62*:239 (2000).

159. M. Borthakur, J. Hazarika, and R.S. Singh, A protocol for micropropagation of *Alpinia galanga*, *Plant Cell Org. Cult. 55*:231 (1999).

160. M. Borthakur and R.S. Singh, Dierct plantlet regeneration from male inflorescence of medicinal yam (*Dioscorea floribunda* Mart & Gal), *In Vitro Cell Dev. Biol. Plant 38*:183 (2002).

161. F. McCuistion and G.W. Elmstrom, Identifying polyploids of various cucurbits, *Proc Fla State Hortic. Soc 106*:155 (1993).

162. M.C. Compton, D.J. Gray, and G.W. Elmstrom, Regeneration of tetraploid plants from cotyledons of diploid watermelon, *Proc. Fla. State Hortic. Soc. 107*:107 (1994).

163. M.C. Compton, Interaction between explant size and cultivar affects shoot organogenic competence of watermelon cotyledons, *HortScience 35*:749 (2000).

164. M.C. Compton, D.J. Gray, and G.W. Elmstrom, Identification of tetraploid regenerants from cotyledons of diploid watermelon cultured in vitro, *Euphytica 87*:165 (1996).

165. M.C. Compton, N. Barnett, and D.J. Gray, Use of fluorescein diacetate (FDA) to determine ploidy of in vitro watermelon shoots, *Plant Cell Tiss. Org. Cult. 58*:199 (1999).

166. M.C. Compton, Dark pretreatment improves adventitious shoot organogenesis from cotyledons of diploid watermelon, *Plant Cell Tiss. Org. Cult. 58*:185 (1999).

167. M.C. Compton and D.J. Gray, Adventitious shoot organogenesis and plant regeneration from cotyledons of tetraploid watermelon, *HortScience 29*:211 (1994).

168. M.C. Compton and D.J. Gray, Shoot organogenesis and plant regeneration from cotyledons of diploid, triploid and tetraploid watermelon, *J.Am.Soc.Hortic. Sci. 118*:151 (1993).

169. M.C. Compton, D.J. Gray, and G.W. Almstrom, A simple protocol for micropropagating diploid and retraploid watermelon using shoot-tip explants, *Plant Cell Tiss. Org. Cult. 33*:211 (1993).

170. M.C. Compton and D.J. Gray, Micropropagation as a means of rapidly propagating triploid and tetraploid watermelon, *Proc. Fla. State Hortic. Soc. 105*:352 (1992).

171. M.C. Compton, B.L. Pierson, and J.E. Staub, Micropropagation for recovery of *Cucumis hystrix*, *Plant Cell Tiss. Org. Cult. 64*:63 (2001).

172. A.C. Cassells, In vitro production of pathogen and contaminant-free plants, *Agricultural Biotechnology* (A. Altman, ed.), Marcel Dekker, New York, 1998, pp. 43–56.

173. G. Staritsky, A.J. Dekkers. N.P. Louwaars, and E.A. Zandvoort, In vitro conservation of aroid germplasm at reduced temperature and under osmotic stress, *Plant Tissue Culture and Its Agricultural Applications* (I. A. Withers and P.G. Anderson, eds.), Butterworth, London, 1986, pp. 277–283.

174. T. Murashighe, Plant propagation through tissue Culture, *Annu. Rev. Plant Physiol. 25*:135 (1974).

175. E.F. George, *Plant Propagation by Tissue Culture*, Exegetics, Basingstoke, 1993, pp. 1–574.

176. M.C. Dias, R. Almeida, and A. Romano, Rapid clonal multiplication of *Lavandula viridis* L'Her through in vitro axillary shoot proliferation. *Plant Cell Tiss. Org. Cult. 68*:99 (2002).

177. A. Romano, S. Barros, and M.A. Martis-Loucao, Micropropagation of the Mediterranean tree *Ceratonia siliqua*, *Plant Cell Tiss. Org.Cult. 68*:35 (2002).

178. A.C. Cassells and R.F. Curry, Oxidative stress and physiological, epigenic and genetic variability in plant tissue culture: implications for micropropagators and genetic engineers. *Plant Cell Tiss. Org. Cult. 64*:145 (2001).

23
Synthetic Seed Biotechnology

I. INTRODUCTION

A. Cellular Totipotency

The cell and molecular biology of the embryogenic development of somatic cell of angiosperms and gymnosperms, known as somatic embryogenesis, has been dealt with extensively in the literature (1–5). In contrast to the rigorously programmed development of the embryo from the zygote (zygotic embryogenesis), is somatic embryogenesis, virtually any somatic cell of the plant body can, under certain experimental conditions, behave like a zygote and faithfully replay a developmental program, producing embryolike structures while remaining innocent of sex. Thus, all plant cells except those that have undergone irreversible differentiation are totipotent and retain the developmental potential to proliferate into an adult plant. Compared to the limited number of embryos arising from gametic fusion, and the difficulty of extracting them from the confines of the ovule, the enormous number of somatic cells potentially capable of developing into embryolike structures by simple experimental manipulation provides an attractive alternative to seed propagation. Although zygotic embryos are identical in appearance and morphogenetic potential to somatic embryos, to emphasize the divergent pathways of their evolution, the latter are generally referred by the term *embryoid* (1).

Plant tissue culture, encompassing protoplast, cell, tissue, and organ culture (anther, embryo), carried out under controlled and aseptic conditions has revealed cells' unique capacity of total genetic potential (totipotency), the inherent potential of the plant cell to give rise to a whole plant. Even a differentiated cell can express totipotency; after it has undergone dedifferentiation, followed by redifferentiation, dedifferentiation often involves cell embryogenesis, leading to callus formation. In vitro culture methods are efficient, safe, and often economic in propagating large quantities of healthy, pathogen-free planting material, bypassing time-consuming requirements of quarantine, and in storing valuable plant germplasms under low-growth conditions. It has now become possible to grow single cells and fuse isolated protoplasts of the unrelated species of crop plants to develop new transgenic plants (see Chapter 24) with higher yields and better qualities. The plant and tissue culture technique has also made it possible to produce artificial/synthetic seeds (synseeds) both through somatic embryogenesis (6–10) and the use of micropropagules other than somatic embryos; for example, shoot buds, axillary buds, embryogenic masses, and protocormlike bodies (11,12).

B. Somatic Seeds Through Embryogenesis

Zygotic embryogenesis refers to division and development of a fertilized egg cell (zygote) into an embryo. The egg may, however, be stimulated to undergo embryogenesis without fertilization. The pollination stimulation alone induces the egg to undergo embryogenic development (parthenogenesis), which can also be effected by the application of growth regulators. Any cell of the same gamatophyte (embryo sac) or of the sporophytic tissue surrounding the embryo sac may give rise to an embryo (6). In many*Citrus* and *Mangifera* species, adventive embryos (embryos from nucellar cells, which mature only if they are pushed into the embryogenic sac at an early stage of development). In nature, there is no instance of ex ovulo embryo development, suggesting that in vivo growth and development of embryos require a special physical and chemical environment to be present in the embryo sac (13).

Investigation of the embryogenic potential of somatic cells has shown that virtually all plant organs can form embryos. It has been established that any diploid cell in which irreversible differentiation has not proceeded too far, if placed in an appropriate medium, can develop in an embryo like way to produce a complete plant. Complex sexual apparatus is thus not an essential prerequisite to remove the effects of aging and to reestablish embryogenic properties (14).

Extensive reviews have been published on in vitro somatic embryogenesis (1,2,5,7,15–22). Somatic embryogenesis was first observed in 1958 (23,24), but it was not until 20 years later that the concept of synthetic seed, that is, a single encapsulated somatic embryo, appeared in print (25). However, the concept of synthetic seed through somatic embryogenesis was mentioned only briefly, and no results appeared in the literature prior to Redenbaugh et al.'s reports (9,26). Ever since in vitro somatic embryogenesis was first observed in *Daucus carota* (23,27), this technique has been widely extended to investigate various aspects of in vitro somatic embryogenesis (28–30) and has been studied in numerous plant species (31–38).

Embryos can be classified as follows on the basis of their formation and evolution (19):

1. *Zygotic embryos*—formed by fertilized egg or the zygote
2. *Nonzygotic embryos*—formed by cells other than the zygote
3. *Somatic embryos*—formed by the sporophytic cells (except the zygote) either in vitro or vivo
4. *Adventive embryos*—somatic embryos arising directly from other embryos or organs (e.g., stem embryos in carrot and buttercup)
5. *Pathenogenic embryos*—formed by the unfertilized egg
6. *Androgenetic embryos*—formed by the male gametophyte (microspores, pollen grains).

The prospect of using plant somatic embryos produced in tissue culture as synthetic/artificial seeds has been a subject of increasing commercial interest (39–41). Somatic embryogenesis has been obtained in more than 150 species of important agricultural crops (20), and is a routine procedure in crops like soybean (42,43), grasses, and cereals (44,45) as well as in conifers, such as European larch (46), Norway spruce (47), and sugar pine (48). The objective of the development of synthetic seeds is to produce a propagule that is genetically, developmentally, and morphologically as close as possible to the seed of the plant species from which it has been derived.

II. ARTIFICIAL SEED BIOTECHNOLOGY

Synthetic/artificial seed biotechnology is thought to have a significant impact on future crop production. Styer (49) envisaged that the major advantage of synthetic seed would be in its potential high production efficiency and clonal nature of the resulting plants, both for vegetatively and seed-propagated crops. For vegetatively propagated crops, like potato and some fruit and nut trees, synthetic seeds would allow direct planting of nongrafted varieties and provide an important alternative to the perpetual maintenance of living specimens for germplasm conservation. The use of elite germplasm would facilitate the production of seed-propagated crops like forest trees (50), date palm, forage grasses, vegetables, oil crops like coconut and oil palm, and a majority of agronomic crops (51). Synthetic seed would also be essential for genetically modified (GM) crop plants that do not breed true owing to the incorporation of meiotically unstable foreign genes. The intentionally introduced meiotic instability then would become an alternative to hybrid seed for commercializing proprietary germplasm. The use of artificial seed would facilitate maintenance of parental inbred lines as an adjunct to conventional breeding programs (52). Outstanding hybrids of high-value crops could be propagated directly without requiring time-consuming procedures to develop parental inbreds.

Gray (51) stressed the need for more research to engineer fragile, rapidly growing somatic embryos to mimic durable, quiescent seeds, which can be handled easily. Methods of tissue culture need to be improved to favor the in vitro development of vigorous mature somatic embryos that would function as well as those of zygotic embryos. Automated, large-scale liquid systems or biorectors may be needed to develop synthetic seeds with a controlled moisture content, nutrients, and appropriate protective encapsulations. Although feasibility has been demonstrated for each step in the overall process of artificial seed production, these steps must be incorporated into an automated apparatus for processing raw material, such as embryogenic cells, nutrients, and seed coat materials into synthetic seeds that could be handled, transported, and stored like orthodox seeds.

A. Induction and Regeneration of Somatic Embryos

Conditions that induce embryogeny in both diploid and haploid somatic tissue have been identified for a number of plant species. The first step in the culture procedure involves the induction of components cells to form proembryogenic structures that either directly or indirectly form somatic embryos. Schenk and Hildebrandt (53) described the technique and basal medium for the induction and growth of embryos of mono- and dicotyledonous plants. The induction medium contained 1 mg of 2,4-dichloropropionic acid and 0.2 mg of kinetin per liter. The responsive genotypes produce somatic embryos from subepidermal cells, and the callus is initiated from rapid cell division in the vascular cambium (54). The auxins differ significantly in their ability to induce embryogenesis, whereas indoleacetic acid (IAA) was ineffective and 4-chlorophenox acetic acid (CPA) and naphthalene acetic acid (NAA) promoted callus formation, 2,4-D and 2,4,5-trichlorophenoxy acetic acid were found to promote both callus and initiation of somatic embryos in alfalfa (55–57). Dudits et al. (58) investigated the time-dose response of auxins and reported that high doses for short periods were effective. Alternatively, the callus can be cultured on a noniductive medium containing NAA and a very high dose of 2,4-D is introduced for 1 hr, followed by thorough washing, to induce embryos in the culture. Shetty and McKersie (59) noticed that including potassium and proline in the medium could modulate the auxin response of alfalfa cells. Stimulation of embryogenesis by auxin has not been resolved fully. It is hypothesized that auxin stimulates the proton pump on the plasmalemma directly (60), and potassium enhances the auxin induction of embryogenesis (59). Changes in solute transport and/or signaling across the plasmalemma are thought to be early events leading to embryogenesis (61). Since proline and purine metabolism are linked together via the NADPH pool (62), it is postulated that stimulation of proline metabolism would consume NADPH and in turn stimulate the pentose phosphate pathway to synthesize a five-carbon compound, ribulose, which is a prerequisite for purine and nucleic acid (DNA) biosynthesis. Since DNA replication precedes cell division, the stimulatory effect of proline on embryogenesis may be through its effect on the cell cycle (61).

Monteiro et al. (63) recently initiated somatic embryogenesis of *Panax ginseng* CA Meyer from suspension aggregates of an embryogenic callus in a liquid medium consisting of half strength of Murashigi and Skong (MS) medium, supplemented with synthetic auxin, benzoselenienyl-3-acetic acid (BSAA). The addition of spermidine to this initiation medium significantly increased the production of somatic embryos. The spermidine supplied to the medium was found to be oxidized by polyamine oxidase and partially metabolized into putrescine. The authors have discussed the role of spermidine and its interaction with auxin in the initiation of the embryogenenic process in *Panax ginseng*.

Somatic embryogenesis was observed in callus initiated from tendril explants of *Vitis vinifera* L. (Fig. 1). cvs. Thompson seedless, Sonaka, and Tas-e-Ganesh on Emershad and Ramming medium supplemented with 1 μM 6-benzylominopurine (BAP). The possible use of tendrils as a novel explant for somatic embryogenesis in grape has been discussed (64). Astarita and Guerra (65) cultured zygotic embryos excised from immature female cones of coniferous *Araucaria angustifolia* to induce embryonal-suspensor masses and investigated conditions for the establishment and multiplication of embryogenic cultures and formation of early somatic embryos. Higher induction rates were obtained with 2,4-D, casein and hydrolysate, BAP, and kinetin depending on the developmental stage of the explants (65).

Mithila et al. (66) reviewed recent advances in *Pelargonium* (geranium) in vitro regeneration systems. The development of modern protocols for in vitro cultures and genetic manipulation has provided new avenues to develop novel varieties of geranium. Optimized techniques of meristem culture have supplemented the culture indexing methods in commercial greenhouse production, resulting in the availability of large-scale pathogen indexed planting material. Currently, technologies are available for the mass in vitro propagation of F_1 hybrid *Pelargonium* through both organogenesis and somatic embryogenesis. The somatic embryogenesis model system has allowed researchers to identity critical factors controlling plant morphogenesis in vitro, such as regulation of regeneration by growth regulators, choice of explant, and characterization of induction and expression phases of morphogenesis in *Pelargonium* (66). Also, optimization of technologies for genetic transformation has opened up the possibilities to develop genotypes with novel characters, including resistance to major diseases.

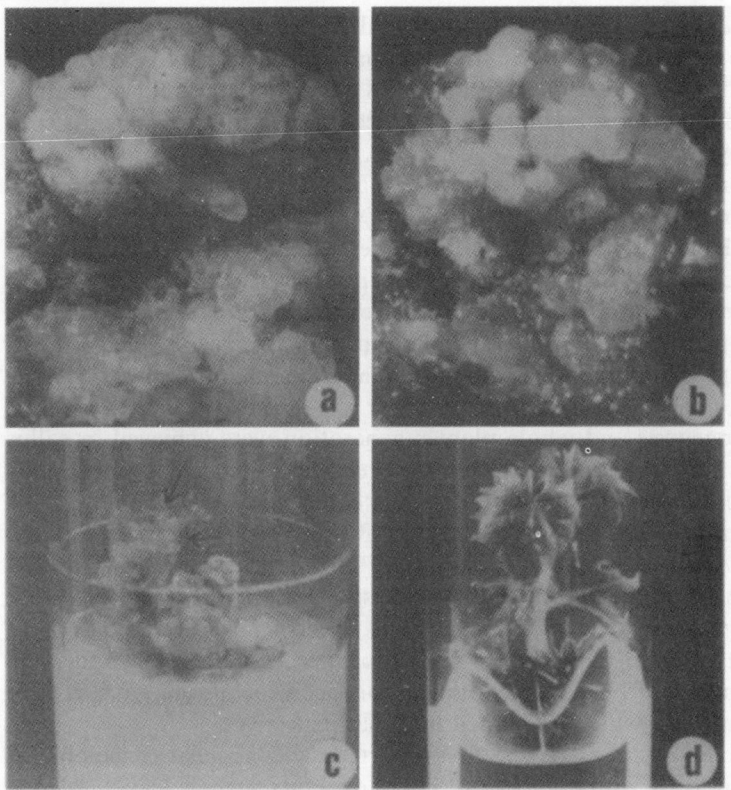

Figure 1 Somatic embryogenesis and plantlet formation from tendril-derived callus of grape. (a) Embryogenic callus, × 10; (b) globular somatic embryos, × 10; (c) shoot development from somatic embryo; (d) complete plantlet on liquid GR medium. (From Ref. 64.)

Regeneration of somatic embryos from plant tissues has been reported in many plant species (1,18,67–69). Since somatic embryos originate from single cells, the problem of producing plants with chimeras can be avoided, which occurs when regeneration has originated from shoots (70,71). The large-scale production of plants using bioreactors (68,72) is more efficient when somatic embryos are used as propagules, because somatic embryos lend themselves to automation and to sorting techniques, such as digital image analysis (73). The rapid multiplication of plants via synthetic seed (10–12,74–76), in which somatic embryos can be encapsulated for direct seeding of crops, will require regeneration over a large range of genotypes and ploidy levels. Somatic embryos can be used with cryopreservation techniques (68,77,78) to alleviate rapid multiplication production difficulties and are a valuable tool for industry. Furthermore, flexibility is available if a propagation protocol has the capability of regenerating viable propagules from several explant sources, such as leaves, roots, and tuber and stem sections. Seabrook and Douglass (79) have recently reported formation of somatic embryos on in vitro cultured stem internodes, leaves, microtubers, and roots of 18 tetraploid potato (*Solanum tuberosum* L.) cultivars, diploid and monoploid germplasm, and three wild *Solanum* species. A two-step protocol with BAP or thidiazuron in the first medium and zeatin, IAA, and GA in the medium produced somatic embryos within 14–28 days. Somatic embryos formed on stem internode section, leaves, and microtuber slices of in vitro–grown plants with clearly evident genotypic differences in regenerative capacity (79). Regeneration of somatic embryos is probably under nuclear control, and the inheritance for regeneration may be quite straightforward (80).

Compton and Koch (81) studied the influence of plant preservative mixture (PPM) on somatic embryogenesis in melon, adventitious shoot orgaogenesis in petunia, and androgenesis in tobacco by culturing explants in regeneration media. The PPM effects on plant regeneration depended on plant species. PPM is a heat-stable, broad-spectrum biocide that reduces microbial contamination in plant tissue cultures.

Somatic embryogenesis and whole plant regeneration were achieved in callus cultures derived from immature zygotic embryos of *Acacia mangium*, an economically important tropical forest species (82). Embryogenic callus was induced on MS medium, containing combinations of 1-phenyl-3- (1,2,3-thiadiazol-5-yl) urea (thiadiazuron, TDZ), and IAA and a mixture of amino acids. This is the first successful report of plant regeneration through somatic embryogenesis in *A. mangium* (82). Mature zygotic embryos of *Liriope platyphylla* Wang et Tang formed embryogenic calluses at 33% frequency when cultured on MS medium, supplemented with 2,4-D and upon transfer to half-strength MS basal medium, embryogenic calluses gave rise to numerous somatic embryos, which developed into plantlets (83).

Somatic embryogenesis in grape was first reported using nucellar tissue of *Vitis vinifera* L. (84), followed by with a variety of grape explants such as leaves, zygotic embryos, ovaries, shoot apices, anthers, and tendrils (85). Salunkhe et al. (85) described a protocol for the induction of somatic embryogenesis in the anther-derived callus of *V. latifolia* and the subsequent conversion of embryos into plantlets.

Compton and Gray (86) described a protocol to obtain watermelon plants from somatic embryos and discussed the effects of 2,4-D, TDZ, and BA, genotype, and explant age on somatic embryogenesis.

B. Synchronization, Development, and Maturation

In vitro embryoids often fail to mature. Cotyledon development is frequently abnormal and embryoids fail to become dessication tolerant. The apical meristem also may be inactive or poorly coordinated (87). Although somatic and zygotic embryos are initiated differently, both embryos start growing with organized cell division, which establishes polarity (61,74). Whereas zygotic embryos are nurtured by the maternal tissue (in many species via the integuments and endosperm), somatic embryos obtain nourishment directly from the growth medium. Again, in contrast to zygotic embryos, which derive protection and control of gas exchange and osmotic potential through their surrounding structure (88), somatic embryos grow and develop in the physicochemical environment regulated by the culture medium and incubator (89). Both somatic and zygotic embryos, however, undergo similar morphological phases of development, including globular, heart, torpedo, and cotyledonary stages (74,90,91). Distinctions in the development of these two types of embryos at the biochemical level showed that although synthesis of storage proteins in somatic embryos of alfalfa is qualitatively similar to that of zygotic embryos, it is low quantitatively (92). In mature alfalfa seeds, proteins are deposited in cotyledons as well as in the seed axes, totaling about 25% of the seed dry weight. The somatic alfalfa embryos are devoid of endosperm and lack the galactomannan reserves of the zygotic seeds. Only about 10% of their final dry weight consists of starch. Because of these important differences, in the accumulation of reserve storage of polysaccharides, the carbohydrate-utilization patterns during the germination of somatic and zygotic embryos are expected to be quite different (61).

The proembryoids are dispersed during the suspension culture phase, followed by formation of new embryogenic clusters. The callus at this stage is transferred to a modified liquid medium, usually containing 2,4-D (93). McKersie et al. (94) observed three predominant cell types in this suspension culture of alfalfa: (a) large clumps of undifferentiated callus and abnormal somatic embryos, (b) small cell clusters composed of nondifferentiated meristematic cells that under appropriate conditions develop into globular embryos, and (c) large banana-shaped single cells. Screening the suspension, first through 500-μM nylon and then a 200-μM screen, separates these cells. The latter screen collects small clusters of proembryonic cells, which are more or less synchronized in their development (61). The suspension loses its embryonic capacity after three subcultures; with each subculturing the single-cell population increases and the embryogenic cell clusters disperse, reducing in proportion of the culture (95). The technique of identification, stabilization, and maintenance of these embryogenic cultures is critical for the scale-up and commercialization of synthetic seed technology (61).

Somatic embryos are further developed by spreading the fraction collected on the 200-μM screen in a thin layer on a hormone-free medium containing 5% sucrose (96). Somatic embryo development in alfalfa can be divided into three phases (61): (a) development, (b) maturation phase I, and (c) maturation phase II, all having different nutritional requirements. In the first developmental phase, the somatic embryos progress through globular and heart-shaped stages to the torpedo stage. About 4 days after screening, globular, heart-shaped embryos appear and develop to the late torpedo stage by the seventh day of sieving. Maturation phase I is a rapidly growing stage between 7 and 14 days after sieving. Most of the storage reserve deposition occurs during this period. Maturation phase II is characterized by the induction of desiccation tolerance, by including abscisic acid (ABA) in the culture medium, or by exposing the culture to sublethal stresses (61). Plating density and light are critical during the development of embryos, and specific plating densities are important to obtain the maximum number of healthy somatic embryos. High light intensities were found to about globular somatic embryos, whereas low light intensity decreased the rate of development and the number of globular embryos elongating to the torpedo stage (93).

Attempts at regulating embryonic culture have taken two approaches (74): (a) physical separation of embryogenic tissue to produce uniform-sized cell masses and favor more uniform embryo development and (b) the use of growth regulators physiologically to synchronize embryo development. Being padded through sieves of defined pore size physically separates embryogenic tissues. Walkar and Sato (87) reported that cell masses from the embryogenic alfalfa callus, passed through a 20- and then a 60-mesh size, measured 234–864 μM in diameter and produced somatic embryos. These cell masses were analogous to isolated preembryonal complexes, which were of a size capable of immediately producing just one embryo, resulting in overall uniformity. The synchrony initiated at such an early stage of embryo development is probably maintained throughout subsequent embryogenesis (74).

ABA has been used physiologically to synchronize embryogenesis. Exogenous-lysupplied ABA has been reported to regulate development of somatic embryos in conifers and carrot (7) but not in grape (74). ABA is thought to interfere with the functions of endogenous growth hormones, probably causing a shift to conditions that favor embryogenesis. The inhibitor may also function as an osmoregulator by decreasing the water content, leading to more synchronous embryo development.

After reaching the torpedo stage, somatic embryos rapidly increase in their fresh weight, accompanied by deposition of major storage reserves in the form of carbohydrates and proteins. This process is enhanced by enriching the culture medium with higher levels of sucrose (50 g/L) and glutamine (50–100 mM) (97–100). Both sucrose and glutamine independently increased the dry weight of somatic embryos of alfalfa, the starch level being increased to about 10% of the dry weight. The absolute amount of alfalfa protein increased twofold with 50 mM glutamine in the phase I maturation medium (54). The sucrose and glutamine responses depend on the embryo density of the maturation medium. higher nutrient levels adversely affected the embryo development, probably by decreasing the osmotic potential of the medium, which could be overcome by frequent (every 2 days) transfer of maturing embryos to the fresh medium (100).

A somatic embryogenesis procedure using a temporary immersion bioreactor was developed for *Coffea arabica*, enabling a mass and virtually synchronous production of germinated somatic embryos without the need for selection before acclimatization (101), with the development of conditions for direct sowing of these somatic embryos on to horticultural soil ex vitro (102). Despite an increased synchronization in the development of the embryos, they still show morphological heterogeneity in the bioreactor and in plant development in the nursery. It is important to improve uniformity of somatic embryos and derived plants for commercial production. Etienne-Barry et al. (103) recently reported on the characterization of the morphological heterogeneity of germinated somatic embryos in a bioreactor and the impact of the heterogeneity observed in vitro on conversion into plantlets after direct sowing in soil and on subsequent plant development in the nursery. The heterogeneity of somatic embryos in the bioreactor affected both the plant conversion efficiency in the soil and plant growth in the nursery, resulting in retarded growth, primarily in plantlets derived from the somatic embryos with small cotyledons (103).

1. Effects of Culture Media

Because somatic embryogenesis (SE) recapitulates the morphological and developmental processes that occur in zygotic embryogenesis, an important factor in SE induction and development is the nutrient composition of the culture medium. Nitrogen is the major element for in vitro morphogenesis (104), and different nitrogen balances and sources in the culture medium can promote SE induction and development in alfalfa (99,100), rice (105), white spruce (106), and cotton (107). Amino acids serve as a primary source of organic nitrogen for the growth of many eukaryotic cells and additionally promote communication between cellsand tissues within multicellular organisms (108). Dal Vesco et al. (109) recently reported the effects of different sources, levels, and balances of nitrogen on both the induction and development of SE from immature and mature zygotic embryos of *Feijoa sellowiana*. Half-strength MS culture medium supplemented with BAP (0.5 μM) enhanced the conversion of seeds to plantlets. Astarita and Guerra (110) investigated biochemical changes in the culture medium during growth of cell suspensions and somatic proembryo formation in *Araucaria angustifolia*.

Fuentes et al. (111) established a more efficient protocol for SE in *Coffea canephora* by studying the effects of silver nitrate and different carbohydrate sources in the culture medium. Silver nitrate, between 30–60 μM, improved embryo yield with higher doses negatively affecting the regenerative capacity. The substitution of maltose, glucose, or fructose for sucrose produced different responses depending upon the five genotypes studied. Whereas fructose significantly increased somatic embryo production in N91 and N128 genotypes, maltose was highly effective for N75 genotype. In addition, more synchronous embryo development occurred in the N91 genotype with glucose as a carbon source (111).

III. DRYING, STORAGE, AND PROTECTIVE ENCAPSULATION

A. Effects of Stresses on Desiccation Tolerance

Synthetic seed technology is envisioned as a way to propagate valuable germplasm clonally via SE and utilize the propagules as an alternative to seeds. In an effort to develop synthetic seeds, hydrated seeds have been delivered to soil in fluid-drilling gel (112,113), encapsulated in alginate (114,115), or dehydrated after encapsulation (8, 116–118). Plants have been obtained from dehydrated somatic embryos (SEs) of orchard grass (*Dactylis glomerata* L.) (119), grape (120), and alfalfa (97,98,121,122). Compton et al. (123) evaluated the potential of synthetic seed technology for maize by examining the effects of various somatic embryo pretreatments on embryo viability: ABA solely or ABA followed by elevated sucrose relative humidity (RH) and the duration of controlled RH dehydration (CRHD) and size of embryos before dehydration. Only embryos sequentially pretreated with ABA and high sucrose remained viable after 2 weeks of dehydration at 70% RH. Up to 34% of SEs survived 2 weeks of dehydration at 70% RH, whereas embryos dehydrated at 50 or 90% RH showed reduced viability (8.7 and 0.8%, respectively). Around 15% of SEs dehydrated at 70% RH developed into plants. Three percent of maize SEs remained viable after 6 weeks of dehydration at 70% RH and 17% developed into plants. Only embryos greater than 5 mm survived 2 weeks of dehydration at 70% RH (123).

Using zygotic embryos of rapeseed, the role of ABA and drying conditions in regulating the maturation process in rapeseed was assessed. The cues regulating the changes occurring during maturation were identified by manipulating culture conditions to produce embryos with higher desiccation tolerance and the capacity to form normal seedlings. The embryos subjected to the maturation treatments (presence of ABA, high osmotic concentrations, or both) were compared with embryos matured in seeds. All treatments conferred some degree of maturity to the embryos, but the most effective treatment was exposure to a high osmotic concentration (124). This effect is consistent with the earlier finding that inhibition of water uptake regulates embryogeny more directly than does ABA (125,126).

Alfalfa SEs acquire desiccation tolerance during maturation phase II (61). Senaratna et al. (121) investigated numerous methods of inducing desiccation tolerance in alfalfa SEs. ABA was found to be critical in inducing desiccation tolerance in these embryos, whether applied directly or induced to accumalate by stress treatments. ABA prevents precocious germination and induces a genetic program in the embryo, which initiates a series of biochemical and physical changes to bind water by protoplasm, increasing its tolerance to water loss (61). ABA also induces the chlorophyll degradation (or degreening), thus minimizing activated oxygen production during the exposure of dry tissue to light (127). ABA, however, does not seem to stimulate the deposition of storage reserves in alfalfa. Also, SEs of alfalfa respond to ABA only after the late torpedo stage of development, and prior to early germination earlier or later applications do not induce desiccation tolerance (61).

Alternatively, sublethal stresses can be employed to induce desiccation tolerance in SEs. According to Anandarajah (127), thermal stress and cold treatment can induce tolerance to dehydration in microspore-derived embryos of *Brassica napus*. In alfalfa, all sublethal stresses (e.g., nutrient deprivation cold stress, thermal treatment, and water stress) induced desiccation tolerance (121). Partial water stress could be applied ·either by removing the seal from the Petri plate containing the medium and embryos for over 23 days or by using polyethylene glycol (PEG-4000) to induce more controlled water stress (128).

Desiccation tolerance can be induced in alfalfa SEs by including 6% sucrose in the maturation medium (97,98). The time of ABA application, exposure to stress treatments, as well as the duration of the exposure are critical in the induction of desiccation tolerance (61). Lecouteux (129) reported that prolonged exposure to ABA or high sucrose beyond 3 days decreased the levels of storage proteins, increased the free amino acid level, and reduced both vigor and desiccation tolerance in alfalfa SEs. Although ABA maintained the embryo in a visually quiescent state, the immobilization of storage reserves and other biochemical changes continued as in precocious germination (129). Desiccation tolerance expresses differently in various situations. In the context of embryo development, McKersie (61) defined desiccation as the loss of cytoplasmic water to less than 0.20 g water per gram dry weight, and desiccation tolerance as the ability of the embryo to regrow after a prolonged period of time in a desiccated state, which may be several weeks.

B. Methods of Drying Somatic Embryos

Drying of SEs requires their transfer from the liquid tissue culture medium to atmospheres of defined RH. The embryos are tranferred sequentially from high humidity to progressively lower humidities with an interval of at least 1 day of equilibrium at each humidity level. Humidity is often regulated on a laboratory scale using desiccators containing saturated solutions of osmoticums, such as PEG-4000. The drying rates vary according to the concentrations and volumes of salt solutions, desiccator size, quantity of embryos, and the atmosphere. Laminar flows are also used for rapid drying of embryos, but results vary more than with controlled drying methods (61).

Two distinct methods are used to dry SEs with varying results: (a) slow drying through a sequence of reducing RHs over a period of week and (b) rapid drying in 1 day at 30–40% RH. Roberts at al. (130,131) noted that SEs of Sitka spruce (*Picea sitchensis* [Bong] Can) and interior spruce (a hybrid of *P. glauca* [Moench.] Voss and *P. engelmannii* Parry) survived drying at 95% RH but not at 81% RH. Attree et al. (128) also reported that when the white spruce (*P. glauca*) SEs were dried through a sequence of decreasing RHs, the survival rates declined significantly below 81% RH, although 10% of the SEs survived drying even at 43% RH. The SEs of celery (*Apium graveolens*) also survived drying at 70–90% RH but not at 50% RH (132). Gray (133) reported that SEs of orchard grass (*Dactylis glomerata*) could survive desiccation at a constant RH of 70–13% moisture content, but their germination capacity decreased within 1 week of storage. The best results to date have been obtained with alfalfa SEs, whether they are dried slowly over a period of 7 days or rapidly in 1 day (94–100). The final moisture content of alfalfa SEs dried by either of the two methods was 15%. Around 90% of visually selected high-quality SEs could survive storage up to 3 weeks in the dried state. These embryos could be stored for 1 year at 15% moisture content with a minimal loss of their viability. The method of drying SEs (at least in the case of alfalfa) is not as critical as the tissue culture technique employed to mature the SEs and induction of the expression of desiccation tolerance (61).

C. Induction of Quiescence/Dormancy

Functional synthetic seeds consist of metabolically active SEs with a nonliving seed coat material, nutrient, and other additives necessary for see germination. Preliminary studies on the induction of quiescence in synthetic seeds were centered on reversibly arresting growth of rapidly developing SEs. Dehydration and rehydration are obvious prospects to control quiescence and germination of SEs, since quiescent seeds resume growth after imbibition.

Gray (133) reported dehydration-induced quiescence in SEs of carrot, grape, and orchard grass. SEs of orchard grass were isolated from the callus and placed in sterile Petri dishes. Unsealed dishes were maintained in the dark at 70% RH and 2.3°C. Dehydration began immediately with a rapid decrease in embryo weight within 24 hr (Fig. 2). Weight loss during the dehydration period varied from 73% moisture content of freshly isolated SEs to 13% of those stored at 70% (133,134). During dehydration, the SEs ceased growing, decreased in size, and became yellowish and brittle and their outer walls collapsed (Fig. 3a–e). Following imbibition of fresh medium, embryos enlarged and were morphologically identical to those not subjected to dehydration

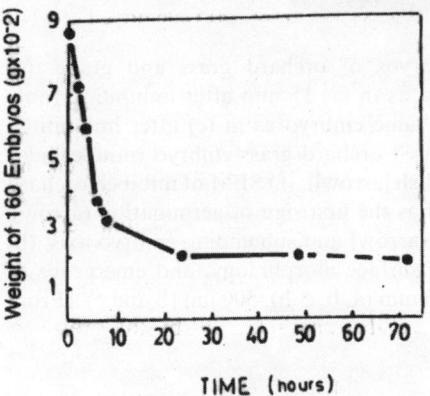

Figure 2 Rate of weight loss of orchard grass somatic embryos during dehydration. After weight stabilization, the dish was subjected to 60°C for 24 hr to remove remaining water (dotted line). (From Ref. 87.)

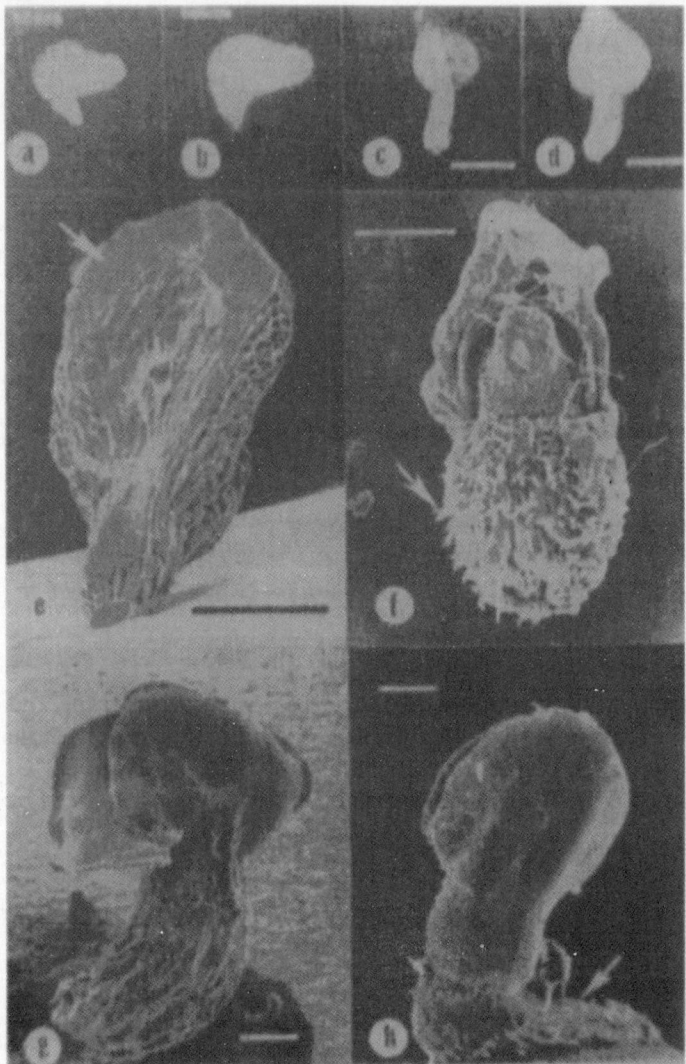

Figure 3 Dehydrated and imbibed somatic embryos of orchard grass and grape. (a) Dehydrated orchard grass embryo; (b) same embryo as in (a) 15 min after imbibition (note increase in size); (c) dehydrated grape embryo; (d) same embryo as in (c) after imbibition; (e) scanning electron micrograph (SEM) of dehydrated orchard grass embryo (note cellular collapse and flattened area due to contact with Petri dish [arrow]); (f) SEM of imbibed orchard grass embryo, with typical development of root hair as the first sign of germination (arrow); (g) SEM of grape embryo with shriveled cotyledons (arrow) and subtending embryo axis; (h) SEM of imbibed grape embryo with restoration of surface morphology, and emergence of roots as sign of germination (arrow). Scale bar = 125 mm (a, b, e–h); 500 μm (b and c). (From Ref. 87.)

(Fig. 3a,f). This change could not be related to metabolic growth, since all embryos grew larger whether or not they subsequently germinated. The root hairs produced from embryo coleorhizal tissue was the first sign of germination (Fig. 3f). Of 300 well-developed embryos, dehydrated and stored for 7 days, only 18% germinated after imbibition, but subsequently died, and 6% produced plants. Only white opaque embryos with well-developed scutellar and coleoptilar tissues (these were also characterized by the presence of starch, suggesting that this storage compound was necessary for survival) survived regardless of their overall size (134). The viability of SEs decreased with increasing storage time (see Table 12 in Chapter 13); 40% of a nondehydrated control group germinated with abbreviated growth and an additional 32% produced viable plants. After 21 days of storage, 8% germinated, but did not grow further, and an additional 4% produced green plants, indicating that SEs could survive dehydration at higher frequencies and longer storage periods than earlier reported. Grape SEs subjected to the same dehydration procedures as orchard grass showed similar changes (Fig. 3c,g). Imbibition rapidly restored the white, opaque nature of the SEs, designated as having survived dehydration. Additionally, SEs producing shoots with green leaves were scored as being viable. Storage studies showed that grape SEs had a higher germination capacity than those of orchard grass and carrot SEs (Table 1). At the 7-day storage period, 63% germinated with no further growth and 28% produced viable plants. At 21 days, only 22% of imbibed SEs remained white and did not respond; 58% germinated with no further growth and 20% produced rooted green plants as compared to only 5% for the control group of SEs not subjected to dehydration (Table 1). After 21 days of dehydrated storage, 34% of SEs from one grape genotype produced plants, followed by imbibition (Table 2). Gray and Conger (134) forwarded the first clear evidence that plants can be produced from dehydrated SEs and that even encapsulated SEs can survive the dehydration procedure (Fig. 4).

Synthetic seed may consist of either a quiescent or nonquiescent SEs with or without any protective covering. The appropriate form of synthetic seed required depends upon the specific application. Naked nonquiescent SEs, germinated in soil plugs, can be used to propagate ornamental crops that are presently micropropa-

Table 1 Effects of Storage Period on Germination Rate of Dehydrated Grape Somatic Embryos[a]

	Dehydrated storage (days)		
Response	0	7	21
No germination	85 (85)[b]	9 (18)	22 (40)
Germination—no further growth	10 (10)	63 (126)	58 (116)
Germination—viable plants	5 (5)	28 (56)	20 (44)
Total	100 (100)	100 (200)	100 (200)

[a] White opaque embryos with well-developed embryo axes and discrete unfused cotyledons were maintained in a dehydrated state at 23°C for 0, 7, or 21 days and then allowed to absorb water. Those that produced roots and/or became yellow to green with no further growth were scored as germinated. Embryos with roots and shoots with green leaves were viable plants.
[b] Figures in parentheses represent total number of somatic embryos.
Source: Ref. 87.

Table 2 Comparison of Dehydration and Benzyladenine (BA) for Inducing Germination in Grape Somatic Embryos[a]

Treatment	Percent germination response[b]			
	Hypotocotyl	Root	Cotyledon	Shoot
Dehydration	77	68	65	34
0.5 μM BA	100	92	98	12
Control	76	88	36	0

[a] Well-developed embryos were either dehydrated for 21 days at 70% RH and 27°C, placed directly on medium with BA, or placed on basal medium (control).
[b] Germination response was based upon either enlargement and greening of hypocotyls and cotyledons or emergence of roots or shoots.
Source: Ref. 108.

gated by tissue culture involving labor (Table 3). Manpower reduction achieved by producing plants by SE, when compared with existing micropropagation, would confer a cost advantage. Dehydrated quiescent SEs without encapsulation can be used for germplasm storage, since they can be hand manipulated and carefully stored in protective containers. The cost of manipulating SEs for germplasm storage is not expected to exceed that of seed. Nonquiescent encapsulated SEs could be used for crops grown in greenhouse before transplanting to the field (e.g., carrot, celery). However, for mass propagation of field crops, protective encapsulation will be necessary (74).

Figure 4 Plants of (a) orchard grass and (b) grape obtained from somatic embryos stored for 7 days in a dehydrated state. (From Ref. 87.)

Table 3 Types of Synthetic Seed and Crop Applications

Synthetic seed type	Relative development cost	Relative cost per seed	Example of crop application
Naked embryo			
Nonquiescent	Low	High	Ornamentals
Quiescent (dried)	Low	High	Germplasm conservation
Encapsulated embryo			
Nonquiescent medium	Medium	High	Vegetable transplants
Quiescent (dried)	High	Low	Agronomic and conifers

Source: Ref. 60.

D. Protective Encapsulation of Somatic Embryos

Kitto and Janick (8,116) investigated the production of synthetic seeds by encapsulating SEs of carrot with synthetic coating material (polyox WSR-N 750, Union Carbide Corp., New York and other parameters that influenced the capacity of asexual embryos to survive dehydration. The method used embryo-rich suspension cultures mixed with polyox, a water-soluble plastic resin, and dried under a sterile airflow in ambient temperature and humidity conditions. Drying, as measured by weight loss, was rapid, reaching equilibrium in about 4 hrs. Only SEs encapsulated in polyox waters survived. The uncoated embryos could not survive drying under any conditions. The protocol used by Kitto and Janick (8) for encapsulating SEs is summarized in Figure 5. The pretreatment with high sucrose (12%), chilling (4°C), ABA (106M), or high inoculum density (0.8 g of suspension culture tissue per 125 mL) increased the survival rate of SEs (116), which decreased rapidly over time in all experiments. The best survival at 26°C was two embryos per 20 wafers after 4 days of storage from a pretreatment combination of ABA with chilling, with the survival time increasing up to 16 days at 4°C (8).

Datta and Potrykus (135) developed an in vitro culture system, which yielded high frequencies of high-quality microspore-derived barley embryos without an intervening callus phase. These SEs were similar to zygotic embryos with regard to their morphology and germination capacity. When encapsulated in sodium alginate to produce individual beads containing one embryo each, these artificial seeds (embryos) germinated well with a root system superior to that of nonencasulated embryos. Germination capacity was maintained for at least 6 months, whereas nonencapsulated embryos did not survive for more than 2 weeks in storage.

Even after satisfactory development of the technique to induce quiescence and dormancy into SEs, the dry, delicate embryos would benefit greatly with the protective seed coating against injury incurred during traditional handling and planting. The protective encapsulation should be designed to provide physical protection as well as essential nutrients, antibiotics, and fungicides to assist SEs during germination and seedling establishment. Two approaches, hydrated and dry encapsulated, have been investigated. The hydrated encapsulation is most satisfactory for nonquiescent SEs that can be planted directly without storage. Calcium alginate has been used as a coating material in hydrated encapsulation with limited success (114,115). Dry encapsulation with synthetic seed coating surrounding SEs, which permits conventional seed storage and handling, is preferred to hydrated encapsulation (136).

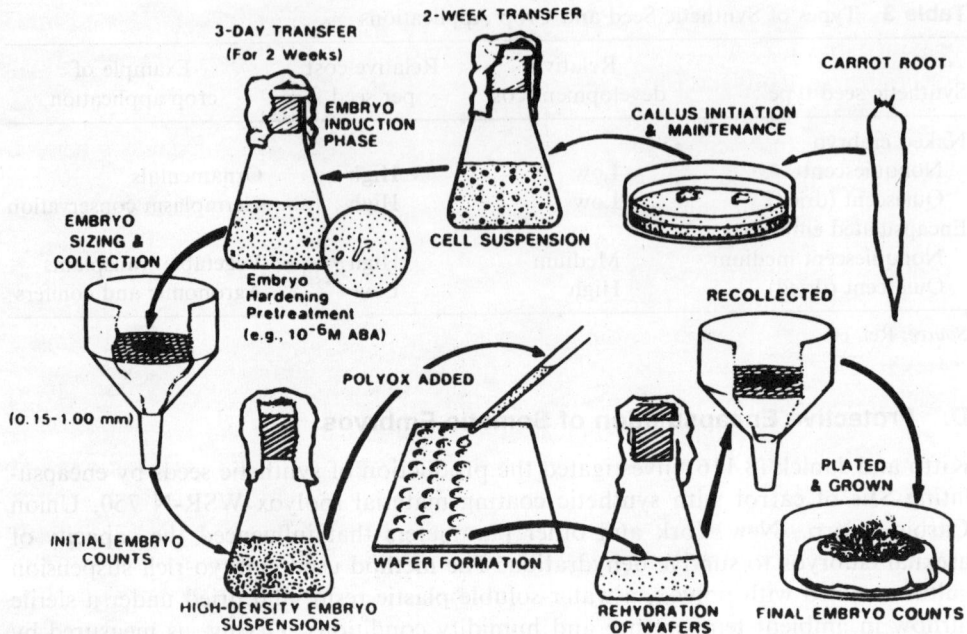

Figure 5 Protocol for carrot embryo encapsulation. The callus induced from secondary phloem of carrot is recultured onto maintenance medium. Cell cultures are initiated from callus and maintained in basal medium containing 2,4-D (0.1 mg/L) and kinetin (0.2 mg/L), and transferred to fresh medium every 2 weeks. Embryos are initiated and grown by placing 0.4 g of cells and cell aggregates (0.15–1.00 mm) per 25 mL of medium without either 2,4-D or kinetin and reculturing every 3 days for 2 weeks. Embryos can be hardened by pretreating with 10^{-6} M ABA during this induction stage. Embryonic suspension ranging in size from 0.15 to 1.00 mm is mixed in equal volumes with 5% polyox to achieve a final concentration of 2.5%. Embryo suspension plus coating compound is dispensed as 0.2-mL drops onto Teflon sheets and dried to wafers in a laminar flow hood. Drying time is based on the ability of wafers to separate from Teflon sheets. The time necessary to form a dried, detachable wafer varies with humidity and temperature, but is not less than 5 hr. Wafers are placed in embryogenic medium to dissolve the coating compound and to dehydrate embryos. Rehydrated embryo suspensions are collected on screens with 0.15-mm spaces, scooped up with a spatula, and dispensed on filter paper supports in Petri dishes containing 3 mL of medium. Petri dishes are sealed with paraffin and placed in a growth room. (From Ref. 89.)

Soneji et al. (137) recently described for the first time a method for the efficient delivery of micropropagated plantlets using tiny (2–5 mm) in vitro shoots of pineapple. The encapsulated shoots represented synthetic seeds that germinated and formed roots in vitro after subculture onto various culture media. One hundred percent germination of synthetic seeds to plantlets occurred after pretreatment of shoots in liquid Pin4 (White's basal medium (138), White's vitamins, 0.56 mM myoinositol, 0.03 M sucrose, 10.8 µM NAA, and 39.4 µM IAA) medium for 12 hr followed by culture of synthetic seeds on Pin2 medium (MS basal medium, MS vitamins, 0.56 mM myoinositol, 0.06 M sucrose, 9.67 µM NAA, 9.84 µM IAA, and 9.29 µM kinetin). Synthetic seeds stored at 4°C remained viable without sprouting for up to 45 days. Plantlets produced in vitro from synthetic seeds were successfully established in soil (Fig. 6). This protocol provided an easy and novel propagation system for pineapple, an otherwise vegetatively propagated fruit crop (137).

Encapsulated in vitro–derived apical buds of M.26 apple rootstock (*Malus pumila* Mill) can be employed to produce synthetic seed (138). Satisfactory levels of conversion (plantlets from synthetic seed) can be achieved with adequate rooting induction treatment, encapsulation protocol, and nutritive and environmental conditions. The largely used sodium alginate for encapsulation is excessively permeable with loss of the nutritive substances (artificial endosperm) and/or dehydration risks during conservation and transport, causing detrimental effects on the synthetic seed conversion and on the plantlet's growth. Simple encapsulation in alginate was compared with double encapsulation and with encapsulation-coating procedures. The presence of a second layer of alginate (double encapsulation) and of a thin external coating layer over the alginate (encapsulation coating) did not show any detrimental effects on viability, sprouting, and regrowth of the encapsulated microcuttings. Satisfactory conversion (70%) was obtained with the encapsulation-coating procedure, whereas the double and simple encapsulation converted less than 40% of the synthetic seed. The addition of an antimicrobial substance, PPM, to the capsule did not compromise the conversion of the encapsulated microcuttings sown in ex vitro nonaseptic

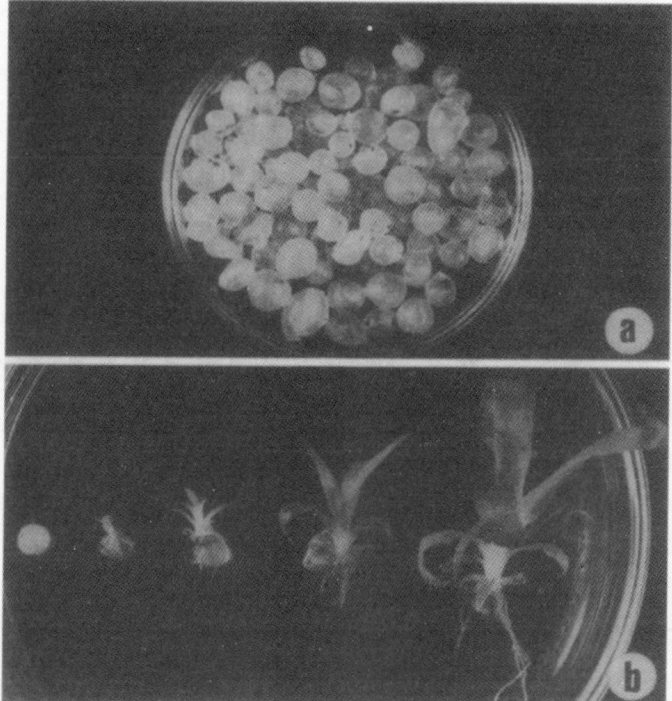

Figure 6 Germination of synthetic seeds of pineapple. (a) Synthetic seeds prepared by encapsulation in 3% sodium alginate in hormone-free MS basal medium (2). MS vitamins, 0.56 mM myoinositol and 0.06 M sucrose (Pin1); (b) germination of a synthetic seed and recovery of a complete plantlet. (From Ref. 137.)

conditions (138). A mechanical protocol for the production of synthetic seed in M.26 apple through encapsulation of differentiating propagules (tissue fragments with shoot primordial) has been devised to achieve a time-saving and hand labor–saving procedure (138a).

Adrinni et al. (139) evaluated the effects of root induction, cold treatments, and addition of sucrose on the conversion of Hayward kiwi fruit synthetic seeds to whole plants following encapsulation of in vitro–derived buds. Encapsulation, although considered to be necessary, depressed the microcuttings' vigor and vegetative activity. Cold treatments boosted bud vigor and subsequent conversion. Increasing sucrose concentration in some steps of the protocol also enhanced conversion, reaching a rate up to 57.5%. Ara et al. (140) described germination and plantlet regeneration from encapsulated SEs of mango with 73.61% germinability and 45.83% plantlet regeneration.

Micropropagated explants of 4 different woody species and 10 genotypes from the genera *Actinidia*, *Malus*, *Olea*, and *Rubus* were used to find the best nutrient formulation to be added to the alginate encapsulation matrix in order to allow the highest viability and regrowth in vitro after 1 month of cold storage (141). The encapsulated *Rubus* microcuttings proved to be usable as synthetic seeds, with a maximum rooting (and conversion) rate of 60.7%. Micheli et al. (142) evaluated the aptitude toward encapsulation in sodium alginate of micropropagated shoot buds of olive, testing the responses of encapsulated explants according to their original position on the shoot, the inductive rooting treatment, the nutrients added to the capsule, and a cold storage treatment. The encapsulated apical buds showed higher

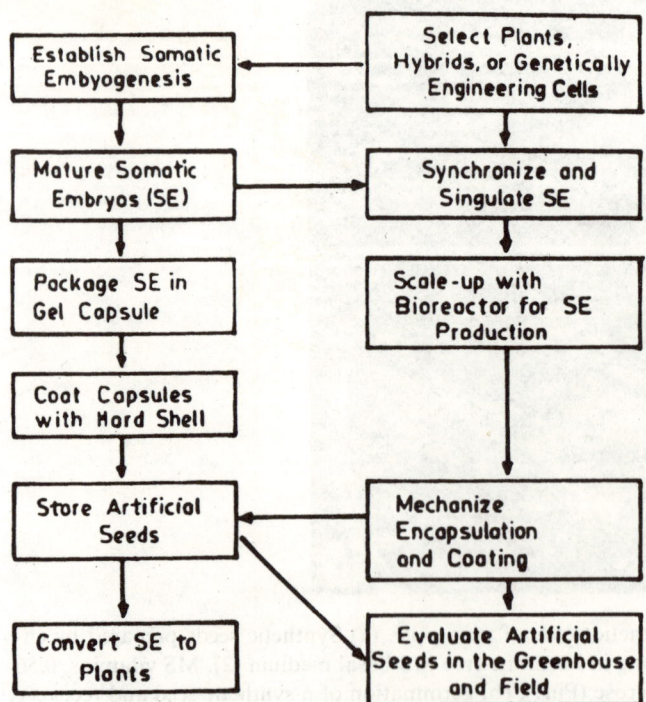

Figure 7 Development of synthetic seeds. (From Ref. 89.)

viability and regrowth ability than the nodal ones. Also, the encapsulated buds can be conveniently stored at 4°C, but not longer than 45 days. Root induction treatments before encapsulation of the buds resulted in low percentages of plantlet regeneration, which was obtained only from apical microcuttings. Modifying the indol-butyric acid (IBA) concentration in the rooting induction phase and/or the composition of the nutrient-enriched capsule and/or the sowing medium usually increased plantlet regeneration rates up to a maximum of 30% (142).

Guerra et al. (143) assessed the effects of genotypes, periods of 2,4-D stock, and different substances supplemented to sodium alginate on the production of synthetic seeds in goiabeira serrana fruit (*Feijoa sellowiana* Berg.). Histological evaluations revealed the direct origin of SEs from the epidemic surface of cotyledons. Synthetic seeds containing MS salts and sucrose resulted in higher rates of contamination than capsules free of these substances (143). Brischia et al. (144) recently evaluated the plantlet-producing capability of encapsulated, machine-ground explants and compared the regeneration capability of such synseeds with those containing hand-cut, unipolar micropropagated explants. The synthetic seeds of M.26 apple rootstock could be produced through organogenesis from machine-processed explants followed by root induction and encapsulation of differentiating propagules.

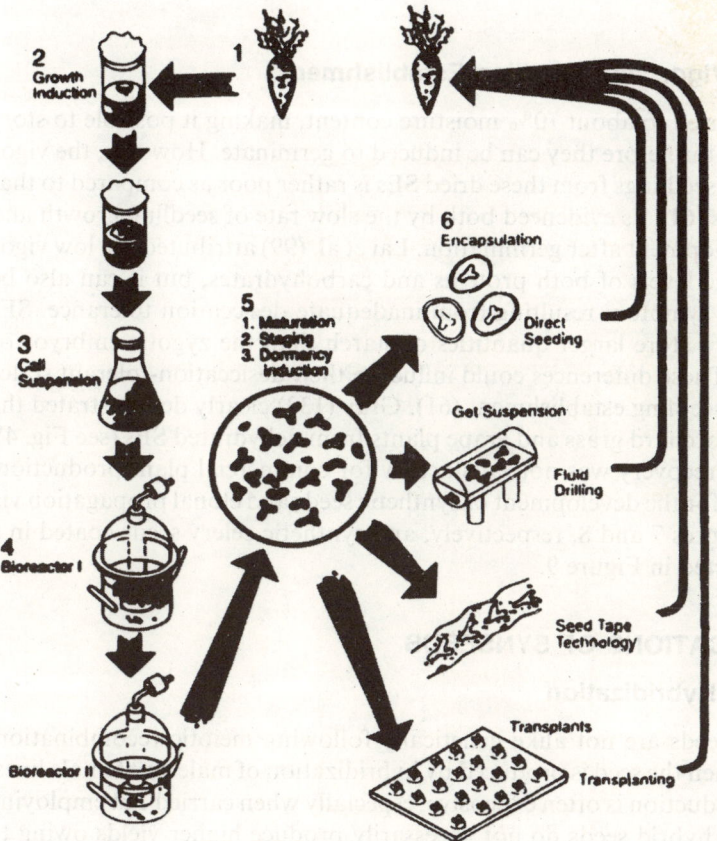

Figure 8 Scheme for vegetative propagation via somatic embryogenesis. (From Ref. 110.)

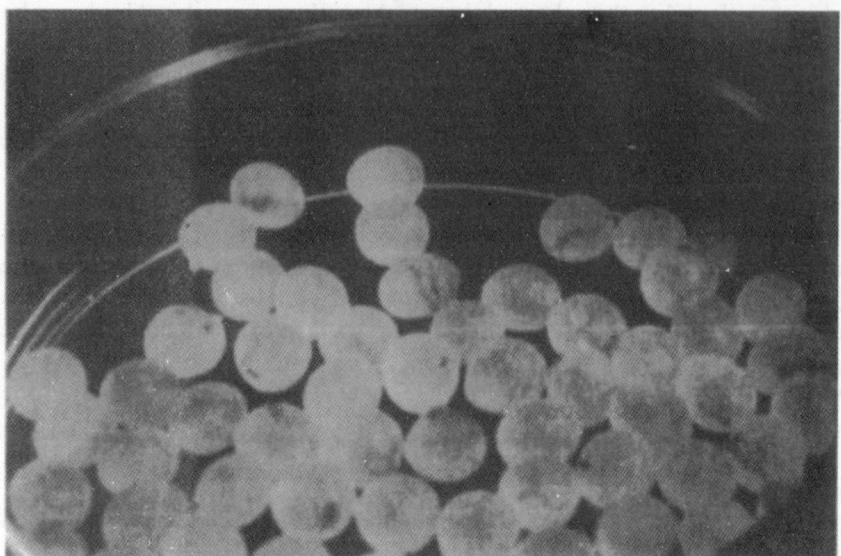

Figure 9 Synthetic celery seeds coated in a calcium gel. (Photograph courtesy of K. Redenbaugh.)

E. Germination, Vigor, and Seedling Establishment

Mature SEs can be dried to about 10% moisture content, making it possible to store them for several months before they can be induced to germinate. However, the vigor and establishment of seedlings from these dried SEs is rather poor as compared to that of conventional seeds (61), as evidenced both by the slow rate of seedling growth and abnormal plant development after germination. Lai et al. (99) attributed the low vigor of SEs to low storage levels of both proteins and carbohydrates, but it can also be considered an injury symptom resulting from inadequate desiccation tolerance. SEs have been reported to store larger quantities of starch than the zygotic embryos or conventional seeds. These differences could influence their desiccation-tolerant states as well as vigor and seedling establishment (61). Gray (133) clearly demonstrated the establishment of the orchard grass and grape plants from dehydrated SEs (see Fig. 4), although the rate of recovery was not satisfactory for commercial plant production. The general schemes for the development of synthetic seeds and clonal propagation via SE are shown in Figures 7 and 8, respectively, and synthetic celery seeds coated in a calcium gel are depicted in Figure 9.

IV. CROP APPLICATIONS OF SYNSEEDS

A. Breeding and Hybridization

Sexually produced seeds are not alike genetically following meiotic recombination. Many yield better when the seed is produced by hybridization of male and female lines. Such hybrid seed production is often expensive, especially when carried out employing manual labor. Also, hybrid seeds do not necessarily produce higher yields owing to unassured hybridization. Moreover, several agricultural and horticultural crops must be propagated vegetatively on account of genetic variability, although these crops may be easily propagated through seeds. These problems in breeding and hybridization have led to the clonal production of synthetic seeds and have become the subject of intensive research activity (10–12,61,136). Synthetic seeds may be of value for several traditional crops (136) (Table 4). Synseeds have a greater potential for crops such as alfalfa and orchard grass in which seeds are not uniform and each plant is a distinct genotype. Synseeds also permit the introduction of specific new genes into single outstanding hybrids, which can then be produced asexually by SEs. The planting efficiency of certain vegetatively propagated fruit and nut trees can be enhanced through synseeds by overcoming the problems of self-incompatibility and longer breeding cycles. The problems of prolonged life cycles of forest conifers (pine and spruce trees) necessary to produce seed can also be resolved by using artificial seeds.

The production of hybrid seeds of crops like cotton and soybean is hindered because flowers are closed at the time of pollination, necessitating hand pollination. SE can circumvent this labor-intensive process. Crops like tomato and watermelon demand high-value hybrid seeds. The production of synthetic seeds of these crops may be cost effective (22,136).

Synseeds can potentially circumvent the need for inbred and male-sterile parental lines in the production of hybrid cereals. The hybrid maize industry relies on inbred parents to produce uniform hybrid seed. Mass hybridization is possible by the use of male-sterile lines as females. Increased production costs over open-pollinated seed incurred by the use of inbred and male-sterile lines outweigh the benefits resulting from yield and quality conferred by hybrid vigor. The development

Table 4 Potential Applications of Synthetic Seed Technology (SST) for Some Selected Crop Species

Crop	Somatic embryo quality[a]	Relative seed cost[b]	Application[c]	Relative need for SST[d]
Alfalfa	h	l	s	m
Corn	p	m	l	m
Cotton	p	m	h	m
Grape	h	na	s,g	m
Loblolly pine	p	h	c	h
Norway spruce	h	—	c	h
Orchard grass	h	l	m	m
Soybean	p	m	h	m
Hybrid tomato	n	v	d	h
Seedless watermelon	n	v	d	h

[a] Relative somatic embryo quality: h = highly developed embryos; p = poorly developed embryos; n = somatic embryos not obtained.

[b] Relative cost of seed: v = seed cost limits planting, h = seed is costly; m = moderate; l = relative inexpensive; na = seed is not used.

[c] Application for synthetic seed: c = circumvent long breeding cycles; d = decrease hybrid seed cost; g = germplasm conservation; h = mass production of hybrids; i = eliminate need for inbreds; s = circumvent self-incompatibility.

[d] Relative need: h = highly useful if implemented; m = existing methods are effective, but implementation should yield improvements.

Source: Ref. 60.

and maintenance of these parental lines also consumes much time, labor, and valuable resources of breeding programs, resulting in slow integration of new germplasm. The use of synthetic seed to propagate new hybrids by eliminating the need for parental inbreds and male-sterile lines is highly intriguing. It would lead to the commercialization of new hybrids, stimulating competition because cultivars could be produced without an existing stock of parental inbreds (74).

The use of synthetic seed is thought to assist in introducing transgenic traits in commercially adapted agricultural crops and their cultivars in a cost-effective manner through clonal multiplication of the transgenic or genetically modified (GM) parent (see Chapter 24). Employing *Agrobacterium* vectors, successful genetic transformation of alfalfa was achieved to introduce useful traits such as herbicide resistance (145) and environmental stress tolerance (146).

The costs of production of double-cross hybrids of crops like alfalfa, even using cytoplasmic male sterility, are enormously high. According to McKersie et al. (94), a double-cross breeding system in which the "cuttings" (of the conventional system) are replaced with synthetic seeds may be desirable, because it reduces high labor costs. However, other problems associated with hybridization, such as the number of parents for maximal hybrid vigor, effective ways of combining the parents, and the need for pollen control must be resolved. Synthetic seed technology is thought to reduce the number of generations of seed production required to produce certified seed that may be more appropriately called reduced-generation synthetic cultivars rather than hybrids (61).

B. Storage of Vegetative Propagules

The progeny of a selected population of hundreds of parents of crops like alfalfa are maintained in nurseries and propagated vegetatively by cuttings. However, this approach is not feasible to maintain a diverse germplasm base or crop plants with a unique combination of genes, which may be required in the future. Transgenic or GM plants with novel genes for various traits, which are easily produced using *Agrobacterium* (145,146), can be stored better using modern biotechnology. Seed propagation of such plants will eradicate the original genotype, with the progeny exhibiting recombination and segregation. These transformants can be preserved in a separate growth nursery (greenhouse) well isolated from other reproductive plants to prevent their inadvertent release into the environment. The use of dry SEs of such plants would be an ideal alternative method of storing vegetative propagules to maintain the valuable germplasm (61).

Etienne et al. (147) recently presented the studies on genetic resources within the complex of *Coffea arabica* L. species—reservoir of resistance genes—and the possible uses of these genes through to the modern tools of genetic improvement of the crop, leading to advances in the micropropagation techniques, indispensable to multiply heterozygous superior genotypes or to develop a genetic transformation program. Economically important genes of the caffeine biosynthetic pathway or genes encoding for seed proteins have been isolated. The high performance already achieved in the in vitro propagation process by SE offers the possibility to mass propagate superior hybrids, pilot productions by SE currently permitting preparation for commercial application. Seed cryopreservation enables a routine use for long-term conservation of coffee genetic resources, and transgenic plants have been obtained for the *C. arabica* and *C. canephora* cultivated species. These authors have discussed the perspectives of application of biotechnology in the long-term improvement of coffee. Etienne et al. (101) developed a method for mass regeneration of *C. arabica* plants from embryogenic cell suspensions. Etienne and Bertrand (148) compared trueness-to-type, growth and yield performance, bean chemical and biochemical characteristics, and cup quality of trees of four *C. arabica* clones produced from embryogenic cell suspension with data obtained for control trees grown from microcuttings. According to Ramanatha Rao and Hodgkin (149), a better understanding of the genetic diversity and its distribution is essential for its conservation and use. This will help to rationalize collections and develop and adopt better protocols for regeneration of germplasm seed. Through improved characterization and development of core collections based on genetic diversity information, it will be possible to exploit the available resources in more valuable ways.

C. Feasibility of Synseeds

The commercial use of cell and tissue culture propagation has been limited to high per-unit value crops, such as ornamentals, flowers, and certain fruit and nut trees. A low-cost, high-volume propagation system if developed could be of significant value to medium per-unit value crops, such as lettuce, celery, and many other vegetable and field crops (Table 5). For these crops, the high-volume propagation potential of SE, combined with the formation of synthetic seeds for low-cost delivery, would open a new field for clonal propagation (10,114). Redenbaugh et al. (114) classified candidate crops for synseed production in to two groups: (a) those having a strong technological basis, for which high-quality SEs are currently being produced, and (b) those with a strong commercial basis for which seed costs are high because of fertility problems, gamete instability, labor-intensive hybrid seed production, or other reasons. Very few crops meet both requirements and are suitable for synthetic seed technology (Group III, Table 5).

The economic feasibility of synseeds, investigated in several crops, showed that the rooted plants of many ornamental and flower crops could be multiplied using current micropropagation technology at a cost of $0.40 to $0.50 per unit (74). The use of producing gel-coated artificial seeds could be $0.068 per unit and that of field plantlets from a greenhouse $0.563 each (22). The cost of desiccated (dry) SEs for use on a commercial scale remains to be established. It also remains to be determined whether hybrid alfalfa seed can be produced economically using synseed technology to multiply vegetatively the parent plants in a double-cross seed-production system (61,95).

Assuming Redenbaugh's cost of production of field plantlets and a plant density of 10,000/ha, the cost of first seed production would be $5630 for alfalfa. Assuming average seed yields of 500 kg/ha and a planting rate of 1 kg/ha, around 1000 tons of certified seed would be possible to obtain over 5000 tons at a retail value of $35 million with an initial investment of $5630 (61,150–153). The production of single-cross hybrids with different economics would require a very high scale-up of the multiplication (of the order of 10^8) of synseeds produced by the tissue culture system using the above assumptions. Bowley et al. (153) predicted that 5000 tons of certified seed would be produced within 2 years over an area of 10,000 ha, that is, 4 years earlier than in the double-crops system, with an added benefit of reduced inbreeding. However, significant improvement in the tissue culture production of artificial seeds through bioreactor scale-ups, as evident by some recent reports, would be needed for this purpose (11,12,151,152). The major limitation of SE technology for crop improvement

Table 5 Potential Crops for Artificial Seeds

Group I	Strong technological basis (for which somatic embryogenesis system already exists): Alfalfa Carawax Carrot Orchard grass *Panicum* *Pennisetum*
Group II	Strong commercial basis (with medium per-unit value): Begonia Broccoli Cauliflower Cotton Cyclamen Geranium Gerbera Impatiens Lettuce Petunia Tobacco Tomato
Group III	Strong technological and commercial basis: Celery Coffee Corn Cotton Oil palm

Source: Ref. 91.

would be the identification of germplasm that would be amenable to multiplication in tissue culture, which would also produce few somatic clonal variants, and would combine well to regenerate into high-performance progeny of the desired crop plant (153). Aly et al. (154) recently reported a rapid and efficient protocol to regenerate plants via SE in *Limonium bellidifolium*, a commercially important species that accumulates beta-alanine betaine (an osmoprotectant), which is a first report of SE in the Plumbaginaceae. Li (155) demonstrated successful production of carrot SEs in a 5 L bioreactor, automated embryo selection using an embryo selector based on flow separation. An instrument was developed to encapsulate carrot SEs, which improved their handling properties. The encapsulation matrix allowed for the incorporation of nutrients and pesticides to improve subsequent performance of the regenerated plants.

V. CONCLUSIONS

A. Challenges and Potential

The major challenges of the production of synthetic seeds through SE include methods to improve and enhance the quality of SEs, asynchronous maturation, and development of SEs (development of extra cotyledons or poorly developed apical meristems) leading to precocious germination and/or reversible induction of quiescence/dormancy, production and application of synthetic coats (encapsulation) for better handling and storage of SEs, and the high scale-up of SEs to produce synthetic seeds on a commercial and economic scale. The concept of direct genetic manipulation of SEs has great potential for the production of superior-performing seeds of a number of crop plants, such as alfalfa, orchard grass, carrot, celery, maize, cotton, oil palm, and many others (114); however, numerous technological and economical hurdles associated with this technique must be resolved to make it a commercially successful venture. Imposition of dormancy and encapsulation to permit the normal handling and storage of SEs, although evolving currently, need to be developed fully. The lack of developmental synchrony in SE is the single most important hurdle to be overcome before advances leading to widespread commercialization of synthetic seed can occur. Asynchronous embryo development makes the harvest of uniformly mature SEs difficult. Attempts to synchronize better somatic embryo development include physical separation of proembryonal cultures to assure uniform callus size and physiological synchronization by adding ABA and other agents to the culture medium (136). ABA causes the cell water (turgor) content to decrease, thereby slowing embryo growth, which inhibits germination of SEs that would tend to germinate precociously (156).

SEs also do not exhibit a quiet resting or dormancy (quiescence) phase, which is a characteristic of orthodox seeds; instead they continue to grow into seedlings or revert back into disorganized callus tissue. This inability of SEs to produce a resting phase in which all embryos are at the same arrested physiological or morphological state is a major challenge to the development of synthetic seed biotechnology. Attempts to induce dormancy on a par with that of orthodox seeds have met with partial success. Some studies (157–160) indicate that dormancy can be successfully introduced into SEs by dehydrating them through various means. Even after successful introduction of desired quiescence into SEs, it would be necessary to provide a protective coating in the form of synthetic encapsulation for easy handling and direct field planting. Such a protective coating can be designed to assist the SEs during their germination by supplying essential nutrients, antibiotics, and fungicides. Of the two approaches used currently, hydrated (calcium alginate) and dry encapsulation, the latter (using a dry, hardened synthetic coat) is preferred (61).

Automation of synthetic seed production would be necessary to reduce the cost (74); the relative need for automation will vary depending upon crop application. The production of artificial seeds can be divided into several steps, each of which is amenable to automation: (a) growth of cell cultures, (b) sorting of SEs, (c) drying and dehydration, and (d) encapsulation. Large numbers of SEs can be induced to form after reaching a suitable mass. The automation of these steps has been described (156). However, there have been few attempts to grow embryogenic cultures in large bioreactors, because it is difficult to produce well-developed, synchronous SEs in a liquid medium (74). Plant recovery from a suspension culture of alfalfa was found to be less successful than from callus, and it decreased further when scale-up to a larger bioreactor was attempted (151). Improvements in bioreactor design and a better understanding of somatic embryo development under liquid culture conditions are needed to make progress in large-scale production. In addition, methods of automating the dehydration and encapsulation of SEs must also be developed. It may be possible to accomplish dehydration and encapsulation simultaneously using an osmotically active synthetic seed coat (133). Ranch (161) has described an experimental model of a large-scale production of soybean SEs, indicating the potential for synthetic seed and addressing other aspects of synseed technology, such as maturation and conversion of SEs, scale-up of embryo proliferation, and plant conversion.

B. Future Research Needs

An ideal synseed consists of a single, well-formed somatic embryo surrounded by a capsule or coating which simulates the endosperm and seed coat. The embryo alone within its capsule can be hydrated or desiccated or the entire encapsulated embryo also may be desiccated (162). For standardized sowing without significant modification of existing planting equipment, the ideal synseed would consist of an embryo desiccated at the same time as its artificial endosperm and coat. The resulting synseed would then be closely analogous to an orthodox seed. However, the desiccated embryos may be encapsulated at sowing time, using a special seed drill like those used in fluid drilling (163). The capsule itself, most often made of alginate, requires research to improve the following qualities (162):

1. Slow embryo imbibition—the desiccated capsule must reabsorb water slowly to avoid ananoxia of excessive duration.
2. Efficient protection of the embryo against mechanical injuries during and after sowing.
3. Well-equilibrated nutritional provision during embryo germination and heterotrophic growth.

Synseeds produced under aseptic conditions as well as embryo-derived seedlings can be protected using appropriate pesticides incorporated in the capsule. Beneficial microorganisms such as *Trichoderma* and *Pseudomonas* can also be introduced to protect seedlings against *Pythium* and *Fusarium*. Synseeds thus can function as a miniecosystem consisting of embryos associated with all their beneficial microflora, protecting seedlings against pathogens. Herbicides may be delivered with seeds or SEs to reduce the overall amount of agricultural chemicals applied to the environment (22).

SEs produced in vitro are often not provided with nutritive reserves, requiring additions of macronutrients and micronutrients, organic compounds (amino acids), sugars, vitamins, and growth regulators. In leguminous crops, the introduction of symbiotic bacteria, *Rhizobia*, into the capsule coating can help overcome certain trophic environmental stresses like iron deficiency and the absence of mycorrhizae on sterilized soils (162). Synseeds with their associated capsule provide mechanical protection to the embryo and enhanced seedling vigor through judicious additives.

Modern agriculture and horticulture require increasing homogeneity in plant growth and development in the field to ensure good harvest at a desired date. Priming and fluid drilling of synseeds have been envisaged to attain improved germination and uniform plant establishment. Priming consists of soaking or hydrating seeds at low temperatures for a few days in PEG or mineral solutions with potassium nitrate, potassium phosphate, potassium chloride, and ammonium or sodium nitrate (approximately 1 M). Aeration (O_2) in the solution improves the priming procedure. Priming can easily be applied to synseeds before or after the coating is applied. If embryos are planted without a capsule, priming can be done a few days before sowing with fluid-drilling equipment (162).

Most synseeds, especiaoly from recalcitrant species, may be required to be sown after seed germination (seedling sowing). Such pregermination may be achieved by soaking seeds in a column filled with water and aerated (75% O_2 and 25% N_2) for several hours at a defined temperature. After the seedlings have germinated and attained roots of sufficient length, they are sorted by density and sown in a fluid hydrogel via a seed drill. This technique, known as fluid drilling, allows a gain in time of plant establishment, synchrony of plant emergence, and uniform harvest, although it is not suitable for all crops (162). Fluid drilling would be suitable for SEs produced in bioreactors, providing their quality is equal to that of zygotic seeds. Numerous additives may be included in the fluid gel.

According to Deunff (162), synseeds are expected to become a reality where vegetative propagation is useful and for applications requiring genetic purity, such as for hybridization. Synseeds can thus ensure propagation and maintenance of genotypes of high value.

Strong allogamy present in woody plants leads to populations of heterogeneous embryos, requiring a prolonged time to select elite individuals of desired agronomic qualities. The prolonged juvenile phase of these crops also delays clonal evaluation. In addition, transformation of woody plants is limited by problems of large genome sizes and differences in regeneration from the transformed cell. Synseeds of woody plants and forest trees can aid in the evaluation of their agronomic qualities in the nursery and during their transformation, cell fusion, and other genetic manipulations.

C. Limitations of Synseeds

The commercial application of synseeds is limited by the following basic and technological constraints (162):

1. Production of High-Quality Embryos

SEs are frequently obtained without an obligatory callus step to produce induced embryogenic determined cells (IEDC), which allows somaclonal variation, which is detrimental to cloning homogeneity. The potential ability to monitor and control genetic variation is necessary for successful production of synseeds on a commercial scale (164). Although of value for inducing variants, somaclonal variation can be a major obstacle for producing uniform plants. In vitro culture affects organelle genomes (chloroplasts and mitochondria) in addition to changes at chromosomal pieces). The cause of these phenomena needs to be resolved to minimize variation.

The nature, concentration, and composition of culture media, which are generally modified Murashige and Skong (MS) media, need to be optimized to better understand the effects of microelements, including inducing auxins like 2,4-D or NAA. Production of SEs on a large scale will require bioreactor utilization. Heterogeneous production of SEs is a problem at both the genetic and physiological levels. It is therefore necessary to evolve processes that would synchronize cell division to achieve desired uniformity. Other methods besides sieving may prove to be useful, such as thermal shock and inhibitors to arrest embryo development at a desired stage (most likely at the globular stage).

According to Deunff (162), the current progress of synthetic seed technology is able to produce synseeds very quickly on a commercial scale for many plant species needing intense vegetative multiplication. Embryo desiccation has been established in some species, and a general method to solve specific problem is available. Also, encapsulation with hydrogels is available, needing only minor modifications for specific applications. Sowing methods depend on the type of synseed produced. For a desiccated embryo with a coating, classic seeders can be used or modified slightly, and for hydrated and desiccated embryos, a hydrated coating is available or a fluid-drilling seeder can be used or modified to limit the number of empty capsules planted or capsules containing two or three embryos.

The synthetic seed technology requires equipment, technology, and education that few developing countries possess. The situation in these countries, already strained by the purchase of seeds produced in classic ways, is likely further to affect when synseeds reach commercialization (162). Two reviews on synthetic seeds (11,12) have focused on the technology developed, its achievements and prospects, as well as limitations resisting the application of the synthetic seed technology.

REFERENCES

1. V. Raghavan, Somatic embryogenesis, *Molecular Embryology of Flowering Plants*, Cambridge University Press, Cambridge, UK, 1997, pp. 467–499.
2. S.A. Merkle, W.A. Parrot, and B.S. Finn, Morphogenic aspects of somatic embryogenesis, *In Vitro Embryogenesis in Plants* (T.A. Thorpe, ed.), Kluwer Academic, Dordrecht, the Netherlands, 1995, pp. 155–203.
3. K. Nomura and A. Komamine, Physiological and biological aspects of somatic embryogenesis, *In Vitro Embryogenesis in Plants* (T.A. Thorpe, ed.), Kluwer Academic, Dordrecht, the Netherlands, 1995, pp. 249–265.
4. D. Dudits, L. Bogre, and J. Gyorgyey, Molecular and cellular approaches to the analysis of plant embryo development from somatic cells in vitro, *J. Cell Sci.* 99:473 (1991).
5. D. Dudits, J. Gyorgyey, L. Bogre, and L. Bako, Molecular biology of somatic embryogenesis, *In Vitro Embryogenesis in Plants* (T.A. Thorpe, ed.), Kluwer Academic, Dordrecht, the Netherlands, 1995, pp. 267–308.
6. S.S. Bhojwani and M.K. Razdan, *Plant Tissue Culture: Theory and Practice*, Elsevier, Amsterdam, 1983.
7. P.V. Ammirato, The regulation of somatic embryo development in plant cell culture: suspension culture techniques and hormone requirements, *Biotechnology* 1:68 (1983).
8. S.L. Kitto and J. Janick, Production of synthetic seeds by encapsulating asexual embryos of carrot, *J. Am. Soc. Hortic. Sci. 110*:277 (1985).
9. K. Redenbaugh, B.O. Paasch, J.W. Nichol, M.E. Kossler, P.R. Viss, and K.A. Walker, Somatic seeds: Encapsulation of asexual plant embryos, *Biotechnology 4*:797 (1986).
10. K. Redenbaugh, ed., *Synseeds: Application of Synthetic Seeds to Crop Improvement*, CRC Press, Boca Raton, FL, 1993.
11. A. Standardi and E. Piccioni, Recent perspectives on synthetic seed technology using nonembryogenic in vitro–derived explants, *Int. J. Plant Sci. 159*:968 (1998).
12. H. Ara, U. Jaiswal and V.S. Jaiswal, Synthetic seed: Prospects and limitations, *Curr. Sci.* 78:1438 (2000).
13. S.S. Bhojwani and S.P. Bhatnagar, *The Embryology of Angiosperms*, Vikas, New Delhi, 1978, p. 280.
14. P.R. Bell, Angiosperm embryology, *Nature (L) 205*:1044 (1965).
15. W. Halperin, Embros of somatic plant cells, *Control Mechanisms in the Expression of Cellular Phenotypes* (H.A. Pdykulla, ed.), Academic Press, New York, 1970. p. 169.
16. B.M. Johri, Embryogenesis in tissue cultures, *Les Cultures de Tissues de Plantes*, Collog. Int. CNRS, Paris, 1971, p. 269.
17. H.E. Street and L. Withers, The anatomy of embryogenesis in culture, *Tissue Culture and Plant Science* (H.E. Street, ed.), Academic Press, London, 1974, p. 71.

18. V. Raghavan, *Experimental Embryology in Vascular Plants*, Academic Press, London, 1976, p. 603.

19. H.W. Kohlenbach, Comparative somatic embryogenesis, *Frontiers of Plant Tissue Culture*, University of Calgary Press, Alberta, Canada, 1978, p. 59.

20. B. Tissert, E.B. Esan, and T. Murashige, Somatic embryogenesis in angiosperms, *Hortic. Rev. 1*:1 (1979).

21. W.R. Sharp, M.R. Sondahl, L.S. Caldas, and L.S. Muraffa, The physiology of in vitro asexual embryogenesis, *Hortic. Rev. 2*:268 (1980).

22. K. Redenbaugh, J.A. Fujii, and D. Slade, Synthetic seed technology, *Cell Culture and Somatic Cell Genetics of Plants*, Vol. 8 (I.K. Vasil, ed.), Academic Press, New York, 1991, p. 35.

23. J. Reinert, Morphogenese and ihre Kontrolle an Gewebekulturen aus Karotten, *Naturwissenschaft 45*:344 (1958).

24. F. Steward, M. Mapes, and K. Mears, Growth and organized development of cultured cells. I. Organization in cultures grown from freely suspended cells, *Am. J. Bot. 45*:705 (1958).

25. T. Murashige, The impact of plant tissue culture on agriculture, *Frontiers of Plant Tissue Culture* (T.A. Thorpe, ed.), Association for Plant Tissue Culture, University of Calgary Press, Alberta, Canada, 1987, p. 15.

26. K. Redenbaugh, J. Nichol, M.E. Kossler, and B. Paasch, Encapsulation of somatic embryos for artificial seed production (Abstract), *In Vitro 20*:256 (1984).

27. F.C. Steward, Growth and development of cultivated cells. III. Interpretations of the growth from free cell to carrot plant, *Am. J. Bot. 45*:709 (1958).

28. F.C. Steward, P.V. Ammarato, and M.O. Mapes, Growth and development of totipotent cells: Some problems, procedures and perspectives, *Am. J. Bot. 34*:761 (1970).

29. L.H. Jones, Factors influencing embryogenesis in carrot culture (*Daucus carota* L.), *Ann. Bot. 38*:1077 (1974).

30. S.M. Smith and H.E. Street, The decline of embryogenic potential as callus and suspension cultures of carrot (*Daucus carota* L.) are serially subcultured, *Ann. Bot. 38*:223 (1974).

31. R.N. Konar and K. Nataraja, Morphogenesis of isolated floral buds of *Ranunculus sceleratus* L. in vitro, *Acta Bot. (Neerl.) 18*:680 (1969).

32. R.N. Konar, E. Thomas, and H.E. Street, Origin and structure of embryods arising from epidermal cells of the stem of *Ranunculus sceleratus* L., *J.Cell Sci. 11*:77 (1972).

33. E. Thomas, R.N. Konar, and H.E. Street, The fine structure of the embryogenic callus of *Ranunculus sceleratus* L., *J.Cell Sci. 11*:95 (1972).

34. H.W. Kohlenbach, Basic aspects of differentiation and plant regeneration from cell and tissue culture, *Plant Tissue Culture and Its Biotechnological Application* (W. Barz, ed.), Springer-Verlag, Berlin, 1977, p. 355.

35. T.S. Rangan, T. Murashige, and W.P. Bitters, In vitro initiation of nucellar embryos in monoembryonate citrus, *HortScience 3*:226 (1968).

36. J. Kochba and P. Spiegel-Roy, Cell and tissue culture for breeding and development studies of citrus, *HortScience 12*:110 (1977).

37. L.C. Monaco, M.R. Sondal, A. Carvalno, O.J. Crocomo, and W.R. Sharp, Application of tissue culture in the improvement of coffee, *Applied and Fundamental Aspects of Plant Cell, Tissue an Organ Culture* (J. Reinert and Y.P.S. Bajaj, eds.), Springer-Verlag, Berlin, 1977, p.109.

38. M.R. Sondahl, D.A. Spahlinger, and W.R. Sharp, A historical study of high frequency induction of somatic embryos in cultured leaf explants of Coffee arabica L, *Z. Pflanzenphiyol. 94*:185 (1979).

39. E.B. Herman, ed., Desiccated somatic embryos: potential synthetic seeds, *Agricell Rep. 5*:21 (1985).

40. McMillon Journals, A germ of an idea, *Nature (L) 317*:664 (1985).
41. M. Rogers, Synthetic seed technology, *Newsweek 102*:111 (1983).
42. U.B. Barwale, H.R. Kerns, and J.M. Widholm, Plant regeneration from callus cultures of several soybean genotypes via embryogenesis and organogenesis, *Planta 167*:473 (1986).
43. J.P. Ranch, L. Cglesby, and A.C. Zielinski, Plant regeneration from embryo-derived tissue cultures of soybeans, *In Vitro Cell Dev. Biol. 21*:653 (1985).
44. D.J. Gray and B.V. Conger, Nonzygotic embryogenesis in tissue cultures of forage grasses, Proceedings of the 40th Southern Pasture and Forage Crop Improvement Conference, April 16–19, 1984, Baton Rouge, LA, p. 18.
45. V. Vasil and I.K. Vasil, Induction and maintenance of embryogenic callus cultures of Gramineae, *Cell Culture and Somatic Cell Genetics of Plants: Laboratory Procedures and Their Applications*, Vol.1 (I.K. Vasil, ed.), Academic Press, New York, 1984, p. 36.
46. R. Nagmani and J.M. Bonga, Embryogenesis in subcultured callus of Larix decidua, *Can. J. For. Res. 15*:1088 (1985).
47. I. Hakman and S. Von Arnold, Plantlet regeneration through somatic embryogenesis in *Picea abies* (Norway spruce), *J. Plant Physiol. 121*:149 (1985).
48. P.K. Gupta and D.J. Durzan, Somatic polyembryogenesis from callus of mature sugar pine embryos, *Biotechnology 4*:643 (1986).
49. D.J. Styer, Biorector technology for plant propagation, *Proceedings of a Symp. on Tissue Culture* (R.R. Henke, K.W. Hughes, M.J. Constantin, and Hollander, eds.), Plenum Press, New York, 1985, p. 117.
50. P. Farnum, R. Timmis, and J.L. Kulp, Biotechnology of forest yield, *Science 219*:694 (1983).
51. D.J. Gray, Introduction to Symposium, Proc. Symp. On Synthetic Seed Technology for Mass Cloning of Crop Plants: Problems and Perspectives, *HortScience 22*:796 (1987).
52. C.P. Meredith and R.H. Lawrence, Report of the vegetable crops round-table, *Env. Exp. Bot. 21*:401 (1981).
53. B.J. Schenk and A.C. Hildebrandt, Medium and techniques for induction and growth of monocotyledonous and dicotyledonous plant cell culture, *Can J.Bot. 50*:199 (1972).
54. C.L. Wenzel and D.C.W. Brown, Histological events leading to somatic embryo formation in cultured petioles of alfalfa, *In Vitro Cell Dev. Biol. 27*:190 (1991).
55. E.T. Bingham, T.J. McCoy, and K.A. Walker, Alfalfa tissue culture, *Alfalfa and Alfalfa Improvement* (A.A. Hanson, D.K. Barnes and R.R. Hills, eds.), American Society of Agronomy, Madison, WI, 1988, p. 903.
56. K. Finstad, Biochemical and Developmental Markers of Induction of Somatic Embryogenesis in Alfalfa Tissue Culture, Ph.D. dissertation, Carleton University, Ottawa, Canada, 1995.
57. D.A. Stuart and C.M. McCall, Induction of somatic embryogenesis using side chain and ring modified forms of phenoxy acid growth regulators, *Plant Physiol. 99*:111 (1992).
58. D. Dudits, L. Bogre, and J. Guorgyey, Molecular and cellular approaches to the analysis of plant embryo development from somatic cells in vitro, *J. Cell Sci. 99*:475 (1991).
59. K. Shetty and B.D. McKersie, Proline, thioproline and potassium-mediated stimulation of somatic embryogenesis in alfalfa (*Medicago sativa* L.), *Plant Sci. 88*:185 (1993).
60. D. Briskin and J.B. Hanson, How does the plant plasma membrane H^+-ATPase pump protons? *J. Exp. Bot. 43*:269 (1992).
61. B.D. McKersie, Somatic embryogenesis in alfalfa: A model for the development of dry artifial seed technology, *Seed Development and Germination* (J. Kigel and G. Galili, eds.), Marcel Dekker, New York, 1995, p. 833.
62. J.M. Phang, The regulatory functions of proline and pyrroline-5-carboxylic acid, *Curr. Topics Cell Reg. 25*:92 (1985).

63. M. Monteiro, C. Kevers, J. Dommes, and T. Gasper, A specific role for spermidine in the initiation phase of somatic embryogenesis in *Panax ginseng* CA Meyer, *Plant Cell Tiss. Org. Cult.* 68:225 (2002).

64. C.K. Salunkhe, P.S. Rao, and M. Mhatre, Induction of somatic embryogenesis and plantlets in tendrils of *Vitis vinifera* L., *Plant Cell Rep.* 17:65 (1997).

65. L.V. Astarita and M.P. Guerra, early somatic embryogenesis in *Araucaria angustifolia*— induction and maintenance of embryonal-suspensor mass cultures. *R. Bras. Fisiol. Veg.* 10:113 (1998).

66. J. Mithila, S.J. Murch, S. Krishna Raj, and P.K. Saxena, Recent advances in *Pelargonium* in vitro regeneration systems, *Plant Cell Tiss. Org. Cult.* 67:1 (2001).

67. V. Raghavan, *Embryogenesis in Angiosperms*, Cambridge University Press, Cambridge, UK, 1986.

68. Y.P.S. Balal, somatic embryogenesis and its application for crop improvement, *Biotechnology in Agriculture and Forestry*, Vol. 30: *Somatic embryogenesis and Synthetic Seed, I* (Y.P.S. Bajaj, ed.), Springer-Verlag, Berlin, 1995, pp. 105–125.

69. D.C.W. Brown, K.I. Finstad, and E.M. Watson, *Somatic embryogenesis in herbaceous dicots, In Vitro Embryogenesis in Plants* (T.A. Thorpe, ed.), Kluwer Academic, Dordrecht, The Netherlands, 1995, pp. 345–415.

70. R. Nagmani, M.R. Becwar, and S.R. Wann, Single-cell origin and development of somatic embryos of *Picea abies* L. Karst. (Norway spruce) and *P. glauca* (Moench) Voss (white spruce), *Plant Cell Rep.* 6:157 (1987).

71. V. Nuti Ronchi and L. Giorgetti, The cell's commitment to somatic embryogenesis, *Biotechnology in Agriculture and Forestry*, Vol. 30: *Somatic Embryogenesis and Synthetic Seed, I.* (Y.P.S. Bajaj, ed.), Springer-Verlag, Berlin, 1995, pp. 3–19.

72. P.D. Denchev and A.I. Atnassov, Micropropagation through somatic embryos, *Biotechnology in Agriculture and Forestry*, Vol. 30: *Somatic Embryogenesis and Synthetic Seed, I.* (Y.P.S. Bajaj, ed.), Springer-Verlag, Berlin, 1995, pp. 193–206.

73. M.A.L. Smith, Machine vision analysis of plant cells and somatic embryos. *Biotechnology in Agriculture and Forestry*, Vol. 30: *Somatic Embryogenesis and Synthetic Seed, I.* (Y.P.S. Bajaj, ed.), Springer-Verlag, Berlin, 1995, pp. 87–101.

74. D.J. Gray and A. Purohit, Somatic embryogenesis and development of synthetic seed technology, *Crit. Rev. Plant Sci.* 10:33 (1991).

75. K. Redenbaugh, J. Fujii, D. Slade, P. Vis, and M. Kossler, Artificial seeds-encapsulated somatic embryos, *Biotechnology in Agriculture and Forestry*, Vol. 30: *Somatic Embryogenesis and Synthetic Seed, I.* (Y.P.S. Bajaj, ed.), Springer-Verlag, Berlin, 1995, pp. 395–416.

76. D.J. Gray, M.E. Compton, R.C. Harrell, and D.J. Cantliffe, Somatic embryogenesis and the technology of synthetic seed, *Biotechnology in Agriculture and Forestry*, Vol. 30: *Somatic Embryogenesis and Synthetic Seed, I.* (Y.P.S. Bajaj, ed.), Springer-Verlag, Berlin, 1995, pp. 126–151.

77. H. Kunitake and M. Mii, Somatic Embryogenesis in *Citrus* species, *Biotechnology in Agriculture and Forestry*, Vol. 30: *Somatic Embryogenesis and Synthetic Seed, I.* (Y.P.S. Bajaj, ed.), Springer-Verlag, Berlin, 1995, pp. 280–298.

78. A.M. Nuutila, U. Kurten, R. Puupponen-Pima, J. Hamalainen, L. Mannonen, and V. Kauppinen, Somatic embryogenesis in birches (*Betula* spp.), *Biotechnology in Agriculture and Forestry*, Vol. 30: *Somatic Embryogenesis and Synthetic Seed, I.* (Y.P.S. Bajaj, ed.), Springer-Verlag, Berlin, 1995, pp. 246–259.

79. J.E.A. Seabrook and L.K. Douglass, Somatic embryogenesis on various potato tissues from a range of genotypes and ploidy levels, *Plant Cell Rep.* 20:175 (2001).

80. J.E.A. Seabrook, L.K. Douglass, and G.C.C. Tai, Segregation for somatic embryogenesis on stem-internode explants from potato seedlings, *Plant Cell Tiss. Org. Cult.* 65:69 (2001).

81. M.E. Compton and J.M. Koch, Influence of plant preservative mixture (PPM) on adventitious organogenesis in melon, petunia and tobacco, *In Vitro Cell Dev. Biol. Plant 37*:259 (2001).

82. D.Y. Xie and Y. Hong, Regeneration of *Acacia mangium* through somatic embryogenesis, *Plant Cell Rep. 20*:34 (2001).

83. S.W. Kim, S.C. Oh, D.S. In, and J.R. Liu, High frequency somatic embryogenesis and plant regeneration in zygotic embryo cultures of *Liriope platyphylla* Wang et Tang, *Plant Cell Tiss. Org. Cult. 63*:227 (2000).

84. M.G. Mullins and C. Srinivasan, Somatic embryos and plantlets from an ancient clone of the grapevine (cv. Cabernet-Sauvignon) by apomixes in vitro, *J.Exp. Bot. 27*:1022 (1976).

85. C.K. Salunkhe, P.S. Rao, and M. Mhatre, Plantlet regeneration via somatic embryogenesis in anther callus of *Vitis latifolia* L., *Plant Cell Rep. 18*:670 (1999).

86. M.E. Compton and D.J. Gray, Somatic embryogenesis and plant regeneration from immature cotyledons of watermelon, *Plant Cell Rep. 12*:61 (1993).

87. K.A. Walker and S.J. Sato, Morphogenesis in callus tissue of *Medicago sativa*: the role of ammonium ion in somatic embryogenesis, *Plant Cell Tiss. Org. Cult. 1*:109 (1981).

88. V. Walbot, Control mechanisms for plant embryogeny, *Dormancy and Developmental Arrest* (M.E. Clutter, ed.), Academic Press, New York, 978. p. 113.

89. J.G. Carman, Embryogenic cells in plant tissue cultures: occurrence and behavior, *In Vitro Cell Dev. Biol. 26*:746 (1990).

90. P.V. Ammarato, Organization events during somatic embryogenesis, *Plant Tissue and Cell Culture* (C.E. Green, D.A. Somers, W.P. Hackett, and D.O. Biesboer, eds.), Liss, New York, 1987, p. 57.

91. N. Ku, K.M. Coulter, J.E. Krochko, and J.D. Bewley, Morphological stages and storage protein accumulation in developing alfalfa (*Medicago sativa* L.) seeds, *Seed Sci. Res. 119*:1 (1991).

92. D.A. Stuart, J. Nelsen, and J.W. Nichol, Expression of 7S and 11S alfalfa seed storage proteins in somatic embryos. *J. Plant Physiol. 132*:134 (1988).

93. K. Anandarajah and B.D. McKersie, The influence of plating density, sucrose and light during development on the germination and vigor of *Medicago sativa* L. embryos after desiccation, *Seed Sci. Res. 2*:133 (1992).

94. B.D. MacKersie, T.Senaratna, S.R. Bowley, D.C.W. Brown, J.E. Krochko, and J.D. Bewley, Application of artificial seed technology in the production of hybrid alfalfa (*Medicago sativa* L.), *In Vitro Cell Dev. Biol. 25*:1183 (1989).

95. F. Lai and B.D. McKersie, Scale-up of somatic embryogenesis in alfalfa (*Medicago sativa* L.), I. Subculture and indirect secondary somatic embryogenesis, *Plant Cell Tiss. Org. Cult. 37*:151 (1994).

96. E.T. Bingham, L.V. Hurley, D.M. Kaatz, and J.N. Sanders, Breeding alfalfa, which regenerates from callus cultures, *Crop Sci. 15*:719 (1975).

97. K. Anandarajah and B.D. McKersie, Enhanced vigor of dry somatic embryos of *Medicago sativa* L. with increased sucrose, *Plant Sci. 71*:261 (1990).

98. K. Anandararah and B.D. McKersie, Manipulating the desiccation tolerance and vigor of dry somatic embryos of *Medicago sativa* L. with sucrose, heat shock and abscisic acid, *Plant Cell Rep. 9*:451 (1990).

99. F. Lai, T. Senaratna and B.D. McKersie, Gutamine enhances storage protein synthesis in *Medicago sativa* L. somatic embryos, *Plant Sci. 87*:69 (1992).

100. F. Lai and B.D. McKersie, Effect of nutrition on maturation of alfalfa (*Medicago sativa* L.) somatic embryos, *Plant Sci. 91*:87 (1993).

101. H. Etienne, B. Bertrand, F. Anthony, F. Cote, and F. Berthouly, Somatic embryogenesis, a tool for coffee genetic improvement, 17th International Scientific Colloquium on Coffee, Nairoby, Kenya, ASIC Publishers, Vvey, Switzerland, 1997, pp. 457–465.

102. D. Etienne-Barry, B. Bertrand, N. Vasquez, and H. Etienne, Direct sowing of *Coffea arabica* somatic embryos mass-produced in a bioreactor and regeneration of plant, *Plant Cell Rep. 19*:111 (1999).

103. D. Etienne-Barry, B. Bertrand, A. Schlonvoigt, and H. Etienne, The morphological variability with in a population of somatic embryos produced in a bioreactor affects the regeneration and the development of plants in the nursery, *Plant Cell Tiss. Org. Cult. 68*:153 (2002).

104. W. Halperin, In vitro embryogenesis: some historical issues and unresolved problems, *In Vitro Embryogenesis in Plants* (T.A. Thorpe, ed.), Kluwer Academic, Dordrecht, The Netherlands, 1995, pp. 1–16.

105. K. Ozawa, D.-H. Ling, and A. Komamine, High- frequency somatic embryogenesis from small suspension-cultured clusters of cell of an interspecific hybrid of *Oryza*, *Plant Cell Tiss. Org. Cult. 46*:157 (1996).

106. J.D. Barrett, Y.S. Park, and J.M. Bonga, The effectiveness of various nitrogen sources on white spruce [*Picea glauca* (Moench) Voss] somatic embryogenesis, *Plant Cell Rep. 16*:411 (1997).

107. M.E. Gonzalez-Benito, J.M.F-C Carvalho, and C. Perez, somatic embryogenesis of an early cotton cultivar, *Pesq. Agrop. Bras. 32*:485 (1997).

108. B.G. Young, D.L. Jack, D.W. Smith, and M.H. Saier, Jr., The amino acid/auxin: proton symport perminease family, *Biochip. Biophys. Acta 1415*:306 (1999).

109. L.L. Dal Vesco and M.P. Guerra, Effectiveness of nitrogen sources in *Feijoa* somatic embryogenesis, *Plant Cell Tiss. Org. Cult. 64*:19 (2001).

110. L.V. Astarita, M.P. Guerra, Conditioning of the culture medium by suspension cells and formation of somatic proembryo in *Araucaria angustifolia* (Coniferae), *In Vitro Cell Dev. Biol. Plant 36*:194 (2000).

111. S.R.L. Fuentes, M.B.P. Calheiros, J. Manetti-Filho, and L.G.E. Vieira, The effects of silver nitrate and different carbohydrate sources on somatic embryogenesis in *Coffea canephora*, *Plant Cell Tiss. Org. Cult. 60*:5 (2000).

112. R.L.K. Drew, The development of carrot (*Daucus carota* L.) embryoids (derived from cell suspension culture) into plantlets on sugar free basal medium, *Hortic Res. 19*:79 (1979).

113. S.L. Kitto, W.G. Pill, and D.M. Molloy, Fluid drilling as a delivery system for somatic embryo-derived plantlets of carrot (*Daucus carota* L.), *Sci. Hortic. 47*:209 (1991).

114. K. Redenbaugh, D.T. Slide, P. Viss, and J.A. Fujii, Encapsulation of somatic embryos in synthetic seed coats, *HortScience 22*:803 (1987).

115. J.A. Fujii, D.T. Slade, K. Redenbaugh, and K.A. Walker, Artificial seeds for plant propagation, *Trends Biotechnol. 5*:335 (1987).

116. S.L. Kitto and J. Janick, Hardening treatments increased survival of synthetically-coated asexual embryos of carrot, *J. Am. Soc. Hortic. Sci. 110*:283 (1985).

117. J. Janick, S.L. Kitto, and Y.H. Kim, Production of synthetic seed by desiccation and encapsulation, *In Vitro Cell Dev. Biol. 25*:1167 (1989).

118. Y.H. Kim and J. Janick, Abscisic acid and proline improve desiccation tolerance and increase fatty acid content of celery somatic embryos, *Plant Cell Tiss. Org. Cult. 24*:83 (1991).

119. D.J. Gray, B.V. Conger, and D.D. Songstad, Desiccated quiescent somatic embryos of orchard grass for use as synthetic seeds, *In Vitro Cell Dev. Biol. 23*:29 (1987).

120. D.J. Gray, Effects of dehydration and exogenous growth regulations on dormancy, quiescence and germination of grape somatic embryos, *In Vitro Cell Dev. Biol. 25*:1173 (1989).

121. T. Senaratna, B.D. McKersie, and S.R. Bowley, Desiccation tolerance of alfalfa (*Meddicago sativa* L.) somatic embryos: influence of abscisic acid, stress pretreatments and drying rates, *Plant Sci. 65*:253 (1989).

122. T. Senaratna, B.D. McKersie, and S.R. Bowley, Artificial seeds of alfalfa (*Medicago sativa* L.): induction of desiccation tolerance in somatic embryos, *In Vitro Cell Dev. Biol. 26*:85 (1990).

123. M.C. Compton, C.M. Benton, D.J. Gray, and D.D. Songstad, Plant recovery from maize somatic embryos subjected to controlled relative humidity dehydration, *In Vitro Cell Dev. Biol. Plant 28*:197 (1992).

124. R.R. Finkelstein and M.L. Crouch, Hormonal and osmotic effects on developmental potential of maturing rapeseed, *HortScience 122*:797 (1987).

125. R.R. Finkelstein, K.M. Tenbarge, J.E. Shumway, and M.L. Crouch, Role of ABA in maturation of rapeseed embryos, *Plant Physiol. 78*:630 (1985).

126. R.R. Finkelstein and M.L. Crouch, Rapeseed embryo development in culture on high osmoticum is similar to that in seeds, *Plant Physiol. 81*:907 (1986).

127. K. Anandarajah, L. Kott, W.D. Beversdorf, and B.D. McKersie, Induction of desiccation tolerance in microspore-derived embryos of *Brassica napus* L. by thermal stress, *Plant Sci. 77*:119 (1991).

128. S.M. Attree, D. Moore, U.K. Sawhney, and L.C. Fowke, Enhanced maturation and desiccation tolerance of white spruce [*Picea glauca* (Moench) Voss] somatic embryos: Effects of a non-plasmolysing water stress and abscisic acid, *Ann. Bot. 68*:519 (1991).

129. C.F. Lecouteux, F. Lai, and B.D. McKersie, Maturation of alfalfa (*Medicago sativa* L.) somatic embryos by abscisci acid, sucrose and chilling, *Plant Sci. 94*:207 (1993).

130. D.R. Roberts, B.C. Sutton, and B.S. Flinn, Synchronous and high frequency germination of interior spruce somatic embryos following partial drying at high relative humidity, *Can. J. Bot. 68*:1086 (1990).

131. D.R. Roberts, WR. Lazaroff, and F.B. Webster, Interaction between maturation and high relative humidity treatments and their effects on germination of Sitka spruce somatic embryos, *J. Plant Physiol. 138*:1 (1991).

132. Y.M. Kim and J. Janick, Synthetic seed technology: Improving desiccation tolerance of somatic embryos of celery, *Acta Hortic. 280*:23 (1990).

133. D.J. Gray, Quiescence in monocotyledonous and dicotyledonous somatic embryos induced by dehydration, *HortScience 22*:810 (1987).

134. D.J. Gray and B.V. Conger, Desiccated quiescent somatic embryos of orchard grass for use as synthetic seed, *In Vitro Cell Dev. Biol. 23*:29 (1987).

135. S.K. Datta and I. Potrykus, Artificial seed in barley: Encapsulation of microspore-derived embryos, *Theorit. Appl. Genet. 77*:820 (1989).

136. L.O. Copeland and M.B. McDonald, *Principles of Seed Science and Technology*, 3rd ed., Chapman and Hall, New York, 1995, p.258.

137. J.R. Soneji, P.S. Rao, and M. Mhatre, Germination of synthetic seeds of pineapple (*Ananas comosus* L. Merr.), *Plant Cell Rep. 20*:891 (2002).

138. M. Micheli, S. Pellgrino, E. Piccioni, and A. Standardi, Effect of double encaplulation and coating on synthetic seed conversion in M.26 apple rootstock, *J. Microencapsulation 19*:347 (2002).

138a. M. Sicurani, E. Piccioni, and A. Standardi, Micropropagation and synthetic seed in M.26 apple rootstocks. I: attepts towards mechanization of organogenesis and encapsulation of adventitious shoot tips in M.26 apple clonal rootstocks, *Plant Cell Tiss. Org. Cult. 66*:207 (2001).

139. M. Adriani, E. Picconi, and A. Standardi, Effect of different treatments on the conversion of 'Hayward' kiwifruit synthetic seeds to whole plants following encapsulation of in vitro-derived buds, *NZ. J. Crop Hortic. Sci. 28*:59 (2000).

140. H. Ara, U. Jaiswal, and V.S. Jaiswal, germination and plantlet regeneration from encapsulated somatic embryos of mango (*Mangifera indica* L.), *Plant Cell Rep. 19*:166 (1999).

141. T. Gardi, E. Picconi, and Standardi, Effect of bead nutrition composition on regrowth of

stored vitro-derived encapsulated microcuttings of different woody species, *J. Micro-encapsulation 16*:13 (1999).

142. M. Micheli, M. Mencuccini, and A, Standardi, Encapsulation of in vitro proliferated buds of olive, *Adv. Hortic. Sci. 12*:163 (1998).

143. M.P. Guerra, L.L. Dal Vesco, J.P.H.J. Ducroquet, R.S. Nodari, and M.S. Dos Reis, somatic embryogenesis in *Goiabeira cerrana*: Genotype response, auxinic shock and synthetic seeds, *R. Bras. Fisiol. Veg. 13*:117 (2001).

144. R. Brischia, E. Piccioni, and A. Standardi, Micropropagation and synthetic seed in M.26 apple rootstock, II: A new protocol for production of encapsulated differentiating propagules, *Plant Cell Tiss. Org. Cult. 68*:137 (2002).

145. K.D. Halluin, J. Botterman, and W. DeGreef, Engineering of herbicide-resistant alfalfa and evaluation under field conditions, *Crop Sci. 30*:866 (1990).

146. B.D. McKersie, Y. Chen, M. deBeus, S.R. Bowley, C. Bowler, D. Inze, K.D'Halluin, and J. Botterman, Superoxide dismutase enhances tolerance of freezing stress in transgenic alfalfa (*Medicago sativa* L.), *Plant Physiol. 103*:1155 (1993).

147. H. Etienne, F. Anthony, S. Dussert, D. Fernandez, P. Lashermes, and B. Bertrandtrand, Biotechnological applications for the improvement of coffee (*Coffea arabica* L.). *In Vitro Cell Dev. Biol. Plant 38*:129 (2002).

148. H. Etienne and B. Bertrand, Trueness-to-type and agronomic characteristics of *Coffea arabica* trees micropropagated by the embryogenic cell suspension technique, *Tree Physiol. 21*:1031 (2001).

149. V. Ramanatha Rao and T. Hodgkin, Genetic diversity and conservation and utilization of plant genetic resources, *Plant Cell Tiss. Org Cult. 8*:1 (2002).

150. B.D. McKersie and S.R. Bowley, Synthetic seeds of alfalfa, *Synseeds: Application of Synthetic Seeds to Crop Improvement* (K. Redenbaugh, ed.), CRC Press, Boca Raton, FL, 1993, p. 231.

151. D.A. Stuart, S.G. Strikland, and K.A. Walker, Bioreactor production of alfalfa somatic embryos, *HortScience 22*:800 (1987).

152. R.J.H. Kessel and A.H. Karr, The effect of dissociated oxygen concentration on growth and differentiation of carrot (*Daucus carota* L.) tissue, *J. Exp. Bot. 23*:996 (1972).

153. S.R. Bowley, G.A. Kielly, K. Anandarajah, B.D. McKersie, and T. Senaratna, Field evaluation following two cycles of back cross transfer of somatic embryogenesis to commercial alfalfa germplasm, *Can. J. Plant Sci. 73*:131 (1993).

154. M.A.M. Aly, B. Rathinasabapathi, and K. Kelley, Somatic embryogenesis in perennial statice, *Limonium bellidifolium*, Plumbaginaceae, *Plant Cell Tiss. Org. Cult. 68*:127 (2002).

155. X-Q Li, Somatic embryogenesis and synthetic seed technology using carrot as a model system, *Synseeds: Application of Synthetic Seeds to Crop Improvement* (K. Redenbaugh, ed.), CRC Press, Boca Raton, FL, 1993, pp. 289–304.

156. D.J. Gray, Synthetic seed for clonal production of crop plants, *Recent Advances in the Development and Germination of Seeds* (R.B. Taylorson, ed.), Plenum Press, New York, 1990, p. 29.

157. B.D. McKersie, S.R. Bowley, T. Senaratna, D.C.W. Brown, and J.D. Bowley, Application of artificial seed technology in the production of alfalfa (*Medicago sativa* L.), *In Vitro Cell Dev. Biol. 24*:71 (1988).

158. J.G. Carman, Improved somatic embryogenesis in wheat by partial simulation of the in-ovulo oxygen, growth regulator and desiccation environments, *Planta 175*:417 (1988).

159. T. Senaratna, S.D. McKersie, and D.C. Brown, Artificial seeds: Desiccated somatic embryos, *In Vitro Cell Dev. Biol. 25*:89 (1989).

160. K. redenbaugh, J.A. Fujii, and D. Slade, Encapsulated plant embryos, *Biotechnology in Agriculture* (A. Mizrani, ed.), Liss, New York, 1988, p. 225.

161. J.P. Ranch, The potential for synthetic soybean seed, *Synseeds: Application of Synthetic*

Seeds to Crop Improvement (K. Redenbaugh, ed.), CRC Press, Boca Raton, FL, 1993, p. 163.

162. Y.L. Deunff, Conclusions and future, *Synseeds: Application of Synthetic Seeds to Crop Improvement* (K. Redenbaugh, ed.), CRC Press, Boca Raton, FL, 1993, pp. 453–461.

163. D. Gray, The effect of sowing pregerminated seeds of lettuce on seedling emergence, *Ann. Appl. Biol. 88*:185 (1978).

164. S. Shohet and P.D.S. Calligari, Variation in somatic embryos: problems and prospects, *Bio/Technology 90*:101 (1990).

24
Transgenic Seeds, Plants, and Crops

I. INTRODUCTION

Genetically modified (GM), or transgenic, seeds, plants, and crops have been the focus of considerable attention of the farmers, agricultural scientists, molecular biologists, the general public, and the media. Barring the one-sided view of the general public to consider GM crops as the latest incarnation of evil biotechnology, most agricultural scientists and biotechnologists are convinced that genetic engineering of plants represents a technology with enormous potential to increase food production in an environmentally benign way. The human population is growing faster than anticipated. According to the recently publised estimates, the world population has been projected to be 9.3 billion in 2050, which is 400 million more than previously estimated. It is an enormous challenge to feed all of these people and thus prevent famine, upheaval, or civil war. Also, more and better food is needed; at least for the majority of people living in the developing countries.

One of the most promising approaches currently available to improve the quality, diversity, and yield of agricultural crops is the evolution of transgenic plants or crops through genetic engineering. GM, or transgenic, seeds, plants, and crops can be produced by manipulating plant cells or organs at the molecular level leading to the introduction, integration, and expression of specific and useful characteristics of foreign genetic material in a host plant. Such genetic manipulations have enabled us to produce new crops and their varieties with useful agronomic traits such as better quality, higher yields, and insect, viral, or herbicidal resistance as well as to generate custom-made male-sterile plants useful in hybrid seed production. In addition, GM crops can be produced with improved postharvest qualities of perishable foods like fruits and vegetables (shelf life enhancement) and crops that synthesize a variety of pharmaceutical chemicals, vaccines, vitamins, and a number of other useful industrial products (1,3).

The subject of genetic engineering of crop plants to produce GM plants or transgenic crops has been dealt with extensively in the literature (1–8). Improvement of agricultural crops so far was based on selection of naturally occurring variation within a given plant species, with induced mutations helping to increase the genetic variation only moderately. Crop improvement from interspecific crosses and hybridization has also met with limited success. Gene-transfer techniques, such as *Agrobacterium*-mediated gene transfer, have changed the situation dramatically. It is now possible (at least theoretically) to use any given character, from any given organism, in any given plant species to do this. In addition to gene transfer, genetic engineering also offers the opportunity to inactivate specific undesirable endogenous genes. However, there

Table 1 Trasgenetic Plants Obtained by Three
Gene-Transfer Techniques

English name	Latin name
I. *Agrobacterium*-Mediated Transformation	
Alfalfa	*Medicago sativa*
Apple	*Malus pumila*
Apricot	*Prunus armeniaca*
Asparagus	*Asparagus officinalis*
Aspen	*Populus tremula*
Birdsfoot trefoil	*Lotus corniculatus*
Black currant	*Ribes nigrum*
Carnation	*Dianthus* spp., *Caryophyllus*
Carrizo citrange	*Citrus* spp.
Carrot	*Daucus carato*
Cauliflower	*Brassica oleracea*
Celery	*Apium graveoleus*
Chickpea	*Cicer arietinum*
Chicory	*Cichorium endivia*
Chrysanthemum	*Chrysanthemum* spp.
Cotton	*Gossypium hirsutum*
Cucumber	*Cucumis sativus*
Flax	*Linum usitatissimum*
Gerbera	*Gerbera hybrida*
Grapevine	*Vitis* spp.
Kiwi fruit	*Actinidia chinensis*
Lettuce	*Lactuca sativa*
Muskmelon	*Cucumis melo*
Mustard	*Brassica juncea*
Papaya	*Carica papaya*
Passion fruit	*Passiflora edulis*
Pea	*Pisum sativum*
Peanut	*Arachis hypogaea*
Pecan	*Carya ovata*
Pepino	*Solanum muricatum*
Plum	*Prunus domestica*
Poplar	*Populas* spp.
Potato	*Solanum tuberosum*
Ramie	*Boehmeria nivea*
Rice	*Oryza sativa*
Soybean	*Glycine max*
Strawberry	*Fragaria x Anannassa*
Stylo	*Stylosanthes humilis*
Subterranean clover	*Trifolium subterraneum*
Sugar beet	*Beta vulgaris*
Sunflower	*Helianthus annuus*
Sweetgum	*Liquidambar styraciflua*
Tamarillo	*Cyphmandra betacea*
Tomatillo	*Physalis ixocarpa*
Tomato	*Lycopersicon esculentum*
Trifoliate orange	*Poncirus trifoliate*
Tobacco	*Nicotiana tabacum*

Table 1 Continued

English name	Latin name
Walnut	*Juglans regia*
Watermelon	*Citrullus vulgaris*
White clover	*Trifolium repens*
White mustard	*Sinapis alba*

II. Ballistic Transformation

Barley	*Hordeum vulgare*
Bean	*Phaseolus vulgaris*
Cotton	*Gossypium hirsutum*
Corn	*Zea mays*
Cranberry	*Vaccinium macrocarpon*
Fescue, red	*Festula rubra*
Fescue tall	*Festula arundinacea*
Oat	*Avena sativa*
Oilseed rape	*Brassica napus*
Orchid	*Dendrobium* spp.
Papaya	*Carica papaya*
Peanut	*Arachis hypogaea*
Poplar	*Populus* spp.
Rice	*Oryza sativa*
Rye	*Secale cereale*
Sorghum	*Sorghum bicolor*
Soybean	*Glycine max*
Sugarcane	*Saccharum officinarum*
Sunflower	*Helianthus annuus*
Tritordeum	*Hordeum x Triticum*
Tobacco	*Nicotiana tabacum*
Bent grass, Redtop	*Agrostis palustris*
Wheat	*Triticum aestium*
White spruce	*Picea glauca*

III. Direct Gene Transfer to Protoplasts

Corn	*Zea mays*
Fescue, red	*Festula rubra*
Fescue, tall	*Festula arundinacea*
Orchard grass	*Dactylis glomerata*
Petunia	*Petunia hybrida*
Potato	*Solanum tuberosum*
Rice	*Oryza sativa*
Tobacco	*Nicotiana tabacum*
Turf grasses: creeping	*Agrostis alba*
Wheat	*Triticum aestivum*

Source: Ref. 1.

are numerous practical problems interfering with the construction of novel miracle crop plants. Genes of desired agronomic characteristics are often not easy to identity and isolate, and gene expression mostly depends on the random integration into the host genome than on the appropriate expression signals (1). Also, gene-transfer techniques yielding thousands of GM plants on a laboratory scale require an enormous investment in terms of manpower and persistence as well as on modern technological equipment to produce a few transgenic plants of commercial interest.

The first transgenic, or GM, "model plant" was produced in 1983 (9). Since then transgenic plants have been recovered from many important crops of commercial importance. Potrykus et al. (1) grouped transgenic plants obtained by gene-transfer techniques into three categories (Table 1). In addition to *Agrobacterium*-mediated, ballistic, and direct gene transfer, there are other techniques that have produced confirmed transgenic plants with the promise for future development; microtargeting, microinjection, electroporation into tissues and whiskers (1). The production of transgenic plants depends on the effectiveness of the gene-transfer technique as well as on the appropriate use of signals to regulate transgene expression and on the vectors used to amplify and transfer the gene. Transformation being a rare event, selectable marker genes or visible reporter genes are normally cotransformed along with the desire gene of agronomic interest, which in turn enables identification of the transgenic clones among large populations of untransformed cells.

II. PLANT GENETIC MANIPULATION

Plant genetic engineering may be simply defined as the isolation, introduction, and expression of foreign DNA in the plant (recombinant DNA technology). However, broadly focusing on the molecular level organization, plant genetic manipulation involves the interfacing of all aspects of cellular and molecular biology and gene-transfer techniques (10). The genetic engineering tools and techniques of tissue culture, somaclonal and gametoclonal variation, cellular selection methods, and recombinant DNA technology either directly or indirectly aim to enhance genetic expression and gene transfer.

The problem of increasing food production without harming the environment will require the concerted use of traditional breeding and organic farming as well as GM crop technology (11). The traditional interspecific hybridization is afflicted with a problem of low probability of obtaining in one individual the desired combination of genes from the parent lines. The speciation process develops reproductive isolation barriers that maintain the integrity of plant species and thus restrict the flow of gens from one to another. According to Cocking (10), it is advantages to pursue efforts to obtain sexual hybrids to their limits before attempting to utilize more sophisticated GM technology. The sexual hybridization of African violet (*Sainpaulia ionantha*) with the alpine wild species of African violet (*S. shumensis*), for example, has successfully produced hybrids that exhibit lower temperature requirements for flowering than *S. ionantha*, whose optimum temperature for flowering is 21°C. This had led to the production of the endurance or cold preference lines of African violet that prefer 5°C cooler cultural temperatures. Also, the "miracle" rices, IR8 and IR36, have been produced by a sophisticated sexual hybridization program (10).

A. Somaclonal and Protoclonal Variation

Somaclonal and gametoclonal variation, arising as a consequence of tissue culture and micropropagation, has been reported in almost all plant species regenerated from in vitro propagation. Among the array of somaclonal variants derived from in vitro culture, the mutants result from changes at preexisting loci, with the evidence for entirely new mutant being only circumstantial. Some genetic events, which occur at a very slow rate spontaneously, occur far more frequently during in vitro cell and tissue culture; somaclonal variation thus may give rise to new genes (12).

Somaclonal variation in plant genetic manipulations has a great potential to create additional genetic variability in coadapted, agronomically useful plant species without the need to resort to hybridization or the production of transgenic plants. Scowcroft and Larkin (12) pointed out that chromosomal rearrangements occur in plants regenerated from tissue culture. Studies on protoclonal variation in the seed progeny of plants regenerated from rice protoplasts showed a wide range of phenotypic characters, and some individuals showing positive variability could be selected preferentially from the population. The assessment of protoclonal variation in rice was greatly facilitated by the availability of a single-cell protoplast system of efficiently regenerating into plantlets through somatic embryogenesis. The development of finely divided fast-growing suspension lines was necessary for efficient protoplast isolation and division and eventual plant regeneration (13). According to Roest and Gilissen (14), cell colonies and calluses were developed from the protoplasts of 212 higher plant species, representing 96 genera of 31 families, which regenerated into embryolike structures, embryoids, or shoots and when cultured on appropriate media developed into whole plants.

B. Somatic Hybridization by Protoplast Fusion

Production of transgenic plants by somatic cell hybridization through protoplast fusion overcomes sexual incompatibility barriers, creating a novel cytoplasmic mix, as organelles of the two fusion partners become compatible to form somatic hybrids. This technique provides a unique opportunity to produce GM plants with new combinations of cytoplasmic genes between the species in one step, thus circumventing time-consuming sexual backcross programs both in sexually compatible and incompatible crosses. Protoplast fusions may be carried out effectively using millions of protoplasts either electrically or using low carbonyl polyethylene glycol (PEG) to obtain a high fusion frequency with better heterokaryon viability and a low level of clumping (15). Fusion products can be isolated directly employing fluorescence-activated cell sorting for rapid isolation of large number of hybrid cells (16). Alternatively, electrofusion of pairs of protoplasts can be observed directly using microfusion methods in droplets of medium through an inverted microscope with suitable electrodes (17). Although technically more demanding and difficult to scale-up, the latter procedure avoids the need to select fusion products (10).

Genetic manipulation has been achieved in numerous plant species by protoplast fusion, especially in the Solanaceae and the Cruciferae, and more recently in the cereals with the availability of efficient procedures for plant regeneration from protoplasts. Finch et al. (18) reported numerous exciting developments from both nuclear gene transfer and cytoplasmic gene transfer in rice, primarily associated with cytoplasmic male sterility (CMS). In these procedures, iodoacetamide was used to inhibit the division of one of the parental rice protoplast lines in the case of nuclear gene transfer, and with gamma-irradiation in the other protoplast system in case of cytoplasmic hybrid rice production (10).

The current interest in the use of fusion technology has shifted from the production of novel hybrids to chromosome transfer and gene introgression (19), although interest in somatic hybridization of some crops like potato continues as an additional tool of crop improvement. In most cases of somatic hybridization, an insufficient number of fusion products were selected and regenerated into whole plants for an adequate evaluation of hybrid fertility; for example, from 21 experiments, only 3 yielded somatic hybrids of *Lycopersicon esculentum* × *L. peruvianum* (2n = 6x = 72), which were hexaploid, fertile, and set seed after self-pollination (20). In the latter case, protoplast fusion resulted in the production of fertile somatic hybrid plants, when sexual crosses (unless produced with difficulty) usually do not produce fertile tomato plants (10).

The possibility of transferring only part of the genome in somatic hybridization to facilitate maintenance of fertility has been thoroughly examined. Research has indicated that irradiation directs the process of chromosome elimination, and the amount of donor DNA undergoing elimination after fusion remains viable and random. More encouragingly the fusion of somatic protoplasts with gametic protoplasts has enabled us to produce triploids with good self- and cross-fertility (21). Pental and Cocking (22) also proposed that triploid plants (ABB) could be produced by fusing protoplasts isolated from microspores at the tetrad stage (n) of species A with the protoplasts isolated from somatic cells (2n) of species B. Production of triploids through gametosomatic fusions enables us to bypass prezygotic and postzygotic sexual incompatibility barriers of the conventional sexual crosses. Pental and Cocking (22) suggested that the ability to produce such interspecific triploids possessing the haploid genome of the alien species may facilitate the transfer of only part of the genome (one or a few chromosomes) from an alien species in to the cultivated species; an alternative may be actual chromosome-mediated transformation (23). *Petunia hybrida* protoplasts, for example, were genetically transformed via microinjection of chromosomes isolated from *P. alpicola*, resulting in changes in the relative activity of enzymes of the flavanoid biosynthesis in the Mendelian manner (10).

C. Genetic Transformation

The production of transgenic plants by genetic transformation in crop plants has been reviewed generally (24) and more specifically in relation to gene transfers in cereals (25). The current limitations in the production of transgenic plants by transformation indicate that it has not yet been possible to utilize *Agrobacterium* to mediate gene transfer into cereals either by direct bacterial interaction with the wounded plant or with explants or by cocultivation with protoplast-derived cells and suspension cultures (10). The latter has led to increased interest in assessing other methods of transformation to produce GM plants, such as the use of DNA-coated microprojectiles, DNA microprojection, using a special single-cell, protoplast injection technique (17), and direct uptake of DNA into protoplasts stimulated either electrically or chemically. Each of these techniques has its own specific limitations. However, it has become evident that the actual delivery of DNA into plant cells either by microprojectiles, microinjection, or direct uptake through an exposed plasma membrane has not remained a major problem. The problem remained to be resolved is the actual competence of the cell receiving the DNA to express that DNA and develop the plant from transformed cells to enable the DNA to become adequately incorporated into the plant germ line to be transmitted to future generations (10).

Christou et al. (26) targeted immature soybean embryos from commercially important cultivated varieties by rapidly accelerated, DNA-coated gold particles. The cell transformation rates in the embryos were of the order of 1 in 10^5. The plants derived from treatment of intact tissues were thought to be chimeras of both transformed and nontransformed cells, complicating the identification and selection of rare transformation events. The recovery of transformation among the self-pollinated progeny of treated plants may be effective if transformation of the germ line can be achieved at practical rates. When particle acceleration was used by electric discharge to introduce DNA-coated free particles into immature soybean seed meristems (27), around 2% of shoots derived from these meristems were chimeric for gene expression, and only one plant was produced from the chimeric plant progeny. The application of a somewhat analogous DNA microinjection approach in cereals for microinjecting selectable marker genes with microspore-derived proembryos of oil seed rape, giving rise to transgenic chimeras, also faced similar problems (28). Potrykus et al. (29) microinjected marker genes into the microspores of several cereals such as wheat, rice, maize, and barley and analyzed the sexual offsprings of regenerated plants for the transgenic properties with no success.

Thus, the only reliable transformation in cereals at present yielding transgenic plants appears to be based on direct gene transfer employing protoplast-based protocols (10). Zhang et al. (30) reported the availability of efficient plant regeneration through somatic embryogenesis to produce fertile transgenic rice plants through electroporation-mediated plasmid uptake into protoplasts. In this single-cell protoplast system, only 1 in 10^5 protoplasts plated became transformed, and it was possible readily to recover transgenic plants, because of the possibility of efficient fertile plant regeneration from protoplasts, similar results having been reported in japonica rice cultivars (31–33), using protoplasts isolated from cell suspension cultures.

Protoplasts are also being used to visualize in RNA expression (10). Video image analysis of tobacco protoplasts electroporoted with luciferase in RNA, for example, demonstrated a wide range in the level of expression of this marker (34). These studies on the production of GM plants are being combined with the construction of restriction fragment length polymorphism (RFLP) linkage maps for major crop plants (35). According to Cocking (10), integration of RFLP techniques into plant breeding programs, including methods leading to the production of transgenic plants, will allow the transfer of novel genes from related wild species to clone genes of complex traits such as disease resistance or stress tolerance.

D. GM Crops with Complex Traits

The ability of legumes and pulses to establish symbiotic relationship with rhizobia and resulting in the formation of nitrogen- fixing root nodules is an example of a complex plant genetic trait. Rhizobia genetically control nodule formation in this complex interactive process. According to Sprent (36), there appears to be no clear reason why nonlegumes like cereals should not nodulate with rhizobia given the appropriate environment and genetic background. It has been reported that treatment to root hairs of clover seedlings with a cell wall–degrading enzyme removed a barrier to *Rhizobium* host specificity (37), indicating that the cell wall at the tip of root hairs plays a key role in controlling rhizobial specificity, and its removal enables bypassing this control. Similarly, the cell wall at the tips of root hairs of both rice and wheat could be removed enzymatically using a cell wall–degrading mixture of cellulase and pectolyse (10). In addition, nodular structures could be induced on rice roots when the roots of 2-day-old seedlings were treated with the cell wall–degrading enzyme mixture, followed by inoculation with rhizobia in the presence of PEG (38). Such nodular structures could also be induced on the roots of similarly treated wheat seedlings (39) and on oilseed rape (10). The presence of *Rhizobium* both within and between the cells of root nodules on these crop plants was confirmed by electron microscopy. The results of these experiments demonstrated that the enzymatic treatment of the root system of non-legumes provided a "port of entry" for rhizobia, probably by exposing the plasma membrane at the tip of root hairs, and identified the cell wall at the tip of root hairs as the target site for genetic manipulation of the complex plant trait that controls nodule formation by rhizobia in both legumes and nonlegumes (10). It has also been shown that *Agrobacterium rhizogenes* can induce tumorlike outgrowths on rice roots when enzyme-PEG-rice seedlings are inoculated with supervirulent A. rhizogenes R1601. Thin sections of these tumorlike outgrowths when examined under the electron microscope showed the presence of agrobacteria between, but not within, some of the cells of the outgrowths (39). While biomolecular and biochemical evidence for genetic transformation is still being awaited, these basic studies have paved a way to a new revolution in the transformation and production of transgenic plants, especially in cereal staple food crops. This hope appears to turn into reality in the light of the observation that a factor inducing *A. tumefaciens vir* gene expression is present in monocotyledonous cereal crops, suggesting that tDNA processing, and possibly its transfer, can take place when the bacterium invades suitable tissues of these crops (10,40).

The techniques of *Agrobacterium*-mediated gene transfer, direct gene transfer to protoplasts, and ballistic gene transfer, the instruments required, and various parameters influencing the transformation as well as the gene-transfer techniques that may be used in the future (microinjection, microtargetting, electroporation, and silicon whiskers) have been discussed extensively (1), along with plant gene expression signals and genetic markers used for transformation.

It is evident from the survey of new hybrid plants produced to date through protoplast fusion that few involve agriculturally important species, which is primarily due to the advanced state and/or ease of in vitro manipulations with many members of the family Solanaceae (41). For this reason, the first new somatic hybrids to be useful agriculturally will most likely be from this family. Recent progress with in vitro culture techniques of important crop plants, such as Crucifereae, has also led to the production of several new hybrid plants that may be useful in hybrid rapeseed production because of their novel combinations of mitochondria and chloroplasts which encode cytoplasmic, male sterility, and herbicidal resistance (see Section III.C), respectively. As protoplast technology and reliable regeneration procedures are further developed, somatic hybridization has been envisaged to play an increasingly significant role in the crop improvement, including the legumes and cereal crops (41).

Predieri (42) reviewed in vitro mutation induction methods in fruits and the in vitro selection procedures available for early screening. Results obtained through in vitro mutation techniques, including somaclonal variation, have been compared with the current achievements and future prospects of transgenic breeding. Plant improvement based on mutations, which changes one or a few specific traits of a cultivar, can contribute to fruit improvement without altering the requirement of the fruit industry. Induced mutations have well-defined limitations in fruit-breeding applications, but their possibilities may be expanded by the use of in vitro techniques. Molecular techniques can provide a better understanding of the potential and limitations of mutation breeding. Mutagenesis in combination with tissue culture appears to be either ineffective or yet to be exploited in fruit, although in vitro mutation induction has a high potential for fruit improvement (42).

Forty-six independently transformed plants were recently regenerated under hygromycin selection from cell suspension–derived protoplasts of *Festuca arundinacea* (Shreb) after PEG-mediated transformation (43). It was suggested that adding nonselected and selected transgenes at a higher molar gene ratio would probably improve the proportion of plants regenerated, which express both transgenes. Olhoft et al. (44) reported that *Agrobacterium*-mediated transformation of soybean cells and the production of fertile transgenic soybean plants using the cotyledonary-node (cot-node) method could be improved by amending the solid cocultivation medium with L-cystine. A protocol has been developed for *Agrobacterium*-mediated genetic transformation of *Acacia mangium* using rejuvenated shoots as explants. The genetic modification of this tropical leguminous tree to confer desirable traits such as modified lignin content and insect resistance, both of which are hard to achieve through traditional breeding, can be achieved through transformation (45).

Ganapathi et al. (46) described the development of embryogenic suspension cultures, *Agrobacterium*-mediated transformation, and plant regeneration of Rasthali (AAB) banana, a preferred Indian cultivar with very sweet pulp. The application of biotechnology for the improvement of this crop would be a useful tool to breeders to introduce traits of interest. This is important, because conventional breeding in banana is complicated by the fact that only a few diploid clones produce viable pollen and that the germplasm of the most important commercial clones is both male and female sterile (47).

Genetic engineering can probably be more successfully applied to maize improvement by identifying and isolating agronomically useful genes and subsequently introducing them directly into the genome of selected elite maize lines. O'Kennedy et al. (48) recently reported transformation of two elite white maize lines, W 506 and M 37W, using a particle flow gun.

III. GENETICALLY MODIFIED TRAITS

A. GM Crops for Insect Resistance

Hilder et al. (49) have described genetic engineering of crops for insect resistance using genes of plant origin. Despite the control measures, around 13% of the total global crop product is lest because of insect pests annually. These losses could be prevented or minimized by incorporating insect resistance into the crop plants (50,51). Hilder et al. (49) pointed out the following advantages of this approach:

1. Continuous crop protection to obtain maximal pest control without the need to predict the occurrence of pest incidence

2. Provision of protection where required, including plant tissues, such as roots, undersides of leaves, and insides of pods, which are difficult to treat with insecticidal chemicals
3. Restricted insecticidal activity only to those insects attacking the plant, thus protecting harmless and beneficial insects such as honeybees and other pollinators
4. Confined problems of environmental pollution. Also, the inherent protection could not "wash off" as a result of rainfall, resulting in additional costs and problems associated with pest application
5. Lower development costs involved compared with producing a novel chemical insecticide

These advantages can be shared with the introduction of genetic resistance to insect pests by a classic breeding program. Genetic engineering, however, offers an opportunity to transfer a single character in one step into a favorable genetic character without cotransfer of possible undesirable characteristics. It also allows transfer of insect resistance genes across the barriers to conventional plant breeding, allowing transfer between species, genera, and even kingdoms (49).

Some of the plant defense mechanisms evolved through millions of years include physical barriers to being eaten by insects—spines, hairs, tough surfaces, and so forth —and an enormous 'armory' of secondary compounds, which may be specific proteins, such as digestive enzyme inhibitors and lectins (52–54). Such single-gene proteins are ideal for genetic engineering, which is well illustrated by cowpea trypsin inhibitor (CpTI).

Of thousands of cowpea lines screened, only one cultivar, designated Tvu2027, showed significant resistance to the larvae of the bruchid beetle, *Callosobruchus maculatus* F (55). This variety provided a potentially useful source of insect resistance genes. It was established that resistance in Tvu2027 was not of a physical barrier type. The seeds of this variety and of susceptible types were then screened for a range of secondary compounds, such as alkaloids, lectins, saponins, nonprotein amino acids, and protease inhibitors (52). Of the various toxic antimetabolic compounds evaluated, the only activities detected were inhibitors of serine proteases, trypsin and chymotrypsin, and bruchid resistance in TVu2027 was found to be associated with high levels of trypsin inhibitor; two- to four-fold higher than the susceptible lines (Table 2). Polyacrylamide gel electrophoresis (PAGE) and isoelectric focusing of the Ti fraction from these varieties indicated that the differences were purely quantitative rather than qualitative (52). The antimetabolic characteristics of CpTI were also confirmed with *C. maculates* bioassay on artificial seeds by incorporating different fractions extracted from cowpea meal. Only those fractions containing active CpTI at levels nearing those naturally occurring in the seeds of TVu2027 were effectively insecticidul to the bruchid beetle (52) (Table 3). The trypsin inhibitors extracted from cowpeas were more effective antimetabolites than those extracted from other legumes (56). The antimetabolite, CpTI, was found to be insecticidal to a wide range of insects in seeding trials on artificial diets, covering insects from orders other than Coleoptera, such as Lepidoptera. Some of these susceptible insects are pests of major economic importance, requiring high expenditure on the use of insecticides; for example, the lepidopterans *Heliothis* (tobacco budworms and corn earworms) and *Spodoptera* (army worms) and the coleopterans *Diabrotica* (corn root worms) and *Anthonomus* (cotton boll weevil).

Table 2 Development of *C. maculates* to Adulthood in Relation to Trypsin Inhibitor (Ti) Content of Seeds from Different Cowpea Lines

Accession emergence	Ti content (% w/w) (%)	Adult
TVu 2027	0.92	0
TVu 4557	0.44	95.1 ± 6.2
TVu 76	0.34	90.0 ± 2.5
TVu 3629	0.30	90.6 ± 9.5
TVu 37	0.26	86.6 ± 4.3
TVu 57	0.25	91.7 ± 6.7
TVu 1109E	0.23	89.0 ± 6.7
TVu 1502-1D	0.19	92.0 ± 6.0

Source: Ref. 52.

It has been proposed that CpTI primarily acts by inhibiting essential digestive proteases, resulting in abnormal development and death of insects due to a deficiency of essential amino acids. It may also affect other insect systems involving proteases. The cite of action being the actual catalytic site of an enzyme, and probably affecting more than one enzyme, the ability of the insects to evolve resistance to CpTI by a single, or even a few, mutational events appears to be minimal. Although CpTI inhibits mammalian trypsin, it is not toxic to mammals. Also, CpTI is degraded by pepsin in the gastric juice before it can encounter the serine proteases of the small intestine in the mammalian gut (49).

Further isolation and characterization of trypsin inhibitors in cowpea showed that they were of the Bowman-Birk type protease inhibitor group (57), which are small polypeptides of around 80 amino acids with a very high degree of disulfide cross linking. They are double-headed inhibitors; that is, each inhibitor molecule can bind to, and thus inhibit, two molecules of enzyme. The cowpea inhibitors comprise a small family of four major isoinhibitors encoded by a larger gene family and consisting of

Table 3 Development of *C. maculates* on Artificial Seeds Supplemented with Various Cowpea Seed Meal Fractions

Seed composition	Adult emergence (% Control)
Control	100
+ 10% albumin	42
+ 10% albumin- CpTI[a]	96
+ 10% globulin	97
+ 0.1% CpTI	102
+ 0.5% CpTI	26
+ 0.8% CpTI	0

[a] CpTI removed from the albumin fraction by affinity chromatography on trypsin-conjugated sepharose.
Source: Ref. 52.

only four active genes. Three of these isoinhibitors are trypsin/trypsin inhibitors and the fourth is trypsin/chymotrypsin type (49).

Hilder and associates determined the protein sequence, which allowed them to synthesize mixed sequence oligonucleotide probes complementary to all the coding possibilities of the short regions of the polypeptide. The latter in turn provided the basis for selecting CpTI-encoding clones from a cowpea cotyledon cDNA library, establishing that these inhibitors represent trypsin/trypsin and trypsin/chymotrypsin types. Using a full-length cDNA clone encoding trypsin/trypsin inhibitor, the selected gene was then transferred to tobacco plants by mobilizing constructs with CpTI encoding sequence into *A. tumefaciens* to transform tobacco leaf discs and regenerating transgenic plants by standard procedures (49).

Hilder et al. (58) studied biomolecular characteristics of the transgenic tobacco plants containing the CpTI gene construct. All the plants investigated revealed multiple, unrearranged copies of the gene construct. Plants expressing high levels of CpTI in the leaves also had insecticidally effective levels of protein expressed in roots, stems, and floral tissues. The insect bioassays with *Heliothis virescens* (a serious pest of tobacco) conclusively established that CpTI-expressing plants had significantly higher resistance to the insect, which extended to a broad range of insect pests that are capable of attacking tobacco, and that the resistance was inherited stably so far up to the seventh generation.

Other published reports on genetically engineered foreign genes into plants to enhance resistance to insects have used modified endotoxin proteins from the bacterium *Bacillus thuringiensis*. Enhanced insect resistance in transgenic tobacco and tomato plants expressing toxic fragments of *B. thuringiensis* has been reported, with comparisons being made between *B. thuringiensis* and CpTI as potential protectants in GM crops (59–61).

Although tomato and tobacco GM plants expressing *B. thuringiensis* display impressive resistance to *B. thuringiensis*–susceptible lepidopteran pests, owing to the *B. thuringiensis*, specificity of activity, the resistance is limited to a small range of insects. Wilcox et al. (62) identified different strains of *B. thuringiensis*, which together cover a wide range of lepidopterans, and against some coleopterans (63). The toxicity of the bacterial protein, *B. thuringiensis* to susceptible insects is considerably higher than that of the plant protein, CpTI. The significance of absolute toxicity of these compounds is often perceived erroneously on the basis of a false analogy with synthetic pesticides, which tend to be generally toxic and persistent as against the types of compounds genetically introduced into plants. Being proteins, the latter are intrinsically biodegradable. The insecticidal levels of CpTI are already present in the consumed part of food crops. The wider toxicological significance of expressing *B. thuringiensis* at effective levels in food crops is not known, but is expected to have an equally low persistence in the environment.

It has been clearly demonstrated through insect feeding trials that effective levels of expression of *B. thuringiensis* and CpTI can be achieved in GM crops. The higher toxicity of *B. thuringiensis* may require much lower levels of expression in transgenic plants than with CpTI. Hilder et al. (49) suggested that the expression of these foreign proteins, especially at the levels required for effectiveness of CpTI, might impose a "yield penalty" on the host plant. These authors have also measured various phenotypic characteristics and "yield" parameters on large CpTI-expressing transenic lines, along with CpTI-nonexpressing transformed plants and untransformed controls. Small differences were noticed between CpTI-expressing and CpTI-nonexpress-ing transformants. Thus, there may be some minor "penalty" resulting from the genetic manipulation process, which could probably be reduced or eliminated in a subsequent breeding program. The expression of 1–2% of a foreign plant protein appears to be well within the plants' "spare" synthetic capacity, without imposing any additional yield penalty (49). Both *B. thuringiensis* and CpTI are promising and novel approaches to provide transgenic protection to crops from insect pests.

It has been envisaged that CpTI will soon be transferred to most of the major crop plants, with the availability of transformation and regeneration systems in these crops (31,33,64,65) as well for many locally important ones. The next step is to identify and obtain other plant genes involved in field resistance to insects, acting on quite different targets within the insects' metabolism. Through these efforts the development of crop plants with inherent, durable resistance to a range of insect pests is expected (49).

B. Virus-Resistant Transgenic Plants

1. Coat Protein–Mediated Protection

The discovery of the expression of a viral coat protein (CP) in transgenic plants, which protects these plants against viral infection, accurred in the late 1930s as a natural extension of the research work reported on crop pollination by Mckinney (66). Cross pollination is simply defined as protecting crop plants against pathogenic invasion due to prior inoculation of plants with another pathogen. This discussion on coat protein–mediated protection against viral infection is based on reviews and other published reports (67–72). Although the phenomenon of crop protection was first observed in related plant viruses, it was also noted for viroids (70,71). Much of the research work on plant viruses and cross protection relates to the provision of viral protection as a result of the presence of the CP, the RNA, or both (67).

Employing tobacco mosaic virus (TMV) strains, which produced either a defective CP (73) or no CP at all (74), it was noticed that protection occurred when plants harboring these strains were challenged with a related virus. Sherwood and Fulton (72), however, demonstrated that the CP of the protecting virus was important for protection. This controversy was resolved by showing that the CP-minus TMV strain used by Sarkar and Smitamana (74) indeed provided protection against superinfection by a TMV strain, although greater protection was provided when a CP-plus strain was used as a protecting virus (75). The protection accorded by CP-minus strain was nonspecific, as shown by the inhibition of infection caused by turnip mosaic virus (TuMV) (poty virus). The protection provided by CP-expressing strain was, however, specific for a related virus, since there was about five-fold greater level of protection against a second TMV strain compared with that observed against TuMV, suggesting that cross protection may have two or more mechanisms of action (76).

In the early 1980s, it was proposed that expression of viral sequences in GM plants could be a way to protect crop plants against viral infections, followed by more refined works regarding specific viral sequences having potential for protection (77–79). It was initially proposed that insertion and expression of the CP gene in plants might result in protection of plants from viruses (72,77), and in addition to the use of a CP gene for the protection, the potential for employing genes that encode antisense viral sequences to provide protection has been discussed (78).

Since the early 1980s, cloning plant viruses and inserting portions of the viral genome into plants began under the control of heterologous promoters to test whether expression of these inserted sequences could provide protection against viral infection. Transgenic tobacco plants that expressed CP of TMV were produced, although the levels of expression were low (around 0.17 µg per gram fresh weight) and no results were presented indicating protection of these plants from viral infection (80). According to Beachy et al. (78), the results of expressing TMV-CP in transgenic tobacco plants were at 10-fold higher levels than those reported by Bevan et al. (80). It was also possible to show that plants are protected after a challenge with TMV (81), and the usefulness of CP gene expression to provide protection in a number of other plant virus combinations (82–86). Nelson et al. (67) reported that of the 28 plant virus families consisting of ssRNA without envelopes, CPs from viruses within seven of the families provided protection against viral infection in transgenic plants. Transgenic tomato plants expressing the TMV-CP gene could be protected in field experiments under conditions of high infection pressure, indicating the potential of this promising technology in practice. Such protection is reflected in the higher fruit yields, 0.25- to 3.0-fold greater than those from inoculated plants not expressing the CP gene (87,88). Lawson et al. (89) also showed that expression of CP from potato virus Y in transgenic potato plants was effective in providing protection against aphid transmission of that virus.

Nelson et al. (67) have summarized the results of research, showing the site of protection and possible mechanisms of CP-mediated protection. The CP-mediated protection is part of cross protection and expression of the entire TMV viral genome in transgenic tobacco plants, closely representing cross protection, and results in a highly effective protection against TMV infection (90). Because these plants are almost completely protected against challenge by TMV-RNA, it is quite possible that its effectiveness in protection is due to more than simply higher levels of CP in the levels prior to challenge inoculation. Thus, for these plants and for those that are cross protected by classic methods, antisense inhibition and/or other mechanism(s) are thought to be involved to a greater extent than in plants protected by CP gene expression. The future research on CP-mediated protection should address how the CP expressed in the transgenic plant prevents or retards viral replication; site-directed mutations of the CP and further research on viral strains have high structural homology but low sequence homology with the transgenic CP to further define the necessary protein conformation for maximal protection. In addition to defining the mechanism of protection at the primary infection site, the mechanism and specific site of protection against the systemic spread of virus need to be elucidated (67).

Recent advances in genetic manipulation have enabled incorporation of agronomically useful traits into recipient transgenic plants. Plant viruses represent an important group of pathogenic entities, causing significant crop losses on a global scale. Some successes in protecting plants from viral diseases have been reported, and the future prospects for manipulating nonconventional plant resistance to viruses look promising. The present strategy adopted has primarily relied on mimicking cross protection against viruses by constructing transgenic plants expressing viral components, such as a coat protein gene or satellite RNAs interfering with viral multiplication (91). According to Covey et al. (92), a further potential novel strategy of viral control would be to exploit those systems where systemic viral infection is restricted because of a hypersensitive response mediated by specific host genes. It is necessary to understand the molecular processes in successful or permissive viral infections rather than in those where establishing a systemic viral infection is prevented by cross protection or a hypersensitive response. Covey et al. (92) have investigated viral infection to characterize the host and viral genes involved in determining viral pathogenesis so as to exploit cauliflower mosaic virus (CaMV) infection of brassica plants.

Many viruses associated with small, extragenomic RNA molecules, often considered to be molecular parasites, are known as satellite RNA (satRNA). They are not required by the virus, but depend on them for all functions necessary for propagation through the infected host plant and transmission from one plant to another. According to Murant and Mayo (93), a significant feature of satRNA is a lack of extensive homology with the helper virus. However, since satRNAs are recognized by replicase and other functions of the helper virus, the features of the secondary or tertiary structures are possibly shared with the helper virus. It has been predicted that the information on the domain structure of satRNA will be used in the development of satRNA for use in crop protection and for understanding the basic mechanism of symptom induction by satRNA. Better understanding of RNA bind to satRNA of cucumber mosaic virus (CMV) would lead to the safe use of satRNA in crop protection (91). Baulcombe et al. (91) have also anticipated that further research on the satRNA pathogenic process will provide information about healthy cells and new methods of engineering disease resistance in crop plants.

C. Herbicide-Resistant Transgenic Plants

Genetically engineered effective herbicidal resistances in crop plants have been developed to a range of potent "knockdown" herbicides, such as glyphosate and the sulfonylureas (94,95). In these cases, the molecular target of the herbicide was characterized and resistances were developed by manipulating or replacing the target gene. *Salmonella typhimurium* is resistant to glyphosate because of a mutation in the 5-enolpyruvylshikimate 3-phosphate synthase (EPSP synthase) gene. The EPSP synthase, the target gene for glyphosate, when expressed in crop plants, this mutant enzyme is substituted for the endogenous plant enzyme that is inactivated by the herbicide (94). Shah et al. (95) selected a petunia cell line that is resistant to glyphosate through the EPSP synthase. This petunia gene was cloned and reintroduced into petunia with a new, much stronger promoter. The new chimeric gene resulted in the overproduction of the enzyme, offering resistance at a whole plant level. Llewellyn et al. (96) cited the work of G. Haughn, J. Smith, B. Mazur, and C. Sommerville, achieving similar results with an acetolactate synthase gene from *Arabidopsis* that confers sulfonylurea resistance on transgenic tobacco plants. Although all of these herbicidal resistance genes are quite effective, they probably have certain deleterious effects, especially in the absence of the herbicide (96).

Another approach that has been used to build herbicidal resistance genes is the engineering of detoxification pathways in susceptible crop plants. Many herbicides are selective (being toxic to some plant species and not to others), the difference in sensitivity often being associated with an ability of the resistant plant to detoxify or metabolize the herbicide before it has a chance to inactivate the susceptible target gene product. A microbial gene encoding an enzyme that detoxifies an herbicide has been modified for expression in plants and introduced into plant species by genetic manipulation (97,98). The novel detoxifying enzymes could be expressed, conferring resistance to relatively high levels of phosphinothricin (Basta) (97) and 3,5-dibromo-4-hydroxybenzonitrile (bromoxynil) (98). According to Llewellyn et al. (96), these artificial systems mimic detoxification systems found in plants and most probably do not place a genetic, physiological, or yield burden on the crop plant either in the presence or absence of the herbicide.

The herbicide, 2,4-D, has a half-life of around 14 days in soil, its breakdown being performed by a range of microorganisms including bacteria, yeasts, and fungi. The most well-characterized organisms are the gram-negative rod bacterium *Alcaligenes eutrophus*, which is found in most aerated soils. These strains have a large 75-kb plasmid, which encodes many enzymes necessary to degrade 2,4-D to simpler metabolites. The 2,4-D degradation pathway in *A. eutrophus* indicated involvement of six plasmic encoded genes. Seed germination experiments carried out in the presence of 2,4-D or its first degradation product, 2,4-dichlorophenol (DCP), showed that, at least in tobacco, DCP was around 50 to 100 times less toxic than 2,4-D, and perhaps only the first enzyme, 2,4-D mono-oxygenase, encoded by the tfdA gene, needs to be expressed in plants to detoxify 2,4-D sufficiently.

Llewellyn et al. (96) have described the genetic engineering of the ifdA gene first into tobacco and subsequently into other crops such as cotton. Transgenic tobacco plants were produced by *Agrobacterium* infection of leaf pieces using standard protocols. The increase in 2,4-D resistance in transgenic plants varied from 10- to 30-fold over the control plants. It was clearly shown that the bacterial 2,4-D mono-oxygenase is effective in protecting transgenic tobacco plants from 2,4-D effects at levels above the normal field application rate.

The herbicidal detoxification system, which has been proved to be extremely effective in at least one plant species (96), has been expected to be useful in other crops, such as cotton, that are sensitive to 2,4-D. The transgenic plants with an engineered resistant gene appear to grow normally, producing viable seeds as nontransformed plants, with no yield penalty in the absence of any applied herbicide. However, when sprayed at relatively high levels of 2,4-D, there is a slight retardation of growth, but plants do recover quickly. Since soil microorganisms degrade most common herbicides, the same principles can be applied to those herbicides to develop quickly herbicide resistance in many commercially important agricultural crops once transformation procedures become routine (96).

D. Higher Photosynthetic Efficiency

Photosynthesis, the basic process leading to biomass accumulation, is intrinsically limited by the low catalytic competence of the enzyme ribulose 1,5-bisphosphate carboxylase/oxygenase (Rubisco) which initiates photosynthesis by carboxylating ribulose 1,5-bisphosphate (RuBP). The enzyme has low specific activity for carboxylase and catalyses several other reactions, including oxygenation of RuBP leading to glycolate production and the subsequent loss of newly assimilated CO_2 from the plant (photorespiration). This oxygenase reaction is the most important metabolic constraint on crop productivity—up to 50% of the carbon fixed in C_3 crops being lost in this way (99). A large amount of Rubisco is required for higher photosynthetic rates due to its low specific activity for carboxylation; thus, it is the enzyme dominating the nitrogen (N) requirements of crop plants (around 25% of leaf N). Even small increases in the CO_2 fixation rates through improved Rubisco efficiency could immensely increase crop production by increasing crop N-use efficiency and water-use efficiency. In addition, enhanced CO_2 fixation from the atmosphere might counteract global warming (100).

Thus, Rubisco and photorespiration are obvious targets for genetic manipulation. Genes can be introduced into the chloroplast to encode a more efficient Rubisco with an improved specificity for CO_2 which can be achieved by exploiting the natural variation in the catalytic properties to Rubisco isolated from different plant species, or by improving the enzyme by site-directed mutagenesis (101). Alternatively, the wasteful photorespiration could be controlled by introducing a CO_2-concentrating mechanism, (already evolved in some plants) or by manipulating photorespiratory metabolism. Genes can be introduced in C_3 plants to modify the photorepiratory pathway to reduce the wasteful consequences of the oxygenase activity; for example, introduction of genes to encode enzymes, glyoxylate carboligase and tartronic semi-aldehyde reductase (normally not found in C_3 plants) together with antisense repression of existing photorespiratory enzymes to prevent the wasteful release of ammonia. Designing of a Rubisco lacking the oxygenase activity would be a more promising speculative approach (100).

Photosynthetic models have suggested that Rubiscos from photosynthetic bacteria *Chromatium vinosum* and *Anacystis nidulans* should out perform most higher plant Rubiscos at elevated CO_2 concentrations and at temperatures of 5–25°C (101). Higher plant Rubisco has eight chloroplast-encoded large subunits and eight nuclear-encoded small subunits. The catalytic properties of holoenzyme are determined by large subunits, with catalytic sites of the enzyme. Parry et al. (100) have described genetic manipulation of photosynthetic metabolism, considering the introduction of *C. vinosum* sequences encoding large subunits of Rubisco into tobacco (102). Despite the expression of higher plant genes of both subunits in *Escherichia coli*, the expressed proteins have failed to assemble into functional enzymes (103). Since the genes for the bacterial forms have been successfully expressed and assembled in *E. coli* and catalytically competent chimeric enzymes with polypeptide subunits from different species have been produced, it is expected that the large subunits of *C. vinosum* get assembled with the small subunits of higher plants to produce holoenzyme, still retaining the catalytic properties of the large subunit (100).

A model system with plants devoid of Rubisco (tobacco lines lacking functional subunits) has been identified. These lines, which cannot photosynthesize, can be maintained in vitro by supplying a carbon source in the form of sucrose and serve as ideal material for photosynthetic metabolism manipulation (104). It has been envisaged that chloroplast transformation would provide an ideal way to introduce a novel Rubisco large-subunit gene (rbcL), although the techniques required are yet to be well established and limited to a single species (105). According to Parry et al. (100), nuclear transformation is routine in many plant species and novel rbcL genes can be integrated into the nuclear genome provided the polypeptide is targeted to the chloroplast by a suitable transit peptide.

E. Modified Proteins/Enzymes/Vaccines

Genes encoding heterogeneous proteins can be introduced into the plant genome for many advantages. The new plant-made proteins are used as a tool in basic biological research. Reporter proteins are used to characterize promoter sequences and other cis-acting sequences. The synthesis of new heterogeneous proteins and enzymes in plants can have several biotechnological applications such as conferring new properties to the plant or transgenic plants serving as a source of important heterologous proteins. To achieve this, the expression profile and accumulation level of proteins in plants need to be stable, reproducible, and economically suitable. De Neve et al. (106) have reviewed the work on screening of transgenic lines with stable and suitable accumulation levels of a heterologous protein.

Most primary transformation shows relatively low expression levels of the transgene (107), and the transmitted expression levels are often not stable in the progeny. Songstad et al. (108) developed different systems to introduce foreign DNA into plant cells, including the one based on the natural gene transfer system of *Agrobacterium*, which has been used extensively in dicotyledonous species (109) and also in some monocotyledonous species (110–112). In these systems, the gene of interest is cloned into *E. coli* between the borders of the transferred DNA (tDNA) and the tDNA construct is then introduced into *Agrobacterium*. The recombinant *Agrobacterium* transfers the tDNA into the plant cell during cocultivation. The transferred DNA is targeted to the nucleus and gets integrated randomly into plant chromosomes (113,114). The transformed cells are selected and regenerated into flowering plants, although this last step has remained a major bottleneck for some plant species (106).

De Neve et al. (106) proposed a protocol which allows identification of lines with suitable accumulation levels of the heterologous protein and stable transmission of their expression level in a homozygous condition through subsequent generations, and described methods for screening of a transgenic *Arabidopsis* or tobacco population obtained by *Agrobacterium*-mediated transformation. The proposed scheme has the following three steps (106):

1. Many primary transformants are screened by a rapid and easy assay for transgene expression to select a limited number of interesting transformants.
2. The number of tDNA copies integrated into each of the selected transformants is determined, preferentially retaining transformants with only one tDNA copy.
3. The transformed lines are sexually propagated, and their progeny are analyzed for the stability of transgenic expression.

According to Pueyo and Hiatt (115), a wide variety of foreign compounds can be obtained from transgenic plants, including proteins and all kinds of modified compounds through the expression of the appropriate enzymes and even catalytic antibodies. The production of plant biomass being inexpensive, the plant cell machinery can provide specific processing that is not available from microbial systems. Also, the isolated plant products are free of bacterial contamination, and the plant seeds can be used as a low-cost storage for these valuable products made by plants using sunlight as a source of energy.

Crop improvement by engineering for specific traits can be achieved through production of enzymes and metabolites encoded by foreign genes. The amino acid composition of seed proteins has been altered, for example, transgenic canola, expressing a Brazil nut albumin, contains around 33% more methionine, an essential amino acid, in the seed protein (116). Yun et al. (117) reported that the production of scopolamine, a medicinal anticholinergic drug, is significantly increased in transgenic

Atropa belladonna plants expressing hyoscyamine 6β-hydroxylase enzyme from *Hyoscyamus niger*, which catalyzes the conversion of the precursor hyoscyamine to scopolamine.

The shelf life of fruits and vegetables after their harvest can be extended significantly. Tomato plants expressing a bacterial ACC deaminase gene, which reduces ethylene synthesis, produced mature fruits that remained firm for at least 6 weeks longer than the nontransgenic controls (118). Transgenic lettuce and tomato have been obtained that express monellin, which is a sweet protein isolated from African berries that increases a flavor and sweetness around 10^4 times that of sucrose on a molar basis (119). Hightower et al. (120) reported an improvement in the freezing properties of harvested tomato fruits produced from plants expressing an antifreeze protein from a polar fish, winter flounder, that inhibits ice recrystallization (121).

A significant amount of toxic cadmium intake in human beings is contributed by vegetable products (over 70%), which can be reduced through genetic transformation. Shoots of transgenic tobacco expressing the mouse metallothioneine I gene have been found to contain a 24% lower cadmium content than nontransgenic plants, because the pollutant metal is sequestered in the roots as cadmium-metallothioneine complexes (122).

Pueyo and Hiatt (115) have reviewed the production of bioactive peptides, human proteins, enzymes, vaccines, various industrial products, and antibodies by transgenic plants. Biologically active pentapeptide leu-enkephalin with opiate activity was produced in *Arabidopsis* and oilseed rape as part of chimeric plant seed storage proteins (123,124). Sijmons et al. (125) reported expression of human serum albumin (HAS), a protein of broad clinical use and of a high commercial value, in transgenic potato. Human interferons (INFs), small defense proteins required to resist viral infection, are high-value proteins that are cloned and produced as recombinant proteins in bacterial and animal cell systems. Edelbaum et al. (126) reported expression of human interferon-beta in transgenic tobacco plants.

Expression of *Bacillus licheniformis* alpha-amylase in transgenic tobacco plants has been reported (127). Transgenic tobacco plants expressing this hydrolytic enzyme could not be distinguished from nontransgenic plants, with leaves containing similar amounts of starch. The latter suggested that the alpha-amylase was located extracellularly, and thus had no access to the intracellular starch for its degradation. The production of transgenic potatoes expressing alpha-amylase is an interesting possibility (115). Pen et al. (128) have also reported expression of *Aspergillus niger* phytase enzyme in transgenic tobacco. Phytase catalyses nutritionally poor phytate (myo-inositolhexaphosphate) into myo-inositol and inorganic phosphate to improve phosphorus utilization in animal feed, especially in monogastric animals like pig and poultry.

Oral vaccines can be produced through genetic manipulation of plants at lower costs by expressing them in edible plant tissues. A vaccine against hepatitis B (hepatitis B surface antigen, HBsAg) has been reported to have been successfully expressed in transgenic tobacco plants (129). Intromuscular injection of HBsAg has been known to immunize and protect healthy individuals from viral infection effectively (130). The maximal levels of HBsAg obtained in transgenic plants were 0.01% of the soluble leaf protein, which is insufficient to produce HBsAg efficiently. This inexpensive plant product could be used in vaccination against hepatitis B in the Third World countries provided the levels of HBsAg can be increased through further research.

F.   Industrial Products/Pharmaceuticals/Antibodies

Transgenic plants can be engineered by introducing genes coding for appropriate enzymes to convert compounds that are normally present in the plant into a range of industrial products and pharmaceuticals of high commercial value (115). Cyclodextrins, cyclic oligosaccharides of 6-α, 7-β or 8-γ, α- 1,4-linked glucopyranoses, with an apolar cavity are capable of forming inclusion complexes with hydrophobic substances and thus provide new properties to the complexed molecule, such as improved stability or higher water solubility. These compounds are used in the pharmaceutical industry as delivery systems and in the food industry for flavor and odor enhancement and to remove certain undesirable compounds like caffeine. Cyelodextrins are normally produced in vitro by the action of the bacterial enzyme cyclodextrin glycosyl transferase (CGT) on prehydrolyzed startch (131). The gene encoding CGT from *Klebsiella pneumoniae* was cloned (132) and expressed in transgenic potato (133). The expressed levels of 2–20 μg of α-cyclodextrin and 2–5 μg of β-cyclodextrin per gram fresh weight or 0.001–0.01% of the total (14%) starch present in potato tuber converted into cyclodextrins are low for the commercial production of these compounds and need to be increased to 1–10% levels (115).

Nonbiodegradable plastics of the modern age have become an environmental nuisance. Efforts have continued to obtain inexpensive biodegradable resins and plastics. Poly-D-(-)-3-hydroxybutyrate (PHB) is a high molecular weight aliphatic polyester found in many bacterial species as a biodegradable thermoplastic storage compound. The production of PHB by bacterial fermentation is not yet economical. The genes encoding three enzymes of PHB synthesis in the *Alcaligene eutrophus* bacterium (3-ketothiolase, acetoacetyl-CoA reductase, and PHB synthase) have been cloned and expressed in *E. coli* leading to PHB production (134). The PHB was found in the hybrid plants when homozygous transgenic plants expressing acetoacetyl-CoA reductase and PHB synthase were cross pollinated. The highest amount of PHB accumulated in leaves was around 100 μg per gram fresh weight (115).

Antibodies offer a wide range of industrial applications. Taken out of the context of an animal immune system, an antibody is a complex protein that binds a single antigen with a high affinity. The antigens can vary from another protein to a synthetic organic molecule. Applications depending solely on the binding affinity of antibodies include pathogenic resistance (insect, fungal, and viral) and metabolic modulations to produce new developmental or nutritional properties (135). Antibodies produced by plant cells on a large scale can be used as therapeutic, diagnostic, or affinity reagents in medicine and basic research in plant biology. Cheaply produced antibodies from the plant world may serve as useful reagents for isolating and processing environmental contaminants and industrial by-products and wastes.

Antibodies possessing catalytic capabilities (enzymes) were isolated from mouse hybridoma cells (136), making it possible to introduce new catalytic capabilities into plants or to produce catalytic antibodies useful in industrial processes. Genetically stable seed stocks of antibody-producing plants can be isolated and stored indefinitely at low cost, and the seed stock can be used to produce any quantity of antibody within one growing season. To date, tobacco has been used principally to initiate antibody studies in plants, although other crops such as forages, alfalfa, soybean, tomato, and potato may be appropriate for antibody production (115).

According to Topfer and Martini (137), strategies developed for genetic engineering are well established and major breakthroughs for engineering metabolic pathways have been achieved. However, the fine-tuning and product developmet and success of the new crops and products on the market are required. In some cases, the plant material used for transformation needs to be preselected for a desired trait, and plant genetic manipulation migh help overcome key steps of a given pathway limiting optimal product formation. Considering industrial traits, it is still difficult to predict which degree of alternation the genetically modified plant can tolerate relative to protein stability and function, changing metabolites and introduction of new pathways. More insight into the biochemistry and interaction and regulation of metabolic pathway will contribute to an improvement of plant products for industrial purposes. The study of partitioning of photosynthate will enable biotechnologists substantially to improve yield characteristics for more productive crop plants. A strategy was developed, for example, to inhibit polygalacturonase (PG), a cell wall–degrading enzyme in tomato, which leads to fruit softening during ripening, so as to be able to harvest tomatoes mechanically. Antisence expression under the control of the cauliflower mosaic virus (CaMV) 35S RNA promoter of a cDNA-encoding tomato PG was found to decrease PG activity significantly without interfering with the normal ripening process (138,139). A reduction of less than 1% of normal tomato PG activity was detected in homozygous lines, thus preventing cell wall degradation and concomitant softening of fruit during ripening (140). Coffee transgenic plants have been obtained through *Agrobacterium*-mediated transformation (141). Etienne et al. (142) recently reviewed biotechnological applications in the improvement of coffee. Economically important genes of the caffeine biosynthetic pathway or genes coding for seed storage proteins have been isolated.

G. Aluminum Tolerance in Transgenic Plants

When solubilized in acid soils, aluminum in the form of Al^{2+}, is toxic to many crops and is the major limiting factor for crop productivity on these soils. Soil acidification accelerated by certain farming practices and by acid rain affects 40% of the arable land worldwide (143,144). Although crop production on acid soil can be sustained by lime application, runoff pollution is an undesirable side effect. Thus, the production of Al-tolerant crop plant varieties either by conventional breeding or genetic engineering appears to be the best solution.

The production of Al-tolerant transgenic plant varieties should be considered as an important alternative to form a part of a crop management strategy to increase agricultural production on acidic soils and protect forests around strongly acidified industrial regions. Of several mechanisms proposed for Al detoxification, the exudation of organic acids that chelate and detoxify Al in the rhizosphere, a common feature of tolerant plants of different species, has been considered to give practical results. The effectiveness of citric acid in alleviating Al toxicity has been demonstrated. Addition of citrate in the nutrient solution has been shown to reverse the inhibition of wheat root growth caused by Al (145).

Transgenic tobacco and papaya plants that overexpress the citrate synthase from *Pseudomonas aeruginosa* in their cytoplasm were produced to assess the impact of elevated levels of citrate on Al tolerance (146). A chimeric gene was constructed to produce these transgenic plants in which the coding sequence of *P. aeruginosa* citrate synthase gene (CSb) was transcriptionally fused to the 35S promoters from the CaMV and the nopaline synthase 3 end sequences (35S CSb). This gene was introduced into the tobacco genome using a Ti plasmid-derived transformation system. Since excretion, rather than intracellular accumulation of citrate is probably responsible for Al tolerance, it was examined whether an increased citrate synthesis leads to a higher rate of its efflux. The amount of citrate exuded through the roots into the media by selected 35S-CSb lines was found to be four-fold higher than the control plants. The analysis of Cbs transgenic tobacco and papaya plants led to the following conclusions (146,147):

1. Oxaloacetate and acetyl-CoA, the substrates of citrate synthase, are readily available in the cytoplasm, and therefore citrate overproduction in transgenic plants can be achieved by cytoplasmic expression of a bacterial citrate synthase.
2. An increased citrate synthesis leads in turn to a higher efflux of citrate, suggesting that active organic acid channels are normally functioning in plant roots.
3. A higher synthesis and excretion of citrate confers aluminum tolerance to higher plants.

Phosphorus (P) is one of the most important plant nutrients limiting agricultural production worldwide. In acid and alkaline soils, which make up over 70% of the world's arable lands, P becomes unavailable for plant use as it forms insoluble phosphates. To ensure plant productivity and reduce P deficiencies, around 30 million tons of fertilizer are applied every year, and up to 80% of the applied P becomes immobile and unavailable for plant uptake. Thus, the development of novel plant varieties more efficient in the use of P represents the best alternative to reduce the use of P fertilizers and achieve a more sustainable agriculture. Lopez-Bucio et al. (148) have recently shown that the ability to use insoluble P compounds can be significantly enhanced by engineering plants to produce more organic acids. The citrate-overproducing transgenic plants have been shown to yield more leaf and fruit biomass as compared to the controls when grown under P-limiting conditions and to require less P fertilizer to achieve optimal growth (148).

Lopez-Buccio et al. (149) have recently reviewed organic acid. metabolism in plants with reference to plant tolerance to environmental stress. Organic acids, in addition to acting as intermediates in carbon metabolism, are key components in mechanisms that some plants use to cope with nutrient deficiencies, toxic metal tolerance, and plant-microbe interactions operating at the root-soil interphase. The physiology and occurrence of organic acids in plants and their special relevance concerning nitrate reduction, phosphorus and iron acquision, aluminum tolerance, and soil ecology have been discussed, and the novel findings in relation to the biotechnological manipulation of organic acids in transgenic models from cell cultures to whole plants have been reviewed (149). It has been envisaged that the novel perspective of organic acid metabolism and its potential manipulation may represent a way to understand the fundamental aspects of plant physiology, leading to new strategies for producing crop varieties that are better adapted to environmental and mineral stress.

IV. CONCLUSIONS

Biotechnological applications, especially transgenic or genetically modified plants, probably hold the most promise in augmenting agricultural production in the first decades of the next millennium. However, the application of these technologies to the agriculture and horticulture of tropical regions where they are most needed remains a major challenge. Herrera-Estrella (150) has discussed some of the important issues that need to be considered to ensure that plant biotechnology, and especially the transgenic (GM plant) biotechnology, is effectively transferred to the developing world. Even if this technology is successfully transferred to developing countries and transgenic varieties are developed for local crops, the problem of getting this technology to the small farmers is still a more important challenge. The national governments need to implement a system for producing and distributing transgenic seeds and any other input, at low or no cost, to the small farmer. Whether the advanced agricultural biotechnology is transferred to the developing countries of the world will depend upon the political will of each national government and the resources needed.

REFERENCES

1. Potrykus, R. Bilang, J. Futterer, C. Sautter, and M. Schrott, Genetic engineering of crop plants, *Agricultural Biotechnology* (A. Altman, ed.), Marcel Dekker, New York, 1998, pp. 119–159.
2. R. Topfer and N. Martini, Engineering of crop plants for industrial traits, *Agricultural Biotechnology* (A. Altman, ed.), Marcel Dekker, New York, 1998, pp. 161–181.
3. J.J. Pueyo and A. Hiatt, Production of foreign compounds in transgenic plants, *Agricultural Biotechnology* (A. Altman, ed.), Marcel Dekker, New York, 1998, pp. 251–261.
4. V. Raghvan, Genetic transformation of embryos, *Molecular Embryology of Flowering Plants*, Cambridge University Press, Cambridge, UK, 1997, pp. 525–531.
5. G.W. Lycett and D. Grierson, eds., *Genetic Engineering of Crop Plants*, Butterworths, London, 1990.
6. S. Shantharam and J.F. Montgomery, eds., *Biotechnology, Biosafety, and Biodiversity: Scientific and Ethical Issues for Sustainable Development*, Oxford & IBH Publishing, New Delhi, 1999.
7. K. Kathiravan, S. Seshadri, S. Prakash, and S. Ignacimuthu, Biotechnological applications in grain legume improvement: a critical review, *Plant Cell Biotechnol. Mol. Biol. 2 (1&2)*: 1 (2001).
8. C. Cunningham and A.J.R. Porter, eds., *Recombinant Proteins from Plants: Production, Isolation and Clinically Useful Compounds, Methos in Biotechnology, Vol. 3*, Humana Press, Totowa, NJ, 1998.
9. R.B. Horsch, R.T. Fraley, S.G. Rogers, P.R. Sanders, A. Lloyd, and N. Hoffman, Inheritance of functional foreign genes in plants, *Science 223*:496 (1984).
10. E.C. Cocking, All sorts of genetic manipulation, *Genetic Engineering of Crop Plants* (G.W. Lycett and D. Grierson, eds.), Butterworths, London, 1990, pp. 1–12.
11. L. Herrera-Esterella and A. Alvarez-Morales, Genetically modified crops: hope for developing countries? *EMBO Rep. 21*:256 (2001).
12. W.R. Scowcroft and P.J. Larkin, Somaclonal variation, *Applications of Plant Cell and Tissue Culture*, CIBA Foundation Symposium 137, Wiley, Chinchester, UK, 1988, pp. 21–35.
13. R. Abdullah, E.C. Cocking, and J.A. Thompson, Efficient plant regeneration from rice protoplasts through somaclonal embryogenesis, *Bio/Technology 4*:1087 (1986).
14. R. Roest and I.J.W. Gilissen, Plant regeneration from protoplasts: literature review, *Acta Bot. Neerl. 38*:1 (1989).
15. A. Kumar and E.C. Cocking, Protoplast fusion: a novel approach to organelle genetics in higher plants, *Am. J. Bot. 74*:1289 (1987).

16. R.G. Alexander, E.C. Cocking, P.J. Jackson, and J.H. Jett, The characterization and isolated of hetero karyons by flow cytometry, *Protoplasma 128*:52 (1985).

17. H.G. Schweiger, J. Dirk, and H.U. Koop, Individual selection, culture and manipulation of higher plant cells, *Theoret. Appl. Genet. 73*:769 (1987).

18. R.P. Finch, P.T. Lynch, J.P. Jotham, and E.C. Cocking, Isolation, culture and genetic manipulation of rice protoplasts, *Biotechnology in Agriculture and Forestry*, Vol. 14 (Y.P.S. Bajaj, ed.), Springer-Verlag, Berlin, 1989.

19. S. Hinnisdaels, I. Negrutui, M. Jacobs, and V. Sidorov, Plant somatic hybridization: evaluations and perspectives, Newsletter, *Int. Assoc. Plant Tiss. Cult. 55*:2 (1988).

20. A. Kinsara, S.N. Patnaik, E.C. Cocking, and J.B. Power, Somatic hybrid plants of *Lycopersicon esculentum* Mill and *Lycopersicon peruvianum* Mill, *J. Plant Physiol. 125*:225 (1986).

21. A. Pirrie and J.B. Power, The production of fertile, triploid somatic hybrid plants (*Nicotiana glutinosa* (n) × *N. tabacum* (2n) via gametic-somatic protoplast fusion, *Theoret. Appl. Genet. 72*:48 (1986).

22. P. Pental and E.C. Cocking, Some theoretical and practical possibilities of plant genetic manipulation of using protoplasts, *Hereditas* (Suppl.) 3:83 (1985).

23. R.J. Griesbach, Chromosome-mediated transformation via microinjection, Plant Sci. 50:69 (1987).

24. K. Weising, J. Schell, and G. Kahl, Foreign genes in plants: transfer, structure, expression and applications, *Annu. Rev. Genet. 22*:421 (1988).

25. E.C. Cocking and M.R. Davey, Gene transfer in cereals, *Science 236*:1259 (1987).

26. P. Christou, D.E. McCabe, and W.F. Swain, Stable transformation of soybean callus by DNA-coated gold particles, *Plant Physiol. 87*:671 (1988).

27. D.E. McCabe, W.F. Swain, B.J. Martinell, and P. Christev, Stable transformation of soybean (*Glycine max*) by particle acceleration, *Bio/Technology 6*:923 (1988).

28. G. Neuhaus, G. Spangenberg, O. Mitteisten-Scheid, and H.G. Schweiger, Transgenic rapeseed plants obtained by the microinjection of DNA into microspore-derived embryoids, *Theoret. Appl. Genet. 74*:30 (1987).

29. I. Potrykus, S.K. Datta, G. Neuhaus, and G. Spangenberg, Approach to cereal transformation via microinjection into microspore-derived and into zygotic proembryos, *J. Cell Biochem.* (Suppl. 13D), UCLA Symp. on Molecular and Cellular Biology (M. 001), 1989, p.228.

30. H.M. Zhang, H. Yang, E.L. Rech, T.J. Golds, A.S. Davis, and B.J. Mulligan, Transgenic rice plants produced by electroporation-mediated plasmid uptake into protoplasts, *Plant Cell Rep. 7*:379 (1988).

31. K. Toriyama, Y. Arimoto, H. Uchimiya, and K. Hinata, Transgenic rice plants after direct gene transfer in to protoplasts, *Bio/Technology 6*: 10 72 (1988).

32. W. Zhang and R. Wu, Efficient regeneration of transgenic plants from rice protoplasts and correctly regulated expression of the foreign gene in the plants, *Theoret. Appl. Genet. 76*: 835 (1988).

33. K. Shimamoto, R. Tereda, T. Izawa, and H. Fujimoto, Fertile transgenic rice plants regenerated from transformed protoplasts, *Nature 338*:274 (1989).

34. D.R. Gallie, W.J. Lucas, and V. Walbot, Visualising mRNA expression in plant protoplasts: factors influencing efficient mRNA uptake and translation, *The Plant Cell 1*:301 (1989).

35. S.D. Tanksley, N.D. Young, A.H. Paterson, and M.W. Bonierbale, RFLP mapping in plant breeding: new tools for an old science, *Bio/Technology 7*: 259 (1989).

36. J.I. Sprent, Transley Review No.15, Which steps are essential for the formation of functional lugume nodules, *New Phytol. 111*:129 (1989).

37. M.K. Al-Mallah, M.R. Davey, and E.C. Cocking, Enzymatic treatment of clover root hairs removes a barrier to Rhizobium-host specificity, *Bio/Technology 5*:1319 (1987).

38. M.K. Al-Mallah, M.R. Davey, and E.C. Cocking, Formation of nodular structures on rice seedlins by rhizobia, *J. Exp. Bot. 40*: 473 (1989).

39. M.K. Al-Mallah, M.R. Davey, and E.C. Cocking, A New approach to the nodulation of non-legumes by rhizobia and the transformation of cereals by agrobacteria using enzymatic treatment of root hairs, *Int. J. Genet. Manipula. Plants 5*: 1 (1989).

40. S. Usami, S. Dkamoto, I. Takebe, and Y. Machida, Factor inducing Agrobacterium tumefaciens vir gene expression is present in monocotyledonous plants, *Proc. Natl. Acad. Sci. USA 85*:3748 (1988).

41. S. Gleddle, W.A. Keller, and G. Stterfield, Production of new hybrid plants through protoplast fusion, *Vegetative Propagation and Biotechnology for Tree Improvement* (K. Kesava Reddy, ed.), Nataraj Publishers, Dehra Dun, India, 1992, pp.231–242.

42. S. Predieri, Mutation induction and tissue culture in improving fruits, *Plant Cell Tiss. Org. Cult. 64*: 185 (2001).

43. A.J.E. Bettany, S.J. Daltan, E. Timms, M.S. Dhanoa, and P. Morris, Effect of selectable gene to reporter gene ratio on the frequency of co-transformation and co-expression of uidA and hpt transgenes in protoplast-derived plants of tall fescue, *Plant Cell Tiss. Org. Cult. 68*: 177 (2002).

44. P.M. Olhoft, K. Lin, J. Galbraith, N.C. Nielsen, and D.A. Somers, The role of thiol compounds in increasing *Agrobacterium*-mediated transformation of soybean cotyledonary-node cells, *Plant Cell Rep. 20*: 731 (2001).

45. D.Y. Xie and Y. Hong, *Agrobacterium*-mediated genetic transformation of Acacia mangium, *Plant Cell Rep. 20*: 917 (2002).

46. T.R. Ganapathi, N.S. Higgs, P.J. Balint-Kurti, C.J. Arntzen, G.D. May, and J.M. Van Eck, *Agrobacterium*-mediated transformation of embryogenic cell suspensions of the banana cultivar, Rasthali (AAB) *Plant Cell Rep. 20*: 157 (2001).

47. F.J. Novak, R. Afza, M. van Duren, Amperes-Dallas, B.V. Conger, and T. Xiaolang, Somatic embryogenesis and plant regeneration in suspension cultures of dessert (AA and AAA) and cooking (ABB) bananas (*Musa* spp.), *Bio/Technology 7*: 147 (1989).

48. M.M. O'Kennedy, J.T. Burger, and D.K. Berger, Transformation of elite white maize using the particle flow gun and detailed analysis of a low-copy integration event, *Plant Cell Rep. 20*: 721 (2001).

49. V.A. Hilder, A.M.R. Gatehouse, and D. Boulter, Genetic engineering of crops for insect resistance using genes of plant origin, *Genetic Engineering of Crop Plants* (G.W. Lycett and D. Grierson, eds.), Butterworths, London, 1990, pp.51–65.

50. A.M.R. Gatehouse and V.A. Hilder, Introduction of genes conferring insect resistance, *Brighton Crop Protection Conference—Pests and Diseases*, Vol. 3, The Lavenham Press, Lavenham, 1988, pp.1245–1254.

51. P.D. Barfoot and R.A. Connett, AGC's cowpea enzyme inhibitor and its potential market opportunity, *Agric. Biotechnol. News Information 1*: 177 (1889).

52. A.M.R. Gatehouse, J.A. Gatehouse, P. Doble, A.M. Kilmister, and D. Boulter, Biochemical basis of insect resistance in Vigna unguiculata, *J. Sci. Food Agric. 30*: 948 (1979).

53. A.M.R. Gatehouse, F.M. Dewey, J. Dove, K.A. Fenton, and A. Pusztai, Effect of seed lectin from *Phaseolus vulgaris* on the development of larvae of Callosobruchus maculates: mechanism of toxicity, *J. Sci. Food Agric. 35*: 373 (1984).

54. D.H. Janzen, C.A. Ryan, I.E. Liener, and G. Pearce, Potentially defensive proteins in mature seeds of 59 species of tropical leguminosae, *J. Chem. Ecol. 12*: 1469 (1986).

55. R.J. Redden, P. Dobie, and A.M.R. Gatehouse, The inheritance of seed resistance to *Callosobruchus maculates* F. in cowpea (*Vigna unguiculata* L. Walp), I. Analysis of parental F_1, F_2, F_3 and backcross seed generation, *Aust. J. Agric. Res. 34*: 681 (1983).

56. A.M.R. Gatehouse and D. Boulter, Assessment of the antimetabolic effects of trypsin inhibitors from cowpea (*Vigna unguiculata*) and other legumes on development of the bruchid beetle, *Callosobruchus maculates*, *J. Sci. Food Agric. 34*: 345 (1983).

57. A.M.R. Gatehouse, J.A. Gatehouse, and D. Boulter, Isolation and characterization of trypsin inhibitors from cowpea (*Vigna unguiculata*), *Phytochemistry 19*: 751 (1980).

58. V.A, Hilder, A.M.R. Gatehouse, S.E. Sheerman, R.F. Barker, and D. Boulter, A novel mechanism of insect resistance engineered into tobacco, *Nature 330*: 160 (1987).

59. K.A. Barton, H.R. Whiteley, and N-S Yang, *Bacillus thuringiensis* delta endotoxin expressed in *Nicotiana tabacum* provides resistance to lepidopteran insects, *Plant Physiol. 85*: 1103 (1987).

60. D.A. Fischhoff, K.S. Bowdish, F.J. Perlak, P.G. Marone, S.M. McCoormick, and J.G. Niedermeyer, Insect tolerant transgenic tomato plants, *Bio/Technology 5*: 807 (1987).

61. M. Vaeck, A. Reynaerts, H. Hofte, S. Jansens, M.D. De Beukleer, and C. Dean, Transgenic plants protected from insect attack, *Nature 328*: 33 (1987).

62. D.R. Wilcox, A.G. Shivakumar, B.E. Melin, M.F. Miller, T.A. Benson, and C.W. Schopp, Genetic engineering of bioinsecticides, *Protein Engineering* (M. Inouye and R. Sama, eds.), Academic Press, New York, 1986, pp.395–413.

63. C. Hernstadt, G.G. Soares, E.R. Wilcox, and D.L. Edwards, A new strain of Bacillus thuringiensis with activity against coleopteran insects, *Bio/Technology 4*: 305 (1986).

64. P. Umbeck, G. Johnston, K. Barton, and W. Swain, Genetically transformed cotton (*Gossypium hirsutum*) plants, *Bio/Technology 5*: 263 (1987).

65. C.A. Rhodes, D.A. Pierce, I.J. Mettler, D. Mascaranhas, and J.T. Detmer, Genetically transformed maize plants from protoplasts, *Science 240*:204 (1988).

66. H.H. McKinney, Mosaic diseases in the Canary Islands, West Africa and Gibraltar, *J.Agric. Res. 39*:557 (1929).

67. R.S. Nelson, P.A. Powell, and R.N. Beachy, Coat protein-mediated protection against virus infection, Genetic Engineering of Crop Plants (G.W. Lycett and D. Grierson, eds.), Butterworths, London, 1990, pp.13–24.

68. R.W. Fulton, Practices and precautions in the use of cross protection for plant virus disease control, *Annu. Rev. Phytopathol. 24*: 67 (1986).

69. F. Ponz and G. Bruening, Mechanism of resistance to plant viruses, *Annu. Rev. Phytopathol. 24*: 355 (1986).

70. K.H. Fernow, Tomato as a test plant for detecting mild strains of potato spindle tuber virus, *Phytopathology 57*: 1347 (1967).

71. C.L. Niblett, E. Dickson, R.H. Fernow, R.K. Horst, and M. Zaitlin, Cross protection among four viroids, *Virology 91*: 198 (1978).

72. J.L. Sherwood and R.W. Fulton, The specific involvement of coat protein in tobacco mosaic virus cross protection, *Virology 119*:150 (1982).

73. M. Zaitlin, Viral cross protection: more understanding is needed, *Phytopathology 66*:382 (1976).

74. S. Sarkar and P. Smitamana, A proteinless mutant of tobacco mosaic virus: evidence against the role of a viral coat protein for interference, *Mol. General Genet. 184*: 158 (1981).

75. J.L. Sherwood, Demonstration of the specific involvement of coat protein in tobacco mosaic virus (TMV) cross protection using a TMV coat protein mutant, *J. Phytopathol. 118*:358 (1987).

76. T.M. Zinnen and R.W. Fulton, Cross protection between sunn-hemp mosaic and tobacco mosaic viruses, *J.Gen. Virol. 67*:1679 (1986).

77. L. Sequeira, cross protection and induced resistance, their potential for plant diseases control, *Trends Biotechnol. 2*:25 (1984).

78. R.N. Beachy, P. Abel, M.J. Oliver, B. De, R.T. Fraley, and S.G. Rogers. Potential for applying genetic transformation to studies of viral pathogenesis and cross-protection, *Biotechnology in Plant Science: Relevance to Agriculture in the Nineteen Eighties* (M. Zaitlin, P. Day and A. Hollander, eds.), Academic Press, New York, 1985, pp. 265–275.

79. J.C. Sanford and S.A. Johnston, The concept of parasite-derived resistance: deriving resistance genes from the parasite's own genome, *J.Theoret. Biol. 113*: 395 (1985).

80. M.W. Bevan, S.E. Mason, and P. Goelet, Expression of tobacco mosaic virus coat protein by a cauliflower mosaic virus promoter in plants transformed by *Agrobactrium*, *EMBO J. 4*:1921(1985).

81. P. Powell-Abel. R.S. Nelson, B. De, N. Hoffman, S.G. Rogers, and R.T. Fraley, Delay of diseases development in transgenic plants that express the tobacco mosaic virus coat protein gene, *Science 232*:838 (1986).

82. L.S. Loesch-Fries, D. Merlo, T. Zinnen, L. Burhop, K. Hill, and K. Karhn, Expression of alfalfa mosaic virus RNA4 in transgenic plants confers virus resistance, *EMBO J. 6*:1845 (1987).

83. N.E. Turner, K.M. O'Connell, R.S. Nelson, P.R. Sanders. R.N. Beachy, and R.T. Fraley, Expression of alfalfa mosaic virus coat protein gene confers cross-protection in transgenic tobacco and tomato plants, *EMBO J. 6*:1181(1987).

84. C.M.P. van Dun, J.F. Bol, and L. van Vloten-Doting, Expression of alfalfa mosaic virus and tobacco rattle virus coat protein genes in transgenic tobacco plants, *Virology 159*:299 (1987).

85. M. Cuozzo, A.M. O'Connell, W. Kaniewski, R, Fang, Nchua, and N.E. Turner, Viral protection in transgenic tobacco plants expressing the cucumber mosaic virus coat protein or its antisense RNA, *Bio/Technology 6*:549(1988).

86. C. Hemenway, R. Fang, W. Kaniewski, N. Chua, and N.E. Turner, Analysis of the mechanism of protection in transgenic plants expressing the potato virus x coat protein or its antisense RNA, *EMBO J. 7*:1273 (1988).

87. R.S. Nelson, S.M. McCormick, X. Delannay, P. Dube, J. Layton, and E.J. Anderson, Virus tolerance, plant growth and field performance of transgenic tomato plants expressing coat protein from tobacco mosaic virus, *Bio/Technology 6*:403 (1988).

88. C.S. Gasser and R.T. Fraley, Genetically engineered plants for crop improvement, *Science 244*:1293 (1989).

89. C. Lawson, W. Kaniewski, L. Haley, R. Rozman, C. Newell, and N. Turner, Expression analysis and field performance of russet Burbank potato genetically engineered for resistance to potato virus X and potato virus Y, Symp. on Horticultural Biotechnology (abst.), 1989, p.42.

90. J. Yamada, M. Yoshioka, T. Meshi, Y. Okada, and T. Ohno, Cross protection in transgenic tobacco plants expressing a mild strain of tobacco mosaic virus, *Mol. Gen. Genet. 215*: 173 (1988).

91. D. Baulcombe, M. Devic, and M. Jaegle, The molecular biology of satellite RNA from cucumber mosaic virus, *Genetic Engineering of Crop Plants* (G.W. Lycett and D. Grieson, eds.), Butterworths, London, 1990, pp.25–32.

92. S.N. Covey, R. Stratford, K. Saunders, D.S. Turner, and A.P. Lucy, Molecular aspects of cauliflower mosaic virus pathogenesis, *Genetic Engineering of Crop Plants* (G.W. Lycett and D. Grieson, eds.), Butterworths, London, 1990, pp33–50.

93. A.F. Murant and M.A. Mayo, Satellites of plant viruses, *Annu.Rev.Phytopathol.20*: 49 (1982).

94. L. Comai, D. Facciotti, W. Hiatt, G. Thompson, R. Rose, and D. Stalker, Expression in plants of a mutant aroA gene from *Salmonella typhimurium* confers tolerance to glyphosate, *Nature 317*: 741 (1985).

95. D. Shah, R. Horsch, H. Klee, G. Kishmore, J. Winter, and N. Tumer, Engineering herbicide tolerance in transgenic plants, *Science 233*:478 (1986).

96. D. Llewellyn, B. Lyon, Y. Cousins, J. Huppatz, E.S. Dennis, and W.J. Peacock, Genetic engineering of plants for resistance to the herbicide, 2,4-D, *Genetic Engineering of Crop Plants* (G.W. Lycett and D. Grieson, eds.), Butterworths, London, 1990, pp.67–77.

97. M. DeBlock, J. Botterman, M. Vandewiele, J. Dockyx, A. Theon, and V. Gossele, En-

gineering herbicide resistance in plants by expression of a detoxifying enzyme, *EMBO J.* 6: 2513 (1987).

98.   D. Stalker, K. McBride, and L. Malyj, Expression in plants of a bromoxynil-specific bacterial nitrilase that confers herbicide resistance, *Genetic Improvement of Agriculturally Important Crops: Progress and Issues* (R.T. Fraley, N.M. Frley and J. Schell, eds.), Cold Spring Harbour Laboratory, Cold Spring Harbour, 1988, pp.37–40.

99.   A.J. Keys, Prospects by increasing photosynthesis by control of photorespiration, *Pesticide Sci. 19*: 313 (1983).

100. M.A.J. Parry, S.P. Colliver, P.J. Madhwick, and M.J. Paul, Manipulation of photosynthetic metabolism, *Recombinant Proteins from Plants: Production, Isolation and Clinically Useful Compounds* (C. Cunnigham and A.J.R. Porter, eds.), *Methods in Biotechnology*, Vol. 3, Humana Press, Totawa, NJ, 1998, pp.229–249.

101. G. Bainbridge, P.J. Madgwick, S. Parmar, R. Mitchell, M.J. Paul, J. Pitts, A.J. Keys, and M.A.J. Parry, Engineering Rubisco to change its catalytic properties, *J.Exp.Bot. 46*: 1269 (1995).

102. A.M. Viale, H. Kobayashi, and T. Akazava, Distinct properties of *Escherichia coli* products of plant-type ribulose-1, 5-bisphosphate carboxylase/oxygenase directed by two sets of genes from the photosynthetic bacterium, Crromatium vinosum, *J.Biol. Chem. 265*:18386 (1990).

103. S. Gutteridge and A.A. Gatenby, Rubisco synthesis, assembly, mechanism and regulation, *Plant Cell 7*:809 (1995).

104. J. Brangeon, A. Nato, and A. Forchioni, Ultrastructural detection of Rubisco and target mRNAs in wild type and holoenzyme-deficient *Nicotiana* using immunogold and in-situ hybridization, *Planta 177*: 151 (1989).

105. Z. Svab and P. Maliga, High frequency plastid transformation in tobacco by selection for a chimaeric aadA gene, *Proc. Natl. Acad. Sci. USA 90*: 913 (1993).

106. M. De Neve, H. Van Houdt, A-M Bruyns, M. Van Montagu, and A. Depicker, Screening for transgenic lines of a heterogenous protein, *Recombinant Proteins from Plants: Production, Isolation and Clinically Useful Compounds* (C. Cunningham and A.J.R. Porter, eds.), *Methods in Biotechnology*, Vol. 3, Humana Press, Totawa, NJ, 1998, pp.203–227.

107. C. Peach and J. Velten, Transgene expression variability (position effect) of CAT and GUS reporter genes driven by linked divergent T-DNA promoters, *Plant Mol. Biol. 17*:49 (1991).

108. D.D. Songstad, D.A. Somers, and R.J. Griesbach, Advances in alternative DNA delivery techniques, *Plant Cell Tiss. Org. Cult. 40*: 1(1995).

109. J.R. Zupan and P. Zambryski, Tranfer of T-DNA from *Agrobacterium* to the plant cell, *Plant Physiol. 107*: 1041 (1995).

110. J. Gold, M. Devey, O. Hasegawa, E.C. Ulian, G. Peterson, and R.H. Smith. Transformation of *Zea mays* L. using *Agrobacterium tumefaciens* and the shoot apex, *Plant Physiol. 95*: 426 (1991).

111. M.T. Chan, T-M Lee, and H-H Chang, Transformation of Indica rice (*Oryza sativa L.*) mediated by *Agrobacterium tumefaciens*, *Plant Cell Physiol. 33*: 577 (1992).

112. Y. Hiei, S. Ohta, T. Komari, and T. Kumashiro, Efficient transformation of rice (*Oryza sativa L.*) mediated by *Agrobacterium* and sequence analysis of the boundaries of the T-DNA, *Plant J. 6*: 271 (1994).

113. P.F. Ambros, A.J.M. Matzke, and M.A. Matzke, Localization of *Agrobacterium rhizogenes* T-DAA in plant chromosomes by in situ hybridization, *EMBO. J. 5*: 2073 (1986).

114. T.P. Robbins, A.G.M. Gerats, H. Fiske, and R.A. Jorgensen, Suppression of recombination in wide hybrids of *Petunia hybrida* as revealed by genetic mapping of marker transgenes, *Theoret. Appl. Genet. 90*: 957 (1995).

115. J.J. Pueyo and A. Hiatti, Production of foreign compounds in transgenic plants, *Agricultural Biotechnology* (A. Altman, ed.), Marcel Dekker, New York, 1998, pp.251–261.

116. S.B. Altenbach, C.C. Kuo, L.C. Staraci, K.W. Pearson, C. Wainwright, A. Georgescu, and J. Townsend, Accumulation of a Brazil nut albumin in seeds of transgenic canola results in enhanced levels of seed protein methionine, *Plant Mol. Biol. 18*:235 (1992).

117. D.J. Yun, T. Hashimoto, and Y. Yamada, Metabolic engineering of medicinal plants: transgenic *Atropa belladonna* with an improved alkaloid composition, *Proc. Natl. Acad. Sci. USA 89*:11799 (1992).

118. H.J. Klee, M.B. Hayford, K.A. Kretzmer, G.F. Barry, and G.M. Kishore, Control of ethylene synthesis by expression of a bacterial enzyme in transgenic tomato plants, *Plant Cell 3*:1187 (1991).

119. L. Penarrubia, R. Kim, J. Giovannoni, S.H. Kim, and R.L. Fisher, Production of the sweet protein monelin in transgenic plants, *Biotechnology 10*:561 (1992).

120. R. Hightower, C. Baden, E. Penzes, P. Lund, and P. Dunsmuir, Expression of antifreeze proteins in transgenic plants, *Plant Mol.Biol. 17*:1013 (1991).

121. K.D. Kenwood, M. Altschuler, D. Hildebrand, and P.L. Davies, Accumulation of type I fish antifreeze protein in transgenic tobacco is cold-specific, *Plant Mol. Biol. 23*:377 (1993).

122. R. Yeargan, I.B. Maiti, M.T. Nielsen, A.G. Hunt, and G.J. Wagner, Tissue partitioning of cadmium to transgenic tobacco seedlings and field grown plants expressing the mouse metallothioneins I gene, *Transgen. Res. 1*: 261 (192).

123. J. Vandekerckhove, J. van Damme, M. van Lijsebettens, J. Botterman, M. de Block, M. Vandewiele, A. de Clercq, J. Leemans, M. van Montagu, and E. Krebbers, Enkephalins produced in transgenic plants using modified 2S seed storage proteins, *Biotechnology 7*:929 (1989).

124. E. Krebbers and J. Vandekerckhole, Production of peptidesin plant seeds, *Trends Biotechnol. 8*:1 (1990).

125. P.J. Sijmons, B.M.M. Dekker, B. Schrammeijer, T.C. Verwoerd, P.J.M. van den Elzen, and A. Hoekema, Production of correctly processed human serum albumin in transgenic plants, *Biotechnology 8*: 217(199).

126. O. Edelbaum, D. Stein, N. Holland, Y. Gafni, O. Livneh, D. Novick, M. Rubinstein, and I. Sela, Expression of active human interferon-beta in transgenic plants, *J.Interferon Res. 12*: 449 (1992).

127. J. Pen, L. Molendijk, W.J. Quax, P.C. Sijmons, A.J.J. van Ooyen, P.J.M. van den Elzen, K. Rietveld, and A. Hoekema, Production of active *Bacillus licheniformis* alpha-amylase in tobacco and its application in starch liquefaction, *Biotechnology 10*:292 (1992).

128. J. Pen, T.C. Verworerd, P.A. van Paridon, R.F. Beudeker, P.J.M. van den Elzen, K. Greerse, J.D. van der Klis, H.A.J. Versteegh, A.J.J. van Ooyen, and A. Hoekema, Phytase-containing transgenic seeds as a novel feed additive for improved phosphorus utilization, *Biotechnology 11*:811 (1993).

129. H.S. Mason, D.M. Lam, and C.J. Arntzen, Expression of hepatitis B surface antigen in transgenic plants, *Proc. Natl. Acad. Sci. USA 89*: 11745 (1992).

130. E.M. Scolnick, A.A. McLean, D.J. West, W.J. McAleer, W.J. Miller, and E.B. Buynak, Clinical evaluation in healthy adult of a hepatitis B vaccine made by recombinant DNA, *J.A.M.A. 251*: 2812 (1984).

131. R.L. Starnes, Industrial potential of cyclodextrin glucosyltransferases, *Cereal Foods World 35*: 1094 (1990).

132. F. Binder, O. Huber, and A. Bock, Cyclodextrin glucosyltransferase from *Klebsiella pneumoniae* M5al: cloning, nucleotide sequence and expression, *Gene 47*: 269 (1986).

133. J.V. Ookes, C.K. Shewmaker, and D.M. Stalker, Production of cyclodextrins, a novel carbohydrate in the tubers of transgenic potato plants, *Biotechnology, 9*: 982 (1991).

134. Y. Poirier, D.E. Dennis, K. Klomparens, and C. Somerville, Polyhydroxybutyrate, a biodegradable thermoplastic, produced in transgenic plants, *Science 256*:529 (1992).

135. A. Hiatt, Antibodies produced in plants, *Nature 344*: 469 (1990).

136. P.G. Schultz, R.A. Lerner, and S.J. Benkovic, Catalytic antibodies, *Chem. Eng. News 68*: 26 (1990).

137. R. Topfer and N. Martini, Engineering crop plants for industrial traits, *Agricultural Biotechnology* (A. Altman, ed.), Marcel Dekker, New York, 1998, pp. 161–181.

138. C.J.S. Smith, C.F. Watson, J. Ray, C.J. Bird, P.C. Morris, W. Schuch, and D. Grierson, Antisense RNA inhibition of polygalacturonase activity in transgenic tomatoes, *Nature 334*: 724 (1988).

139. R.E. Sheehy, M. Krammer, and W.R. Hiatti, Reduction of polygalacturonase activity in tomato fruit by antisense RNA, *Proc. Natl. Acad. Sci. USA 85*: 8805 (1988).

140. C.J.S. Smith, C.F. Watson, P.C. Morris, C.J. Bird, G.B. Seymour, J.E. Gray, C. Arnold, G.A. Trucker, W. Scuch, S.E. Harding, and D. Grierson, Inheritance and effects of ripening of antisense polygalacturonase genes in transgenic tomatoes, *Plant Mol. Biol. 14*: 369 (1990).

141. T. Hatanaka, Y.E. Choi, T. Kusano, and H. Sano, Transgenic plants of coffee (*Coffea canephora*) from embryogenic callus via *Agrobacterium tumefaciens*–mediated transformation, *Plant Cell. Rep. 19*: 106 (1999).

142. H. Etienne, F. Anthony, S. Dussert, D. Fernandez, P. Lashermes, and B. Bertrand, Biotechnological applications for the improvement of coffee (*Coffea arabica* L.), *In Vitro Cell Dev. Biol. Plant 38*: 129 (2002).

143. C.D. Foy, R.L. Chaney, and M.C. White, The physiology of metal toxicity in plants, *Annu. Rev. Plant Physiol. 29*: 511 (1978).

144. L.V. Kochian, Cellular mechanisms of aluminum toxicity and resistance in plants, *Annu. Rev. Plant Physiol. Plant Mol. Biol. 46*: 237 (1995).

145. J.D. Ownby and H.R. Popham, Citrate reverses the inhibition of wheat growth caused by aluminum, *J.Plant Physiol. 135*: 588 (1989).

146. J.M. De la Fuente, V. Ramirez-Rodriguez, J.L. Cabrera-Ponce, and L. Herrera-Estrella, Aluminum-tolerance in transgenic plants by alteration of citrate synthesis, *Science 276*: 1566 (1997).

147. J.M. De la Fuente-Martinez, and L. Herrera-Estrella, Advantages in the understanding of aluminum-tolerant transgenic plants, *Adv. Agron. 66*: 103 (1999).

148. J. Lopez-Bucio, O.M. de la Vega, A. Guevara-Garcia, and L. Herrera-Estrella, Enhanced phosphorus uptake in transgenic tobacco plants that overproduce citrate, *Nat. Biotechnol. 18*: 450 (2000).

149. J. Lopez-Bucio, M.F. Nieto-Jacob, V. Ramirez-Rodriguez, and L. Herrera-Estrella, Organic acid metabolism in plants: from adaptive physiology to transgenic varieties for cultivation in extreme soils, *Plant Sci. 160*: 1 (2000).

150. L. Herrera-Estrella, Transgenic plants for tropical regions: Some considerations about their development and their transfer to the small farmer, *Proc. Natl. Acad. Sci. USA 96*: 5678 (1999).

25
Loss Reduction Biotechnology of Seeds

I. INTRODUCTION

Whether crops are raised for food or seed purposes, a bulk of the produce is lost during production, harvesting, threshing, drying, processing, storage, marketing, and distribution. One conservative estimate of food crop loss is 30–40% (1). May (2) estimated preharvest and postharvest losses of foods at 48% on a worldwide basis. Seed loss varies greatly and is a function of crop species and variety, pests and diseases, climate, harvesting system, processing, storage, handling, marketing practices and methods, and the existing food/seed laws. Most of the information available on seed loss emphasizes quantitative losses caused by disease, insects, and rodents during storage. Besides these quantitative losses, enormous losses also occur in the nutritional quality and viability of seeds during postharvest handling, storage, and processing. Although reliable estimates for quantitative and qualitative losses of seeds during harvesting, threshing, drying, transportation, or processing are not available, it is clear that significant losses do occur and that their elimination or reduction warrants serious efforts at the farm, processor, trader, and consumer levels (3,4).

II. MAGNITUDE OF SEED CROP LOSSES

According to the National Academy of Sciences of the United States (5), the magnitude of postharvest losses of food grains in various developing countries ranges from 2 to 40% for rice, 1 to 100% for maize, 2 to 42% for wheat, and 1 to 68% for legumes. Quantitative postharvest food grain losses worldwide (Table 1) indicate that the extent of food grain losses in developing countries is much higher than estimated.

Adams (6) distinguished between estimates of losses obtained in laboratory studies or small field trials and those derived from investigations of normal postharvest practices. Among the food crops, cereals have received more attention than legumes, which may be partly due to their relative importance as food and the greater prevalence of grading standards. Also, more attention has been given to quantitative losses in weight and market quality rather than to nutritional or seed viability losses. An examination of the available field information on the causes of food losses has shown that emphasis has been on losses by insect infestation, followed by those caused by vertebrates, fungi, and physical factors. Most work has been done on losses of cereals at the farm level rather than at the trader or other large-scale storage levels. Adams (6) concludes that most attempts to evaluate quantitative postharvest grain losses were based on experience in Asia and Africa, specifically on cereal crops, and

Table 1   Postharvest Losses of Food Grains

Country	Estimated losses (%)	Remarks
World	5–50	Total postharvest losses
India	15–25	Preharvest losses
	0.2–7	Harvesting, threshing, transportation, processing, and milling losses
	1.7–30	Storage losses
	40–61	Total postharvest losses
Bangladesh	5	Storage losses by rodents
Philippines	2–3	Storage losses by rodents
Brazil	4–8	Storage losses by rodents
Tropical Africa	30	Storage losses
Developing countries	50	Total postharvest losses
United States	0.5–1	Transport and grading losses

Source: Ref. 3.

that these were caused by insects at the farm level. Other agents such as rodents, microorganisms, and chemical reactions also cause significant losses.

The magnitude of leguminous seed losses varies according to the type of legume and the storage and environmental conditions (Table 2). One conservative estimate indicated that a minimum of 107 million tons of food was lost in 1976 (5). The total amount of food lost in cereal and leguminous grains alone would provide more than the annual minimum caloric requirements for 169 million people (3). It has been suggested that controlling postharvest losses could improve the food supply to the

Table 2   Losses of Food Legumes Within Postharvest Systems in Developing Nations

Nation	Legume	Total loss (%)
Ghana	Shelled bean	7.45
Nigeria	Cowpea	5.4
	Groundnut	4.5
Kenya	Pulses	30
Sudan	Groundnut	4–27
Uganda	Groundnut	9–18
Zambia	Cowpea	40
India	Pulses	8.5
Pakistan	Pulses	5–10
Thailand	Soybean	10–30
Brazil	Drybean	15–25
Honduras	Drybean	20–25
Nicaragua	Soybean	10–35
Paraguay	Soybean	15

Source: Ref. 4.

extent that the amount of food produced today would be enough to feed the next few generations (7).

Efficient use of water and nitrogenous fertilizers and control of pests and diseases constitute the most important constraints in the production of seed crops. World crop losses to the three major pest groups—insects, pathogens, and weeds—have been estimated at about 35% (8,9). Losses to mammals and birds are more severe in the tropics and subtropics than in temperate regions, but are still low compared with the losses to these three pest groups. Representative data on losses of world cereal crops due to insects and pathogens are summarized in Table 3, and the average losses of food crops caused by insects, diseases, and weeds in different regions are shown in Table 4. The losses caused by pests at different stages of production and storage in developed and developing countries have been estimated to vary from about 25 to 43% annually. According to Deshpande and Deshpande (10), crop losses may sometimes be negligible and at other times total, especially as a result of sporadic outbreaks of nonendemic pests and diseases.

Pimentel (8) estimated U.S. preharvest losses of food plants to pests to be about 37% even with the use of modern pest control technology. Insects accounted for about 13% of these losses; plant pathogens, 12%; and weeds, 12% U.S. postharvest losses were estimated to be about 9%. Thus, total crop losses to pests in the United States alone are more than 40%. It is possible that losses are even higher in developing countries, where advanced pest control technology is neither available nor economically feasible (11). Nutritional and germinative losses of seeds, which may be combination of loss of quantity and quality, are rather difficult to assess and measure.

Untimely and frequent rains in India (29.0 mm in April, 100.2 mm in May, and 105.2 mm in June 1997) at the time of seed maturity in pea cv. Bonnevile, onion cv. Pb. Naroya, carrot cv. Selection 21, radish cv. Pb. Safed, turnip cv. L-1 and cauliflower cv. Pb. Giant-26 caused heavy losses due to rotting and discoloration of the seeds. For each of the species, data are tabulated on frequency of seven fungal species after 0, 2, and 4 months of storage under ambient conditions beginning in June 1997. Pea seeds suffered the greatest damage during storage and their viability declined from 86.5 to 28.5%. The major damage in stored seed was caused by *Aspergillus flavus, A. niger, A. fumigatus, Mucor* spp., *Rhizopus* spp., and *Penicillium* and *Fusarium* spp. After storage, the seeds were treated with 0.2% Thiram 75WP, Captan 75WP, and Bavistin

Table 3 Estimated Losses (%) in World Cereal Crops Due to Insects and Diseases

Cereal	Insects	Diseases	Total
Wheat	5	9	14
Rice	27	9	36
Corn	12	9	21
Barley	4	8	12
Oats	8	9	17
Sorghum and millet	10	11	21
Rye	2	3	5

Source: Ref. 12.

Table 4 Regional Crop Losses (%) Due to Pests

Region	Insects	Diseases	Weeds	Total
Europe	5	13	7	25
North and Central America	9	11	8	28
South America	10	15	8	33
Africa	13	13	16	42
Asia	21	11	11	43

Source: Refs. 5 and 13.

[carbendazim] 50WP before sowing. Treatment with Bavistin in pea and with Thiram and Captan in the other crops significantly improved seed germination (11a).

Solid matrix priming offers an effective means of raising seed performance in sweet corn (*Zea mays*) carrying the shrunken-2 *sh-2* gene. Nevertheless, the storability of primed *sh-2* seeds is still unknown. Chang and Sung (11b) investigated the emergence performance of vermiculite-primed *sh-2* seeds of two hybrids during 12 months of dry storage. Several physiological traits in relation to lipid peroxidation and peroxide-scavenging were also examined. The vermiculite priming improved seedling emergence, decreased initial seed leakage, reduced lipid peroxidation, and enhanced activities of several peroxide-scavenging enzymes. However, the benefits of priming were diminished and the longevity was shortened when primed seeds were stored for 12 months, especially under 25°C storage conditions. The loss of viability might be linked to the increased seed leakage (after 12 months of storage) and lipid peroxidation, marked changes in peroxide scavenging activities and the shift of glutathione redox status toward a more oxidized form. Significant differences also existed between hybrids in emergence percentage and related physiological activities, although the responses expressed as a function of storage duration are rather similar. The priming effects on *sh-2* sweet corn seeds could be maintained after storage for 6 months under 10°C. However, the primed *sh-2* seeds deteriorated more rapidly than nonprimed *sh-2* seeds when stored at 25°C for 6 months.

III. CAUSES OF SEED CROP LOSSES

A. Preharvest Factors

General causes of seed crop losses include preharvest factors such as the genetic characteristics of the seed variety and species, which greatly influence both preharvest and postharvest crop losses. Traditional varieties of seed crops are generally well adapted to both their usual environment and to postharvest handling. The seeds or grains that survive storage and are used in subsequent seasons have evolved characteristics that favor their survival. The lower moisture content in the ripe seed that dries more readily and a thicker seed coat for repelling insects and rodents will, for example, have better storability. Introduction of newer, high-yielding seed crop varieties has resulted in greater postharvest losses, where the new varieties are not as well adapted to the postharvest conditions as the traditional varieties. Damage to the growing seed crop affects its postharvest characteristics, as does the crop-protection treatment prior to harvest. Insect infestation of a maturing seed crop may increase its vulnerability to loss after harvest, although residual insecticide may decrease the extent of postharvest insect damage to seeds (5).

The time at which a seed crop is harvested has an important effect on the subsequent storage quality of the seed. Typically, the harvest may be begun before the grains are fully ripe and may extend until mold and insect damage are prevalent and shattering has occurred. Seeds or grains that are not fully developed contain more moisture and will deteriorate more quickly than mature grains, because enzyme systems are still active. If the grain remains in the field after maturity, repeated wetting from rain and dew at night along with drying by the hot sun by day may cause seed to crack (e.g., long-grain paddy) and may also increase insect damage, especially in maize, paddy, and pulses. Seed crops standing in the field after maturity become more liable to harvest losses. Ripened seed is more likely to be shattered onto the ground during harvesting. Maize loss may result from the loosening of the husk after it is ripe, and subsequent mold infection or insect attack insect infestation in the field is also likely to increase if the seed crop stands too long, in addition to losses caused by rodents, grain-eating birds, and other vertebrates.

B. Biological Factors

1. Insects and Nematodes

The variety of organisms associated with damage to cultivated seed crops ranges from the smallest viruses, bacteria, fungi, nematodes, and insects to higher vertebrate animals such as birds and rodents. The principal biological agents of seed deterioration during storage are insects, mites, fungi, and rodents. Insect pests are a greater problem in the regions of high relative humidity, but temperature is the overriding factor influencing insect multiplication. At about 32°C, insects can multiply at a monthly compounded rate of 50-fold. Thus, it is theoretically possible that 50 insects at the harvest of a seed crop could multiply to become more than 312 million within a period of 4 months of storage. Insects often attack the most valuable part of seeds. For example, four important pests of maize attack the embryo, rejecting the starchy endosperm, thus removing the most nutritious part of the grain as well as destroying its ability to germinate.

Leguminous seeds are particularly susceptible to insect attack and provide a nutritious and palatable substance for a large and extremely diverse range of pests worldwide (14). In nature, in the absence external influences, the number of insects tends to fluctuate around an equilibrium, known as the general equilibrium level, which is defined as the average density of the population over a period in the absence of permanent environmental changes (13). Any organism causing 5% loss in the yield of a crop is described as an economic pest and the amount of crop damage that would justify control measures is known as economic damage. The lowest pest population density causing economic damage is called the economic injury level. This level varies considerably with the insect pest, crop, season, and locality. The effects of different numbers of pests on the yields of seed crop is very important and must be estimated in order to justify control measures (10).

According to Youdeowei and Adeniji (13), the economic threshold forms the basis for making decisions about when to apply control measures. It is defined as the pest population density or damage level at which control measures must be applied to prevent a pest population from reaching the economic injury level. Several natural enemies play an important role in preventing insect pests from causing economic damage (15), which is demonstrated by using chemicals either to remove these natural enemies or to reduce their effectiveness. One consequence of pesticide use may be a resurgence of pest populations to densities far greater than those in untreated areas (16). In addition, previously minor pests may become serious when chemicals kill off their natural enemies. Pesticide use must therefore be restricted to selected application of chemicals that are less harmful to important beneficial species but are effective in keeping pest populations under control (17).

Insects, nematodes, and spider mites are the most troublesome pests of leguminous seed crops. Most pests are specific to the part of the plant they attack; for example, leaves, stems, flowers, fruits, seeds, or roots. Pests can be classified according to their feeding habits. The four insect orders that most threaten food legumes are bugs (Hemiptera), moths and butterflies (Lepidoptera), beetles (Coleoptera), and grasshoppers and crickets (Orthoptera). The sucking insects that pierce the plant and feed on its juices form the most abundant class of insect pests affecting seed crops, including the pulses. Many are pod suckers, causing substantial yield and quality losses. Many of the pod-chewing insects also feed on foliage, blooms, petioles, and stems. Most economic losses, however, occur because of direct feeding on pods. Coleoptera, Lepidoptera, and Thysanoptera (thrips) species feed on foliage and also chew into the pods. Insects having biting and chewing mouth parts tunnel into the stems of crops and remain inside (stem eaters/stem borers), consuming large quantities of tissue and quickly killing the plants. Tennites (stem eaters) actually devour stem tissues and are serious economic pests of leguminous seed crops in the tropics. Stem borers include both lepidopteran and dipteran larvae as well as beetle larvae.

Nematodes are serious root pests of leguminous seed crops in the tropics and belong to following five classes:

1. Root-knot nematodes (genus *Meloidogyne*)
2. Lesion nematodes (genus *Pratylenchus*)
3. Cyst nematodes (genus *Heterodora*)
4. Sting nematodes (genus *Belonolaimus*)
5. Reniform nematodes (genus *Rotylenchus*)

2. Fungi

Fungi, including *Pythium*, *Fusarium*, and *Rhizoctonia*, are commonly responsible for the seed and seedling diseases of a wide range of seed crops. Blight is characterized by sudden and serious damage to all aerial plant parts (*Phytophthora*, *Ascochyta*, and *Rhizoctonia*). *Cercospora* causes leaf spot disease in various seed crops.

3. Viruses

Seedborne infection and insect vectors play major roles in the transmission of viral diseases. Seedborne infections are important for the carryover of virus between seasons, and vectors are vital agents of secondary spread. The incidence of seed transmission depends on the virus and its strain as well as the host plant, its cultivar, and the environment (18). Although natural infection of a susceptible host usually leads to the systemic spread of virus throughout the plant, seeds of infected plants do not invariably become virus infected, and some plant viruses are not even seedborne. Seed-transmitted viruses include cowpea mild mottle, tobacco ringspot, and bean common mosaic. Bos (19) has reviewed the mechanism of seed transmission of viruses.

Embryonic infection leads to direct infection of progeny. Although viruses can be carried outside the embryo in or on the seed coat, very few viruses are seedborne in this manner. Bean common mosaic and cowpea aphidborne mosaic viruses are examples of internal seedborne viruses. The rate of seed transmission depends on the time of infection. Plants infected after flowering do not produce infected seeds, and viral transmission can take place only before cytoplasmic separation of the developing embryo from maternal tissue (19). Aphids are the most important group of viral vectors that transmit virus in a nonpersistent manner (soybean dwarf, peanut rosette, and pea leaf roll). Beetles and white flies also are important vectors of viral diseases (10).

4. Bacteria

Bacteria are disseminated primarily through water, insects, seeds, plant parts, machinery, tools, or any method of moving soil from one place to another. Free water on the plant surface is necessary for the motility of bacterial cells, which enter the plant through a natural opening or wound. There the bacterial cell multiplies and establishes an infection under favorable environmental conditions. The symptoms of bacterial diseases vary from galls on roots, discoloration of xylem and other tissues, to decomposition characterized by wet, slimy, or smelly rots. Aboveground plant parts may have water-soaked spots or chlorotic, yellow spots, tan spots, or streaks (20). The entire plant may wilt, showing grayish green foliage and often a discoloration of seeds (11).

The principal bacterial pathogens of food legumes belong to the genera *Pseudomonas*, *Xanthomonas*, and *Corynebacterium*. Unlike fungi and viruses, there are comparatively few bacterial diseases of leguminous seed crops. However, bacterial pathogens induce a range of blights, cankers, wilt, leaf spots, and pustules in susceptible hosts leading to significant economic crop losses. Deshpande and Deshpande (10) have described the important bacterial diseases of food legumes.

5. Rodents

Rodent damage to seed crops can occur in more than one way. In addition to consuming seeds, rodents foul stored seeds with their excretions, which may carry pathogenic microorganisms. Damage by rodents to maize seeds is characteristic in that the embryo is usually first removed, rendering the seed incapable of germination. Rodents also gnaw holes in containers, resulting in leakages and wastage of seeds, and they paw into and scatter seeds while they eat. This scattered seed, along with that leaking from gnawed holes, is subject to contamination by microorganisms and admixture with impurities. Damage to seeds stored in bulk is usually less than to seed stored in bags.

Three main species of rodents causing losses to seed lots are (a) the Norvey, common, or brown rat (*Rattus norvegicus*), (b) the roof, ship, or black rat (*Rattus rattus*), and (3) the house mouse (*Mus musculus*). Bandicoot rats (*Bandicota bengalensis*) are important pest in isolated areas.

C. Postharvest Factors

1. Threshing and Shelling

Traditional methods of threshing to separate seeds from the plant, such as use of animals to trample sheaves in the threshing yard or the modern equivalent using tractor wheels, may result in loss of seed grain. This method also allows impurities to become mixed with the seed, causing subsequent storage problems. The use of flails to beat the seed from the stalk may damage the grains or kernels and is not always effective. Threshing and shelling will contribute to seed losses if carried out in a manner that results in cracking of seeds. Modern threshing and shelling devices may be used incorrectly or for seed crops for which they are not intended, with excessive breakage of seeds.

2. Drying Losses

Drying is a particularly vital operation in the chain of seed handling, since the moisture content of seeds is the most important factor determining whether and to what extent they will be liable to deterioration during subsequent storage. Drying is also used temporarily to inhibit the germination of seeds and to reduce the moisture content to a level that prevents the growth of microorganisms, especially fungi and bacteria. It also retards attacks on the seed by insects and mites.

Sun and air are commonly used to dry seeds in developing countries, although supplemental heat is sometimes employed. In many cases, seeds are treated separately from food grains and with greater care. Drying is a complex process requiring considerable skill and effort on the part of seedsmen, and the success with which the seed is preserved over shorter or longer periods depends to a great extent on the care and attention given to the drying and subsequent storage. Artificial and quicker drying methods may be required to dry seed crops harvested during wet seasons.

Overdrying, which can easily occur in arid regions or after excessive exposure to sun or other heat, can cause breakage, damage to the seed coat, bleaching, scorching, discoloration, and loss of germinability and nutritional quality. Too rapid drying of seed crops with a high moisture content may also damage seeds, causing bursting or "case-hardening," in which the surface of the seed dries out rapidly, sealing moisture within the inner layers. Similarly, under drying or slow drying, problems frequently encountered in humid regions, may result in seed deterioration due to fungi and bacteria, which in extreme cases may lead to total loss. Solar technology for artificial drying has received increasing attention owing to its negligible running costs. However, solar devices do not operate effectively when they are needed most to dry seeds during a wet spell or the rainy season. Clearly, methods for drying seed crops must be selected to suit the particular climate and economic and social circumstances in which they will be used.

3. Storage Losses

The extent to which seed deterioration and loss occurs during storage depends on physical and production factors, the storage environment, and biological factors. Physical damage to the seed crop during harvest and subsequent operations may also affect storage losses. Even undamaged seeds in their pods (cowpea), shells (groundnut), or husks (paddy) require protection from infestation by insect species and rodents.

The rate of deterioration during storage is influenced by environmental conditions. High temperatures and relative humidities encourage development of mold and rapid growth of insect populations. Seed deterioration is minimal in cool and dry climates and more marked in hot and dry ones; high in cool and damp conditions and very high in hot and damp climates. Climatic conditions during and after harvest affect the ease with which natural drying may be carried out and dictate the need for artificial drying. Seasonal and diurnal temperature differences between stored seeds and the surrounding environment can result in moisture translocation or migration among the quantities of bulk- or bag-stored seeds or in condensation of moisture on the seeds. Some climates lessen the residual activity of certain pesticides and can even reduce the effective life of storage containers and structures. Seed deterioration during storage is also closely related to the method of storage and management. Properly designed open-sided cribs allow relatively rapid drying of unhusked ears of maize and reduce losses due to mold. Traditional pest control methods like admixture of pulses with sand to fill the intergranular spaces effectively inhibit the development of bruchid beetles. The admixture or overlay of wood ashes or dried animal dung is another method affording protection against insect attack (5).

D. Seed Movement/Transportation

Seed losses can occur during transportation of field stalks to the threshing yard, from the threshing yard to the place of storage at the farm, or from the farm to the market and other distribution centers. The movement of seeds from the field to the final user involves a variety of transportation systems. The extent of seed losses depends upon the transportation system and the climatic conditions prevailing at the time of transit. Ultimately, seed losses during transportation are mainly due to spillage.

Mwasha et al. (20a) investigated the effect of the storage environment on seed deterioration in five clones of cashew (*Anacardium occidentale*). The seeds survived desiccation to 6% moisture content and short-term (up to 70 days) storage at cool temperatures (0 and $-20°C$). In addition, one sample was stored hermetically at 4.3% moisture content and $-20°C$ for 2 years and showed 100% normal germination. However, deterioration was detected when seeds were stored at 40°C. The duration of hermetic storage influenced seed germination ($P < .01$). There was a negative logarithmic relation ($P < .005$) between the seed moisture content and longevity (defined as the standard deviation of the frequency distribution of seed deaths in time) at 40°C. There was no significant difference among the clones in this relation; the negative slope (Cw of the seed viability equation) was 2.877 (s.c. 0.239). Thus, cashew shows orthodox seed storage behavior. A provisional model of the effect of seed storage moisture content and temperature on cashew seed longevity is provided.

IV. LOSS REDUCTION BIOTECHNOLOGY

A. Efficient Harvesting

The time of harvest of seed crops depends on the method of harvest used (manual or mechanical) and on the cultivar. Harvesting can be delayed somewhat longer with manual harvesting than with mechanical harvesting. In the case of leguminous seeds, harvesting should begin before the lower pods become dry enough to shatter. When the beans are harvested by combine, the crop should be fully mature and as dry as possible. In most varieties, the attachment point of the bean seeds to the pod turns black when the crop is ripe. The moisture content of the combine-harvested beans is usually higher than that of cereal crops. For combined-harvested beans, 15–20% is a low moisture content, 21–25% medium, and 26–30% high (3,4).

Single-row harvesting machines move down a row removing leaves, stalk, and pods. They are elevated to a position where the leaves and stalks are blown out of the rear, leaving the pods to be collected in sacks or on a platform. This is similar to cereal grain collection on a combine harvester. For cereals, manual harvesting is better suited to smaller fields, wherein the harvested stocks are immediately taken to the threshing yard. The moisture content of cereal crops such as paddy is the best index for determining the optimum time of harvest irrespective of varieties or dates of flowering. The seed moisture content, lodging, and type of harvester are important factors in the mechanical harvesting of seed crops. Maize should be harvested when cobs can be safely cribbed or dried to prevent spoilage. The nonlodging cultivars are better suited to mechanical harvesting.

B. Careful Threshing, Shelling, and Dehusking

When properly dried and stored, threshed seeds have been observed to show better keeping quality than seeds stored on the earheads. Unthreshed grains suffer greater losses because of insects and molds. The heads of cereal crops should be spread on mats or cement blocks, the seeds sun dried to 10–12% moisture content, and then threshed immediately to reduce exposure to birds, rats, and rain. Locally manufactured gasoline or electric power threshers are quite suitable in less developed countries. The treadle-operated drum thresher is semiportable and widely used in India and Japan to thresh cereals like sorghum. Threshing is achieved by holding a bundle of produce against a rotating drum, which is furnished with wire loops to strike the grain from the stalks. Winnowing is carried out in a separate operation (3).

Maize is shelled or dehusked manually in developing countries. Although this method is labor intensive and costly, it is efficient in stripping the cob and minimizing grain damage. It also permits removal of infested and damaged seeds from sound ones. Other traditional methods include beating bagged cobs with a stick. Losses occur because of incomplete shelling and mechanical damage to the grains, which may become susceptible to subsequent infestation by microorganisms during storage. Mechanical shelling is relatively more efficient for maize seeds with a low moisture content. Care must be taken to ensure that seed is not damaged during threshing, shelling, or dehusking.

C. Effective Drying

Cereal seeds may be sun dried or mechanically dried. Traditional sun drying of spread out grain is an effective means of reducing the moisture content of seeds in the dry season. Wet-season crops require forced-air drying with heated or ambient temperature air. Commercial driers are designed either to dry seed in batches in deep or shallow beds in which grain may be stationary or mechanically circulated or continuously in stages as the grain flows through the drier. The physical design and operating characteristics, including factors such as air to grain volume ratio and air temperature, influence the rate of drying and the quality of the dried seed. Mechanical characteristics of driers vary for different kinds of grain, and great care must be exercised in using a drier for a grain other than one for which it is designed. Other characteristics of the drying equipment to be considered include the time required to dry seed from a high moisture content and the peak volumes that must be handled during the harvest season. Mechanical driers based on medium- or high-temperature forced-air drying, such as bin driers, batch driers, or continuous airflow driers, are commonly used for paddy drying.

The type of drying system used depends mainly on the relative humidity of the atmosphere. At relative humidities below 50%, the equilibrium moisture content of wheat seeds would be below 14%, so that natural air drying can be effectively employed. However, when the relative humidity is above 60%, heated air must be used to reduce grain moisture to a safe level. Drying temperatures depend on the end use of the crop. Muckle and Stirling (21) recommended a maximum of 49°C for drying seed-purpose wheat, whereas wheat for milling could be dried at 66°C. Salunkhe et al. (4) recommended maximum air temperatures for drying and moisture contents for the safe storage of some legumes (Table 5). The moisture content of seeds varies according to the grain type, chemical composition, moisture at harvest, harvesting methods, relative humidity of the atmosphere, and seasonal fluctuations. The optimum or safe storage moisture content of seeds is normally defined as the amount of water present at which the rate of respiration is low enough to prevent seed germination and the consequent deterioration of their quality (4).

A distinct optimum seed moisture content of 8.4% was found to be necessary for their ability to germinate during 10.5 years of hermetic seed storage of arabica coffee seeds cv. Caturra at 20°C (21a). There was no evidence that the optimum moisture content for seed storage longevity was increased, the cooler the storage temperature, despite a marked discontinuity in relation between storage temperature and longevity. Seed longevity at −20°C was greater than at 15°C, but longevity at 15°C was considerably greater than at 0°C. This ability to survive medium-term storage at −20°C is exceptional.

D. Improved Storage Technology

Seeds must be stored properly to preserve them for the next planting season. Seeds may be stored at the farm, trader, market, government, or retailer level. The mode of storage may be traditional, intermediate, or improved (e.g., modern steel or concrete elevators).

More than 70% of the cereal grains produced in developing countries are stored at the farm in traditional storage structures. Losses in vitality and seedling vigor occur during the storage of seeds as a result of cellular, biochemical, and cytogenetical changes in the environment.

Table 5 Recommended Maximum Air Temperature for Drying and Moisture Contents for Safe Storage of Some Legumes

Legume	Maximum air temperature (0°C)	Moisture content (% on wet basis)
Cowpea	38	15
Dry beans (animal feed)	45	15
Dry beans (seeds)	38	15
Lentil	38	15
Peanut	37	7
Peas	38	14
Soybean	38	11

Source: Ref. 4.

Bagged seeds absorb more moisture and suffer greater losses owing to infestation than do bulk lots (22). Hence, bulk storage of seeds in bins at the farm level and in silos at the commercial level is recommended. Seeds in concrete bins also pick up moisture from the atmosphere during the rainy season. Such seeds are more often spoiled owing to heat damage, development of acidity (oil seeds), and loss of viability than seeds stored in aluminum bins (23).

Different storage structures have been designed to prevent seed loss during storage. Storage structures suitable for farm, urban, and commercial storage of grains and seeds in tropical countries have been described by Salunkhe et al. (3,4), Deshpande and Deshpande (10), Pingale (24), and Birewar et al. (25).

1. Farm Storage Structures

The circular steel bin has an opening at the top for filling and a spout at the bottom for removing the seed or grain. The bottom is made of plain sheets of mild steel. This structure can be easily assembled on site and can be removed when not needed. It has a built-in arrangement for preventing the development of uneven temperatures, which may lead to moisture migration and deterioration of seed quality. Grains stored in the bin can he fumigated whenever required.

The plastic bin is a low-cost structure suitable for indoor storage of seeds or grains and has a tube-shaped metal base with a provision for placing bamboo sticks vertically around the side of the metal drum. A cylindrical rubberized fabric is hung inside and, into which the seeds are placed. The seeds can be removed from the top or through a sliding door provided at the bottom of the metal base. This bin is a compact and stable storage structure and can easily be dismantled when not in use.

The prefabricated steel pin with hopper bottom is a strong and durable outdoor structure with a sloping roof, a manhole for filling, and a hopper bottom with a sliding door for discharging seeds. The bin provides for natural aeration to prevent build-up of a temperature gradient and the subsequent problems of translocation of moisture and deterioration of seed quality. The bin stands on a firm support creating a clearance space of 60 cm below the hopper. Both the manhole and the hopper gates have locking arrangements, and a metallic ladder and pulley are provided to facilitate filling of the bin.

The aluminum bin is an outdoor storage structure consisting of a cylindrical body built of several corrugated aluminum curved sheets and a conical roof of flat aluminum sheets. The bin is built on a 60-cm–high platform and has a spout embedded in the platform and a manhole in the roof, both provided with locking arrangements. The hopper gates have locking arrangements.

The reinforced cement concrete (RCC) bin is circular, sturdy, weatherproof, and suitable for both indoor and outdoor storage. Its construction is similar to that of the storage structures described above.

The PUSA bin is a traditional mud bin using a polythene sheet to make the structure airproof and moistureproof. The PUSA bin is simple in design, easy to construct, and inexpensive to suit the needs of small fanners. When filled to capacity and properly closed, the bin becomes a airtight, dampproof structure that prevents thermal conductivity.

2. Urban Storage Structures

In urban areas, Pingale (24) recommended a rectangular, rodent- and moistureproof steel structure with an arrangement for controlled aeration as well as fumigation for the storage of cereals and legumes in one place. These structures are made of galvanized iron sheets and are usually partitioned into four or five compartments of different sizes.

Commercial large-scale storage structures are used to store large quantities of seed by the government agencies who have adequate economic and technical support to use more sophisticated structures for both bag and bulk storage of seeds.

A well-planned *bag storage warehouse* should ideally have waterproof walls, roof, and floors and a sealable opening for controlled ventilation and fumigation. The opening may be used for both natural and fan-controlled aeration. The metal roof should be insulated to minimize temperature build-up. The seed bags are placed on wooden planks erected above the floor to provide insulation against floor moisture and temperature. The building must be rodent-, bird-, and insectproof.

A *bulk storage warehouse* should meet the basic requirements of the bag storage structure. However, sidewall insulation is more important in a bulk storage structure, because the grains are in direct contact with the bin walls. White paint on the metal skins of these bins serves as a good reflector and radiator of low-temperature heat energy. Depending upon the requirements, bulk storage structures may be small or large. Large elevators may be flat or upright, whereas upright storage bins are usually large cylindrical and silo types. A pointed metal or concrete silo can be built singly or into a row of bins.

Santana-Buzzy et al. (25a) evaluated mature seeds of *Coffea arabica* L. cvs. Caturra and Catimor stored for 0–18 months for germination and growth in a greenhouse. Zygotic embryos were isolated, germinated in vitro, and after germination grown in a greenhouse. High in vitro germination (75–199%) was observed with time versus a sharp decrease in germination (100–25%) in the greenhouse after a few months of storage. The low germination percentage in the greenhouse after 18 months of storage could not be attributed to the loss of embryo viability or environmental conditions; it may have been due to endosperm tissue deterioration.

Modified atmosphere (MA) and controlled atmosphere (CA) storage are alternatives to toxic, residue-building chemical fumigants to protect stored seed from insect pest infestations (25b). Being storaged in CO_2-rich atmospheres might have a beneficial or adverse effect on seed quality; that is, viability and vigor. Lower concentrations (5, 10, 20% CO_2) and longer exposure periods (2, 4, 6 months) were tested to study the effects of CO_2 on insect infestation and the seed germinability, vigor, and biochemical parameters of quality of wheat seed. A concentration of 20% CO_2 completly controlled insect population build-up and insect damage to seed during 2 months' storage. Germinability and vigor of wheat seed remained unaffected under all CO_2 concentrations. Biochemical parameters like dehydrogenase and mealondialdehyde activity were not modified by any of the CO_2 concentrations tested (25b).

Patil and Nagaraja (25c) conducted a study to determine the viability of chili seeds in pods and of extracted seeds kept in polythene bags of 700 gauge, cloth bags, earthen pots, or gunny bags. The study revealed that the fruits stored in polythene bags retained higher seed viability. Seeds of cv. Byadagi kaddi maintained certifiable limits of germination even after 6 months' storage, whereas those of cv. Dyavanoor local could be kept only for 3 months to be certifiable.

Seeds of aubergine cultivars Emi and Long Negro were produced over three seasons between 1988 and 1990 and stored at ambient (25 ± 5°C) or low (5 ± 2°C) temperature. In the absence of humidity control, seeds stored satisfactorily for 4 years at 25°C, whereas at 5°C a reduction in germination was observed after 1–2 years of storage. When the storage time was extended to 7 years, there was a significant drop in germination capacity, although in some seed lots, the percentage germination was still over 70%. Under Mediterranean conditions and in the absence of humidity control, long-term (>1–2 years) storage of aubergine seeds should be carried out at ambient (25 ± 5°C) rather than at low (5 ± 2°C) temperature (25d).

Seed germination in nine medicinal plants of India was studied after storage in liquid nitrogen for 1 week. Seeds of *Andrographis paniculata*, *Abrus precatorius*, *Coleus forskohlii*, *Dipteracanthus patulus*, *Hemidesmus indicus*, *Ocimum gratissimum*, and *Tylophora indica* contained 7–15% moisture at seed shedding. After desiccation to 4–7% moisture, they exhibited 80–95% germination, which was not affected significantly by 1 week of storage in liquid nitrogen. Seeds isolated from ripe berries of *Rauwolfia micrantha* and *Embelia ribes* contained 40% moisture and germinated to 88% and 78%, respectively. The seeds of *R. micrantha* survived desiccation to a 5–7% moisture level and cryopreservation with germination of 73% comparable to that of the desiccation control. Seeds of *E. ribes*, which were desiccated down to 4% moisture, germinated well (77%) under ambient conditions but failed to germinate after storage in liquid nitrogen. The study reveals a possible link between seed moisture content and liquid nitrogen tolerance in some, but not all, Indian medicinal plants (25e).

Stumpf et al. (25f) examined the storage potential of onion seeds stored in hermetic cans with a low moisture content under environmental temperature conditions at Pelotas, Brazil. The Ellis and Roberts seed viability equation was also used to determine whether it could be used to predict changes in onion seed germination in the variable temperatures encountered during warehouse storage. Moisture content, germination, and vigor tests were determined on 154 commercial seed lots stored for 1–10 years. Each seed lot consisted of two cans of 0.5 kg, which were left over from larger lots after commercialization. The seed lots were analyzed according to the storage period and seed moisture content intervals of <5.5%, 5.5–6.5%, 6.5–7.0%, and 7.0–7.5%. For onion seeds with a minimum of 90% germination, hermetically packaged, with no more than a 7.5% moisture content, and at a annual average temperature of 17.6–18.3°C, it was concluded that (a) as the moisture content and storage period increases, the viability equation becomes less accurate, being practically useful up to 3 years; (b) physiological quality determined by the germination test only is not enough to predict the storage potential of the onion seed lots, and possible alternatives are discussed; and (c) some onion seed lots can be stored for 10 years and retain a high physiological quality, although some may lose their quality after 2 years of storage.

B. juncea cv. RH-30 seeds were treated with 0.25% thiram or untreated, and were stored in cloth or polyethylene bags for up to 30 months. Seed germination declined more rapidly in cloth bags, but retained the minimum certification standards until the growing season after harvest. Untreated seeds stored in polyethylene bags maintained 93.5% germination after 30 months' storage compared with 74% in treated seeds (25g).

E. Seed Treatment

Various seed treatments have been successfully employed to reduce losses during storage and distribution. Taylor and Hannan (26) reviewed the use of seed treatments to protect sown seeds from soil fungi, discussing technologies such as seed coating, seed pelleting, physiological seed treatments (e.g., drilling and seed priming), and seed treatments with beneficial microorganisms (amendments that enhance biological seed treatments). Recently, a novel application of the agar diffusion test has been reported for the control of seed treatments with pesticides. Inhibition zone formation of pesticides and microbial strains is used to assess the sensitivity of seeds to pesticides (27). Shah and Mariappan (28) observed that after 9 months of storage, sorghum seeds (cv. Co 19) treated with 0.2% thiram, 0.2% Vitavax, 0.4% Panotine (guazatine acetate), and 0.4% Panoram (fenfuram) maintained 66–73.6% seed viability compared with 45% for untreated seeds. Gulyaev (29) described problems in automating grain treatment and storage, including centralized control of production lines for grain treatment, automatic control of grain drying, cleaning, and sorting, forced ventilation and storage, automatic control equipment for grain treatment and storage, and the economic efficiency of all of these processes. The most commonly used current seed treatments include coating with pesticides or beneficial microorganisms, pelleting, or physiological treatments, such as osmotic conditioning or priming (30). Uppar et al. (31) showed that wheat seeds treated with iodine and stored in polythene bags showed higher values for germination, root length, shoot length, seedling length, and vigor index as compared to seeds stored in clothe bags. Also, electrical conductivity of seeds stored in polythene bags was slower than those stored in clothe bags. According to Ramphal et al. (32), green gram seeds treated with the insecticide deltamethrine (4 mg/kg) and the fungicide thiram (2.5 g/kg) and packed in polythene bags stored best as indicated by higher germination (82%), seedling length (19 cm), and seed vigor index (1569.69).

Xalxo and Yadav (33) observed the incidence of mycoflora on 4 Desi and 10 Kabuli cultivars of chickpea using the blotting method. It was recommended that chickpea seeds should essentially be treated with a suitable fungicide, such as thiram (3g/kg) alone or in combination with turmeric and white lime (4 g/kg), to prevent mycoflora on seeds, seed and seedling rot during the standard germination test, and improve field emergence in both Desi and Kabuli chickpea seeds, especially in large-seed types. Varshney et al. (34) reported that irrespective of the age of maize hybrids and their parent lines after seed treatment with thiram + bavistin and thiram, and having packed them in 700-gauge polythene bags under ambient conditions of storage for 180 days, maintained not only a higher level of germination and seedling dry weight but also reduced the seed mycoflora drastically as compared to clothe bags, thereby improving the seed quality, including seedling emergence in the field.

REFERENCES

1. R. J. Miller, Proceedings of National Food Losses Conference (M. V. Zaehringer and J. O. Early, eds.), College of Agriculture, University of Idaho, Moscow, 1976.
2. R. M. May, Foods lost of pests, *Nature* 267:669 (1977).
3. D. K. Salunkhe, J. K. Chavan, and S. S. Kadam, *Postharvest Biotechnology of Cereals*, CRC Press, Boca Raton, FL, 1985.
4. D. K. Salunkhe, S. S. Kadam, and J. K. Chavan. *Postharvest Biotechnology of Food Legumes*, CRC Press, Boca Raton, FL, 1985.
5. *Postharvest Food Losses in Developing Countries*, National Academy of Sciences, Washington, DC, 1978.
6. J. M. Adams, A review of the literature concerning losses in stored cereals and pulses published since 1964, *Trop. Sci. 19*:1 (1977).
7. H. A. B. Parpia, Role of food technology in improving the level of nutrition in developing countries, Paper presented at the 7th International Congress of Nutrition, Humburg, Germany, 1966.

8. D. Pimentel, *Handbook of Pest Management in Agriculture*, Vol. I, CRC Press, Boca Raton, FL, 1981.

9. R. H. Davidson and W. F. Lyon, *Insect Pests of Farm, Garden and Orchard*, Wiley, New York, 1987.

10. U. S. Deshpande and S. S. Deshpande, Legumes, *Foods of Plant Origin: Production, Technology and Human Nutrition* (D. K. Salunkhe and S. S. Deshpande, eds.), Van Nostrand Reinhold, New York, 1994, p. 137.

11. S. S. Deshpande, B. Singh, and U. Singh, Cereals, *Foods of Plant Origin: Production, Technology and Human Nutrition* (D. K. Salunkhe and S. S. Deshpande, eds.), Van Nostrand Reinhold, New York, 1991, p. 6.

11a. R. C. Sharma, Neelu Kohli, S. S. Gill, B. S. Gill, and N. Kohli, Deterioration of rain affected vegetable seeds by storage fungi, *Seed Res. 28*: 229(2000).

11b. S. M. Chang and J. M. Sung, Deteriorative changes in primed sweet corn seeds during storage, *Seed Sci. Technol. 26*: 613(1998).

12. F. O. McEven, Food production: The challenge for pesticides, *BioScience 28*:773 (1978).

13. A. Youdeowei and M. O. Adeniji, Crop production, *Introduction to Tropical Agriculture* (A. Youdeowei, F. O. C. Ezedinma, and O. C. Onazi, eds.), Longman, New York, 1986, p. 132.

14. H. F. Van Emden, Insects and mites of legume crops, *Advances in Legume Science* (R. J. Summerfield and A. H. Bunting, eds.), Royal Botanical Gardens, Kew, UK, 1981, p. 187.

15. L. D. Newsom, Progress in integrated pest management of soybean pests, *Pest Control Strategies* (E. H. Smith and D. Pimentel, eds.), Academic Press, New York, 1978, p. 157.

16. M. Shepard, G. R. Carner, and S. G. Turnip seed, Colonization and resurgence of insect pests of soybeans in response to insecticides and field isolation, *Environ. Entomol. 6*:50 I (1977).

17. S. G. Turnipseed, A survey of viral diseases of pulse crops in Uttar Pradesh, G. B. Pant University of Agriculture and Technology, Pantnagar, India, 1972.

18. R. M. Goodman and J. Bird, Bean golden mosaic virus. CMI/AAB descriptions of plant viruses No. 192, Commonwealth Mycological Institute and Association of Applied Biologists, Kew, UK, 1978.

19. L. Bos, Seedborne viruses, *Plant Health and Quarantine in Transfer of Genetic Resources* (W. B. Hewitt and L. Chiarappa, eds.), CRC Press, Florida, 1977.

20. R. F. Nyvall, *Field Crops Diseases Handbook*, Van Nostrand Reinhold, New York, 1989.

20a. A. J. Mwasha, R. H. Ellis, and T. D. Hong, The effect of desiccation on the subsequent survival of seeds of cashew (*Anacardium occidentale* L.), *Seed Sci. Technol. 25*: 115 (1997).

21. T. B. Muckle and H. G. Stirling, Review of drying cereals and legumes in the tropics, *Trop. Stored Prod. Res. 22*:11 (1971).

21a. T. D. Hong and R. H. Hills, Optimum moisture status for the exceptional survival of seeds of arabica coffee (*C. arabica* L.) in medium-term storage at −20°C, *Seed Sci. Technol. 30*: 131 (2002).

22. J. N. Sarid, L. Rai, K. Krishnamurthy, and S. V. Pingale, Studies on large-scale storage of food grains in India, Part I, Studies on places, *Bull. Grain Technol. 3*:87 (1965).

23. J. N. Sarid, L. Rai, K. Krishnamurthy, and S. V. Pingale, Studies on large-scale storage of food grains in India. Part II. Studies on relative stability of cement concrete and aluminum bins for storing wheat, *Bull. Grain Technol. 5*:3 (1965).

24. S. V. Pingale, Storage and *Handling of Food Grains*, Indian Council of Agricultural Research, New Delhi, 1976.

25. B. R. Birewar, B. K. Verma, C. P. Raman, and S. C. Kanjilal, *Traditional Storage Structures in India and Their Improvements*, Indian Grain Storage Institute, Hapur, India, 1980.

25a. N. Santana-Buzzy, V. M. Loyla-Vargas, M. Valcarcel, M. L. Barzaga, M. M. Hernandez, M. E. Gonzalez, F. Barahona, and J. Mijangos-Cortes, The effect of in vitro germination in maintaining germination levels over time in storage for two cultivars of *Coffea arabica* L., *Seed Sci. Technol. 30*: 119 (2002).

25b. A. Bera, N. C. Singhal, S. N. Sinha, R. K. Paul and Srivastava, Studies on carbo-di-oxide as wheat seed protectant, XI National Seminar on 'Quality Seed to Enhance Agricultural Productivity,' UAS, Dharwar, January 18–20, 2002, *Seed Tech News 32* (A. Gaur, A. K. Vari, J. L. Varshney, and K. Kant, eds.), Indian Society Seed Technology, Division of Seed Science and Technology, IARI, New Delhi, March 2002, p.229.

25c. K. N. Patil and A. Nagaraja, Accelerated aging as a tool for predicting storability of chilli seeds, *Karnataka J.Agric Sci. 12*: 189 (1999).

25d. H. C. Passam, K. Akoumianakis, and E. M. Khah, Long-term storage of aubergine seed following production under high ambient temperature, *Seed Sci. Technol. 27*: 977 (1999).

25e. S. W. Decruse, S. Seeni, and P. Pushpangandan, Effects of cryopreservation on seed germination of selected rare medicinal plants of India, *Seed Sci. Technol. 27*: 501(1999).

25f. C. L. Stumpf, S. T. Peske, and L. Baudet, Storage potential of onion seeds hermetically packaged at low moisture content, *Seed Sci. Technol. 25*: 25(1997).

25g. R. C. Sharma, S. S. Gill, H. R. Kaur, and A. S. Dhatt, Effect of fungicidal treatment and containers on storage life and field performance of raya (*Brassica juncea* L.) seed, *Seed Res. 23*: 55 (1995).

26. A. G. Taylor and G. E. Harman, Concepts and technologies of selected seed treatments, *Ann. Rev. Phytopathol. 28*:321 (1990).

27. G. Kovas and K. Moller, Novel application of the agar diffusion test for control of seed treatments, *Seed Sci. Technol. 19*(1): 175 (1991).

28. S. E. Shah and V. Mariappan, Effect of seed dressing fungicides on the storage and viability of sorghum seeds, *Madras Agric. J. 77*(7–8): 278 (1990).

29. G. A. Gulyaev, *Automation of the Processes of Grain Postharvest Treatment and Storage*, Agropromizdat, Moscow, 1990.

30. A. Tarquis and J. M. Duran, Pre-treated seeds: concept of vigor, *Agric. Rev. Agropecu. 59*: 130 (1990).

31. D. S. Uppar, M. B. Kurdikeri, M. B. Chetti, R. R. Hanchinal, and A. S. Nalini, Influence of halogen and storage containers on storability of *Dicoccum* wheat, XI National Seminar on 'Quality Seed to Enhance Agricultural Productivity,' UAS, Dharwar, January 18–20, 2002, *Seed Tech News 32* (A. Gaur, A. K. Vari, J. L. Varshney, and K. Kant, eds.), Indian Society Seed Technology, Division of Seed Science and Technology, IARI, New Delhi, March 2002, p.124.

32. Ramphal, A. A. Khan, V. P. Kanaujia, C. B. Singh, and A. L. Jadhav, Effect of abiotic and biotic factors on storability of green gram (*Vigna radiata*. Wilzek) seeds, XI National Seminar on 'Quality Seed to Enhance Agricultural Productivity,' UAS, Dharwar, January 18–20, 2002, *Seed Tech News 32* (A. Gaur, A. K. Vari, J. L. Varshney, and K. Kant, eds), Indian Society Seed Technology, Division of Seed Science and Technology, IARI, New Delhi, March 2002, p.125.

33. P. Xalxo and S. P. Yadav, Incidence of mycoflora on desi and kabuli chickpea, before and after storage and its eradication, XI National Seminar on 'Quality Seed to Enhance Agricultural Productivity,' UAS, Dharwar, January 18–20, 2002, *Seed Tech News 32* (A. Gaur, A. K. Vari, J. L. Varshney, and K. Kant, eds), Indian Society Seed Technology, Division of Seed Science and Technology, IARI, New Delhi, March 2002, p.237.

34. J. L. Varshney, S. Mahajan, S. P. Sharma and A. Gaur, Effect of fungicidal treatment on seed quality of two maize hybrids and their parental lines during storage, *Seed Technology and Seed Pathology* (T. Singh and K. Agrawal, eds.), Pointer Publishers, Jaipur, India, 2002, pp.182–199.

26

World Food Security: Future Challenges and Research Needs

I. INTRODUCTION

Agriculture has evolved with human culture and civilization, first as a means of food security to the family and then as a source of family income and profitability (1). Plants and animals found in the wild were domesticated with gradual long-term improvement in their quality and quantity leading to higher productivity and what is now collectively termed "agriculture" (2). Agriculture thus replaced the former nomadic habit of food collection for immediate consumption (3). Human efforts continued to improve plant and animal yield, as has been documented in the ancient scriptures of many civilizations of the Old and New World (4). Plant and animal domestication associated with food storage probably coincided with the growth of microorganisms (5), giving rise to classic food fermentation and the use of microorganisms to produce fermented food products, the earliest known application of biotechnology (2), which was used for beer brewing, wine production, and bread baking. Domestication of an increasing number of plant crops and animal species from the wild with gradual improvement in agricultural techniques, step-by-step selection of better-performing and more-adaptive genotypes, along with intuitive breeding continued at a slow pace until Mendel's discoveries of the laws of genetics in the 1860s revolutioned plant-breeding programs through planned and controlled experiments (1,2). Plant and animal breeding along with the "old" biotechnology was harnessed fully in the later part of the twentieth century resulting in improved crops and farm animals. The rapid scientific development significantly improved agricultural yields and product quality, better supporting the increasing human demands for a variety of foods. The evolution of short-stalk wheats and isolation of high-yield rice formed the basis of successful achievements of the "Green Revolution" of the 1960s (2,6,7), which was followed by more productive genotypes of cereals like corn, sorghum, rice, and food legumes as well as cattle and poultry. Soon the old biotechnology entered into another revolution with the cloning of DNA in 1973 (1,4) and the development of genetically modified organisms (GMOs) through recombinant DNA technology by the end of 1970s, giving rise to a "new" agricultural biotechnology (2,8).

II. WORLD FOOD HUNGER

Despite the successes of the Green Revolution, the battle to ensure food security for hundreds of millions of poor people has been far from won. Mushrooming populations, changing demographics, and inadequate poverty intervention programs have eroded many of the gains of the Green Revolution (9). Thirty year ago, in his acceptance speech for the Nobel Peace Prize, Norman E. Borlaug said that the Green Revolution had won a temporary success in man's war against hunger which if fully implemented, could provide sufficient food for the mankind through the end of the twentieth century. But he warned that unles the frightening power of human reproduction was curbed, the success of the Green Revolution would only be ephemeral.

According to Chrispeels (10), 800 million people on earth are poor and malnourished, living on less than a dollar a day. They cannot be sure that their fields will yield enough food or that they will earn enough money to buy food. Around 40,000 people die each day of malnutrition, one-half of them children. The doubling of food production enabled by the Green Revolution has not solved the problems of malnutrition and hunger of the Third World countries. There were around a billion hungry people some 40 years ago, and population projections indicate that there may be still be 600 million poor people by 2025, when the earth's population will have grown to about 8 billion. The Green Revolution cannot wipe out poverty. Not enough jobs have been created in either the rural areas or the cities to generate the purchasing power that provides farmers with the incentive to grow more food. Ironically, the world food hunger still persists while the prices for agricultural commodities are at an all-time low (10).

The significant role of recombinant DNA technology and GMOs in the agricultural productivity of the twenty-first century has bee discussed extensively (11–14). Continued improvement in crop and animal quality and productivity are crucial to feed the world of 10 billion people that will come into existence sometime after the middle of the current century (11).

Until the advent of the Green Revolution after 1960, the major contributor to increase in the world food supply was the extension of arable land (13), with the increase in food production since then having been obtained primarily through rising yields, a feature that has began to show some signs of slowing down (11). According to Evans (13), most, or even all, of the arable land on the earth is already under cultivation, suggesting there is no more land available for that purpose. Although worldwide this is not true, the actual situation is complex. There is a lot of potentially arable land that is currently not under cultivation but much of it is undisturbed forest and wetland, whereas other land is arable but marginal. Also, arable land is increasingly being lost all the time to urbanization and being replaced with previously uncultivated land, keeping the total roughly constant. Evans (13) has dealt with the complexities of the issues related to producing enough food to keep up with population growth.

The relationship between food production and population growth can be viewed from two angles: The one view of Thomas Malthus states that the food supply is the driving variable influencing the population growth, and the other view of Ester Boserup sees it the other way around; that is, the population growth is the driving force for the agricultural production. This issue cannot be resolved easily. According to Siedow (11), the correct answer, if one truly exists, would have a large bearing on the eventual acceptance of genetically modified crops, especially in the developing countries.

Smil (14) has discussed an ecological perspective of the challenge of feeding the world in the twenty-first century, addressing the issue of sustainable agriculture. He has dealt with the often-quoted limit of 4 billion people that can be sustained if nitrogen were only applied following the principles of organic farming, and practical approaches to achieving a sustainability that can support 10 billion people. To achieve long-term agricultural sustainability, the food chain from crop productivity through postharvest losses of food crops (15–18) on to food production distribution, consumption, and human nutrition (19) needs to be improvised. According to Smil (14), there is considerable slack in the current system, and the prospects for more efficient use of existing resources at all levels are very large and real. An omnivorous world can be sustained only if it done in an intelligent way; for example, with more chicken and much less beef. There is a need to achieve increased efficiency with the existing technologies and knowledge bases, with the GMOs being just one part of the solution to feeding a 10 billion–person world (11,14).

Conway (12) has provided a detailed picture of the world agricultural scene, including the poverty and many socioeonomic factors that contribute to the existence of significant numbers of underfed people in a world of sufficient food supplies. The opponents of GMOs often use this fact, pointing to poverty as the main issue and not a lack of food. However true the latter may be, alleviating poverty is not a practical or workable solution and will not address the future need to feed 60–70% more people than exist at present (12). Conway has envisaged the next Green Revolution as being "doubly green" to achieve a goal of a more sustainable form of agriculture through newer biotechnologies, especially the plant and animal GMOs. It is necessary to have the vision of the potential gains from the application of GMOs, which may far outweigh their perceived risks at this stage. This is especially true when GMOs are envisioned as being an important way out of the cycle of the large-scale application of pesticides associated with the first Green Revolution. There is a need for far a more broad-ranging partnership between the industrial companies acting as stewards of the agricultural biotechnology and the public agricultural research centers in the developing countries, and for empowering and including local farmers in the new partnership for the their feedback. A comprehensive agricultural revolution, one that includes the technological, the ecological, and the sociological issues, is called for. Although this is not an easy task to accomplish, the cost of failure to act may be extremely high (12).

Despite the different outlooks of plant biologists on the GMO debate, it is generally agreed that feeding a world of 10 billion inhabitants will not be accomplished without making significant changes, especially in the developing world, that ran throughout the food chain from agricultural quality and productivity to socioeconomic issues. However, the battle over the application of GMO technology to help feed the earth's growing population presently rests in the hands of the developed world, whereas most of the people that will need to be fed are located in the developing Third World. Ironically, owing to the modern agricultural technologies, the population of the developed countries has access to the most abundant, healthiest, and cheapest supply of food in the history of the human race. The GMO opponents do not need to prove whether their claims about the possible dangers of GMOs are true or not. Being in a position to be picky about what they choose to or not to eat, they can decide not to eat GMO food. Such an easy decision comes with no apparent consequence for the cost, availability, or quality of the food eaten, and this luxury is not afforded to someone in a country where food is nowhere near as cheap and available (11).

The application of technology to the problems of poor in the developing countries will not be straightforward, and the models from the developed countries will probably not be directly applicable (10). Agriculture in developing countries does not need to be "modernized," although it does need to be improved. The developing countries can hopefully skip the high-input unsustainable phase through which agriculture is now passing in developed countries, proceeding immediately to more sustainable practices. To enable agricultural research from the bottom up, and not from the top down, crops must be created that fit not only in the agroecology of the poorest regions, often characterized by marginal and heterogeneous environments, but the crops must fit into the socioeconomic systems. The major objective has to be the productivity and profitability of small-holder farms with synergy between food crops, cash crops, livestock, agroforestry, and aquaculture with integrated management of soil, water, and nutrients (20). This goal and the process for achieving it are more important than the introduction of GM crops (10).

Crispeels (10) has dealt with many aspects of providing food for the poor that are well beyond the control of either biotechnologists or agricultural advisors in the field. The national governments of the poor countries must realize that agriculture can be an important engine of economic growth and therefore must invest more in agricultural research. Agricultural development should be encouraged to create a rural infrastructure that will permit crop surpluses to be marketed. Cheap food policies that favor the urban poor often discourage development of food production capacity, leading to a transfer of wealth from the agricultural sector to the industrial sector. The developing countries may strive for self-sufficiency at all costs (21), but they should look at the entire package—food production, rural development, job creation, land reform, and lending institutions—and enact enabling policies. These policies should also benefit the small farmers who might be engaged in more sustainable practices. Biotechnology is a potentially important tool in the struggle to reduce and improve the living conditions of both the rural and the urban poor (22).

Funding for agricultural research has declined 50% on a worldwide basis. The intrusion of intellectual property rights does as not seem to be benefiting the agricultural research in the developing countries. There are no serious health issues at stake in the consumption of GM foods. However, the environmental issues of GM crops that are still unresolved and the ethical considerations of transgenic crops pale in comparison to the environmental impact of rural populations that practice low-yield agriculture on marginal lands and the ethical considerations of not improving the lives of the poorer populations in the Third World countries (10).

III. PROBLEMS OF TRADITIONAL AGRICULTURE

Traditional agriculture suffers from several serious limitations in view of the present changes in the world trade and international markets. The world transition to a global village and the increased information flow has also changed the market conditions (6–8,23). Local pricing policies and growers' profitability are no longer valid and effective except in certain "closed" communities, and they are increasingly affected by the volume of world trade and international prices of food and other agricultural products (2,8,23).

Agriculture, which strongly depends on the availability of natural resources, is also facing deteriorating conditions of reduced land and water availability. It is further aggravated by increased soil, water, and air quality reduction owing to global climatic changes, decertification, and industrial pollution (2).

A. Resistance to Diseases, Insect Pests, and Environmental Stresses

Classic breeding methods have not yet exploited their potential to improve plant and animal productivity fully. Classic breeding and selection for pest resistance has resulted in better yields, and genotypes that are more tolerant to extreme climates, water and soil conditions are being selected (1,2,4,7). Although plant breeders have succeeded in breeding resistance in many crops to pathological diseases and insect pests and a range of physiological and environmental stresses (drought, salinity, heat, cold), the biochemical basis of plant resistance and the molecular mechanisms involved therein need to be elucidated completely.

There is no intrinsic reason why some plant diseases should not be practically eradicated. Wart disease of potato, for example, could be reduced to a level of unimportance by banning cultivars susceptible to prevailing races. Many crops with soft seeds (e.g., club kafir sorghum) suffer from seed rots and seedling blights, producing poor crop stands. The potential yields of seeds of peas, beans, and other pulse legumes can be increased most effectively by breeding varieties that are resistant to diseases and insect pests.

In seed-propagated crops, species hybridization has played an important role in crop improvement (e.g., roses, dahlias, gladioli, and amaryllis). The usefulness of this method is limited by the percentage of seed set and the viability of such seed. Viable seedlings can be obtained from sterile crosses using embryo culture wherein the young embryos from developing seeds are removed and transferred to an appropriate nutrient media to breed new varieties (e.g., *Iris* species).

Molecular biology has resulted in the generation of revolutionary powerful and precise breeding tools. There are many examples of successful integration of new biotechnologies in classic breeding, such as microscopic visualization of chromosomes in tissue culture, the fluorescent in situ staining by hybridization (FISH) for fluorescent labeling of chromosome markers in cells, restriction fragment length polymerization (RFLP) to isolate DNA markers, development of the simple-sequence repeat (SSR) DNA markers to isolate important plant traits, and the use of Southern and Western blot techniques to validate successful trait isolation (2). All of these genetic techniques constitute an integral part of classic breeding, contributing successfully to shorten the breeding and selection cycles and significant accelerate classic breeding for plant and animal improvement (7,8,23,24). Newer biotechnologies of genetic analysis and gene transfer to produce GM crops (see Chapter 24), newer in vitro clonal propagation methods, and production of artificial/synthetic seeds through somatic embryogenesis (see Chapters 22 and 23) are thought to make a significant contribution to crop improvement in the future. The RFLP and randomly amplified polymorphic DNA (RADP) techniques are now being used most successfully to develop genetic maps of cultivated and diploid crop species through tissue culture and direct manipulation of DNA (25).

B. Germination, Dormancy, and Seed Viability

More research and analysis would be needed before broad generalizations can accurately relate to various aspects of seed quality, including longevity for an integrated consideration of seed biochemistry, physiology, taxonomy, and crop economy. For example, seeds of some lettuce varieties require light for germination; seeds covered with soil and those buried deep in soil remain dormant. This light requirement can be eliminated by breeding or the use of growth regulators. Not much is known about what chain of biochemical events is set in motion by a flash of light as short as one thousandth of a second in these light-sensitive seeds. It is also not known why some seeds require alternating temperatures in order to grow, whereas others do not. How do some seeds live for decades and scores of years, whereas others, apparently as well protected, die in a short period of time? Why do some small plants produce seeds that are much larger than the seeds of some larger plants? Why do some seeds require years to develop, whereas others are ready in a few days? How is it that seeds of some crops so astonishingly different from each other quite evidently evolve to accomplish exactly the same thing? These questions can be answered only through further research. Our dependence on agricultural crops developed through hard work and research efforts with seeds behooves us to learn more about their origin, biology, structure, and function.

Not much is known about the nature of germination inhibitors in seeds, including that of sugar beet (26), or about the factors that influence inhibitor concentrations in the fruits. Meager research has been carried out on how the conditions in the field and concentrations of germination inhibitors in seeds are related to reduce emergence, contributing to significant variations in plant development. The relationship between the seed constituents and bolting tendency in sugar beets needs to be established fully.

Although seed dormancy is valuable in moderation, it can be disastrous in either extreme. Temporary dormancy is advantageous in crops like oat, sorghum, sweet corn, and others that are subject to field sprouting in the shock (stook) or on the standing stalk before the ripe seed is threshed. On the other hand, seed dormancy becomes a handicap in northern regions where wheat must be sown after its harvest.

C. Seed Size, Hardness, Color, and Vigor

The soft starchy seeds of maize and sorghum are subject to mechanical injury during harvesting and postharvest operations, resulting in poor storability. The size and vigor of seedlings are closely associated with the seed size. There is a direct relationship between the seed weight and dry weight of the seedling. Large seeds permit a seedling to emerge from deeper in the soil. Larger seeds are particularly important in small-seed crops like grasses and forage legumes. Such seeds must be shallowly sown for better emergence and are thus subject to drying in the upper soil layers, losing vitality. Large seeds can be sown deep enough to permit the seedling roots to reach moist soil before the upper soil layer is dry. However, large seeds in some food crops such as tomato, okra, eggplant, watermelon, and cucurbits are not acceptable to consumers. Few attempts may be made to breed varieties of such vegetable crops with larger seeds to enhance their planting value. Research efforts to cross newly introduced large-seeded annual species of sweet clover with the domesticated species have met with partial success.

The presence of hard seeds that do not germinate promptly (e.g., alfalfa, sweet clover, true clover) poses difficulty in crop establishment. Also, the hard or corneous seeds of cotton, corn, and sorghum are more resistant to seed rots and seedling blights than the softer-seeded varieties of these crops. Selection for an increased percentage of hard seed of legumes in Canada has occasionally reduced plant vigor (27).

Seed color is important too. Colored seeds generally have better germination and vigor than white seeds, which canners prefer. Green cotyledon in garden peas and lima beans is preferred to yellow cotyledon. The degree of maturity of green seeds is difficult to detect by visual inspection, making it difficult to judge appropriate harvest maturity. The inheritance of seed color and its pattern is often complicated. The expression of anthocyanin colors depends on a few "basic" genes, which can be suppressed by a dominant inhibitor. Tenderness of the seed coat is important for all seeds that are directly consumed. Tenderness is generally associated with thin seed coats, whose thickness increases with crop maturity. More appropriate tests must be developed to judge the correct harvest maturity for tenderness as well as for sweetness consistency, and flavor to select seeds, (e.g., sweet corn) for improved quality.

The relationship between the seed coat color and seedling quality was assessed in *Emblica officinalis* (*Phyllanthus emblica*) cv. BSR 1 seeds collected from Bhavanisagar, Tamil Nadu, India, in 1995. Bulk seeds with different color shades were categorized into three color grades: yellowish brown, chestnut brown, blackish brown. Seeds with a chestnut brown color recorded the highest values for recovery (60%), 100–seed weight (2.188 g), germination percentage (70.63%), seedling length (12.6 cm), and vigor index (1064). These results suggest that the seed coat color can be used as a parameter to assess the seed quality of *E. officinalis* (27a).

Five cowpea cultivars differing in seed coat pigmentation were stored under simulated Ghanaian conditions (30°C, 75.5% RH) for 6 months and accelerated aging conditions (40°C, 100% RH) for 6 days. Pigmented cultivars retained high germinations throughout the storage period, whereas those of the unpigmented cultivars declined. All cultivars showed increased seed leachate conductivity and reduced staining of the cotyledons with tetrazolium chloride as storage time increased, although in both storage conditions, the unpigmented cultivars deteriorated far more rapidly, possibly due to the greater increase observed in their seed moisture content when held at the high relative humidity. Unpigmented cultivars also deteriorated more rapidly during controlled deterioration at a constant moisture content (20%) and 40°C, indicating that they are also genetically more predisposed to rapid deterioration than are pigmented cultivars. Imhibition damage did not occur in unstored seeds whereas the reduced vital staining of the cotyledons after rapid imhibition of stored, that is, aged, seeds revealed the incidence of imhibition damage. In pigmented seeds, the incidence of imhibition damage in seeds that had aged in storage could have been partly due to the increase in the rate of imhibition observed after storage. However, the rate of imhibition of the unpigmented cultivars remained unchanged, suggesting an increased predisposition of the aged seeds to imhibition damage. Asiedu and Powell (27b) have discussed the interaction of seed aging during storage and imhibition damage in their influence on seed quality.

Seed longevity is an important consideration in the selection of desired lines. Certain homozygous recessive characteristics reduce the vitality seeds within a short time. A higher proportion of hard seeds adversely affects germination, resulting in poor cooking quality. Split seed coats and deformity in cotyledons of lima beans, snap beans, and soybeans can be minimized by breeding and selection. These defects contribute significantly to threshing injury, leading to poor storability (27).

D. Bolting Resistance

Certain biennial crops such as beets, cabbage, celery, lettuce, onion, and spinach tend to produce seed during the crop season. Such plants use food and energy to produce a seed stalk, making the root or top worthless for harvest. It is essential to breed slow-bolting varieties of such crops. However, the plants must be able to bolt during the seed-producing season. The genetic make-up of the plant determines its response to light and temperature, but treatment with growth regulators like gibberellins may induce plants to bolt. More research is required to understand the biochemistry and physiology of bolting in crop plants (26).

E. Prevention of Seed Shattering

Most seed crops shatter heavily during harvesting and handling, and also take a long time to mature, making the harvest difficult. The yields of forage crops like birdsfoot trefoil, lupines, and reed canary grass are low mainly because of heavy shattering losses before all seeds on the plants are mature. Attempts to reduce seed shattering in grasses like phalaris by selection for more compact and inflexible inflorescences have been successful, as has the development of special harvesting machinery and the use of sprays to bind inflorescences and prevent seed drop (28). The most successful approach, however, has been the identification and selection of nonshedding natural mutants possessing a strong, nonbrittle rachilla, which anchors the seed in the spikelet at maturity. In terms of genetic control, mutations resemble ones that have occurred in major cereals, distinguishing them from their wild ancestors. It is possible that mutations of type found in phalaris that prevent seed shattering may be present in many other grass species and crops, including peas and beans. However, because these mutations are at a selective disadvantage in the wild, the frequency of genes controlling them in natural populations will be low.

Nevertheless, it should be possible to locate such natural mutants in other important crops to overcome one of the major problems inhibiting the domestication of these crop species for agriculture.

F. Newer Plant Growth Regulators and their Novel Uses

Seed shattering at harvest time often causes a 25% or greater loss of productivity in rice, beans, oilseeds, alfalfa, clovers, and other crops. Seed-bearing organs, pods, and capsules develop well-defined thin-walled layers of cell or abscission zones as they mature. Even a brief delay in cell separation through the use of new growth-regulating chemicals (abscission inhibitors) would reduce losses from seed shattering and increase seed yields significantly.

Biochemical manipulation of crop growth and development is an exceedingly powerful tool with which the seed yields of agricultural crops can be increased substantially. The high cost of such chemicals, however, precludes their application to herbage seed crops on a commercial scale (e.g., ancymidol in grasses). Depressed yields of herbage seed crops are frequently associated with adverse weather conditions at anthesis and harvest (29). If lodging of forage and other crops can be prevented using growth regulators such as cycocel (CCC) and other growth inhibitors, it may be possible to increase the yields of these crops for long-term economy through greater yield stability. If a suitable growth-regulating chemical is to be developed, efforts must also be made to establish other related aspects. Management factors such as the amount and time of application and the time and method of harvesting will need reevaluation (30).

The soon to be realized gains from the control of the internal metabolism of crops by the use of plant growth regulators have not been fully appreciated even by many agricultural scientists. Many researchers feel that this approach, in its broad aspects, may result in the most important quantitative gains yet to be achieved in agriculture. Plant growth regulators have already been found to promote rooting and plant propagation; initiate or terminate dormancy of seeds, buds, and tubers; promote or delay flowering and fruit ripening; induce or prevent leaf, fruit, or seed drop (abscission); control fruit set and development; control plant or organ size; prune plants chemically; increase plant resistance to pests and unfavorable environmental factors such as temperature, water stress, and air pollution; prevent postharvest spoilage; enhance mineral uptake from the soil; alter the timing of crop growth and development; and even regulate the chemical composition and color of plant products (31).

Newer plant growth–regulating substances are being synthesized in laboratories and other new growth regulators are being obtained directly from plants. Through further research with these compounds, additional novel ways may be found so that they can be used safely and even more effectively in the production of high-quality seeds and agricultural crops. However, further research is needed to understand (a) the mechanism(s) of action of growth regulators at the molecular and genic levels; (b) their effects on enzyme activities (activation and/or synthesis or inhibition); (c) the role of ethylene in major plant hormone activity and fruit ripening; (d) the role of auxins, gibberellins, cytokinins, and inhibitors in seed germination and dormancy, ripening and senescence, and abscission; and (e) their use in tissue and in vitro cultures, breeding and production of synseeds, and transgenic plants (13).

G. Effective Weed and Pollination Control

The evaluation of newer herbicides for weed control in crop production has proved to be beneficial. New herbicides have been derived largely in an arable context. In the future, seed growers will have a closer affinity with grassland farmers regarding the choice of weed treatment. Although herbicides are important, alternative measures must not be neglected. Little research work has been reported on cultural weed control methods in seed crops of grasses such as *Lolium* spp. (32). The importance of crop rotation with the benefits of weed control for cereal crops cannot be overemphasized, nor can the need for good establishment and care of the seed crop after sowing to reduce the problem of grass weed invasion be overstressed (33).

The seed-setting potential of autotetraploid crops such as red clover can be greatly improved using suitable selection techniques. However, a full exploitation of the seed-setting potential will require a marked improvement in pollination conditions. Bumblebees are not reliable as pollinators because of their highly variable and decreasing numbers. Efforts must be made to improve pollination by honeybees either by breeding plants for improved attractiveness or by inducing the bees to visit larger forage crop areas.

H. Preservation of Seed Quality

Inexpensive low-quality seed often results in a poor crop, and seed can be produced cheaply when a high-yielding variety is sown. The concept of seed quality consists of several components, such as analytical, genetic or species purity, cultivar purity, germination capacity, vigor, size, uniformity, moisture content, and freedom from weeds and diseases (34).

Curtis (35) stressed three important aspects of seed quality: (a) achieving seed quality, (b) maintaining seed quality, and (c) measuring seed quality. Inevitably, the most attention has been focused on the maintenance and measurement of seed quality, and although these are undoubtedly important, it is sometimes forgotten that seed quality is initially determined before harvest. Harvesting and all postharvest operations are concerned only with the preservation of seed quality.

Fortunately, cultivation practices that result in maximum yield usually produce high-quality seed. High seed yields are associated with early sowing and timely harvesting of large well-defined, disease-free ears in most crops, such as maize. The recent switch from double and three-way crosses to single and modified single crosses in maize has resulted in lower yields of commercial seed. It has been predicted that increases in seed yields will stem from the current trend among seed companies to emphasize research into seed-production methods. In a competitive seed industry, agronomic research in a broad sense cannot be concerned solely with the development of superior hybrids, and it must also discover newer methods of producing high-quality seed more efficiently.

Despite a significant improvement in the quality of cereal seed available to the farmer today—with the help of excellent legislation concerning seed certification, both in the field and in the laboratory, coupled with significant developments made in germination and vigor-testing techniques—the cereal seed industry must not be complacent. There are numerous areas of investigation yet to be undertaken with regard to cereal seed quality. Very little is currently known about cereal seed itself. It is necessary to be more methodical in the collection of data concerning seed quality, particularly with respect to purity and germination. Seed must be treated with respect, since it is often the last factor to be given credit for a successful crop. Seed can prove to be the most costly factor of a crop if it is of poor quality (36).

IV. EXPECTATIONS FROM NEWER BIOTECHNOLOGIES

A. Future Prospects: The Promise of Biodiversity

The new biotechnology refers to the use of recombinant DNA and in vitro biological techniques, including micropropagation, synthetic seed technology, and production of genetically modified or transgenic plants. These techniques will be useful in three major areas as (2): (a) powerful tools in classic breeding methods, (b) a means of generating transgenic plants, animals, and other useful organisms, and (c) a means of integrating beneficial microorganisms into various agricultural production systems (4,6,23,37), such as agricultural and industrial waste management (38). The new biotechnology that has been adapted to agriculture during the last couple of decades indeed has opened new vistas along the way. Its achievements thus far have already exceeded previous expectations. However, continued success will depend upon further research development supported by the national governments, economic growth, favorable regulatory climate, and public support (4,39).

Environmental concerns (see Sec. IV.B) and biodiversity are the central guidelines in agricultural biotechnology (2). The basic need to save natural germplasm variability and to protect it against possible extinction and the economic value of biodiversity in itself has attracted the attention of industrial and financial organizations (37). The World Monetary Fund (WMF) announced at the World Congress on Biodiversity held in Buenos Aires, Argentina, in 1996, the allocation of 100 million U.S. dollars for 3 years toward the preservation of biodiversity, which was the obvious outcome of the Rio de Janeiro convention held in 1992, where 162 states signed an agreement to save the biodiversity of germplasms, species, and ecosystems (23). Central issues discussed in the convention were the importance of biodiversity as a natural resource of the countries involved and its use of new product development and protection of oceans and specific ecological regions, such as the Amazon, which represent the frontier of discovery of new genotypes, products, medicines, pharmaceuticals, and alternative food sources (2,4,23,37).

Biodiversity provides developing countries a unique opportunity to establish profitable industries based on newly discovered compounds and rare genotypes that are specific to those regions. For example, the neem tree was used in India for many years for fertility control in agriculture, which is now being industrially processed by developed countries for the same applications. Palclitaxel (Taxol), extracted from trees of the rain forest in South America and used as an anticancer drug (40), is another example of the potential benefit of biodiversity to human health. Important legal issues are involved touching on how to sustain the rights of the developed and developing countries to determine the compensatory agreement between the sources of biodiversity and the accessors that develop these sources into an economic product (23,37). Thus, developing countries will be acknowledged for the natural resources of their regions, and industrial companies will establish, in the developing countries, new ventures for turning the basic knowledge into a commercially viable product (2). Ramanatha Rao and Hodgkin (41) have recently reviewed the available literature on the genetic diversity and conservation and utilization of plant genetic resources and identified the future research needs.

B. Environmental Considerations

There is much basic information available about how various environmental factors, both singly and interacting, affect the production and germination of seeds. Biochemical mechanisms that sense environmental conditions (e.g., phytochrome) and regulate the plant's response are also understood to some extent. Genetic and hormonal control mechanisms are beginning to be understood, as are the synthesis and degradation of specific macromolecules in seeds and enzymes and their role in cellular metabolism.

However, there is still much more to learn about the genetics, biochemistry, and physiology of seed, regardless of the environment, before wider ecological implications are considered. A newer approach is needed requiring a thorough knowledge of the ecology of plant species to choose them on the basis of their natural distribution. Wherever possible, comparative studies involving species with varying degrees of tolerance to environmental factors should be attempted. This fundamental knowledge will untimately benefit all ecologists to enable them to have some control over the growth and development of seeds and seedlings of crops.

In many countries like India, agriculture still depends to a large extent on the vagaries of monsoon rains, with the resultant lower crop productivity. No much progress has been made in terms of droughtproofing. Breeding crops for tolerance to abiotic stresses like drought, salinity, alkalinity, frost, cold, and heat through new biotechnology are the intriguing tasks ahead. Changes in the global temperature, CO_2, and ozone layer are other challenges of the future agricultural scenario. Weather extremes like excessive rains and severe droughts, along with a changing crop complex, may create new insect pests and abiotic problems. Appropriate strategies and research and developmental efforts to be undertaken by the public and private sectors are needed, especially in view of the legal, technical, economical, and moral issues involved in granting intellectual property rights to plant breeders. Finding and developing better seeds for agricultural crops through newer biotechnological tools, such as tissue culture and in vitro micropropagation, synseeds, and transgenic technology, that can resist drought, heat, and cold as well as the threat of diseases and insect pests is becoming increasingly important to ensure adequate food for all populations. Agricultural crop plants are the key for continuance of human life and seed is the prime ingredient in this effort.

1. Biosafety

A debt of gratitude is owed to the environmental movement that has taken place over the past 40 years, which has led to legislation to improve air and water quality, protect wildlife, control the disposal of toxic wastes, protect the soils, and reduce the loss of biodiversity (9). According to Shantharam (42) of the Animal and Plant Health Inspection Service (APHIS), U.S. Department of Agriculture (USDA), four funda-mental questions surrounding the biosafety and environmental impacts of GMOs must be addressed in any good developmental practice (GDP) for product develop-ment: (a) Is the introduced gene stably inherited in the transgenic plants and its off springs? (b) What are the routes by which the introduced genes might possibly escape from the GMOs? (c) What are the consequences of such a gene escape? (d) Do the GMOs have the potential to become weeds and aggressive colonizers? These are the most straightforward biosafety questions which most of the scientific and regulatory scientists can defend. Ticklish questions relate to the long-tem effects of growing these plants in monoculture, the associated changes that might occur in agricultural practices, and the socioeconomic benefits of adopting GMOs for large-scale cultiva-tion and their impact on biodiversity. At the present time, there are no useful working models that are quantitative in nature with any sense of predictive value to base the environment risk assessments on, although the potential impacts or consequences on the environment can be estimated by using qualittive questions and the knowledge of the reproductive nature of a given organism. Using this process, USDA, APHIS that has allowed for thousands of field tests since 1987, has gained a considerable amount of experience in this area, thereby establishing global leadership in regulating biotechnology. This has allowed for the safe deployment of GMOs, leading to the successful commercialization of more than 20 transgenic crop plants in the United States. These years of experience has also allowed, APHIS to pare down the regulatory oversight by introducing a notification system for almost all crops (42).

The agricultural environment in most tropical countries is clearly of a different nature, many of them being the centers of origin of the plants in question. Biodiversity is under threat in almost all parts of the world owing to human activity. Modern agriculture has contributed its share, and GMOs will cause additional concern. Monoculture farming has allowed domesticating only a select group of about five food crops, covering millions of hectares of land being cultivated by just a few varieties of crop plants, purely because of their yield potential. As a consequence of the uncontrolled human population, biodiversity is being destroyed, resulting in cutting down of forests. Transgenic crop plants are expected to be cultivated in the same areas as those being cultivated with conventionally bred varieties of crop plants. Large-scale cultivation of genetically engineered crop plants is not expected to escape from well-tended cultivation plots and take over the countryside and overrun the wilderness of biodiversity. There is no scientific reason or evidence for the aggressive colonization of GMOs. The possibility of gene escape has been demonstrated in the case of crop plants whose wild and weedy relatives are found in the surrounding areas of cultivation. But whether the escaped gene introgresses into the wild and weedy relatives and whether there will be sufficient pressure for the selection and propagation of such genes in the wild remain a enigmatic questions. The evidence from conventionally bred crop plants for similar traits does not show any record to have spread in the wild. Only a long-term monitoring of freely cultivated GM crop plants for 10 or more years might shed some light on this question. This exercise of an academic research problem in plant ecology should not prevent field testing and carefully monitored commercialization of GMOs (42).

Ellstrand (43) has recently quoted two reports (44,45), illustrating how easy it is to lose track of transgenes. Without careful checking, there are plenty of opportunities for them to move from one variety to another. The field release of "third-generation" transgenic crops that are grown to produce pharmaceutical and other industrial biochemicals may pose special challenges for containment if these compounds are not expected to appear in the human food supply (43).

Three species of insect-resistant transgenic crops with δ-endotoxin genes from *Bacillus thuringiensis* were released to farmers in the United States in 1996, with several other *B. thuringiensis* crops likely to be released in many countries over the next few years. Insecticide products containing *B. thuringiensis* have been widely used in agriculture for than three decades, with an excellent safety record (46). However, researchable questions remain regarding the environmental impact of *B. thuringiensis* crops, differing for particular crop species and geographical areas, including the indirect effects on the management of nontarget pest species by nature's biological control, the fate and persistence of δ-endotoxins in the soil, the consequences of outcrossing of δ-endotoxin genes to wild and weedy crop relatives, and the evolution of pest resistance to δ-endotoxins. Risks associated with the use of *B. thuringiensis* crops have to be balanced against potential benefits; notably a reduction in the use of chemical insecticides in some crop systems (46).

Since viruses constitute a major threat to global food security, the use of virus-resistant crop plants is the most effective and durable approach to manage viral diseases. Incorporating defined viral sequence(s) into plants may confer resistance to challenge infection by both homologous and heterologous viruses through genetic engineering technology. Field tests carried out under natural conditions in several countries have shown promising results in reducing the impact of viral infections in many crops (47). However, concerns have been raised about the possibility of generation of new viruses as a result of interactions between the transgene and the challenge virus through mechanisms such as recombination and transcapsidation. Although both of these events have resulted in the production of recombinant virus from a transgenic plant under experimental condition this should be viewed in the context of the frequency of such genetic event(s) and the level of selection pressure exerted in nature. Approaches to reduce the frequency and/or the effects of such events include (47): (a) use of untranslatable RNA sequences of viral origin, preventing further expression of viral gene, (b) expressing defective/truncated forms of viral gene, (c) avoiding viral sequences that might initiate recombination, and (d) selecting transgenic plant lines that express minimal levels of the transgene. The potential risks of using virus-resistant GMO plants should be weighed against the benefits such as increased yield, a better-quality product, and a significant reduction in the use of pesticides.

Transgenic plants with novel herbicidal tolerances may become a significant part of agricultural system within the next few years, with an anticipation of further regulatory reviews of plant lines (48). Ultimately, the marketplace will decide how popular these lines will be to the growers compared to currently available varieties. Passing rigorous regulatory reviews can minimize any negative effects on the environment and biodiversity. Signatories to the Convention on Biological Diversity, introduced at the 1992 "Rio Summit," are currently considering a Biosafety Protocol, providing a worldwide regulatory mechanism for the transboundry movement of living GMOs (48).

2. Threat of Antiscience Zealotry

Although we now have the agricultural biotechnology available to feed the 8.3 billion people anticipated in the next quarter of a century (9), the more pertinent question today is whether farmers will be permitted to use that technology. For many people in the developed countries, GM crops have become the latest incarnation of evil biotechnology that sacrifices both humans and the environment. On one side of heated discourse are people who firmly believe that GM crops pose a threat to human health and biodiversity, whereas the other side mainly comprises scientists who are convinced that GMOs represent a technology with an enormous potential for increasing food, feed, and fiber production in an environmentally benign way. Extremists in the environmental movement, largely from the developed rich nations and/or the privileged societies in the poor nations, have encouraged the antibiotechnology zealots to wage their campaigns of propaganda and vandalism. Even some scientists, many of whom should or do know better, have jumped on this extremist environmental bandwagon (9).

There is no such thing as safe food, and there has never been. This is not to suggest that all our foods are dangerous, only an acknowledgment that trace levels of such contaminants as toxins and carcinogens are present in everything human beings eat. A primary rule of toxicity refers to the importance of dosage: "Every substance is a poison, but it is the dosage that makes it poisonous" (49).

Plants have evolved to produce an array of chemicals to protect themselves against pests, diseases, and herbivores (50), and human beings consume roughly 5000–10,000 natural toxins daily (51). For instance, roasted coffee has over 1000 chemicals, of which 27 have been tested and 19 of them were found to be rodent carcinogens (52). The fat-soluble neurotoxins solanine and chaconine are present in potatoes and can be detected in the blood stream of all potato eaters (53). Even when the crops are bred for resistance to pests and diseases by transfering genes through conventional breeding methods, the resistance is often accompanied by an increase in such toxic compounds (51).

Problems did exist with conventionally bred varieties of crops. Any crop variety found to pose a real health risk was promptly removed from the food chain, but those varieties, in contrast to GM crops, were never routinely tested for safety. One pest-resistant variety that produced rashes in agricultural workers was found to contain 6200 ppb of carcinogenic psoralens compared to 800 ppb in the control celery (54). This celery was removed from cultivation, as was also the case with the potato variety Lenape, which contained very high levels of the toxic alkaloid solanine.

Humans have learned from trial and error with all innovations. Similarly, crop-improvement practices have also evolved over time with continued refinement. It is common in human nature to generate an exaggerated fear of new innovations, whereas perceiving the older or so-called "natural" products as always being more benign. Huber (55) has discussed this double standard in the larger context of risk regulation; that is, being lenient toward existing known and greater hazards and failing to recognized and "exorcise" much larger older risks, whereas creating "gatekeepers" to minimize new risks.

Human beings must eat to live and therefore have pursued the most ecologically demanding endeavor of organized food production. Agricultural expansion over the millennia has destroyed millions of acres of forestland around the world. Alien plant species have been introduced through traditional breeding into nonnative environments to provide food, feed, fiber, and timber, and as a result has disrupted local fauna and flora. Certain aspects of modern farming also has had a negative impact on the biodiversity of crop plants and on air, soil, and water quality; nevertheless, it sustains and nurtures most of the world's six billion people with affordable food (51).

The potential environmental concerns of GM crops can be addressed in the context of our experience with traditional crop variety deployment. Genes for pest and disease resistance have been continuously introduced through conventional breeding into most of our agricultural crops, including traits such as stress tolerance and herbicidal tolerance with altered growth habits of every crop. The risks of gene flow to weedy relatives have always existed, with the possible occurrence of "gene flow." It is comforting to recognize that no major "superweeds" have developed since the advent of modern plant breeding. Noxious weeds like kudzu, water hyacinth, and parthenium have resulted from the introduction of semidomestication of wild plants into nonnative environments without the checks nd balances of their native pests. Yet, there are probably no dwarf plants among the wild *Oryza* spp. and *Triticum* spp. populations in the Middle East or Asia despite the fact that diminutive rice and wheat varieties have been grown for decades.

The risk of gene transfer to weeds is similar with both conventional and GM crops, with it not being contingent on how these genes have been introduced into plants. It is essential to be vigilant to ensure that weeds do not become noxious as a result of any new crop variety. The current case-by-case testing and monitoring approach with biotech crops is a good regimen for the future, whereas the past experience with conventional crops assures that such risks will be minimal and manageable (51).

The potential positive impacts of GM crops on the environment are (51) (a) decreasing agricultural expansion to preserve wild ecosystems; (b) improving air, soil, and water quality by prompting reduced tillage, and reduced use of chemical pesticides and fuel; (c) improving biodiversity through resuscitation of older varieties and promotion of beneficial insects; and (d) cleaning up contaminated soil and air through pytoremediation.

Human beings and crops have been and will always be mutually dependent on each other's survival and will continue to guide the evolution of crops, which will increasingly be more knowledge based and responsible. An appreciation of agricultural development may provide with a useful road map to develop appropriate strategies for informing and rationalizing societal responses to continued crop improvement. Paraphrasing the American philosopher George Santayana, "ignoring history may condemn us to repeat it, but an understanding of the past may as well lead us to an enlightened future" (51).

According to Buchanan (56), the problem of genetic engineering and the allergy issue is transitory, and once appropriate allergen testing capability is in place, health concerns will abate and the development of transgenic foods will continue apace. An animal model needs to be developed to identify transgenic plant proteins that either or have become allergens in humans. This will enable formulation of a safe and reasonable testing policy to allay consumer concerns.

The Strategic Alliance for Biotechnology Research in African Development (SABRAD) workshop held at Accra, Ghana in November 2000, resolved to achieve food self-sufficiency and improved living conditions of resource-poor farmers identified as the target recipients for products generated from biotechnology applications. The SABRAD initiative is one step in the right direction that deserves support from all those who want to help African scientists and farmers to feed their own people (57).

3. Conclusions

Borlaug (9) has reiterated that the world has technology that is either available or well advanced in the research pipeline to feed a population of 10 billion people. Agricultural scientists and national leaders have a moral obligation to warn the political, educational, and religious leaders about the magnitude and seriousness of the arable land, food, and population problems that lie ahead even with breakthroughs in biotechnology. Failing to do this, the world food situation may be condemned to face the pending chaos of incalculable millions of deaths by starvation.

REFERENCES

1. R. Budd, The uses of life, *A History of Biotechnology*, Cambridge University Press, New York, 1993.
2. H. Meiri and A. Altman, Agriculture and agricultural biotechnology: development trends towards 21st century, *Agricultural Biotechnology* (A. Altman, ed.), Marcel Dekker, New York, 1998, pp. 1–17.
3. R.A. Colwell and A. Sasson, Biotechnology and development, World Science Report, 1994, pp. 253–268.
4. B.D. Davis, *The Genetic Revolution: Scientific Prospects and Public Perceptions*, Johns Hopkins University Press, Baltimore, 1991.
5. B.A. Law, Biotechnology in food manufacture, *Chem. Ind.* 1994, pp. 502–505.
6. Office of Technology Assessment, Biotechnology on a Global Economy, US Committee Report, 1993.
7. J.E. Smith, Biotechnology, *Studies in Biology*, 3rd ed., Cambridge University Press, Cambridge, UK, 1996, p. 236.
8. The New Biotechnology, World Report of 1994, Paris, 1994, pp. 1–57.
9. N.E. Borlaug, Ending world hunger: the promise of biotechnology and the threat of antiscience zealotry, *Plant Physiol. 124*:487 (2000).
10. M.J. Chrispeels, Biotechnology and the poor, *Plant Physiol.* 124:3 (2000).
11. J.N. Siedow, Feeding ten billion people: three views, *Plant Physiol. 126*:20 (2001).
12. C. Conway, *The Doubly Green Revolution: Food for All in the Twenty First Century*, Cornell University Press, Ithaca, NY, 1997.
13. L.T. Evans, *Feeding the Ten Billion*, Cambridge University Press, Cambridge, UK, 1998.
14. V. Smil, *Feeding the World: A Challenge for the 21st Century*, MIT Press, Cambridge, MA, 2000.

15. D.K. Salunkhe and B.B. Desai, *Posthrvest Biotechnology of Fruits*, Vols. 1&2, CRC Press Boca Raton, FL, 1984.

16. D.K. Salunkhe and B.B. Desai, *Posthrvest Biotechnology of Vegetables*, Vols. 1&2, CRC Press Boca Raton, FL, 1984.

17. D.K. Salunkhe and B.B. Desai, *Posthrvest Biotechnology of oilseeds*, CRC Press Boca Raton, FL, 1986.

18. D.K. Salunkhe and B.B. Desai, *Posthrvest Biotechnology of Sugar Crops*, CRC Press Boca Raton, FL, 1988.

19. B.B. Desai, *Handbook of Diet and Nutrition*, Marcel Dekker, New York, 2000.

20. I. Serageldin, Biotechnology and food security in the 21st century, *Science 285*:387 (1999).

21. C.F. Runge and B. Senauer, A removable feast, *Foreign Affairs 79*:39 (2000).

22. G.J. Persley, Agricultural Biotechnology and the Poor. Promethean Science (http://www.cgiar.org./biotech/repo100/persley.pdf). Cited by Chrispeels (10).

23. Biotechnology and Economic Development: Forecast and Potential Report, Paris, 1995, pp. 1–64.

24. C.N. Stewart, Jr., Monitoring transgenic plants using in vitro markers, *Nature Bio/Technology 14*:682 (1996).

25. A.K. Weissenger, Biotechnology for improvement of groundnut (*Arachis hypogaea*), *Groundnut- A Global Perspective*, Proceedings of an International Workshop, November 25–29, 1991,ICRISAT Center, New Delhi, India, 1992, p. 317.

26. K. Lexander, Seed composition in connection with germination and bolting of *Beta vulgaris* L. (sugar beet), *Seed Production* (P.D. Hebbelethwaite, ed.), Butterworths, London, 1980, p. 271.

27. J.H. Martin and S.H. Yarnell, Problems and rewards of improving seeds, Seeds, *The Yearbook of Agriculture*, US Department of Agriculture, Washington, DC, 1961, p. 113.

27a. T.V. Karivarhadaraaju, P. Srimathi, and K. Malarkodi, Seed coat color as a parameter for seedling quality of amla (*Emblica offcinalis* L. Gaertn), *Adv. Plant Sci. 14*:271 (2001).

27b. E.A. Asiedu and A.A. Powell, Comparison of the storage potential of cultivars of cowpeas (Vigna unguiculata) differing in seed coat pigmentation, Seed Sci. Technol. *26*:211 (1998).

28. J.R. McWilliam, The development and significance of seed retention in grasses, *Seed Production* (P.D. Hebbelethwaite, ed.), Butterworths, London, 1980, p. 51.

29. D.J. Griffiths, H.M. Roberts, and J. Lewis, The seed yield potential of grasses, Welsh Plant Breeding Station, Annual Report, 1973, p. 117.

30. P.D. Hebbelethwaite, D. Wright, and A. Noble, Some physiological aspects of seed yields in *Lolium perenne* L. (perennial rye grass), *Seed Production* (P.D. Hebbelethwaite, ed.), Butterworth, London, 1980, p. 71.

31. B.B. Desai, G.K. Zende, and D.K. Salunkhe, Future decade of research in agricultural chemistry, *Indi. J. Agric. Chem. 16* (1):1 (1983).

32. A.A. Oswald and R.J. Hagger, Weed control in ryegrass grown for seed, *Seed Production*, (P.D. Hebbelethwaite, ed.), Butterworths, London, 1980, p. 121.

33. B.A. Dennis, Breeding for improved seed production in autotetraploid red clover, *Seed Production* (P.D. Hebbelethwaite, ed.), Butterworths, London, 1980, p. 229.

34. J.R. Thomson, *An Introduction to Seed Technology*, Leonard Hill, 1979, p.1.

35. D.L. Curtis, Some aspects of *Zea mays* L. (corn), *Seed Production* (P.D. Hebbelethwaite, ed.), Butterworths, London, 1980, p. 389.

36. M. Carver, The production of quality cereal seed, *Seed Production* (P.D. Hebbelethwaite, ed.), Butterworths, London, 1980, p. 295.

37. C. Brenner and J. Conner, International Initiative in Biotechnology for Developing Country Agriculture: Promises and Problems, OECD, Paris, 1994.

38. Anonymous. *Waste Management and Recycling International*, Sterling, London, 1994.

39. Task Group on Public Perceptions of Biotechnology. The application of human genetic research, Briefing Paper 3, European Federation of Biotechnology, 1995.
40. S.M. Edginton, Taxol out of woods, *Biotechnology 9*:933–935, 1012–1013 (1991).
41. V. Ramanath Rao and T. Hodgkin, Genetic diversity and conservation and utilization of plant genetic resources, *Plant Cell Tiss. Org. Cult. 68*:1–19 (2002).
42. S. Shantharam, biotechnology and development: Moving technology forward using environmental assessment, *Biotechnology, Biosafety, and Biodiversity: Scientific and Ethical Issues for Sustainable Development* (S. Shantharam and J.F. Mongomery, eds.), Oxford & IBH Publishing, Co., New Delhi, 1999, pp. 1–4.
43. N.C. Ellstrand, When transgenes wander, should we worry, *Plant Physiol. 125*:1543 (2001).
44. M. MacArthur, Triple-resistant canola weeds found in Alberta. The WesternProducer Http//www.producer.com/articles/20000210/news/200000210newssol.html(Feb., 10, 2000). Cited by Ellstrand (43).
45. P. Callaban, Genetically altered protein is found in still more corn, *Wall Street J. 236*: B5 (2000). Cited by Ellstrand (43).
46. M.B. Cohen, Environmental impact of crops transformed with genes from Bacillus thuringiensis (Bt) for insect resistance, *Biotechnology, Biosafety, and Biodiversity: Scientific and Ethical Issues for Sustainable Development* (S. Shantharam and J.F. Montgomery, eds.), Oxford & IBH Publishing, New Delhi, 1999, pp. 31–39.
47. H.R. Pappu, Biosafety issues of genetically engineered virus-resistant plants, *Biotechnology, Biosafety, and Biodiversity: Scientific and Ethical Issues for Sustainable Development* (S. Shantharam and J.F. Montgomery, eds.), Oxford & IBH Publishing, New Delhi, 1999, pp.52–64.
48. S. Yarrow, Biosafety issues of genetically engineered Canada's perspectives, Biosafety, and Biodiversity: Scientific and Ethical Issues for Sustainable Development (S. Shantharam and J.F. Montgomery, eds.), Oxford & IBH Publishing, New Delhi, 1999, pp. 42-45.
49. A. Poole and G.B. Leslie, *A Practical Approach to Toxicological Investigations*, Cambridge University Press, Cambridge, UK, 1989.
50. B.N. Ames, M. Profet, and L.S. Gold, Dietary pesticides (99.99 percent all natural), *Proc. Natl. Acad. Sci. USA 87*:7777 (1990).
51. C.S. Prakash, The genetically modified crop debate in the context of agricultural evolution, *Plant Physiol. 126*:8 (2001).
52. B.N. Ames and L.S. Gold, Pollution pesticides and cancer misconceptions, what Risk (R. Bate, ed.), Butterworth-Heinemann, Boston, 1997, pp. 173–190.
53. B.N. Ames, M. Profet, and L.S. Gold Nature's chemicals synthetic chemicals comparative toxicology, *Proc. Natl. Acad. Sci. USA 87*:77782 (1990).
54. B.N. Ames and L.S. Gold, Chemical carcinogens: too many rodent carcinogens, *Proc. Natl. Acad. Sci. USA 87*:1 (1990).
55. P. Huber, Exorcists vs. gatekeepers in risk regulation, *Regulation 7*:23 (1983).
56. B.B. Buchanan, Genetic engineering and the allergy issue, *Plant Physiol 126*:5 (2001).
57. J. Machuka, Agricultural biotechnology for Africa. Africa scientists and farmers must feed their own people, *Plant Physiol. 126*:16 (2001).

Index